# COMMUNICATION SYSTEMS
## Analog and Digital

# COMMUNICATION SYSTEMS
## Analog and Digital

**R P Singh**
**SD Sapre**
*Department of Electronics Engineering*
*Maulana Azad College of Technology*
*(A Regional Engineering College)*
*Bhopal,*

**Tata McGraw-Hill Publishing Company Limited**
New Delhi

*McGraw-Hill Offices*
**New Delhi**  New York  St Louis  San Francisco  Auckland  Bogota  Guatemala
Hamburg  Lisbon  London  Madrid  Mexico  Milan  Montreal  Panama
Paris  San Juan  Sao Paulo  Singapore  Sydney  Tokyo  Toronto

**Tata McGraw-Hill**
*A Division of The **McGraw·Hill** Companies*

© 1995, Tata McGraw-Hill Publishing Company Limited

Sixteenth reprint 2003
RQLLCRXDRCRYY

No part of this publication can be reproduced in any form or by any means without the prior written permission of the publishers

This edition can be exported from India only by the publishers,
Tata McGraw-Hill Publishing Company Limited

ISBN 0-07-460339-6

Published by Tata McGraw-Hill Publishing Company Limited,
7 West Patel Nagar, New Delhi 110 008, and printed at
Sheel Print-N-Pack.

*In the sacred memories of*
*Late Shri Daya Narayan Singh & Late Smt. Sonvarsha Devi*
*parents of Rajendra Prasad Singh*
AND
*Late Dr. Damodar Krushna Sapre & Late Smt. Laxmibai Sapre*
*parents of Sadanand Damodar Sapre*

# PREFACE

Communication Systems is a core subject for undergraduate courses in Electronics & Communication, and Computer Engineering. It is also offered as a special subject for undergraduate courses in Electrical and Instrumentation Engineering, and for postgraduate courses in Physics, Electronics and Computer Science. This book is primarily intended to serve as a textbook for undergraduate courses in Electronics & Communication Engineering. The first course on Analog Communication Systems is generally offered in the pre-final year, and the second course on Digital Communication Systems in the final year. Accordingly, the material covered in the book has been divided into two main parts. The first part, Chapters 1, 2, 4, 5 and 6 cover Analog Communication Systems, and the second part consisting of Chapters 3, 7, 8, 9, 10 and 11, covers Digital Communication Systems.

Chapters 1 and 2 discuss Signals and Systems. Classical topics like Fourier series, Fourier transforms, linear systems, energy-power correlation, and autocorrelation for analog signals have been reviewed in detail.

Chapter 3 reviews the Probability Theory and the Random Signal Theory. These are important topics, as the signals to be communicated are normally probabilistic in nature. Chapter 4 covers topics such as Noise Signal Generation, Analysis and Measurements.

Analog communication techniques have been covered in Chapters 5 and 6, which discuss Amplitude Modulation (AM) and Angle Modulation. Angle Modulation includes Frequency Modulation (FM) and Phase Modulation (PM). Analysis of AM, FM, and PM signals, their generation and detection, their relative performance from various points of view including noise considerations have been described in detail. Many practical circuits for AM and FM generation and detection have also been incorporated.

Chapter 7 explains various Pulse Modulation Systems with practical circuits. The next chapter on Pulse Code Modulation (PCM) also discusses other systems such as Delta Modulation and Adaptive Delta Modulation. Chapter 9 is on Data Transmission and explains Amplitude Shift Keying, Frequency Shift Keying and Phase Shift Keying. Chapters 10 and 11 cover topics on Information Theory and Coding Theory respectively.

The book includes many solved and unsolved problems. Objective type questions are given at the end of the book.

Lecture notes of the authors on the subject proved to be very useful in preparing the manuscript. Also, while teaching the subject, the authors encountered the difficulties of the students. The questions asked by them helped the authors in identifying the topics that needed special and careful explanation. This proved to be of great help in writing the book, and the authors wish to thank their students for their inquisitiveness.

The authors hope that the book will fulfil the pressing need of interested readers, and would welcome suggestions to make it even more useful.

R P Singh
S D Sapre

# INTRODUCTION

Communication is the process of conveying message at a distance. If the distance involved is beyond the direct communication, then communication engineering comes into the picture. The branch of engineering which deals with communication systems is known as Telecommunication Engineering.

In telecommunication, the physical message, such as sound, words, pictures, etc., is converted into an electrical message called *signal* and this electrical signal is conveyed at the distant place, where it is reconverted into the physical message through some media. Thus, a communication system has three basic components (Fig. I.1):

Fig. I.1 Block diagram of electrical communication system

(i) transmitter, (ii) transmission media, and (iii) receiver.

A transmitter is the equipment which converts a physical message into an electrical signal. The *receiver*, on the other hand, is an equipment which converts the electrical signal back to the physical signal. The electrical signal from transmitter is conveyed to the receiver through the transmission media. Telecommunication engineering is divided into the following two categories depending on the transmission media used.

(i) line communication, and (ii) radio communication.

## LINE COMMUNICATION

In line communication, the media of transmission is a pair of conductors called transmission line. Each transmission line can normally convey only one message at a time, and is known as the line channel. The installation and maintenance of a transmission line is not only costly and complex, but also overcrowds the open space. Apart from this, its message transmission capability is also limited. Hence, in many applications, it is worthwhile to dispense with the transmission lines and use a wireless communication. A wireless message is transmitted through open space by electromagnetic waves called *Radio Waves*. This mode of communication is known as radio communication, where radiowaves are radiated from the transmitter in open space through a device called antenna. A receiving antenna intercepts the radio waves at the receiver.

## RADIO COMMUNICATION

In radio communication, signals from various sources are transmitted through a common media– the open space. This causes interference among various signals, and, as such, no useful message is received by the receiver. The problem of interference is solved by translating the message signals to different radio frequency spectra. This is done by the transmitter by a process known as *modulation*. Each radio frequency spectrum behaves as a separate *frequency channel* and thus, avoids interference. The receiver selects the desired radio frequency and produces the useful message, by another process known as *demodulation*.

The entire radio frequency ($rf$) range is classified into several bands as given below :

| Class | rf range | Wave length ($\lambda$) |
|---|---|---|
| Very Low Frequency (VLF) | 10 kHz – 30 kHz | $3 \times 10^4 - 10^4$ m |
| Low Frequency (LF) | 30 kHz – 300 kHz | $10^4 - 10^3$ m |
| Medium Frequency (MF) | 300 kHz – 3 MHz | $10^3 - 10^2$ m |
| High Frequency (HF) | 3 MHz – 30 MHz | $10^2 - 10$ m |
| Very High Frequency (VHF) | 30 MHz – 300 MHz | $10 - 1$ m |
| Ultra High Frequency (UHF) and Microwaves | above 300 MHz | below 1 m |

The VLF and LF radio frequencies are also referred to as long waves, their length being comparatively longer. On similar basis, HF frequencies are called short waves. The medium wave range used for broadcasting purposes lies between 550 kHz and 1600 kHz and is known as *broadcast band*.

The main objective of a communication system designer is to transmit messages as speedily as possible, with least probability of error. Fast communication is possible by : (i) reducing the time of each message; but this, in turn, increases the bandwidth and (ii) simultaneous transmission of several messages over a single physical channel. This process is known as *multiplexing*. An error in a received message is due to the occurrence of various types of distortions. Another cause of error is the *noise* which is mingled with the desired signal during the various communication processes.

The noise is a randomly varying signal, and is one of the most serious problems in a communication system. The noise intermingles with the desired signal, and corrupts the latter, with the result that the message received becomes erroneous.

There are many reasons for distortion in the received signal. The signal may be distorted mainly due to following reasons :
 (i) insufficient channel bandwidth,
 (ii) random variations in the channel characteristics,
 (iii) external interference, and
 (iv) noise.

It has been observed that if an effort is made to reduce the probability of error in the received message, the speed of transmission suffers, and vice-versa. A compromise, therefore, has to be made in order to get an optimum system. Communication system, as a subject, covers the study of all aspects of message transmission with particular emphasis on the following :
 (i) reliability of the system,
 (ii) accuracy (least error),
 (iii) speed of transmission,
 (iv) bandwidth of the signal and system,
 (v) complexity, and
 (vi) cost.

The first point considers reliability with an eye on the requirements and restrictions given by points (ii) to (vi). The second, third and fourth points are essentially requirements of the system and they are decided by the restrictions of the fifth and sixth points.

# CONTENTS

*Preface*   vii
*Introduction*   ix

1. **Signal Analysis**   1
   - 1.1 Fourier Series  *1*
   - 1.2 Complex Fourier Spectrum (Discrete Spectrum or Line Spectrum)  *9*
   - 1.3 The Fourier Transform  *20*
   - 1.4 Continuous Spectrum  *23*
   - 1.5 Fourier Transform Involving Impulse Function  *27*
   - 1.6 Properties of Fourier Transform  *40*
   - 1.7 Fourier Transform of Periodic Functions  *62*
   - 1.8 Convolution  *68*
   - 1.9 Sampling Theorem  *75*
   - *Problems*  *84*

2. **Linear Systems**   87
   - Introduction  *87*
   - 2.1 The System Function (Transfer Function)  *89*
   - 2.2 Distortionless Transmission  *91*
   - 2.3 Paley-Wiener Criterion  *93*
   - 2.4 Energy Signals and Power Signals  *104*
   - 2.5 Correlation  *118*
   - 2.6 Autocorrelation  *122*
   - *Problems*  *125*

3. **Probability and Random Signal Theory**   128
   - Introduction  *128*
   - 3.1 Set Theory  *128*
   - 3.2 Introduction to Probability  *132*
   - 3.3 Conditional Probability and Statistical Independence  *139*
   - 3.4 Baye's Theorem  *146*
   - 3.5 Random Variables  *150*
   - 3.6 Discrete Random Variables  *150*

- 3.7 Continuous Random Variables *154*
- 3.8 Joint Distributions *160*
- 3.9 Characteristics of Random Variables *169*
- 3.10 Binomial, Poisson and Normal Distributions *181*
- 3.11 Uniform and Other Distributions *186*
- 3.12 Random Processes *187*
- 3.13 Markov Processes *189*
  *Problems 190*

## 4. Noise 193

Introduction *193*
- 4.1 Sources of Noise *193*
- 4.2 Shot Noise *194*
- 4.3 Resistor Noise *198*
- 4.4 Calculation of Noise in Linear Systems *203*
- 4.5 Noise Bandwidth *209*
- 4.6 Available Power *211*
- 4.7 Noise Temperature *214*
- 4.8 Noise in Two-Port Networks *215*
- 4.9 Noise-Figure *219*
- 4.10 Cascaded Stages *225*
- 4.11 Measurement of Noise-figure *231*
- 4.12 Signal in Presence of Noise *233*
- 4.13 Narrow Band Noise *233*
  *Problems 237*

## 5. Amplitude Modulation Systems 240

Introduction *240*
- 5.1 Suppressed Carrier Systems (DSB-SC) *242*
- 5.2 Single Sideband Modulation (SSB) *258*
- 5.3 Vestigial Sideband Modulation (VSM) *277*
- 5.4 Amplitude Modulation with Large Carrier (AM) *284*
- 5.5 Generation of AM Waves *304*
- 5.6 Demodulation of AM Waves *312*
- 5.7 AM Transmitters and Receivers *318*
- 5.8 Noise in Amplitude Modulated Systems *330*
- 5.9 Comparison of Various AM Systems *342*
- 5.10 Frequency Division Multiplexing (FDM) *344*
  *Problems 345*

## 6. Angle Modulation Systems 348

Introduction *348*
- 6.1 Definition *348*
- 6.2 Narrowband FM *355*

6.3  Wideband FM  *361*
6.4  Some Remarks about Phase Modulation  *369*
6.5  Multiple-Tone Wideband FM (Non-linear Modulation)  *372*
6.6  FM Modulators and Transmitters  *373*
6.7  FM Demodulators and Receivers  *388*
6.8  Noise in Angle Modulated Systems  *398*
6.9  Comparison between AM and FM  *418*
6.10 Threshold Improvement in Discriminators  *419*
6.11 Comparisons between AM and FM  *427*
    *Problems  436*

## 7. Pulse Modulation Systems  439

   Introduction  *439*
7.1  Pulse Amplitude Modulation  *439*
7.2  Pulse Time Modulation  *454*
    *Problems  465*

## 8. Pulse Code Modulation  466

   Introduction  *466*
8.1  PCM System  *468*
8.2  Intersymbol Interference  *469*
8.3  Eye Patterns  *469*
8.4  Equalization  *470*
8.5  Companding  *471*
8.6  Synchronous Time Division Multiplexing  *471*
8.7  Asynchronous Time Division Multiplexing—Pulse Stuffing  *472*
8.8  Bandwidth of the PCM System  *472*
8.9  Noise in PCM System  *474*
8.10 Delta Modulation  *477*
8.11 Limitations of DM  *478*
8.12 Adaptive Delta Modulation  *479*
8.13 Comparison between PCM and DM  *480*
8.14 Delta or Differential PCM (DPCM)  *480*
8.15 S-ary System  *481*
    *Problems  481*

## 9. Data Transmission  482

   Introduction  *482*
9.1  Amplitude Shift Keying (ASK)  *482*
9.2  Frequency Shift Keying (FSK)  *482*
9.3  Phase Shift Keying (PSK)  *484*
9.4  Differential Phase Shift Keying (DPSK)  *486*
9.5  Baseband Signal Receiver  *488*
9.6  Probability of Error  *490*

- 9.7 The Optimum Filter *492*
- 9.8 The Matched Filter *495*
- 9.9 Correlator *497*
- 9.10 Probability of Error in ASK *499*
- 9.11 Probability of Error in FSK *500*
- 9.12 Probability PSK *501*
- 9.13 Probability DPSK *502*
- 9.14 Comments on $P_e$ of Different Methods *502*
  Problems *502*

## 10. Information Theory — 503

  Introduction *503*
- 10.1 Unit of Information *503*
- 10.2 Entropy *505*
- 10.3 Rate of Information *508*
- 10.4 Joint Entropy and Conditional Entropy *509*
- 10.5 Mutual Information *516*
- 10.6 Channel Capacity *523*
- 10.7 Shannon's Theorem *532*
- 10.8 Continuous Channel *533*
- 10.9 Capacity of a Gaussian Channel: Shannon-Hartley Theorem *536*
- 10.10 Bandwidth $S/N$ Trade-off *539*
  Problems *541*

## 11. Coding — 545

  Introduction *545*
- 11.1 Coding Efficiency *546*
- 11.2 Shannon-Fano Coding *549*
- 11.3 Huffman Coding *552*
- 11.4 Error-Control Coding *555*
- 11.5 Block Codes *555*
- 11.6 Convolutional Codes *566*
  Problems *571*

*Answers to Selected Unsolved Problems* — 573

*References* — 587

*Index* — 588

# ONE
# SIGNAL ANALYSIS

The most fundamental function in electrical communications is the sinusoidal signal because,
1. The response of a sine wave to a linear time invariant system is also sinusoidal, i.e., if a sinusoidal signal is applied at the input of a linear system, the output waveform is also sinusoidal with the same frequency as the input, except for a constant transmission delay and gain (or attenuation).
2. The sinusoidal analysis of electrical networks is even more simple and convenient.

(The terms signal and function are interchangeably used as there is no difference between the two. A function is defined as a set of rules which substitute one number for another. A "signal" is a function of time; and therefore, "signal" may be defined as a set of rules to assign a number $f(t)$ for every number $t$, a time domain.)

However, communication systems involve waveforms that are complex in nature, and it is desirable to resolve them in terms of sinusoidal functions. Signal analysis is a tool for achieving this aim. Signal analysis may be defined as the method of resolving a function in terms of an infinite series. The method suggested by J.B.J. Fourier for signal analysis provides the origin of the theory of *Fourier series* and *Fourier transform*.

## 1.1 FOURIER SERIES

A generalised Fourier series is the representation of a function $f(t)$ by the linear combination of elements of a *closed set* of infinite *mutually orthogonal* functions.

### Mutually Orthogonal Functions

Two real time functions are said to be "mutually orthogonal" over an interval between $t_1$ and $t_2$, if the integral of their product over this interval is zero; i.e., $f(t)$ is orthogonal to $h(t)$, if

$$\int_{t_1}^{t_2} f(t)h(t) = 0 \qquad (1.1.1)$$

Therefore a set of time functions $\{g_1(t), g_2(t), \cdots\}$ forms an orthogonal set over the interval between $t_1$ and $t_2$, if

$$\int_{t_1}^{t_2} g_j(t)g_k(t)dt = 0, (j \neq k) \qquad (1.1.2)$$

The restriction $j \neq k$ assures that the product of two *different* members of the set is integrated.

The terms orthogonality (defined as relating to right angles) of vectors is applied to their dot product operation; i.e., Two vectors are said to be orthogonal if their dot product is zero; which means that one vector has zero component in the direction of the other; and that they have nothing in common. This concept of orthogonality can be extended to signals. If a function $f_1(t)$ does not contain any component that is of the form of any other function $f_2(t)$; then, the functions are mutually orthogonal. For example, rectangular function $f(t)$ has a component 'sin($t$)', and hence they are not mutually orthogonal functions. The examples of orthogonal functions are: Legendre polynomials, Jacobi polynomials, trigonometric and exponential functions. Only the last two will be discussed here because of their significance in communication systems.

## Orthogonality in Complex Functions

So far, we have considered only real functions. If $f_1(t)$ and $f_2(t)$ are complex functions of the real variable $t$, then the two functions are orthogonal over the interval between $t_1$ and $t_2$, if

$$\int_{t_1}^{t_2} f_1(t) f_2^*(t) dt = \int_{t_1}^{t_2} f_1^*(t) f_2(t) dt = 0 \tag{1.1.3}$$

where $f_1^*(t)$ and $f_2^*(t)$ are complex conjugates of $f_1(t)$ and $f_2(t)$, respectively. The complex functions $e^{jn\omega_0 t}$ $n = 0, \pm 1, \pm 2, \cdots$ form a set of orthogonal functions.

## Representation of a Function by a Closed Set of Mutually Orthogonal Functions

A function $f(t)$ can be exactly represented by a linear combination of orthogonal functions provided they form a complete (closed) set.

A set of functions $\{g_1(t), g_2(t) \ldots g_r(t) \ldots\}$ mutually orthogonal over the interval between $t_1$ and $t_2$, is said to be a complete, or closed, set if there exists no function $m(t)$, such that

$$\int_{t_1}^{t_2} m(t) g_k(t) dt = 0; \quad \text{for } k = 1, 2 \ldots$$

If there exists a function $m(t)$ satisfying the above integral, then obviously $m(t)$ is a member of the mutually orthogonal set $\{g_r(t)\}$ and this set cannot be a complete set without $m(t)$.

## Dirichlet's Conditions

A function $f(t)$ can be represented within the interval $(t_1, t_2)$ by a complete set of orthogonal functions, and this representation is called Fourier series. There are sufficient conditions that need to be satisfied by a function $f(t)$ for its Fourier series representation within the interval $(t_1, t_2)$. These are as follows:
1. The function $f(t)$ is a single-valued function of the variable $t$ within the interval $(t_1, t_2)$.
2. The function $f(t)$ has a finite number of discontinuities in the interval $(t_1, t_2)$.
3. The function $f(t)$ has a finite number of maxima and minima in the interval $(t_1, t_2)$.
4. The function $f(t)$ is absolutely integrable, i.e.,

$$\int_{t_1}^{t_2} |f(t)| dt < \infty$$

These conditions are known as Dirichlet's conditions.

## Trigonometric Fourier Series

A set of harmonically related sine and cosine functions, i.e., $\cos n\omega_0 t$ and $\sin n\omega_0 t$ ($n = 0, 1, 2, ...$) forms a complete orthogonal set over the interval $t_0 \le t \le t_0 + \frac{2\pi}{\omega_0}$ (Prob. 1.1). Therefore, a function $f(t)$ can be represented by a Fourier series comprising the following sine and cosine functions:

$$\begin{aligned} f(t) &= a_0 + a_1 \cos \omega_0 t + a_2 \cos 2\omega_0 t + ... + a_n \cos n\omega_0 t + ... \\ &\quad + b_1 \sin \omega_0 t + b_2 \sin 2\omega_0 t + ... b_n \sin n\omega_0 t + ... \\ &= a_0 + \sum_{n=1}^{\infty} (a_n \cos n\omega_0 t + b_n \sin n\omega_0 t), \ (t_0 \le t \le t_0 + T) \end{aligned} \quad (1.1.4)$$

Where
$$T = \frac{2\pi}{\omega_0}$$

Equation 1.1.4 is the *trigonometric Fourier series* representation of $f(t)$ over an interval $(t_0, t_0 + T)$. The constant $a_0$ corresponds to $n = 0$, and is given by

$$a_0 = \frac{1}{T} \int_{t_0}^{t_0+T} f(t) dt \quad (1.1.5a)$$

It is evident that $a_0$ is the average value; i.e., the d.c. component of $f(t)$ over the interval $(t_0, t_0 + T)$. Also that constant $b_0 = 0$, , as $\sin n\omega_0 t$ is zero for $n = 0$.

The other constants corresponding to $a_n$ and $b_n$ are given by

$$a_n = \frac{2}{T} \int_{t_0}^{t_0+T} f(t) \cos n\omega_0 t \, dt \quad (1.1.5b)$$

$$b_n = \frac{2}{T} \int_{t_0}^{t_0+T} f(t) \sin n\omega_0 t \, dt \quad (1.1.5c)$$

The trigonometric series may also be represented in a simple form as given below. This is obtained by an easy trigonometric manipulation of Eq. (1.1.4)

$$f(t) = c_0 + \sum_{n=1}^{\infty} c_n \cos(n\omega_0 t - \phi_n) \quad (1.1.6)$$

Where $c_0, c_n$ and $\phi_n$ are related to $a_0, a_n$ and $b_n$ by the equations.

$$c_0 = a_0 \quad (1.1.7a)$$

$$c_n = \sqrt{a_n^2 + b_n^2} \quad (1.1.7b)$$

and
$$\phi_n = \tan^{-1}\left(\frac{b_n}{a_n}\right) \quad (1.1.7c)$$

The coefficients $c_n$ are called *spectral amplitudes* i.e. $c_n$ is the amplitude of the spectral component $c_n \cos(n\omega_0 t - \phi_n)$ having a frequency $nf_0$ whereas $\phi_n$ specifies the phase information of the spectral component $nf_0$.

## Complex Exponential Fourier Series

As the exponential form of fourier series is simpler and more compact, it has extensive application in communication theory. It is apparent that a set of complex exponential functions $\{e^{jn\omega_0 t}\}$ ($n = 0, \pm 1, \pm 2, ..$) forms a closed orthogonal set over an interval $\left(t_0, t_0 + \frac{2\pi}{\omega_0}\right)$ for any value of $t_0$ (Prob. 1.2) and therefore, it can be used to represent a Fourier series. An arbitrary function $f(t)$ can be represented by a complex exponential series over an interval $\left(t_0, t_0 + \frac{2\pi}{\omega_0}\right)$ as given below:

$$\begin{aligned}f(t) &= F_0 + F_1 e^{j\omega_0 t} + F_2 e^{j2\omega_0 t} + ... + F_n e^{jn\omega_0 t} + ... \\ &\quad + F_{-1} e^{-j\omega_0 t} + F_{-2} e^{-j2\omega_0 t} + ... + F_{-n} e^{-jn\omega_0 t} + ... \\ &= \sum_{n=-\infty}^{\infty} F_n e^{jn\omega_0 t} \quad (t_0 \leq t \leq t_0 + T)\end{aligned} \quad (1.1.8)$$

where $T = \frac{2\pi}{\omega_0}$, and the summation is for the integral values of $n$ from $-\infty$ to $\infty$, including zero. The coefficients $F_n$ are given by

$$F_n = \frac{1}{T}\int_{t_0}^{t_0+T} F(t) e^{-jn\omega_0 t} \, dt \quad (1.1.9)$$

The trigonometric series and the complex exponential series are two different ways of representing the same series; and one series can be derived from the other, using interrelations between trigonometric and complex exponential functions. The complex function $e^{jn\omega_0 t}$ can be seen as a vector of unit length and angle, $n\omega_0 t$, in the complex two-dimensional plane; i.e. $e^{jn\omega_0 t} = \cos(n\omega_0 t) + j\sin(n\omega_0 t)$. Similarly, the complex exponential $e^{-jn\omega_0 t}$ can be viewed as a vector of unit length and angle $-n\omega_0 t$ i.e.

$$e^{-jn\omega_0 t} = \cos n\omega_0 t - j\sin n\omega_0 t$$

The functions $e^{jn\omega_0 t}$ and $e^{-jn\omega_0 t}$ may be taken as two vectors rotating in opposite directions, and when added they yield a *real* function of time (sine or cosine function, Thus

$$e^{jn\omega_0 t} + e^{-jn\omega_0 t} = 2\cos(n\omega_0 t)$$

and

$$e^{jn\omega_0 t} - e^{-jn\omega_0 t} = 2j\sin(n\omega_0 t)$$

In other words, the functions $e^{jn\omega_0 t}$ and $e^{-jn\omega_0 t}$ are the signals of frequencies $n\omega_0$ and $-n\omega_0$, respectively; and they combine to yield a real function of time. Here $-n\omega_0$ represents *negative frequencies* that are actually fictitious.

## Concept of Negative Frequency

The exponential form of Fourier series represents an arbitrary function $f(t)$ over the interval $(t_0, t_0 + T)$ by a linear combination of complex exponential functions of frequencies, $0, \pm \omega_0, \pm 2\omega_0, \pm 3\omega_0 \cdots$. The presence of negative frequencies is simply because the mathematical model of a signal requires the use of negative frequencies. In fact, a real function $f(t)$ cannot be represented by the complex exponential phasors $e^{jn\omega_0 t}$ without associating them with their opposite rotating counterpart $e^{-jn\omega_0 t}$. The signal $e^{-jn\omega_0 t}$ of frequency $-n\omega_0 t$ (negative frequency) represents the counterpart of the signal $e^{jn\omega_0 t}$ (positive frequency), and when combined they yield a real function of time. Indeed, the complex-valued basic functions, (viz. $e^{jn\omega_0 t}$ and $e^{-jn\omega_0 t}$) have no physical meaning separately, but yield real functions ($\cos n\omega_0 t$ or $\sin n\omega_0 t$) when combined together i.e., negative frequency signals are not physical signals, rather they are only a mathematical concept required to give a compact mathematical description of a *real signal* by a combination of complex exponentials of positive and negative frequencies.

## Derivation of Trigonometric Fourier Series from Complex Exponential Series

The preceding discussions show that both the series are not different. Indeed, one series can be derived from the other. Let us consider the following exponential series:

$$f(t) = F_0 + F_1 e^{j\omega_0 t} + F_2 e^{j2\omega_0 t} + \cdots + F_n e^{jn\omega_0 t} + \cdots$$
$$+ F_{-1} e^{-j\omega_0 t} + F_{-2} e^{-j2\omega_0 t} + \cdots + F_{-n} e^{-jn\omega_0 t} + \cdots$$

Replacing the exponential functions by their trigonometric identities, the series becomes

$$f(t) = F_0 + F_1(\cos\omega_0 t + j\sin\omega_0 t) + \cdots + F_n(\cos n\omega_0 t + j\sin n\omega_0 t) + \cdots$$
$$+ F_{-1}(\cos\omega_0 - j\sin n\omega_0 t) + \cdots + F_{-n}(\cos n\omega_0 t - j\sin n\omega_0 t) + \cdots$$

Rearranging the right-hand side, we have

$$f(t) = F_0 + (F_1 + F_{-1})\cos\omega_0 t + \cdots + (F_n + F_{-n})\cos n\omega_0 t + \cdots$$
$$+ j(F_1 - F_{-1})\sin\omega_0 t + \cdots + j(F_n - F_{-n})\sin n\omega_0 t + \cdots$$

which is a trigonometric series representation of $f(t)$ derived from exponential series. Comparing this derived trigonometric series with actual form of trigonometric series (Eq. 1.1.4). The following identities between coefficients of exponential and trigonometric series can be achieved

$$a_0 = F_0 \tag{1.1.10a}$$
$$a_n = F_n + F_{-n} \tag{1.1.10b}$$
$$b_n = j(F_n - F_{-n}) \tag{1.1.10c}$$

Similarly, an exponential series can be derived from the trigonometric series by replacing the sine and cosine functions with their exponential equivalents (Prob. 1.3).

The Eqs 1.1.10 a,b and c may be written as

$$F_0 = a_0 \tag{1.1.11a}$$

$$F_n = \frac{1}{2}(a_n - jb_n) \tag{1.1.11b}$$

$$F_{-n} = \frac{1}{2}(a_n + jb_n) \tag{1.1.11c}$$

which represent the coefficients of exponential series in terms of coefficients of trigonometric series. The trigonometric series has no negative frequencies as against the exponential series which contains negative frequencies.

## Fourier Series Representation of a Periodic Function over the Entire Interval ($-\infty < t < \infty$)

So far, we have considered the Fourier series representation of an arbitrary function over a finite interval between $t_0$ and $(t_0 + T)$. The function and its Fourier series may not be equal outside this interval. A periodic function which repeats after every $T$ seconds is expected to have the same Fourier series for the entire interval $(-\infty, \infty)$. A periodic function has an identical Fourier series for the entire interval $(-\infty, \infty)$ as for the interval $(t_0, t_0 + T)$ *since* the same function repeats after every $T$ seconds. Thus for a periodic function $f(t)$ we have,

$$f(t) = \sum_{n=-\infty}^{\infty} F_n e^{jn\omega_0 t}, \quad (-\infty < t < \infty) \tag{1.1.12}$$

$$F_n = \frac{1}{T} \int_{t_0}^{t_0 + T} f(t) e^{-jn\omega_0 t} \, dt, \quad \omega_0 = \frac{2\pi}{T} \tag{1.1.13}$$

The choice of $t_0$ is immaterial for a periodic function; however, it is significant for the series expansion of a function in a finite interval; and such a series expansion is therefore not unique; whereas, the Fourier series expansion of a periodic function is unique, irrespective of the location of $t_0$ in the waveform.

### Example 1.1.1

Expand a function $f(t)$ shown in Fig. 1.1.1 by trigonometric Fourier series over the interval (0, 1).

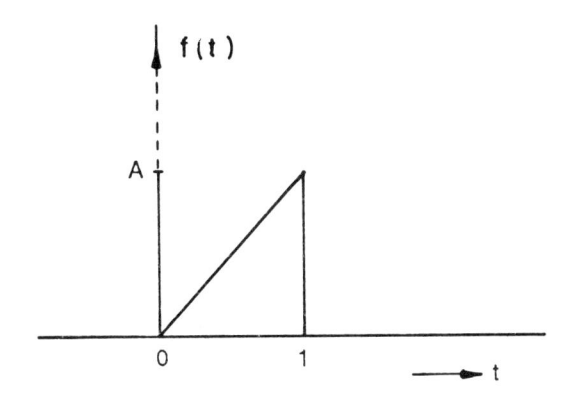

Fig. 1.1.1  A Triangular Pulse

*Solution*

It is obvious from the Fig. 1.1.1 that $f(t) = At, (0 \leq t \leq 1)$

Take $t_0$ at zero, then

$$\omega_0 = \frac{2\pi}{T} = 2\pi \quad \text{(as } T = 1\text{)}$$

The Fourier series is given by

$$f(t) = a_0 + \sum_{n=1}^{\infty} (a_n \cos 2\pi nt + b_n \sin 2\pi nt)$$

$$= a_0 + a_1 \cos 2\pi t + a_2 \cos 4\pi t + a_3 \cos 6\pi t + \cdots$$
$$+ b_1 \sin 2\pi t + b_2 \sin 4\pi t + b_3 \sin 6\pi t + \cdots$$

where the coefficients are evaluated as follows:

$$a_0 = \frac{1}{T} \int_0^T f(t) dt = 1 \int_0^1 At \, dt = \frac{A}{2}$$

$$a_n = \frac{2}{T} \int_0^1 At \cos 2\pi nt \, dt$$

$$= \frac{A}{2\pi^2 n^2} [\cos 2\pi nt + 2\pi nt \sin 2\pi nt]_0^1 = 0$$

$$b_n = \frac{2}{T} \int_0^1 At \sin 2\pi nt \, dt$$

$$= \frac{A}{2\pi^2 n^2} [\sin 2\pi nt - 2\pi nt \cos 2\pi nt]_0^1 = -\frac{A}{\pi n}$$

# 8 COMMUNICATION SYSTEMS: Analog and Digital

Therefore, the trigonometric series will be as follows :

$$f(t) = \frac{A}{2} - \frac{A}{\pi}\sin 2\pi t - \frac{A}{2\pi}\sin 4\pi t - \frac{A}{3\pi}\sin 6\pi t \cdots$$

$$= \frac{A}{2} - \frac{A}{\pi}\sum_{n=1}^{\infty}\frac{\sin 2\pi n t}{n} \quad (0 < t < 1) \tag{1.1.14}$$

Hence, the function $f(t)$ has a d.c. component $A/2$, and sinusoidal components of frequencies $2\pi, 4\pi, \cdots$ with magnitudes of $-\frac{A}{\pi}, -\frac{A}{2\pi}, \cdots$ respectively.

We should note the following points:

1. The series expansion given by Eq. 1.1.14 is valid for any other function identical to $f(t)$ over the interval (0,1), irrespective of its waveform outside this interval.
2. The Fourier series representation for a function $f(t)$ by Eq.1.1.14 is not restricted to the interval (0,1), provided the function repeats itself after every 1 second. Thus, the periodic waveform $f_p(t)$ in Fig. 1.1.2 will have the same Fourier series representation as that of $f(t)$ given by Eq.1.1.14, and it is valid for the entire interval $(-\infty, \infty)$. The Fourier series will be the same irrespective of the choice of $t_0$ at zero, or elsewhere. Therefore, Fourier series representation is unique for periodic function $f_p(t)$.

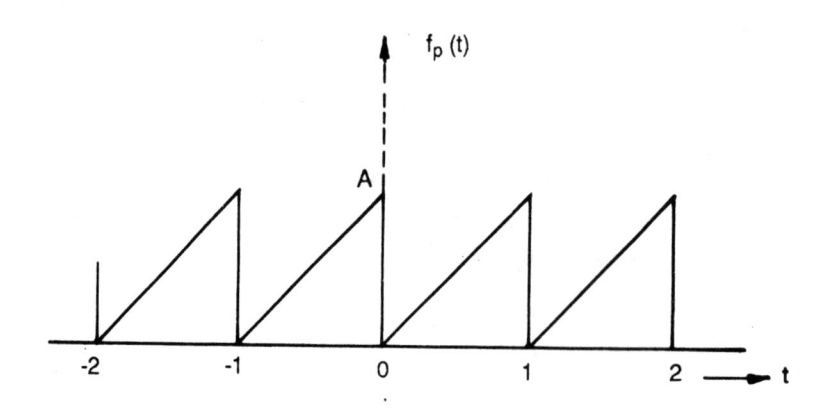

Fig. 1.1.2  A Periodic Triangular Wave

### Example 1.1.2

Derive the exponential Fourier series from the trigonometric Fourier series for the waveform shown earlier in Fig.1.1.1.

**Solution**

The coefficients of trigonometric series for the waveform shown in Fig.1.1.1 are listed below which are obtained by Eq.1.1.14

$$a_0 = \frac{A}{2}; a_n = 0, b_n = -\frac{A}{\pi n}$$

The coefficients of corresponding exponential series may be evaluated by using Eq.1.1.11

$$F_0 = a_0 = \frac{A}{2}; \; F_n = \frac{1}{2}(a_n - jb_n) = \frac{jA}{2\pi n}$$

and
$$F_{-n} = \frac{1}{2}(a_n + jb_n) = -\frac{jA}{2\pi n}$$

Thus, the exponential Fourier series is given by

$$f(t) = \frac{A}{2} + \frac{jA}{2\pi}e^{j2\pi t} + \frac{jA}{4\pi}e^{j4\pi t} + \cdots$$
$$- \frac{jA}{2\pi}e^{-j2\pi t} - \frac{jA}{4\pi}e^{-j4\pi t} \cdots$$

$$= \frac{A}{2} + \frac{jA}{2\pi} \sum_{n=-\infty}^{\infty} \frac{1}{n} e^{j2\pi nt} \qquad (1.1.15)$$

## 1.2 COMPLEX FOURIER SPECTRUM (DISCRETE SPECTRUM OR LINE SPECTRUM)

The complex Fourier Series representation of a time function $f(t)$ is equivalent to resolving the function in terms of harmonically related frequency components of a fundamental frequency $f_0\left(=\frac{1}{T}\right)$. A complex weighting factor $F_n$ (for exponential series), or $C_n$ (for trigonometric series) is assigned to each frequency component. The weighting factor $F_n$, or $C_n$ is called spectral amplitude and represents the amplitude of $n^{th}$ harmonic. Graphical representation of spectral amplitude alongwith the spectral phase is called complex frequency spectrum. An amplitude spectrum without the phase information does not specify the waveform; because in general the weighting factor $F_n$ is a complex number. Therefore such spectrum is called a complex frequency spectrum. However, if $F_n$ is purely real or purely imaginary, we can disregard the phase spectrum. In a typical amplitude spectrum shown in Fig.1.2 a vertical line has been drawn at each harmonic frequency, choosing a suitable scale so that the height of a line represents the amplitude of the corresponding harmonic frequency. This spectrum called a *Discrete spectrum* exist only at discrete frequencies that are harmonically related.

Figure 1.2 (a) represents the spectrum of a trigonometric series extending from 0 to $\infty$, producing a one-sided spectrum because negative frequencies do not exist in this type of series whereas, Fig.1.2 (b) represents a two-sided spectrum corresponding to an exponential series having negative frequencies. The phase spectrum is not shown.

The exponential series is more popular; and as such, this series will be used in most of the discussions to follow. It is obvious from Fig.1.2 (b) that the amplitude spectrum corresponding to the exponential series is symmetrical about vertical axis. This is true of all real periodic functions and is proved as follows:

$$F_n = \frac{1}{T}\int_{-T/2}^{T/2} f(t)\, e^{jn\omega_0 t}\, dt$$

Placing $-n$ for $n$ in the above equation,

$$F_{-n} = \frac{1}{T}\int_{-T/2}^{T/2} f(t)\, e^{-jn\omega_0 t}\, dt$$

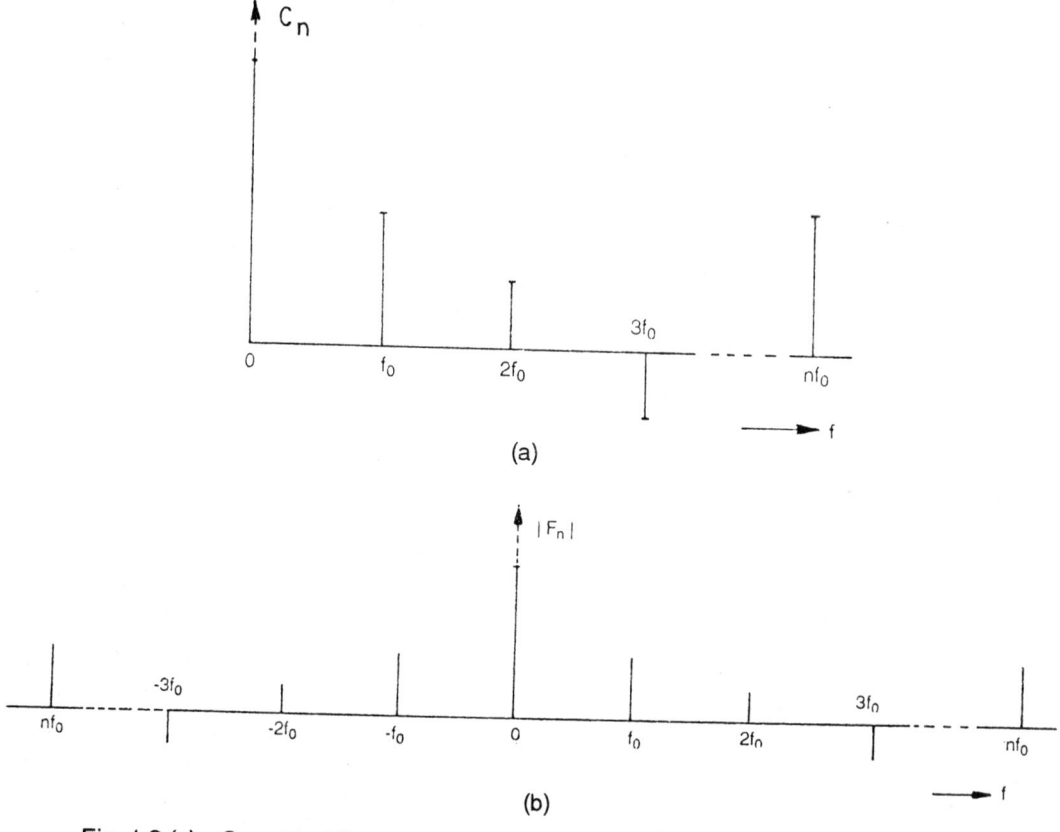

Fig. 1.2 (a) One-sided Spectrum of Trigonometric Series  (b) Two-sided Spectrum of Corresponding Exponential Series

According to the above equations $F_n$ and $F_{-n}$ are complex conjugate of each other:
$$F_{-n} = F_n{}^*$$
If $F_n$ is a general complex number, then
$$F_n = |F_n|e^{j\theta n}$$
and
$$F_{-n} = |F_n|e^{-j\theta n}$$
so
$$|F_n| = |F_{-n}|$$

It is obvious that the magnitude spectrum is symmetrical about the vertical axis passing through the origin; and thus it is, an even function of $\omega$. It is called the *even symmetry* of magnitude spectrum.

Also $\theta_n$ is the phase of $F_n$, and $-\theta_n$ is the phase of $F_{-n}$ which means that the phase spectrum is antisymmetrical about the vertical axis. It is called *odd symmetry* of phase spectrum. Accordingly, for a real valued periodic function, the magnitude spectrum is symmetrical (an even function of $\omega_n$) and the phase spectrum is antisymmetrical (an odd function of $\omega_n$) about the vertical axis passing through the origin.

## Time Domain and Frequency Domain Representation of a Signal

A signal $f(t)$ can be represented in terms of relative amplitudes of various frequency components present in signal. This is possible by using exponential Fourier series. This is a frequency domain representation of the signal. The time domain representation specifies a signal at each instant of time. This means that a signal $f(t)$ can be specified in two equivalent ways:
 (i) Time domain representation; where $f(t)$ is represented as a function of time. Graphical time domain representation is termed as *wave-form*.
 ii) The frequency domain representation; where the signal is defined in terms of its spectrum.

Any of the above two representations uniquely specifies the function, i.e., if the signal $f(t)$ is specified in time domain, we can determine its spectrum. Conversely, if the spectrum is specified, we can determine the corresponding time domain of a signal. In order to determine the function, it is necessary that both amplitude spectrum and phase spectrum are specified. However, in many cases, the spectrum is either real or imaginary, as such, only an amplitude plot is enough as all frequency components have identical phase relation.

### Example 1.2.1

 (i) Evaluate the trigonometric Fourier series expansion of a full wave rectified cosine function shown in Fig. 1.2.1.

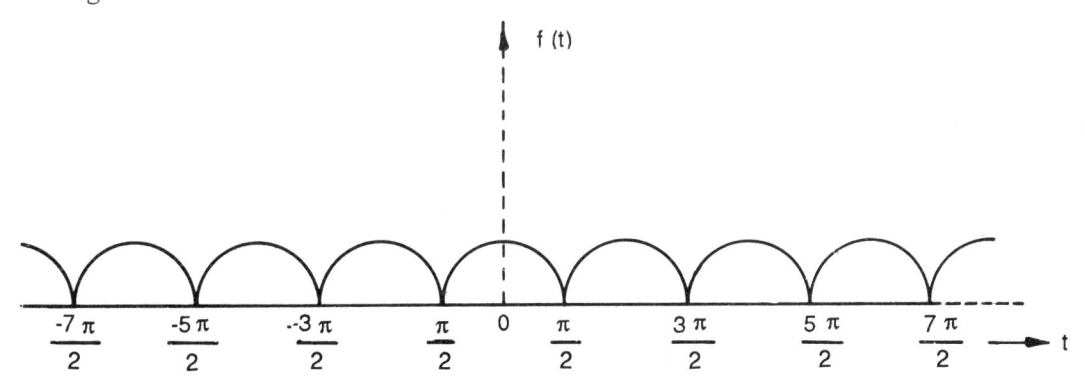

Fig. 1.2.1 A Rectified cosine Wave

 (ii) Derive the corresponding exponential Fourier series.
 (iii) Draw the complex Fourier spectrum.

*Solution*
 (i) The trigonometric Fourier series:

It is obvious from the waveform that the time period $T = \pi$, and $\omega_0 = \dfrac{2\pi}{T} = 2$. The series is, therefore, of the form

$$f(t) = a_0 + \sum_{n=1}^{\infty} \left[ a_n \cos 2nt + b_n \sin 2nt \right] \quad (1.2.1)$$

$$a_0 = \frac{1}{T} \int_{-\frac{T}{2}}^{\frac{T}{2}} f(t)\,dt = \frac{1}{\pi} \int_{-\frac{\pi}{2}}^{\frac{\pi}{2}} \cos t\, dt = \frac{2}{\pi}$$

and

$$a_n = \frac{2}{T} \int_{-\frac{T}{2}}^{\frac{T}{2}} f(t)\cos 2nt\, dt = \frac{2}{\pi} \int_{-\frac{\pi}{2}}^{\frac{\pi}{2}} \cos t \cos 2nt\, dt$$

$$= \frac{1}{\pi} \int_{-\frac{\pi}{2}}^{\frac{\pi}{2}} [\cos(2n-1)t + \cos(2n+1)t]\, dt$$

$$= \frac{1}{\pi}\left[ \frac{\sin(2n-1)t}{2n-1}\bigg|_{-\frac{\pi}{2}}^{\frac{\pi}{2}} + \frac{\sin(2n+1)t}{2n+1}\bigg|_{-\frac{\pi}{2}}^{\frac{\pi}{2}} \right]$$

since

$$\sin\left\{(2n-1)\frac{\pi}{2}\right\} = (-1)^{n+1}$$

$$\sin\left\{(2n+1)\frac{\pi}{2}\right\} = (-1)^n$$

Therefore,

$$a_n = \frac{2}{\pi}\left[ \frac{(-1)^{n+1}}{2n-1} + \frac{(-1)^n}{2n+1} \right] \qquad (1.2.2)$$

Similarly,

$$b_n = \frac{2}{T} \int_{-\frac{T}{2}}^{\frac{T}{2}} f(t)\sin 2nt\, dt$$

Since $f(t) \sin 2nt$ is an odd function, the integral for $b_n$ will yield zero, i.e.

$$b_n = 0$$

So the series is given by

$$f(t) = \frac{2}{\pi} + \sum_{n=1}^{\infty}\left\{ \frac{2}{\pi}\left[ \frac{(-1)^{n+1}}{2n-1} + \frac{(-1)^n}{2n+1} \right] \cos 2nt \right\} \qquad (1.2.3)$$

The first few terms of the series may be given as

$$f(t) = \frac{2}{\pi} + \left[1 + \frac{2}{3}\cos 2t - \frac{2}{15}\cos 4t + \frac{2}{35}\cos 6t \cdots \right]$$

(ii) Derivation of exponential series:
The corresponding exponential series is given by

$$f(t) = \sum_{n=-\infty}^{\infty} F_n e^{jn\omega_0 t} = \sum_{n=-\infty}^{\infty} F_n e^{j2nt} \qquad (1.2.4)$$

The coefficient $F_n$ is given as (using Eq. 1.1.11)

$$F_n = \frac{1}{2}(a_n - jb_n)$$

Therefore,

$$F_n = \frac{1}{2}\left\{\frac{2}{\pi}\left[\frac{(-1)^{n+1}}{2n-1} + \frac{(-1)^n}{2n+1}\right]\right\}$$

$$= \frac{1}{\pi}\left[\frac{(-1)^{n+1}}{2n-1} + \frac{(-1)^n}{2n+1}\right] \qquad (1.2.5)$$

The first few terms of $F_n$ are:

$$F_0 = \frac{2}{\pi} \qquad \text{(for } n = 0\text{)}$$

$$F_1 = F_{-1} = \frac{1}{2}a_1 = \frac{2}{3\pi} \qquad \text{(for } n = \pm 1\text{)}$$

$$F_2 = F_{-2} = = \frac{1}{2}a_2 = \frac{-2}{15\pi} \qquad \text{(for } n = \pm 2\text{)}$$

Thus, the series is of the form

$$f(t) = \frac{2}{\pi} + \frac{2}{3\pi}e^{j2t} - \frac{2}{15\pi}e^{j4t} + \cdots$$

$$+ \frac{2}{3\pi}e^{-j2t} - \frac{2}{15\pi}e^{-j4t} + \cdots$$

(iii) The complex Fourier spectrum is just a sketch of $F_n V_s n\omega_0$ as shown in Fig.1.2.2.
Since $F_n$ is real, only amplitude plot is required. The spectrum is symmetrical about the vertical axis passing through the origin, as in every periodic *real* function. The negative amplitude signifies the phase opposition of the corresponding harmonic.

### Example 1.2.2

For the periodic *Gate function* shown in Fig.1.2.3;
(a) Find the trigonometric Fourier series.
(b) Derive the corresponding exponential Fourier series.
(c) Find the exponential series directly; and show that the result is in agreement with (b).

**14** COMMUNICATION SYSTEMS: Analog and Digital

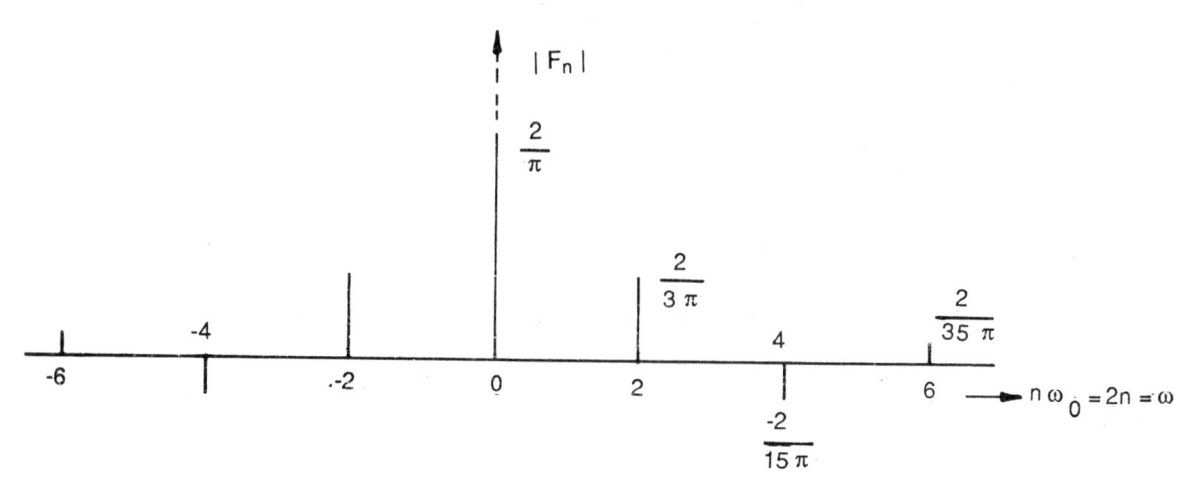

Fig. 1.2.2  Spectrum of Fig. 1.2.1

(d) Draw the spectrum for (c).

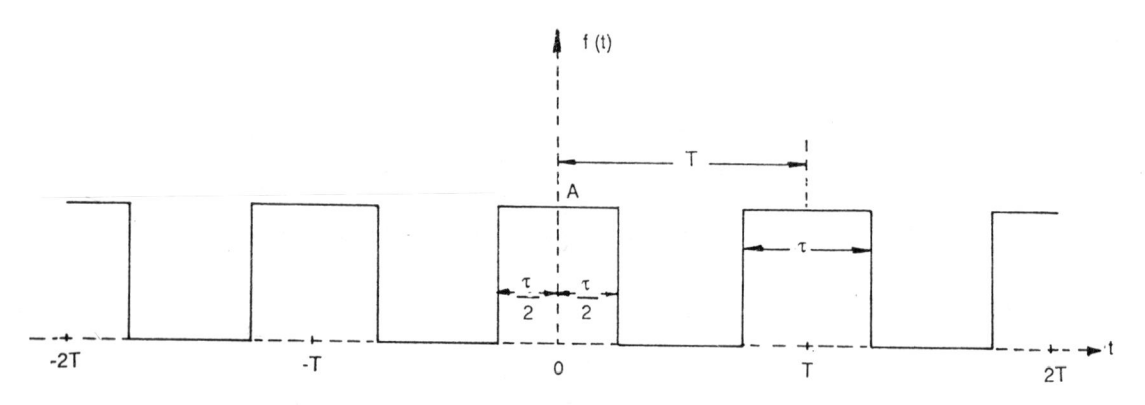

Fig. 1.2.3  A Periodic Gate Function

*Solution*

The periodic Gate function appears like a periodic pulse with pulsewidth $\tau$ and repetition time $T$. Clearly, from its shape, a Gate function can be described over one period as follows:

$$f(t) = \begin{cases} a & \dfrac{-\tau}{2} < t < \dfrac{\tau}{2} \\ 0 & \dfrac{\tau}{2} < t < T - \dfrac{\tau}{2} \end{cases}$$

For convenience, let us choose the origin to coincide with the center of a pulse.
  (a) Trigonometric Fourier series
  The coefficients of the series are found as follows:

$$a_0 = \frac{1}{T}\int_{-\tau/2}^{\tau/2} f(t)dt = \frac{A\tau}{T} \tag{1.2.6a}$$

$$a_n = \frac{2}{T}\int_{-\tau/2}^{\tau/2} f(t)\cos n\omega_0 t\, dt; \quad \left(\omega_0 = \frac{2\pi}{T}\right)$$

$$= \frac{2}{T}\int_{-\tau/2}^{\tau/2} A\cos n\omega_0 t\, dt = \frac{2A\tau}{T}\frac{\sin\left(n\omega_0\frac{\tau}{2}\right)}{\left(n\omega_0\frac{\tau}{2}\right)} \tag{1.2.6b}$$

$$b_n = 0 \tag{1.2.6c}$$

Therefore, the series is given by

$$f(t) = \frac{A\tau}{T} + \frac{2A\tau}{T}\sum_{n=1}^{\infty}\frac{\sin\left(n\omega_0\frac{\tau}{2}\right)}{\left(n\omega_0\frac{\tau}{2}\right)}\cos n\omega_0 t \tag{1.2.6d}$$

(b) Derivation of exponential series from the trigonometric series:
We know the relation,

$$F_n = \frac{1}{2}(a_n - jbn)$$

Hence, the coefficient of corresponding exponential series is

$$F_n = \frac{A\tau}{T}\frac{\sin\left(n\omega_0\frac{\tau}{2}\right)}{\left(n\omega_0\frac{\tau}{2}\right)}$$

The exponential Fourier series is given by

$$f(t) = \frac{A\tau}{T}\sum_{n=-\infty}^{\infty}\frac{\sin\left(n\omega_0\frac{\tau}{2}\right)}{\left(n\omega_0\frac{\tau}{2}\right)}e^{jn\omega_0 t} \tag{1.2.7}$$

(c) Direct determination of exponential Fourier series:
The exponential Fourier series is defined as

$$f(t) = \sum_{n=-\infty}^{\infty} F_n e^{jn\omega_0 t}$$

We can directly determine the coefficient $F_n$ by using the integral

$$F_n = \frac{1}{T}\int_{-\tau/2}^{\tau/2} f(t)e^{-jn\omega_0 t}\, dt = \frac{1}{T}\int f(t)e^{-jn\omega_0 t}\, dt = \frac{1}{T}\int_{-\tau/2}^{\tau/2} A e^{-jn\omega_0 t}\, dt$$

# 16 COMMUNICATION SYSTEMS: Analog and Digital

$$= -\frac{A}{jn\omega_0 T}e^{-jn\omega_0 t}\Big|_{-\frac{\tau}{2}}^{\frac{\tau}{2}} = \frac{2A}{n\omega_0 T}\frac{\left(e^{jn\omega_0\tau/2} - e^{-jn\omega_0\tau/2}\right)}{2j}$$

$$= \frac{2A}{n\omega_0 T}\sin\left(n\omega_0\frac{\tau}{2}\right)$$

which may be written as

$$F_n = \frac{A\tau}{T}\frac{\sin(n\omega_0\tau/2)}{n\omega_0\tau/2} \tag{1.2.8}$$

and the series is given by

$$f(t) = \frac{A\tau}{T}\sum_{n=-\infty}^{\infty}\frac{\sin(n\omega_0\tau/2)}{n\omega_0\tau/2}e^{jn\omega_0 t} \tag{1.2.9}$$

Equations 1.2.9 and 1.2.7 are identical; hence the directly determined series is in agreement with the one derived from trigonometric series.

(d) Spectrum:

The spectrum is the plot of $F_n$ as function of $n\omega_0$. Since $F_n$ is real, only amplitude plot is required. Let us rewrite the expression for $F_n$ given by Eq.1.2.8.

$$F_n = \frac{A\tau}{T}\frac{\sin(n\omega_0\tau/2)}{n\omega_0\tau/2} \tag{1.2.10a}$$

If we define a normalized and dimensionless variable $x = \frac{n\omega_0\tau}{2}$,

then

$$F_n = \frac{A\tau}{T}\left[\frac{\sin x}{x}\right] \tag{1.2.10b}$$

The function $\frac{\sin x}{x}$, known as *sampling function* plays an important role in communication theory. We will further study this function in detail as this will appear in many of the problems to be discussed.

## The Sampling Function and the Sinc Function

The sampling function is defined by

$$Sa(x) = \frac{\sin x}{x} \tag{1.2.11}$$

Closely related to the sampling function is the sinc function defined by

$$\operatorname{Sinc} x = \frac{\sin \pi x}{\pi x} \tag{1.2.12}$$

The fourier series of periodic gate function as given in Eq.1.2.9 may be expressed in terms of sampling function as follows:

$$f(t) = \frac{A\tau}{T}\sum_{n=-\infty}^{\infty} Sa\left(\frac{n\omega_0\tau}{2}\right)e^{jn\omega_0 t} \tag{1.2.13}$$

Figure 1.2.4 shows a sampling function. The function has its maximum value of unity at $x = 0$, and it approaches zero as $x$ approaches infinity. The function is oscillatory in nature with period of $2\pi$. It has a decaying amplitude in either direction of $x$, and has zeros at $x = \pm\pi, \pm 2\pi, \pm 3\pi, \cdots$
We may rewrite Eq.1.2.10 in terms of sampling function as follows

$$F_n = \frac{A\tau}{T} Sa\left(n\omega_0 \frac{\tau}{2}\right)$$

Taking $\omega_0 = \frac{2\pi}{T}$, we get

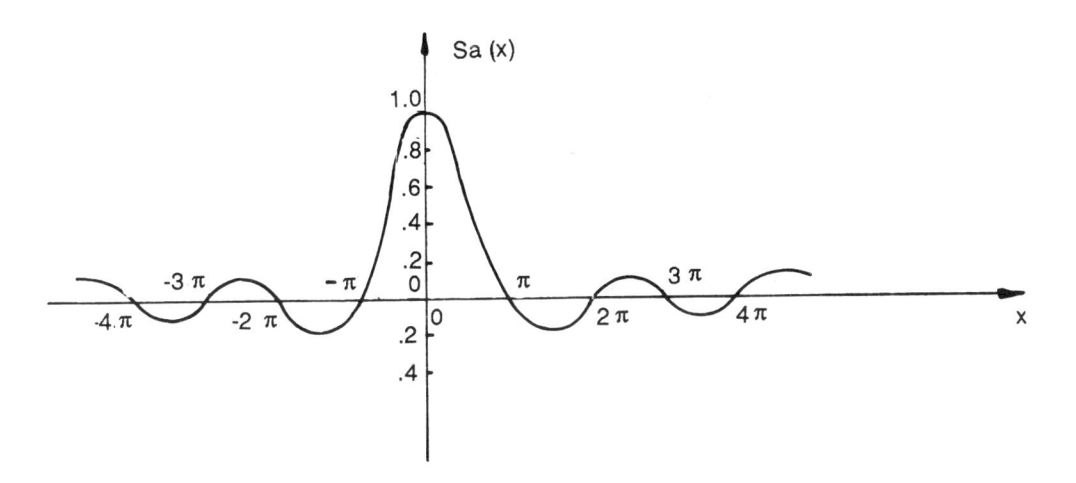

Fig 1.2.4 The Sampling Function

$$F_n = \frac{A\tau}{T} Sa\left(\frac{n\pi\tau}{T}\right) \tag{1.2.14}$$

Note that $|F_n| = |F_{-n}|$ because $Sa(x)$ is an even function of $x$.

The amplitude plot of $F_n$ is a discrete spectrum existing at $\omega = 0, \pm\omega_0, \pm 2\omega_0, \pm 3\omega_0, \cdots$ having amplitudes $\frac{A\tau}{T}, \frac{A\tau}{T} Sa\left(\frac{\pi\tau}{T}\right), \frac{A\tau}{T} Sa\left(\frac{2\pi\tau}{T}\right), \cdots$ etc. respectively. $\frac{\tau}{T}$ is called the duty cycles of periodic waveform.

The tendencies of both amplitude and phase spectrum are shown in Fig. 1.2.5. The phase spectrum here is shown for a fuller understanding, although it is not necessary. The phase spectrum obviously lies between 0 and $-180°$ for positive frequencies. By antisymmetry, the spectrum lies between 0 and $+180°$ for negative frequencies.

The envelope of the amplitude plot is similar to the sampling function curve shown in Fig.1.2.4. The spacing between successive lines is $\omega_0 = \frac{2\pi}{T}$. Zero crossings occur in the envelope of the amplitude plot when

18 COMMUNICATION SYSTEMS: Analog and Digital

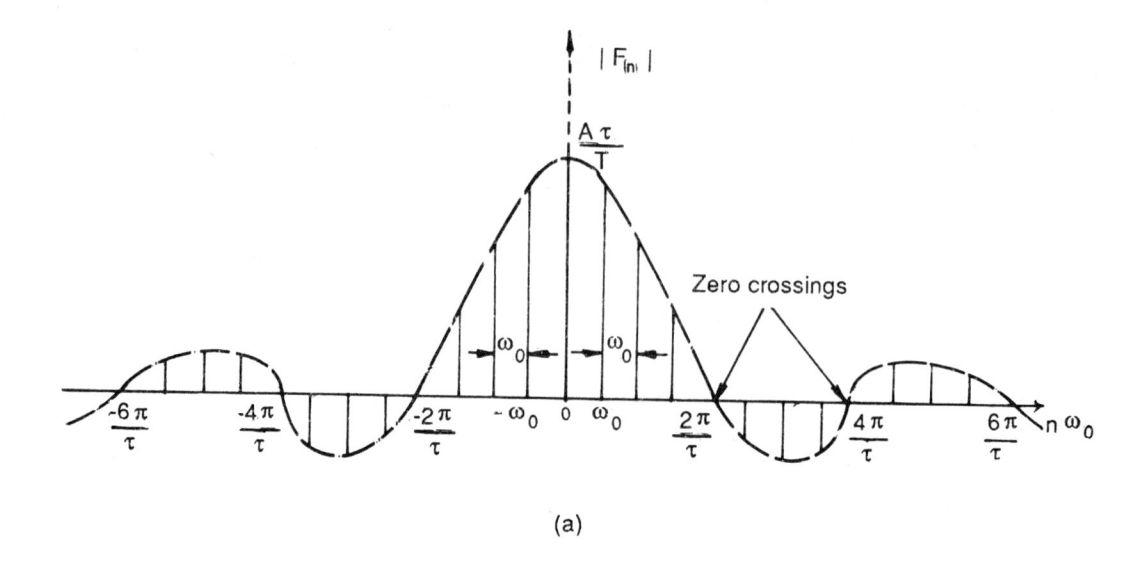

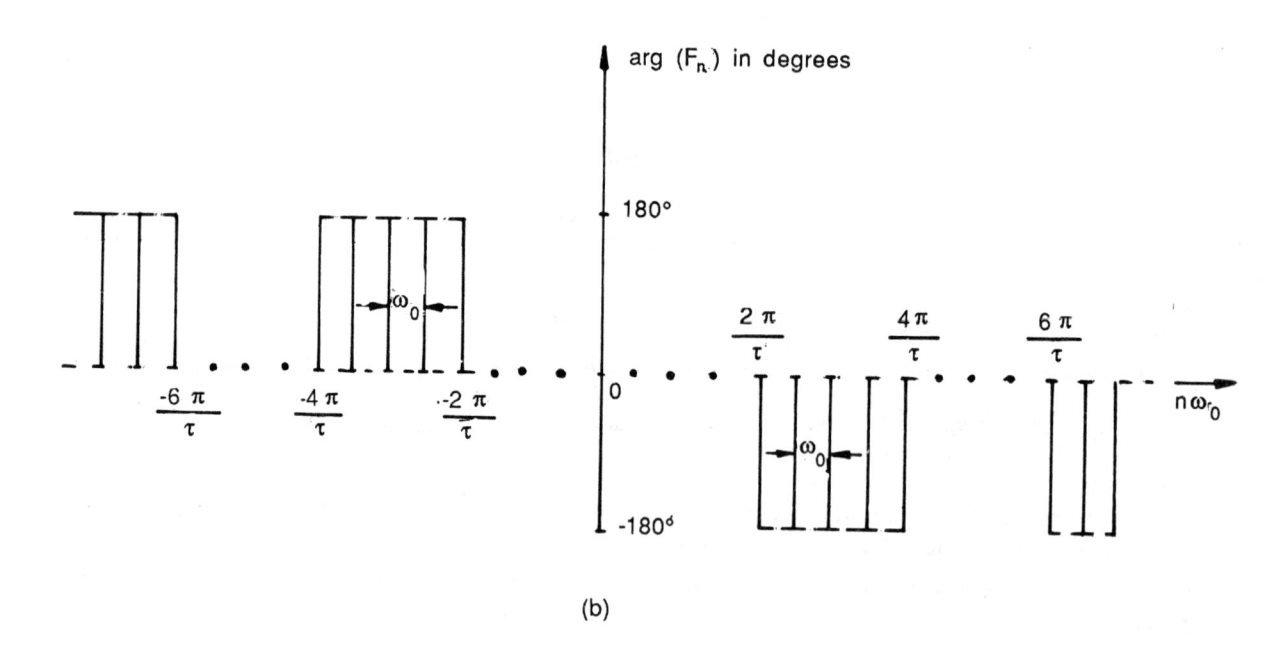

Fig. 1.2.5 (a) Amplitude Spectrum (Symmetrical)  (b) Phase-spectrum (Antisymmetrical)

$$\frac{n\omega_0 \tau}{2} = \pm n\pi \quad (n = 1, 2, 3, \cdots)$$

First zero crossing corresponding to $n = 1$ will occur when $\omega$ (i.e. $n\omega_0$) is given by

$$\omega = \frac{2\pi}{\tau}$$

The practical bandwidth of periodic pulse signal (Gate function) corresponds to the first zero crossing, because the band beyond this has a small amplitude. Hence, the practical bandwidth depends on $\tau$, and is equal to $\frac{2\pi}{\tau}$ radians or $\frac{1}{\tau}$ Hz.

## Example 1.2.3

For the periodic gate function shown in earlier Fig.1.2.3; draw the spectra for the following cases

(a) $\tau = \frac{1}{10}$ s, period $T = \frac{1}{2}$ s

(b) $\tau = \frac{1}{10}$ s, period $T = 1$ s

*Solution*

The spectrum of a periodic function is the plot of $F_n$ with frequency $\omega$. From Eq. 1.2.14, we have

$$F_n = \frac{A\tau}{T} Sa\left(\frac{n\pi\tau}{T}\right)$$

(a) Substituting $\tau = \frac{1}{10}$ and $T = \frac{1}{2}$

$$F_n = \frac{A}{5} Sa\left(\frac{n\pi}{5}\right)$$

The fundamental frequency $\omega_0 = \frac{2\pi}{T} = 4\pi$ radians/sec. Therefore, the lines of the spectrum exist at $\omega = 0, \pm 4\pi, \pm 8\pi, \cdots$ etc., corresponding to $n = 0, \pm 1, \pm 2, \cdots$ etc. The first zero of the envelope will occur when

$$\frac{n\pi}{5} = \pm \pi$$

or

$$n = \pm 5$$

which means the first zero of the envelope will occur at fifth harmonic corresponding to $\omega = n\omega_0 = 5 \times 4\pi = 20\pi$. This is same as $\frac{2\pi}{\tau}$ and satisfies the fact that zero crossings of the envelope occur at frequencies that are multiple of $\frac{2\pi}{\tau}$. The maximum amplitude of the spectrum is A/5. The spectrum is shown in Fig.1.2.6a.

(b) As in the case (a) substituting $\tau = \frac{1}{10}$ s and $T = 1$ s

$$F_n = \frac{A}{10} Sa\left(\frac{n\pi}{10}\right)$$

The fundamental frequency
$$\omega_0 = \frac{2\pi}{1} = 2\pi \text{ radians/s}.$$

The first zero will occur when
$$\frac{n\pi}{10} = \pm \pi$$

or $n = \pm 10$, i.e., at the 10th harmonic. This corresponds to frequency $\omega_0 = 10 \times 2\pi = 20\pi$, same as in case (a) because $\tau$ is same in both the cases. The spectrum is shown in Fig.1.2.6 (b)

**Interesting Phenomena**

Figure 1.2.6 emphasizes some interesting phenomena given as follows:

(i) As pulse width $\tau$ decreases, $\frac{2\pi}{\tau}$ increases which means that the frequency content of the signal extends over a larger frequency range. There is, thus, an inverse relationship between the pulse width $\tau$, and the frequency spread of the spectrum. The frequency spread decides the bandwidth of the signal. The practical bandwidth $B$ is given by the frequency range of the signal from zero to the first zero crossing i.e., $B = \frac{1}{\tau}$ which is 10 Hz in this example. As the period becomes larger, the fundamental frequency $\omega_0$ becomes smaller, generating more frequency components in a given range of frequency; and therefore the spectrum becomes denser. However, the amplitudes of the frequency components become smaller.

In the limit $T$ goes to infinity, we are left with a single pulse of width $\tau$ in the time domain as the next pulse comes after infinite interval. The fundamental frequency $\omega_0$, of this waveform, approaches zero i.e., no spacing is left between two line-components; the spectrum becomes continuous and exists at all frequencies rather than only at the discrete frequencies. However, there is no change in the shape of the envelope of the spectrum, and this is independent of period $T$.

The continuous spectrum described above corresponds to a single non-repetitive pulse; i.e., a non-periodic function existing over the entire interval $(-\infty < t < \infty)$ has a continuous spectrum. Thus, we have arrived at the spectrum of a non-periodic function, taking it as a special case of periodic function with period $T$ approaching infinity. This approach will be discussed in detail in the next section to analyse a non-periodic waveform.

## 1.3 THE FOURIER TRANSFORM

So far, the discussion was confined to the use of Fourier series in the analysis of the following cases of wave forms:

 (i) an arbitrary wave form over a finite interval
 (ii) a periodic waveform over an entire interval $(-\infty, \infty)$.

However, it is desirable to analyse any general waveform, periodic or not, over an entire interval $(-\infty, \infty)$ because a vast majority of interesting signals extend for all time $(-\infty, \infty)$ and are non-periodic in nature.

SIGNAL ANALYSIS 21

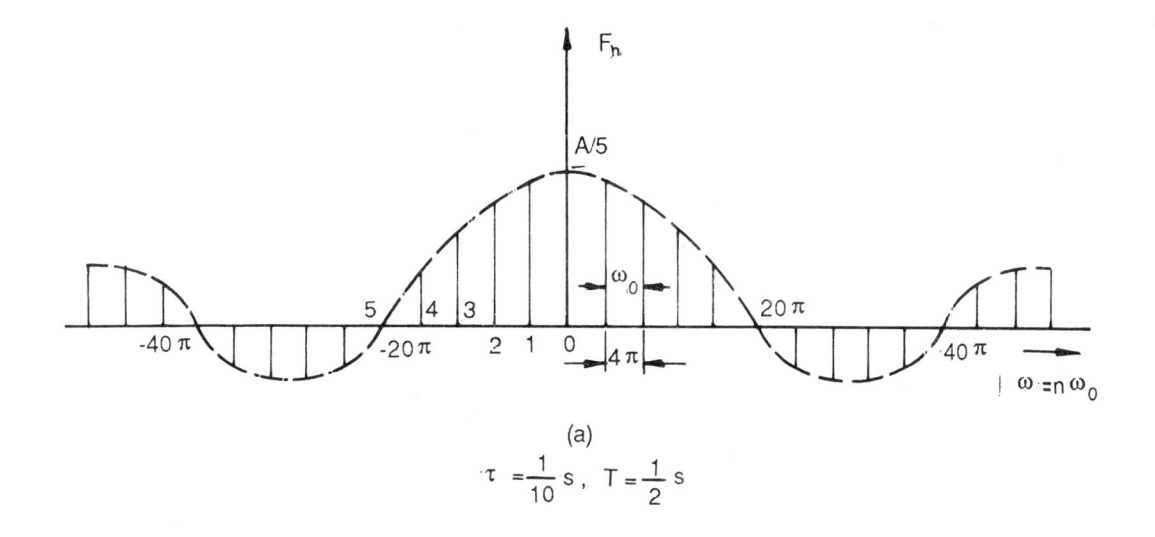

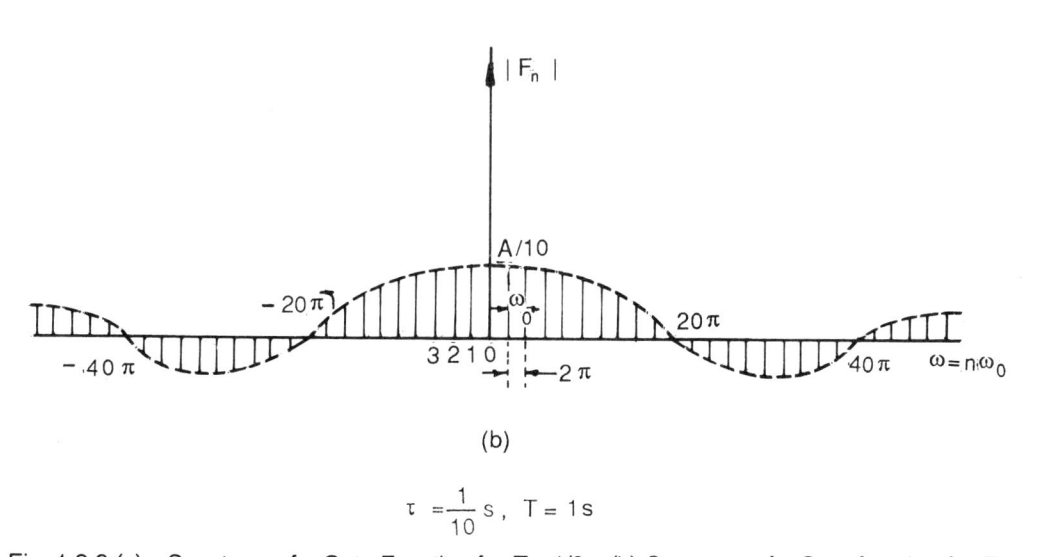

Fig. 1.2.6 (a) Spectrum of a Gate Function for $T = 1/2$ s (b) Spectrum of a Gate function for $T = 1$ s

## Analysis of a Non-periodic Function over Entire Interval

As discussed in Ex.1.2.3, a non-periodic signal can be viewed as a limiting case of a periodic signal; where the period of the signal approaches infinity. This approach will be used to develop the frequency domain representation of a non-periodic signal over an entire interval.

Consider a periodic function $f(t)$ with period $T$. The Fourier series representation of this function is

$$f(t) = \frac{1}{T} \sum_{n=-\infty}^{\infty} F_n e^{jn\omega_0 t} \, ; \, \omega_0 = \frac{2\pi}{T} \tag{1.3.1a}$$

where

$$F_n = \int_{-T/2}^{T/2} f(t)e^{-jn\omega_0 t} dt \qquad (1.3.1b)$$

A typical discrete spectrum plot would look like Fig.1.3.1.

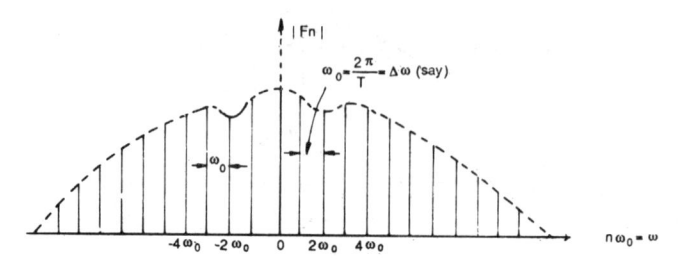

Fig. 1.3.1. A Typical Discrete Spectrum

The spacing between successive harmonics is

$$\omega_0 = \frac{2\pi}{T} = \Delta\omega \text{ (say)}$$

$$\frac{1}{T} = \left(\frac{\Delta\omega}{2\pi}\right)$$

Equation (1.3.1a), then, may be written as

$$f(t) = \frac{1}{2\pi} \sum_{n=-\infty}^{\infty} F_n e^{jn\omega_0 t}(\Delta\omega) \qquad (1.3.2)$$

Now let us consider the limiting case as $T \to \infty$. Then $\Delta\omega \to 0$ and the discrete line in the spectrum merges to form a continuous spectrum. The infinite sum in Eq. 1.3.2 becomes the ordinary Riemann integral. $F_n$ is now defined for all frequencies, rather only the integral multiples of $\omega_0 = \frac{2\pi}{T}$. The integer $n$ has no significance now and in the limit as $T \to \infty$; $F_n$ becomes a continuous function $F(\omega)$ given as,

$$F(\omega) = \lim_{T \to \infty} F_n$$

or
$$= \int_{-\infty}^{\infty} f(t)e^{-j\omega t} dt$$

The Fourier series given in Eq.1.3.2 for a periodic signal now becomes the Fourier integral representation for a non-periodic function over entire interval, given by

$$f(t) = \frac{1}{2\pi} \int_{-\infty}^{\infty} F(\omega)e^{j\omega t} d\omega \qquad (1.3.3a)$$

$$F(\omega) = \int_{-\infty}^{\infty} f(t)e^{-j\omega t} dt \qquad (1.3.3b)$$

In general, $F(\omega)$ is a complex function of $\omega$ and may be written as

$$F(\omega) = |F(\omega)|e^{j\theta(\omega)} \qquad (1.3.4)$$

$F(\omega)$ represented by Eq. 1.3.3b is called Fourier transform of $f(t)$. Similarly $f(t)$ represented by Eq. 1.3.3a is referred as inverse Fourier transform of $F(\omega)$. The functions $f(t)$ and $F(\omega)$ constitute a *Fourier transform pair* and one is called the mate of the other. Symbolically, these transforms are written as

$$F(\omega) = F[f(t)]$$

and

$$f(t) = F^{-1}[F(\omega)]$$

We will generally use lower case letter for time functions, and upper case letter for corresponding transforms.

## Merits of Fourier Transform

A transform is a set of rules substituting one function for another. A function $f(t)$ can have a variety of transforms. Fourier transform is most useful for analysing signals involved in communication systems. Some of the main advantages of this transform are:
  (i) The original time functions can be uniquely recovered from it.
  (ii) It has a property analogous to common logarithm that helps in evaluating convolution integrals which will be discussed later.
  (iii) Although Laplace transforms is extensively used for solving the transient problems of electrical systems, Fourier transform is much more useful in communication system because here the phase and amplitude characteristics are readily known. Laplace transform, on the other hand, is more useful in electrical systems, where network transfer function as a ratio of polynomials in $s(s = \sigma + j\omega)$ is readily specified and the analysis is based on poles and zeros.

## Limitations of Fourier Transform

The Fourier transform $(s = j\omega)$ is closely related with Laplace transform. In Fourier transform, damping factor $\sigma = 0$, and hence the Fourier transform may not converge for many time-functions. In other words, there are many time functions for which Fourier transform does not exist. Such functions are not *absolutely integrable* i.e., their Fourier integral does not converge in the limit $T \to \infty$. The Laplace transform for such function may exist because damping factor $\sigma$ makes the Laplace integral to converge.

## Existence of the Fourier Transform

For a function $f(t)$ to be Fourier transformable, it is sufficient that $f(t)$ satisfies Dirichlet's conditions given below:
  1. The function $f(t)$ *is* a single valued with a finite number of maxima and minima; and a finite number of discontinuities in any finite time interval.
  2. The function $f(t)$ is absolutely integrable, i.e.,

$$\int_{-\infty}^{\infty} |f(t)| dt < \infty$$

## 1.4 CONTINUOUS SPECTRUM

The spectrum of a non-periodic function is continuous. Consider the amplitude spectrum plot shown in Fig.1.3.1. This is a discrete line spectrum corresponding to a periodic function with period $T$. Successive

## 24 COMMUNICATION SYSTEMS: Analog and Digital

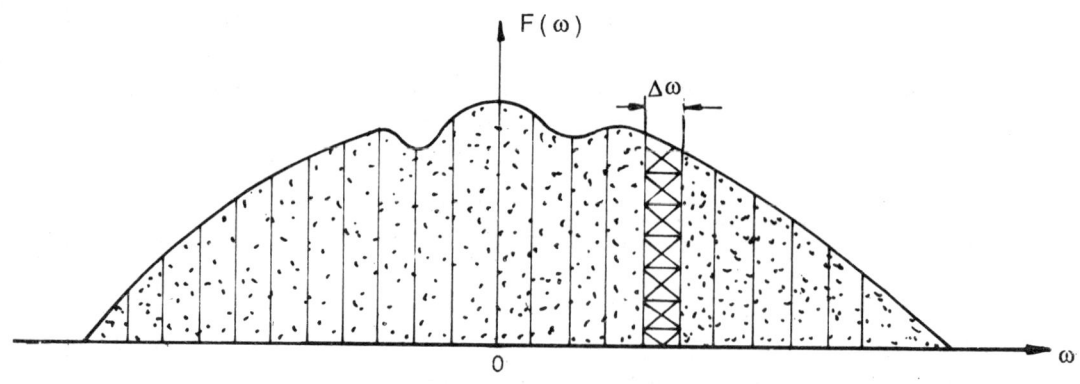

Fig. 1.4.1. A Continuous Spectrum

harmonics are spaced by $(\Delta\omega) = \omega_0 = \dfrac{2\pi}{T}$. The spectrum can be extended to a corresponding non-periodic function by considering the limiting case as $T \to \infty$. Then, $\Delta\omega \to 0$; the discrete lines in the spectrum merge and we get a continuous frequency spectrum as shown in Fig.1.4.1. The amplitude of a single frequency component becomes infinitesimal. However, the amplitude for a range of frequency does exist, and is proportional to $F(\omega)$. The Fourier transform $F(\omega)$ is also known as the spectral density function because $F(\omega) \cdot d(\omega)$ gives the amplitude of frequency components lying in the range '$d(\omega)$'.

The concept of continuous spectrum is confusing because it does not provide the amplitude of any single frequency component; rather, it specifies density of the spectrum by which an amplitude for a range of frequency components $d\omega$ can be evaluated. The concept becomes clearer by considering the analogous phenomenon of a loaded beam. The discrete spectrum is analogous to a discrete loaded beam. In this case the spectrum exists only at discrete points, and is absent at others. On the other hand, continuous spectrum is analogous to a continuously loaded beam. Here the spectrum exists at every point, but at any one frequency the spectrum will have zero magnitude. The symmetry conditions for continuous spectrum and discrete spectrum are the same. For a real functions $f(t)$, the continuous spectrum has a symmetrical magnitude and antisymmetrical phase spectrum. (Prob 1.22)

### Example 1.4.1

Find the Fourier transform of a single-sided exponential function $e^{-bt} u(t)$ shown in Fig.1.4.2 and draw the spectrum. $u(t)$ is the unit step function.

**Solution**

$$f(t) = e^{-bt} u(t)$$

$$F(\omega) = \int_{-\infty}^{\infty} f(t) e^{-j\omega t} dt$$

$$= \int_{-\infty}^{\infty} e^{-bt} u(t) e^{-j\omega t} dt$$

Since the unit step function exists for $t > 0$, $u(t)$ can be removed if $-\infty$ in the lower limit is replaced by zero, i.e.,

$$F(\omega) = \int_0^{\infty} e^{-bt} e^{-j\omega t} dt = \int_0^{\infty} e^{-(b+j\omega)t} dt$$

$$= \frac{1}{b+j\omega} = \frac{(b-j\omega)}{(b+j\omega)(b-j\omega)} = \frac{b}{b^2+\omega^2} - j\frac{\omega}{b^2+\omega^2}$$

$$= \frac{1}{\sqrt{b^2+\omega^2}} e^{-j\tan^{-1}\left(\frac{\omega}{b}\right)}$$

The magnitude spectrum $|F(\omega)|$, and phase spectrum $\theta(\omega)$, are given as

$$|F(\omega)| = \frac{1}{\sqrt{b^2+\omega^2}}$$

and

$$\theta(\omega) = -\tan^{-1}\left(\frac{\omega}{b}\right)$$

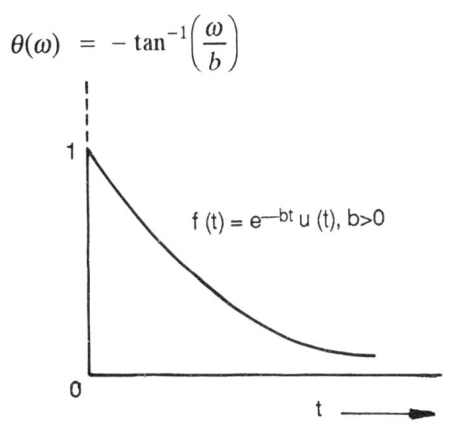

Fig. 1.4.2 A Single-sided Exponential Function

*Spectrum*: The magnitude and phase spectrum are shown in Fig.1.4.3. Note that Fourier transform exists only for the positive value of $b$, and it is not absolutely integrable for the negative value of $b$.

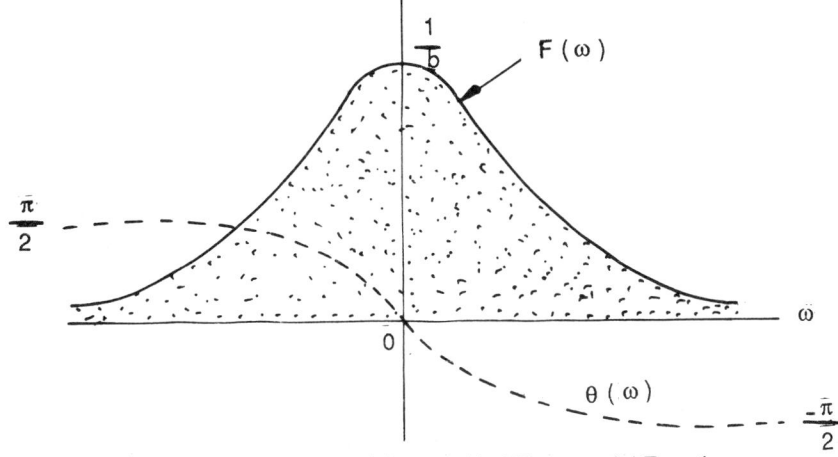

Fig. 1.4.3 Spectrum of Single-sided Exponential Function

# 26 COMMUNICATION SYSTEMS: Analog and Digital

The magnitude spectrum has a value $\frac{1}{b}$ at $\omega = 0$, it decreases as $\omega$ increases. Similarly the phase spectrum varies between $-\frac{\pi}{2}$ to $\frac{\pi}{2}$, depending on the values of $-\tan^{-1}\left(\frac{\omega}{b}\right)$ at different frequencies. Since Fourier transform is complex, both plots are necessary.

## Example 1.4.2

Find the Fourier transform of a double-sided exponential signal $e^{-bt}$ shown in Fig.1.4.4 (a) and draw the spectrum.

**Solution**

$$f(t) = e^{-bt}$$

$$F(\omega) = \int_{-\infty}^{\infty} e^{-bt} e^{-j\omega t} dt$$

For convenience, the integral can be split into two parts, depending upon its being positive, or negative.
(i) For negative $t$, the limit will be $-\infty$ to $0$
(ii) For positive $t$, the limit will be $0$ to $\infty$

Thus,

$$F(\omega) = \int_{-\infty}^{0} e^{-b(-t)} e^{-j\omega t} dt + \int_{0}^{\infty} e^{-bt} e^{-j\omega t} dt$$

Note that the limits of integral do not affect $e^{-j\omega t}$

$$F(\omega) = \int_{-\infty}^{0} e^{(b-j\omega)t} dt + \int_{0}^{\infty} e^{-bt} e^{-j\omega t} dt$$

$$= \frac{1}{b - j\omega} + \frac{1}{b + j\omega} = \frac{2b}{b^2 + \omega^2}$$

As the spectrum is real, only magnitude plot is desired. The phase spectrum $\theta(\omega) = 0$ which means all the frequency components are in the same phase. The amplitude spectrum is shown in Fig.1.4.4b.

## Example 1.4.3

Find the Fourier transform of a Gate function of amplitude $K$ and width $\tau$ as shown in Fig.1.4.5(a)

**Solution**

A gate function $G_\tau(t)$ is a rectangular pulse defined by

$$f(t) = G_\tau(t) = \begin{cases} K & |t| < \frac{\tau}{2} \\ 0 & \text{elsewhere} \end{cases} \quad |t| \text{ stands for positive as well as negative value of } t.$$

The Fourier transform of this function is obtained as follows:

$$F(\omega) = \int_{-\tau/2}^{\tau/2} K e^{-j\omega t} dt = \frac{K}{j\omega}\left(e^{\frac{j\omega\tau}{2}} - e^{\frac{-j\omega\tau}{2}}\right)$$

SIGNAL ANALYSIS 27

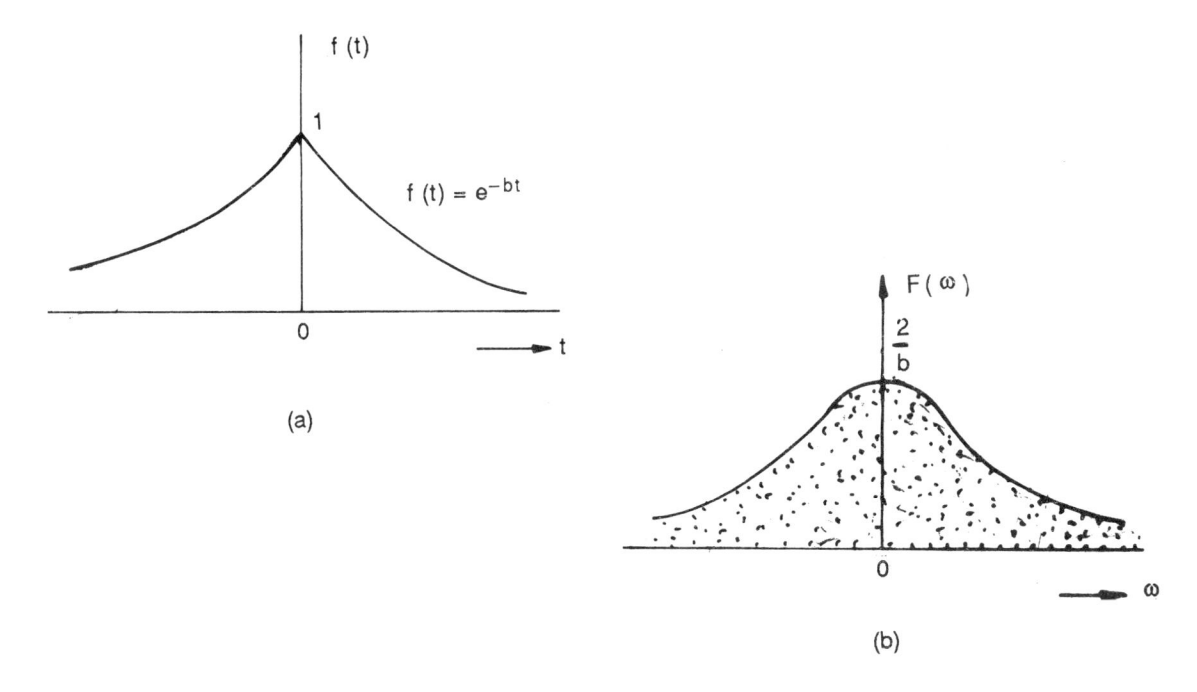

Fig. 1.4.4 (a) Time Domain of Double-sided Exponential Function (b) Frequency Domain of Fig. 1.4.4a

$$= K\tau \frac{\sin\left(\omega\frac{\tau}{2}\right)}{\left(\omega\frac{\tau}{2}\right)} = K\tau\, Sa\left(\frac{\omega\tau}{2}\right)$$

The continuous amplitude and phase spectrum are shown in Figs 1.4.5(b) and (c), respectively. Here, $F(\omega)$ is real function and hence only amplitude plot is sufficient. However, for a clearer understanding the phase plot is also shown.

The practical bandwidth of a Gate function corresponds to the first zero crossing. Therefore, the practical bandwidth of the pulse is $\frac{2\pi}{\tau}$ radians/s or $\frac{1}{\tau}$ Hz.

Thus, we can write

Bandwidth of a Gate function $= \dfrac{1}{\text{pulsewidth}}$.

## 1.5 FOURIER TRANSFORM INVOLVING IMPULSE FUNCTION

The sufficient condition required for the Fourier transform of a function to exist is that the function $f(t)$ should be absolutely integrable i.e.,

28  COMMUNICATION SYSTEMS: Analog and Digital

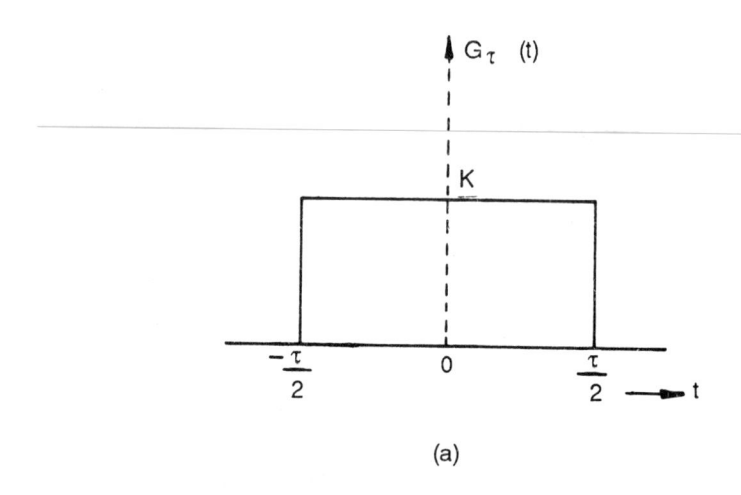

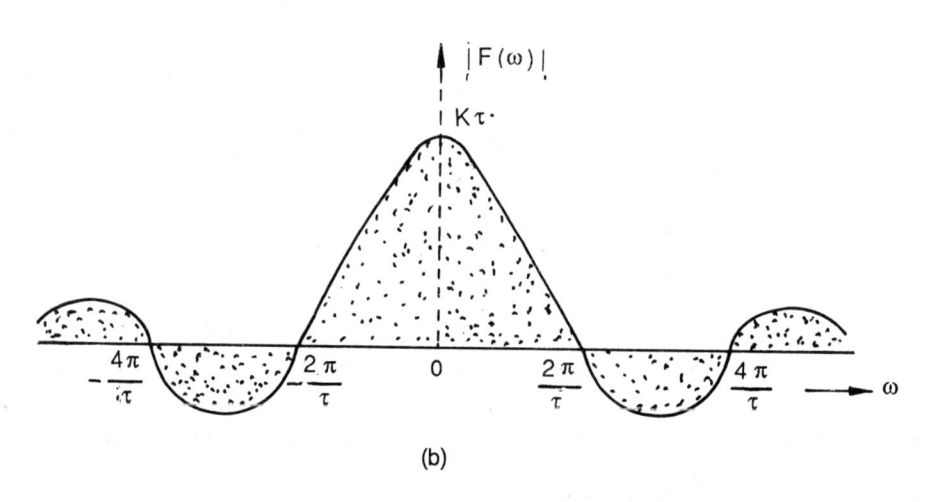

Fig. 1.4.5  (a) A Gate Pulse;  (b) Amplitude Spectrum

$$\int_{-\infty}^{\infty} f(t)\,dt < \infty$$

But this is not the necessary condition for existence of the Fourier transform. A number of useful signals in communication systems are not absolutely integrable (e.g. a d.c. signal), but their Fourier transform exists in the limiting condition. The Fourier transform for these signals are obtained by using the concept of impulse function or Delta function denoted by $\delta(t)$. The impulse function $\delta(t)$ falls under the category of singularity functions.

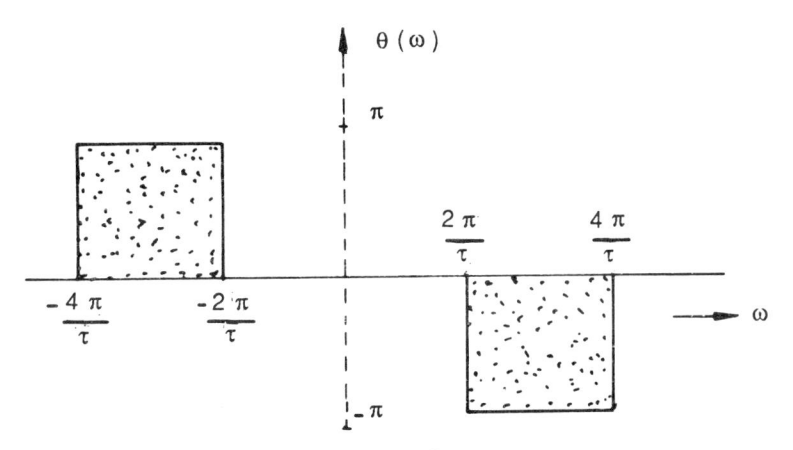

Fig 1.4.5 (c) Phase Spectrum

## Singularity Function

Singularity functions are discontinuous functions, or have discontinuous derivatives. These functions may have continuous derivatives only up to finite order.

Let us now examine the significance of singularity functions in signal analysis. When a unit step voltage is applied to a capacitor, as shown in Fig. 1.5.1 (a), the current $i(t)$ flowing through the circuit is

$$i(t) = \frac{du(t)}{dt}$$

The idealized unit step function, shown in Fig. 1.5.1 b is unity for $t > 0$, but discontinuous at $t = 0$. Hence, the current $i(t)$ is zero everywhere except at $t = 0$, where the derivative of the step function is undefined.

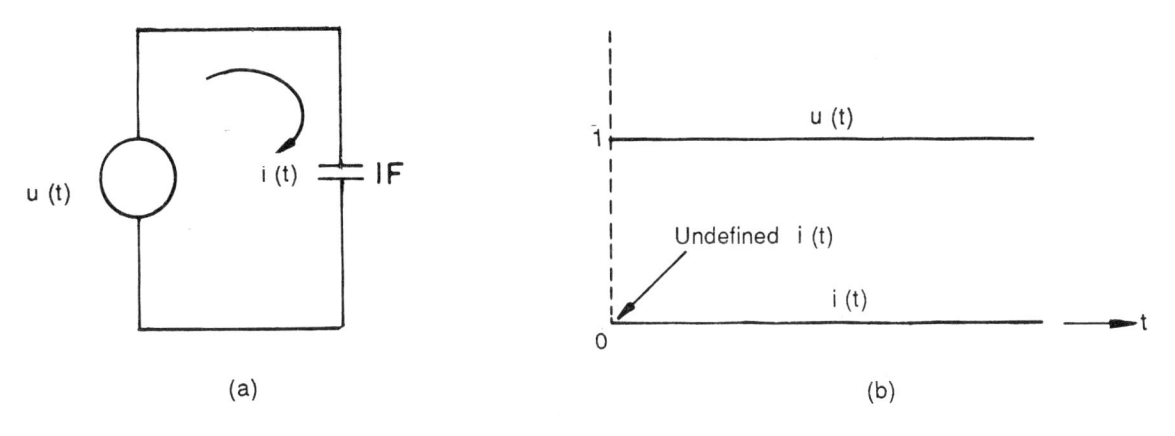

Fig. 1.5.1

The unit impulse function, which is mathematically defined as derivative of the unit step function, is not a true function in the mathematical sense where a function should be defined at every value of $t$. The unit impulse function, however, can be defined as a 'limiting case of derivatives of unidealized step functions'. The concept is made clear by the following example:

Consider an unidealized voltage function $u_b(t)$ shown in Fig.1.5.2(a). This voltage function becomes a unit step function if the limit $b$ goes to zero, i.e.

$$\lim_{b \to 0} u_b(t) = u(t)$$

when this voltage $u_b(t)$ is applied to a capacitor, the current $i = \dfrac{du_b(t)}{dt}$ will be a rectangular pulse of height $\dfrac{1}{b}$, and width $b$ as shown in Fig. 1.5.2 (b).

(a)    (b)

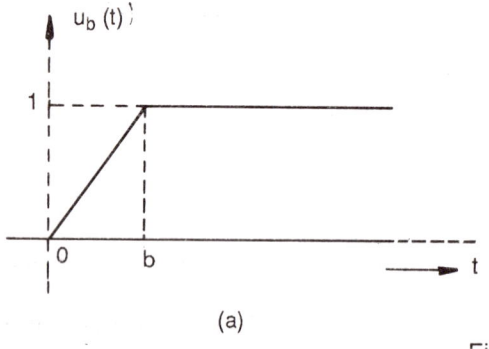

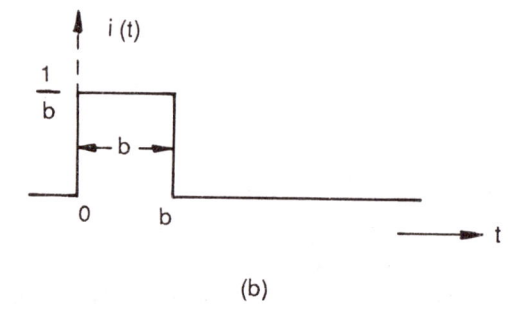

Fig. 1.5.2

Let us now analyse same situations for various decreasing values of $b$. For $b = b_1, b_2, b_3,$ and $b_4$, the function $u_b(t)$ and the resulting respective currents $i_1, i_2, i_3,$ and $i_4$, are shown in Fig. 1.5.3. The current pulse corresponding to $b = b_1$ has the width $b_1$ and height $\dfrac{1}{b_1}$, with the resulting area of unity. The current pulse corresponding to $b = b_2$ has reduced in width, but increased in height to result in the same area of unity. Same is the case for $b = b_3$ and $b = b_4$. Thus as the value of $b$ reduces, the height of the pulse sequences increases, and the area under the current pulse curve is always unity, i.e., a constant.

In the limit when $b$ goes to zero, the width of the pulse becomes zero, and the height of the pulse approaches infinity, maintaining the area under the pulse-curve unity. In the limit $b$ goes to zero, $u_b(t)$ becomes a unit step function $u(t)$, and the corresponding current pulse becomes an impulse function $\delta(t)$, given by

$$\delta(t) = \lim_{b \to 0} \frac{d}{dt}\{u_b(t)\}$$

It is clear that the derivative of an unidealized step function can define a unit impulse function, inspite of the fact that the derivative of an idealized unit step function does not exist. Hence, we can define the unit impulse function as the limit of the sequence of derivatives of unidealized function $u_b(t)$ as $b$ tends to zero. In the limit $b \to 0$, the function $\delta(t)$ *takes* the form of a pulse having the following properties:
 (i) The width of the pulse is zero: i.e. pulse exists only at $t = 0$.
 (ii) The height of the pulse goes to infinity.
 (iii) The area under the pulse-curve is always unity: i.e., a constant

Mathematically, the above mentioned properties define the impulse function as follows :

## SIGNAL ANALYSIS

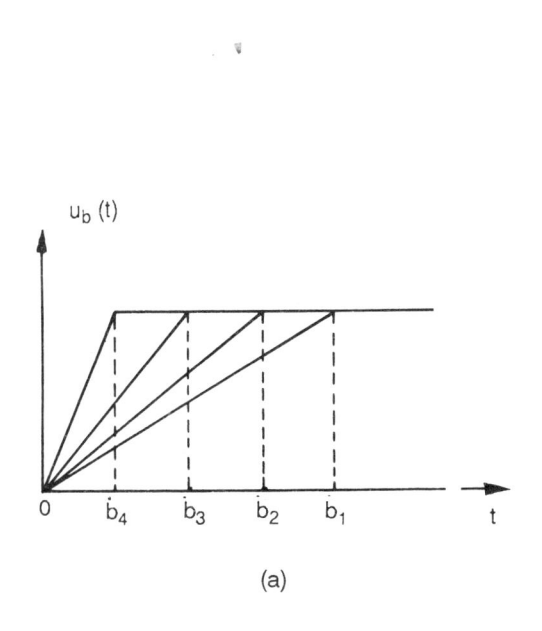

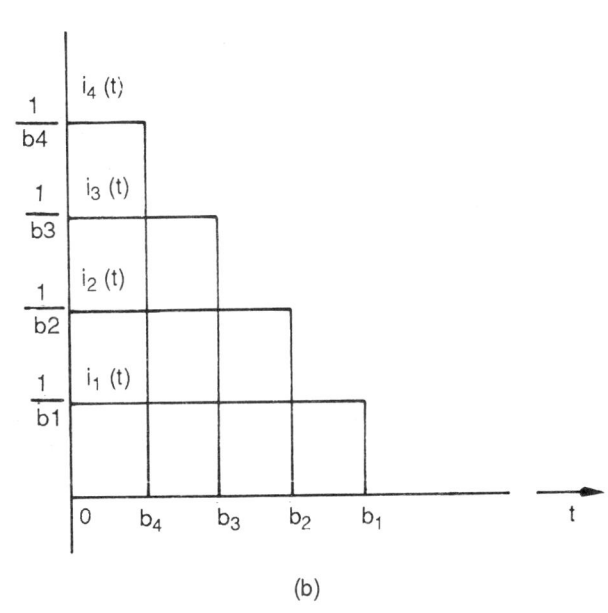

Fig. 1.5.3

$$\delta(t) = 0, \quad t \neq 0 \tag{1.5.1a}$$

$$\int_{-\infty}^{\infty} \delta(t)\,dt = 1 \tag{1.5.1b}$$

P.A.M. Dirac gave the above definition, and hence the impulse function is also called *Dirac delta function*.

The delta function was developed by considering the limiting case of the sequence of "Derivatives of Unit step functions (unidealized)". Hence one can observe that the integral i.e., antiderivative of $\delta(t)$ is $u(t)$ the unit step function

$$\int_{-\infty}^{\infty} \delta(\tau)\,d\tau = u(t) = \begin{cases} 1 & t > 0 \\ 0 & t < 0 \end{cases}$$

which conversely defines the unit impulse function: unit step function is an integral of delta function.

The delta function may also be defined as a "limiting case of sequences of regular functions". A number of regular functions defining impulse function by their sequences under the limiting case, are shown in Fig.1.5.4.

The definition of the impulse function is given alongwith the corresponding figure. Sequence of sampling functions are shown in Fig. 1.5.5. In the limit, area under the curves is unity.

## Sampling Square Function

Sequence of the square of sampling function may also lead to delta function in the limiting case.

$$\delta(t) = \lim_{b \to \infty} \frac{b}{\pi} Sa^2(bt) \tag{1.5.2}$$

Hence, the area under the sampling square function is unity, i.e.,

$$\int_{-\infty}^{\infty} \frac{b}{\pi} Sa^2(bx)\,dx = 1 \tag{1.5.3}$$

**32** COMMUNICATION SYSTEMS: Analog and Digital

$$\delta(t) = \lim_{\tau \to 0} \frac{1}{\tau} e^{(-\pi t^2/\tau^2)}$$

(a)

$$\delta(t) = \begin{cases} \lim_{\tau \to 0} \frac{1}{\tau}\left[1 - \frac{|t|}{\tau}\right]; |t| < \tau \\ 0, \text{ elsewhere} \end{cases}$$

(b)

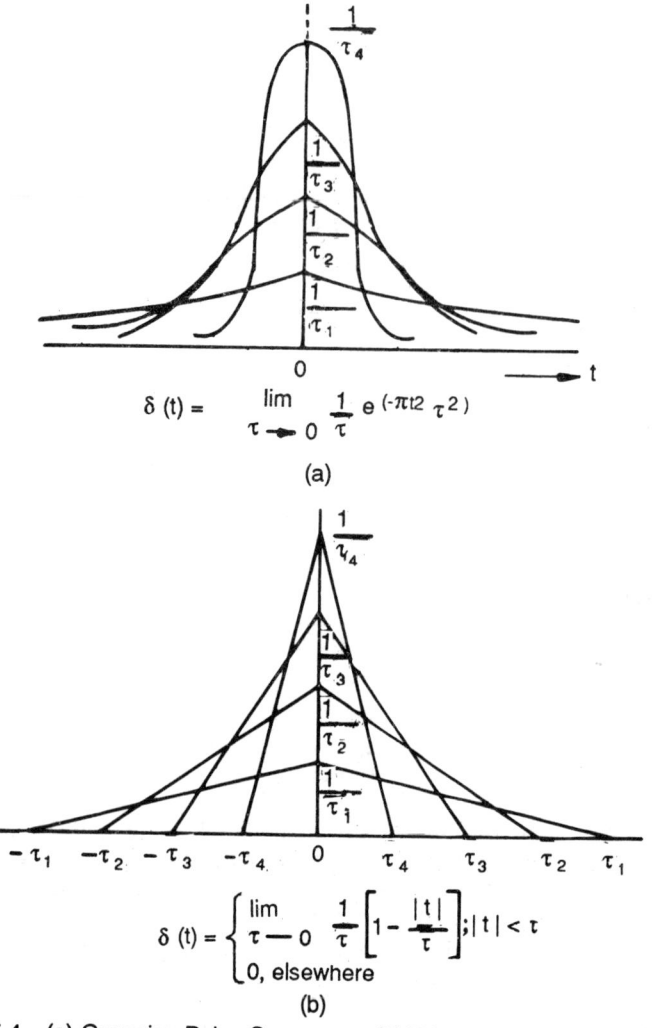

Fig. 1.5.4 (a) Gaussian Pulse Sequence; (b) Triangular Pulse Sequence

## Shifting Property of the Delta Function

Let us take the integral of the product of $\delta(t)$, and any function $f(t)$ that is continuous at $t = 0$. Since $\delta(t)$ exists only at $t = 0$, the product yields the function $f(t)$ existing only at $t = 0$, i.e.,

$$\int_{-\infty}^{\infty} f(t)\delta(t)\,dt = f(0)\int_{-\infty}^{\infty}\delta(t)\,dt = f(0).1 = f(0) \qquad (1.5.4)$$

where; $f(0)$ is the value of the function $f(t)$ at $t = 0$. This is very significant result, and is appropriately known as *shifting or sampling property* of the delta function because the impulse shifts the value of $f(t)$ at $t = 0$; equivalently, the value of $f(t)$ has been sampled at $t = 0$.

The sampling or shifting can be done at any instant $t = t_0$ (rather than only at $t = 0$) by defining the impulse function at the instant $t_0$. Figure 1.5.6. shows an impulse function centered at $t = 0$, and another one centered

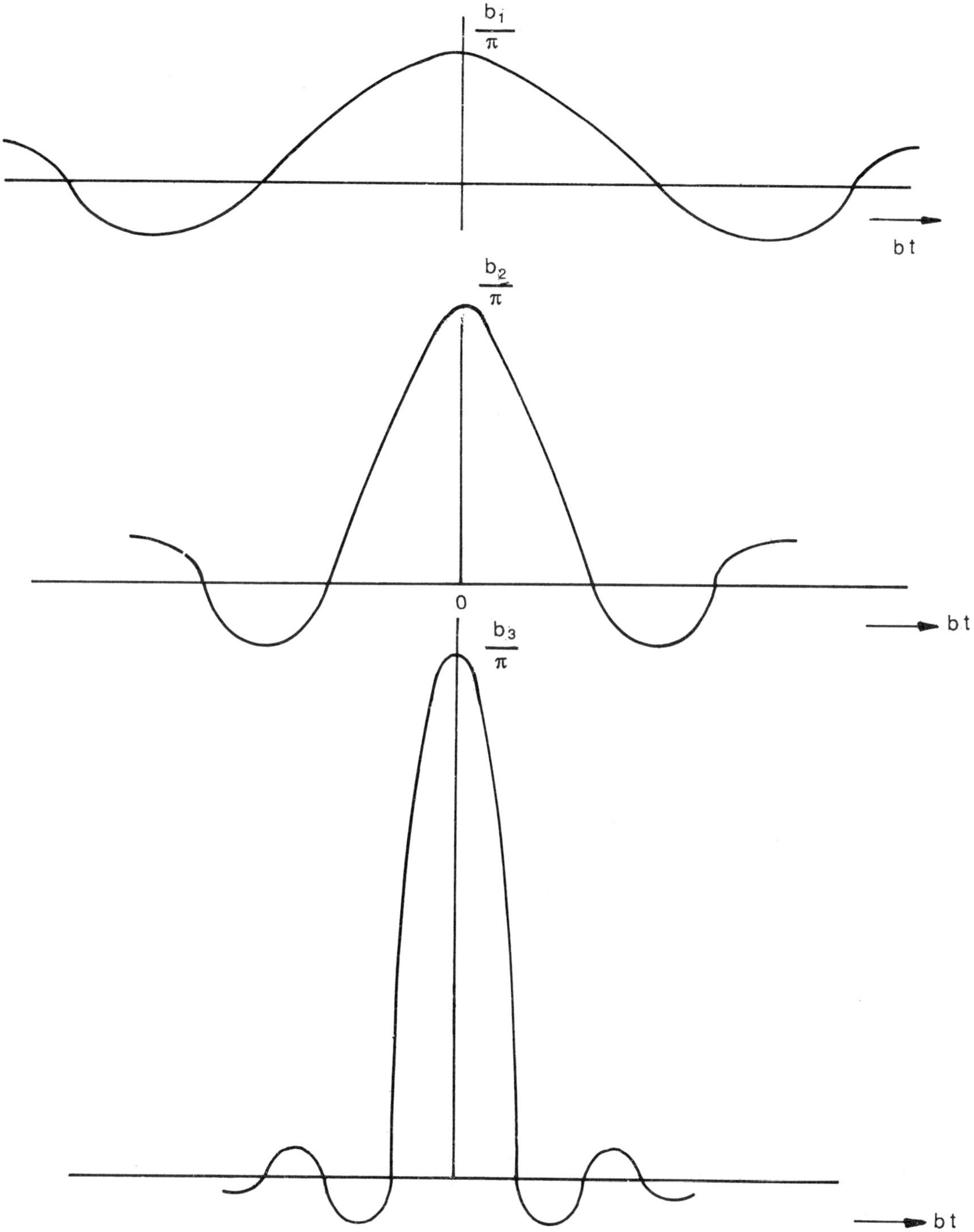

SIGNAL ANALYSIS 33

$$\delta(t) = \lim_{k \to \infty} \frac{b}{\pi} \, Sa(bt)$$

Fig. 1.5.5 Sampling Function Sequence

**34** COMMUNICATION SYSTEMS: Analog and Digital

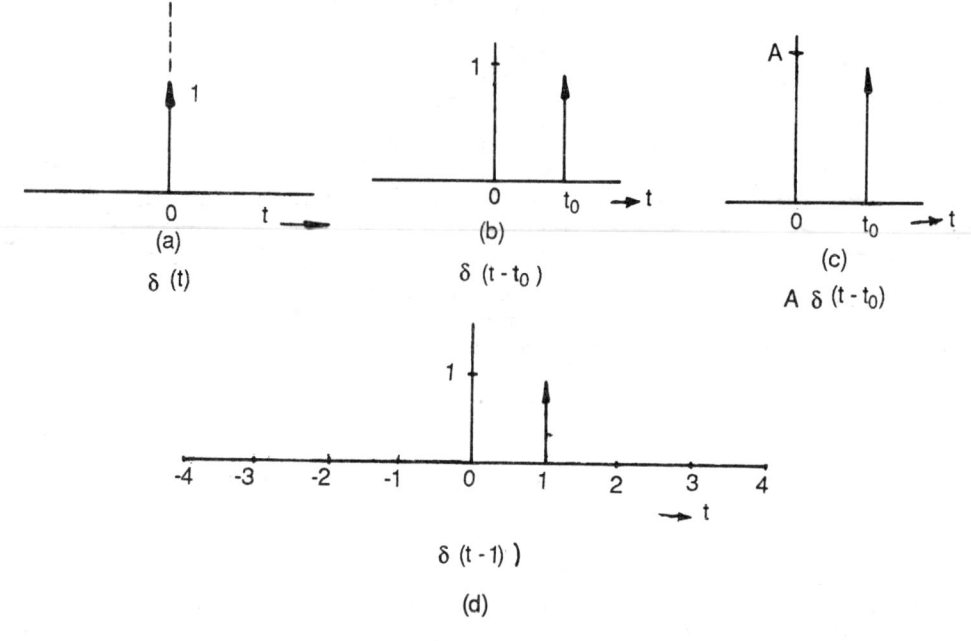

**Fig. 1.5.6**

at $t = t_0$, denoted by $\delta(t)$ and $\delta(t - t_0)$ respectively. The symbols are explained below:
(i) The upward pointing arrow indicates the actual value of infinity.
(ii) The height of the arrow indicates the total area under the impulse. This area is unity for unit impulse function; as shown in Figs 1.5.6(a) and (b). These represent delta functions of the unit area centered at origin and $t_0$ respectively.
(iii) The delta function shown in Fig.1.5.6(c) has the area $A$ centered at $t = t_0$. This impulse written as $A \delta(t - t_0)$ where $A$ indicates the total area under the pulse. Similarly, Fig.1.5.6(d) represents an impulse function at $t = 1$ and having unity area.

The total area under the pulse is also called the *strength* of the impulse. Thus, the impulse in Figs 1.5.6 (a) and (b) has unity strength and in Fig. 1.5.6 (c) has a strength of $A$. With this definition of an impulse function, the shifting property, defined by Eq.1.5.4, can be extended for shifting to any arbitrary instant $t_0$

$$\int_{-\infty}^{\infty} f(t) \, \delta(t - t_0) \, dt = f(t_0) \tag{1.5.5}$$

which states that the multiplication of a continuous function $f(t)$ with an impulse function $\delta(t - t_0)$ yields the sampled value of $f(t)$ at $t = t_0$, i.e. $f(t_0)$.

Impulse function is frequently used to represent point mass, point charge, point source and other similar physical entities used in science and engineering. The step function and the impulse function with all their higher derivatives are known as *singular function*.

### Example 1.5.1

Evaluate the following integrals:

(a) $$\int_{-\infty}^{\infty} \left[t^2 + 1\right] \delta(t) \, dt$$

(b) $$\int_{-1}^{1} \left[t^4 + 1\right] \delta(t-1) \, dt$$

(c) $$\int_{3}^{5} \left[t^3 + 4t + 2\right] \delta(t-1) \, dt$$

(d) $$\int_{-\infty}^{\infty} \left[t^4 - 2\right] \delta(1-t) \, dt$$

*Solutions*

(a) Using shifting property of the impulse function, the integral is given as :

$$\int_{-\infty}^{\infty} \delta(t) t^2 \, dt + \int_{-\infty}^{\infty} \delta(t)(1) \, dt = \left[t^2\right]_{t=0} + \text{area under } \delta(t) = 0 + 1 = 1$$

(b) $$\int_{-1}^{2} (t^4 + 1) \delta(t-1) \, dt$$

Figure 1.5.6d shows that the impulse at $t = 1$ which falls within the range of integration i.e., $-1$ to 2. Since impulse exists only at $t = 1$, using shifting property the integral becomes

$$\int_{-1}^{2} t^4 \, \delta(t-1) \, dt + \int_{-1}^{2} \delta(t-1) \, dt = \left[t^4\right]_{t=1} + \left[\text{Area under the impulse at } t = 1\right] = 1 + 1 = 2$$

(c) $$\int_{3}^{5} \left[t^3 + 4t + 2\right] \delta(t-1) \, dt$$

The impulse function exists at $t = 1$ which is outside the range of integration (3 to 5), as shown in Fig.1.5.6d. Since the impulse function is absent in the range of integration, the integral vanishes. Therefore

$$\int_{3}^{5} \left[t^3 + 4t + 2\right] \delta(t-1) \, dt = 0$$

(d) $\delta(1-t)$ represents unimpulse function at an instant for which $1-t = 0$, i.e., at $t = 1$. Therefore;

$$\delta(1-t) = \delta(t-1)$$

Both represent a delta function at $t = 1$.
Using shifting property

$$\int_{-\infty}^{\infty} \delta(t-1) \left[t^4 - 2\right] dt = \int_{-\infty}^{\infty} \delta(t-1) t^4 \, dt - \int_{-\infty}^{\infty} 2\delta(t-1) \, dt$$

$$= \left[t^4\right]_{t=1} - 2 \times (\text{area under impulse curve}) = 1 - 2 = -1$$

## Fourier Transform of Functions Involving Impulse Function

Now, we will find the Fourier transform of some useful functions which can be easily determined using delta functions, but their Fourier transform does not exist otherwise, as they are not absolutely integrable.

### Example 1.5.2

Find the Fourier transform of an impulse function

$$f(t) = \delta(t)$$

## Solution

The Fourier transform from the basic definition is

$$F(\omega) = \int_{-\infty}^{\infty} f(t) e^{-j\omega t} dt = \int_{-\infty}^{\infty} \delta(t) e^{-j\omega t} dt$$

Using shifting property

$$F(\omega) = \left[e^{-j\omega t}\right]_{t=0} = 1$$

A symbol of double ended arrow (↔) is used to represent a time function, and its Fourier transform.

$$\delta(t) \leftrightarrow 1 \qquad (1.5.6a)$$

The Fourier transform of an impulse function is unity.

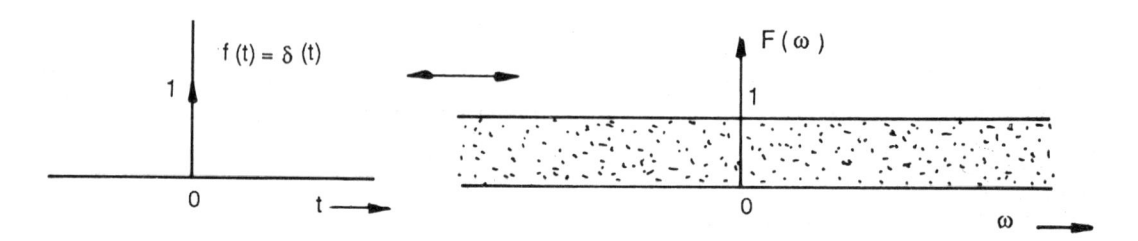

Fig. 1.5.7 An Impulse Function and its Spectrum

Figure 1.5.7 shows an impulse function and it's Fourier transform. The figure clearly shows that an impulse function contains the entire frequency component ($-\infty$ to $\infty$) of identical magnitude, and same phase, i.e. the bandwidth of the delta function **is infinite**. The phase plot is not required as $F(\omega)$ is real. The result can be extended to an impulse function of an arbitrary strength $A$.

$$A\delta(t) \leftrightarrow A \qquad (1.5.6b)$$

## Example 1.5.3

Find the Fourier transform of a constant function

$$f(t) = A$$

## Solution

A function is constant if it has a fixed value over an entire interval $(-\infty, \infty)$, as shown in Fig.1.5.8. This function is not absolutely integrable, but its Fourier transform can be determined using the concept of delta function.

The constant function can be taken as a Gate function of height $A$; and width $\tau$, in the limit $\tau$ approaching infinity. The Fourier transform of the Gate function is given as

$$G_\tau(t) \leftrightarrow A\tau \, Sa\left(\frac{\omega\tau}{2}\right)$$

Therefore, the Fourier transform of a constant function is,

SIGNAL ANALYSIS 37

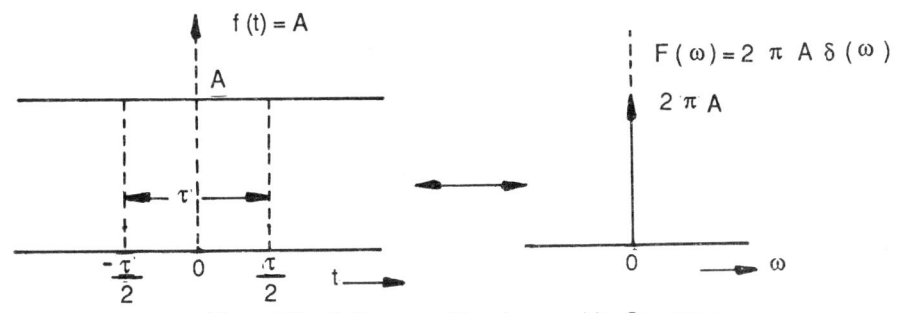

Fig. 1.5.3 A Constant Function and its Spectrum

$$F[A] = \lim_{\tau \to \infty} F\{G_\tau(t)\} = \lim_{\tau \to \infty} A\tau Sa\left(\frac{\omega\tau}{2}\right) = 2\pi A \lim_{\tau \to \infty} \frac{\tau}{2\pi} Sa\left(\frac{\omega\tau}{2}\right)$$

As in the limit $\tau \to \infty$, sampling function approaches a delta function (Fig.1.5.5),

$$F[A] = 2\pi A\delta(\omega) \tag{1.5.7}$$

The Fourier transform plotted in Fig.1.5.8 shows that the delta function contains a single frequency component at $\omega=0$ which is a d.c. This is quite logical, since a constant function over an entire interval is a pure d.c. signal, and has no other frequency component. When plotted as a function of angular frequency $\omega$, the magnitude of d.c. signal is $2\pi A$. Considering it with respect to linear frequency $f$, the magnitude would be $A$.

### Example 1.5.4

Find the Fourier transform of a signum function denoted by *sgn* (*t*).

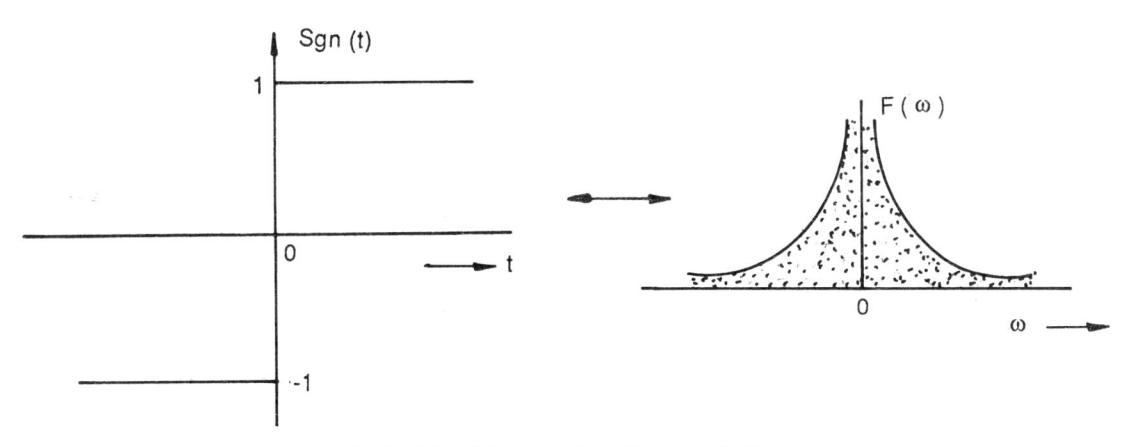

Fig.1.5.9(a) A Signum Function and its Spectrum

*Solution*

The signum function shown in Fig.1.5.9a is expressed as

$$sgn(t) = \begin{cases} 1 & t > 0 \\ -1 & t < 0 \end{cases} \tag{1.5.8}$$

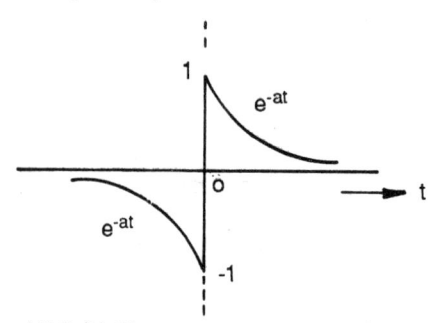

Fig. 1.5.9 (b) Signum Function as a Limiting Case

The function can be defined in terms of unit step function as

$$sgn(t) = 2u(t) - 1 \qquad (1.5.9)$$

For $t > 0$, $2u(t) = 2$; and for $t < 0$, $2u(t) = 0$, therefore Eq.1.5.9 satisfies the definition given in Eq.1.5.8. Keeping following facts in view,

(i) $\quad \lim_{a \to 0} e^{-at} = 1, \quad t > 0$

(ii) $\quad \lim_{a \to 0} e^{at} = 1, \quad t < 0$ (i.e. negative values of $t$)

(iii) $\quad u(t) = 1, \quad t > 0$

(iv) $\quad u(-t) = 1 \quad t < 0$

the positive half of the signum function can be expressed as $\lim_{a \to 0} e^{-at} u(t)$, and the negative half can be expressed as $\lim_{a \to 0} -e^{at} u(-t)$, as shown in Fig.1.5.9b. The total function can be expressed as,

$$sgn(t) = \lim_{a \to 0} \left[ e^{-at} u(t) - e^{at} u(-t) \right]$$

The above form of expression of the signum function in easily evaluating the Fourier integral. The Fourier transform is given as

$$F[sgn(t)] = \lim_{a \to 0} \left[ \int_0^\infty e^{-at} e^{-j\omega t} dt - \int_{-\infty}^0 e^{at} e^{-j\omega t} dt \right]$$

$$= \lim_{a \to 0} \left[ \frac{-2j\omega}{a^2 + \omega^2} \right]$$

$$= \frac{2}{j\omega} \qquad (1.5.10)$$

This being purely imaginary, only amplitude plot is required which is shown in Fig.1.5.9(a).

### Example 1.5.5

Find the Fourier transform of a unit step function shown in Fig. 1.5.10.

SIGNAL ANALYSIS 39

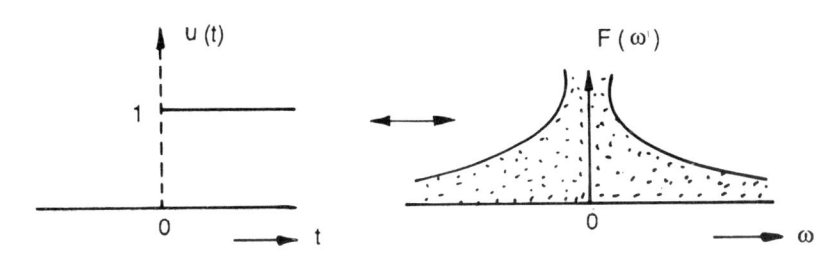

Fig.1.5.10  A Unit Step Function and its Spectrum

*Solution*

The step function shown in Fig.1.5.10 is expressed as

$$u(t) = 1, t > 0$$

The Fourier transform of $u(t)$ can be easily determined using the spectrum of signum function. The step function can be taken as the sum of a signum and a constant function, as shown in Fig.1.5.11. Hence

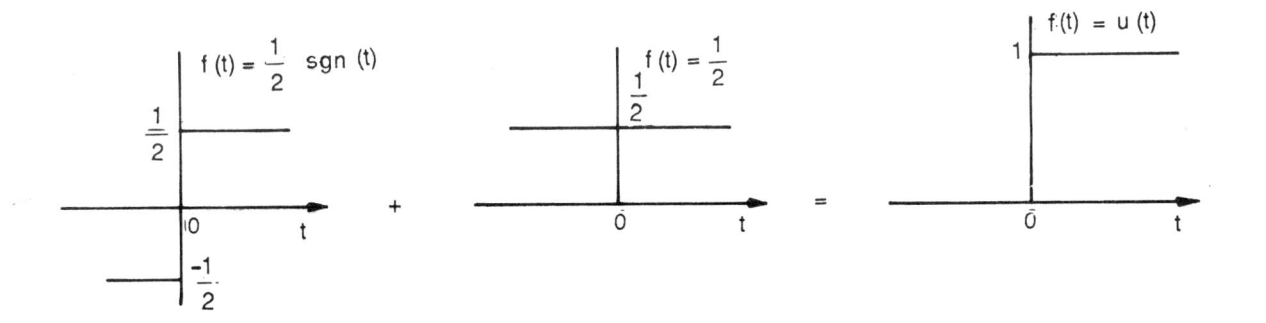

Fig.1.5.11

$$u(t) = \frac{1}{2}[1 + sgn(t)]$$

The Fourier transform is given by

$$F\{u(t)\} = \frac{1}{2}F[1] + \frac{1}{2}F[sgn(t)]$$

Since

$$F[1] = [F\{A\}]_{A=1} = [2\pi A\delta(\omega)]_{A=1} = 2\pi\delta(\omega)$$

and

$$F[sgn(t)] = \frac{2}{j\omega}$$

Hence

$$F[u(t)] = \pi\delta(\omega) + \frac{1}{j\omega}$$

Figure 1.5.10 shows the spectrum which consists of a frequency component at $\omega = 0$ (equivalent to d.c. signal), along with other frequency components extending up to infinity. The unit step function appears to be a d.c. signal, for $t > 0$; and hence the spectrum has an impulse at $\omega = 0$. However, it is not a pure d.c. signal because it is zero for $t < 0$ and abruptly discontinues at $t = 0$, which is the source of other high frequency components in the spectrum. A pure d.c. signal has a fixed value over the entire interval $(-\infty, \infty)$ and its consists only one frequency component, as show in Ex.1.5.3.

## 1.6 PROPERTIES OF FOURIER TRANSFORM

Fourier transform has many important properties. Apart from giving simple solutions of complicated Fourier transform, these properties also help in finding the effect of various time domain operations on frequency domain.

### (1) Linearity Property
Let
$$f_1(t) \leftrightarrow F_1(\omega)$$
$$f_2(t) \leftrightarrow F_2(\omega)$$
$$f_3(t) \leftrightarrow F_3(\omega)$$
$$f_n(t) \leftrightarrow F_n(\omega)$$

Then
$$a_1 f_1(t) + a_2 f_2(t) + a_3 f_3(t) + \cdots + a_n f_n(t) \leftrightarrow a_1 F_1(\omega) + a_2 F_2(\omega) + \cdots + a_n F_n(\omega) \quad (1.6.1)$$

where $a_1, a_2, a_3, \cdots a_n$ are arbitrary constants.

This property is proved easily by linearity property of integrals used in defining Fourier transform.

### (2) Time-scaling Property
Let
$$f(t) \leftrightarrow F(\omega)$$
then
$$f(bt) \leftrightarrow \frac{1}{|b|} F\left(\frac{\omega}{b}\right) \quad (1.6.2)$$

where $b$ is a real constant.

### Proof

$$F[f(bt)] = \int_{-\infty}^{\infty} f(bt) e^{-j\omega t} dt$$

set
$$bt = x$$

so that
$$dt = \frac{dx}{b}$$

## Case 1

When $b > 0$ ($b$ becomes a positive constant)

$$F\{f(bt)\} = \frac{1}{b}\int_{-\infty}^{\infty} f(x) e^{-j\left(\frac{\omega}{b}\right)x} dx = \frac{1}{b}F\left(\frac{\omega}{b}\right)$$

## Case 2

When $b < 0$ ($b$ becomes a negative constant). As in Case 1,

$$f(bt) \leftrightarrow \frac{1}{|b|}F\left(\frac{\omega}{b}\right)$$

Combined, the two cases are expressed as,

$$f(bt) \leftrightarrow \frac{1}{|b|}F\left(\frac{\omega}{b}\right)$$

### Significance

The function $f(bt)$ represents compression of $f(t)$ in the time domain, i.e., the period of the function is reduced by a factor $b$. The function $F\left(\frac{\omega}{b}\right)$ represents expansion of $F(\omega)$ by the same factor. Thus, the scaling property states that "compression of a function in time domain by a factor $b$ is equivalent to expansion of its frequency domain by the same factor $b$ or vice-versa".

Compression in the time domain means reducing the period of the wave. Since frequency is reciprocal of the period, the range of frequency components present in the waveform will increase; i.e., the spectrum is expanded. Let us take the example of a function $\sin \omega_0 t$, a signal of frequency $\omega_0$. When compressed in the time domain by a factor 2, it becomes $\sin 2\omega_0 t$. The frequency now is $2\omega_0$, which is double as compared to the previous case. Thus, compression of a sine wave in the time domain (reducing the period) is equivalent to expansion in the frequency domain (increasing the frequency).

In the special case when $b = -1$, the property yields,

$$f(-t) \leftrightarrow F(-\omega) \tag{1.6.3}$$

The theorem is also known as "scaling uncertainty principle" as it implies that both the signal and its Fourier transform cannot be of short duration. A signal of shorter duration yields a wider spectrum.

### (3) Duality or Symmetry Property

Let $$f(t) \leftrightarrow F(\omega)$$

then $$F(t) \leftrightarrow 2\pi f(-\omega)$$

or $$F(t) \leftrightarrow f(-f)$$

For example, for the Gate function in Fig.1.6.1, the Fourier transform is a sampling function. According to symmetry property, a sampling function has a Fourier transform which is a Gate function as in Fig.1.6.2.

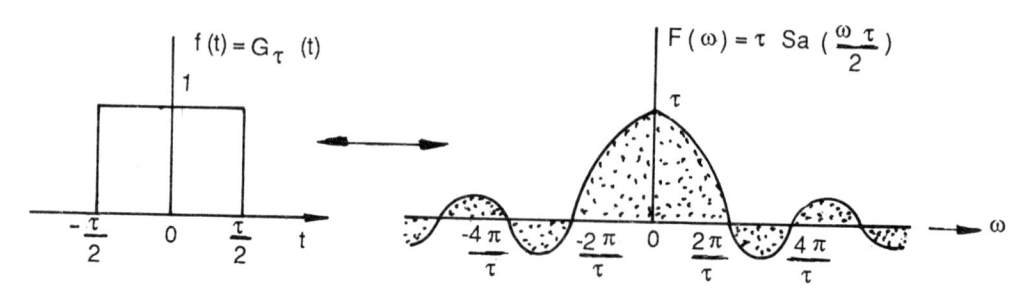

Fig. 1.6.1  A Gate Function and its Spectrum

## Proof

The inverse Fourier transform of $F(\omega)$ is given by

$$f(t) = \frac{1}{2\pi}\int_{-\infty}^{\infty} F(\omega)e^{j\omega t}\,d\omega$$

Therefore

$$2\pi f(-t) = \int_{-\infty}^{\infty} F(\omega)e^{-j\omega t}\,d\omega$$

Since, $\omega$ is a dummy variable, we can interchange the variable $t$ and $\omega$, i.e.

$$2\pi f(-\omega) = \int_{-\infty}^{\infty} F(t)e^{-j\omega t}\,dt = F[F(t)]$$

Thus, the Fourier transform of the time function $F(t)$ is $2\pi f(-\omega)$,

$$F(t) \leftrightarrow 2\pi f(-\omega)$$

For an even function

$$f(-\omega) = f(\omega)$$

and hence

$$F(t) \leftrightarrow 2\pi f(\omega)$$

This is called a *perfect symmetry*.

## Example 1.6.1

Find the inverse Fourier transform of
(a) $sgn(\omega)$;   (b) $u(\omega)$

*Solution*

(a) The Fourier transform of a signum function has already been determined and given as

$$F[sgn(t)] = \frac{2}{j\omega}$$

By applying the duality property, we get

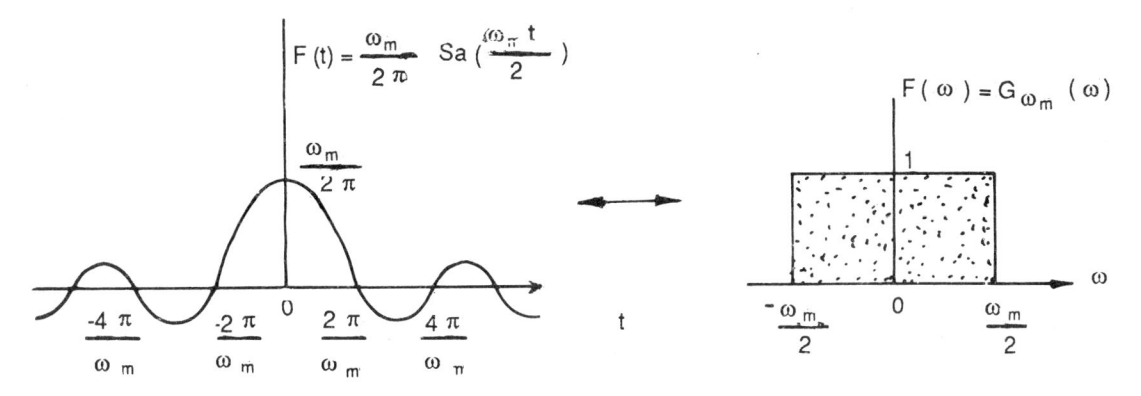

Fig. 1.6.2  A Sampling Function and its Spectrum

$$F\left[\frac{2}{jt}\right] = 2\pi \, sgn\,(-\omega)$$

Since $sgn\,(\omega)$ is an odd function i.e.; $sgn\,(-\omega) = -sgn\,(\omega)$
hence

$$\frac{2}{jt} \leftrightarrow -2\pi \, sgn\,(\omega)$$

or

$$-\frac{1}{jt} \leftrightarrow \pi \, sgn\,(\omega)$$

or

$$\frac{j}{\pi t} \leftrightarrow sgn\,(\omega)$$

Thus, the inverse Fourier transform of $sgn\,(\omega)$ is purely imaginary function $\frac{j}{\pi t}$.

(b)
$$u\,(\omega) = \frac{1}{2} + \frac{1}{2} sgn\,(\omega)$$

By linearity property

$$F^{-1}[u\,(\omega)] = F^{-1}\left[\frac{1}{2}\right] + F^{-1}\left[\frac{1}{2} sgn\,(\omega)\right]$$

$$= \frac{1}{2} \delta\,(t) + \frac{j}{2\pi t}$$

### (4) Time Shifting Property
Let
$$f\,(t) \leftrightarrow F\,(\omega)$$

## 44 COMMUNICATION SYSTEMS: Analog and Digital

then

$$f(t-b) \leftrightarrow F(\omega)e^{-j\omega b} \qquad (1.6.6)$$

### Statement

This theorem states that a shift in the time domain by an amount $b$ is equivalent to multiplication by $e^{-j\omega b}$ in the frequency domain, i.e. magnitude $F(\omega)$ spectrum remains unchanged, but phase spectrum is changed by $-\omega b$. A shift in time domain does not change the magnitude of a frequency component. A shift of $b$ at a frequency $\omega$ is equivalent to a phase shift of $\omega b$

### Proof

$$f(t-b) \leftrightarrow \int_{-\infty}^{\infty} f(t-b) e^{-j\omega t} dt$$

put $(t-b) = y$, so that $dt = dy$
then

$$f(t-b) \leftrightarrow \int_{-\infty}^{\infty} f(y) e^{-j\omega(b+y)} dy = \int_{-\infty}^{\infty} f(y) e^{-j\omega y} e^{-j\omega b} dy$$

Since $y$ is a dummy variable

$$f(t-b) \leftrightarrow F(\omega) e^{-j\omega b}$$

### Example 1.6.2

Find the Fourier transform of the pulse function $f(t)$ shown in Fig.1.6.3a.

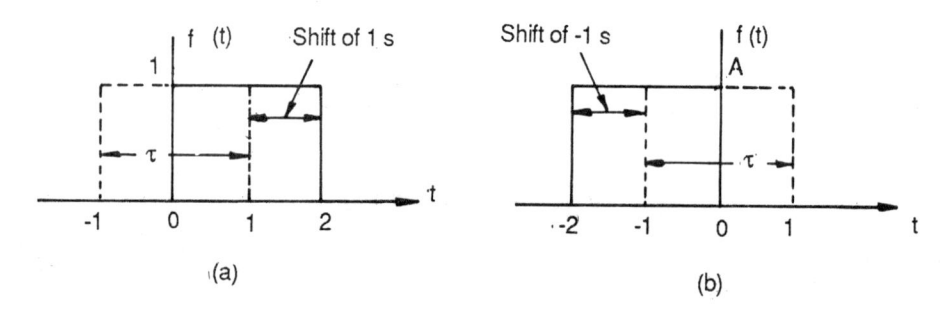

Fig. 1.6.3

### Solution

The function $f(t)$ shown in Fig.1.6.3a is a Gate function $G_\tau(t)$s shifted by $+1$ s. Fourier transfer of $G_\tau(t)$ is given by, ($G_\tau(t)$ is identified by dotted lines)

$$G_\tau(t) \leftrightarrow \tau Sa\left(\frac{\omega \tau}{2}\right)$$

Here

$$\tau = 2$$

hence

$$G_\tau(t) \leftrightarrow 2\, Sa(\omega)$$

Since the function $f(t)$ is $G_\tau(t)$ shifted by 1 s. $(b = 1)$, using time shifting theorem, we have

$$f(t) \leftrightarrow F[G_\tau(t)]e^{-j\omega b} = 2e^{-j\omega} Sa(\omega)$$

## Example 1.6.3

Find the Fourier transform of the pulse function $f(t)$ shown by continuous lines in Fig.1.6.3b.

*Solution*

The function $f(t)$ is a Gate function $G_\tau(t)$ shifted by $-1$ s. The Fourier transform of the Gate function is

$$G_\tau(t) \leftrightarrow A\tau\, Sa\left(\frac{\omega\tau}{2}\right)$$

putting $\tau = 2$ sec.

$$G_\tau(t) \leftrightarrow 2A\, Sa(\omega)$$

The Fourier transform of $f(t)$ is the spectrum of $G_\tau(t)$ shifted by $-1$ sec. Thus

$$f(t) \leftrightarrow 2A\, e^{j\omega} Sa(\omega)$$

## (5) Frequency Shifting Property

If
$$f(t) \leftrightarrow F(\omega)$$
then
$$e^{j\omega_c t} f(t) \leftrightarrow F(\omega - \omega_c) \tag{1.6.7}$$

*Proof*

The Fourier transform $F(\omega)$ is given as

$$F(\omega) = \int_{-\infty}^{\infty} f(t)\, e^{-j\omega t}\, dt$$

Therefore,

$$f(t)\, e^{j\omega_c t} \leftrightarrow \int_{-\infty}^{\infty} e^{j\omega_c t} f(t) e^{-j\omega t}\, dt = \int_{-\infty}^{\infty} f(t) e^{-j(\omega - \omega_c)t}\, dt$$

or

$$F\left[f(t)\, e^{j\omega_c t}\right] = F(\omega - \omega_c)$$

*Significance*

According to this property, multiplication of a function $f(t)$ by $e^{j\omega_c t}$ is equivalent to shifting its Fourier transform $F(\omega)$ in the positive direction by an amount $\omega_c$ i.e., spectrum $F(\omega)$ is translated by an amount $\omega_c$. Therefore, this theorem is also known as *Frequency translation theorem*.

Translation of a spectrum is a very important phenomenon in communication system. This translation helps in simultaneous transmission of messages using frequency division multiplexing and is achieved by a process known as *modulation*. This process can be performed by multiplying the known signal $f(t)$ by a sinusoidal signal. This is because a sinusoidal function can be expressed as a sum of exponentials as follows:

**46** COMMUNICATION SYSTEMS: Analog and Digital

$$f(t)\cos\omega_c t = \frac{1}{2}\left[f(t)e^{j\omega_c t} + f(t)e^{-j\omega_c t}\right]$$

Therefore,

$$F[f(t)\cos\omega_c t] = \frac{1}{2}[F(\omega+\omega_c) + F(\omega-\omega_c)] \tag{1.6.8a}$$

Similarly,

$$F[f(t)\sin\omega_c t] = \frac{j}{2}[F(\omega+\omega_c) - F(\omega-\omega_c)] \tag{1.6.8b}$$

The multiplication of a time function $f(t)$ with a sinusoidal function translates the whole spectrum $F(\omega)$ to $\pm\omega_c$. We can obtain a similar result by the process of modulation. This result, also known as modulation theorem, states that the multiplication of a time function $f(t)$ with a sinusoidal function of frequency $\omega_c$ translates the spectrum $F(\omega)$ by an amount $\omega_c$. It should be noted that $e^{j\omega_c t}$ can also provide frequency translation, but it is not a real signal, whereas, sinusoidal function is a real signal. Hence, sinusoidal, function is used in practical modulation systems.

### Example 1.6.4
Find the Fourier transform of Eternal sinusoids (i) $\cos\omega_c t$ and, (ii) $\sin\omega_c t$ shown in Fig. 1.6.4 (a) and (c) respectively.
*Solution*
(i) Using Euler's identiy to express the cosine function,

$$\cos\omega_c t = \frac{e^{j\omega_c t} + e^{-j\omega_c t}}{2}$$

Note that, *Eternal sinusoid exists forever* $(-\infty, \infty)$. Using the linearity property of the Fourier transform we have

$$F[\cos\omega_c t] = F\left[\frac{1}{2}e^{j\omega_c t}\right] + F\left[\frac{1}{2}e^{-j\omega_c t}\right]$$

The Fourier transform of a constant $\frac{1}{2}$ is $\pi\delta(\omega)$. Therefore, using frequency shifting property,

$$F[\cos\omega_c t] = \pi\delta(\omega-\omega_c) + \pi\delta(\omega+\omega_c) \tag{1.6.9a}$$

The spectrum as been shown in Fig. 1.6.4b, and has two impulses located at $\pm\omega_c$

(ii) 
$$\sin\omega_c t = \frac{1}{2j}\left[e^{j\omega_c t} - e^{-j\omega_c t}\right]$$

Using linearity property,

$$F[\sin\omega_c t] = F\left[\frac{1}{2j}e^{j\omega_c t}\right] - F\left[\frac{1}{2j}e^{-j\omega_c t}\right]$$

Using frequency shifting property, as in case (i)

$$F[\sin \omega_c t] = \frac{1}{j}[\pi\delta(\omega - \omega_c) - \pi\delta(\omega + \omega_c)]$$

$$= j\pi[\delta(\omega + \omega_c) - \delta(\omega - \omega_c)] \quad (1.6.9b)$$

The spectrum has two impulses located at $\pm \omega_c$ as shown in Fig.1.6.4 (d). Since transform is purely imaginary, only amplitude plot is required. The spectrum of a sinusoidal function exists only at single frequency $\pm \omega_c$. This is true because a pure sinusoidal function has only the fundamental frequency component.

### Example 1.6.5

Find the Fourier transform of an Eternal exponential $e^{j\omega_0 t}$

*Solution*

The Eternal exponential exists for entire period $(-\infty, \infty)$.

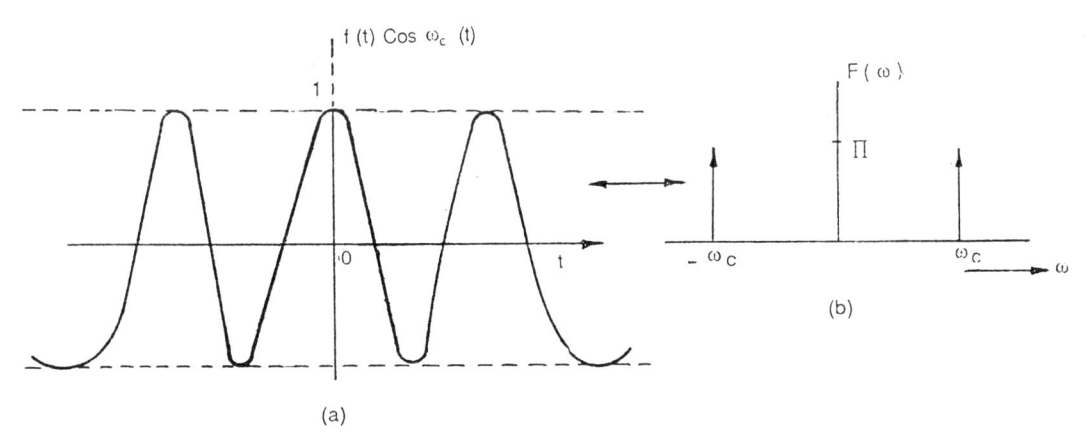

Fig. 1.6.4

$$F\left[e^{j\omega_0 t}\right] = F\left[1 e^{j\omega_0 t}\right]$$

The Fourier transform of a constant '1' is known to be $2\pi\delta(\omega)$. Hence, using frequency shifting property, we have

$$F\left[e^{j\omega_0 t}\right] = 2\pi\delta(\omega - \omega_0) \quad (1.6.10)$$

The spectrum is shown in Fig. 1.6.5. Interestingly the spectrum has a single impulse at $\omega = \omega_0$, and there is no associated negative frequency component and hence the spectrum has no even symmetry. This is because $e^{j\omega_0 t}$ is an imaginary function and not a real function of time.

48  COMMUNICATION SYSTEMS: Analog and Digital

(c)

(d)

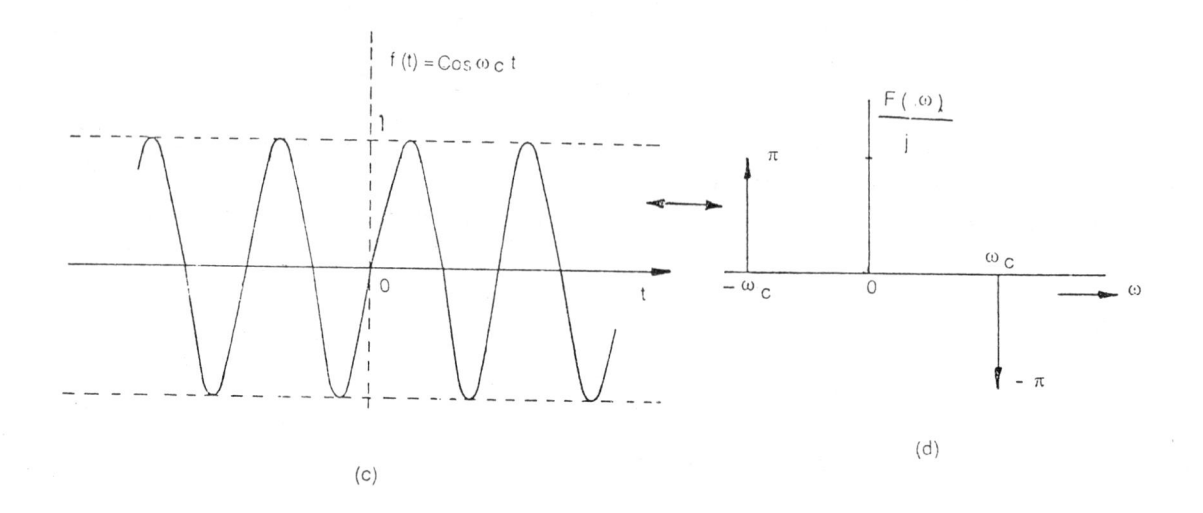

Fig. 1.6.4

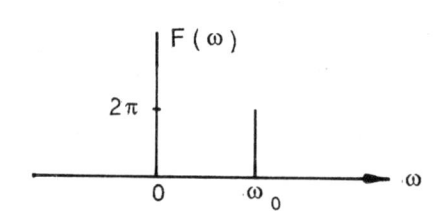

Fig. 1.6.5  Spectrum of an Eternal Exponential

## Example 1.6.6

Find the Fourier transform of a radio frequency pulse shown in Fig.1.6.6(a)

*Solution*

The function consists of a sinusoidal wave of frequency $\omega_0$ extending from $\frac{-\tau}{2}$ to $\frac{\tau}{2}$. The function $f(t)$ may be considered as multiplication of an Eternal sinusoid $\cos \omega_0 t$ and Gate function $G_\tau(t)$ as in Fig.1.6.6 b. This may be expressed Mathematically as,

$$f(t) = G_\tau(t) \cos \omega_0 t = G_\tau(t) \left\{ \frac{1}{2} \left( e^{j\omega_0 t} + e^{-j\omega_0 t} \right) \right\}$$

$$= \frac{1}{2} \left[ G_\tau(t) e^{j\omega_0 t} + G_\tau(t) e^{-j\omega_0 t} \right]$$

SIGNAL ANALYSIS 49

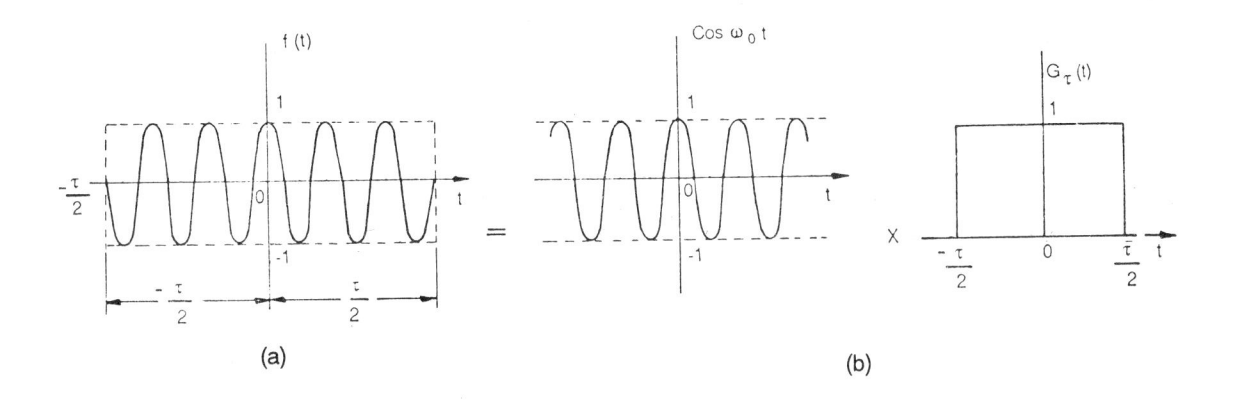

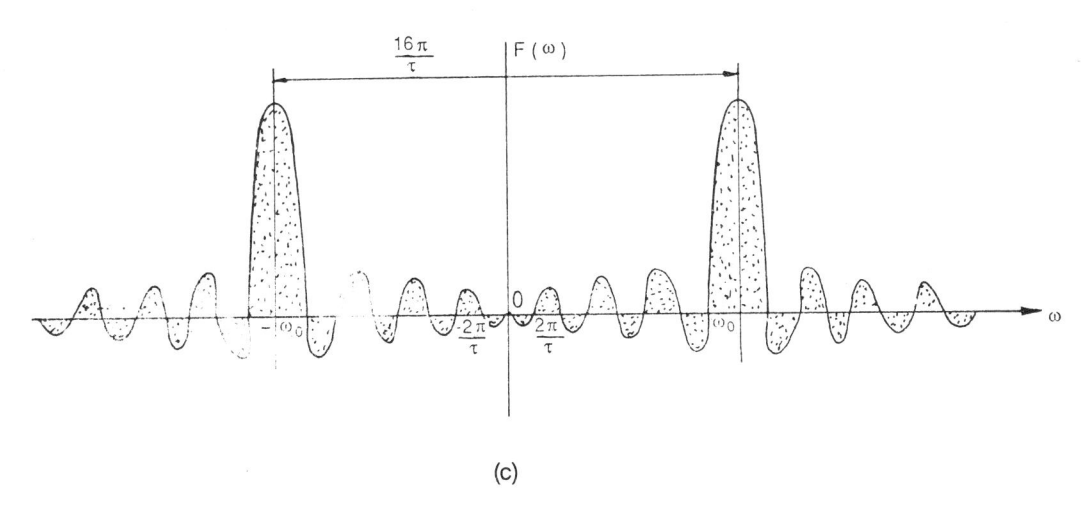

Fig. 1.6.6 (a) and (b) An RF Pulse, (c) Spectrum

The Fourier transform of a Gate function $G_\tau(t)$ is known to be sampling function, $\tau\, Sa\left(\dfrac{\omega\tau}{2}\right)$. Hence, using frequency shifting property;

$$F[f(t)] = \frac{\tau}{2}\left[Sa\left\{\frac{(\omega-\omega_0)\tau}{2}\right\} + Sa\left\{\frac{(\omega+\omega_0)\tau}{2}\right\}\right] \qquad (1.6.11)$$

The spectrum is shown in Fig. 1.6.6 (c) for $\tau = \dfrac{8}{f_0}$ corresponding to a function $\cos\omega_0 t$ existing only for 8 cycles i.e.,

**50** COMMUNICATION SYSTEMS: Analog and Digital

$$f(t) = \begin{cases} \cos\omega_0 t, & |t| < \dfrac{8}{f_0} \quad \left(\text{i.e. up to 8th cycle, each cycle of duration } \dfrac{1}{f_0} \text{ sec.}\right) \\ 0, & |t| > \dfrac{8}{f_0} \quad \text{(i.e., beyond 8th cycle)} \end{cases}$$

where $f_0 = \dfrac{\omega_0}{2\pi}$

It should be observed that most of the energy is concentrated near $\pm\omega_0$, and this concentration increases as $\tau$ is increased. In the limit $\tau \to \infty$, the entire energy is concentrated at $\pm\omega_0$ providing a spectrum with only two impulses at $\pm\omega_0$ which is identical to the Eternal sinusoid $\cos\omega_0 t$. Indeed as $\tau \to \infty$, the radio pulse becomes an Eternal sinusoid. Further, this example illustrates the modulation theorem. The spectrum of the Gate function has been translated to $\pm\omega_0$ as is obvious from Fig.1.6.6c.

### Example 1.6.7

An audio oscillator is switched on at $t = 0$. Find the spectrum of the generated (i) cosine waveform; (ii) sine waveform both of frequency $\omega_c$.

**Solution**

The generated waveforms start at $t = 0$ as shown in Fig.1.6.7(a). The waveforms appear to be pure sinusoid (single frequency sinusoid) and are expected to have impulse only at $\pm\omega_0$ in their respective spectrum. But a pure sinusoid exists forever $(-\infty < t < \infty)$, rather than only for $t \geq 0$. An oscillator switched on at $t = 0$ will exist only for $0 < t < \infty$. It is associated with transients and will contain other frequency components alongwith the frequency $\omega_c$. The spectrum discussed below makes it clear. The sinusoid may be expressed as a product of an Eternal sinusoid and a unit step function,

$$u(t)\cos\omega_c t = \frac{1}{2}u(t)\left[e^{j\omega_c t} + e^{-j\omega_c t}\right]$$

$$u(t)\sin\omega_c t = \frac{1}{2j}u(t)\left[e^{j\omega_c t} - e^{-j\omega_c t}\right]$$

(i)
$$F[u(t)\cos\omega_c t] = F\left[\frac{1}{2}u(t)e^{j\omega_c t} + \frac{1}{2}u(t)e^{-j\omega_c t}\right]$$

Since, Fourier transform of $u(t)$ is known to be,

$$u(t) \leftrightarrow \pi\delta(\omega) + \frac{1}{j\omega}$$

Therefore, using, frequency shifting property

$$F[u(t)\cos\omega_c t] = \frac{\pi}{2}\delta(\omega - \omega_c) + \frac{1}{2j(\omega - \omega_c)} + \frac{\pi}{2}\delta(\omega + \omega_c) + \frac{1}{2j(\omega + \omega_c)}$$

$$= \frac{\pi}{2}[\delta(\omega - \omega_c) + \delta(\omega + \omega_c)] + \frac{1}{2j(\omega - \omega_c)} + \frac{1}{2j(\omega + \omega_c)}$$

or
$$u(t)\cos\omega_c t = \frac{\pi}{2}[\delta(\omega-\omega_c)+\delta(\omega+\omega_c)]+\frac{j\omega}{\omega_c^2-\omega^2} \qquad (1.6.12a)$$

(ii) Similar to case (i)

$$F[u(t)\sin\omega_c t] \leftrightarrow F\left[\frac{1}{2j}u(t)e^{j\omega_c t}-\frac{1}{2j}u(t)e^{-j\omega_c t}\right]$$

using frequency shifting property,

$$u(t)\sin\omega_c t \leftrightarrow \frac{\pi}{2j}[\delta(\omega-\omega_c)-\delta(\omega+\omega_c)]+\frac{\omega_c}{\omega_c^2-\omega^2} \qquad (1.6.12b)$$

The spectra shown in Fig.1.6.7(b) consist of impulses at $\pm\omega_c$ and other frequency components concentrated about $\pm\omega_c$ which die down as we move away from $\pm\omega_c$.

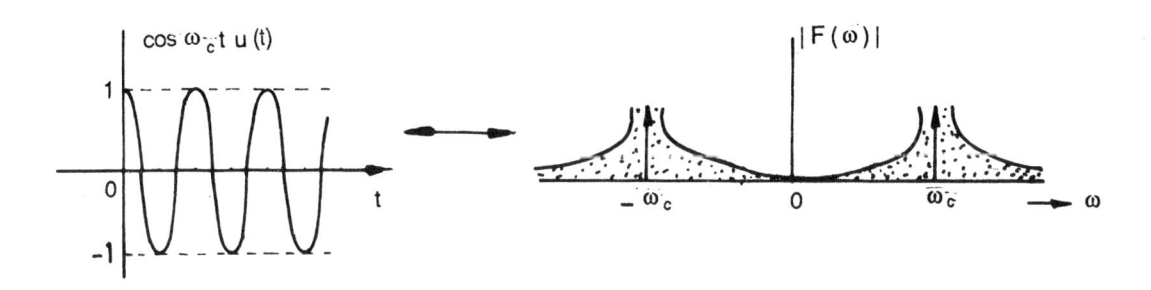

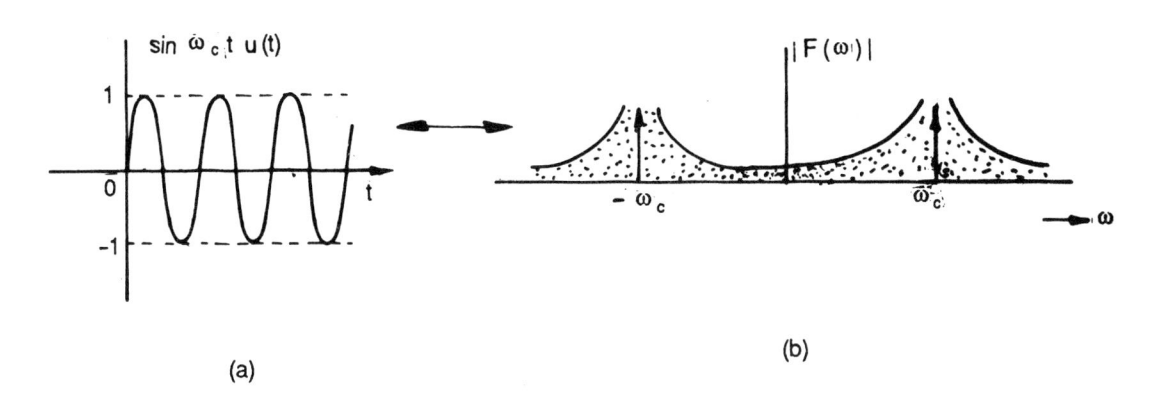

Fig. 1.6.7

## Example 1.6.8

Find the Fourier transform of a damped sinusoidal waveform of frequency $\omega_0$ shown in Fig. 1.6.8 (a)

52 COMMUNICATION SYSTEMS: Analog and Digital

(a)

(b)

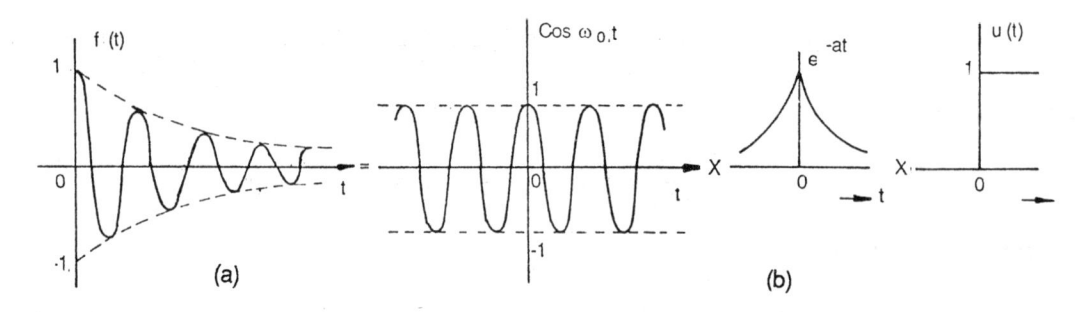

Fig. 1.6.8 Damped Sinusoid

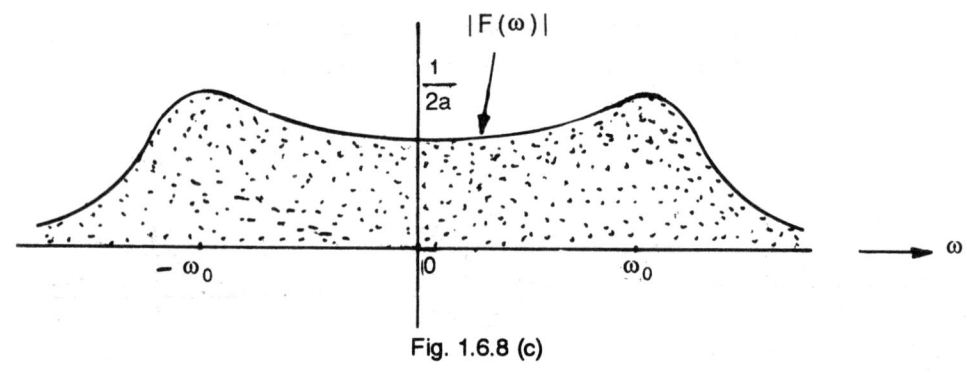

Fig. 1.6.8 (c)

*Solution*

The waveform in Fig.1.6.8(a) may be considered as the product of three waveforms: viz., an Eternal cosine function, a double sided exponential, and a unit step function, as shown in Fig.1.6.8 (b). The function can be expressed as

$$f(t) = e^{-at} \cos \omega_0 t \, u(t)$$

The function $e^{-at} u(t)$ is a single-sided exponential; its Fourier transform has already been determined

$$e^{-at} u(t) \leftrightarrow \frac{1}{a + j\omega}$$

Therefore, using frequency shifting property, we have

$$F\left[e^{-at} \cos \omega_0 t \, u(t)\right] = F\left[\frac{1}{2} e^{-at} u(t) e^{j\omega_0 t} + \frac{1}{2} e^{-at} e^{-j\omega_0 t}\right]$$

$$= \frac{1}{2}\left[\frac{1}{a + j(\omega - \omega_0)} + \frac{1}{a + j(\omega + \omega_0)}\right]$$

$$= \frac{(a + j\omega)}{(a + j\omega)^2 - (j\omega_0)^2}$$

$$= \frac{(a + j\omega)}{(a + j\omega)^2 - (a + j\omega)^2}$$

The spectrum is shown in Fig.1.6.8c.

## Example 1.6.9

Find the Fourier transform of $f(t)$ given by,

$$f(t) = \begin{cases} e^{jt}, & |t| < 1 \\ 0, & \text{otherwise} \end{cases}$$

*Solution*

The function $f(t)$ may be written as

$$f(t) = \begin{cases} 1\, e^{jt}, & |t| < 1 \\ 0, & \text{elsewhere} \end{cases}$$

If $e^{jt}$ is ignored it will be a Gate function having sampling function as its Fourier transform; as shown in Fig. 1.6.9

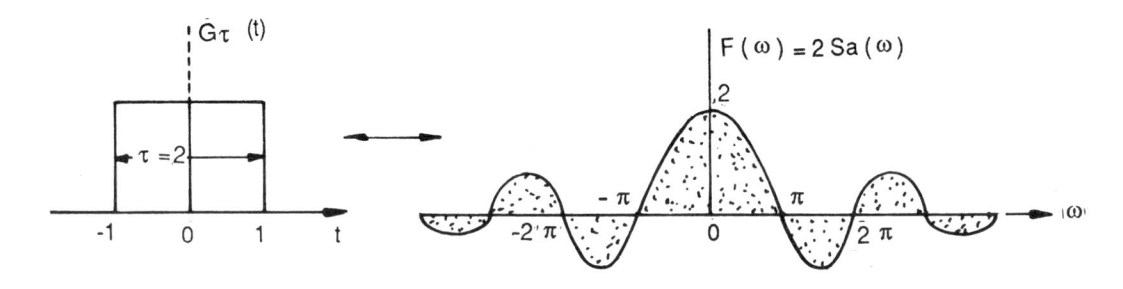

Fig. 1.6.9

This can be expressed as ($\tau = 2$)

$$G_\tau(t) \leftrightarrow 2\, Sa(\omega)$$

The function $f(t)$ is the Gate function multiplied by $e^{j.1.t}$ (i.e., $e^{j\omega_0 t}$ with $\omega_0 = 1$). Using frequency shifting property

$$f(t) \leftrightarrow 2\, Sa(\omega - 1)$$

The Fourier transform of the function $f(t)$ is shown in Fig. 1.6.10. The spectrum is obtained by shifting the spectrum of Fig. 1.6.9 by 1 second.

## Example 1.6.10

Find the Fourier transform of the waveform shown in Fig. 1.6.11 (a).

*Solution*

The waveforms representing $f(t)$ may be expressed as the sum of two waveforms shown in Fig. 1.6.11 (b), and (c). The Fourier transform of the Gate function in Fig. 1.6.11 b is $2\, Sa(\omega)$, and the Fourier transform of shifted

54 COMMUNICATION SYSTEMS: Analog and Digital

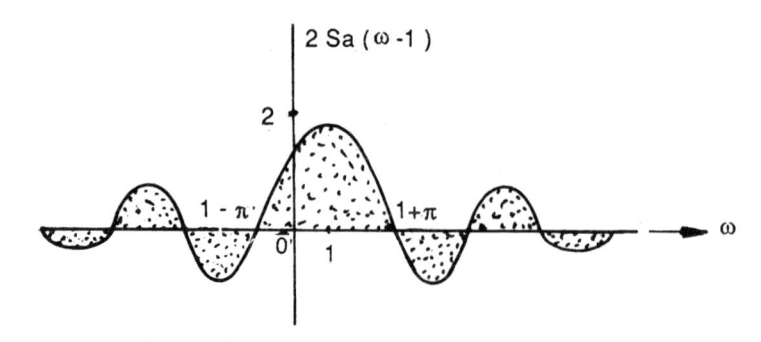

Fig. 1.6.10

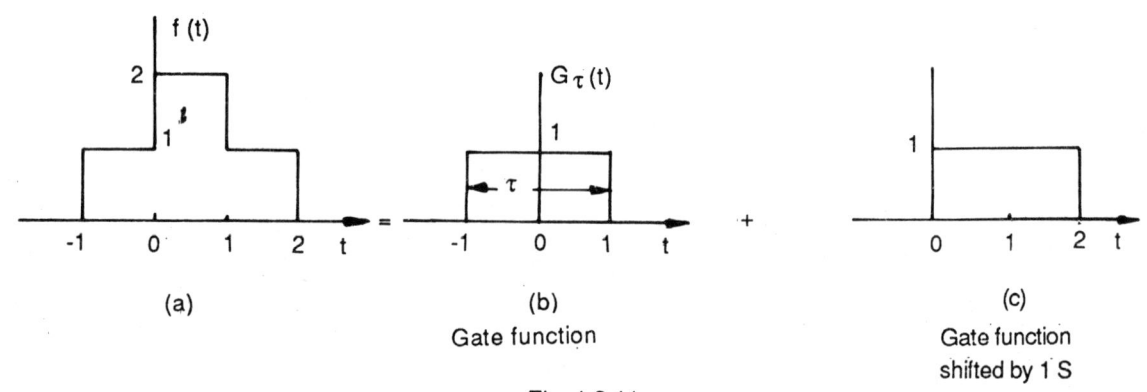

Fig. 1.6.11

(a)

(b) Gate function

(c) Gate function shifted by 1 S

Gate function in Fig.1.6.11 (c) is $2Sa(\omega)e^{-j(\omega)}$ (using time shifting property). Hence the fourier transform of $f(t)$ may be obtained using linearity property of Fourier transform.

$$F[f(t)] = 2Sa(\omega) + 2Sa(\omega)e^{-j\omega}$$

or

$$f(t) \leftrightarrow 2Sa(\omega)\left[1 + e^{-j\omega}\right]$$

## (6) Area under the Curve

(i) The area under a function $f(t)$ is equal to the value of its Fourier transform $F(\omega)$ at $\omega = 0$, i.e., if
$$f(t) \leftrightarrow F(\omega)$$
then

$$\int_{-\infty}^{\infty} f(t)\,dt = \frac{1}{2\pi} F(0) \qquad 1.6.13a$$

This can be proved by putting $\omega = 0$ in the expression of $F(\omega)$.

(ii) The area under the Fourier transform $F(\omega)$ of a function $f(t)$ is equal to the value of the function $f(t)$ at $t = 0$, i.e.,

$$f(0) = \int_{-\infty}^{\infty} F(\omega)\, d\omega \qquad (1.6.13b)$$

The result can be proved by taking $t = 0$ in the expression of inverse Fourier transform of $F(\omega)$.

## (7) Time Differentiation and Integration

(i) Differentiation of a function $f(t)$ in the time domain is equivalent to multiplication of its Fourier transform by a factor $j\omega$,

$$\frac{df(t)}{dt} \leftrightarrow j\omega F(\omega) \qquad (1.6.14a)$$

### Proof

The result can be proved by taking the first derivative of both the sides of the expression for inverse Fourier transform,

$$f(t) = \frac{1}{2\pi} \int_{-\infty}^{\infty} F(\omega) e^{j\omega t}\, d\omega$$

$$\frac{df(t)}{dt} = \frac{1}{2\pi} \frac{d}{dt}\left[\int_{-\infty}^{\infty} F(\omega) e^{j\omega t}\, d\omega\right]$$

Changing the order of differentiation and integration, we get

$$\frac{df(t)}{dt} = \frac{1}{2\pi} \int_{-\infty}^{\infty} \frac{d}{dt}\left\{F(\omega) e^{j\omega t}\right\} d\omega$$

$$= \frac{1}{2\pi} \int_{-\infty}^{\infty} j\omega F(\omega) e^{j\omega t}\, d\omega$$

$$= F^{-1}[j\omega F(\omega)]$$

Hence,

$$F\left[\frac{df(t)}{dt}\right] = j\omega F(\omega)$$

or

$$\frac{df(t)}{dt} \leftrightarrow j\omega F(\omega)$$

The theorem is logical because Function $f(t)$ is expressed as the continuous sum of exponential function $e^{j\omega t}$. Hence the differentiation of $f(t)$ yields $j\omega$ in its Fourier transform since time derivative of $e^{j\omega t}$ is $j\omega e^{j\omega t}$.

In general,

$$\frac{d^n f(t)}{dt^n} = (j\omega)^n F(\omega) \qquad (1.6.14b)$$

(ii) Integration of a function $f(t)$ in time domain is equivalent to division of its Fourier transform by the factor $j\omega$, as integration of $e^{j\omega t}$ yields $\frac{1}{j\omega} e^{j\omega t}$. The theorem is expressed as,

56  COMMUNICATION SYSTEMS: Analog and Digital

$$\int_{-\infty}^{t} f(\tau)d\tau \leftrightarrow \frac{1}{j\omega} F(\omega) \tag{1.6.14c}$$

Provided

$$F(0) = 0$$

The theorem can be proved with an approach similar to the time differentiation theorem (Prob.1.8)

### (8) Conjugate Functions

If

$$f(t) \leftrightarrow F(\omega)$$

then

$$f^*(t) \leftrightarrow F^*(-\omega) \tag{1.6.15}$$

where, the asterisk denotes the complex conjugate operation. The theorem can be proved by taking the complex conjugates of both sides of the equation for inverse Fourier transform of $F(\omega)$ and then by replacing $f$ with $-f$ (Prob.1.8)

### (9) Frequency Differentiation (Dual of time differentiation)

If

$$f(t) \leftrightarrow F(\omega)$$

then

$$-jtf(t) \leftrightarrow \frac{dF}{d\omega} \tag{1.6.16}$$

The theorem can be proved by taking frequency derivative of $F(\omega)$, and then changing the order of integration and differentiation (Prob. 1.8).

### Example 1.6.11

Determine the Fourier transform of a *trapezoidal function* $f(t)$ shown in Fig. 1.6.12a.

*Solution*

The first order derivative of function $f(t)$ gives the pulses of height $\left|\frac{A}{n-m}\right|$, as shown in Fig.1.6.12(b), and second order derivative yields impulses of strength $\left|\frac{A}{n-m}\right|$ located at $t = -n, -m, m$ and $n$, as shown in Fig.1.6.12(c). Thus, we have from Fig.1.6.12c.

$$\frac{d^2 f(t)}{dt^2} = \frac{A}{n-m}[\delta(t+n) + \delta(t-n) - \delta(t+m) - \delta(t-m)]$$

Therefore, applying time differentiation and time shifting property of Fourier transform we get

SIGNAL ANALYSIS 57

(a)

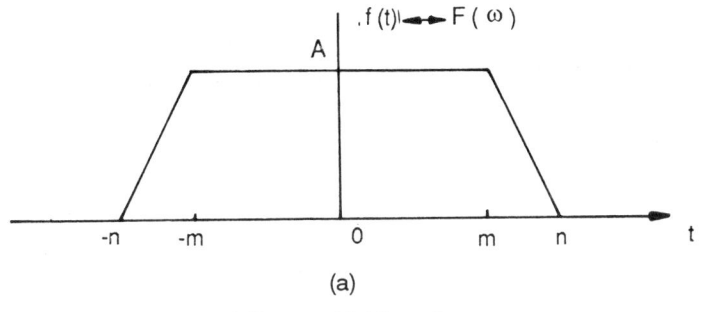

A Trapezoidal Function

(b)

(c)

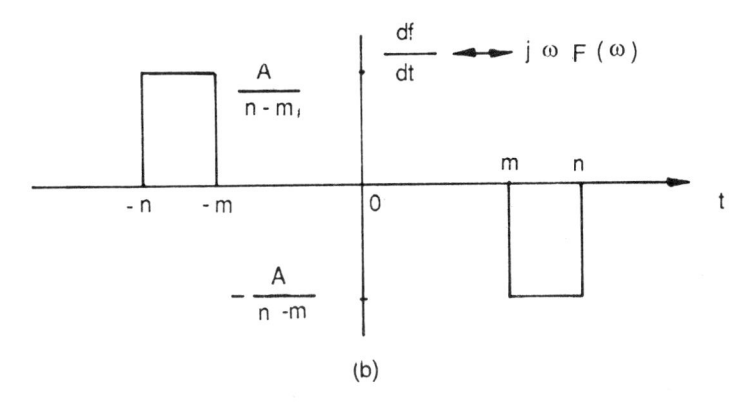

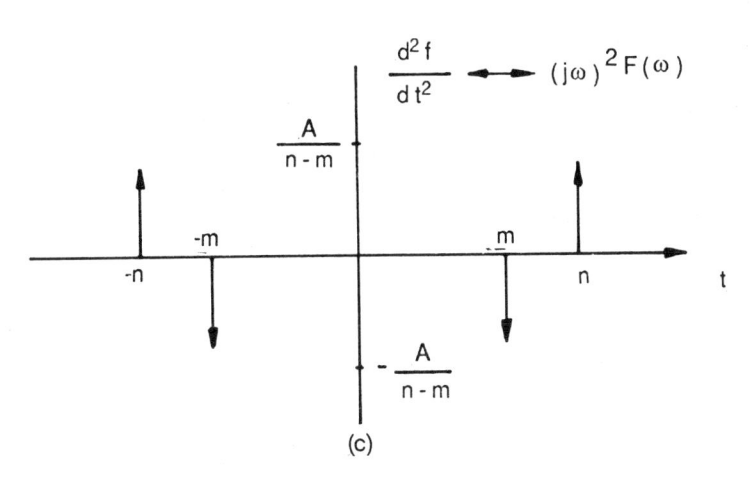

Fig. 1.6.12

$$(j\omega)^2 F(\omega) = \frac{A}{n-m}(e^{j\omega n} + e^{-j\omega n} - e^{j\omega m} - e^{-j\omega m})$$

which yields

$$F(\omega) = \frac{2A}{(n-m)}\left(\frac{\cos m\omega - \cos n\omega}{\omega^2}\right)$$

## 58 COMMUNICATION SYSTEMS: Analog and Digital

### Numerical Method

Example 1.6.11 provides a tool for evaluating the Fourier transform of an arbitrary waveshape. The waveshape may be approximated by line segments as in Fig.1.6.13. The more the number of segments, the less is the error in approximation.

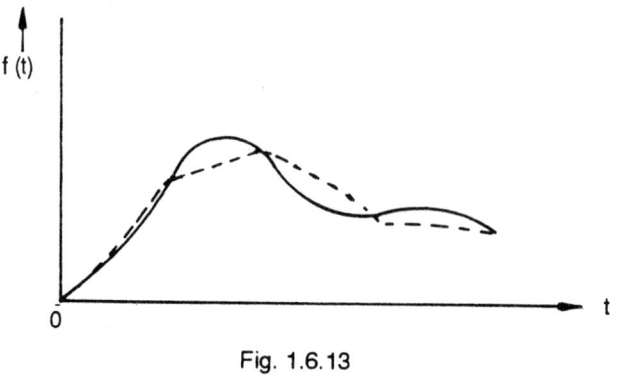

Fig. 1.6.13

The second derivative of $f(t)$ yields train of impulses depending on the nature of the line segments, the Fourier transform of these can be easily determined. This is know as the numerical method of signal analysis.

### Example 1.6.12

Find the Fourier transform of a triangular RF pulse shown in Fig.1.6.14 a, and draw its spectrum. The radio frequency is $\omega_0$.

*Solution*

The given pulse may be considered as a triangular pulse multiplied by $\cos\omega_0 t$ at every instant as shown in Fig. 1.6.14 b. Since $\cos\omega_0 t$ has a negative excursion, the multiplied signal has also a negative excursion. According to modulation theorem the Fourier transform of Fig. 1.6.14a is the Fourier transform of a triangular pulse shifted by $\pm \omega_0$ (recall Eq. 1.6.8).

The main problem is in finding the Fourier transform of the triangular pulse redrawn in Fig.1.6.15 a. The differentiation of the pulse yields a doublet pulse $G(t)$ shown in Fig.1.6.15 b. The Fourier transform of this pulse can be obtained by using time shifting property.

The function $G(t)$ in Fig. 1.6.15b is split in two parts; $G_1(t)$; and $G_2(t)$, as in Fig.1.6.15 (d) with the Fourier transforms as given below (refer to time shift theorem).

$$G_1(t) \leftrightarrow \tau \times \frac{1}{\tau} Sa\left(\frac{\omega\tau}{2}\right) e^{j\omega\frac{\tau}{2}}$$

$$G_2(t) \leftrightarrow -\tau \times \frac{1}{\tau} Sa\left(\frac{\omega\tau}{2}\right) e^{-j\omega\frac{\tau}{2}}$$

The Fourier transform of doublet pulse $G(t)$ in Fig. 1.6.15b is sum of the Fourier transforms of $G_1(t)$ and $G_2(t)$ (using linearity property) i.e.,

SIGNAL ANALYSIS 59

$$f_1(t) = \begin{cases} 1 - \left|\dfrac{t}{\tau}\right|, & |t| < \tau \\ 0, & \text{elsewhere} \end{cases}$$

$$f_2(t) = \cos \omega_0 t$$

(a)  (b)

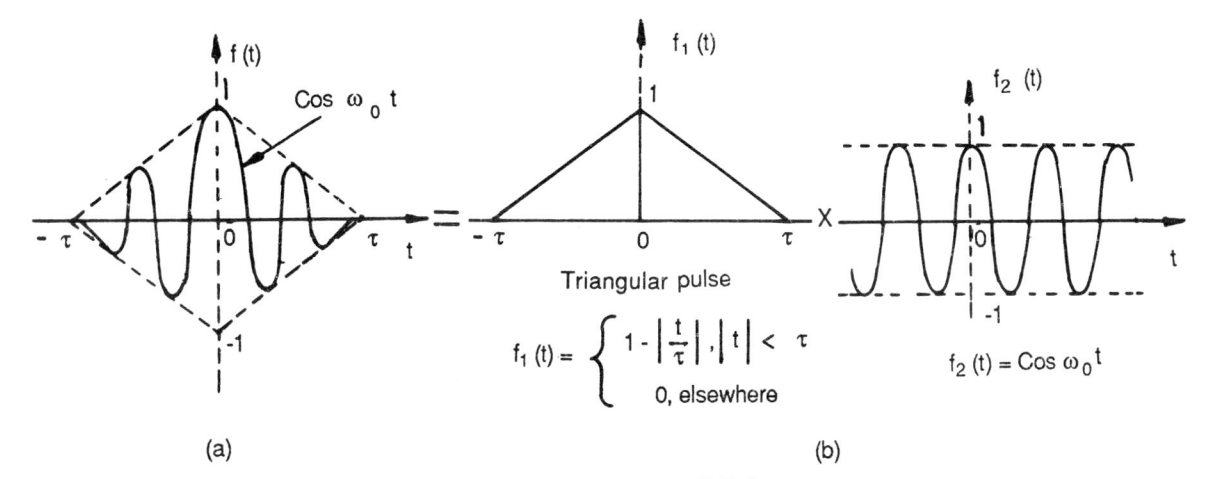

Fig. 1.6.14 A Triangular RF Pulse

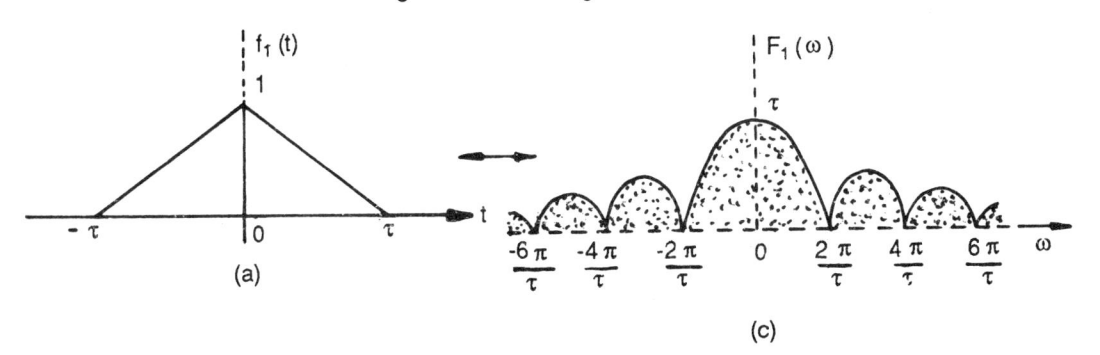

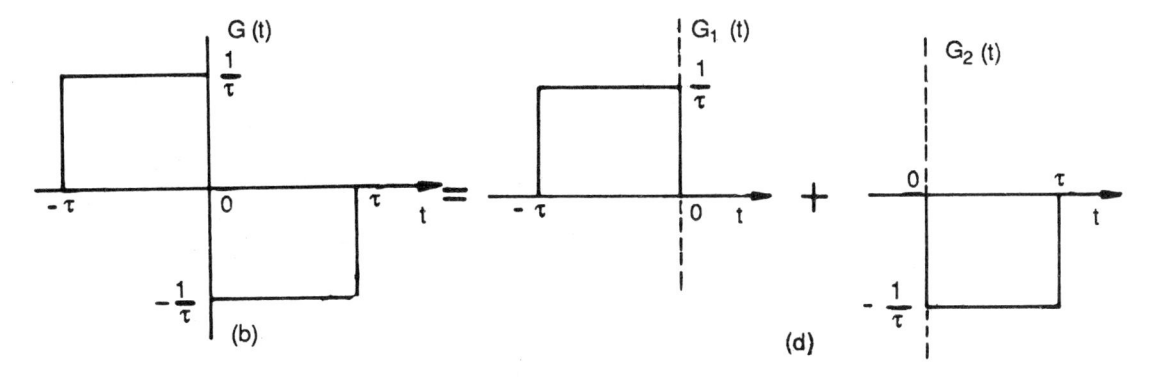

(a) (c)

Fig. 1.6.15

$$G(t) \leftrightarrow Sa\left(\frac{\omega\tau}{2}\right)\left[e^{j\frac{\omega\tau}{2}} - e^{-j\frac{\omega\tau}{2}}\right] = Sa\left(\frac{\omega\tau}{2}\right)(2j)\frac{e^{j\frac{\omega\tau}{2}} - e^{-j\frac{\omega\tau}{2}}}{2j}$$

or

$$G(t) \leftrightarrow 2j\, Sa\left(\frac{\omega\tau}{2}\right) \sin\left(\frac{\omega\tau}{2}\right)$$

Since, differentiation of triangular pulse $f_1(t)$ yields $G(t)$, we have

$$F[f_1(t)] \times j\omega = F[G(t)]$$

Therefore, Fourier transform of the triangular pulse $f_1(t)$ is given by

$$f_1(t) \leftrightarrow \frac{1}{j\omega}(2j)\, Sa\left(\frac{\omega\tau}{2}\right)\sin\left(\frac{\omega\tau}{2}\right) = \tau\, Sa\left(\frac{\omega\tau}{2}\right)\frac{\sin\left(\frac{\omega\tau}{2}\right)}{\left(\frac{\omega\tau}{2}\right)}$$

or

$$f_1(t) \leftrightarrow \tau\, Sa^2\left(\frac{\omega\tau}{2}\right) \qquad (1.6.17)$$

The spectrum of the triangular pulse $f_1(t)$ is shown in Fig.1.6.15(c) which is a sampling square function.

The Fourier transform of $f(t)$ is the Fourier transform of $f_1(t)$ shifted by $\pm\omega_0$ (using frequency shifting theorem),

$$f(t) \leftrightarrow \frac{\tau}{2} Sa^2\left[(\omega - \omega_0)\frac{\tau}{2}\right] + \frac{\tau}{2} Sa^2\left[(\omega + \omega_0)\frac{\tau}{2}\right] \qquad (1.6.18)$$

The spectrum is plotted in Fig. 1.6.16.

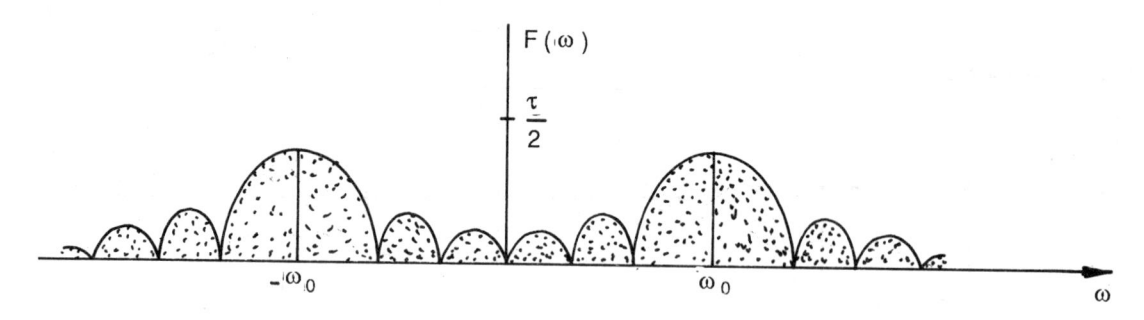

Fig.1.6.16 Spectrum of Triangular RF Pulse

### Example 1.6.13

Show that a normalised Gaussian pulse is its own Fourier transform.
*Solution*
A pulse is said to be normalized when its central ordinate is equal to the area under the curve. Let us consider a Gaussian pulse which is similar to a Gaussian probability density function.

SIGNAL ANALYSIS 61

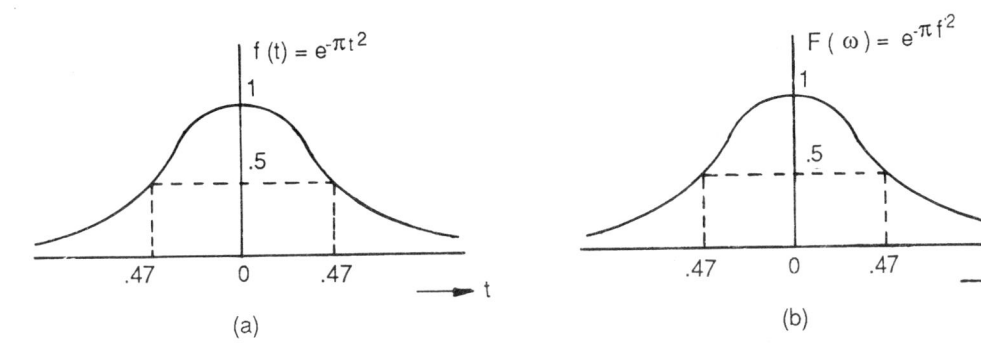

Fig. 1.6.17 (a) Gaussian Pulse and; (b) its Spectrum

The pulse is shown in Fig.1.6.17a and may be expressed as

$$f(t) = e^{-\pi t^2} \tag{1.6.19}$$

Differentiating $f(t)$ we get

$$\frac{df(t)}{dt} = -2\pi t e^{-\pi t^2}$$

Taking

$$e^{-\pi t^2} = f(t), \text{ we have}$$

$$\frac{df(t)}{dt} = -2\pi t f(t)$$

$$= 2\pi j \{jt f(t)\}$$

The Fourier transform of both sides yields

$$F\left[\frac{df(t)}{dt}\right] = 2\pi j F[jt f(t)]$$

By using time differentiation theorem on the left-hand side, and frequency differentiation theorem on right-hand side of the above equation, we get

$$j\omega F(\omega) = -j2\pi \frac{dF(\omega)}{d\omega}$$

which can be written as,

$$\frac{dF(\omega)}{d\omega} = -\frac{\omega}{2\pi} F(\omega)$$

The above differential equation yields the solution

$$F(\omega) = e^{\frac{-\omega^2}{4\pi}} = e^{\frac{-(2\pi f)^2}{4\pi}}$$

or

$$F(\omega) = e^{-\pi f^2}$$

which is also a Gaussian function as shown in Fig. 1.6.17b.

Thus, the Fourier transform of a Gaussian function is also Gaussian. The Gaussian function $e^{-\pi t^2}$ is a normalized function, i.e., its control ordinate and *area under the curve,* are same as unity. The area under the pulse can be obtained by using the property (i.e., area under the curve)

$$\int_{-\infty}^{\infty} f(t)dt = F(0) = \left| e^{-\pi f^2} \right|_{f=0} = 1$$

which is same as the central ordinate of the curve $e^{-\pi t^2}$. Thus the Gaussian pulse is a normalised pulse, and has its own Fourier transform.

## 1.7  FOURIER TRANSFORM OF PERIODIC FUNCTIONS

We conclude from the discussions made so far that a function can be analysed over an entire interval by:
(i) using Fourier series, if the function is periodic, and by
(ii) using Fourier transform, if the function is non periodic.

However, once we determine the Fourier transform of a periodic function, the Fourier transform will provide us a unified tool for analysing both periodic and non-periodic waveforms over the entire interval. Fortunately, this can be done using the concept of delta function. Fourier transform of a periodic function can be determined in limiting cases, as was done for sinusoidal function, inspite of the fact that periodic function fails to satisfy the condition of absolute integrability. Thus, the Fourier transform provides a tool for generalized analysis of arbitrary waveforms over the entire interval.

Let us now find the Fourier transform of a periodic function $f(t)$ with periodic $T$. The function can be expressed in terms of complex Fourier series.

$$f(t) = \sum_{n=-\infty}^{\infty} F_n e^{jn\omega_0 t}, \quad \omega_0 = \frac{2\pi}{T}$$

The Fourier transform of both sides yields

$$F[f(t)] = F\left[\sum_{n=-\infty}^{\infty} F_n e^{jn\omega_0 t}\right] = \sum_{n=-\infty}^{\infty} F_n F\left[1.e^{jn\omega_0 t}\right]$$

Using frequency-shifting theorem, we have

$$F\left[1.e^{jn\omega_0 t}\right] = 2\pi\delta(\omega - n\omega_0)$$

Therefore,

$$F[f(t)] = 2\pi \sum_{n=-\infty}^{\infty} F_n \delta(\omega - n\omega_0) \tag{1.7.1}$$

Thus, the Fourier transform of a periodic function consists of a train of equally spaced impulses. These are located at the harmonic frequencies of the signal, and the strength (area) of each impulse is given as $2\pi F_n$. This is logical, as the Fourier series analysis of periodic functions yields the discrete lines located at harmonic frequencies. Here we have reached a discrete spectrum starting with Fourier transform, which normally has a continuous spectrum.

### Example 1.7.1

(i) Prove that a Dirac comb is its own Fourier transform, (ii) find the exponential Fourier series of this Dirac comb and, (iii) show that the result is similar to its Fourier transform.

## Solution

(i) A Dirac comb is a comb-like waveform consisting of a sequence of equidistant impulses shown in Fig.1.7.1. The sequence of unit impulses are spaced by $T$ seconds.

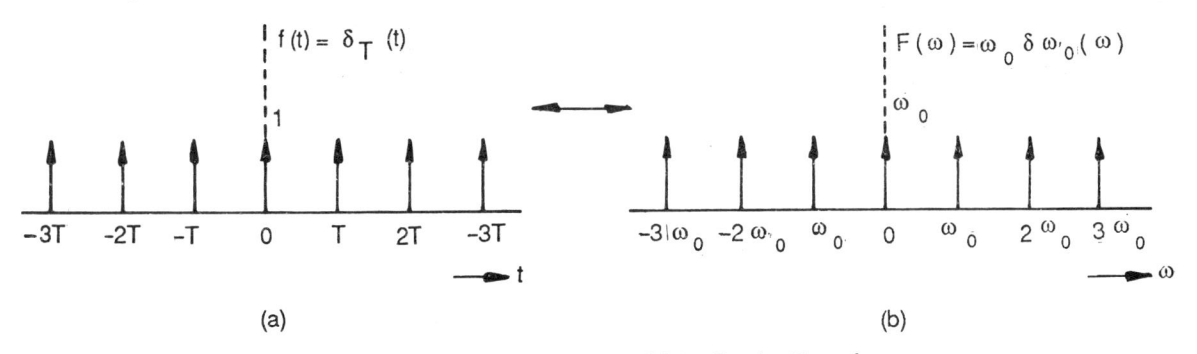

Fig. 1.7.1 (a) Dirac comb, and (b) Its Fourier Transform

The function may be expressed as

$$\delta_T(t) = \delta(t) + \delta(t-T) + \delta(t-2T) + \cdots \delta(t-nT) + \cdots + \delta(t+T) + \delta(t+2T) + \cdots + \delta(t+nT) + \cdots$$

The Fourier transform of the Dirac comb may be obtained by Eq.1.7.1. The value of $F_n$ is evaluated as usual,

$$F_n = \frac{1}{T} \int_{-\frac{T}{2}}^{\frac{T}{2}} \delta_T(t) e^{-jn\omega_0 t} dt$$

Function $\delta_T(t)$ has only one impulse in the interval $\left(-\frac{T}{2}, \frac{T}{2}\right)$, and hence is equivalent to $\delta(t)$. The $F_n$ can be written as

$$F_n = \frac{1}{T} \int_{-\frac{T}{2}}^{\frac{T}{2}} \delta(t) e^{-jn\omega_0 t} dt$$

Using sampling property of unit impulse function, the integral has a value of unity.

Therefore, $$F_n = \frac{1}{T} \qquad (1.7.2)$$

The Fourier transform of $\delta_T(t)$ is obtained using Eq. 1.7.1

$$F[\delta_T(t)] = 2\pi \sum_{n=-\infty}^{\infty} \frac{1}{T} \delta(\omega - n\omega_0) = \omega_0 \sum_{n=-\infty}^{\infty} \delta(\omega - n\omega_0) \qquad (1.7.3)$$

or $$\delta_T(t) \leftrightarrow \delta_{\omega_0}(\omega)$$

The spectrum is shown in Fig.1.7.1. It consists of a sequence of impulses located at harmonics of $\omega_0$ and each impulse has constant strength of $\frac{2\pi}{T} = \omega_0$. Thus, the Fourier transform of the periodic unit impulse function is also a periodic unit impulse function. In other words, the periodic unit impulse function (Dirac comb) has its own transform.

(ii) Fourier series of periodic impulse function (Dirac comb):
The Fourier series of a function is given by

$$f(t) = \sum_{n=-\infty}^{\infty} F_n e^{jn\omega_0 t}$$

Where $F_n$ is same as given in Eq.1.7.2, i.e. $F_n = \frac{1}{T}$.

The discrete (line) spectrum of a periodic unit impulse function ($F_n$ $V_s$ $\omega$) plotted in Fig.1.7.2 is similar to Fourier transform spectrum in Fig. 1.7.1b, however, the magnitude of each impulse is $\frac{1}{T}$ in place of $\frac{2\pi}{T}$. The results detained are in agreement with Fourier transform in the sense that strength of impulse in Fourier transform = $2\pi \times$ (corresponding coefficient of Fourier series i.e., $F_n$).

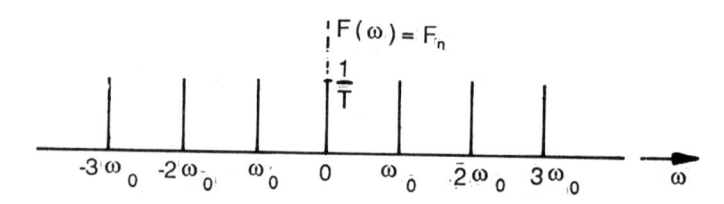

Fig. 1.7.2 Line Spectrum of Dirac Comb

## Example 1.7.2

Find the Fourier transform of a periodic Gate function with period $T = \frac{1}{2}$ and; width $\tau = \frac{1}{20}$.

*Solution*

The periodic Gate function with its Fourier transform is shown in Fig. 1.7.3. The value of $F_n$ for periodic Gate function has been evaluated earlier, and is given as

$$F_n = \frac{A\tau}{T} Sa\left(n\pi \frac{\tau}{T}\right)$$

For $\tau = \frac{1}{20}$; and $T = \frac{1}{2}$, the duty cycle $\frac{\tau}{T} = \frac{1}{10}$

Hence,

$$F_n = \frac{A}{10} Sa\left(\frac{n\pi}{10}\right)$$

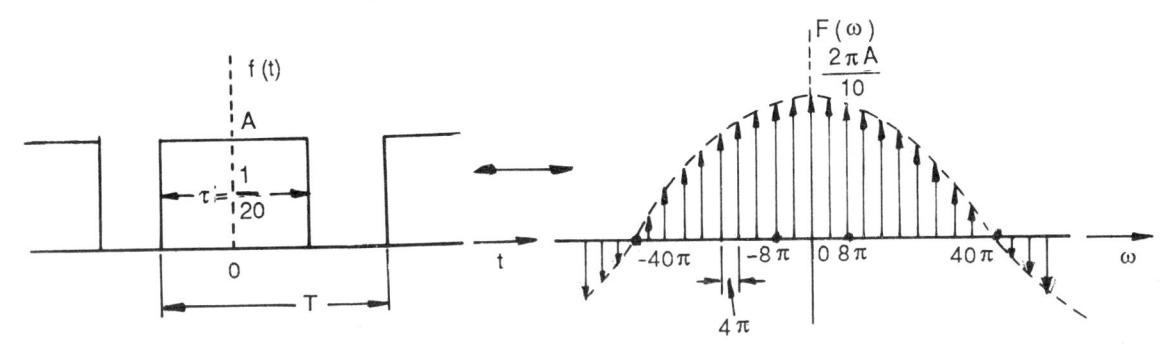

Fig. 1.7.3 Periodic Gate Pulse and its Fourier Transform

The Fourier transform is obtained by substituting this value of $F_n$ in Eq.1.7.1,

$$F(\omega) = \frac{2\pi A}{10} \sum_{n=-\infty}^{\infty} Sa\left(\frac{n\pi}{10}\right) \delta(\omega - n\omega_0)$$

therefore the transform, consists of impulses, located at $\omega = 0, \pm \omega_0, \pm 2\omega_0, \pm n\omega_0$. The magnitude of the impulse at $\omega = n\omega_0$ is given by $\frac{2\pi A}{10} Sa\left(\frac{n\pi}{10}\right)$

The first zero of the envelope will occur at 10th harmonic $\left(\frac{n\pi}{10} = \pm \pi \text{ or } n = \pm 10\right)$ Each harmonic being spaced by $\frac{2\pi}{T} = 4\pi$, the first zero will occur at $\omega = 40\pi$. The maximum magnitude of the envelope is $\frac{2\pi A}{10}$. The spectrum again is similar to the Fourier series spectrum except that the magnitude is $2\pi \times$ i.e., (corresponding Fourier coefficient $F_n$).

## Example 1.7.3

Find the Fourier transform of a periodic train of triangular pulses with period $T$, base width of $2\tau$ and altitude $A$.

*Solution*

The periodic triangular pulse alongwith its first and second derivatives is shown in Fig.1.7.4. The second derivative shown in Fig.1.7.4 (c) can be considered as sum of three Dirac combs, i.e., three *train of impulse* each with periodic $T$. The first comb corresponds to impulse at $t = -\tau$, the second comb corresponds to impulse at $t = 0$; and the third corresponds to impulse $t = \tau$. If we find the Fourier transform of each comb and add them, we get the transform of Fig.1.7.4c.

The Fourier transform of a comb $\delta_T(t)$ is already found to be $2\pi \sum_{n=-\infty}^{\infty} F_n \delta(\omega - n\omega_0)$. In this case, $F_n$ is the sum of $F_{n1}, F_{n2}$, and $F_{n3}$, corresponding to first, second and third comb respectively. Assume $\omega = n\omega_0$. Then for the first comb, having an amplitude $\frac{A}{\tau}$, we have

$$F_{n1} = \frac{A}{\tau T} \int_{-T/2}^{T/2} \delta(t+\tau) e^{-j\omega t} dt = \frac{A}{\tau T} \left| e^{-j\omega t} \right|_{t=-\tau} = \frac{A}{\tau T} e^{j\omega \tau}$$

and, for second comb having an amplitude $-\dfrac{2A}{\tau}$, we have

$$F_{n2} = -\frac{2A}{\tau T} \int_{-T/2}^{T/2} \delta(t) dt = -\frac{2A}{\tau T}$$

Similarly, for third comb having an amplitude $\dfrac{A}{\tau}$,

$$F_{n3} = \frac{A}{\tau T} e^{-j\omega \tau}$$

Therefore, $\qquad F_n = F_{n1} + F_{n2} + F_{n3}$

$$= \frac{A}{\tau T}\left[e^{j\omega\tau} - 2 + e^{-j\omega\tau}\right] = \frac{A}{\tau T}[2\cos\omega\tau - 2]$$

$$= -\frac{2A}{\tau T} 2\sin^2\left(\frac{\omega\tau}{2}\right) = -\frac{4A}{\tau T}\sin^2\left(\frac{\omega\tau}{2}\right)$$

The Fourier transform of $\dfrac{d^2 f_T(t)}{dt^2}$ in Fig.1.7.4 c is given by

$$\frac{d^2 f_T(t)}{dt^2} \leftrightarrow 2\pi \sum_{n=-\infty}^{\infty} F_n \delta(\omega - n\omega_0)$$

Using time differentiation property on L.H.S. and substituting the value of $F_n$ on R.H.S. we get

$$(j\omega)^2 F\{f_T(t)\} = -\frac{8\pi A}{\tau T}\sin^2\left(\frac{\omega\tau}{2}\right) \sum_{n=-\infty}^{\infty} \delta(\omega - n\omega_0)$$

$$F\{f_T(t)\} = \frac{8\pi A}{\tau T \omega^2}\sin^2\left(\frac{\omega\tau}{2}\right) \sum_{n=-\infty}^{\infty} \delta(\omega - n\omega_0)$$

$$= \frac{2\pi A \tau}{T} \frac{\sin^2\left(\dfrac{\omega\tau}{2}\right)}{\left(\dfrac{\omega\tau}{2}\right)^2} \sum_{n=-\infty}^{\infty} \delta(\omega - n\omega_0)$$

SIGNAL ANALYSIS 67

(a) Periodic triangular pulse train

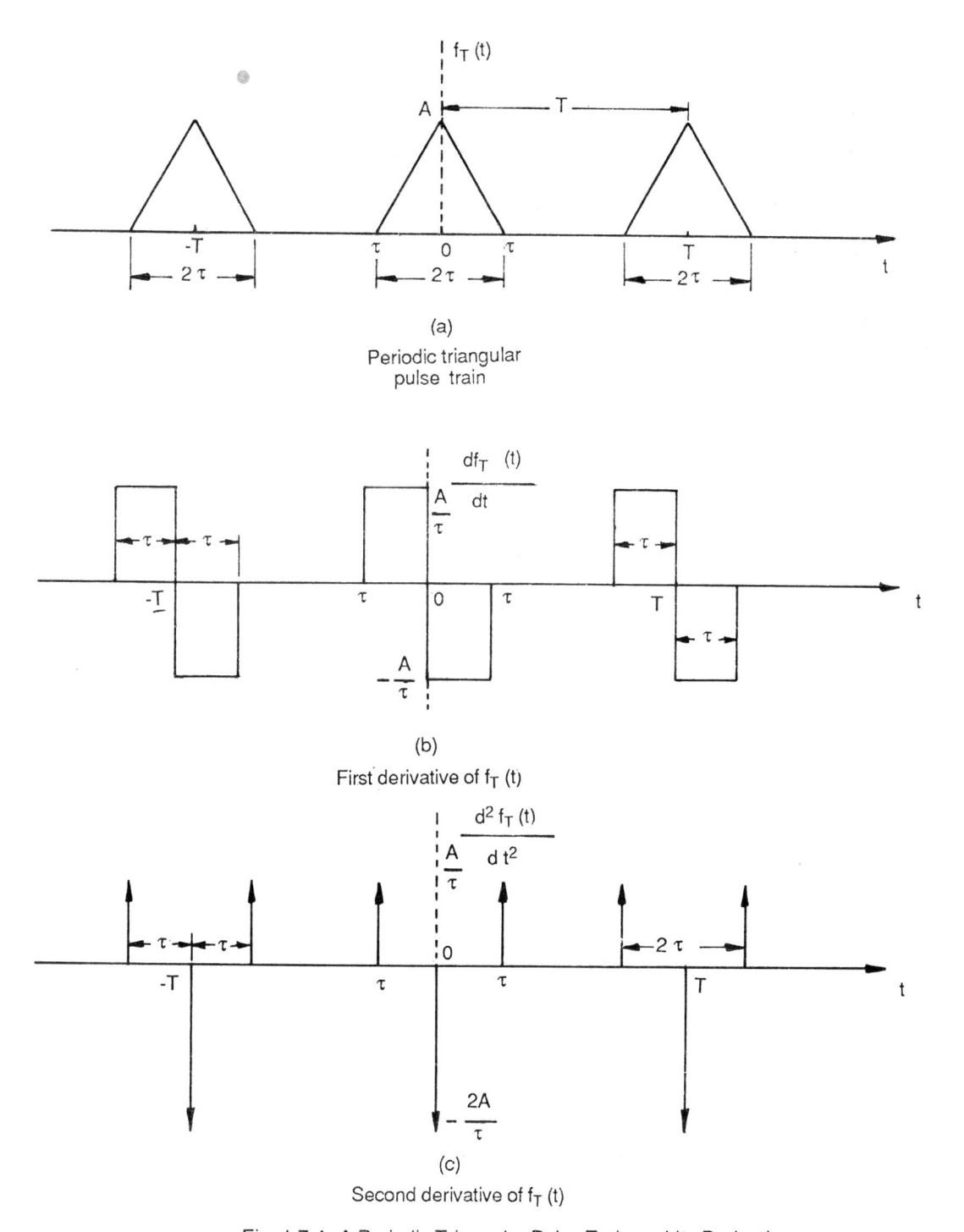

(b) First derivative of $f_T(t)$

(c) Second derivative of $f_T(t)$

Fig. 1.7.4 A Periodic Triangular Pulse Train and Its Derivatives

$$= \frac{2\pi A\tau}{T} Sa^2\left(\frac{\omega\tau}{2}\right) \sum_{n=-\infty}^{\infty} \delta(\omega - n\omega_0)$$

Replacing $\omega = n\omega_0$, in the sampling square function and putting it within summation, we get

$$F\{f_T(t)\} = \frac{2\pi A\tau}{T} \sum_{n=-\infty}^{\infty} Sa^2\left(\frac{n\omega_0\tau}{2}\right) \delta(\omega - n\omega_0)$$

## 1.8 CONVOLUTION

Convolution is a mathematical operation and is useful for describing the input/output relationship in a linear time invariant system. It is an important analytical tool for the communication engineer.

The convolution $f(t)$ of two time functions $f_1(t)$, and $f_2(t)$, is defined by the following integral

$$f(t) = \int_{-\infty}^{\infty} f_1(\tau) f_2(t - \tau) d\tau \qquad (1.8.1)$$

An encircled multiplication $\otimes$ notation will be used to denote the representation of a convolution. Thus, $f_1(t) \otimes f_2(t)$ is read as $f_1(t)$ convolved with $f_2(t)$.

### Convolution by Inspection (Graphical Convolution)

Normally, an analytical evaluation of convolution becomes complicated for many functions. However, the convolution can be estimated by inspection, and certain observations about convolution can be interpreted without actually performing the detailed calculation. In many applications of communication system the inspection procedure provides the information needed without complicated calculations. The inspection procedure is known as graphical convolution.

### Graphical Convolution Technique

Let us examine the expression of a convolution integral

$$f_1(t) \otimes f_2(t) = \int_{-\infty}^{\infty} f_1(\tau) f_2(t - \tau) d\tau$$

Here

(i) $f_1(\tau)$ is the first function, where, an independent variable $(t)$ is replaced by a dummy variable $\tau$.
(ii) $f_2(-\tau)$ is the mirror image of $f_2(\tau)$, which means $f_2(\tau)$ is flipped around the y-axis.
(iii) $f_2(t-\tau)$ represents the function $f_2(-\tau)$ shifted to right by, "$t$" seconds.
(iv) For a particular value of $t=b$, integration of the product $f_1(\tau)f_2(b-\tau)$ represents the area under the product curve (common area). This common area represents the convolution of $f_1(t)$ and $f_2(t)$ for a shift of, $t=b$ i.e.,

$$\int_{-\infty}^{\infty} f_1(\tau) f_2(b - \tau) d\tau = [f_1(t) \otimes f_2(t)]_{t=b} \qquad (1.8.2)$$

(v) The procedure is repeated for different values of $t$ to evaluate the convolution. For a negative value of $t$, function $f_2, (-\tau)$ is shifted to left by $t$ seconds. The function $f_2, (-\tau)$ may be considered as a rigid frame progressed along $\tau$ – axis.

(vi) The value of convolution obtained at different values of $t$ (both positive as well as negative) may be plotted on a graph.

### Example 1.8.1

Find the convolution of the functions $f_1(t)$; and $f_2(\tau)$ shown in Fig. 1.8.1 (a).

*Solution*

(i) Fig.1.8.1 b shows the functions $f_1(\tau)$ and $f_2(-\tau)$ i.e., $f_2(\tau)$ flipped around y-axis (mirror image).
(ii) Fig.1.8.1 c represents the functions $f_1(\tau)$ and $f_2(b-\tau)$ i.e., $f_2(-\tau)$ shifted by $b$ seconds towards right.
(iii) The shaded portion in Fig.1.8.1 d represents the product (common area) for $b$ = 2 s.
(iv) Figs 1.8.1e and f show respectively the common area for $b$ = 3 s and 3.5.
(v) Note that the common area is maximum for a shift of $b$ = 3 s towards right.
(vi) Fig. 1.8.1g shows the convolution for zero shift.
(vii) Similarly, we can evaluate for negative shifts.
(viii) The common are is zero beyond $-1 < t < 5$.

The convolution is plotted in Fig.1.8.1 h.

## Laws of Convolution Relationship

The convolution relationship follows the laws of ordinary multiplication stated as follows :

### (1) Commutative Law

$$f_1(t) \otimes f_2(t) = f_2(t) \otimes f_1(t) \qquad (1.8.3)$$

The convolution of $f_1(t)$ and $f_2(\tau)$ is same as that of $f_2(\tau)$ and $f_1(t)$. Either of the two functions may be taken as rigid frame for evaluating convolution.

### (2) Associative Law

$$f_1(t) \otimes \{f_2(t) \otimes f_3(t)\} = \{f_1(t) \otimes f_2(t)\} \otimes f_3(t) \qquad (1.8.4)$$

### (3) Distributive Law

$$f_1(t) \otimes \{f_2(t) \otimes f_3(t)\} = f_1(t) \otimes f_2(t) + f_1(t) \otimes f_3(t) \qquad (1.8.5)$$

## The Convolution Theorems

Let us study the convolution theorems associated with Fourier transform. There are two theorems: time-convolution and the frequency-convolution.

## Time-Convolution Theorem

This theorem states that convolution in time domain is equivalent to multiplication of their spectra in frequency domain; i.e., if

70 COMMUNICATION SYSTEMS: Analog and Digital

(a)

(b)

(c)

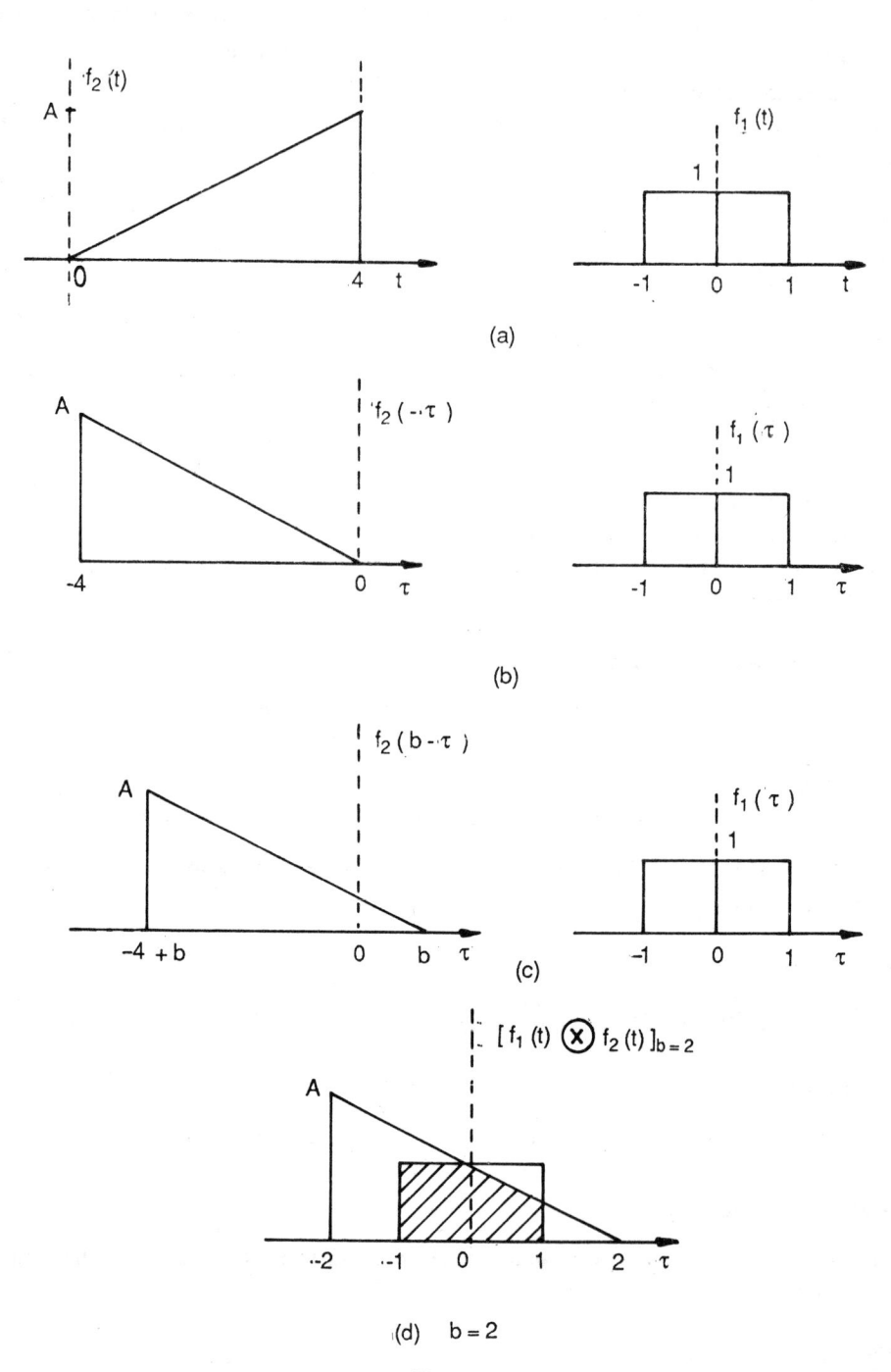

(d) b = 2

Fig. 1.8.1

SIGNAL ANALYSIS 71

**Fig. 1.8.1**

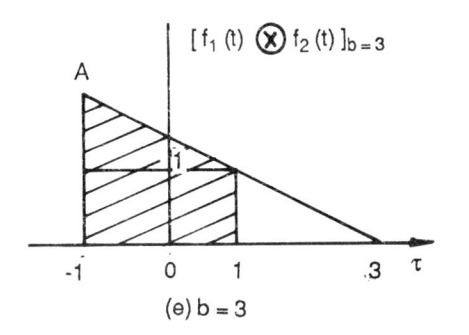

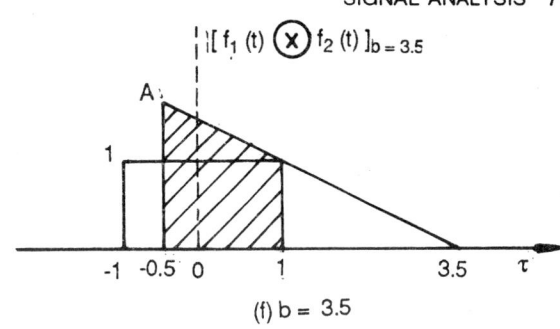

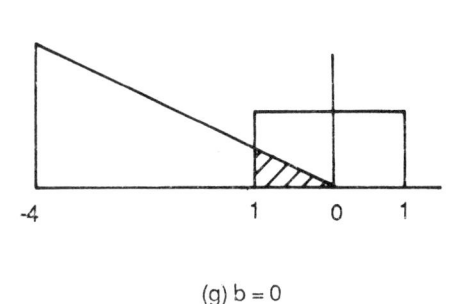

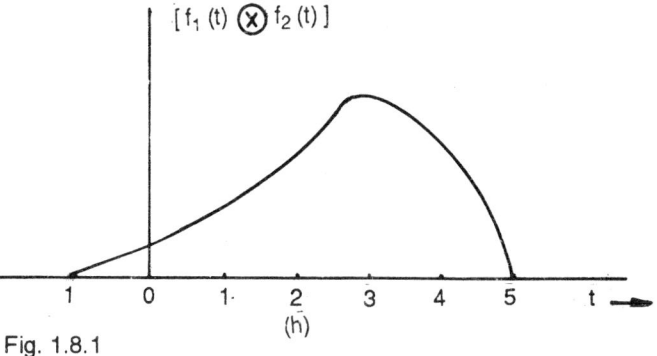

(e) b = 3
(f) b = 3.5
(g) b = 0
(h)

$$f_1(t) \leftrightarrow F_1(\omega)$$

and

$$f_2(t) \leftrightarrow F_2(\omega)$$

then

$$f_1(t) \otimes f_2(t) \leftrightarrow F_1(\omega) F_2(\omega) \qquad (1.8.6)$$

**Proof**

$$F[f_1(t) \otimes f_2(t)] = \int_{-\infty}^{\infty} \left[ \int_{-\infty}^{\infty} f_1(\tau) f_2(t-\tau) d\tau \right] e^{-j\omega t} dt$$

After a simple manipulation, we have

$$[f_1(t) \otimes f_2(t)] \leftrightarrow \int_{-\infty}^{\infty} f_1(\tau) \left[ \int_{-\infty}^{\infty} e^{-j\omega t} f_2(t-\tau) dt \right] d\tau$$

using time shifting theorem, the term under bracket on RHS is given as

$$\int_{-\infty}^{\infty} e^{-j\omega t} f_2(t-\tau) dt = F_2(\omega) e^{-j\omega \tau}$$

Hence

$$f_1(t) \otimes f_2(t) \leftrightarrow \left\{ \int_{-\infty}^{\infty} f_1(\tau) e^{-j\omega \tau} d\tau \right\} \cdot F_2(\omega) = F_1(\omega) F_2(\omega)$$

## Frequency Convolution Theorem

According to this theorem multiplication of two functions in time domain is equivalent to convolution of their spectra in the frequency domain, i.e., if

$$f_1(t) \leftrightarrow F_1(\omega)$$
$$f_2(t) \leftrightarrow F_2(\omega)$$

then
$$2\pi f_1(t) f_2(t) \leftrightarrow F_1(\omega) \otimes F_2(\omega) \qquad (1.8.7)$$

The proof is similar to time-convolution theorem. (Prob 1.13).

### Example 1.8.2

Evaluate the following integral using the convolution theorem.

$$I = \int_{-\infty}^{\infty} \frac{\sin 4x}{x} \cdot \frac{\sin(t-x)}{(t-x)} dx$$

*Solution*

It can be seen that a above integral is convolution of two time functions

$$I = \frac{\sin 4t}{t} \otimes \frac{\sin t}{t}$$

Let 
$$f_1(t) = \frac{\sin 4t}{t};$$

and 
$$f_2(t) = \frac{\sin t}{t}$$

Then according to the time-convolution theorem, convolution of $f_1(t)$, and $f_2(t)$ can be determined by evaluating inverse Fourier transform of the Product $[F_1(\omega) \cdot F_2(\omega)]$, i.e.,

$$I = F^{-1}[F_1(\omega) F_2(\omega)] = f_1(t) \otimes f_2(t)$$

The functions $f(t)$ and $f_2(t)$ are sampling functions. Hence, the Fourier transforms $F_1(\omega), F_2(\omega)$ and their product will be Gate functions, as shown in Fig.1.8.2.

The inverse Fourier transform of product function (i.e. a Gate function), is a sampling function given by

$$F^{-1}[F_1(\omega) F_2(\omega)] \leftrightarrow \frac{\pi \sin t}{t}$$

Hence, integral $I = \dfrac{\pi \sin t}{t}$

### Example 1.8.3

Find the convolution with itself of a rectangular pulse shown in Fig.1.8.3 (a): (i) graphically, (ii) using time-convolution theorem

SIGNAL ANALYSIS 73

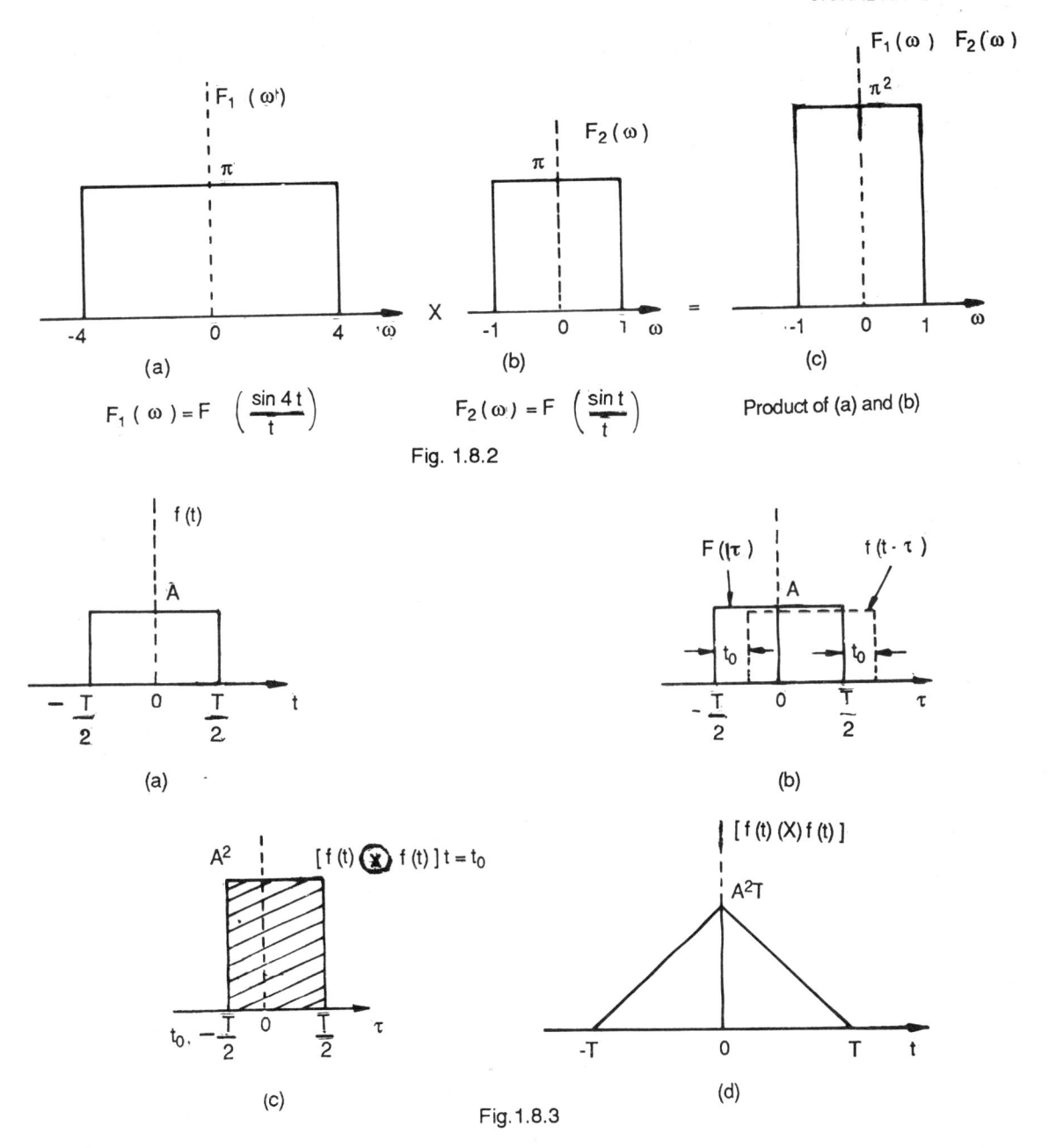

Fig. 1.8.2

(a) $F_1(\omega) = F\left(\dfrac{\sin 4t}{t}\right)$

(b) $F_2(\omega) = F\left(\dfrac{\sin t}{t}\right)$

(c) Product of (a) and (b)

Fig. 1.8.3

*Solution*
(i) Since the rectangular pulse is an even function of time $t$, its mirror image results in the same function. The dotted curve in Fig.1.8.3(b) shows the rectangular pulse $f(t)$ shifted by $t_0$ s. The shaded portion in Fig.1.8.3 (c) represents the product of the original and shifted pulse, and provides the convolution for $t = t_0$. Similarly, the convolution is evaluated for different values of $t$. The result is plotted in Fig.1.8.3 (d).
(ii) According to the time-convolution theorem

Hence
$$f_1(t) \otimes f_2(t) \leftrightarrow F(\omega).F(\omega)$$

$$f_1(t) \otimes f_2(t) = F^{-1}\left[\{F(\omega)\}^2\right]$$

The Fourier transform $F(\omega)$ for a rectangular pulse is a sampling function given as

$$F(\omega) = AT\, Sa\left(\frac{\omega T}{2}\right)$$

so

$$|F(\omega)|^2 = A^2 T^2 Sa^2\left(\frac{\omega T}{2}\right)$$

The inverse Fourier transform of $A^2 T^2 Sa^2\left(\frac{\omega T}{2}\right)$ is a triangular function shown in Fig.1.8.3 (d), which is obvious from Ex. 1.6.12. Hence, convolution of a rectangular pulse $f(t)$ with itself is a triangular pulse shown in Fig.1.8.3(d). Thus, the result is in agreement with the one obtained graphically in (i).

### Example 1.8.4

Show that the convolution of a function $f(t)$ with a unit impulse function results the function itself.
*Solution*
The convolution of a function $f(t)$ with unit impulse function is given as

$$f(t) \otimes \delta(t) = \int_{-\infty}^{\infty} f(\tau)\, \delta(t - \tau)\, d\tau$$

Using the sampling property of impulse function, the right-hand side yields the function $|f(\tau)|_{\tau=t}$ i.e., $f(t)$. Hence, the convolution of $f(t)$ and $\delta(t)$ is $f(t)$ itself.

The result can also be proved by using the time convolution theorem, according to which

$$f(t) \otimes \delta(t) = F^{-1}[F\{f(t)\}.F\{\delta(t)\}]$$

It is known that

$$F\{f(t)\} = F(\omega) \text{ and }. F\{\delta(t)\} = 1$$

Hence,

$$f(t) \otimes \delta(t) = F^{-1}[F(\omega)] = f(t) \qquad (1.8.8)$$

### Example 1.8.5

Prove the following relations:

(i) $$f(t) \otimes \delta(t-b) = f(t-b) \qquad (1.8.9a)$$
(ii) $$f(t-b) \otimes \delta(t-c) = f(t-b-c) \qquad (1.8.9b)$$
(iii) $$f(t-b) \otimes \delta(t-c) = f(t-b-c) \qquad (1.8.9c)$$

*Solution*
(i) It is known that

$$f(t) \leftrightarrow F(\omega); \delta(t-b) = 1.e^{j\omega b} = e^{-j\omega b}$$

From time convolution theorem

$$f(t) \otimes \delta(t-b) = F^{-1}\left[F(\omega)e^{-j\omega b}\right] = f(t-b)$$

(ii) Using time shifting property of Fourier transform

$$f(t-b) \leftrightarrow F(\omega)e^{-j\omega b}; \delta(t-c) \leftrightarrow 1.e^{-j\omega c}$$

From time convolution theorem,

$$f(t-b) \otimes \delta(t-c) = F^{-1}\left[F(\omega)e^{-j\omega b}e^{-j\omega c}\right] = F^{-1}\left[F(\omega)e^{-j\omega(b+c)}\right]$$

Again, using time shifting property of Fourier transform

$$F^{-1}\left[F(\omega)e^{-j\omega(b+c)}\right] = f(t-b-c)$$

Hence, $\quad f(t-b) \otimes f(t-c) = f(t-b-c)$

(iii) Using time shifting property of Fourier transform

$$\delta(t-b) \leftrightarrow e^{-j\omega b}; \delta(t-c) = e^{-j\omega c}$$

From time convolution theorem,

$$\delta(t-b) \otimes \delta(t-c) = F^{-1}\left[e^{-j\omega b}e^{-j\omega c}\right] = F^{-1}\left[e^{-j\omega(b+c)}\right]$$
$$= \delta(t-b-c) \qquad \text{(Using time shifting property)}$$

## 1.9 SAMPLING THEOREM

The sampling theorem is significant in communication systems because it provides the basis for transmitting analog signals by use of digital techniques. The sampling theorem may be stated in two equivalent ways as under:

(1) A band limited signal having no frequency components higher than $f_m$ Hz is completely described by its sample values at uniform intervals less than or equal to $\dfrac{1}{2f_m}$ s apart. This is a frequency domain statement.

(2) A band limited signal having no frequency components higher than $f_m$ Hz may be completely recovered from the knowledge of its samples taken at the rate of at least $2f_m$ samples per second. This is a time domain statement.

The above stated theorem is called uniform sampling theorem, since the samples are taken at uniform intervals. A band limited signal has zero value of Fourier transform beyond the frequency $f_m$ Hz. No useful signal is band limited in the true mathematical sense, since its Fourier transform extends from $-\infty$ to $\infty$. But, after a particular frequency, the magnitude of the transform diminishes to such an extent that it can be neglected, and the Fourier transform provides a finite bandwidth. Such signals are considered as band limited for all practical purposes.

The two statements of sampling theorem stated above can be proved by using the convolution theorem. Let us consider a band limited signal $f(t)$ having no frequency components beyond $f_m$ Hz i.e., $F(\omega)$ is zero for $|\omega| > \omega_m$ ($\omega_m = 2\pi f_m$). When this signal is multiplied by a periodic impulse function $\delta_T(t)$; the product yields a sequence of impulses located at uniform intervals of $T$ seconds. The strength of resulting impulses

**76** COMMUNICATION SYSTEMS: Analog and Digital

is equal to the value of $f(t)$ at the corresponding instants. Figure 1.9.1 a shows the function $f(t)$ and its Fourier transform $F(\omega)$. The periodic impulse function $\delta_T(t)$ and its Fourier transform $\omega_0 \delta_{\omega 0}(\omega)$ is shown in Fig. 1.9.1 b. The product of $f(t)$, and $\delta_T(t)$ is shown in Fig.1.9.1 (c). This product of $f(t)$, and $\delta_T(t)$ represents the samples of $f(t)$, located at uniform interval of $T$. Let us denote this sampled version of $f(t)$ as $f_s(t)$. Then

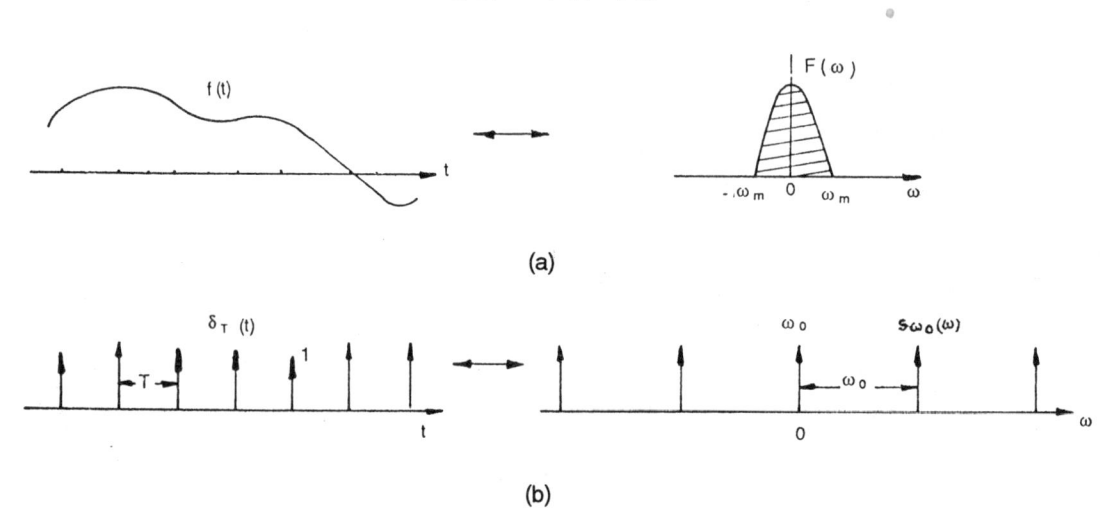

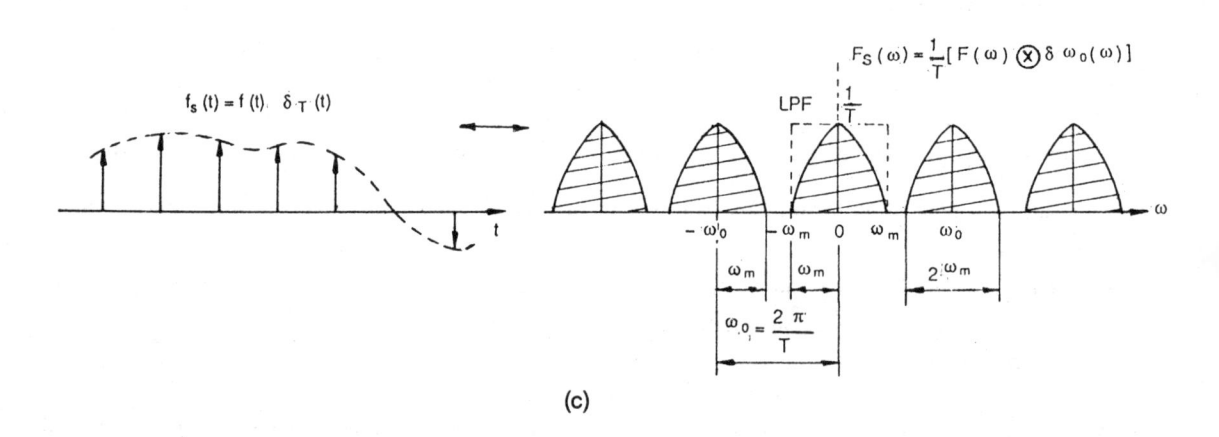

$$f_s(t) = f(t)\, \delta_T(t)$$

Fig. 1.9.1 (a) A function $f(t)$ and its Spectrum, (b) Periodic Delta function and its Spectrum; and (c) Periodic impulse function and its Spectrum

The spectrum (Fourier transform) of $f_s(t)$ can be obtained by using frequency-convolution theorem. Accordingly,

$$f_s(t) \leftrightarrow \frac{1}{2\pi}\left\{F(\omega) \otimes \omega_0 \delta_{\omega_0}(\omega)\right\}$$

putting $\dfrac{\omega_0}{2\pi} = \dfrac{1}{T}$ in the above equation we get

$$f_s(t) \leftrightarrow \dfrac{1}{T}\{F(\omega) \otimes \delta_{\omega_0}(\omega)\} \qquad (1.9.1)$$

Thus, the spectrum of $f_s(t)$ can be obtained by convolving $F(\omega)$ and $\delta_{\omega_0}(\omega)$. The convolution can be obtained graphically by flipping the function $\delta_T(t)$ around the vertical axis. Since, $\delta_T(t)$ is an even function of time, flipping yields the same function $\delta_T(t)$. The operation of convolution yields $F(\omega)$, repeating itself every $\omega_0$ radians per second, as in Fig.1.9.1 (c). $F(\omega)$, periodically repeating, forms the spectrum of $f_s(t)$, and is denoted by $F_s(\omega)$.

The spectrum $F_s(\omega)$ can also be achieved analytically. The periodic function $\delta_{\omega_0}(\omega)$ can be written as the sum of impulses located at $\omega = 0, \pm \omega_0, + 2\omega_0, \cdots + m\omega_0$

$$\begin{aligned}\delta_{\omega_0}(\omega) &= \delta(\omega) + \delta(\omega - \omega_0) + \cdots + \delta(\omega - m\omega_0) + \cdots \\ &\quad + \delta(\omega + \omega_0) + \cdots + \delta(\omega + m\omega_0) + \cdots \\ &= \sum_{m=-\infty}^{\infty} \delta(\omega - m\omega_0) \qquad (m = 1, 2, 3, \ldots)\end{aligned}$$

From Eq. 1.9.1. we have

$$f_s(t) \leftrightarrow \dfrac{1}{T}\left[F(\omega) \otimes \sum_{m=-\infty}^{\infty}\delta(\omega - m\omega_0)\right]$$

or

$$F_s(\omega) = \dfrac{1}{T}\sum_{m=-\infty}^{\infty} F(\omega) \otimes \delta(\omega - m\omega_0)$$

By using sampling property of a delta function

$$F_s(\omega) = \dfrac{1}{T}\sum_{m=-\infty}^{\infty} F(\omega - m\omega_0) \qquad (1.9.2)$$

The summation represents $F(\omega)$ repeating every $\omega_0$ radians per second, as the one obtained by graphical convolution.

It is obvious from $F_s(\omega)$ shown in Fig.1.9.1 (c), that $F(\omega)$ will repeat periodically without overlapping, provided

$$\omega_0 \geq 2\omega_m$$

or

$$\dfrac{2\pi}{T} \geq 2(2\pi f_m)$$

# 78 COMMUNICATION SYSTEMS: Analog and Digital

Or
$$\frac{1}{T} \geq 2f_m$$

and
$$T \leq \frac{1}{2f_m} \text{ s} \tag{1.9.3 a}$$

Where $T$ is the uniform sampling interval. A related term *sampling rate* or sampling frequency $f_0 = \frac{1}{T}$, should meet the following condition

$$f_0 \geq 2 \; f_m \text{ samples/sec} \tag{1.9.3 b}$$

Eq. 1.9.3 a and b proves the sampling them

## Recovery of Original Spectrum F(ω)

Thus, we find that as long as signal is sampled at interval $T$ as given in Eq. (1.9.3 a) or at a sampling rate $f_0$ satisfying relation in Eq. 1.9.3 b, the spectrum $F(\omega)$ will repeat periodically without overlap. The spectrum extends up to infinity and the ideal bandwidth of sampled signal is infinite. *However, the desired spectrum $F(\omega)$ centered at $\omega = 0$ can be recovered by passing the sampled signal with spectrum $F_s(\omega)$ through a low pass filter with cut off frequency $\omega_m$.* The transfer function of an ideal law pass filter is shown in the spectrum $F_s(\omega)$ of Fig. 1.9.1c. The sampled signal after filtering yields all the frequency components present in desired signal $f(t)$ having spectrum $F(\omega)$.

## Nyquist Interval

It is evident from Eq. 1.9.3 a, that maximum sampling interval is given as

$$T = \frac{1}{2f_m} \text{ s}$$

This maximum sampling interval is known as Nyquist sampling interval. Similarly minimum sampling rate $f_0 = (1/2 f_m)$ is called as Nyquist sampling rate. When the bandlimited signal is sampled at Nyquist sampling interval (i.e. Nyquist rate), the spectrum $F_s(\omega)$ will contain non-overlapping $F(\omega)$ repeating periodically, however, each spectrum $F(\omega)$ will be touching the neighbouring ones as shown is Fig. 1.9.2 a. The original spectrum $F(\omega)$ can be recovered from this sampled spectrum also by using a low pass filter with cut off frequency $f_m$.

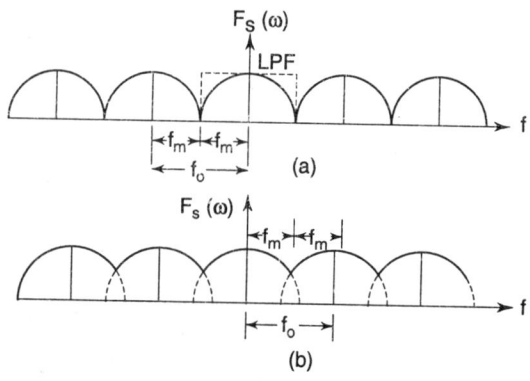

Fig. 1.9.2 Spectrum of f(t) sampled at (a) Nyquist rate (b) Slower than Nyquist rate

## Aliasing

When a band limited signal is sampled at rate lower than Nyquist rate, $f_0 < 2f_m$ or sampling interval is higher than Nyquist interval ($T > 1/2f_m$), then periodically repeating $F(\omega)$ in the spectrum of sampled signal overlap with neighbouring ones as shown Fig. 1.9.2.b. The signal is under-sampled in this case and some aliasing is produced in this under-sampling process. Aliasing refers to the phenomenon in which a high frequency component in the spectrum of the signal seemingly taking on the identity of a lower frequency in the spectrum of its sampled version. Because of the overlap in Fig. 1.9.2 b due to aliasing phenomenon, it is no longer possible to recover $f(t)$ from $f_s(t)$ by low pass filtering since the spectral components in the overlap regions add and therefore the signal is distorted. The aliasing spectrum is shown by solid curve in the Fig. 1.9.2 b. In order to combat the effect of aliasing we may use prior to sampling a low pass anti-aliasing filter, and then filtered signal is sampled at a rate slightly higher than the Nyquist rate.

number of samples per second, i.e., the sampling rate $f_s = \frac{1}{T}$, is given as

$$f_s \geq 2f_m$$

Hence, the sampling rate should atleast be equal to twice the maximum frequency component present in the signal $f(t)$. This means that at least two samples per cycle are needed for a complete recovery of the signal from $f_s(t)$. Thus, the minimum sampling rate is given by

$$f_s = 2f_m$$

This minimum rate of sampling is known as *Nyquist sampling rate*.

The foregoing discussion substantiates the first statement of the sampling theorem. The second statement is proved by considering the recovery of the signal $f(t)$ from its sampled version $f_s(t)$.

## Signal Recovery from its Sampled Version

Earlier we had seen that an original signal $f(t)$ can be recovered in frequency domain by passing its sampled version through a low pass filter with a cut-off frequency $f_m$. Now, we will see that the function $f(t)$ can be reconstructed in time domain from its sampled version.

For a function $f(t)$; sampled at Nyquist rate $\omega_0$, we have

$$\omega_0 = 2\omega_m$$

Substituting this value of $\omega_0$ in 1.9.2, we get

$$F_s(\omega) = \frac{1}{T} \sum_{m=-\infty}^{\infty} F(\omega - 2m\omega_m)$$

The schematic diagram for the recovery of a function $f(t)$ from its sampled version $f_s(t)$ is shown in Fig. 1.9.3a.

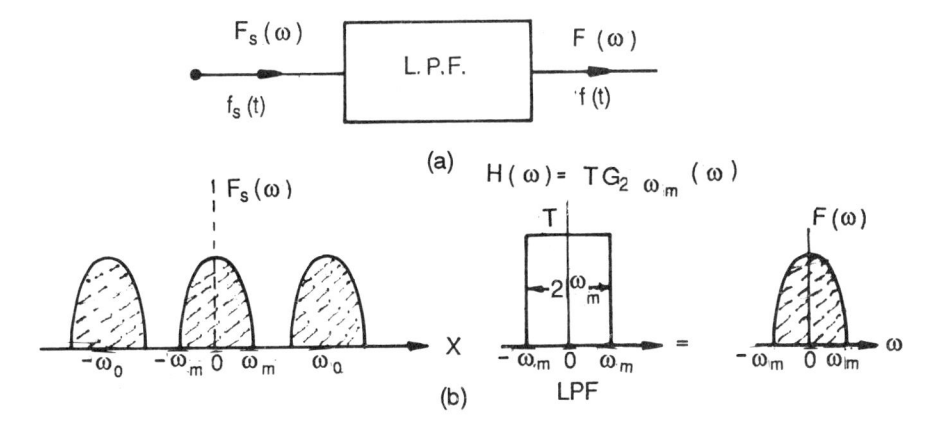

Fig. 1.9.3 Recovery of a Function from its Sampled Version

The low pass filter (LPF) has a cut-off frequency of $\omega_m$ radians. The transfer function of a LPF is a Gate function. Hence the baseband spectrum $F(\omega)$ can be recovered from $F_s(\omega)$ by multiplying the latter with a Gate function. This is illustrated in Fig.1.9.3b. We have from this figure,

## 80 COMMUNICATION SYSTEMS: Analog and Digital

$$F_s(\omega) \cdot G_{2\omega_m}(\omega) = \frac{1}{T} F(\omega)$$

or

$$F(\omega) = F_s(\omega) \cdot T G_{2\omega_m}(\omega) \tag{1.9.4}$$

The Gate function $G_{2\omega_m}(\omega)$ is representing a low pass filter with a cut-off frequency $\omega_m$. In other words, the action of low pass filtering is equivalent to multiplying the sampled signal $F_s(\omega)$ with a Gate function $T G_{2\omega_m}$ and this action yields the baseband spectrum $F(\omega)$. The Gate function has height $T$, and width $2\omega_m$.

The function $f(t)$ can be obtained by evaluating time domain equivalent (inverse Fourier transform) of $F(\omega)$ in Eq.1.9.4. The time domain equivalent of right hand side of Eq.1.9.4. can be evaluated using time convolution theorem i.e.,

$$f(t) = f_s(t) \otimes T \frac{\omega_m}{\pi} Sa(\omega_m t)$$

$$= f_s(t) \otimes T \frac{\omega_0}{2\pi} Sa(\omega_m t) \quad \text{(since } 2\omega_m = \omega_0\text{)}$$

$$= f_s(t) \otimes Sa(\omega_m t) \quad \left(\text{as } T = \frac{2\pi}{\omega_0}\right) \tag{1.9.5}$$

The sampled function $f_s(t)$ can be considered as a sum of impulses located at sampling instants $nT$, having strength equal to sample value $f_n$ at that instant as shown in Fig.1.9.4 a. Hence, the sampled function $f_s(t)$ can be expressed as

$$f_s(t) = \sum_n f_n \, \delta(t - nT) \tag{1.9.6}$$

$$f_s(t) = \sum_n f_n \, \delta(t - nT)$$

where $f_n$ is the $n^{th}$ sample of $f(t)$.

Substituting $f_s(t)$ from Eq.1.9.6 into Eq.1.9.5, the function $f(t)$ is given as,

$$f(t) = \sum_n f_n \, \delta(t - nT) \otimes Sa(\omega_m t)$$

Using sampling property of delta function, we get

$$f(t) = \sum_n f_n \, Sa[\omega_m(t - nT)] \tag{1.9.7a}$$

putting $\omega_m T = \pi$, we get

$$f(t) = \sum_n f_n \, Sa(\omega_m t - n\pi) \tag{1.9.7b}$$

Equation 1.9.7 represents that the function $f(t)$ can be constructed by multiplying its samples ($f_n$) by a sampling function $Sa(\omega_m t - n\pi)$, and adding the multiplied values. Note that $Sa[\omega_m(t - nT)]$ represents the sampling functions at sampling instants $t = nT$. The construction of $f(t)$ in this way is shown in Fig.1.9.4 b.

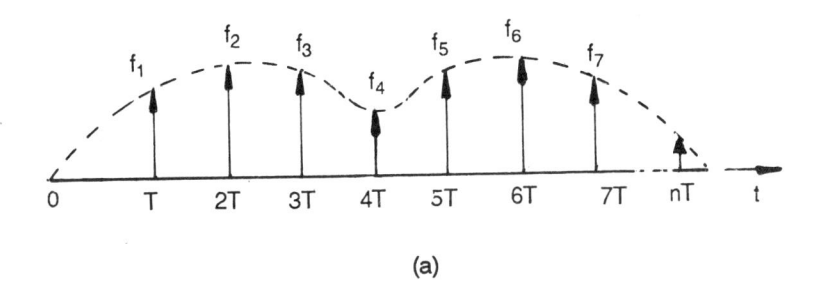

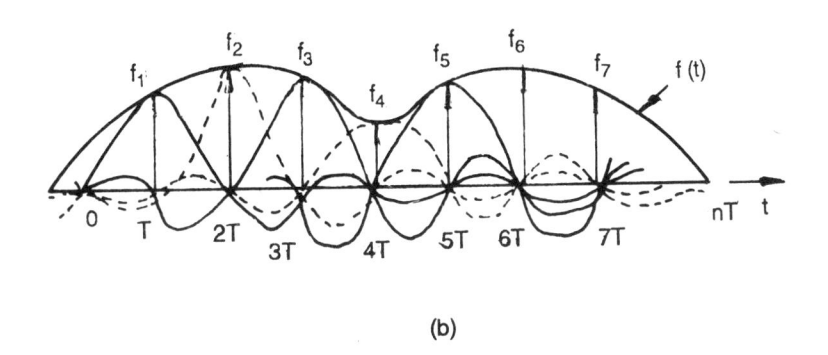

Fig. 1.9.4 (a) Sampled function and (b) Reconstruction of $f(t)$ in Time Domain

## Sampling Theorem (Frequency Domain)

The sampling theorem states that a time limited signal which does not exist beyond $T$ is uniquely determined by the samples of its frequency spectrum at uniform frequency intervals less than $1/2T$ Hz apart. Mathematically, it is stated as

$$F(\omega) = \sum_{m=-\infty}^{\infty} F\left(\frac{m\pi}{T}\right) Sa(\omega T - m\pi) \qquad (1.9.8)$$

This theorem is dual of time domain sampling theorem and can be proved in a similar way (Prob.1.15).

## Example 1.9.1

A function $f(t) = \sin \omega_c t$ is sampled at Nyquist rate. If the sampling starts from a zero crossing;
 (a) can we recover $f(t)$ from its samples?
 (b) what do you interpret from (a) for unambiguous recovery from sampled signal.

*Solution*

(a) The function $f(t) = \sin \omega_c t$ is shown by continuous lines in Fig.1.9.5. The function is a pure sinusoid having a single frequency component $\omega_c$ i.e., maximum frequency component present is $\omega_c$. The Nyquist sampling interval is given as $T_s = \dfrac{1}{2f_c}$ which means two samples per cycle are taken, and the samples are separated from each other by half cycles. Now, when the sampling starts from a zero crossing, it can be seen from Fig.1.9.5 that all the samples lie at zero crossings, because after each half cycle a zero crossing exists. Thus, the value of all the samples are zero, and no signal is received at the receiver. The signal $f(t)$ cannot be recovered although the sampling has been done at Nyquist rate.

(b) We can interpret that the sampling theorem does not provide unambiguous recovery of signals if sampled exactly at Nyquist rate. However, in the above example, if the sampling rate is greater than Nyquist rate, then all the samples would not lie on zero crossings and the signal recovery can be done. This is shown in Fig.1.9.5, where the sampling is done twice the Nyquist rate. In this case, alternate samples lie at zero crossing, but the other samples fall at maxima and minima shown by arrows in Fig.1.9.5 and signal $f(t)$ can be very well recovered.

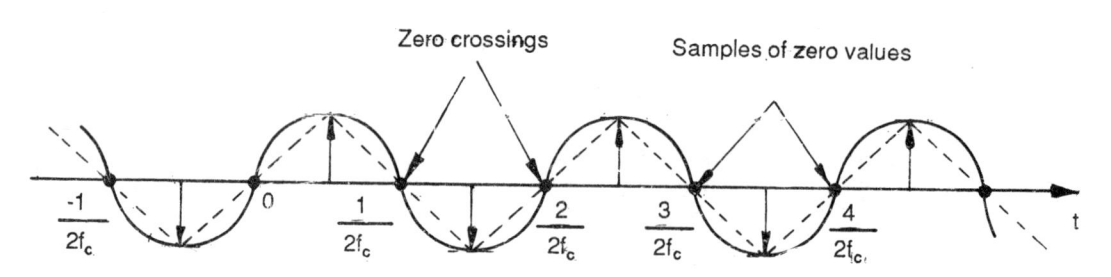

Fig. 1.9.5 Sampling at Zero Crossings of a Sine Wave

Thus, an unambiguous definition of sampling states that inequality in Eq.1.9.3. must mean *strictly less than* and *not less than, or equal to*,

$$T < \dfrac{1}{2f_m} \text{ and not } T \leq \dfrac{1}{2f_m}$$

and

$$f_s > 2f_m \text{ and not } f_s \geq 2f_m$$

Although the upper limit of sampling interval $T$ is $\dfrac{1}{2f_m}$ for an unambiguous recovery it should be less than $\dfrac{1}{2f_m}$.

### Example 1.9.2

A function $f(t) = \sin\omega_c t$ is sampled at twice the minimum rate, as shown in Fig.1.9.5. Justify that the samples correspond to the function $\sin\omega_c t$ and not to a saw-tooth type of waveform shown by dotted lines in Fig.1.9.5.

**Solution**

Although it appears that samples may represent both the waveforms, the saw-tooth waveform is not bandlimited to $\omega_c$, rather, theoretically, it possesses infinite frequencies. On the other hand, $\sin\omega_c t$ is bandlimited to $\omega_c$ and hence samples represent the function $\sin\omega_c t$, satisfying the necessary condition of band, limitation for sampling theorem.

## Sampling of Bandpass Signals

The sampling theorem discussed above applies to bandlimited signals centered at origin, i.e., low pass signals. Let us consider a more general case of bandpass signal with $\omega_H$ as upper cut-off frequency and $\omega_L$ as lower cut-off frequency, centered at $\omega_c$ as in Fig. 1.9.6. Obviously, the low pass signals are the special case of this band pass signal when $\omega_c = 0$.

It is evident from Fig. 1.9.6 that maximum frequency component present in the signal is $\omega_H = \omega_c + \omega_m$ and hence, minimum sampling rate is expected to be $2(\omega_c + \omega_m)$, but actually it is much less than this value. The rate is specified as follows:

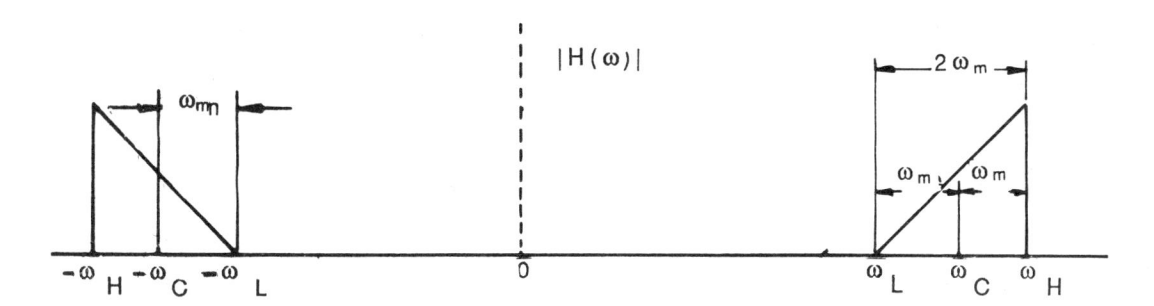

Fig. 1.9.6 Sampling of Bandpass Signals

*Case 1* If either $\omega_H$ or $\omega_L$ is a harmonic of sampling frequency $\omega_s$, the minimum sampling rate is given as

$$\omega_s = 2(\omega_H - \omega_L) = 2 \times 2\omega_M = 4\omega_m$$

Let us consider that $\omega_L$ is integral multiple (harmonic) of $\omega_s$ (i.e., $\omega_L = n\omega_s$). Then the spectrum of the corresponding bandpass signal may be shown to contain the entire passband, and the original bandpass signal can be recovered from its sampled version by passing the sampled signal through a bandpass filter with arbitrarily sharp cut-off, and having passband from to $\omega_L$ to $\omega_H$.

*Case 2* If $\omega_L$ or $\omega_H$ is not harmonic of sampling frequency $\omega_s$, then, a more general sampling condition is given as follows. The minimum sampling rate is given by

$$\omega_s = \frac{2(\omega_c + \omega_m)}{m}$$

where $m$ is the largest integer not exceeding $\dfrac{\omega_c + \omega_m}{2\omega_m}$

Actually case 1 is a special case of case (2) where the frequency band occupied by signal is located between adjacent multiplies of $2\omega_m$, i.e., $m = \dfrac{f_c + f_m}{2f_m}$. In this case, the minimum sampling rate is twice the bandwidth, i.e., $2 \times 2\omega_m$. This is same as case (1), where minimum sampling rate was shown to be $2(f_L - f_H)$ or $2 \times 2 f_m$.

### Example 1.9.3

The spectral range of a function extends from 10.0 MHz to 10.2 MHz. Find the minimum sampling rate, the maximum sampling time.

*Solution*
Let us find the sampling rate using *Case* 1
$$f_s = 2(f_H - f_L) = 2(10.2 - 10) = 0.4 \text{ MHZ}$$
since $f_H$ (i.e., 10 MHz) is $25^{th}$ harmonic of $f_s$ (i.e. 0.4 MHz), hence the necessary condition needed for Case 1 is satisfied. Therefore, the sampling rate of 0.4 MHz is the desired answer.
The corresponding sampling time $T_s$ is given as,
$$T_s = \frac{1}{0.4 \text{ MHz}} = \frac{10^6}{0.4}$$
$$= 2.5 \mu s$$

## PROBLEMS

1.1 Show that a set of function $\cos n\omega_0 t$ and $\sin n\omega_0 t$ ($n = 0, 1, 2, \cdots$) form a complete orthogonal set over the interval $t_0 \leq t \leq t_0 + T$ for any choice of $t_0$. $\left(T = \frac{2\pi}{\omega_0}\right)$.

1.2 Show that a set of functions $e^{jn\omega_0 t}$ ($n = 0, \pm 1, \pm 2, \pm 3, \cdots$). is a closed orthogonal set over the interval $(t_0, t_0 + T)$.

1.3 Derive an exponential series from a trigonometric series by replacing the sine and cosine functions with their exponential equivalents, and show that the co-efficients are related as follows:
$$F_0 = a_0$$
$$F_n = \frac{1}{2}(a_n - jb_n)$$
$$F_{-n} = \frac{1}{2}(a_n + jb_n)$$

1.4 Starting from fundamental, find the exponential Fourier series of Fig.1.1.1 and derive corresponding trigonometric series.

1.5 Starting from fundamental, derive the trigonometric series of a half-rectified sine wave, with time period $T$ and amplitude $A$. Derive the corresponding exponential Fourier series and draw the spectrum. Show how you can use this wave for obtaining a d.c. signal by using a low-pass-filter. Also, show that magnitude of d.c in a full-rectified wave is higher than the one in a half-rectified wave.

1.6 Find the Fourier series of a periodic impulse function with period $T$ and strength $A$. Draw the spectrum.

1.7 Determine the Fourier transform of a radio frequency pulse shown in Fig.1.6.6a by using the convolution theorem, and show that the result is identical to Ex.1.6.6.

1.8 Prove the following properties of Fourier transform.
(i) Conjugate function.
(ii) Frequency differentiation theorem.
(iii) Time integration theorem.

1.9 Find the Fourier transform of a triangular R.F. pulse shown in Fig.1.6.14a by using double differentiation property, and show that the result is identical with Ex. 1.6.12.

1.10 Find the Fourier transform of the function $f(t)$ given below

  (a) $f(t) = te^{-at}u(t)$;

  (b) $f(t) = |t|$

1.11 Find the Fourier transform of a Gaussian pulse given by

$$f(t) = e^{\left(\frac{-t^2}{2\sigma^2}\right)}$$

1.12 Prove the following convolution laws:
  (i) Commutative law,
  (ii) Associative Law, and
  (iii) Distributive law

1.13 State and prove the frequency convolution theorem.

1.14 Plot the convolution of a function $f(t)$ with a pair of impulses.

1.15 State and prove sampling theorem (frequency domain).

1.16 The triangular pulse shown in Fig. Prob.1.16 is multiplied by a sinusoidal signal $\cos 2\pi.10^6 t$. Find the expression for the spectrum of the product signals and plot it.

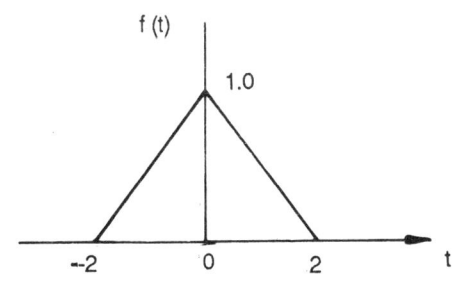

Fig. Prob. 1.16

1.17 Find Fourier transform and draw the spectrum of a Gate function with period $T = \frac{1}{2}$ s, and duration $\tau = \frac{1}{40}$ s. Assume unity amplitude.

1.18 An A.M. signal is given by

$$f(t) = 10 \sin(2\pi.10^6 t) + \left[8 \cos(2\pi.10^3 t) + 6 \sin(2\pi.10^2.t)\right] \sin(2\pi.10^6.t)$$

Find the Fourier transform $F(\omega)$. Draw the spectrum.

1.19 Find the specrum of a 5V-pulsed carrier of 2 μs width at 100 Hz pulse repetition rate. Carrier frequency is 1 MHz. Draw the spectrum.

1.20 Find the Fourier transform of the half cosine pulse shown in Fig. Prob. 1.20.

1.21 Find the amplitude and phase of the following signals

  (a) $\dfrac{\sqrt{3}}{2} \cos \omega t + \dfrac{1}{2} \sin \omega t$

  (b) $\dfrac{\sqrt{3}}{2} \cos \omega t + \dfrac{1}{2} \sin \omega t$

f(t)

A

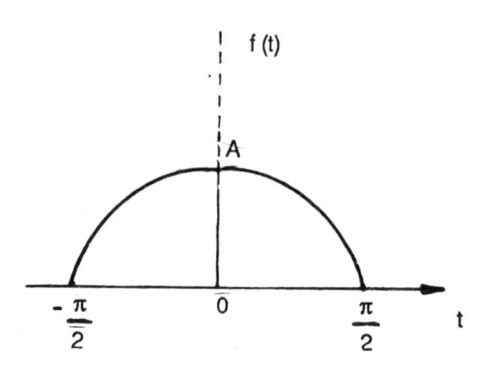

$-\frac{\pi}{2}$    0    $\frac{\pi}{2}$    t

Fig. Prob. 1.20

1.22 Show that the magnitude and phase spectrum of the Fourier transform $F(\omega)$ are respectively symmetrical and antisymmatrical about the vertical axis.

# TWO
# LINEAR SYSTEMS

## INTRODUCTION

In communication systems the transfer of information may be treated as a transfer of *signal* through a set of inter-connected functional blocks called *system*. Let us define some common terms related to signals and systems.

## Signal

For our purpose, a signal may be defined as any ordinary function of time. The value of a signal at any time may be real or complex.

## *Deterministic and Non-deterministic (Random) Signals*

Signals that can be modelled by explicit mathematical expressions are called deterministic signals, e.g. $f(t) = 20 \cos 50\,t$. On the other hand, a *random signal* is one about which there is some degree of uncertainty before it actually occurs, e.g., noise.

## Classification of Signals

### *Periodic and Non-periodic Signals*

A signal $f(t)$ is called periodic if there exists constant $T$, such that

$$f(t) = f(t + T), \qquad -\infty < t < \infty$$

$T$ is called the period of the signal, and signal $f(t)$ repeats after every $T$ s. Those signals that do not satisfy the above conditions are called *non-periodic* or *aperiodic* signals.

### *Energy and Power Signals*

These signals will be defined in Sec. 2.4.

### *Impulse Signals*

These signals are described by singularity functions already discussed in Sec. 1.5.

## 88 COMMUNICATION SYSTEMS: Analog and Digital

### Causal Signals

The signals having zero value for $t<0$ are called causal signals. Otherwise they are non-causal.

### System

A system is defined as a set of rules that associates an *output* time function to ever input time function. This is represented by the block-diagram shown in Fig.2.1.

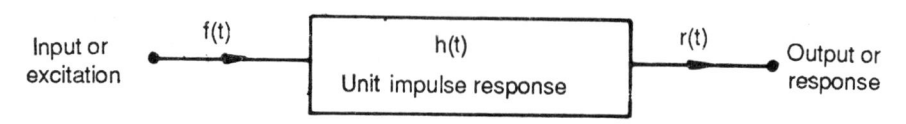

**Fig.2.1 Block Diagram of a System**

In the Fig.2.1, $f(t)$ is the *input* signal, also known as excitation or *source signal* or *driving function*; $r(t)$ is the output signal also known as *response signal*; $h(t)$ is the response of the system when input is a unit impulse function. The function $h(t)$ is known as *unit impulse response* of the system. The actual physical structures of the system determines the exact relationship between $f(t)$ and $r(t)$ and specifies $r(t)$ for every possible input $f(t)$.

Symbolically, input and response are represented as $f(t) \to r(t)$ and read as input $f(t)$ causes a response $r(t)$

### Classification of systems

#### Linear and Non linear System

Linearity is closely related with the concept of superposition. A system is said to obey superposition when the output obtained due to a sum of inputs is equal to the sum of the outputs caused by individual inputs, i.e., if

$$f_1(t) \to r_1(t), \text{ and } f_2(t) \to r_2(t)$$

then

$$f_1(t) + f_2(t) \to r_1(t) + r_2(t)$$

A system is said to be linear if the following relationship holds for all values of the constants $a_1$, and $a_2$,

$$a_1 f_1(t) + a_2 f_2(t) \to a_1 r_1(t) + a_2 r_2(t)$$

In general

$$a_1 f_1(t) + a_2 f_2(t) + a_3 f_3(t) + \cdots + a_n f_n(t) \to a_1 r_1(t) + a_2 r_2(t) + a_3 r_3(t) + \cdots + a_n r_n(t);$$

otherwise, the system is non linear. Thus the linear systems are characterized by superposition theorem and homogeneity.

#### Time-invariant and Time-varying Systems

A system is said to be time-invariant if the response due to an input is independent of the actual time of occurrence of the input, i.e., a time shift in the input results in an equal time shift in the output. Symbolically, if

$$f(t) \to r(t)$$

then, for a time-invariant system
$$f(t - t_0) \rightarrow r(t - t_0)$$
Systems that do not meet the above requirements are called time-varying systems. A sufficient condition for an electrical network to be time-invariant is that its component elements (resistance, capacitance, inductances etc.), do not change with time.

## Causal and Non-causal Systems

A system where the response does not begin before the input function is applied is known as a causal system. In other words, the value of the output, $r(t)$ at any instant $t = t_0$ depends only on the values of the input $f(t)$ for $t \leq t_0$ i.e.,
$$r(t_0) = f[f(t); t \leq t_0] \qquad -\infty < t, t_0 < \infty$$
The unit impulse response $h(t)$ of a causal system is zero for $t < 0$, i.e.,
$$h(t) = 0, t < 0$$
as unit impulse function $\delta(t)$ exists only at $t = 0$.

This means the system is operating in real time. Non-causal systems do not satisfy the above mentioned conditions. We will have more discussion on causal system in Sec. 2.3 where we will see that causal systems are physically realizable and non-causal systems do not exist in the real world.

## 2.1 THE SYSTEM FUNCTION (TRANSFER FUNCTION)

Let us characterize a linear time invariant system. The response $r(t)$ of a linear system to a given input $f(t)$ can be determined by taking advantage of the superposition theorem. The following steps can be adopted to determine the response using superposition theorem:

(i) Resolve the input function $f(t)$ in terms of simpler functions such as exponential, or impulse function, for which response can be easily evaluated.
(ii) Determine individually the response of a linear system for the simple input functions.
(iii) Find the sum of the individual responses which will become the overall response, $r(t)$ of function $f(t)$. The sum may be discrete or continuous depending on whether $f(t)$ has been resolved by a discrete or continuous sum of the simple functions.

Let us use the impulse function to represent $f(t)$, i.e., we shall first represent input function $f(t)$ as a continuous sum of impulse functions then find the continuous sum of the response to these impulses.

### Representation of a Function $f(t)$ as Continuous Sum of Impulse Function

Let us consider an arbitrary excitation $f(t)$ as shown in Fig.2.1.1, where the function has been represented in terms of element areas. In the limit $\Delta t \rightarrow 0$, $n^{th}$ element area may be considered as a rectangle of width $\Delta t$ and height $f(n \Delta t)$. For example, the shaded portion in Fig. 2.1.1 has width $\Delta t$, and height $f(2\Delta t)$ and the area of this shaded portion will be "$f(2\Delta t).(\Delta t)$". In the limit $\Delta t \rightarrow 0$, this element area approaches a delta function of strength $f(2\Delta t)(\Delta t)$ located at $t = 2\Delta t$. This delta function is symbolically represented as $f(2 \Delta t)(\Delta t) \delta(t - 2 \Delta t)$. The function $f(t)$ is continuous sum of such impulse functions,

$$f(t) = \lim_{\Delta t \rightarrow 0} \sum_{n = -\infty}^{\infty} f(n \Delta t) (\Delta t) \delta(t - n \Delta t) \qquad (2.1.1)$$

# 90 COMMUNICATION SYSTEMS: Analog and Digital

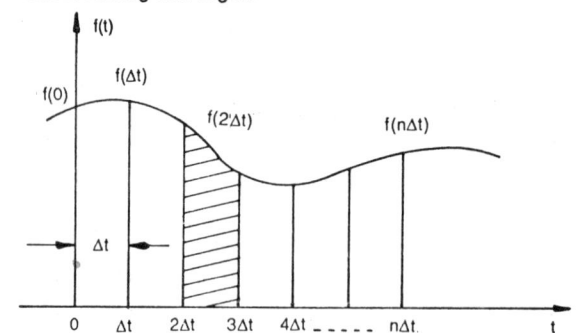

Fig. 2.1.1 Approximation of a Driving Function $f(t)$

## Superposition Theorem to Determine Response $r(t)$

Let $h(t)$ be the unit impulse response of a linear system i.e., $h(t)$ is the response of linear system when input function is a unit impulse function $\delta(t)$ located at $t = 0$, and of unit strength. Obviously, the response of the system for an impulse function of strength $f(n\Delta t)(\Delta t)$ located at $t = n\Delta t$ will be given as $f(n\Delta t)(\Delta t)h(t - n\Delta t)$. According to superposition theorem, the response $r(t)$ of the system for input function $f(t)$ as expressed by Eq.2.1.1 will be given by

$$r(t) = \lim_{\Delta t \to 0} \sum_{n=-\infty}^{\infty} f(n\Delta t)h(t - n\Delta t)(\Delta t)$$

Smaller the $\Delta t$ better is the approximation, and in the limit $\Delta t \to 0$, the summation becomes integration yielding $r(t)$ as, ($n$ has no significance for integration)

$$r(t) = \int_{-\infty}^{\infty} f(\tau)h(t-\tau)d\tau \qquad (2.1.2a)$$

$$= f(t) \otimes h(t) = h(t) \otimes f(t) \qquad (2.1.2b)$$

The relation in Eq. 2.1.2 is the convolution integral. Therefore, the output of any general input can found by convolving the input function $f(t)$ with the system's unit impulse response $h(t)$.

Thus, if the response of a linear system is known for an impulse function, then the response to any other function $f(t)$ can be obtained by Eq. 2.1.2. Therefore, the unit impulse function may be considered as *test function*, and can be used to characterize a system. It should be recalled that $\delta(t)$ contains all frequency components, and hence deserves to be called a test function.

By using the time convolution theorem, Eq. 2.1.2 can be represented in frequency domain as follows:

$$R(\omega) = F(\omega) H(\omega)$$

Where $r(t) \leftrightarrow R(\omega)$, $f(t) \leftrightarrow F(\omega)$, and $h(t) \leftrightarrow H(\omega)$.

Therefore, the output in frequency domain is equal to the product of the input function in frequency domain and unit impulse response in frequency domain. $H(\omega)$ characterizes the linear system, and hence is known as *system function*. $H(\omega)$ is also called a *transfer function*, as it represents the ratio of output (frequency domain) to input (frequency domain).

## Definition of Transfer Function or System Function

In Eq. 2.1.2. $H(\omega)$ is equal to $R(\omega)$, if $F(\omega) = 1$. Since $F(\omega) = 1$ corresponds to a unit impulse function in the time domain we can say that $H(\omega)$ is response (frequency domain) of the system when its input is a unit impulse function. Therefore, we can define *transfer function* $H(\omega)$ as follows:

"The transfer function of a linear system is transform of its response when the input is a unit impulse function" with zero initial conditions. $H(\omega)$ in general, is complex and may be expressed in the form

$$H(\omega) = |H(\omega)| e^{j\theta\omega} \quad (2.1.3)$$

where $|H(\omega)|$ is known as the amplitude response and $\theta(\omega)$, the phase response. $H(\omega)$ exhibits conjugate symmerty for a linear system, which means the amplitude response is symmetrical, and phase response is antisymmetrical about the vertical axis i.e.,

$$|H(\omega)| = |H(-\omega)| \quad (2.1.4a)$$
$$\theta(\omega) = -\theta(-\omega) \quad (2.1.4b)$$

Sometimes, it is preferable to work on the logarithmic scale. Taking logarithm on both sides of Eq.2.1.3.

$$\ln H(\omega) = \ln |H(\omega)| + j\theta(\omega) \quad (2.1.5a)$$
$$= \alpha(\omega) + j\theta(\omega) \quad (2.1.5b)$$

The function $\alpha(\omega)$ is called the gain of the system and $\theta(\omega)$ is called the *phase shift* introduced by the system. $\alpha(\omega)$ is measured in neper or decibels depending on the base of the logarithm, i.e.,

$$\alpha(\omega) = \ln |H(\omega)| \quad \text{nepers}$$
$$= 20 \log_{10} |H(\omega)| \quad \text{decibel or dB}$$

$\theta(\omega)$ is measured in radians. Note that gain has positive value if the response is greater than excitation, and has zero value if response is same as excitation. On the other hand, if $\alpha(\omega)$ is negative in dB, the response will be less than the excitation, and the system will produce a loss. This negative value of gain $\alpha(\omega)$ for which the system produces a loss, is known as attenuation. The neper and decibel (dB) is related to each other as follow:

$$1 \text{ neper} = 8.69 \text{ dB}$$

## 2.2 DISTORTIONLESS TRANSMISSION

Transmission of a signal through a system is said to be distortionless if the output signal is an exact replica of the input signal. However, a constant change in magnitude and a constant time delay in the output replica is not considered as distortion. The magnitude of the output replica may increase or decrease during transmission depending on whether the system offers a gain or attenuation. Therefore, an input function $f(t)$ is said to be transmitted without distortion if the output signal $r(t)$ is defined by

$$r(t) = K f(t - d) \quad (2.2.1)$$

where constant $K$ represents the change in magnitude (gain or attenuation), and constant $d$ accounts for a constant time delay.

Using time shifting property, the frequency domain representation of Eq. 2.2.1 yields

$$R(\omega) = K F(\omega) e^{-j\omega d} \quad (2.2.2 \text{ a})$$

The response $R(\omega)$ is related with input $F(\omega)$ by its transfer function i.e.,

$$R(\omega) = H(\omega) F(\omega) \qquad (2.2.2b)$$

Comparing Eq. 2.2.2a with Eq. 2.2.2b, we get

$$H(\omega) = K e^{-j\omega d} \qquad (2.2.3)$$

The magnitude and the phase angle of this transfer function $H(\omega)$ is given by,

$$|H(\omega)| = K \qquad (2.2.4a)$$

$$\theta(\omega) = -\omega d$$

or more generally

$$\theta(\omega) = n\pi - \omega d \ (n \ is \ + ve \ or \ - ve \ integer) \qquad (2.2.4b)$$

Equation 2.2.4 reveals that magnitude $|H(\omega)|$ should be constant for a distortionless transmission. It means that all the frequency components present in $f(t)$ are equally attenuated or amplified, and system bandwidth is infinite. The phase spectrum $\theta(\omega)$ should be directly proportional to the frequency, as shown in Fig. 2.2.1 The reason is clear. A time delay of $d$ introduced by the system is equivalent to a phase shift of $\omega d$, which is directly proportional to $\omega$. For example, if a signal $\sin \omega_0 t$ is delayed by $d$, the resulting signal is $\sin \omega (t - d)$, which can be expressed as $\sin(\omega t - \omega d)$. The phase shift in the new signal is $\omega d$, which is proportionate to $\omega$. No system has infinite bandwidth in practice and, hence, distortionless conditions are never met exactly. The following distortions may occur during transmission of signals through linear systems:

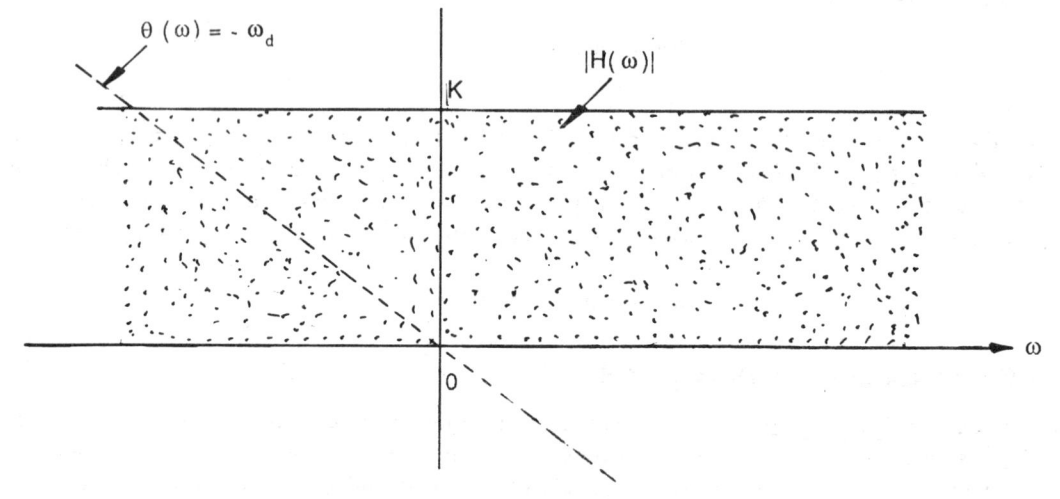

Fig. 2.2.1 A Distortionless Transmission System

(i) *Amplitude Distortion*: This distortion occurs when $|H(\omega)|$ is not constant within the frequency band of interest and frequency components present in the input signal are transmitted with different gain or attenuation. Some frequency components may remain unchanged, some may be greatly attenuated, or amplified; and others may be slightly attenuated, or amplified.

(ii) *Phase Distortion*: Phase distortion occurs when $\theta(\omega)$ is not linearly changing with time, and the different frequency components in the input are subjected to different time delays during transmission.

The above distortions change the shape of the output waveform as compared to input waveform.

## Bandwidth of a System

For distortionless transmission, the system bandwidth should be infinite, but physical systems are limited to finite bandwidth. The system with finite bandwidth can provide distortionless transmission for a bandlimited signal, provided the system function $|H(\omega)|$ is constant over the bandwidth of the input signal. The system response is constant over a band, if $|H(\omega)|$ is constant for the band. This constancy of the magnitude function $|H(\omega)|$, or gain $\alpha(\omega)$, over a band of frequency is specified by a parameter called *bandwidth* of the system. The bandwidth is defined as the interval of frequencies over which $|H(\omega)|$ remains within $\frac{1}{\sqrt{2}}$ time of its midband value. A typical $|H(\omega)|$ plot having bandwidth of $(\omega_H - \omega_L)$ is shown in Fig. 2.2.2.

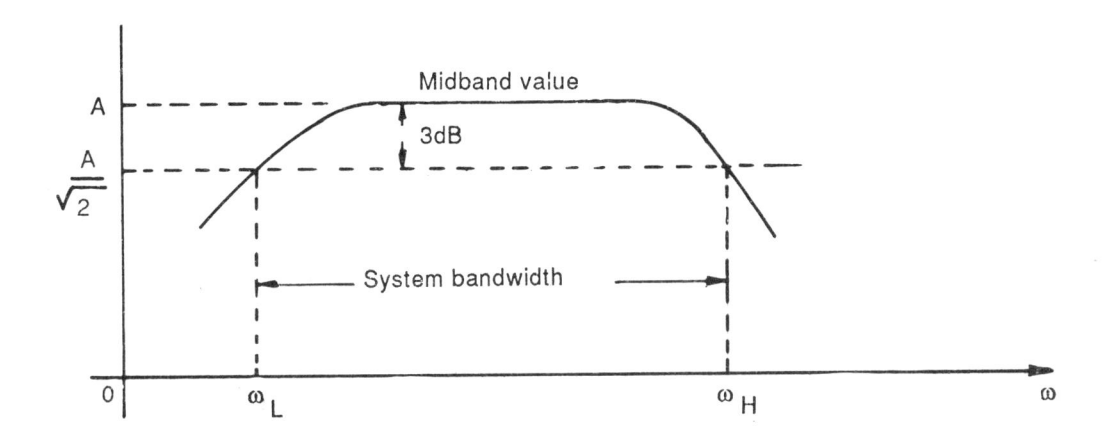

Fig. 2.2.2

In most practical applications, input signals are approximated as bandlimited signals, because magnitude reduces appreciably after a certain frequency component. The bandlimited signals can be transmitted without distortion if the *system bandwidth is at least equal to the signal bandwith*.

## 2.3 PALEY-WIENER CRITERION

*Minimum Phase System*: Let us consider a linear time invariant system in which phase and gain components of the transfer function $H(\omega)$ are uniquely related to each other by the following equations:

$$\theta(f) = \frac{1}{\pi} \int_{-\infty}^{\infty} \frac{\alpha(\lambda)}{\lambda - f} d\lambda \qquad (2.3.1a)$$

and

$$\alpha(f) = \alpha(\infty) - \frac{1}{\pi} \int_{-\infty}^{\infty} \frac{\theta(\lambda)}{\lambda - f} d\lambda \qquad (2.3.1b)$$

where $\lambda$ is a dummy variable, and $\alpha(\infty)$ is the value of gain at infinite frequency. The unique relations given above represent that if gain $\alpha$ is known for the entire frequency, then $\theta$ can be completely described, and vice-versa, except the constant $\alpha(\infty)$. In such a system, for a known gain characteristic, the transfer function has the minimum phase shift possible. This system is known as *minimum phase system*. Any change in phase characteristic of such systems causes a corresponding change in the gain characteristic and vice-versa.

## Paley-Wiener Criterion

Let us consider a system with gain $\alpha(\omega)$, and phase $\theta(\omega)$. The necessary and sufficient condition for the existance of the minimum phase $\theta(\omega)$ for a given $\alpha(\omega)$ is that

$$\int_{-\infty}^{\infty} \frac{|\alpha(\omega)|}{1+\omega^2} d\omega < \infty$$

$$\int_{-\infty}^{\infty} \frac{|\ln|H(\omega)||}{1+\omega^2} d\omega < \infty \qquad (2.3.2)$$

provided $|H(\omega)|$ is square integrably i.e., $\int_{-\infty}^{\infty} |H(\omega)|^2 d\omega < \infty$

The condition given by Eq. 2.3.2 is known as *Paley Wiener Criterion*.

## Causality Condition and Physical Realizability

The Paley Wiener criterion stated in Eq. 2.3.2 is frequency domain equivalent of the causality condition for a system. Therefore, the time domain equivalent of Paley Wiener criterion is stated as :
(i) A system violating the Paley Wiener criterion has a non-casual unit impulse response. It means that response is present prior to excitation. Obviously, such systems are not physically realizable as no physical system can produce response before excitation.
(ii) On the other hand, a system satisfying the Paley Wiener criterion has a causal unit impulse response, i.e.

$$h(t) = 0, t < 0$$

Such systems are physically realizable. The above statements represent the time domain conditions for the physical realizability of a system.

The frequency domain conditions for the physical realizability of a system can be interpreted from the frequency domain Paley Wiener criterion given by Eq. 2.3.2. The following frequency domain interpretations may be drawn from Eq. 2.3.2.
(a) Magnitude function $|H(\omega)|$ of a physically realizable system may be zero for some discrete frequencies but it can never be zero for a finite band of frequency, otherwise, integral in Eq. 2.3.2 would tend to be infinite. For example, a system having a transfer function $H(\omega)$ as shown in Fig. 2.3.1a, is not physically realizable, because its $|H(\omega)|$ has zero value for a band of frequencies beyond $\omega_m$. Such systems are called *ideal* low pass filters. On the other hand, a system, as represented by Fig. 2.3.1 b, is physically realizable even for a small value of $a$ as it has non zero value anywhere for a band of frequency, although some discrete frequencies have zero magnitude.

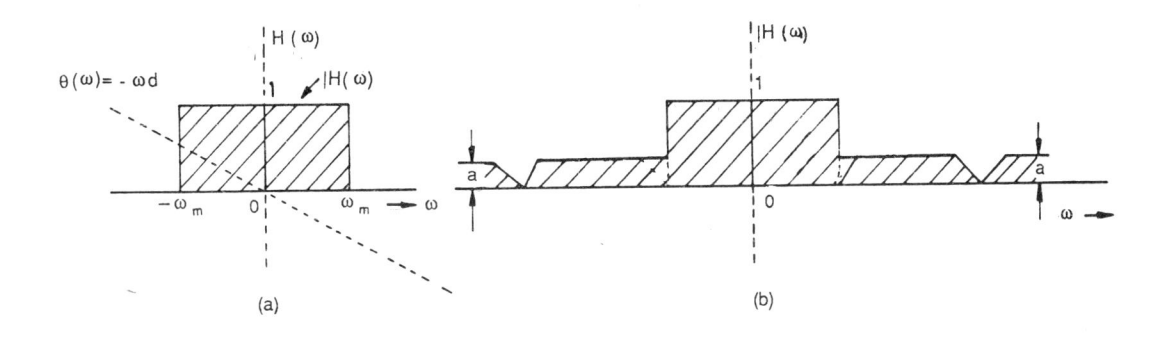

Fig. 2.3.1 Frequency Domain Causality Condition (a) Unrealizable System; (b) Realizable System

(b) It can be seen from Eq. 2.3.2 that the magnitude function $|H(\omega)|$ for a realizable system cannot decay faster than a function of exponential order. Thus $H(\omega) = Ce^{-\alpha|\omega|}$ corresponds to a realizable system, but $H(\omega) = Ce^{-\alpha\omega^2}$ is not realizable.

## Filters

A filter is a frequency selective network that permits a band of frequency to pass through it with little attenuation. The attenuation increases severely for the remaining frequencies. The band allowed is called *Passband* and the band that is severely attenuated and not allowed to pass through it is called *Stopband*. Figure 2.3.2 shows the transfer functions of a high pass and bandpass filter. The low pass filter is already shown in Fig. 2.3.1 (a)

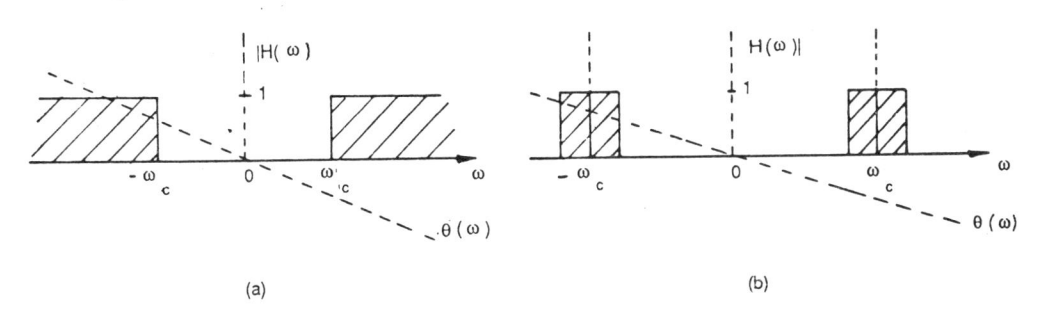

Fig. 2.3.2 Transfer Function of Filters (a) High pass filter; (b) Bandpass Filter

Linear systems may behave like filters depending on the characteristics of the function $H(\omega)$. Since the response of a linear system is related to the input function by the relation $R(\omega) = F(\omega) H(\omega)$, the spectrum of the response is modified depending on the nature of $H(\omega)$. If for a band of frequencies, $H(\omega)$ is zero, then the response will also be zero, and that particular portion of the band will not be allowed to pass. The shape (time domain) of the output (response) is also modified, if certain band of frequency is not allowed to pass, as shown in Fig. 2.3.3.

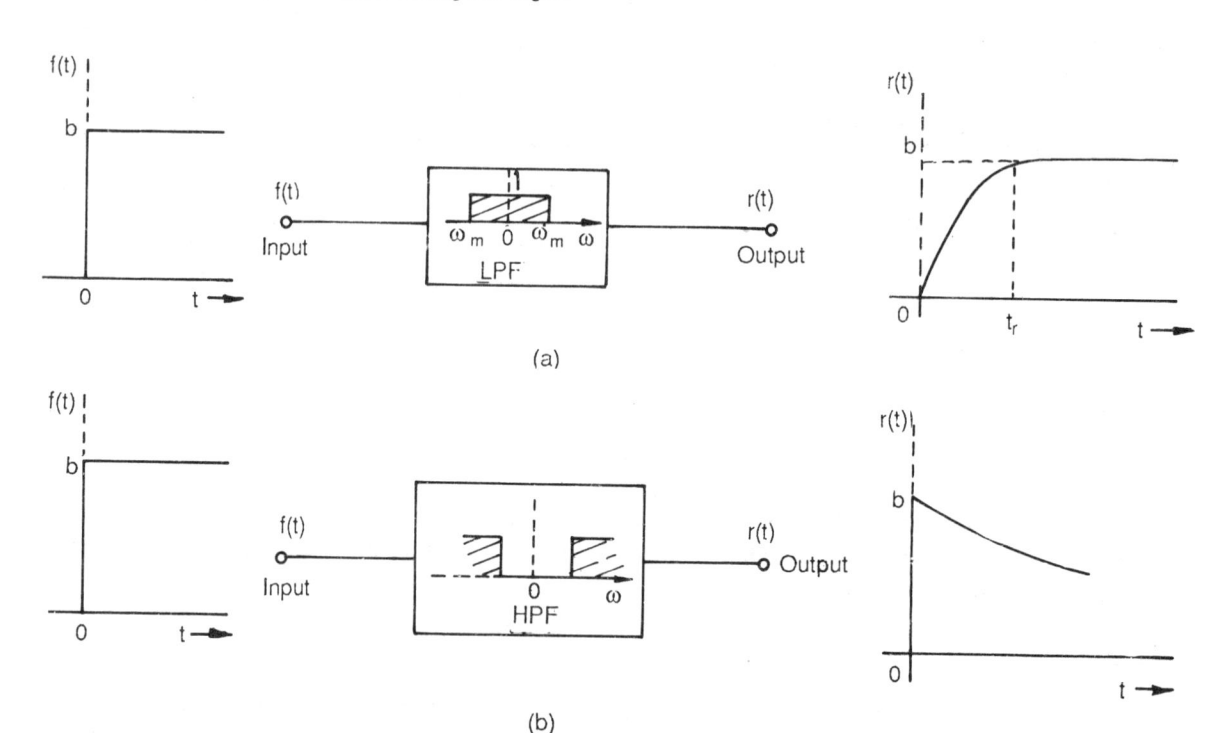

Fig. 2.3.3 Time Domain Response of Filters (a) LPF; (b) HPF

The output of a low pass filter (LPF) is shown in Fig. 2.3.3 a. Since the LPF attenuates the high-frequency components, the rise of the pulse is not instantaneous, rather, it is delayed. The sharp rise of the pulse causes the presence of high frequencies in the input signal which are not allowed to pass, and the waveshape of the output is modified. Similarly, a tilt is produced in the output of a high pass filter due to the absence of low frequencies.

## Ideal Filters

An ideal filter passes the pass band frequency without any distortion, and completely attenuates the stop band frequencies. Thus, an ideal low pass filter (Fig.2.3.1a) passes the frequency band from 0 to $\omega_m$ radians without any distortion, and completely attenuates the frequencies above $\omega_m$. Figure 2.3.2 illustrates the transfer function of some ideal filters. The ideal filters are characterized by sharp rise of $|H(\omega)|$ at cut of frequency. It is obvious from Fig. 2.3.2 that ideal filters have zero value of $H(\omega)$ for a band of frequency and hence, they are not physically realizable. The time domain response $h(t)$ of such system is non-causal. This will be illustrated in Ex. 2.3.1.

## Example 2.3.1

Show that unit impulse response of an ideal low pass filter is non-casual.

*Solution*

The magnitude function $|H(\omega)|$ for an ideal low pass filter is shown in Fig. 2.3.1a. The cut-off frequency is $\omega_m$.

The transfer function can be expressed as
$$H(\omega) = |H(\omega)|e^{j\theta(\omega)}$$
As shown in Fig.2.3.1a. $|H(\omega)|$ is a Gate function $G(\omega)$. The phase function is $\theta(\omega) = \omega d$, hence,
$$H(\omega) = G(\omega)e^{-j\omega d}$$
Taking the inverse Fourier transform
$$h(t) = F^{-1}\left[G(\omega)e^{-j\omega d}\right]$$
By using time shifting property, and recalling that a Gate function is a Fourier transform of a sampling function, the unit impulse response $h(t)$ is given as,
$$h(t) = \frac{\omega_m}{2\pi} Sa[\omega_m(t-d)] \qquad (2.3.3)$$
This is shown in Fig. 2.3.4.

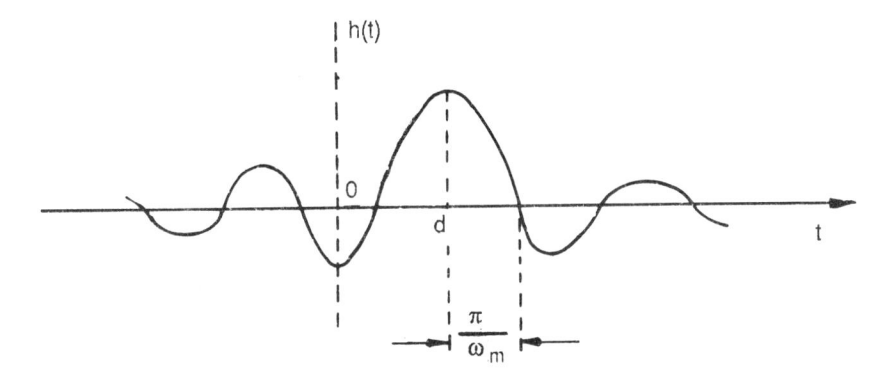

Fig. 2.3.4  Unit Impulse Response of a LPF

It is clear that unit impulse response exists even for negative values of $t$. Hence, the ideal low pass filter is a non-causal system, hence physically unrealizable.

Similarly, it can be seen that ideal high pass filters, and band pass filters have a non-causal impulse response, and hence cannot be physically realized. However, practical filters can be designed to have characteristics very close to ideal filters as illustrated in Ex. 2.3.2

## Example 2.3.2

Show that the practical filter circuit shown in Fig. 2.3.5a has a transfer function close to ideal low pass filter and also that its unit impulse response is causal.

*Solution*

The transfer function of the circuit is
$$H(\omega) = \frac{R(\omega)}{F(\omega)}$$

Here $\quad F(\omega) = E_i(\omega)$, and $R(\omega) = E_o(\omega)$, hence

$$H(\omega) = \frac{E_0(\omega)}{E_i(\omega)}$$

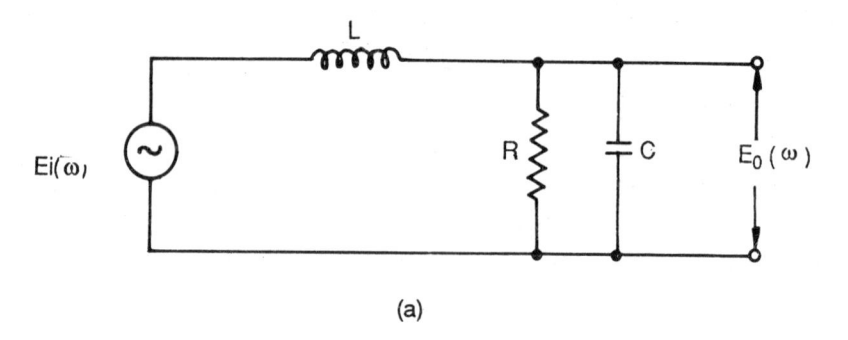

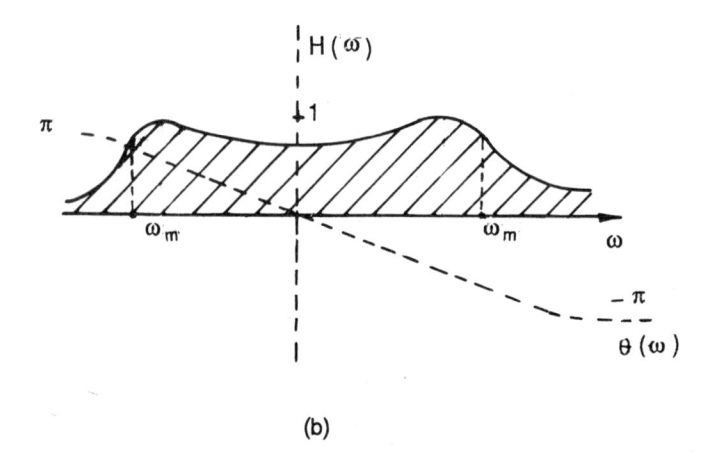

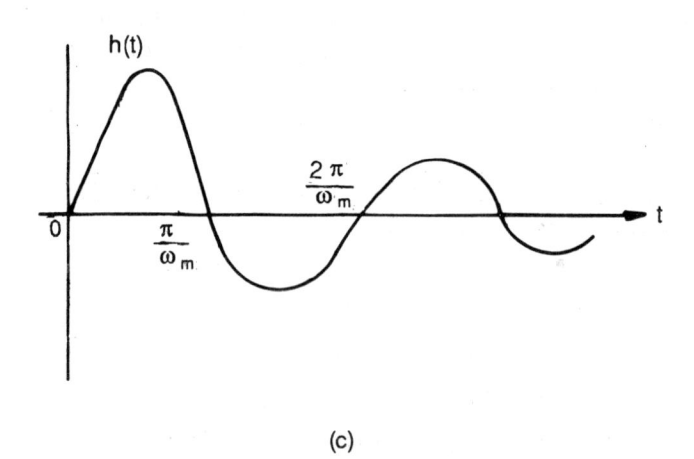

Fig. 2.3.5 A Practical LPF (a) Circuit; (b) Transfer function; (c) Unit impulse response

For the network shown in Fig. 2.3.5a

$$H(\omega) = \frac{\left[\dfrac{1}{(1/R + j\omega c)}\right]}{j\omega L + \dfrac{1}{\dfrac{1}{R} + j\omega c}} = \frac{1}{1 - \omega^2 LC + j(\omega L/R)}$$

Let $\dfrac{1}{\sqrt{LC}} = \omega_m$ and $\sqrt{\dfrac{L}{C}} = R$, then

$$H(\omega) = \frac{2\omega_m}{\sqrt{3}} \frac{\dfrac{\sqrt{3}}{2}\omega_m}{\left(\dfrac{\omega_m}{2} + j\omega\right)^2 + \left(\dfrac{\sqrt{3}}{2}\omega_m\right)^2}$$

The unit impulse response is given as

$$h(t) = F^{-1}[H(\omega)] = \frac{2\omega_m}{\sqrt{3}} e^{-(\omega_m t/2)} \sin\left(\frac{\sqrt{3}}{2}\omega_m t\right)$$

The plots of $H(\omega)$ and unit impulse response are shown in Fig. 2.3.5b and c respectively, which reveal that the magnitude, and phase response of $H(\omega)$ are very close to ideal low pass filter. The impulse response is identical to that of ideal low pass filter except that it is causal, i.e., $h(t)$ is zero for the negative values of $t$. Thus, it satisfies the Paley Wiener criterion, both in frequency and time domains.

Similarly, it can be shown that the ideal high pass and ideal bandpass filters are physically un-realizable, however, filters having characteristics close to ideal ones may be realized. Bandpass filter is analysed by using Hilbert transform. This analysis is quite complicated, and as such, will not be treated in this text.

## Active Filters

Practical filters which are realized by using passive components, e.g., inductors, capacitors and resistors are called passive filters. They have a disadvantage in that their output power is always less than the input power, i.e., they attenuate the signal because such filters absorb and store the energy passing through them. Also the inductors used are bulky and costly.

Active filters on the other hand, provide gain and therefore, output power is greater than the input, because they are realized by using active elements called operational Amplifiers (OPAMP) as basic building blocks. The OPAMPs add energy to the signal passing through the filter. The advantages of active filters may be stated as follows:

   (i) They provide gain, since the active component (OPAMP) adds energy.
   (ii) Inductors are not used, rather, they are realized by OPAMPs, and hence, such filters are light and cheap.
   (iii) They are versatile and simple to design.
   (iv) Arbitrary-causal transfer function can be achieved by these filters.

## Rise Time and System Bandwidth

It was seen earlier that the response of a filter is modified depending on the range of frequency components allowed to pass through the filter (i.e., system bandwidth). The *system bandwidth* is referred to as the range of frequencies that a system can handle. The system bandwidth is considered important in communication systems, and can be determined by evaluating the system function $H(\omega)$. As this involves a lengthy process, it is convenient to find the system bandwidth by estimating $h(t)$ or $r(t)$. It will be seen in the following discussion that the system bandwidth can be related to a time function parameter called *rise time*.

Let us consider a case in which a unit step function $u(t)$ is applied to an ideal low pass filter as shown in Fig. 2.3.1a. The response $r(t)$ does not rise sharply; rather, it shows a gradual rise. The time taken by the function to rise to its final value depends on the cut-off frequency of the filter i.e., system bandwidth. We will find the relation between filter cut-off frequency $\omega_m$ and the rise time $t_r$.

### Rise-time

The rise time $(t_r)$ may be defined as the time required for a signal to reach the final value from its initial value. This is one of the several definitions of rise time.

The transfer function $H(\omega)$ of the low pass filter is equivalent to a Gate function given by

$$H(\omega) = G(\omega) e^{-j\omega d}$$

The Fourier transform of the unit step function (input) is given by

$$F[u(t)] = \pi\delta(\omega) + \frac{1}{j\omega}$$

The Fourier transform of response $R(\omega)$ is obtained by multiplying $H(\omega)$ with the Fourier transform of $u(t)$ i.e.,

$$R(\omega) = \left[\pi\delta(\omega) + \frac{1}{j\omega}\right] H(\omega)$$

$$= \pi\delta(\omega) H(\omega) + \frac{1}{j\omega} H(\omega)$$

Since $\delta(\omega)$ exists at $\omega = 0$, and $|H(\omega)|_{\omega=0} = 1$, hence $\delta(\omega) H(\omega) = \delta(\omega)$

Therefore, $R(\omega)$ becomes

$$R(\omega) = \pi\delta(\omega) + \frac{1}{j\omega} H(\omega)$$

The time domain of response is given as,

$$r(t) = F^{-1}\left[\pi\delta(\omega) + \frac{1}{j\omega} G(\omega) e^{-j\omega d}\right]$$

Since, inverse Fourier transform of $\pi\delta(\omega) = \frac{1}{2}$, hence

$$r(t) = \frac{1}{2} + F^{-1}\left[\frac{1}{j\omega} G(\omega) e^{-j\omega d}\right]$$

$$= \frac{1}{2} + \frac{1}{2\pi} \int_{-\infty}^{\infty} \frac{1}{j\omega} G(\omega) e^{-j\omega d} e^{j\omega t} d\omega$$

Since, Gate function $G(\omega)=1$ for $|\omega|\leq \omega_m$ and zero elsewhere,

$$r(t) = \frac{1}{2} + \frac{1}{2\pi} \int_{-\omega_m}^{\omega_m} \frac{e^{j\omega(t-d)}}{j\omega} d\omega$$

Putting the complex exponential function in terms of cosine and sine, we get

$$r(t) = \frac{1}{2} + \frac{1}{2\pi} \int_{-\omega_m}^{\omega_m} \frac{\cos[\omega(t-d)]}{j\omega} d\omega + \frac{1}{2\pi} \int_{-\omega_m}^{\omega_m} \frac{\sin[\omega(t-d)]}{\omega} d\omega$$

The first integral vanishes as its solution yields an odd cosine function. The integrand of second integral is an even function of $\omega$.
Hence,

$$r(t) = \frac{1}{2} + \frac{1}{\pi} \int_{0}^{\omega_m} \frac{\sin[\omega(t-d)]}{\omega} d\omega$$

put $\omega(t-d) = y$, so that $d\omega(t-d) = dy$, then

$$r(t) = \frac{1}{2} + \frac{1}{\pi} \int_{0}^{\omega_m(t-d)} \frac{\sin y}{y} dy = \frac{1}{2} + \frac{1}{\pi} \int_{0}^{\omega_m(t-d)} Sa(y) dy$$

The integral of a sampling function is a standard integral known as sine integral $Si(y)$ given by

$$Si(y) = \int_{0}^{y} Sa(\tau) d\tau \qquad (2.3.4)$$

Hence, $r(t)$ becomes

$$r(t) = \frac{1}{2} + \frac{1}{\pi} Si[\omega_m(t-d)] \qquad (2.3.5)$$

The sine integral $Si(t)$, unit step input function $u(t)$ and its response $r(t)$ are shown in Fig. 2.3.6. It is obvious from Fig. 2.3.6 (c) that as cut-off frequency $\omega_m$ reduces, the response $r(t)$ rises more slowly.

Assuming the minimum value as the initial value and maximum value as final value, the rise time is defined as,

$$t_r = \frac{2\pi}{\omega_m} = \frac{1}{B}.$$

where $B$, denotes the bandwidth of the filter in Hz i.e., $B = \frac{\omega_m}{2\pi}$. Obviously, the rise-time is inversely proportional to the system bandwidth.

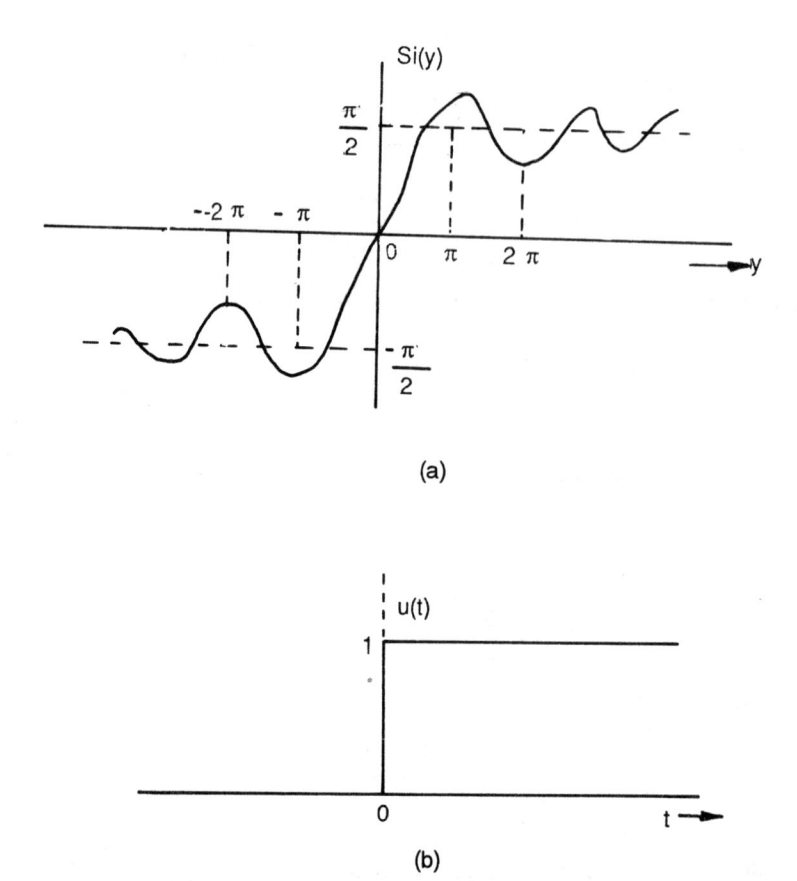

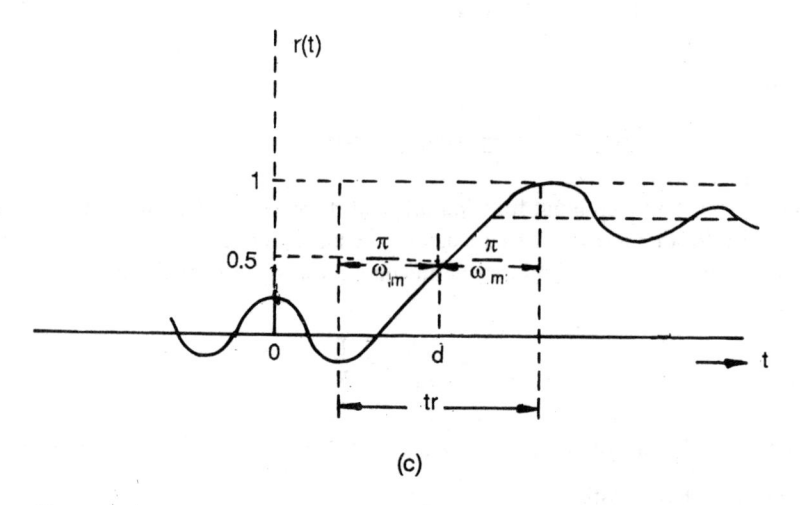

Fig. 2.3.6 (a) Sine integral; (b) Unit step function; (c) Responce, r (t)

## Example 2.3.3

Find the response of a low pass filter to a rectangular Gate pulse $G(t)$ shown in Fig. 2.3.7a and also show that dispersion occurs in the response due to low pass filtering. Assume $\omega_m$ as the cut-off frequency of LPF

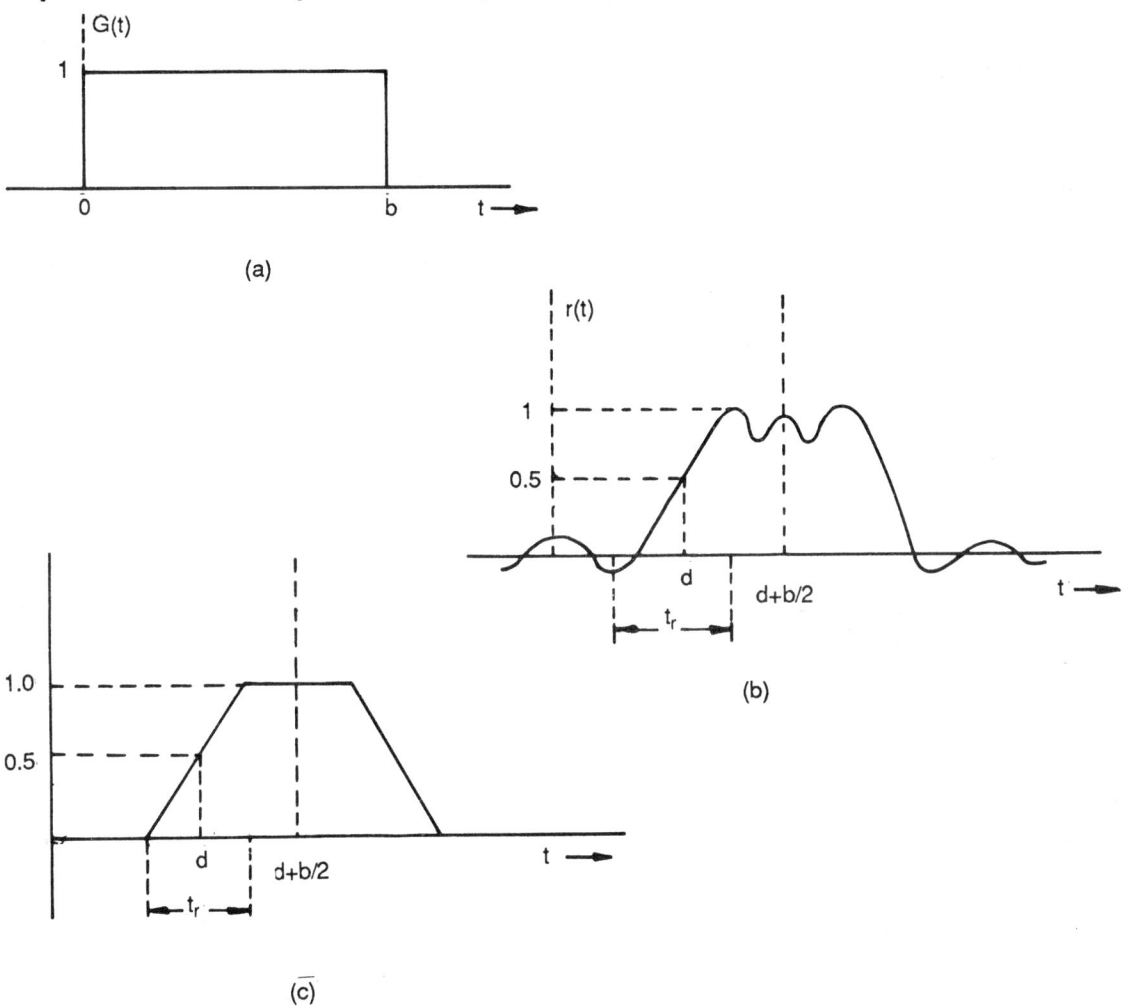

Fig. 2.3.7 (a) A Rectangular Pulse; (b) Response of LPF to Pulse in (a); (c) Approximation of Figure (b)

*Solution*

The rectangular pulse can be considered as a unit step function $u(t)$ from which another unit step function at $t = b$ i.e., $u(t - b)$ has been subtracted, thus,

$$G(t) = u(t) - u(t - b)$$

Therefore, the response $r(t)$ can be obtained by superposition theorem i.e., by subtracting the response corresponding to $u(t - b)$ from the response of $u(t)$. This is given as (using Eq. 2.3.5.)

$$r(t) = \frac{1}{\pi}\{Si[\omega_m(t - d)] - Si[\omega_m(t - d - b)]\}$$

The response is plotted in Fig. 2.3.7b. An approximation of r(t) is shown in Fig. 2.3.7c, which reveals that the transmission of a rectangular pulse from a low pass filter causes dispersion (spreadout) in the response.

## 2.4 ENERGY SIGNALS AND POWER SIGNALS

A primary goal of the communication system is to transmit more signal power as against noise power to achieve greater signal to noise ratio. Noise, being random in nature, cannot be expressed as a time function, like deterministic waveforms. It is represented by its power. Hence, for evaluation of signal to noise power ratio, it is necessary to evolve a method for calculating the power content of a signal.

### The Energy Signal

Let us consider a voltage $f(t)$ applied across a 1-ohm resistor, or a current $f(t)$ passing through a 1-ohm resistor. The energy dissipated as heat is given by

$$E = \int_{-\infty}^{\infty} f^2(t) \, dt \qquad (2.4.1)$$

This energy $E$, dissipated across 1-ohm resistor by a signal $f(t)$ is called *normalized energy* of the signal. The term energy appearing in our discussions would mean normalized energy unless specified otherwise. The unit of energy is Joule.

The energy of a signal exists only if the integral in Eq. 2.4.1 is finite. The signals for which Eq. 2.4.1 is finite are called *energy signals*. Aperiodic signals are examples of energy signals.

The energy may be infinite for many signals, and, for such signals, we have to define another useful parameter called *average power* of the signal. Average power is the time average of energy, or average time derivative of the energy (rate of energy flow). Average power of a signal may exist even if its energy is infinite. Such signals with finite average power are called *power signals*. Periodic signals, unit step function, Dirac Delta function etc., are examples of power signals.

### Parseval's Theorem for Energy Signals

The Parseval's theorem defines energy of a signal in terms of its Fourier transform. The theorem is very useful as it helps in evaluating the energy of a signal without knowing its time domain. When the Fourier transform of a signal is known, its energy can be evaluated without finding the inverse Fourier transform i.e., (time domain of signal).

Let us consider a function $f(t)$ with its Fourier transform as $F(\omega)$. The energy $E$ of $f(t)$ is given by

$$E = \int_{-\infty}^{\infty} f^2(t) \, dt = \int_{-\infty}^{\infty} f(t) . f(t) \, dt$$

Replacing second $f(t)$ in terms of the inverse Fourier transform of $F(\omega)$, we get

$$E = \int_{-\infty}^{\infty} f(t) \left\{ \frac{1}{2\pi} \int_{-\infty}^{\infty} F(\omega) e^{j\omega t} \, d\omega \right\} dt$$

By interchanging the order of integration,

$$E = \frac{1}{2\pi} \int_{-\infty}^{\infty} F(\omega) \left[ \int_{-\infty}^{\infty} f(t) e^{j\omega t} dt \right] d\omega$$

The integral under (bracket is $F(-\omega)$, hence

$$E = \frac{1}{2\pi} \int_{-\infty}^{\infty} F(\omega) F(-\omega) d\omega$$

For a real function $f(t)$, the Fourier transform $F(\omega)$ and $F(-\omega)$ are complex conjugates, i.e., $F(-\omega) = F^*(\omega)$ and for such functions

$$F(\omega) F(-\omega) = |F(\omega)|^2$$

as magnitude of $F(\omega)$ and $F^*(\omega)$ is same. Hence

$$E = \int_{-\infty}^{\infty} f^2(t) dt = \frac{1}{2\pi} \int_{-\infty}^{\infty} |F(\omega)|^2 d\omega \qquad (2.4.2a)$$

$$= \frac{1}{2\pi} \left[ \text{Area under } |F(\omega)|^2 \text{ curve} \right]$$

The factor $2\pi$ can be eliminated if integrated with respect to the linear frequency $f$ rather than '$\omega$' i.e.

$$E = \int_{-\infty}^{\infty} |F(\omega)|^2 df \qquad (2.4.2b)$$

Equation 2.4.2 is called Parseval's theorem for energy (periodic) signals. Equation 2.4.2 is also referred to as Rayleigh's energy theorem, according to which the energy of a signal is obtained by the knowledge of only magnitude spectrum of Fourier transform. The phase spectrum has no effect on energy.

## Energy Spectral Density

Consider a signal $f(t)$ applied to an ideal low pass filter as shown in Fig. 2.4.1a. The transfer function of the low pass filter $H(\omega)$ is shown in Fig. 2.4.1b.

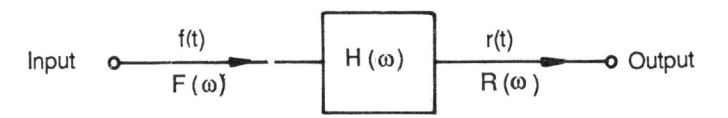

Fig. 2.4.1(a) LPF

The response $r(t)$ of the system is given by the relation,

$$R(\omega) = H(\omega) F(\omega)$$

where $F(\omega)$ is Fourier transform of $f(t)$ and $R(\omega)$ is Fourier transform of $r(t)$.

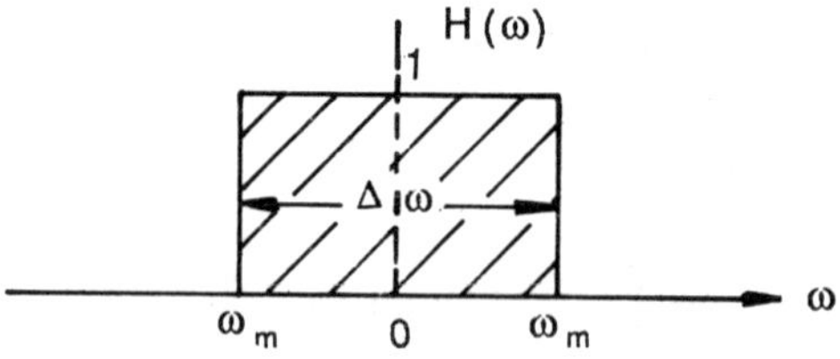

[Figure: Transfer function H(ω) shown as a rectangular LPF of width Δω centered at 0, extending from -ω_m to ω_m, with amplitude 1.]

Fig. 2.4.1 (b)  Transfer Function of LPF

The energy $E_0$ of the output signal is given by

$$E_0 = \frac{1}{2\pi} \int_{-\infty}^{\infty} |R(\omega)|^2 \, d\omega$$

$$= \frac{1}{2\pi} \int_{-\infty}^{\infty} |H(\omega) F(\omega)|^2 \, d\omega$$

It is obvious from Fig. 2.4.1b that $H(\omega) = 0$, except for the narrow band $-\omega_m$ to $\omega_m$ for which it is unity. $F(\omega)$ is constant with frequency for a narrowband filter ($\Delta \omega \to 0$). Hence, energy over this narrowband $\Delta \omega = 2\omega_m$ is given as

$$E_0 = \frac{1}{2\pi} \int_{-\omega_m}^{\omega_m} |F(\omega)|^2 \, d\omega = \frac{1}{2\pi} |F(\omega)|^2 (2\omega_m)$$

putting $\quad 2\omega_m = \Delta \omega$

$$E_0 = \frac{1}{2\pi} |F(\omega)|^2 (\Delta \omega) = |F(\omega)|^2 (\Delta f)$$

Therefore, $E_0$ represents the contribution of energy due to bandwidth ($\Delta \omega$) of the signal (including negative frequencies). Hence, energy contribution per unit bandwidth is given by

$$\frac{E_0}{\Delta f} = |F(\omega)|^2$$

Thus, $|F(\omega)|^2$ represents energy per unit bandwidth and is called *Energy Spectral Density or Energy Density spectrum*, denoted by, $\psi(\omega)$ It is obvious that $\psi(\omega)$ is the frequency derivative of the energy $E_0$. The total energy of the signal is obtained by integrating this $\psi(\omega)$ over the bandwidth of the signal. Energy contribution includes both by negative and positive frequencies, and the contribution is equal as

$$|F(\omega)|^2 = |F(-\omega)|^2$$

From the above discussion, energy density spectrum is defined as,

$$\psi(\omega) = |F(\omega)|^2 \tag{2.4.3}$$

The total energy $E$ as given by Eq. 2.4.2a is

$$E = \frac{1}{2\pi} \int_{-\infty}^{\infty} |F(\omega)|^2 d\omega$$

$$= \frac{1}{2\pi} \int_{-\infty}^{\infty} \psi(\omega) d\omega \qquad (2.4.4)$$

For real functions, $|F(\omega)|^2 = |F(-\omega)|^2$ i.e., spectrum is symmetrical, and the contribution by negative and positive frequencies is the same.
Hence,

$$E = \frac{1}{\pi} \int_{0}^{\infty} \psi(\omega) d\omega \qquad (2.4.5)$$

The relation between energy densities of input and response can be derived as follows:
since

$$R(\omega) = H(\omega) F(\omega)$$

Hence

$$|R(\omega)|^2 = |H(\omega) F(\omega)|^2 = |H(\omega)|^2 |F(\omega)|^2$$

Let $\psi_r(\omega)$ denote the energy density spectrum of response $r(t)$, and $\psi_f(\omega)$ as energy density spectrum of excitation $f(t)$, then

$$\psi_r(\omega) = |H(\omega)|^2 \psi_f(\omega) \qquad (2.4.6)$$

where

$$\psi_r(\omega) = |R(\omega)|^2 \text{ and } \psi_f(\omega) = |F(\omega)|^2$$

### Example 2.4.1

Find the energy density spectrum of a Gate function of width $\tau$, and amplitude $A$.

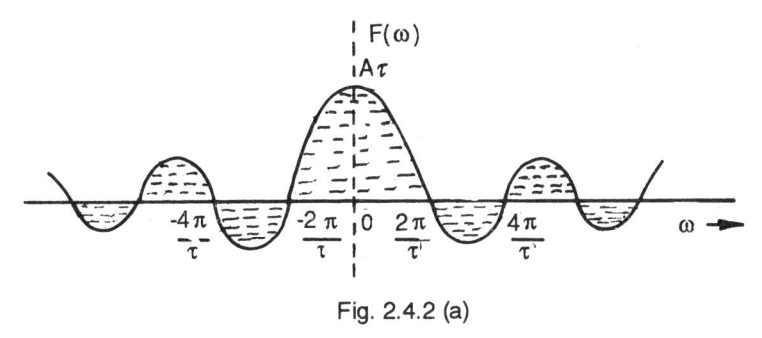

Fig. 2.4.2 (a)

*Solution*
The Fourier transform of the Gate function is a sampling function given as

$$F(\omega) = A\tau \, Sa\left(\frac{\omega\tau}{2}\right)$$

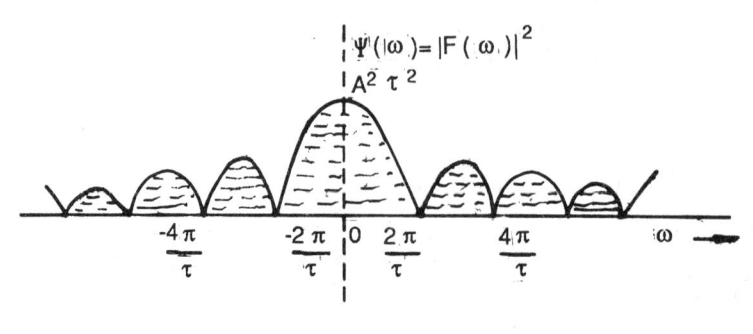

Fig. 2.4.2(b)

Hence, the energy density spectrum is,

$$\psi(\omega) = |F(\omega)|^2 = A^2\tau^2 Sa^2\left(\frac{\omega\tau}{2}\right)$$

The Fourier transform $F(\omega)$ and energy density spectrum $\psi(\omega)$, are shown in Fig. 2.4.2a and b, respectively. $\psi(\omega)$ being a square function, it has no negative value anywhere in the spectrum.

### Example 2.4.2

A signal $e^{-3t} u(t)$ is passed through an ideal low pass filter with cut-off frequency of 1 rad per second (a) test whether the input is an energy signal, and (b) find the input and output energy.

*Solution*

(a) The input signal $f(t)$ is given by

$$f(t) = e^{-3t} u(t)$$

The energy of the input signal is given as (unit step function vanishes, for $t < 0$, hence $t = 0$ is taken as lower limit of the integral).

$$E_i = \int_{-\infty}^{\infty} f^2(t)\,dt = \int_0^{\infty} (e^{-3t})^2\,dt = \int_0^{\infty} e^{-6t}\,dt = \frac{1}{6}$$

Since the energy is finite, the input signal is an energy signal.

(b) The energy density spectrum of the output $r(t)$ is given as

$$\psi_r(\omega) = \psi_f(\omega)|H(\omega)|^2$$

Here,

$$\psi_f(\omega) = |F(\omega)|^2$$

and

$$F(\omega) = F\left[e^{-3t} u(t)\right] = \frac{1}{j\omega + 3}$$

Hence,
$$\psi_f(\omega) = \left|\frac{1}{(j\omega+3)}\right|^2 = \left|\frac{3-j\omega}{\omega^2+9}\right|^2 = \frac{1}{\omega^2+9}$$

The square of the transfer function of low pass filter is given by
$$|H(\omega)|^2 = \begin{cases} 1 & |\omega|<1 \\ 0 & \text{otherwise} \end{cases}$$

which gives
$$\psi_r(\omega) = \psi_f(\omega)|H(\omega)|^2 = \begin{cases} \dfrac{1}{\omega^2+9} & |\omega|<1 \\ 0 & \text{elsewhere} \end{cases}$$

The total energy contained in output is given by
$$E_0 = \frac{1}{\pi}\int_0^\infty \psi_r(\omega)\,d\omega = \frac{1}{\pi}\int_0^1 \frac{1}{\omega^2+9}\,d\omega$$

$$= \frac{1}{3\pi}\tan^{-1}\left(\frac{1}{3}\right)$$

## The Power Signal

Signals having infinite energy, but finite average power, are called power signals. For such signals, the time average of energy (i.e., average power) of the signal is a meaningful parameter.

### Average Power

The average power dissipated by a voltage $f(t)$ applied across a 1-ohm resistor is defined as the average power or simply power of the signal $f(t)$. This is same as the power dissipated by a current $f(t)$ passing through a 1-ohm resistor. The average power $P$, therefore, may be expressed as

$$P = \lim_{T\to\infty}\frac{1}{T}\int_{-\frac{T}{2}}^{\frac{T}{2}}|f(t)|^2\,dt \tag{2.4.7}$$

The Eq. 2.4.7 represents the mean square value of the signal $f(t)$ and hence, the average power $P$ is same as the mean square value (*ms* value) of $f(t)$. The unit of power is watt. The ratio of power $P_1$, and $P_2$, is defined in terms of decibel i.e.,

$$\alpha = 10\log_{10}\left(\frac{P_2}{P_1}\right) = 20\log_{10}\left(\frac{V_2}{V_1}\right)$$

where $V_2$ and $V_1$ are rms voltage corresponding to power $P_2$ and $P_1$ respectively.

### Parseval's Power Theorem

This theorem is similar to Parseval's theorem for energy signals. The theorem defines the power of a signal in terms of its Fourier series coefficients, i.e., in terms of amplitudes of the harmonic components present in the signal.

Let us consider a function $f(t)$. We know that

$$|f(t)|^2 = f(t) f^*(t)$$

where $f(t)$ is the complex conjugate of the function $f(t)$. The power of the signal $f(t)$ over a cycle is given by,

$$P = \frac{1}{T} \int_{-\frac{T}{2}}^{\frac{T}{2}} |f(t)|^2 dt = \frac{1}{T} \int_{-\frac{T}{2}}^{\frac{T}{2}} f(t) f^*(t) dt$$

Replacing $f(t)$ by its exponential Fourier series,

$$P = \frac{1}{T} \int_{-\frac{T}{2}}^{\frac{T}{2}} f^*(t) \left[ \sum_{n=-\infty}^{\infty} F_n e^{jn\omega_0 t} \right] dt, \text{ where } \omega_0 = \frac{2\pi}{T}$$

Interchanging the order of integration and summation,

$$P = \frac{1}{T} \sum_{n=-\infty}^{\infty} F_n \int_{-\frac{T}{2}}^{\frac{T}{2}} f^*(t) e^{jn\omega_0 t} dt$$

The integral in the above expression is equal to $T F_n^*$. Hence we may write

$$P = \sum_{n=-\infty}^{\infty} F_n F_n^* = \sum_{n=-\infty}^{\infty} |F_n|^2 \qquad (2.4.8)$$

Equation 2.4.8 is known as Parseval's Power Theorem. The equation defines that the power of the signal is equal to the sum of the square of the magnitudes of various harmonics present in a discrete spectrum. This is a special case of Parseval's theorem defined earlier for energy signals having continuous spectrum. $|F_n|^2$ is referred as the discrete power spectrum of the signal $f(t)$.

### Example 2.4.3

Find the power of a sinusoidal signal $A \cos \omega_0 t$.

*Solution*

The exponential Fourier series of the cosine function exists only at $\pm \omega_0$. The amplitude of the two components located at $\pm \omega_0$ is $\frac{A}{2}$. Hence, by using Eq. 2.4.8, the power is given as

$$P = \left(\frac{A}{2}\right)^2 + \left(\frac{A}{2}\right)^2 = \frac{A^2}{2}$$

This is true because the mean square value of the cosine function is

$$\frac{1}{T}\int_{-\frac{\tau}{2}}^{\frac{\tau}{2}} A^2 \cos^2 \omega_0 t \, dt = \frac{A^2}{2}$$

### Example 2.4.4

Find the power of the signal $a + f(t)$, where $a$ is a constant, and $f(t)$ is a power signal with zero mean value.

*Solution*

The power corresponding to constant $a$ is $a^2$, and the power of $f(t)$ is its mean square value $\overline{f^2(t)}$. Hence, the total power of the signal is $a^2 + \overline{f^2(t)}$.

This can be proved otherwise also by evaluating, the mean square value of the entire signal $a + f(t)$, i.e.,

$$P = \lim_{T \to \infty} \frac{1}{T} \int_{-\frac{\tau}{2}}^{\frac{\tau}{2}} [a + f(t)]^2 \, dt$$

$$= \lim_{T \to \infty} \frac{1}{T} \left[ \int_{-\frac{\tau}{2}}^{\frac{\tau}{2}} a^2 \, dt + \int_{-\frac{\tau}{2}}^{\frac{\tau}{2}} f^2(t) \, dt + 2a \int_{-\frac{\tau}{2}}^{\frac{\tau}{2}} f(t) \, dt \right]$$

$$= a^2 + \overline{f^2(t)} + 0 \text{ (mean value)}$$

$$= a^2 + \overline{f^2(t)}$$

### The Power Density Spectrum

We will derive an expression for the power density spectrum (or spectral power density) assuming the power signal as a limiting case of an energy signal. Consider a power signal $f(t)$ (extending to infinity) as shown in Fig. 2.4.3a. Let us truncate this signal so that it is zero outside the interval $\pm \frac{\tau}{2}$, as shown 2.4.3 b. Let us call

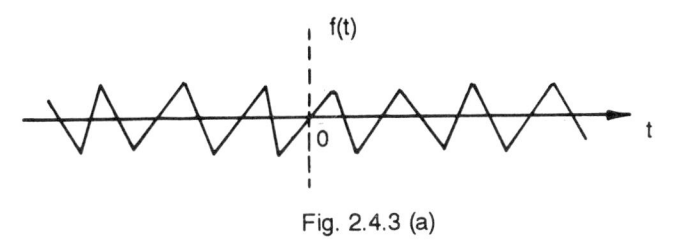

Fig. 2.4.3 (a)

this truncated signal as $f_\tau(t)$ which can be expressed as

## 112 COMMUNICATION SYSTEMS: Analog and Digital

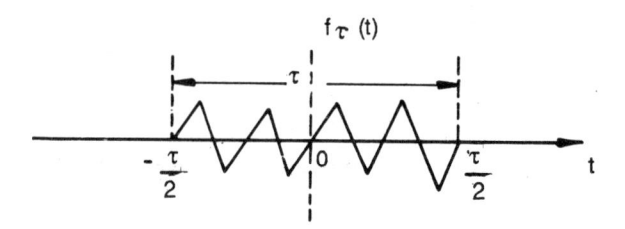

Fig. 2.4.3 (b)

$$f_\tau(t) = \begin{cases} f(t) & |t| < \dfrac{\tau}{2} \\ 0 & \text{otherwise} \end{cases}$$

Signal $f_\tau(t)$ is of finite duration $\tau$, and hence it is an energy signal with energy $E_\tau$ given by

$$E_\tau = \int_{-\infty}^{\infty} |f_\tau(t)|^2 \, dt = \int_{-\infty}^{\infty} |F_\tau(\omega)|^2 \, df$$

where $F_\tau(\omega)$ is Fourier transform of $f_\tau(t)$. Since $f(t)$ over the interval $\left(-\dfrac{\tau}{2}, \dfrac{\tau}{2}\right)$ is same as $f_\tau(t)$ over the interval $(-\infty, \infty)$ we have

$$\int_{-\infty}^{\infty} |f_\tau(t)|^2 \, dt = \int_{-\tau/2}^{\tau/2} |f(t)|^2 \, dt$$

Hence,

$$\frac{1}{\tau} \int_{-\tau/2}^{\tau/2} |f(t)|^2 \, dt = \frac{1}{\tau} \int_{-\infty}^{\infty} |F_\tau(\omega)|^2 \, df$$

In the limit $\tau \to \infty$, the left-hand side represents the average power $P$ of the function $f(t)$. Hence

$$P = \int_{-\infty}^{\infty} \lim_{\tau \to \infty} \frac{|F_\tau(\omega)|^2}{\tau} \, df \qquad (2.4.9a)$$

In the limit $\tau \to \infty$, the ratio $\dfrac{|F_\tau(\omega)|^2}{\tau}$ may approach a finite value. Let us denote this finite value by $S(\omega)$,

i.e.,
$$S(\omega) = \lim_{\tau \to \infty} \frac{|F_\tau(\omega)|^2}{\tau} \qquad (2.4.9b)$$

From Eq. 2.4.9a, average power $P$ of the function $f(t)$ is given as

$$P = \overline{f^2(t)} = \int_{-\infty}^{\infty} S(\omega) df = \frac{1}{2\pi} \int_{-\infty}^{\infty} S(\omega) d\omega \qquad (2.4.10)$$

where $\overline{f^2(t)}$ represents the mean square (ms value) of $f(t)$. According to expression 2.4.10, the total power is obtained by multiplying $S(\omega)$ with bandwidth $\Delta\omega$ and integrating over the entire bandwidth. Therefore, $S(\omega)$, may be considered as average power per unit bandwidth, and hence known as *power density spectrum, or spectral power density*. It is obvious that $S(\omega)$ is the frequency derivative of the power P defined by Eq. 2.4.9a. The total power can be obtained by multiplying $S(\omega)$ with bandwidth involved (including negative frequencies), because

$$|F_\tau(\omega)|^2 = |F_\tau(-\omega)|^2$$

That is, contribution to power by positive and negative frequencies is identical. Hence, the average power may be written as

$$P = 2\int_0^{\infty} S(\omega) df = \frac{1}{\pi} \int_0^{\infty} S(\omega) d\omega \qquad (2.4.11)$$

Expression for power density spectrum given by Eq.2.4.9b, encounters only the magnitude of $F_\tau(\omega)$, and not the phase, i.e., phase information is lost in power density spectrum. Hence, it can be stated that :
(i) A given signal has a unique power density spectrum. (ii) But the converse is not true, i.e., a power density spectrum, may correspond to a large number of signals with identical frequency spectrum, but with different phase functions. This means that a signal with the same amplitude, but with different phase functions, will have identical power density spectrum.

## Frequency Shifting Theorem for Power Density Spectrum

Just as we did for the frequency shifting property of Fourier transform, we can define frequency shifting theorem for power density spectrum. This theorem states that multiplication of a signal by a sinusoidal function of frequency $\omega_c$ shifts its power density spectrum by $\omega_c$. If a function $f(t)$ has power density spectrum $S_f(\omega)$, then the power density spectrum of $\Phi = f(t) \cos \omega_c t$ is given as

$$S_f(\omega) = \frac{1}{4}[S_f(\omega + \omega_c) + S_f(\omega - \omega_c)] \qquad (2.4.12)$$

### *Proof*

Let $f(t) \leftrightarrow F(\omega)$
From frequency shifting property of Fourier transform

$$\Phi(t) = f(t) \cos \omega_c t \leftrightarrow \frac{1}{2}[F(\omega + \omega_c) + F(\omega - \omega_c)]$$

**114  COMMUNICATION SYSTEMS: Analog and Digital**

Since, the power density spectrum is related with square of the Fourier transform Hence,

$$S_f(\omega) = \frac{1}{4} \lim_{\tau \to \infty} \left[ \frac{|F_\tau(\omega + \omega_c)|^2}{\tau} + \frac{|F_\tau(\omega - \omega_c)|^2}{\tau} \right]$$

$$= \frac{1}{4}[S_f(\omega + \omega_c) + S_f(\omega - \omega_c)]$$

where subscript $\tau$ corresponds to truncated function. The cross product term $F_\tau(\omega + \omega_c).F_\tau(\omega - \omega_c)$ is taken to be zero as the two spectra are non-overlapping.

Similarly, we can see that the power density spectrum of $f(t) \sin \omega_c t$ is exactly same as Eq.2.4.12. The power density spectrum remains the same whether a signal is multiplied by $\sin \omega_c t$ or $\cos \omega_c t$. This is because phase information is lost in power density spectrum.

### Example 2.4.5

(a) Show that the power density spectrum of a periodic function $f(t)$ with period $T$ is given by

$$S_f(\omega) = 2\pi \sum_{n=-\infty}^{\infty} |F_n|^2 \delta(\omega - n\omega_0); \quad \omega_0 = \frac{2\pi}{T} \quad (2.4.13a)$$

(b) Determine the power density spectrum of the function

$$f(t) = A\cos(\omega_c t + \theta)$$

**Solution**

Let $f_T(t)$ represent the truncated portion of a periodic function $f(t)$ over a period of $T$. This truncated function $f_T(t)$ can be considered as function $f(t)$ multiplied by a Gate function $G_T(t)$,

$$f_T(t) = f(t) G_T(t)$$

The Fourier transform of $F_T(t)$ can be obtained using frequency convolution theorem,

$$F_T(\omega) = \frac{1}{2\pi} T.Sa\left(\frac{\omega T}{2}\right) * F(\omega)$$

The Fourier transform of a periodic function $f(t)$ is given as

$$F(\omega) = 2\pi \sum_{n=-\infty}^{\infty} F_n \delta(\omega - n\omega_0)$$

Substituting this value of $F(\omega)$, we get

$$F_T(\omega) = T \sum_{n=-\infty}^{\infty} F_n Sa\left(\frac{\omega T}{2}\right) \otimes \delta(\omega - n\omega_0)$$

Using sampling property of impulse function,

$$F_T(\omega) = T \sum_{n=-\infty}^{\infty} F_n Sa\left[\frac{(\omega - n\omega_0)T}{2}\right]$$

The power density spectrum is given as,

$$S_f(\omega) = \lim_{T \to \infty} \frac{|F_T(\omega)|^2}{T}$$

Putting the value of $F_T(\omega)$, we get

$$S_f(\omega) = \lim_{T \to \infty} T \sum_{n=-\infty}^{\infty} |F_n|^2 Sa^2\left[\frac{(\omega - n\omega_0)T}{2}\right]$$

In the limit $T \to \infty$, the sampling square function is reduced to delta function $2\pi\, \delta(\omega - n\omega_0)$. Hence

$$S_f(\omega) = 2\pi \sum_{n=-\infty}^{\infty} |F_n|^2\, \delta(\omega - n\omega_0)$$

Thus, if the spectral density function $F_n$ is known for a periodic function, its power density can be obtained by Eq.2.4.13 which will be a sequence of periodic impulses or Dirac comb with intensities $2\pi F_n^2$. A typical $F_n$ and corresponding $S_f(\omega)$ is shown in Fig. 2.4.4.

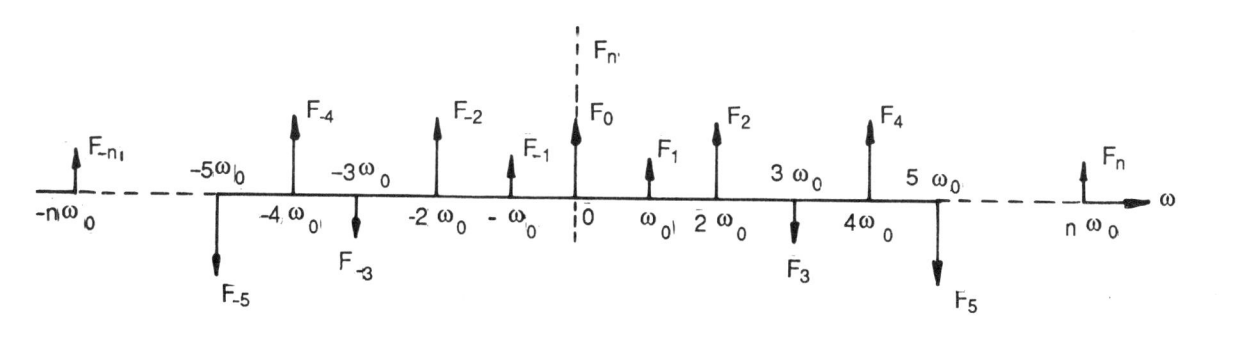

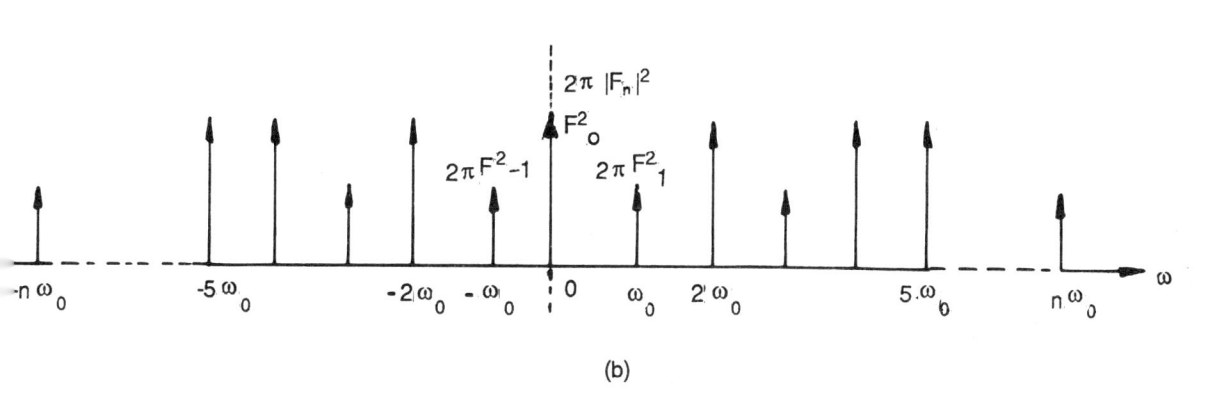

Fig. 2.4.4 (a) $F_n$ of a Typical Function (b) Power Density Spectrum of the Function

## Spectral Power Densities of Input and the Response

Consider a signal $f(t)$ having response $r(t)$ to a system $h(t)$. Let $f_T(t)$ and $R_T(t)$, represent the truncated version of $f(t)$ and $r(t)$ respectively over $\left(-\frac{T}{2}, \frac{T}{2}\right)$. Now, if the signal $f_T(t)$ is applied to the system, the response may not be restricted within $t \pm \frac{T}{2}$ rather, it may extend beyond $t = \pm \frac{T}{2}$. That is, in general, response may not be $r_T(t)$ inspite of the fact that input is absent beyond $t = \pm \frac{T}{2}$. But for a stable system, response beyond $t = \pm \frac{T}{2}$ decays very rapidly as $T$ increases, and in the limit $T \to \infty$, the magnitude of the response beyond $t = \pm \frac{T}{2}$ is negligible and it may be considered that response to $f_T(t)$ is $r_T(t)$.
Hence,

$$\lim_{T \to \infty} f_T(t) \to r_T(t)$$

Therefore,

$$\lim_{T \to \infty} R_T(\omega) = F_T(\omega) H(\omega)$$

where

$$r_T(t) \leftrightarrow R_T(\omega); h(t) \leftrightarrow H(\omega), \text{ and } f_T(t) \leftrightarrow F_T(\omega)$$

The power density spectrum of the response $r_T(t)$ is given as

$$S_r(\omega) = \lim_{T \to \infty} \frac{1}{T} |R_T(\omega)|^2$$

Putting $R_T(\omega) = F_T(\omega) H(\omega)$, we get

$$S_r(\omega) = \lim_{T \to \infty} \frac{1}{T} |H(\omega) F_T(\omega)|^2$$

$$= |H(\omega)|^2 \left\{ \lim_{T \to \infty} \frac{1}{T} |F_T(\omega)|^2 \right\}$$

The term within brackets is $S_f(\omega)$, hence

$$S_r(\omega) = |H(\omega)|^2 S_f(\omega) \tag{2.4.14}$$

This expression is similar to that of energy density spectrum.

### Example. 2.4.6

A function $f(t)$ has power density spectrum $S_f(\omega)$. Find the power density spectrum of (i) **integral of $f(t)$** (ii) time derivative of $f(t)$.
*Solution*
(i) Let $f(t) \leftrightarrow F(\omega)$, then for integral of $f(t)$, we have

$$\int_{-\infty}^{\infty} f(\tau) d\tau \leftrightarrow \frac{1}{j\omega} F(\omega)$$

Therefore if a function $f(t)$ is applied at the input of an ideal integrator; then Fourier transform of the output $r(t)$ will be $\dfrac{1}{j\omega}F(\omega)$. Hence, the transfer function of the integrator circuit is

$$H(\omega) = \frac{R(\omega)}{F(\omega)} = \frac{(1/j\omega)\,F(\omega)}{F(\omega)} = \frac{1}{j\omega}$$

and

$$|H(\omega)|^2 = \left|\frac{1}{j\omega}\right|^2 = \frac{1}{\omega^2}$$

Therefore, the power density spectrum of integrator output is

$$S_r(\omega) = |H(\omega)|^2 S_f(\omega) = \frac{1}{\omega^2} S_f(\omega)$$

(ii) Again, if $f(t) \leftrightarrow F(\omega)$ then derivative of $f(t)$ has

$$\frac{df(t)}{dt} \leftrightarrow j\omega\, F(\omega)$$

Accordingly, ideal differentiator can be considered to have transfer function $H(\omega) = j\omega$ and

$$|H(\omega)|^2 = \omega^2$$

$$S_r(\omega) = |H(\omega)|^2 S_f(\omega) = \omega^2 S_f(\omega)$$

### Example 2.4.7

A signal with power spectral density

$$D_f(\omega) = A\pi[\delta(\omega - 1) + \delta(\omega + 1)]$$

is applied to the input of a linear system shown in Fig. 2.4.5. Find the ratio of average power of the response and excitation.

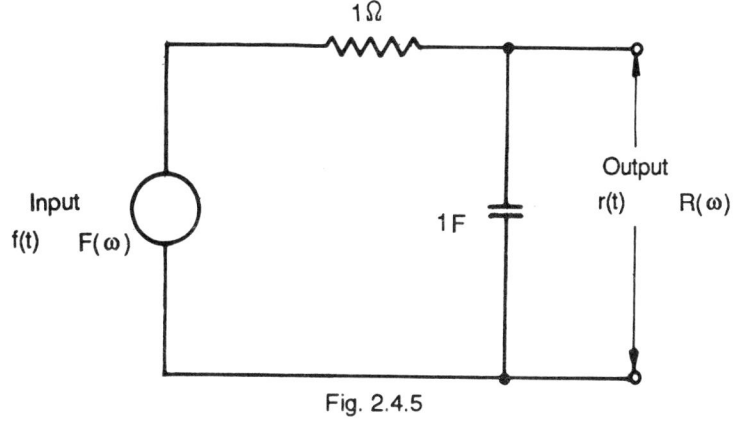

Fig. 2.4.5

### Solution
The transfer function of the network is

$$H(\omega) = \frac{R(\omega)}{F(\omega)} = \frac{(1/j\omega.1)}{1 + (1/j\omega.1)} = \frac{1}{1 + j\omega} = \frac{1 - j\omega}{1 + \omega^2}$$

The magnitude of $H(\omega)$ is

$$|H(\omega)| = \frac{1}{1+\omega^2} \times \sqrt{1+\omega^2} = \frac{1}{\sqrt{1+\omega^2}}$$

$$|H(\omega)|^2 = \frac{1}{1+\omega^2}$$

The power density spectrum of the output is

$$S_r(\omega) = |H(\omega)|^2 S_f(\omega)$$

$$= \frac{1}{1+\omega^2}[A\pi\{\delta(\omega-1)+\delta(\omega+1)\}]$$

$$= \frac{A\pi}{1+\omega^2}[\delta(\omega-1)+\delta(\omega+1)]$$

The average output power is given as

$$P_o = \frac{1}{\pi}\int_0^\infty S_r(\omega).d\omega = A\int_0^\infty \frac{1}{1+\omega^2}[\delta(\omega-1)+\delta(\omega+1)]d\omega$$

$$= \int_0^\infty \frac{A}{1+\omega^2}\delta(\omega-1)d\omega + \int_0^\infty \frac{A}{1+\omega^2}\delta(\omega+1)d\omega$$

The impulse functions in first and second integrals exists at $\omega = 1$ and $\omega = -1$, respectively. The second integral vanishes for $\omega = -1$, because the limit of integration starts form '0'. Hence

$$P_o = \left.\frac{A}{1+\omega^2}\right|_{\omega=1} = \frac{A}{2}$$

The input power is given as

$$P_i = \frac{1}{\pi}\int_0^\infty S_f(\omega)d\omega = A\int_0^\infty [\delta(\omega-1)+\delta(\omega+1)]d\omega = A$$

The power ratio is

$$\frac{P_o}{P_i} = \frac{1}{2}$$

## 2.5 CORRELATION

The correlation, or more precisely cross-correlations between two waveforms is the measure of similarity between one waveform, and time delayed version of the other waveform. This expresses how much one waveform is related to the time delayed version of the other waveform when scanned over time axis.

The expression for correlation is very close to convolution. Consider two general complex functions $f_1(t)$ and $f_2(t)$ which may or may not be periodic, and not restricted, to finite interval. The cross-correlation, or simply correlation $R_{1,2}(\tau)$ between two functions is defined as follows:

$$R_{1,2}(\tau) = \lim_{T \to \infty} \int_{-T/2}^{T/2} f_1(t) f_2^*(t+\tau) dt \qquad (2.5.1a)$$

This represent the shift of function $f_2(t)$ by an amount $-\tau$ (i.e. towards left). A similar effect can be obtained by shifting $f_1(t)$ by an amount $+\tau$ (i.e., towards right). Hence, correlation may also be defined in an equivalent way, as

$$R_{1,2}(\tau) = \lim_{T \to \infty} \int_{-T/2}^{T/2} f_1(t-\tau) f_2^*(t) dt \qquad (2.5.1b)$$

Let us define the correlation for two cases, (i) energy (non periodic) functions, and (ii) power (periodic) functions. In the definition of correlation, limits of integration may be taken as infinite for energy signals,

$$R_{1,2}(\tau) = \int_{-\infty}^{\infty} f_1(t) f_2^*(t+\tau) dt = \int_{-\infty}^{\infty} f_1(t-\tau) f_2^*(t) dt \qquad (2.5.2a)$$

For power signals of period $T_0$, the definition in Eq. 2.5.2a may not converge. Hence, average correlation over a period $T_0$ is defined as

$$R_{1,2}(\tau) = \frac{1}{T_0} \int_{-T_0/2}^{T_0/2} f_1(t) f_2^*(t+\tau) dt = \frac{1}{T_0} \int_{-T_0/2}^{T_0/2} f_1(t-\tau) f_2^*(t) dt \qquad (2.5.2b)$$

The correlation definition represents the overlapping area between the two functions. In all the above definitions of correlation, the *conjugate symbol \*, is removed if the functions are real.*

### Searching Parameter ($\tau$) :

The time delay ($\tau$) introduced in the expressions of correlation is known as *searching* or *scanning* parameters. The time $t$ is a dummy variable, and the correlation $R_{1,2}(\tau)$ a function of the delay parameter $\tau$. The parameter $\tau$ is needed to find the maximum possible correlation between waveforms. Actually, two waveforms having no correlation when $\tau = 0$ (i.e., no delay between them) may have significant correlation by adjusting the suitable value of parameter $\tau$. This can be seen from the waveforms shown in Fig. 2.5.1. The waveform appear to have considerable similarity, but the correlation expression defined by Eq. 2.6.2 provides zero value of correlation for $\tau = 0$, because they are non-overlapping. However, when one signal is delayed, correlation is non-zero. The correlation increases as the delay parameter $\tau$ increases, and it is maximum for $\tau = \tau_0$. Thus, $\tau$ is a scanning parameter which corresponds to the maximum possible correlation between two waveforms.

### Uncorrelated (Incoherent) Signal

Functions for which $R_{1,2}(\tau)$ is zero for all values of $\tau$ are called *uncorrelated* or *incoherent* functions. Correlation is also known as *coherence*.

During the process of scanning, it is essential to specify which function is being shifted. In general, $R_{1,2}(\tau)$, which is obtained by shifting $f_2(t)$ in one direction is not the same as $R_{1,2}(\tau)$, which is obtained by shifting

$f_1(t)$ in the same direction. It can be seen that (Prob. 2.3), for real function $f_1(t)$, and $f_2(t)$,

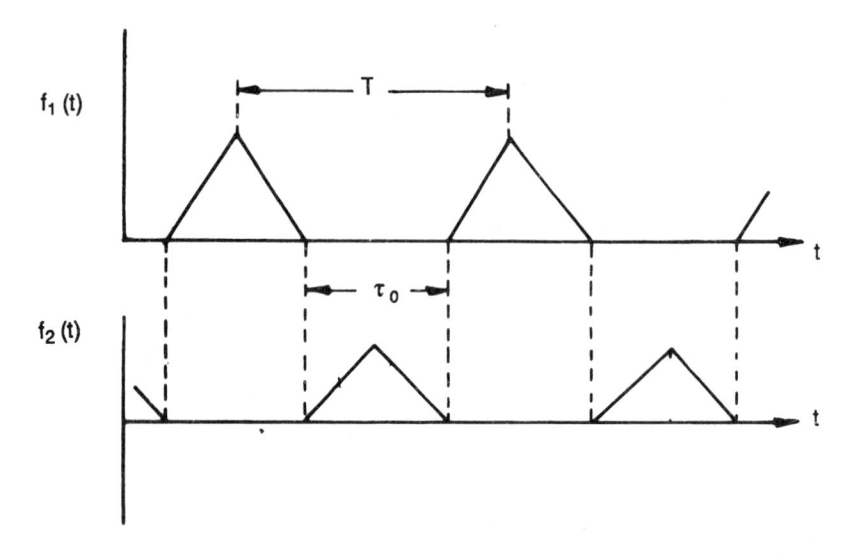

Fig. 2.5.1

$$R_{2,1}(\tau) = \frac{1}{T_0} \int_{-T_0/2}^{T_0/2} f_1(t+\tau) f_2(t) dt = R_{1,2}(-\tau) \tag{2.5.3a}$$

where $R_{1,2}(-\tau)$ is defined as

$$R_{1,2}(-\tau) = \frac{1}{T_0} \int_{-T_0/2}^{T_0/2} f_1(t) f_2(t-\tau) dt$$

Thus, cross correlation function is non-commutative. This is true, as shifting of one function in one direction is equivalent to shifting the other in the opposite direction. Accordingly, Eq. 2.5.2a can be written for $-\tau$ as

$$R_{1,2}(-\tau) = \int_{-\infty}^{\infty} f_1(t+\tau) f_2^*(t) dt \tag{2.5.3b}$$

## Difference between Correlation and Convolution

As stated earlier the two are closely related, with the following minor differences:
(i) In correlation physical time $t$ plays the role of a dummy variable and it disappears after solution of an integral. But, in convolution, delay $\tau$ plays the role of a dummy variable, and (ii) Correlation $R_{1,2}(\tau)$ is a function of the delay parameter $\tau$, whereas, convolution is a function of time $t$. Let us replace the dummy variable $t$ in Eq. 2.5.1a by a new variable $-\sigma$. Then we get

$$R_{1,2}(\tau) = \int_{-\infty}^{\infty} f_1(-\sigma) f_2^*(\tau-\sigma) d\sigma$$

Comparing this expression with the definition of convolution, we find that the correlation can be obtained by convolving $f_1(-t)$, and $f_2^*(t)$. Hence, it can be seen that convolution does not depend on which function is being shifted, whereas, correlation does i.e., convolution is commutative. Using convolution theorem, $R_{1,2}(\tau)$ will have the Fourier transform which is a multiplication of frequency domain of $f_1(t)$ and $f_2^*(t)$ i.e.,

$$R_{1,2}(\tau) \leftrightarrow F_1(\omega) F_2^*(\omega)$$

$R_{1,2}$ is the inverse Fourier transform of R.H.S., as given below (for energy signals),

$$R_{1,2}(\tau) = \frac{1}{2\pi} \int_{-\infty}^{\infty} F_1(\omega) F_2^*(\omega) e^{j\omega\tau} d\omega \qquad (2.5.4a)$$

Similarly, $R_{1,2}(\tau)$ for power signals (periodic signal), can be written as

$$R_{1,2}(\tau) = \frac{1}{T} \int_{-T/2}^{T/2} F_1(\omega) F_2^*(\omega) e^{j\omega\tau} d\omega \qquad (2.5.4b)$$

## *Power of Correlated Functions*

Let us consider a function $f_1(t)$ with power $P_1$, and another function $f_2(t)$ with power $P_2$. We have to find the power of the combination of the two functions, i.e., $f_1(t) + f_2(t)$. Since, the time delay does not affect the normalized power, and power of $f_2(t)$ and, $f_2(t+\tau)$ is the same, we can consider a more general case of combined functions $f_1(t) + f_2(t+\tau)$. The normalized power (m.s. value), $P_{1,2}$ of the combined functions is given by,

$$P_{1,2} = \frac{1}{T} \int_{-T/2}^{T/2} [f_1(t) + f_2(t+\tau)]^2 dt$$

$$= \frac{1}{T} \int_{-T/2}^{T/2} f_1^2(t) dt + \frac{1}{T} \int_{-T/2}^{T/2} f_2^2(t+\tau) dt + \frac{2}{T} \int_{-T/2}^{T/2} f_1(t) f_2(t+\tau) dt$$

$$P_{1,2} = P_1 + P_2 + 2R_{1,2}(\tau) \qquad (2.5.5)$$

The following conclusions are drawn from Eq. 2.5.5.

(i) The power of two correlated functions is equal to the sum of powers of each individual function plus twice the cross-correlation between them.
(ii) If the functions $f_1(t)$ and $f_2(t)$ are uncorrelated, i.e., $R_{1,2}(\tau) = 0$, then the powers of the combined functions is equal to the sum of the powers of each individual function, and
(iii) Functions correlated by dc components are considered as uncorrelated, because dc components are rarely of any interest for correlation consideration.

Consider two uncorrelated functions $f_1(t)$ and $f_2(t)$. If the dc components $F_1$ and $F_2$ are added to them, then the functions become $f_1(t) = F_1 + f_1(t)$ and $f_2(t) = F_2 + f_2(t)$. It can be seen that the correlation between the functions is $R_{1,2}(\tau) = F_1 F_2$. This correlation is due to dc components and hence the functions may be considered as uncorrelated.

## 2.6 AUTOCORRELATION

Autocorrelation is a special form of cross-correlation. It is defined as the correlation of a function with itself. Thus, if $f_1(t) = f_2(t) = f(t)$, then Eq. 2.5.1 provides the expression for autocorrelation as given below:

$$R(\tau) = \lim_{T \to \infty} \frac{1}{T} \int_{-T/2}^{T/2} f(t) f^*(t + \tau) dt \qquad (2.6.1a)$$

which is equivalent to

$$R(\tau) = \lim_{T \to \infty} \frac{1}{T} \int_{-T/2}^{T/2} f(t - \tau) f^*(t) dt \qquad (2.6.1b)$$

It is obvious that autocorrelation function is the measure of similarity of a function with its delayed replica. An analogy case may be stated as "comparison of your present photograph and the photograph taken five years back".

### Properties of Autocorrelation $R(\tau)$

(a) The autocorrelation for $\tau = 0$ is average power $P$ of the signal, i.e.,

$$R(0) = \lim_{T \to \infty} \int_{-T/2}^{T/2} f(t) f^*(t) dt = \lim_{T \to \infty} \frac{1}{T} \int_{-T/2}^{T/2} |f(t)|^2 dt = P \qquad (2.6.2)$$

For energy signals $R(0)$ is the energy $E$ of the signal

$$R(0) = \int_{-\infty}^{\infty} |f(t)|^2 dt = E \qquad (2.6.3)$$

(b) The autocorrelation function exhibits conjugate symmetry, i.e.,
$$R(\tau) = R^*(-\tau) \qquad (2.6.4)$$
For real functions,
$$R(\tau) = R(-\tau) \qquad (2.6.5)$$
In other words, the real part of $R(\tau)$ is an even function of $\tau$ and the imaginary part is an odd function of $\tau$. This property follows directly from the definition of autocorrelation function. Recall that shifting the function towards right or left is equivalent when integration extends from $-\infty$ to $\infty$, i.e.,
$$R(\tau) = R(-\tau)$$

(c) The maximum value of autocorrelation function $R(\tau)$ occurs at origin,
$$R(0) \geq R(\tau) \text{ for all } \tau \qquad (2.6.6)$$
That is, $R(0)$ is the maximum value of $R(\tau)$.

### Proof

Consider an integral $I$, given by

$$I = \lim_{T \to \infty} \frac{1}{T} \int_{-T/2}^{T/2} [f(t) - f(t+\tau)]^2 \, dt$$

$$= \lim_{T \to \infty} \frac{1}{T} \int_{-T/2}^{T/2} f^2(t) \, dt + \lim_{T \to \infty} \frac{1}{T} \int_{-T/2}^{T/2} |f(t+\tau)|^2 \, dt - 2 \lim_{T \to \infty} \int_{-T/2}^{T/2} f(t) f(t-\tau) \, dt$$

$$= P + P - 2R(\tau)$$

$$= 2P - 2R(\tau)$$

From the property (a), power $P = R(0)$. Hence $I = 2R(0) - 2R(\tau) = 2[R(0) - R(\tau)]$
Since the integral $I$ is always positive, as it is an integral of a square quantity,

$$I \geq 0$$

$$2[R(0) - R(\tau)] \geq 0$$

Therefore
$$R(0) \geq R(\tau)$$

(d) (i) Autocorrelation $R(\tau)$ of a periodic waveform is also periodic with same period.
(ii) Power spectral density function $S(\omega)$ and autocorrelation function of a power signal are Fourier transform pair,

$$R(\tau) \leftrightarrow S(\omega)$$

*Proof*

(i) A periodic function $f(t)$ is expressed in frequency domain as

$$f(t) = \sum_{n=-\infty}^{\infty} F_n e^{jn\omega_0 t}$$

The autocorrelation function of a periodic real function (power signal) $f(t)$ is given by

$$R(\tau) = \frac{1}{T} \int_{-T/2}^{T/2} f(t) f(t+\tau) \, dt$$

Substituting the frequency domain versions we get

$$R(\tau) = \frac{1}{T} \int_{-T/2}^{T/2} \left[ \sum_{m=-\infty}^{\infty} F_m e^{jm\omega_0 t} \right] \left[ \sum_{n=-\infty}^{\infty} F_n e^{jn\omega_0(t+\tau)} \right] dt$$

Changing the order of integration and summation

$$R(\tau) = \sum_{m=-\infty}^{\infty} \sum_{n=-\infty}^{\infty} \frac{1}{T} e^{jn\omega_0\tau} \int_{-T/2}^{T/2} F_m F_n e^{j(m+n)\omega_0 t} dt$$

$$= \sum_{m=-\infty}^{\infty} \sum_{n=-\infty}^{\infty} F_m F_n e^{jn\omega_0\tau} \left[ \frac{\sin \pi(m+n)}{\pi(m+n)} \right] \tag{2.6.7}$$

Since $m$ and $n$ are integers, the expression given above is zero except for $m = -n$ (i.e., $m+n=0$). For the case $m = -n$ the solution of Eq. 2.6.7 yields

$$R(\tau) = \sum_{n=-\infty}^{\infty} F_n F_{-n} e^{jn\omega_0\tau} = \sum_{n=-\infty}^{\infty} |F_n|^2 e^{jn\omega_0\tau} \tag{2.6.8a}$$

$$= |F_0|^2 + 2 \sum_{n=1}^{\infty} |F_n|^2 \cos n\omega_0\tau; \, \omega_0 = \frac{2\pi}{T} \tag{2.6.8b}$$

The expression for $R(\tau)$ given by Eq.2.6.8b is a periodic function with period $T$. Hence, it is proved that a periodic function has a periodic autocorrelation function with the same period. Also, it is obvious that $R(\tau) = R(-\tau)$.

(ii) Taking Fourier transform of Eq. 2.6.8a, we get

$$F[R(\tau)] = \int_{-\infty}^{\infty} \left[ \sum_{n=-\infty}^{\infty} |F_n|^2 e^{jn\omega_0\tau} \right] e^{-j\omega\tau} d\tau$$

Interchanging the order of integration and summation,

$$f[R(\tau)] = \sum_{n=-\infty}^{\infty} |F_n|^2 \int_{-\infty}^{\infty} e^{-j\tau(\omega - n\omega_0)} d\tau$$

$$= 2\pi \sum_{n=-\infty}^{\infty} |F_n|^2 \delta(\omega - n\omega_0) \tag{2.6.9}$$

R.H.S. is the power spectral density function $\delta(\omega)$ of the periodic functions $f(t)$ Hence it is obvious that autocorrelation and the power spectral density of a power signal are a Fourier transform pair, i.e.,

$$f[R(\tau)] = D(\omega) \tag{2.6.10}$$

$$f^{-1}[D(\omega)] = R(\tau) \tag{2.6.11}$$

(e) Autocorrelation function of a pulse type signal (energy signal), and its energy density spectrum $\psi(\omega)$ form a Fourier transform pair, i.e.,

$$R(\tau) \leftrightarrow \psi(\omega)$$

## Proof

The cross-correlation of two energy functions $f_1(t)$, and $f_2(t)$, is given by (Eq. 2.5.4a)

$$R_{1,2}(\tau) = \frac{1}{2\pi} \int_{-\infty}^{\infty} F_1(\omega) F_2'(\omega) e^{j\omega\tau} d\omega$$

If both the functions are same, i.e., $f_1(t) = f_2(t) = f(t)$, then

$$R(\tau) = \frac{1}{2\pi} \int_{-\infty}^{\infty} F(\omega) F^*(\omega) e^{j\omega\tau} d\omega$$

$$= \frac{1}{2\pi} \int_{-\infty}^{\infty} |F(\omega)|^2 e^{j\omega\tau} d\omega$$

$$= F^{-1}\left[|F(\omega)|^2\right]$$

$$F^{-1}[\psi(\omega)] \qquad (2.6.12a)$$

Equivalently,

$$F[R(\tau)] = \psi(\omega) \qquad (2.6.12b)$$

Hence, autocorrelation and the energy density spectrum are a Fourier transform pair.

## PROBLEMS

2.1 Show that the power of an amplitude-modulated signal $f(t)\cos\omega_0 t$ is half the power of the signal $f(t)$.

2.2 Show that the mean square value of a band limited signal in terms of its sample values is given by

$$\overline{f^2(t)} = \overline{f_k^2}$$

where $f_k$ is the $k^{th}$ sample.

2.3 Show that the following relation regarding cross-correlation function is true

$$R_{1,2}(\tau) = R^*_{2,1}(-\tau)$$

2.4 Show that the cross-correlation of two functions $f_1(t)$, and $f_2(t)$, shown in Fig. Prob. 2.4 (a) and (b) is Fig. (c).

126  COMMUNICATION SYSTEMS: Analog and Digital

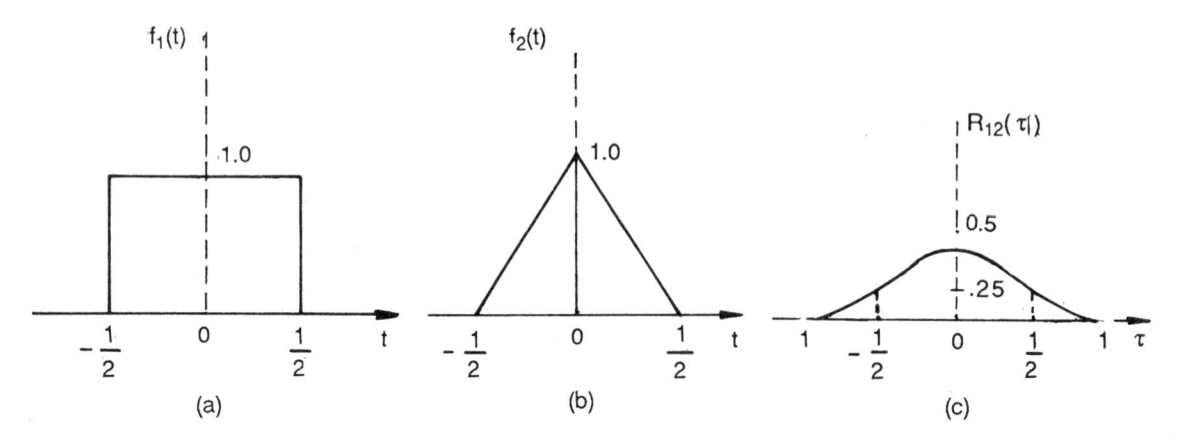

Fig. Prob 2.4

2.5 Plot the cross-correlation of functions $\sin \omega t$ and $\cos \omega t$
2.6 If $f(t) = \sin \omega_0 t$,
   (a) Plot its autocorrelation $R(\tau)$
   (b) Find the power density spectrum directly, and show that it is same as Fourier transform of $R(\tau)$.
2.7 Find the power of the following signal.
   (a) $A \cos 300 \pi t \cos 3000 \pi t$
   (b) $[B + \sin 400 \pi t] \cos 4000 \pi t$
2.8 An amplifier has a gain characteristic (transfer function) as given below:

$$H(\omega) = \frac{-B}{\omega_0 \left( \dfrac{j\omega}{\omega_0} + 1 \right)}$$

   (a) Plot the magnitude and phase characteristics of the amplifier.
   (b) Determine the delay time introduced in the transmission of signals.
2.9 Determine the unit impulse reponse, and the unit step response of the ideal high-pass filter whose transfer function is given by

$$H(\omega) = \left[1 - G_{2\omega_m}(\omega)\right] e^{-j\omega t_0}$$

2.10 A signal $f(t) = 3e^{-t} u(t)$ is passed through an ideal low pass filter with cut-off frequency of 2 radians per second. Find the energy density spectrum of the output of the filter. Determine the energies of the input and output signals.
2.11 If $f(t) = A$, is a power signal, find the power of the signal.
2.12 Find the highest value of autocorrelation of a function $f(t)$, where $f(t) = 2 \cos t + 3 \cos 3t + 4 \sin 4t$

2.13 If
$$f(t) = \begin{cases} e^{-bt} & t \geq 0 \\ 0 & \text{elsewhere} \end{cases}$$

and

$$h(t) = \begin{cases} \dfrac{1}{T} e^{-t/T} & t \geq 0 \\ 0 & \text{elsewhere} \end{cases}$$

Derive the response $r(t)$ of the linear system $h(t)$ when it is excited by the input signal $f(t)$.

2.14 The transfer function of an ideal bandpass filter is given by

$$H(\omega) = K\left[G_{\omega_m}(\omega - \omega_c) + G_{\omega_m}(\omega + \omega_c) e^{-j\omega t_0}\right]$$

(i) Sketch the magnitude and phase plot of this transfer function.
(ii) Derive the impulse response of this filter.
(iii) Is this filter physically realizable?

2.15 Find the average power of the signal

$$2 + 3 \sin \omega t + 8 \cos \omega t$$

2.16 An AM signal, $10 \sin 10^6 t + (8 \cos 100 t + 6 \sin 1000 t) \sin 10^6 t$ is applied across a 1-ohm resistor. Determine the average power dissipated.

2.17 Using Parseval's theorem for power signals, evaluate

$$\int_{-\infty}^{\infty} e^{-2t} u(t) dt$$

2.18 A signal $e^{-t/\lambda} u(t)$ is passed through a high pass R C filter with time constant $\lambda$. Find the energy density at the output of the circuit and show that it is half of the energy density of the input signal.

# THREE
# PROBABILITY AND RANDOM SIGNAL THEORY

## INTRODUCTION

Deterministic signals are the signals that can be modelled as completely specified functions of time, and are easy to analyse. There is another class of signals known as non-deterministic signals or random signals or statistical signals. The precise values of these signals cannot be predicted in advance. Many signals, and noise, in a communication system are of this nature. These signals can be described in terms of their statistical properties. Probability theory is a mathematical tool to deal with these signals.

## 3.1 SET THEORY

The concept of set theory is very useful in understanding the probability theory.

A set is a collection of objects called elements, or members, of the set. Synonyms for the word set are class, aggregate, and collection. In general, unless otherwise specified, a set is denoted by upper-case letters, such as $A$, $B$, $C$, and an element by lower-case letters such as $a, b, c$.

If '$a$' is an element of the set '$A$', then this relationship is expressed by

$$a \in A \qquad \text{'}a\text{' belong to '}A\text{'}$$

If '$a$' is not a member of '$A$' then it is written as

$$a \notin A \qquad \text{'}a\text{' does not belong to '}A\text{'}$$

For example, if '$A$' is the set of all even numbers less than or equal to 10, then

$$4 \in A, \ \sqrt{6} \notin A, \ 7 \notin A, \ 12 \notin A$$

A set can be defined either by giving all its elements (roster method), or by defining some property held by all members and no nonmembers (property method).
The above set $A$ may be defined in both the ways
  (i) $A = \{2, 4, 6, 8, 10\}$ — roster method.
  (ii) Let $I$ be the set of all integers, then
$$A = \{2a \in I \,|\, 1 \leq a \leq 5\} \text{ — property method.}$$

PROBABILITY AND RANDOM SIGNAL THEORY 129

If every element of $B$ is also an element of $A$, then it is said that $B$ is a subset of $A$, written as $B \subset A$ or $A \supset B$, and read as "$B$ is contained in $A$" or "$A$ contains $B$", respectively. It follows that for all sets $A$, $A \subset A$, or $A \supset A$.

The $A$ and $B$ are said to be equal or identical sets (written as $A = B$) if $A \subset B$ and $B \subset A$. In this case the elements in the two sets are the same. If the elements in the two sets $A$ and $B$ are not the same, then it is said that $A$ and $B$ are distant sets (written as $A \neq B$).

If $A \subset B$ but $A \neq B$, then $A$ is said to be a proper subset of $B$.

In many cases, we may have in mind a particular class of objects. Then the set constituted by these objects is known as *universal set U*, or *sample space S*. All other sets are formed from the elements of $U$ and are subsets of $U$.

It is useful to consider a set having no elements. This is called a *null set* or *empty set* and it is denoted by $\phi$. It is a subset of any set. Then, for every set $A$, we have

$$\phi \subset A \subset U$$

and, for every element $a$, we have

$$a \in U \quad \text{and} \quad a \notin \phi$$

### 3.1.1 Venn Diagram

It is sometimes convenient to represent a universal set, sets and subsets with the help of geometric diagrams known as *Venn diagrams*. The universal set is indicated by a rectangle and subsets of the universe are indicated

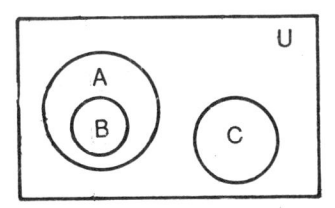

Fig. 3.1.1.1 Venn Diagram

by circles within the rectangle as in Fig. 3.1.1.1. Here A, B, and C are subsets of universal set $U$ and B is a proper subset of A, so that

$$B \subset A$$

### 3.1.2 Set Operations

1. *Union or sum*: The set of all elements in $A$, or $B$, or both is known as union, or sum of $A$ and $B$, and is denoted by $A \cup B$ or $A + B$ (shaded area in Fig. 3.1.2.1a).

2. *Intersection or product*: The set of elements common to both A and B is known as intersection or products of $A$ and $B$, and is denoted by $A \cap B$ or $A.B$ or $AB$ (shaded area in Fig. 3.1.2.1b).

If $AB = \phi$, i.e., if $A$ and $B$ have no common elements, then they are known as disjoint or mutually exclusive sets. In Fig. 3.1.1.1, $A$ and $C$ are disjoint sets, and so are $B$ and $C$.

3. *Difference*: The set consisting of all elements of $A$ which do not belong to $B$ is called the difference of $A$ and $B$, and is denoted by $A - B$ (shaded area in Fig. 3.1.2.1c).

4. *Complement*: The set of all elements not present in $A$ is known as complement of $A$, and is denoted by $A^c$ or $A'$ or $\overline{A}$ (shaded area in Fig. 3.1.2.1d).

If $A \subset B$, then $(B - A)$ is called the complement of $A$ relative to $B$, and is denoted by $A'_B$ or $A^C_B$ or $\overline{A}_B$ (shaded area in Fig.3.1.2.1e).

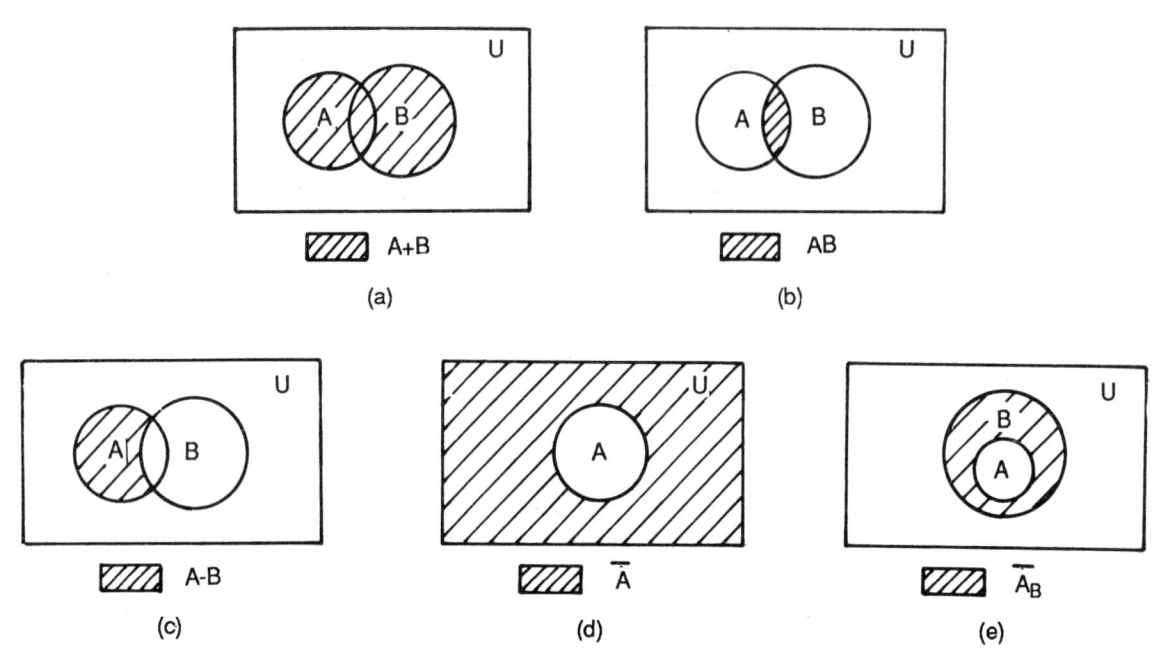

Fig. 3.1.2.1 (a) (b) (c) (d) and (e) Set Operations

It can be seen that

$$\overline{A} = U - A$$

and

$$\overline{A}_B = B - A \quad \text{if } A \subset B$$

### 3.1.3 Some Laws Involving Sets

**Idempotent laws:**
- 1a. $A + A = A$
- 1b. $AA = A$

**Identity laws:**
- 2a. $A + \phi = A$
- 2b. $A\phi = \phi$
- 3a. $A + U = U$
- 3b. $AU = A$

**Complement laws:**
- 4a. $A + \overline{A} = U$
- 4b. $A\overline{A} = \phi$
- 5a. $\overline{(\overline{A})} = A$
- 5b. $\overline{U} = \phi; \overline{\phi} = U$

**Commutative laws:**
- 6a. $A + B = B + A$
- 6b. $AB = BA$

Associative laws:
7a.  $(A + B) + C = A + (B + C) = A + B + C$   7b.  $(AB)C = A(BC) = ABC$

Distributive laws:
8a.  $A(B + C) = AB + AC$   8b.  $A + (BC) = (A + B)(A + C)$

DeMorgan's laws:
9a.  $\overline{(A + B)} = \overline{A}\,\overline{B}$   9b.  $\overline{(AB)} = \overline{A} + \overline{B}$

Difference laws:
10a.  $(AB) + (A - B) = A$   10b.  $(AB)(A - B) = \phi$
11a.  $A - B = A\overline{B}$   11b.  $B - A = \overline{A}B$

The above laws can easily be proved by using Venn diagrams. The proofs are left as exercise.

## Example 3.1.3.1

Verify the following relation:
$$(A + B) - AB = A\overline{B} + \overline{A}B$$

*Solution*
$$(A + B) - AB = (A + B)(\overline{AB}) \text{ using law (11a)}$$
$$= (A + B)(\overline{A} + \overline{B}) \text{ using law (9b)}$$
$$= (A + B)\overline{A} + (A + B)\overline{B} \text{ using law (8a)}$$
$$= A\overline{A} + B\overline{A} + A\overline{B} + B\overline{B} \text{ using law (8a)}$$
$$= \phi + B\overline{A} + A\overline{B} + \phi \text{ using law(4b)}$$
$$= B\overline{A} + A\overline{B} \text{ using law (2a)}$$
$$= A\overline{B} + B\overline{A} \text{ using law (6a)}$$
$$= A\overline{B} + \overline{A}B \text{ using law (6b)}$$

The same problem can be solved by using Venn diagrams as follows:

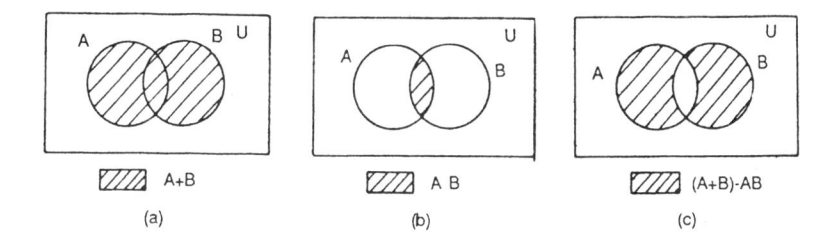

Fig. 3.1.3.1 (a); (b) and (c) LHS of Equation in Example 3.1.3.1

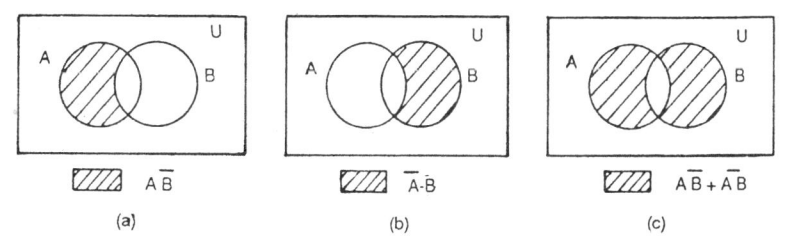

Fig. 3.1.3.2 (a); (b) and (c) RHS of Equation in Example 3.1.3.1

Figure 3.1.3.1(c) is derived from Fig. 3.1.3.1(a) and Fig. 3.1.3.1(b). Similarly, Fig. 3.1.3.2(c) is derived from Fig. 3.1.3.2(a) and Fig. 3.1.3.2(b). It is observed that the shaded areas in Fig. 3.1.3.1(c) and Fig. 3.1.3.2(c) are same. Hence $(A+B) - AB = A\bar{B} - \bar{A}B$

### Example 3.1.3.2

Find out the set composed of the shaded region of Fig. 3.1.3.3 in terms of specified sets.

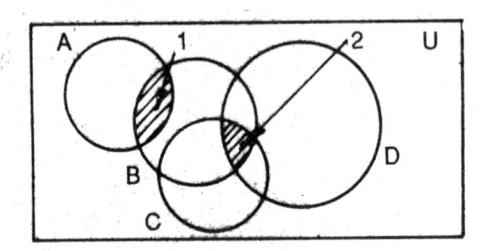

Fig. 3.1.3.3 Venn Diagram for Example 3.1.3.2

*Solution*

The shaded area 1 = $AB$

The shaded area 2 = $BCD$

Hence, total shaded area = $AB + BCD = B(A + CD)$

## 3.2 INTRODUCTION TO PROBABILITY

Probability is the study of random experiments. In any random experiment, there is always an uncertainty as to whether a particular event will occur or not. As a measure of chance or probability of occurrence of an event, a number between 0 and 1 is assigned. If we are sure that the event will occur, then we say that its probability is 100% or 1. On the other hand, if we are sure that the event will not occur, then we say that its probability is 0% or 0. If we are not sure whether the event will occur or not, then its probability is between 0 and 1. For example, the probability of occurrence of 28 February in a year is 1, as it is certain to occur every year; whereas, the probability of occurrence of 30 February in a year is 0, as it never occurs. On the other hand, the probability of occurrence of 29 February in a year is neither 0 nor 1. It is between 0 and 1 (in fact it is $\frac{1}{4}$ as it occurs every leap year, i.e., once in four years).

There are three approaches to understand probability:

### 1. Classical or a priori Approach

If an event $A$ can occur in $s$ different ways out of a total of $n$ equally likely ways, then the probability $P$ of an event $A$ is given by

$$p = P(A) = \frac{s}{n}$$

For example, in tossing of a die, an even number can occur in 3 ways out of 6 equally likely ways. Hence,

$$p = \frac{3}{6} = \frac{1}{2}$$

## 2. Frequency or a porteriori Approach

If in $n$ repetitions of an experiment, where $n$ is very large, an event $A$ occurs $s$ times, then the probability of the event is given by

$$p = P(A) = \frac{s}{n}$$

It is observed that the ratio $\frac{s}{n}$, called the relative frequency, becomes stable as $n$ increases. For example, in tossing of a fair coin, if $n = 1$, then the probability of occurrence of head $P(H)$ is either 1 or 0. But as $n$ becomes very large, say 1000, then both head and tail will occur around 500 times (since the coin is fair), and $P(H)$ will be very near to 0.5.

The classical approach deals with the experiments whose outcomes are equally likely. In practice, however, the probabilities of outcomes of an experiment are arbitrary, and the classical definition of probability fails. On the other hand, the phrase, *very large* of frequency approach is vague and hence the probability given by frequency approach is not precise. These limitations of the above two approaches lead to another approach based on set theory, known as axiomatic approach.

## 3. Axiomatic Approach

Let $S$ be the sample space consisting of all possible outcomes of an experiment. The events $A, B, C...$ are subset of the space. A function $P(.)$ is defined which associates with each event $A$ a real number called the probability of $A$. This function $P(.)$ has to satisfy the following axioms:

*Axiom* 1. For every event $A$, $0 \leq P(A) \leq 1$.

*Axiom* 2. For a sure or certain event, $P(S) = 1$.

*Axiom* 3. If $A$ an $B$ are mutually exclusive events, i.e. if $AB = 0$, then

$$P(A + B) = P(A) + P(B)$$

In general, if $A_1, A_2, A_3, \cdots$ is a sequence of mutually exclusive events, then

$$P(A_1 + A_2 + A_3 + \cdots) = P(A_1) + P(A_2) + P(A_3) + \cdots$$

We can now prove various theorems on probability using the above axioms.

### Theorem 1

If $\phi$ is an empty set, then $P(\phi) = 0$

### Proof

Let $A$ be any set. Then $A$ and $\phi$ are mutually exclusive, and $A + \phi = A$. Then by Axiom 3, $P(A + \phi) = P(A)$ or $P(A) + P(\phi) = P(A)$.
Subtracting $P(A)$ from both sides gives $P(\phi) = 0$

## Theorem 2

$$P(\overline{A}) = 1 - P(A)$$

### Proof

The sample space $S$ can be divided into two mutually exclusive events $A$ and $\overline{A}$, as shown in Fig. 3.2.1. Then by Axioms 2 and 3,

$$1 = P(s) = P(A + \overline{A}) = P(A) + P(\overline{A})$$

or $\qquad P(\overline{A}) = 1 - P(A)$

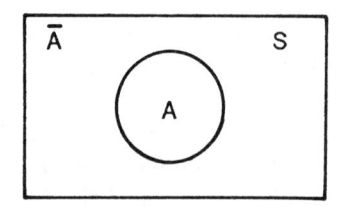

Fig. 3.2.1 Venn Diagram for Theorem 2 on Probability

## Theorem 3

If $A \subset B$, then $P(A) \leq P(B)$

### Proof

If $A \subset B$, then $B$ can be divided into two mutually exclusive events $A$ and $(B - A)$ as shown in Fig. 3.2.2. Thus, $P(B) = P(A) + P(B - A)$

Now, $\qquad\qquad P(B - A) \geq 0 \qquad\qquad$ Axiom 1

Hence,
$$P(A) \leq P(B)$$

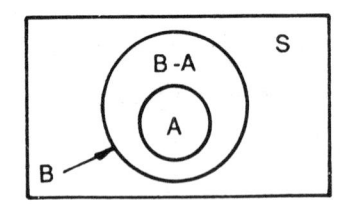

Fig. 3.2.2 Venn Diagram for Theorem 3 on Probability

## Theorem 4

If $A$ and $B$ are two events, then

$$P(A - B) = P(A) - P(AB)$$

## Proof

The event $A$ can be divided into two mutually exclusive events $(A - B)$ and $AB$ as shown in Fig.3.2.3.

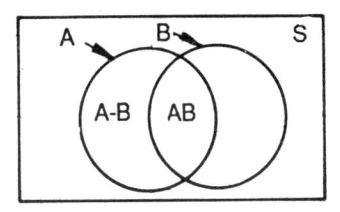

Fig. 3.2.3 Venn Diagram for Theorem 4 on Probability

Thus, $\qquad P(A) = P(A - B) + P(AB)$

Hence $\qquad P(A - B) = P(A) - P(AB)$

## Theorem 5

If A and B are any two events, then $P(A + B) = P(A) + P(B) - P(AB)$

## Proof

The event $(A + B)$ can be divided into two mutually exclusive events $(A - B)$ and $B$ as shown in Fig.3.2.4.

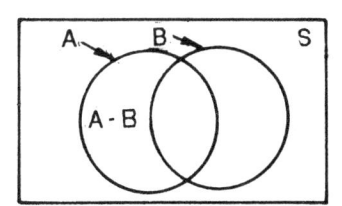

Fig. 3.2.4 Venn Diagram for Theorem 5 on Probability

Thus $\qquad P(A + B) = P(A - B) + P(B)$

$\qquad\qquad\qquad = P(A) - P(AB) + P(B) \quad$ Theorem 4

$\qquad\qquad\qquad = P(A) + P(B) - P(AB)$

## Example 3.2.1

A box contains 3 White, 4 Red, and 5 Black balls. A ball is drawn at random. Find the probability that it is (a) Red, (b) not Black, and (c) Black or White.

*Solution*

Let $W$, $R$ and $B$ be the events of drawing a White, a Red, and a Black ball respectively.

(a)
$$P(R) = \frac{\text{Ways of choosing a red ball}}{\text{total ways of choosing a ball}}$$

$$= \frac{4}{3+4+5}$$

$$= \frac{1}{3}$$

(b) We have to find $P(\bar{B})$
Now,
$$P(B) = \frac{5}{3+4+5}$$

$$= \frac{5}{12}$$

Hence,
$$P(\bar{B}) = 1 - P(B)$$

$$= 1 - \frac{5}{12}$$

$$= \frac{7}{12}$$

(c) First method

$P(\text{black or white})$
$$= P(B+W)$$
$$= P(B) + P(W) \quad \text{(as } B \text{ and } W \text{ are mutually exclusive events)}$$
$$= \frac{5}{12} + \frac{3}{12}$$
$$= \frac{2}{3}$$

Second method
$P(\text{Black or White}) = P(\text{not Red})$
$$= P(\bar{R})$$
$$= 1 - P(R)$$
$$= 1 - \frac{1}{3}$$
$$= \frac{2}{3}$$

## Example 3.2.2

Two dice are thrown. What is the probability of getting a 5?
*Solution*
First method: Each die has six faces. Hence a total of $(6 \times 6 = 36)$ outcomes are possible. Each outcome has

an equal probability of occurrence. Thus probability of any outcome is (1/36). There are 4 different face combinations that give a 5. They are 4,1; 3,2; 2,3; 1,4. The probability of the overall event is then

$$P(5) = \frac{1}{36} + \frac{1}{36} + \frac{1}{36} + \frac{1}{36} = \frac{1}{9}$$

*Second method*: Of the 36 possible outcomes, 4 are favourable ones corresponding to the occurrence of 5. Then we can call the occurrence of a 5 as the desired event rather than any single combination that gives a 5. The probability of the desired event 5 is then

$$P(5) = \frac{4}{36} = \frac{1}{9}$$

[Note: As a check, it can be seen that

$$P(2) + P(3) + \cdots + P(11) + P(12) = 1]$$

## Example 3.2.3

A card is drawn at random from an ordinary deck of 52 playing cards. Find the probability of its being (a) an Ace, (b) a Six, or a Heart, (c) neither a Nine, nor a Spade.

**Solution**

Let us use the letters $C, D, H$ and $S$ to indicate Club, Diamond, Heart and Spade respectively, and, the numbers 1,2,3...12,13 for Ace, Two, Three...Queen, King, respectively. Then $3D$ means 3 of Diamond, and $(12+H)$ means Queen or Heart. Since the events are equally likely, the probability of drawing any card is $\frac{1}{52}$.

(a) As there are four Aces in a pack of cards,

$$P(1) = \frac{\text{ways of choosing an Ace}}{\text{total ways of choosing a card}}$$

$$= \frac{4}{52}$$

$$= \frac{1}{13}$$

(b) Since 6 and $H$ are not mutually exclusive events (because a 6 of Heart is an event 6 as well as an event Heart)

$$P(6 + H) = P(6) + P(H) - P(6H)$$

$$= \frac{4}{52} + \frac{13}{52} - \frac{1}{52}$$

$$= \frac{4}{13}$$

(c) First method: The probability of getting neither a 9, nor a Spade is given by $P(\overline{9}\overline{S})$. Now $(\overline{9}\overline{S}) = (\overline{9+S})$ (DeMorgan's theorem). Hence,

$$P(\overline{9}\overline{S}) = P(\overline{9+S})$$

$$= 1 - P(9+S)$$

$$= 1 - [P(9) + P(S) - P(9S)]$$

$$= 1 - \left[\frac{4}{52} + \frac{13}{52} - \frac{1}{52}\right]$$

$$= \frac{9}{13}$$

Second method: Figure 3.2.5 gives the Venn diagram of complement of the desired event drawn over the

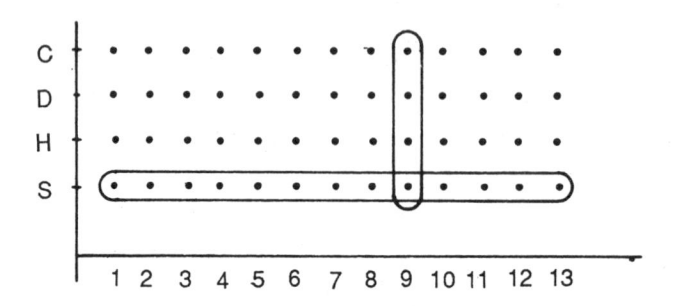

Fig. 3.2.5 Venn Diagram for Example 3.2.3

sample space of 52 points. Since this complement has 16 sample points, the desired event has $52 - 16 = 36$ sample points in it. Each sample point is assigned probability $\frac{1}{52}$. Hence the required probability is

$$P(\overline{9}\overline{S}) = \frac{36}{52}$$

$$= \frac{9}{13}$$

(Note: Part (a) and (b) of this example can also be solved by using the sample space method.)

### Example 3.2.4

Two cards are drawn at random from an ordinary deck of 52 playing cards. Find the probability $P$ that one is a Diamond and the other is a Heart.

*Solution*

There are $\binom{52}{2} = \frac{52 \times 51}{1 \times 2} = 1326$ ways of drawing 2 cards from a deck of 52 cards. Since there are 13 Diamonds and 13 Hearts, there are $13 \times 13 = 169$ ways of drawing a Diamond and a Heart. Hence,

$$P = \frac{\text{the number of ways of drawing a Diamond and a Heart}}{\text{the total number of ways of drawing two cards}}$$

$$= \frac{169}{1326}$$

$$= \frac{13}{102}$$

## 3.3 CONDITIONAL PROBABILITY AND STATISTICAL INDEPENDENCE

The problem of Example 3.2.4 can be looked at from a different angle. Instead of considering drawing of two cards as a single event, we may consider drawing of each card as a separate event. Let drawing of a Diamond be event $D$, and drawing of a Heart be event $H$. Then $P(DH)$ is the probability of drawing first a Diamond, and then a Heart, and $P(HD)$ is the probability of drawing first a Heart and then a Diamond. $P(DH)$ and $P(HD)$ are known as *joint probabilities*. As we are interested in a Diamond-Heart combination, irrespective of the order in which they occur, the desired probability is

$$P = P(DH) + P(HD)$$

Now, the probability of drawing a Diamond as the first card is

$$P(D) = \frac{13}{52} = \frac{1}{4}$$

since 13 out of the 52 cars are Diamonds. The probability of Drawing a Heart as the second card is $\frac{13}{51}$, as only 51 cards are left, out of which 13 are Hearts. Thus the second drawing is dependent upon, or conditioned by, the first drawing. The probability corresponding to the second drawing is designated by the symbol $P(H/D)$: the probability of event $H$ occurring, it being known that event $D$ has occurred. $P(H/D)$ is known as *conditional probability*.

*Note*: If the first card is replaced before the second drawing, then the probability of second drawing is independent of first drawing, and it is $P(H) = \frac{13}{52} = \frac{1}{4}$

The probability of drawing first a Diamond, and then a Heart is

$$P(DH) = P(D) P(H/D)$$
$$= \frac{13}{52} \cdot \frac{13}{51}$$

and the probability of drawing first a Heart, and then a Diamond is

$$P(HD) = P(H) P(D/H) = \frac{13}{52} \cdot \frac{13}{51}$$

Hence the desired probability is

$$P = P(DH) + P(HD) = \left(\frac{13}{52} \cdot \frac{13}{51}\right) + \left(\frac{13}{52} \cdot \frac{13}{51}\right) = \frac{13}{102}$$

which is in agreement with the answer of Example 3.2.4.

We can now generalize the concepts of joint probability and conditional probability. Let $n$ be the number of times an experiment is performed. Let $n$ be a large number. The outcome $A$ appears $n_A$ times, the outcome $B$ appears $n_B$ times, the combination $AB$ appears $n_{AB}$ times and the combination $BA$ appears $n_{BA}$ times. Then since $n$ is a large number, the joint probability of first $A$, and then $B$ occurring is

$$P(AB) = \frac{n_{AB}}{n} \qquad (3.3.1)$$

(*Note*: The probabilities of Example 3.2.3 with two terms in bracket such as $P(9S)$ and, $P(\overline{9S})$ should not be mistaken as joint probability, since $9S$ as well as $\overline{9S}$ are single event. The first event ($9S$) is a Nine of Spades and the Second event ($\overline{9S}$) is a card which is neither a Nine, nor a Spade. In contrast, $P(AB)$ of Eq.3.3.1 is a joint probability as $A$ and $B$ are separate events.)

Now $n_A$, must include $n_{AB}$, since some of the times when $A$ appears, it is followed by $B$. If $n_A$ is a large number, then the relative frequency of occurrence of $B$ preceded by $A$, i.e., the conditional probability $P(B/A)$ is given by

$$P(B/A) = \frac{n_{AB}}{n_A} \qquad (3.3.2)$$

Also, the probabilities of events $A$, and $B$, respectively, are given by

$$P(A) = \frac{n_A}{n}, P(B) = \frac{n_B}{n} \qquad (3.3.3)$$

Equation 3.3.2 can be rewritten as

$$P(B/A) = \frac{n_{AB}/n}{n_A/n} = \frac{P(AB)}{P(A)} \qquad (3.3.4)$$

or

$$P(AB) = P(A) P(B/A) \qquad (3.3.5)$$

Similarly, the joint probability of first $B$ and then $A$ occurring is

$$P(BA) = \frac{n_{BA}}{n} \qquad (3.3.6)$$

Now, $n_B$ must include $n_{BA}$, since some of the times when $B$ appears, it is followed by $A$. If $n_B$ is a large number, then the relative frequency of occurrence of $A$ preceded by $B$, i.e., the conditional probability $P(A/B)$ is given by

$$P(A/B) = \frac{n_{BA}}{n_B} \qquad (3.3.7)$$

$$= \frac{n_{BA}/n}{n_B/n}$$

$$= \frac{P(BA)}{P(B)} \qquad (3.3.8)$$

or

$$P(BA) = P(B) P(A/B) \qquad (3.3.9)$$

Equations 3.3.4 and 3.3.8 are not valid when $P(A)$ and $P(B)$, respectively, are zero. Hence the formal definition of conditional probability takes the following form:

The conditional probability $P(B/A)$ of event $B$ based on the hypothesis that event $A$ has occurred is defined as the ratio of the probability of occurrence of the joint event $AB$ to the probability of occurrence of the hypothesis event $A$, under the condition that the latter is non-zero. Thus,

$$P(B/A) = \frac{P(AB)}{P(A)}, P(A) \neq 0 \qquad (3.3.10)$$

Similarly,

$$P(A/B) = \frac{P(BA)}{P(B)}, P(B) \neq 0 \qquad (3.3.11)$$

The following example explains the concepts of joint and conditional probabilities:

An experiment is performed 50 times, and the outcomes appear as given by the sequence below :
BAABBBBABAAABBAAAABABBABAABABAAABBABBAABABBABABBAA
It is seen that $n = 50$, $n_A = 26$, $n_B = 24$, $n_{AB} = 14$, $n_{BA} = 15$.
Thus the different probabilities are as follows:

$$P(A) = \frac{n_A}{n} = \frac{26}{50} \qquad P(B) = \frac{n_B}{n} = \frac{24}{50}$$

$$P(AB) = \frac{n_{AB}}{n} = \frac{14}{50} \qquad P(BA) = \frac{n_{BA}}{n} = \frac{15}{50}$$

$$P(B/A) = \frac{n_{AB}}{n_A} = \frac{14}{26} \qquad P(A/B) = \frac{n_{BA}}{n_B} = \frac{15}{24}$$

Equations 3.3.5 and 3.3.9 are verified by substituting these values of probabilities
$$P(AB) = P(A)\,P(B/A) \text{ gives}$$
$$\frac{14}{50} = \frac{26}{50} \times \frac{14}{26} \text{ hence verified}$$

and

$$P(BA) = P(B)\,P(A/B) \text{ gives}$$
$$\frac{15}{50} = \frac{24}{50} \times \frac{15}{24} \text{ hence verified}$$

Now, let us consider a situation where the probability of the event $B$ occurring is independent of the event $A$. Such a situation would be true in the two-card problem if the first card were immediately replaced after having been drawn. In this case then

$$P(B/A) = P(B) \qquad (3.3.12)$$

implying that the probability of event $B$ is independent of event $A$, and Eq. 3.3.5 becomes

$$P(AB) = P(A)\,P(B) \qquad (3.3.13)$$

Similarly, when the probability of event $A$ occurring is independent of event $B$, then

$$P(A/B) = P(A) \qquad (3.3.14)$$

implying that the probability of event $A$ is independent of event $B$, and the Eq. 3.3.9 becomes

$$P(BA) = P(B)\,P(A) \qquad (3.3.15)$$

From Eqs 3.3.13 and 3.3.15, we get

$$P(AB) = P(BA) = P(A)\,P(B) \qquad (3.3.16)$$

The two events $A$ and $B$ are said to be *Statistically Independent* if their probabilities satisfy the Eqs 3.3.12, 3.3.14 and 3.3.16.

Equation 3.3.5 can also be written as

$$P(DC) = P(D)\,P(C/D) \qquad (3.3.17)$$

simply by replacing $A$ and $B$ by $D$ and $C$, respectively.
Now let $D$ be the joint event $AB$. Hence Eq. 3.3.17 becomes

$$P(ABC) = P(AB)\,P(C/AB)$$
$$= P(A)\,P(B/A)\,P(C/AB) \text{ using Eq.3.3.5}$$

The same concept may be extended to $n$ events, thus giving

$$P(A_1\,A_2\,A_3\cdots A_n) = P(A_1)\,P(A_2/A_1)\,P(A_3/A_1 A_2)\cdots P(A_n/A_1 A_2\cdots A_{n-1}) \qquad (3.3.18)$$

Equation 3.3.18 is known as the *Multiplication Theorem.*

## Example 3.3.1

Urn A contains two White and three Black balls, and Urn B contains three White and four Black balls. One of the urns is selected at random and a ball is chosen from it. What is the probability of drawing a White ball?

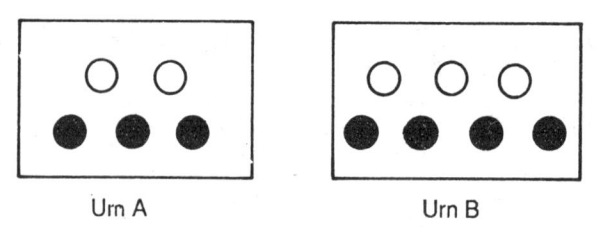

Fig.3.3.1 Two Urn Problem of Example 3.3.1

*Solution*

The situation is explained in Fig.3.3.1. The White ball may either be taken from urn A or from urn B. The corresponding probabilities are $P(AW)$ and $P(BW)$. Since these are mutually exclusive events, the probability of drawing a White ball is given by

$$P(W) = P(AW + BW) = P(AW) + P(BW)$$

Now

$$P(AW) = P(A) P(W/A)$$

where $P(A)$ is the probability of selecting urn A

$\left[ P(A) = P(B) = \dfrac{1}{2}, \text{ since the urns are selected at random} \right]$ and $P(W/A) = \dfrac{2}{5}$ is the probability of drawing a White ball from urn A.

Hence,

$$P(AW) = \frac{1}{2} \times \frac{2}{5} = \frac{1}{5}$$

Similarly,

$$P(BW) = P(B) P(W/B)$$
$$= \frac{1}{2} \times \frac{3}{7}$$
$$= \frac{3}{14}$$

Thus,

$$P(W) = P(AW) + P(BW)$$
$$= \frac{1}{5} + \frac{3}{14}$$
$$= \frac{29}{70}$$

## Example 3.3.2

An urn contains 4 White and 3 Black balls. Three balls are drawn from the urn successive. What is the probability that the first two balls are White and third is Black?

*Solution*

The probability of the first ball being a white ball is $\frac{4}{7}$, i.e., $P(A_1) = \frac{4}{7}$, as there are four White balls out of a total of 7 balls. The probability of the second ball being also white is, $\frac{3}{6}$, i.e., $P(A_2 / A_1) = \frac{3}{6}$, as after the first drawing there remain 3 white balls out of a total of 6 balls. The probability of the third ball being black is $\frac{3}{5}$, i.e., $P(A_3 / A_1 A_2) = \frac{3}{5}$, as after the second drawing there are 3 black balls out of a total of 5 balls. The desired probability is then found by using Multiplication Theorem as follows:

$$P = P(A_1 A_2 A_3) = P(A_1) P(A_2 / A_1) P(A_3 / A_1 A_2)$$

$$= \frac{4}{7} \times \frac{3}{6} \times \frac{3}{5}$$

$$= \frac{6}{35}$$

## Example 3.3.3

Each letter of the word ATTRACT is written on a separate card. The cards are then thoroughly shuffled, and four of them are drawn in succession. What is the probability of getting result a TACT?

*Solution*

The probability of getting $T$ as the first letter is $\frac{3}{7}$, since there are three $T$s in seven letters. The probability of $A$ as the second letter is $\frac{2}{6}$ as there are two $A$'s in the remaining six letters. The probability of $C$ as the third letter is $\frac{1}{5}$ as there is one $C$ in the remaining five letters. The probability of $T$ as the fourth letter is $\frac{2}{4}$ as there are two $T$'s (one of the three $T$s having been drawn first time) in the remaining four letters. The desired probability is then found by using Multiplication Theorem as follows:

$$P = \frac{3}{7} \times \frac{2}{6} \times \frac{1}{5} \times \frac{2}{4} = \frac{1}{70}$$

## Example 3.3.4

An urn contains five White, three Red, and two Black balls. If three balls are drawn in succession, what is the probability that they will be of different colours?

*Solution*

Let $W$, $R$, and $B$ be the events of drawing White, Red, and Black balls, respectively. There are six possible ways of getting the balls of different colours They are:

WRB, WBR, RWB, RBW, BWR and BRW

The probability of first combination WRB is given by the Multiplication Theorem as

$$P(WRB) = \frac{5}{10} \times \frac{3}{9} \times \frac{2}{8}$$

Where $\frac{5}{10}$ is the probability of drawing first a white ball, $\frac{3}{9}$ is the probability of drawing second a red ball, and $\frac{2}{8}$ is probability of drawing third a black ball. In the same way, the probabilities of other combinations can be found out. As the six events as mutually exclusive, the desired probability is given by

$$\begin{aligned}P &= P(WRB + WBR + RWB + RBW + BWR + BRW)\\&= P(WRB) + P(WBR) + P(RWB) + P(RBW) + P(BWR) + P(BRW)\\&= \left(\frac{5}{10} \times \frac{3}{9} \times \frac{2}{8}\right) + \left(\frac{5}{10} \times \frac{2}{9} \times \frac{3}{8}\right) + \left(\frac{3}{10} \times \frac{5}{9} \times \frac{2}{8}\right) + \left(\frac{3}{10} \times \frac{2}{9} \times \frac{5}{8}\right)\\&\quad + \left(\frac{2}{10} \times \frac{5}{9} \times \frac{3}{8}\right) + \left(\frac{2}{10} \times \frac{3}{9} \times \frac{5}{8}\right)\\&= \frac{1}{4}\end{aligned}$$

### Example 3.3.5

40% of the population of a town are voters, 50% are educated, and 20% are educated voter. A person is chosen at random. (a) If he is educated, what is the probability that he is a voter? (b) If he is a voter, what is the probability that he is not educated? (c) What is the probability that he is neither a voter nor educated?

**Solution**

Let $A$ and $B$ be the events that the person is voter, and educated, respectively. Then $AB$ is the event that the person is an educated voter. The situation is explained in the Venn diagram of Fig. 3.3.2.

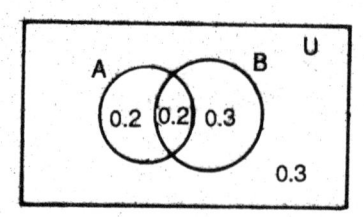

Fig. 3.3.2  Venn Diagram for Example 3.3.5

The probabilities are $P(A) = 0.4$, $P(B) = 0.5$ and $P(AB) = 0.2$.

Hence
$$P(A + B) = P(A) + P(B) - P(AB)$$
$$= 0.4 + 0.5 - 0.2$$
$$= 0.7$$

(a) The desired probability is
$$P(A/B) = \frac{P(AB)}{P(B)} = \frac{0.2}{0.5} = 0.4$$

(b) The desired probability is
$$P(\bar{B}/A) = \frac{P(\bar{B}A)}{P(A)} = \frac{0.2}{0.4} = 0.5$$

(c) The desired probability is
$$P(\overline{A}\overline{B}) = P(\overline{A+B}) \quad \text{using DeMorgan's theorem}$$
$$= 1 - P(A+B)$$
$$= 1 - 0.7$$
$$= 0.3$$

## Example 3.3.6

Three men, A, B and C hit a target with the respective probabilities 1/3, 1/4 and 1/5. Each one of them shoots once at the target. (a) What is the probability that one of them misses the target? (b) If only one misses the target, then what is the probability that it is the second man?

*Solution*
Given
$$P(A) = \frac{1}{3}, P(B) = \frac{1}{4}, P(C) = \frac{1}{5}$$
Hence
$$P(\bar{A}) = \frac{2}{3}, P(\bar{B}) = \frac{3}{4}, P(\bar{C}) = \frac{4}{5}$$

(a) There are three ways by which one of them can miss the target. They are (i) $A$ misses, while $B$ and $C$ hit, (ii) $B$ misses while $A$ and $C$ hit, (iii) $C$ misses while $A$ and $B$ hit. Since these are mutually exclusive events, the desired probability of one of them missing the target is
$$P(M) = P(\bar{A}BC + A\bar{B}C + AB\bar{C})$$
$$= P(\bar{A}BC) + P(A\bar{B}C) + P(AB\bar{C})$$

Also, individual hitting or missing are statistically independent events.
Thus,
$$P(\bar{A}BC) = P(\bar{A}) P(B) P(C) \text{ etc.} \cdots$$

Hence,

$$P(M) = [P(\bar{A})P(B)P(C)] + [P(A)P(\bar{B})P(C)] + [P(A)P(B)P(\bar{C})]$$

$$= \left(\frac{2}{3} \times \frac{1}{4} \times \frac{1}{5}\right) + \left(\frac{1}{3} \times \frac{3}{4} \times \frac{1}{5}\right) + \left(\frac{1}{3} \times \frac{1}{4} \times \frac{4}{5}\right)$$

$$= \frac{3}{20}$$

(b) The desired probability is

$$P(B/M) = \frac{P(BM)}{P(M)} = \frac{P(A\bar{B}C)}{P(M)}$$

$$= \frac{P(A)P(\bar{B})P(C)}{P(M)} = \frac{\frac{1}{3} \times \frac{3}{4} \times \frac{1}{5}}{\frac{3}{20}} = \frac{1}{3}$$

## 3.4 BAYES' THEOREM

Let the events $A_1, A_2, \cdots, A_n$ form a partition sample space $S$, i.e., $A_i$ are mutually exclusive and their union is $S$. Let there be another event $B$. (See Fig.3.4.1). Then

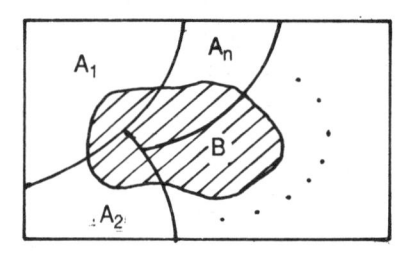

Fig. 3.4.1  Bayes' Theorem

$$B = SB$$
$$= (A_1 + A_2 + \cdots + A_n)B$$
$$= A_1B + A_2B + \cdots + A_nB$$

Since $A_i$ are mutually exclusive, $A_iB$ are also mutually exclusive. Hence

$$P(B) = P(A_1B + A_2B + \cdots + A_nB)$$
$$= P(A_1B) + P(A_2B) + \cdots + P(A_nB)$$
$$= P(A_1)P(B/A_1) + P(A_2)P(B/A_2) + \cdots + P(A_n)P(B/A_n)$$

or

$$P(B) = \sum_{i=1}^{n} P(A_i) P(B/A_i) \qquad (3.4.1)$$

Also, we have

$$P(A_i/B) = \frac{P(A_i B)}{P(B)} = \frac{P(A_i) P(B/A_i)}{P(B)}$$

Therefore

$$P(B) = \frac{P(A_i) P(B/A_i)}{P(A_i/B)} \qquad (3.4.2)$$

Eqs 3.4.1 and 3.4.2 give

$$\frac{P(A_i) P(B/A_i)}{P(A_i/B)} = \sum_{i=1}^{n} P(A_i) P(B/A_i)$$

or

$$P(A_i/B) = \frac{P(A_i) P(B/A_i)}{\sum_{i=1}^{n} P(A_i) P(B/A_i)} \qquad (3.4.3)$$

Equation 3.4.3 is known as Bayes' Theorem.

### Example 3.4.1

In a factory, four machines $A_1, A_2, A_3$ and $A_4$ produce 10%, 20%, 30% and 40% of the items respectively. The percentage of defective items produced by them is 5%, 4%, 3%, and 2% respectively. An item selected at random is found to be defective. What is the probability that it was produced by the machine $A_2$?

*Solution*

The situation is explained in Fig.3.4.2.

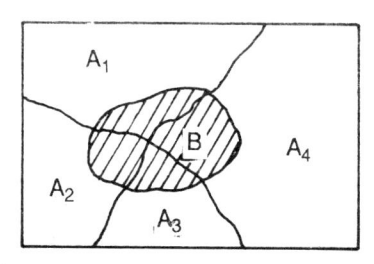

Fig. 3.4.2 Venn Diagram for Example 3.4.1

Let $B$ be the event that the item is defective. The desired probability is given by Bayes' Theorem as

$$P(A_2/B) = \frac{P(A_2) P(B/A_2)}{\sum_{i=1}^{4} P(A_i) P(B/A_i)}$$

$$= \frac{0.2 \times 0.04}{(0.1 \times 0.05) + (0.2 \times 0.04) + (0.3 \times 0.03) + (0.4 \times 0.02)}$$

$$= \frac{4}{15}$$

The same problem can be solved using a *tree diagram* of Fig. 3.4.3. $D$ stands for defective item, and $N$ stands for nondefective item. Each branch of the tree gives the corresponding probability. For example, the

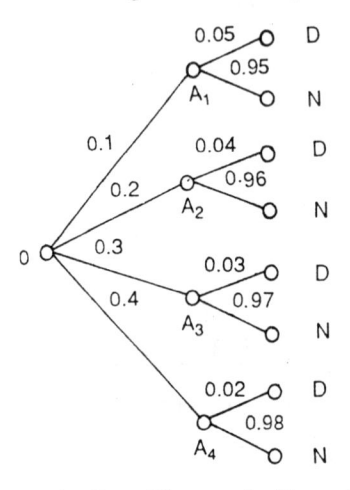

Fig. 3.4.3 Tree Diagram for Example 3.4.1

probability of selecting machine $A_2$ and its producing a defective item is given by path $OA_2 D$, whose probability is

$$P(A_2) P(B/A_2) = 0.2 \times 0.04 = 0.08$$

Now,
$$\begin{aligned} P(B) &= \text{Path } OA_1 D + \text{Path } OA_2 D + \text{Path } OA_3 D + \text{Path } OA_4 D \\ &= (0.1 \times 0.05) + (0.2 \times 0.4) + (0.3 \times 0.03) + (0.4 \times 0.02) \\ &= 0.3 \end{aligned}$$

Hence, that probability of defective item coming from $A_2$ is

$$P(A_2/B) = \frac{\text{Path } OA_2 D}{P(B)}$$

$$= \frac{0.08}{0.3}$$

$$= \frac{4}{15}$$

### Example 3.4.2

A town has a population of 10,000 people. Of these, 6,000 are males, and 4000 are females. Also 300 males and 400 females of this population are unemployed. An unemployed person is chosen at random. What is the probability that he is a male?

## Solution
First method:

Let $A_1$ and $A_2$ represent male, and female, respectively. Hence,

$$P(A_1) = \frac{6000}{6000 + 4000} = 0.6$$

and

$$P(A_2) = \frac{4000}{6000 + 4000} = 0.4$$

Let $B$ represent an unemployed person. Then,

$$P(B/A_1) = \frac{300}{6000} = 0.05$$

$$P(B/A_2) = \frac{400}{4000} = 0.1$$

The desired probability is then found by Bayes' Theorem as

$$P(A_1/B) = \frac{P(A_1)\,P(B/A_1)}{P(A_1)\,P(B/A_1) + P(A_2)\,P(B/A_2)}$$

$$= \frac{0.6 \times 0.05}{(0.6 \times 0.05) + (0.4 \times 0.1)}$$

$$= \frac{3}{7}$$

Second Method:

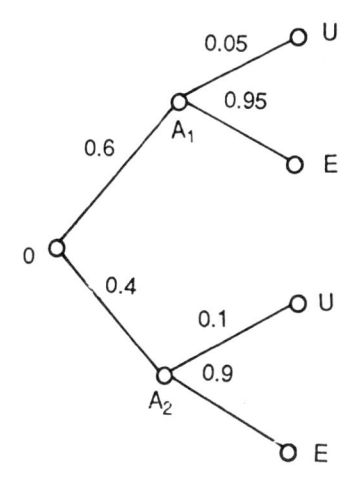

Fig. 3.4.4  Tree Diagram for Example 3.4.2

The tree diagram for the problem is given in Fig. 3.4.4. $U$ stands for unemployed person and $E$ stands for employed person. Then, the probability of unemployed person is

150  COMMUNICATION SYSTEMS: Analog and Digital

$$P(B) = \text{Path } OA_1U + \text{Path } OA_2U$$
$$= (0.6 \times 0.05) + (0.4 \times 0.1)$$
$$= 0.07$$

Hence, the probability of unemployed person being a male is

$$P(A_1/B) = \frac{\text{Path } OA_1U}{P(B)} = \frac{0.6 \times 0.05}{0.07} = \frac{3}{7}$$

## 3.5  RANDOM VARIABLES

In previous sections, we have used terms like random experiment, sample space, event, etc. Some of them were defined and some were not. For convenience, they are all defined in this section.

The word *random* stresses the fundamental fact that we are dealing with experiments governed by laws of chance, rather than by any deterministic laws.

An experiment whose outcome cannot be predicted exactly, and hence is random, is called a *random experiment*. (e.g., tossing of a coin, drawing of a card from a deck of playing cards etc.)

The collective outcomes of a random experiment form a *sample space*. A particular outcome is called a *sample point* or *sample*. Collection of outcomes is called an *event*. Thus an event is a subset of sample space.

A *random variable* is a real valued functions defined over the sample space of random experiment. It is also known as *stochastic variable*, or *random function* or *stochastic function*. The random variables are denoted by upper-case letters such as $X$, $Y$ etc and the values assumed by them are denoted by lower-case letters with subscripts such as $x_1, x_2, y_1, y_2$, etc.

## 3.6  DISCRETE RANDOM VARIABLES

A random variable that takes on a finite number of values is known as a *discrete random variable*.

Let us take for example, the experiment of three tosses of a fair coin. There are eight possible outcomes of this experiment. These will constitute the sample space. Let the number of heads be the random variable $X$. The sample space $S$, and the random variable $X$ are as shown below:

$$S = [\ HHH \quad HHT \quad HTH \quad THH \quad HTT \quad THT \quad TTH \quad TTT\ ]$$
$$X = [\ x_1 \quad x_2 \quad x_3 \quad x_4 \quad x_5 \quad x_6 \quad x_7 \quad x_8\ ]$$
$$= [\ 3 \quad\ \ 2 \quad\ \ 2 \quad\ \ 2 \quad\ \ 1 \quad\ \ 1 \quad\ \ 1 \quad\ \ 0\ ]$$

Many other random variables can be defined over the sample space, such as, number of tails, square of the number of heads, difference of number of heads and number of tails etc.

### 3.6.1  Discrete Probability Distributions

Let $X$ be a discrete random variable, and also let $x_1, x_2, x_3, \cdots$ be the values that $X$ can assume in increasing order of magnitude.
Let

$$P(X = x_j) = f(x_j)\ j = 1, 2, 3 \ldots \tag{3.6.1.1}$$

be the probability of $x_j$.

Let there be a function $f(x)$ such that

1. $$f(x) \geq 0 \qquad (3.6.1.2)$$
2. $$\sum_x f(x) = 1 \qquad (3.6.1.3)$$

Then $f(x)$ is known as *probability function* or *probability distribution* of the discrete random variable.

## Example 3.6.1.1

Find the probability function for the coin tossing experiment of Sec. 3.6.
*Solution*

Since the coin is fair, the probability of each of the 8 outcomes is $\frac{1}{8}$. Then,

$$P(X = 0) = P(x_8) = \frac{1}{8}$$

$$P(X = 1) = P(x_5) + P(x_6) + P(x_7) = \frac{1}{8} + \frac{1}{8} + \frac{1}{8} = \frac{3}{8}$$

$$P(X = 2) = P(x_2) + P(x_3) + P(x_4) = \frac{1}{8} + \frac{1}{8} + \frac{1}{8} = \frac{3}{8}$$

$$P(X = 3) = P(x_1) = \frac{1}{8}$$

The probability function is then as given below:

| $X$ | 0 | 1 | 2 | 3 |
|---|---|---|---|---|
| $f(x)$ | 1/8 | 3/8 | 3/8 | 1/8 |

Figures 3.6.1.1 and 3.6.1.2 show the bar chart, and the histogram for example 3.6.1.1, respectively. It may be noted that the sum of the ordinates of the bar chart is 1, and the sum of the area of the histogram (shaded portion shown in Fig. 3.6.1.2) is 1.

## Example 3.6.1.2

An urn contains 4 White and 3 Black balls. Two balls are drawn successively with $X$ denoting the number of Black balls. (a) Find the probability function of $X$ (b). Draw the bar chart and histogram.
*Solution*
(a) The sample space $S$ and the random variable $X$, are as shown below:

$$S = [\ WW \quad WB \quad BW \quad BB\ ]$$
$$X = [\ 0 \quad\quad 1 \quad\quad 1 \quad\quad 2\ ]$$

Now,

$$P(X = 0) = P(WW) = \frac{4}{7} \times \frac{3}{6} = \frac{2}{7}$$

$$P(X = 1) = P(WB) + P(BW) = \left(\frac{4}{7} \times \frac{3}{6}\right) + \left(\frac{3}{7} \times \frac{4}{6}\right) = \frac{4}{7}$$

$$P(X = 2) = P(BB) = \frac{3}{7} \times \frac{2}{6} = \frac{1}{7}$$

The probability function is

| $X$ | 0 | 1 | 2 |
|---|---|---|---|
| $f(x)$ | 2/7 | 4/7 | 1/7 |

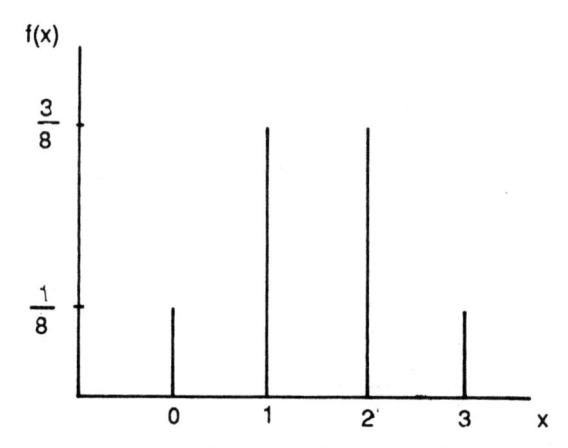

Fig. 3.6.1.1 Bar Chart for Example 3.6.1.1

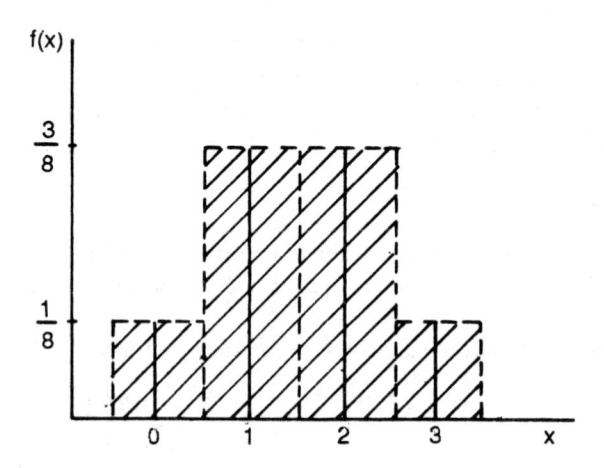

Fig. 3.6.1.2 Histogram for Example 3.6.1.1

The bar chart and histogram are shown in Figs 3.6.1.3, and 3.6.1.4, respectively.

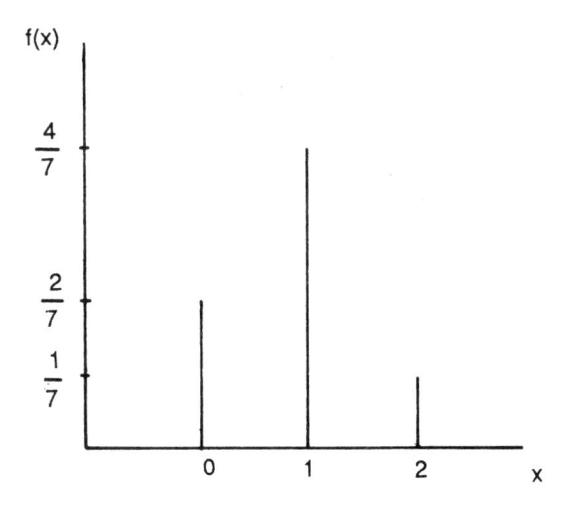

Fig. 3.6.1.3 Bar Chart for Example 3.6.1.2

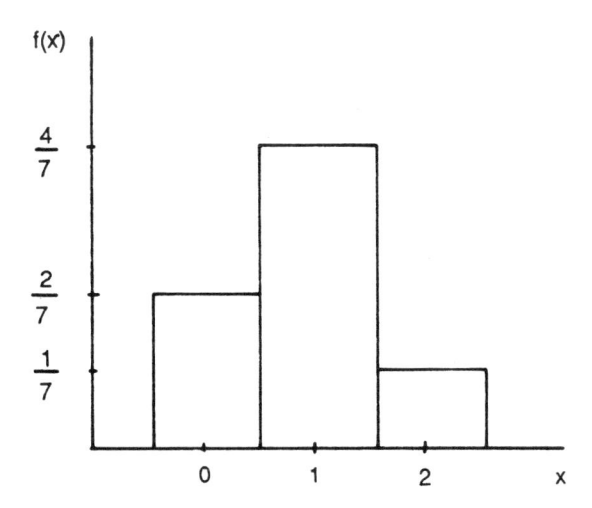

Fig. 3.6.1.4 Histogram for Example 3.6.1.2

## 3.6.2 Cumulative Distribution Function for a Discrete Random Variable

The *cumulative distribution function* or *distribution function* for a discrete random variable is defined as

$$F(X) = P(X \leq x) = \sum_{u \leq x} f(u) \qquad -\infty < x < \infty \qquad (3.6.2.1)$$

If $X$ can take on the values $x_1, x_2, x_3, \ldots, x_n$, then the distribution function is given by

$$F(x) = \begin{cases} 0 & -\infty \le x < x_1 \\ f(x_1) & x_1 \le x < x_2 \\ f(x_1) + f(x_2) & x_2 \le x < x_3 \\ \vdots \\ f(x_1) + f(x_2) + \cdots + f(x_n) & x_n \le x < \infty \end{cases} \qquad (3.6.2.2)$$

### Example 3.6.2.1

Find the distribution function for the random variable $X$ of Example 3.6.1.1.
*Solution*
The distribution function is found by using Eq. 3.6.2.2.

$$F(x) = \begin{cases} 0 & -\infty \le x < 0 \\ 1/8 & 0 \le x < 1 \\ 4/8 & 1 \le x < 2 \\ 7/8 & 2 \le x < 3 \\ 1 & 3 \le x < \infty \end{cases}$$

Figure 3.6.2.1 shows the graph for $F(x)$. The value of $F(x)$ at an integer is obtained at the higher step. Thus $F(x)$ at $x = 1$ is $4/8$ and not $1/8$. It can be seen that the jumps at 0,1,2, and 3 are 1/8, 3/8, 3/8, and 1/8 respectively,

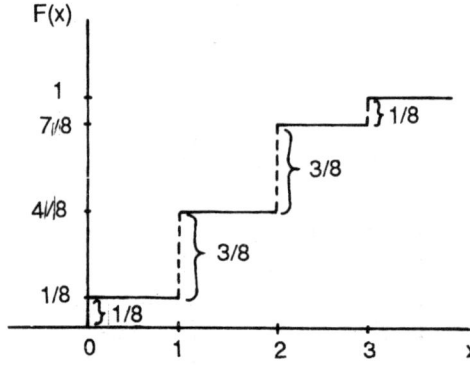

**Fig. 3.6.2.1** Distribution Function for Random Variable of Example 3.6.1.1

which are precisely the values of ordinates of Fig 3.6.1.1. Thus the probability function can be obtained from the distribution function and vice versa. It can be seen that $F(x)$ remains the same or increases as $X$ increases. Hence $F(x)$ is said to be a monotonically increasing function.

## 3.7 CONTINUOUS RANDOM VARIABLES

A random variable that takes on an infinite number of values is known as a *continuous random variable*.

As there are infinite possible values of $X$, the probability that it takes on any particular value is $\frac{1}{\infty}$, or 0. Hence probability function in this case cannot be defined as in the discrete case. In a continuous case, the probability that $X$ lies between two different values is non-zero. For example, if $X$ represents the height of a person, then the probability that it is exactly 160 cm would be zero, but the probability that it is between 155 cm and 165 cm would be non-zero.

### 3.7.1 Continuous Probability Distributions

With the ideas of Sec. 3.7, we can define probability function for a continuous case.
Let there be a function $f(x)$ such that

1. $$f(x) \geq 0$$

2. $$\int_{-\infty}^{\infty} f(x)\, dx = 1$$

Here the function $f(x)$ is known as *probability function* or probability distribution, for a continuous random variable, but it is more popularly known as *Probability Density Function* (PDF) or *density function*.

The probability of $X$ lying between $a$ and $b$ is defined by

$$P(a < X < b) = \int_a^b f(x)\, dx \qquad (3.7.1.1)$$

For a continuous case, the probability of $X$ being equal to any particular value is zero. Hence, either or both the signs $<$ in Eq. 3.7.1.1 can be replaced by the sign $\leq$. Thus,

$$P(a < X < b) = P(a \leq X < b) = P(a < X \leq b) = P(a \leq X \leq b) \qquad (3.7.1.2)$$

### Example 3.7.1.1

(a) Find the constant $C$ so that the function

$$f(x) = \begin{cases} C(x-1) & 1 < x < 4 \\ 0 & \text{otherwise} \end{cases}$$

is a density function. (b) Find $P(2 < X < 3)$,

**Solution**

(a) The property 2 gives

$$1 = \int_{-\infty}^{\infty} f(x)\, dx = \int_1^4 C(x-1)\, dx = C\left[\frac{x^2}{2} - x\right]_1^4 = \frac{9C}{2}$$

Hence

$$C = \frac{2}{9}$$

(b) $$P(2 < X < 3) = \int_2^3 \frac{2}{9}(x-1)\, dx = \frac{2}{9}\left[\frac{x^2}{2} - x\right]_2^3 = \frac{1}{3}$$

### 3.7.2 Distribution Function for Continuous Random Variable

The distribution function for a continuous random variable is defined as

156  COMMUNICATION SYSTEMS: Analog and Digital

$$F(x) = P(X \le x) = P(-\infty < X \le x) = \int_{-\infty}^{x} f(u)\,du \qquad (3.7.2.1)$$

At points of continuity of $f(x)$, the sign $\le$ in Eq. 3.7.2.1 can be replaced by the sign $<$.

The plots of $f(x)$ and $F(x)$ are shown in Fig. 3.7.2.1 and Fig. 3.7.2.2 respectively. Since $f(x) \ge 0$, the curve in Fig. 3.7.2.1 cannot fall below the x-axis. The total area under the curve in Fig. 3.7.2.1 must be 1, and the

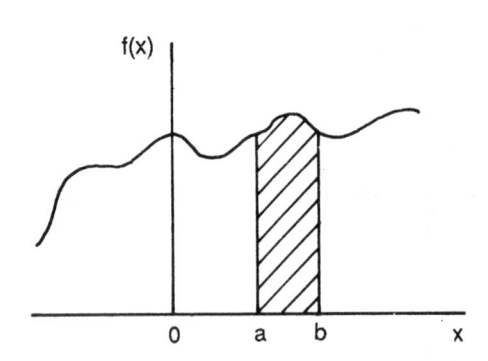

Fig. 3.7.2.1 Probability Density Function of Continuous Random Variable

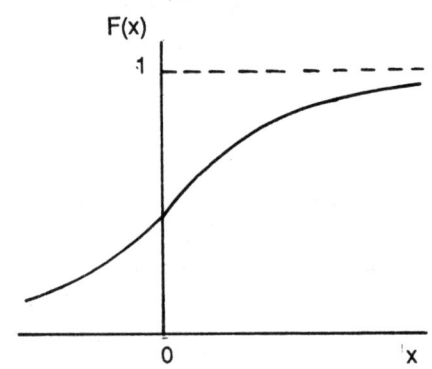

Fig. 3.7.2.2 Distribution Function of Continuous Random Variable

shaded area in it gives the probability that $x$ lies between $a$, and $b$, i.e., $P(a < X < b)$. The distribution function $F(x)$ is a monotonically increasing function that increases from 0 to 1 as shown in Fig. 3.7.2.2.

### Example 3.7.2.1

(a) Find the distribution function for the random variable of Example 3.7.1.1 (b). Find $P(2 < X < 3)$ using the result of (a).

*Solution*

(a) For $\quad X < 1, \quad F(x) = \int_{-\infty}^{x} f(u)\,du = \int_{-\infty}^{x} (0)\,du = 0$

For $\quad 1 < X < 4, \quad F(x) = \int_{1}^{x} f(u)\,du = \int_{1}^{x} \frac{2(u-1)}{9}\,du = \frac{(x-1)^2}{9}$

For $\quad 4 < X, \; F(x) = \int_{1}^{4} f(u)\,du + \int_{4}^{x} f(u)\,du$

$$= \int_{1}^{4} \frac{2(u-1)}{9}\,du + \int_{4}^{x} (0)\,du = 1$$

Thus,
$$F(x) = \begin{cases} 0 & X < 1 \\ (x-1)^2/9 & 1 < X < 4 \\ 1 & 4 < X \end{cases}$$

(b) We have
$$P(2 < X < 3) = P(X < 3) - P(X < 2)$$
$$= \frac{(3-1)^2}{9} - \frac{(2-1)^2}{9}$$
$$= \frac{1}{3}$$

## Example 3.7.2.2

For the PDF shown in Fig. 3.7.2.3, find
(a) the relationship between $a$ and $b$.

(b) $P\left(X > \dfrac{a}{2}\right)$

*Solution*

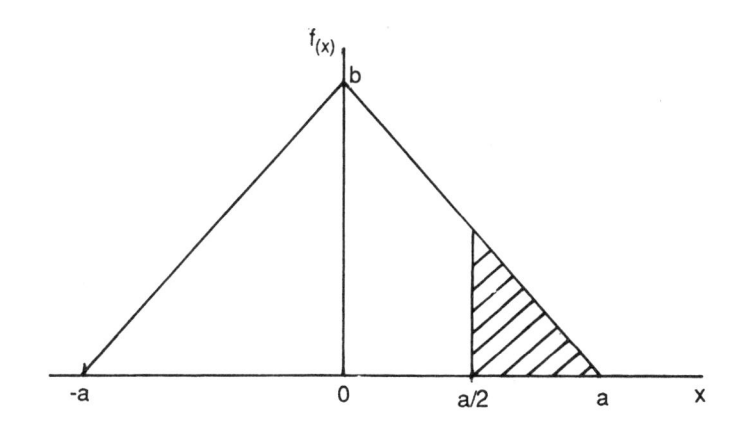

Fig. 3.7.2.3  Probability Density Function for Example 3.7.2.2

The PDF for random variable of Fig. 3.7.2.3 is

$$f(x) = \begin{cases} \dfrac{b}{a}x + b & x < 0 \\ -\dfrac{b}{a}x + b & x > 0 \end{cases}$$

or

$$f(x) = -\frac{b}{a}|x| + b$$

(a)
$$1 = \int_{-\infty}^{\infty} f(x)\,dx$$

$$= \int_{-a}^{0}\left(\frac{b}{a}x + b\right)dx + \int_{0}^{a}\left(-\frac{b}{a}x + b\right)dx$$

$$= \left[\frac{b}{a}\frac{x^2}{2} + bx\right]_{-a}^{0} + \left[-\frac{b}{a}\frac{x^2}{2} + bx\right]_{0}^{a}$$

$$= ab$$

Hence,
$$b = \frac{1}{a}$$

(b)
$$P\left(X > \frac{a}{2}\right) = \int_{a/2}^{\infty} f(x)\,dx$$

$$= \int_{a/2}^{a}\left(-\frac{b}{a}x + b\right)dx$$

$$= \left[-\frac{b}{a}\frac{x^2}{2} + bx\right]_{\frac{a}{2}}^{a}$$

$$= \frac{ab}{8}$$

$$= \frac{1}{8}$$

### Example 3.7.2.3

A random variable has an exponential PDF given by $f(x) = ae^{-b|x|}$, where $a$ and $b$ are constants. Find (a) the relationship between $a$ and $b$. (b) The distribution function of $x$.

*Solution*

(a)
$$1 = \int_{-\infty}^{\infty} f(x)\,dx$$

$$= \int_{-\infty}^{\infty} ae^{-b|x|}\,dx$$

$$= \int_{-\infty}^{0} ae^{bx}\,dx + \int_{0}^{\infty} ae^{-bx}\,dx$$

$$= \left[\frac{a}{b}e^{bx}\right]_{-\infty}^{0} + \left[-\frac{a}{b}e^{-bx}\right]_{0}^{\infty}$$

$$= \left[\frac{a}{b}(1-0)\right] + \left[-\frac{a}{b}(0-1)\right]$$

$$= \frac{2a}{b}$$

Hence,
$$b = 2a$$

(b) For $X < 0$
$$F(x) = \int_{-\infty}^{x} f(u)\,du$$

$$= \int_{-\infty}^{x} ae^{bu}\,du$$

$$= \left[\frac{a}{b}e^{bu}\right]_{-\infty}^{x}$$

$$= \frac{a}{b}(e^{bx} - 0)$$

$$= \frac{1}{2}e^{bx}$$

For $X > 0$,
$$F(x) = \int_{-\infty}^{x} f(u)\,du$$

$$= \int_{-\infty}^{x} ae^{-b|u|}\,du$$

$$= \int_{-\infty}^{0} ae^{bu} \, du + \int_{0}^{x} ae^{-bu} \, du$$

$$= \left[\frac{a}{b}e^{bu}\right]_{-\infty}^{0} + \left[-\frac{a}{b}e^{-bu}\right]_{0}^{x}$$

$$= \left[\frac{a}{b}(1-0)\right] + \left[-\frac{a}{b}(e^{-bx}-1)\right]$$

$$= 1 - \frac{1}{2}e^{-bx}$$

Thus,

$$F(x) = \begin{cases} \frac{1}{2}e^{bx} & X < 0 \\ 1 - \frac{1}{2}e^{-bx} & X > 0 \end{cases}$$

## 3.8 JOINT DISTRIBUTIONS

So far we have considered the case of a single random variable. The idea of the previous sections involving single random variable can easily be extended to two or more random variables. Random variables may either be all discrete or all continuous, or some discrete and some continuous. A typical case is of two random variables of the same type. This is even more important as, in the study of a communication system, we have to study the behaviour of the transmitter and receiver — each being assigned a random variable. Moreover, communication systems involve either both discrete or both continuous random variables. Hence, we will study only these two cases. However, the results obtained in these cases can easily be extended to more complex situations.

### 3.8.1 Discrete Case

Let $X$, and $Y$, be two discrete random variables. The *joint probability function* of $X$ and $Y$ is given by

$$f(x, y) = P(X = x, Y = y) \tag{3.8.1.1}$$

where $f(x, y)$ satisfies following properties

(1) $$f(x, y) \geq 0$$

(2) $$\sum_{x}\sum_{y} f(x, y) = 1$$

Now, let

$$[X] = [\, x_1 \quad x_2 \quad \ldots \quad x_m \,]$$

and

$$[Y] = [\, y_1 \quad y_2 \quad \ldots \quad y_n \,]$$

PROBABILITY AND RANDOM SIGNAL THEORY  161

Then,

$$[XY] = \begin{bmatrix} x_1 y_1 & x_1 y_2 & \cdots & x_1 y_n \\ x_2 y_1 & x_2 y_2 & \cdots & x_2 y_n \\ \cdots & \cdots & \cdots & \cdots \\ x_m y_1 & x_m y_2 & \cdots & x_m y_n \end{bmatrix}$$

The joint probability function $f(x, y)$ can then be represented by a joint probability table as shown in Table 3.8.1.1.

Table 3.8.1.1.

| X \ Y | $y_1$ | $y_2$ | ... | $y_n$ | Totals |
|---|---|---|---|---|---|
| $x_1$ | $f(x_1, y_1)$ | $f(x_1, y_2)$ | ... | $f(x_1, y_n)$ | $f_1(x_1)$ |
| $x_2$ | $f(x_2, y_1)$ | $f(x_2, y_2)$ | ... | $f(x_2, y_n)$ | $f_1(x_2)$ |
| ... | ... | ... | ... | ... | ... |
| $x_m$ | $f(x_m, y_1)$ | $f(x_m, y_2)$ | ... | $f(x_m, y_n)$ | $f_1(x_m)$ |
| Totals | $f_2(y_1)$ | $f_2(y_2)$ | ... | $f_2(y_n)$ | Grand Total=1 |

The probability that $X = x_j$ is known as *marginal probability function* of $X$, and is obtained by adding all the entries in the row corresponding to $x_j$. Thus,

$$f_1(x_j) = P(X = x_j) = \sum_{k=1}^{n} f(x_j, y_k)$$

Similarly, the probability that $Y = y_k$ is known as *marginal probability function* of $Y$ and is obtained by adding all the entries in the column corresponding to $y_k$. Thus,

$$f_2(y_k) = P(Y = y_k) = \sum_{j=1}^{m} (x_j, y_k)$$

It may be noted that

$$\sum_{j=1}^{m} f_1(x_j) = 1$$

$$\sum_{k=1}^{n} f_2(y_k) = 1$$

and

$$\sum_{j=1}^{m} \sum_{k=1}^{n} f(x_j, y_k) = 1$$

## 162 COMMUNICATION SYSTEMS: Analog and Digital

The *joint distribution function* of $X$ and $Y$ is defined as

$$F(x, y) = P(X \leq x, Y \leq y) = \sum_{u \leq x}\sum_{v \leq y} f(u, v)$$

The 'marginal distribution function' or simply 'distribution function' of $X$ and $Y$ are defined as

$$F_1(x) = P(X \leq x) = \sum_{u \leq x}\sum_{v} f(u,v)$$

and

$$F_2(y) = P(Y \leq y) = \sum_{u}\sum_{v \leq y} f(u,v)$$

### Example 3.8.1.1

The joint probability function of two random variables $X$ and $Y$ is given by

$$f(x, y) = \begin{cases} c(x^2 + 2y) & x = 0, 1, 2 \quad y = 1, 2, 3, 4 \\ 0 & \text{otherwise} \end{cases}$$

Find (a) the value of $c$, (b) $P(X=2, Y=3)$, (c) $P(X \leq 1, Y > 2)$, and (d) marginal probability functions of $X$ and $Y$.

**Solution**

Table 3.8.1.2

| x \ y | 1 | 2 | 3 | 4 | |
|---|---|---|---|---|---|
| 0 | 2c | 4c | 6c | 8c | 20c |
| 1 | 3c | 5c | 7c | 9c | 24c |
| 2 | 6c | 8c | 10c | 12c | 36c |
|   | 11c | 17c | 23c | 29c | 80c |

(a) Table 3.8.1.2 gives $f(x,y)$. The grand total, $80C$, must be 1. So we get

$$80C = 1$$

or

$$C = \frac{1}{80}$$

(b) The desired probability is given by the entry corresponding to $X = 2$, and $Y = 3$. (Shown by the dotted circle in Table 3.8.1.2.) Thus,

$$P(X = 2, Y = 3) = 10C = \frac{10}{80} = \frac{1}{8}$$

(c) The desired probability is given by adding the entries in dotted rectangle of Table 3.8.1.2. Thus,

$$P(X \leq 1, Y > 2) = (6C + 8C + 7C + 9C) = 30C = \frac{30}{80} = \frac{3}{8}$$

(d) The marginal probability functions of $X$ and $Y$ are obtained by adding the entries of the rows, and columns, respectively. Thus,

PROBABILITY AND RANDOM SIGNAL THEORY    163

$$f_1(x) = P(X = x) = \begin{cases} 20C = \dfrac{1}{4} & x = 0 \\ 24C = \dfrac{3}{10} & x = 1 \\ 36C = \dfrac{9}{20} & x = 2 \end{cases}$$

$$f_2(y) = P(Y = y) = \begin{cases} 11C = \dfrac{11}{80} & y = 1 \\ 17C = \dfrac{17}{80} & y = 2 \\ 23C = \dfrac{23}{80} & y = 3 \\ 29C = \dfrac{29}{80} & y = 4 \end{cases}$$

### 3.8.2 Continuous Case

Let $X$ and $Y$ be two continuous random variables. The *joint probability function* of $X$ and $Y$ is defined by

1.  $$f(x, y) \geq 0$$

2.  $$\int_{-\infty}^{\infty} \int_{-\infty}^{\infty} f(x, y) \, dx \, dy = 1$$

$f(x,y)$ is more commonly known as *joint density function* of $X$ and $Y$.

As in the case of discrete random variables, the *marginal density functions* or simply *density functions* of $X$, and $Y$, are given by

$$f_1(x) = \int_{v = -\infty}^{\infty} f(x, v) \, dv$$

and

$$f_2(y) = \int_{u = -\infty}^{\infty} f(u, y) \, du$$

The joint distribution function of $X$, and $Y$, is given by

$$F(x, y) = P(X \leq x, Y \leq y) = \int_{u = -\infty}^{x} \int_{v = -\infty}^{y} f(u, v) \, du \, dv$$

The marginal distribution functions or simply distribution functions of $X$ and $Y$ are given by

$$F_1(x) = P(X \leq x) = \int_{u = -\infty}^{x} \int_{v = -\infty}^{\infty} f(u, v) \, du \, dv$$

and

$$F_2(y) = P(Y \leq y) = \int_{u = -\infty}^{\infty} \int_{v = -\infty}^{y} f(u, v) \, du \, dv$$

## 3.8.3 Independent Random Variables

Let $X$ and $Y$ be discrete random variables. If the events $X = x$ and $Y = y$ are independent events for all $x$ and $y$, then $X$ and $Y$ are independent random variables. In such case, we have

$$P(X = x, Y = y) = P(X = x) P(Y = y)$$

or

$$f(x, y) = f_1(x) f_2(y)$$

Note that if for all values of $x$ and $y$, $f(x,y)$ can be expressed as the product of a function of $x$ alone and a function of $y$ alone, then $X$ and $Y$ are independent random variables.

Let $X$ and $Y$ be continuous random variables. If the events $X \leq x$, and $Y \leq y$, are independent events for all $x$ and $y$, then $X$ and $Y$ are independent random variables. In such a case, we have

$$P(X \leq x, Y = y) = P(X \leq x) P(Y \leq y)$$

or

$$F(x, y) = F_1(x) F_2(y)$$

Note that if for all values of $x$ and $y$, $F(x,y)$ can be expressed as a product of a function of $x$ alone and a function of $y$ alone, then $X$ and $Y$ are independent random variables.

## 3.8.4 Conditional Distributions

If $X$ and $Y$ are discrete random variables, then the conditional probability function of $Y$ given $X$ is defined as

$$f(y/x) = P(Y = y/X = x) = \frac{f(x, y)}{f_1(x)}$$

Similarly, the conditional probability function of $X$, given $Y$, is defined as

$$f(x/y) = P(X = x/Y = y) = \frac{f(x, y)}{f_2(y)}$$

If $X$ and $Y$ are continuous random variables, then the *conditional density function of $Y$ given $X$* is defined as

$$f(y/x) = \frac{f(x, y)}{f_1(x)}$$

Similarly, the conditional density function of $X$ given $Y$ is defined as

$$f(x/y) = \frac{f(x, y)}{f_2(y)}$$

### Example 3.8.4.1

In Example 3.8.1.1, find whether the random variables are independent or not.

## Solution
*First method*

For $X$ and $Y$ to be independent, we must have $P(X = x, Y = y) = P(X = x) P(Y = y)$ for all $x$ and $y$. Let us check this for $x = 2$, and $y = 3$. It is found that

$$P(X = 2, Y = 3) = \frac{1}{8}, \quad P(X = 2) = \frac{9}{20}, \quad P(Y = 3) = \frac{23}{80}$$

Now,

$$\frac{1}{8} \neq \frac{9}{20} \times \frac{23}{80}$$

Hence $X$ and $Y$ are not independent.

*Second Method*

As $f(x, y) = C(x^2 + 2y)$ cannot be expressed as the product of a function of $x$ alone and a function of $y$ alone, the random varaibles are not independent.

### Example 3.8.4.2

Find (a) $f(y/1)$, (b) $f(x/2)$, (c) $P(Y = 3/X = 2)$, (d) $P(X = 0/Y = 4)$, (e) $f(2/y)$ and, (f) $f(3/x)$ in Ex.3.8.1.1.

*Solution*

(a) 
$$f(y/x) = \frac{f(x, y)}{f_1(x)} = \frac{(x^2 + 2y)/80}{f_1(x)}$$

For $x = 1$

$$f(y/1) = \frac{(1 + 2y)/80}{3/10} = \frac{1 + 2y}{24}$$

(b) 
$$f(x/y) = \frac{f(x, y)}{f_2(y)} = \frac{(x^2 + 2y)/80}{f_2(y)}$$

For $y = 2$

$$f(x/2) = \frac{(x^2 + 4)/80}{17/80} = \frac{x^2 + 4}{17}$$

(c) 
$$P(Y = 3/X = 2) = f(3/2) = \frac{(4 + 6)/80}{9/20} = \frac{5}{18}$$

(d) 
$$P(X = 0/Y = 4) = f(0/4) = \frac{(0 + 8)/80}{29/80} = \frac{8}{29}$$

(e) 
$$f(2/y) = f(x/y)|_{x=2} = \frac{(4 + 2y)/80}{f_2(y)}$$

Hence.

$$f(2/y) = \begin{cases} \dfrac{(4+2y)/80}{11/80} = \dfrac{4+2y}{11} = \dfrac{6}{11} & \text{for } y = 1 \\[6pt] \dfrac{(4+2y)/80}{17/80} = \dfrac{4+2y}{17} = \dfrac{8}{17} & \text{for } y = 2 \\[6pt] \dfrac{(4+2y)/80}{23/80} = \dfrac{4+2y}{23} = \dfrac{10}{23} & \text{for } y = 3 \\[6pt] \dfrac{(4+2y)/80}{29/80} = \dfrac{4+2y}{29} = \dfrac{12}{29} & \text{for } y = 4 \end{cases}$$

(f)
$$f(3/y) = f(y/x)\big|_{y=3} = \dfrac{(x^2+6)/80}{f_1(x)}$$

Hence,

$$(3/x) = \begin{cases} \dfrac{(x^2+6)/80}{1/4} = \dfrac{x^2+6}{20} = \dfrac{3}{10} & \text{for } x = 0 \\[6pt] \dfrac{(x^2+6)/80}{3/10} = \dfrac{x^2+6}{24} = \dfrac{7}{24} & \text{for } x = 1 \\[6pt] \dfrac{(x^2+6)/80}{9/20} = \dfrac{x^2+6}{36} = \dfrac{5}{18} & \text{for } x = 2 \end{cases}$$

### Example 3.8.4.3

The joint density function of two continuous random variables $X$ and $Y$ is given by

$$f(x, y) = \begin{cases} 2 & \text{for } 0 < x < 1,\ 0 < y < x \\ 0 & \text{otherwise} \end{cases}$$

Find (a) the marginal density functions, and
(b) the conditional density functions.

*Solution*

(a)
$$f_1(x) = \int_{v=-\infty}^{\infty} f(x, v)\, dv = \int_0^x 2\, dv = 2x \quad 0 < x < 1$$

$$f_1(x) = 0 \qquad \qquad \text{otherwise}$$

$$f_2(y) = \int_{u=-\infty}^{\infty} f(u, y)\, du = \int_y^1 2\, du = 2(1-y) \quad 0 < y < 1$$

$$f_2(y) = 0 \qquad \qquad \text{otherwise}$$

(b) $$f(y/x) = \frac{f(x,y)}{f_1(x)} = \frac{2}{2x} = \frac{1}{x} \quad 0 < x < 1$$

$f(y/x)$ is not defined elsewhere, as outside $0 < x < 1$, $f_1(x)$ is zero.

$$f(x/y) = \frac{f(x,y)}{f_2(y)} = \frac{2}{2(1-y)} = \frac{1}{1-y} \quad 0 < y < 1$$

$f(x/y)$ is not defined elsewhere, as outside $0 < y < 1$, $f_2(y)$ is zero.

### Example 3.8.4.4

The joint density function of two continuous random variables is

$$f(x, y) = \begin{cases} Cxy & 0 < x < 2, \ 1 < y < 3 \\ 0 & \text{otherwise} \end{cases}$$

Find (a) $C$, (b) $P(0 < X < 1, 1 < Y < 2)$, (c) $P(X < 1, Y > 2)$, (d) marginal distribution functions of $X$ and $Y$, (e) joint distribution functions of $X$ and $Y$ and, (f) $P[(X+Y) < 3]$.

**Solution**

(a) We must have

$$\int_{-\infty}^{\infty}\int_{-\infty}^{\infty} f(x,y)\,dx\,dy = 1$$

Hence,

$$\int_{x=0}^{2}\int_{y=1}^{3} Cxy\,dx\,dy = 1$$

This gives $C = \frac{1}{8}$

(b) $$P(0 < X < 1, 1 < Y < 2) = \int_{x=0}^{1}\int_{y=1}^{2} \frac{xy}{8}\,dx\,dy = \frac{3}{32}$$

(c) $$P(X < 1, Y > 2) = \int_{x=0}^{1}\int_{y=2}^{3} \frac{xy}{8}\,dx\,dy = \frac{5}{32}$$

(d) For $x < 0$, $F_1(x) = 0$

For $0 < x < 2$,

$$F_1(x) = \int_{u=-\infty}^{x}\int_{v=-\infty}^{\infty} f(u,v)\,du\,dv$$

$$= \frac{1}{8}\int_{u=0}^{x}\int_{v=1}^{3} uv\,du\,dv = \frac{x^2}{4}$$

For $x > 2$,
$$F_1(x) = 1$$

Hence,
$$F_1(x) = \begin{cases} 0 & x < 0 \\ x^2/4 & 0 < x < 2 \\ 1 & x > 2 \end{cases}$$

For $y < 1$,
$$F_2(y) = 0$$

For $1 < y < 3$
$$F_2(y) = \int_{u=0}^{\infty} \int_{v=-\infty}^{y} f(u,v)\, du\, dv$$
$$= \frac{1}{8} \int_{u=0}^{2} \int_{v=1}^{y} uv\, du\, dv = \frac{y^2 - 1}{8}$$

For $y > 3$
$$F_2(y) = 1$$

Hence,
$$F_2(y) = \begin{cases} 0 & y < 1 \\ (y^2 - 1)/8 & 1 < y < 3 \\ 1 & y > 3 \end{cases}$$

(e) $f(x,y)$ can be written as the product of a function of $x$ alone and a function of $y$ alone, as given below:
$$f(x, y) = cxy = (c_1 x)(c_2 y) \quad \text{where } c_1 c_2 = c$$

Hence,
$$F(x, y) = F_1(x) F_2(y)$$

Thus $F(x, y)$ for different values of $x$ and $y$ is as shown in Fig. 3.8.4.1.

(f) Figure 3.8.4.2 shows the region $0 < x < 2$, $1 < y < 3$ within which $f(x,y)$ is nonzero. $f(x, y)$ when integrated over entire rectangular area, is 1. The probability $P[(X+Y) < 3]$ is obtained by integrating $f(x, y)$ over the shaded area $A$ of Fig. 3.8.4.2 for which $f(x, y)$ is nonzero and $(x+y) < 3$. Hence,

$$P[(X+Y) < 3] = \iint_A f(x,y)\, dx\, dy = \int_{x=0}^{2} \int_{y=1}^{3-x} \frac{xy}{8} dx\, dy = \frac{1}{4}$$

or

$$P[(X+Y) < 3] = \iint_A f(x,y)\, dx\, dy = \int_{x=0}^{3-y} \int_{y=1}^{3} \frac{xy}{8} dx\, dy = \frac{1}{4}$$

PROBABILITY AND RANDOM SIGNAL THEORY 169

| $F(x,y)=0$ | $F(x,y)=\dfrac{x^2}{4}$ | $F(x,y)=1$ |
|---|---|---|
| y=3 | | |
| $F(x,y)=0$ | $F(x,y)=\dfrac{x^2(y^2-1)}{4\times 8}$ | $F(x,y)=\dfrac{y^2-1}{8}$ |
| y=1 | | |
| $F(x,y)=0$ | $F(x,y)=0$ | $F(x,y)=0$ |
| | x=0 | x=2 |

Fig. 3.8.4.1 Joint Distribution Function of X and Y for Example 3.8.4.4

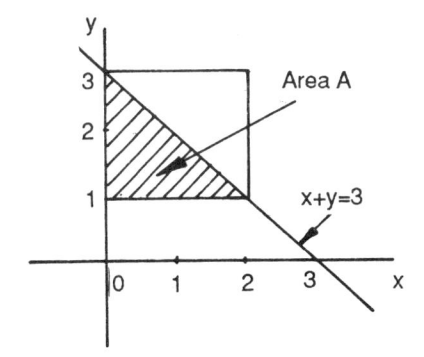

Fig. 3.8.4.2.2 $P[(X+Y)<3]$ for Example 3.8.4.4

## 3.9 CHARACTERISTICS OF RANDOM VARIABLES

In this section we will deal with some important characteristics of random variables.

### 3.9.1 Expectation

Let $X$ be a discrete random variable such that

$$[X] = [x_1\ x_2\ \cdots\ x_m]$$

The *expectation* or *expected value* or *mean* of $X$ is defined as

$$\mu_x = E(x) = \sum_{j=1}^{m} x_j\, f(x_j) \qquad (3.9.1.1)$$

where $f(x_j)$ is the probability function.

If all the probabilities are equal, then

$$f(x_1) = f(x_2) = \cdots = f(x_m) = \frac{1}{m}$$

Hence Eq. 3.9.1.1 becomes

$$\mu_x = E(x) = \frac{\sum_{j=i}^{m} x_j}{m} = \frac{x_1 + x_2 + \cdots + x_m}{m}$$

Thus in this case

Expectation = arithmetic mean.

For a continuous random variable $X$ having a density function $f(x)$, the expectation is defined as

$$\mu_x = E(x) = \int_{-\infty}^{\infty} x f(x) dx \qquad (3.9.1.2)$$

### Example 3.9.1.1

In a gambling game, if face 1, 2, 3, 4, 5, 6 of a die turns up, then the player is paid Rs. 15, 10, 5, −5, −10, −15, respectively. Find the average value of profit per throw if (a) the die is fair, (b) the probabilities of faces 1, 2, 3, 4, 5, 6 are 2/15, 1/6, 1/6, 1/6, 1/6, 3/15 respectively.

*Solution*

Table 3.9.1.1 explains the problem.

Table 3.9.1.1

|     | Dice throw | 1 | 2 | 3 | 4 | 5 | 6 |
|-----|------------|---|---|---|---|---|---|
|     | $x_j$ | 15 | 10 | 5 | −5 | −10 | −15 |
| (a) | $f(x_j)$ | $\frac{1}{6}$ | $\frac{1}{6}$ | $\frac{1}{6}$ | $\frac{1}{6}$ | $\frac{1}{6}$ | $\frac{1}{6}$ |
| (b) | $f(x_j)$ | $\frac{2}{12}$ | $\frac{1}{6}$ | $\frac{1}{6}$ | $\frac{1}{6}$ | $\frac{1}{6}$ | $\frac{3}{15}$ |

(a) In this case,

$$E(X) = (15 \times 1/6) + (10 \times 1/6) + (5 \times 1/6) - (5 \times 1/6) - (10 \times 1/6) - (15 \times 1/6) = 0$$

Thus the average profit per throw is zero.

(b) In this case,

$$E(X) = (15 \times 2/15) + (10 \times 1/6) + (5 \times 1/6) - (5 \times 1/6) - (10 \times 1/6) - (15 \times 3/15) = -1$$

Thus the average profit per throw is Re −1 i.e., average loss per throw is Re.1.

### Example 3.9.1.2

The density function of a continuous random variable is

$$f(x) = \begin{cases} x/8 & 0 < x < 4 \\ 0 & \text{otherwise} \end{cases}$$

Find $E(X)$.

**Solution**

$$E(x) = \int_{-\infty}^{\infty} x f(x) dx$$

$$= \int_0^4 x(x/8) dx$$

$$= \frac{8}{3}$$

### 3.9.2 Functions of Random Variables

If $X$ is a random variable, then $Y = g(x)$ is also a random variable. Hence,

$$E(Y) = E[g(x)] = \sum_x g(x) f(x) \qquad \text{discrete case}$$

and

$$E(Y) = E[g(x)] = \int_{-\infty}^{\infty} g(x) f(x) dx \qquad \text{continuous case}$$

The above results can easily be generalized for functions of two, or more, random variables. Thus,

$$E[g(x, y)] = \sum_x \sum_y g(x, y) f(x, y) \qquad \text{discrete case}$$

and

$$E[g(x, y)] = \int_{-\infty}^{\infty} \int_{-\infty}^{\infty} g(x, y) f(x, y) dx\, dy \qquad \text{continuous case}$$

**Example 3.9.2.1**

The joint probability function of two discrete random variables $X$ and $Y$ is given by

$$f(x, y) = \begin{cases} (x + 2y)/14 & x = 0, 1 \ y = 1, 2 \\ 0 & \text{Otherwise} \end{cases}$$

Find (a) $E(X)$, (b) $E(Y)$ and (c) $E(2X + 3Y)$.

**Solution**

(a) $$E(X) = \sum_x \sum_y x f(x, y) = \sum_{x=0}^{1} \sum_{y=1}^{2} x(x + 2y)/14 = \frac{8}{14}$$

**172** COMMUNICATION SYSTEMS: Analog and Digital

(b) $$E(Y) = \sum_x \sum_y y f(x, y) = \sum_{x=0}^{1} \sum_{y=1}^{2} y(x + 2y)/14 = \frac{23}{14}$$

(c) $$E(2X + 3Y) = \sum_x \sum_y (2x + 3y) f(x, y)$$

$$= \sum_{x=0}^{1} \sum_{y=1}^{2} (2x + 3y)(x + 2y)/14 = \frac{85}{14}$$

### Example 3.9.2.2

The joint density function of two continuous random variables is given by

$$f(x, y) = \begin{cases} xy/8 & 0 < x < 2, 1 < y < 3 \\ 0 & \text{Otherwise} \end{cases}$$

Find (a) $E(X)$, (b) $E(Y)$ and (c) $E(2X + 3Y)$.

**Solution**

(a) $$E(X) = \int_{-\infty}^{\infty} \int_{-\infty}^{\infty} x f(x, y) dx\, dy = \int_{x=0}^{2} \int_{y=1}^{3} x(xy/8) dx\, dy = \frac{4}{3}$$

(b) $$E(Y) = \int_{-\infty}^{\infty} \int_{-\infty}^{\infty} y f(x, y) dx\, dy = \int_{x=0}^{2} \int_{y=1}^{3} y(xy/8) dx\, dy = \frac{13}{6}$$

(c) $$E(2X + 3Y) = \int_{-\infty}^{\infty} \int_{-\infty}^{\infty} (2x+3y) dx\, dy = \int_{x=0}^{2} \int_{y=1}^{3} (2x + 3y)(xy/8) dx\, dy = \frac{55}{6}$$

Following are the important theorems on expectation.

**Theorem 1** If $c$ is any constant, then

$$E(cX) = cE(X) \qquad (3.9.2.1)$$

**Theorem 2** If $X$ and $Y$ are any random variables, then

$$E(X + Y) = E(X) + E(Y) \qquad (3.9.2.2)$$

**Theorem 3** If $X$ and $Y$ are independent random variables, then

$$E(XY) = E(X) E(Y) \qquad (3.9.2.3)$$

**Proof of Theorem 3**

If $X$ and $Y$ are independent random variables, then

$$f(x, y) = f_1(x) f_2(y)$$

Thus,

$$E(XY) = \sum_x \sum_y xy\, f(x, y)$$

$$= \sum_x \sum_y xy\, f_1(x) f_2(y)$$

$$= \sum_y \left[ y f_2(y) \sum_x x f_1(x) \right]$$

$$= \sum_y [y f_2(y) \, E(X)]$$

$$= E(X) \sum_y y f_2(y)$$

$$= E(X) \, E(Y)$$

Proof for the continuous case can be obtained by replacing summations with appropriate integrations Proof of Theorems 1 and 2 is straightforward, and is left as an exercise.

Part (C) of Example 3.9.2.1 and Example 3.9.2.2 can be solved by using the above theorems.

In the first problem

$$E(2X + 3Y) = E(2X) + E(3Y) = 2E(X) + 3E(Y)$$
$$= \left(2 \times \frac{8}{14}\right) + \left(3 \times \frac{23}{14}\right) = \frac{85}{14}$$

In the second problem,

$$E(2X + 3Y) = E(2X) + E(3Y) = 2E(X) + 3E(Y)$$
$$= \left(2 \times \frac{4}{3}\right) + \left(3 \times \frac{13}{6}\right) = \frac{55}{6}$$

### 3.9.3 Variance and Standard Deviation

Let us consider two random variables whose density functions are given by curves I and II of Fig. 3.9.3.1.

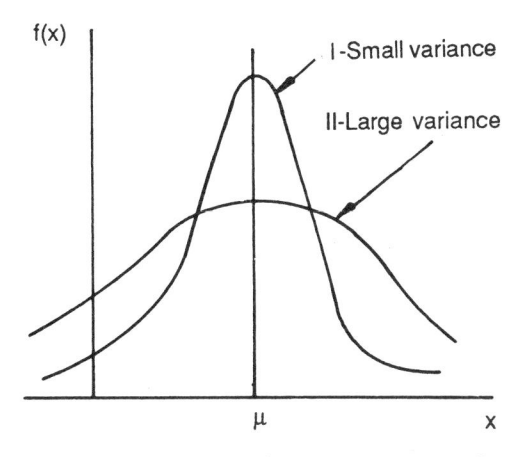

Fig. 3.9.3.1 Random Variables of same Mean and Different Variance

It may be seen that both of them have the same mean ($\mu$). Yet both the random variables differ in some way. Thus we see that the mean alone does not characterise a random variable. To characterise a random variable, we must also know how it varies or deviates, from its mean. We may be inclined to take $E(X - \mu)$ as another characteristic of a random variable to indicate its deviation, or dispersion, about its mean.

Let us consider the curves I and II, to be symmetric about $\mu$. Then $(X - \mu)$ is positive for $X > \mu$, and negative for $X < \mu$. Because of the symmetry of the curve about $\mu$, the two differences cancel each other, resulting in $E(X - \mu) = 0$ in both the cases. Thus $E(X - \mu)$ does not serve our purpose. If a function treats both positive and negative differences identically, our purpose is served. $E(|X - \mu|)$, and $E[(X - \mu)^2]$ are two such functions. However, it is found that $E[(X - \mu)^2]$ is more useful function of the two. It can be seen that $E[(X - \mu)^2]$ is small for curve I, and relatively large for curve II. $E[(X - \mu)^2]$ is known as *variance* of X. The unit of variance is a square unit. To have the same unit as that of X, we must take the square root of variance, and this is known as *standard deviation* of X.

We can formally define Variance and Standard Deviation as
Variance of a random variable $X$ is defined as

$$\text{Var}(X) = \sigma_x^2 = E[(X - \mu)^2], \quad \text{where } \mu = E(X) \tag{3.9.3.1}$$

and

Standard deviation of a random variable $X$ is defined as the positive square root of variance. Thus,

$$\sigma_x = \sqrt{\text{Var}(X)} = \sqrt{E[(X - \mu)^2]}, \quad \text{where } \mu = E(X) \tag{3.9.3.2}$$

When we interpret $f(x)$ as mass density on X-axis, the mean is the centre of gravity of mass. Variance, equals the moment of inertia of the probability masses and gives some notion of their concentration near the mean.

Following are important theorems on variance:

### Theorem 1

$$\sigma^2 = E[(X - \mu)^2] = E(X^2) - \mu^2 \tag{3.9.3.3}$$

**Proof**

$$\sigma^2 = E[(X - \mu)^2]$$
$$= E[(X^2 - 2X\mu + \mu^2)]$$
$$= E(X^2) - 2\mu E(X) + E(\mu^2)$$
$$= E(X^2) - 2\mu(\mu) + \mu^2$$
$$= E(X^2) - \mu^2$$

### Theorem 2

If $C$ is any constant, then

$$\text{Var}(CX) = C^2 \text{Var}(X) \tag{3.9.3.4}$$

## Proof

$$\text{Var}(CX) = E[(cx - c\mu)^2]$$
$$= E[c^2(X - \mu)^2]$$
$$= c^2 E[(X - \mu)^2]$$
$$= c^2 \text{Var}(X)$$

## Theorem 3

If $X$ and $Y$ are independent random variables, then

(a) $\quad$ Var $(X+Y)$ = Var $(X)$ + Var $(Y)$ $\quad\quad$ (3.9.3.5)

(b) $\quad$ Var $(X-Y)$ = Var $(X)$ + Var $(Y)$ $\quad\quad$ (3.9.3.6)

## Proof of (a)

$$\text{Var}(X + Y) = E\left[\{(X + Y) - (\mu_X + \mu_Y)\}^2\right]$$
$$= E\left[\{(X - \mu_X) + (Y - \mu_Y)\}^2\right]$$
$$= E\left[(X - \mu_X)^2 + 2(X - \mu_X)(Y - \mu_Y) + (Y - \mu_Y)^2\right]$$
$$= E\left[(X - \mu_X)^2\right] + 2E[(X - \mu_X)(Y - \mu_Y)] + E\left[(Y - \mu_Y)^2\right]$$
$$= \text{Var}(X) + 2E[(X - \mu_X)(Y - \mu_Y)] + \text{Var}(Y)$$
$$= \text{Var}(X) + 2E[(X - \mu_X)] E[(Y - \mu_Y)] + \text{Var}(Y)$$

(since $X$ and $Y$ are independent).

Now, $E(X - \mu_X) = 0$ and $E(Y - \mu_Y) = 0$

Hence, Var $(X+Y)$ = Var $(X)$ + Var $(Y)$

## Proof of (b)

It can be proved directly like theorem 3(a) or indirectly as follows

$$\text{Var}(X - Y) = \text{Var}[X + (-Y)]$$
$$= \text{Var}(X) + \text{Var}(-Y) \quad\quad \text{theorem 3(a)}$$
$$= \text{Var}(X) + (-1)^2 \text{Var}(Y) \quad\quad \text{theorem 2}$$
$$= \text{Var}(X) + \text{Var}(Y)$$

## Example 3.9.3.1

The rms value (including signal and internal noise) of output of a unity gain amplifier is 10 volts. When the signal is doubled, the rms value of output becomes 19.3 volts. Calculate the rms value of internal noise. Assume that both signal and internal noise have zero mean, and they are independent.

## Solution
As the mean of signal and internal noise is zero,
$$\text{Var}(S) = \sigma_s^2 = E[(S - \mu_S)^2] = E(S^2)$$
and
$$\text{Var}(N) = \sigma_N^2 = E[(N - \mu_N)^2] = E(N^2)$$
Now,
$$\text{rms value of } S = \sqrt{E(S^2)} = \sqrt{\sigma_s^2} = \sigma_S$$
and
$$\text{rms value of } N = \sqrt{E(N^2)} = \sqrt{\sigma_N^2} = \sigma_N$$
As $S$ and $N$ are independent,
$$\text{rms value of } (S + N) = \sigma_{S+N} = \sqrt{\text{Var}(S+N)}$$
$$= \sqrt{\text{Var}(S) + \text{Var}(N)}$$
$$= \sqrt{\sigma_S^2 + \sigma_N^2}$$
$$= 10 \tag{3.9.3.7}$$
When the signal is doubled,
We have,
$$\sigma_{s'} = 2\sigma_s$$
Hence
$$(\sigma_{s'})^2 = (2\sigma_s)^2 = 4\sigma_s^2$$
or
$$\text{Var}(S') = 4\text{Var}(S)$$
Hence the rms value of $(S' + N)$ is
$$\sigma(S' + N) = \sqrt{\text{Var}(S' + N)}$$
$$= \sqrt{\text{Var}(S') + \text{Var}(N)}$$
$$= \sqrt{4\text{Var}(S) + \text{Var}(N)}$$
$$= \sqrt{4\sigma_S^2 + \sigma_N^2}$$
$$= 19.3 \tag{3.9.3.8}$$
Solving Eqs 3.9.3.7 and 3.9.3.8 yields
$$\sigma_s = 9.45 \text{ V and } \sigma_N = 3\text{V}$$
Thus, the rms value of internal noise = 3 V.

### 3.9.4 Standardized Random Variables

If $\mu$ and $\sigma(\sigma > 0)$ are the respective mean and standard deviation of random variable $X$, then an associated *standardized random variable* or *normalized random variable* is defined as

PROBABILITY AND RANDOM SIGNAL THEORY 177

$$X^* = \frac{X - \mu}{\sigma} \qquad (3.9.4.1)$$

$X^*$ has following properties:
(1) $\qquad E(X^*) = 0$
(2) $\qquad \text{Var}(X^*) = 1$

Proof of these properties is left as an exercise.

From Eq. 3.9.4.1 it can be seen that $X^*$ is a dimensionless quantity as X, $\mu$ and $\sigma$ have the same units Standardized random variables are useful for the comparison of different distributions.

### 3.9.5 Moments

The $r^{th}$ moment of a random variable $X$ about the origin is defined as

$$\mu'_r = E[(X)^r] \qquad r = 0, 1, 2, \cdots \qquad (3.9.5.1)$$

The $r^{th}$ moment of a random variable $X$ about the mean $\mu$ is defined as

$$\mu_r = E[(X - \mu)^r] \, r = 0, 1, 2 \qquad (3.9.5.2)$$

$\mu_r$ is also known as $r^{th}$ central moment.
It can be seen that

$$\mu'_0 = 1, \; \mu'_1 = \text{mean} = \mu, \; \mu'_2 = \mu_2 + \mu^2$$
$$\mu_0 = 1, \; \mu_1 = 0, \; \mu_2 = \text{Var}(X)$$

In discrete case,

$$\mu'_r = \sum_m x_m^r f(x_m)$$

and

$$\mu_r = \sum_m (x_m - \mu)^r f(x_m)$$

In continuous case,

$$\mu'_r = \int_{-\infty}^{\infty} x^r f(x) dx$$

and

$$\mu_r = \int_{-\infty}^{\infty} (x - \mu)^r f(x) dx$$

### Example 3.9.5.1

Establish a relationship between the moments about mean and origin.
*Solution*
From the Binomial expansion, we have

$$(X - \mu)^r = \sum_{m=0}^{r} (-1)^m \binom{r}{m} X^{r-m} \mu^m$$

Hence,

$$\mu_r = E[(X - \mu)^r]$$

$$= \sum_{m=0}^{r} (-1)^m \binom{r}{m} \mu^m E[X^{r-m}]$$

$$= \sum_{m=0}^{r} (-1)^m \binom{r}{m} \mu^m \mu'_{r-m} \qquad (3.9.5.3)$$

Equation 3.9.5.3 gives the desired relationship.

### 3.9.6 Moment Generating Function and Characteristic Function

The moment generating function of a random variable $X$ is defined as

$$M(t) = E[e^{tX}] \qquad (3.9.6.1)$$

Now,

$$e^{tX} = 1 + tX + \frac{t^2 X^2}{2!} + \cdots + \frac{t^m X^m}{m!} + \cdots$$

Hence,

$$M(t) = E[e^{tX}] = E\left[1 + tX + \frac{t^2 X^2}{2!} + \cdots + \frac{t^m X^m}{m!} + \cdots\right]$$

$$= 1 + t E(X) + \frac{t^2}{2!} E(X^2) + \cdots + \frac{t^m}{m!} E(X^m) + \cdots$$

$$= 1 + t\mu'_1 + \frac{t^2}{2!}\mu'_2 + \cdots + \frac{t^m}{m!}\mu'_m + \cdots \qquad (3.9.6.2)$$

Differentiating Eq. 3.9.6.2 with respect to $t$, and then putting $t = 0$, we get

$$\left.\frac{dM(t)}{dt}\right|_{t=0} = \mu'_1$$

Similarly, differentiating $m$ times with respect to $t$, and then putting $t = 0$, we get

$$\left.\frac{d^m M(t)}{dt^m}\right|_{t=0} = \mu'_m \qquad (3.9.6.3)$$

Moments about origin can be found out from the moment generating function by using Eq. 3.9.6.3. The moments about mean can then be found out by using Eq. 3.9.5.3.

The characteristic function of a random variable $X$ is defined as

$$\phi(\omega) = M(i\omega) = E[e^{i\omega x}], i = \sqrt{-1} \qquad (3.9.6.4)$$

By expanding Eq. 3.9.6.4, as in the case of a moment generating function, it can be shown that

$$\left.\frac{d^m \phi(\omega)}{d\omega^m}\right|_{\omega=0} = (i)^m \mu'_m \qquad (3.9.6.5)$$

Moments about origin can be found out from the characteristics function using Eq. 3.9.6.5. The moments about mean can then be found out by using Eq. 3.9.5.3.

## 3.9.7 Tchebycheff's Inequality

It was seen in Sec. 3.9.3 that variance is closely related to the idea of deviation, or dispersion, of the distribution about the mean. Thus small variance indicates that large deviations from the mean are improbable. The Tchebycheff's inequality gives a more precise relation concerning the above statement.

Let $X$ be a random variable with mean $\mu$ and let $C$ be any real number. Then if $E\left[(X-c)^2\right]$ is finite and $\epsilon$ is any positive number, then

$$P[|X - c| \geq \epsilon] \leq \frac{E\left[(X-c)^2\right]}{\epsilon^2} \quad (3.9.7.1)$$

Equation 3.9.7.1 is known as Tchebycheff's inequality. The complementary statement of Eq. 3.9.7.1 is

$$P[|X - c| < \epsilon] \geq \left[1 - \frac{E\left[(X-c)^2\right]}{\epsilon^2}\right] \quad (3.9.7.2)$$

For $c = \mu$, Eq. 3.9.7.1 becomes

$$P[|X - \mu| \geq \epsilon] \leq \frac{\sigma^2}{\epsilon^2} \quad (3.9.7.3)$$

If $\epsilon = k\sigma$, then Eq. 3.9.7.1 becomes

$$P[|X - \mu| \geq k\sigma] \leq \frac{1}{k^2} \quad (3.9.7.4)$$

### Proof

We will prove Eq. 3.9.7.1 as it is the most general form of Tchebycheff's inequality. We have

$$P[|X - c| \geq \epsilon] = \int_{|X-c| \geq \epsilon} f(x)\,dx \quad (3.9.7.5)$$

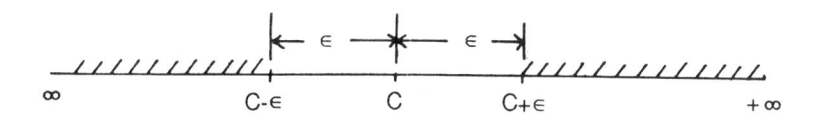

Fig. 3.9.7.1 Tchebycheff's Inequality

The meaning of integration in the above equation can be understood from Fig. 3.9.7.1. The integral is integrated over $|X - c| \geq \epsilon$, i.e., from $-\infty$ to $(c - \epsilon)$ and from $(c + \epsilon)$ to $+\infty$ as shown by the shaded region in Fig. 3.9.7.1. Thus

**180** COMMUNICATION SYSTEMS: Analog and Digital

$$\int_{|x-c|\geq \epsilon} f(x)\,dx = \int_{-\infty}^{c-\epsilon} f(x)\,dx + \int_{c+\epsilon}^{+\infty} f(x)\,dx$$

Now,

$$|x-c| \geq \epsilon \text{ can be written as}$$

$$(x-c)^2 \geq \epsilon^2 \quad \text{or} \quad \frac{(x-c)^2}{\epsilon^2} \geq 1$$

Hence,

$$\int_{|x-c|\geq \epsilon} f(x)\,dx \leq \int_{|x-c|\geq \epsilon} \frac{(x-c)^2}{\epsilon^2} f(x)\,dx \qquad (3.9.7.6)$$

Also,

$$\int_{|x-c|\geq \epsilon} \frac{(x-c)^2}{\epsilon^2} f(x)\,dx \leq \int_{-\infty}^{\infty} \frac{(x-c)^2}{\epsilon^2} f(x)\,dx \qquad (3.9.7.7)$$

as RHS is integrated over entire region, whereas LHS is integrated over the shaded region of Fig. 3.9.7.1. Hence, Eqs. (3.9.7.6) and (3.9.7.7) give

$$\int_{|x-c|\geq \epsilon} f(x)\,dx \leq \int_{-\infty}^{\infty} \frac{(x-c)^2}{\epsilon^2} f(x)\,dx$$

But

$$\int_{-\infty}^{\infty} (x-c)^2 f(x)\,dx = E\big[(x-c)^2\big]$$

Hence,

$$\int_{|x-c|\geq \epsilon} f(x)\,dx \leq \frac{E\big[(x-c)^2\big]}{\epsilon^2}$$

Using Eq. 3.9.7.5, this becomes

$$P\big[|X-c|\geq \epsilon\big] \leq \frac{E\big[(x-c)^2\big]}{\epsilon^2}$$

The physical significance of Tchebycheff's inequality can be understood from Fig. 3.9.7.2.

The total area under the curve $f(x)$ of Fig. 3.9.7.2 is 1. The equality sign of Eq. 3.9.7.1 gives the shaded area under the curve $f(x)$ and the equality sign of Eq. 3.9.7.2 gives the unshaded area under the curve $f(x)$.

If we know the probability distribution of a random variable $X$, then we may compute its mean and variance. But if we know the mean and variance of a random variable, then we cannot know its probability distribution.

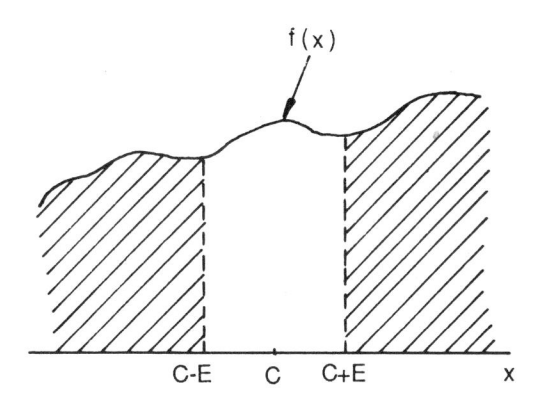

Fig. 3.9.7.2 Physical Significance of Tchebycheff's Inequality

Hence we cannot find the probabilities like $P[|X - c| \geq \epsilon]$ or $P[|X - c| < \epsilon]$ etc. But with the help of Tchefycheff's inequality, we can get upper or lower bounds to such probabilities which serve our purpose to a great extent.

## 3.10 BINOMIAL, POISSON AND NORMAL DISTRIBUTIONS

Three most important probability distributions are Binomial, Poisson, and Normal. Binomial and Poisson distributions are for discrete random variables, whereas, the Normal distribution is for continuous random variables.

### 3.10.1 Binomial Distribution

Let us consider an experiment with only two possible outcomes. (Such an experiment is also known as *Bernoulli trial*.) One outcome is called success and other is called failure. Let the experiment be repeated a number of times. Let us consider that the probability of success p in each trial is the same, i.e. the trials are independent. Then the probability of failure in each trial is $q = (1 - p)$. The probability of $x$ successes in $n$ trials is given by a probability function known as *Binomial distribution*:

$$f(x) = P(X = x) = \binom{n}{x} P^x q^{n-x} \tag{3.10.1.1}$$

The properties of Binomial distribution are

$$\text{mean} = np, \text{ variance} = npq$$

### Example 3.10.1.1

A fair die is tossed 5 times. A toss is called a success if face 1 or 6 appears. Find (a) the probability of 2 successes, (b) the mean and standard deviation for the number of successes.

*Solution*

(a) $\qquad n = 5, p = \dfrac{2}{6} = \dfrac{1}{3}, q = 1 - p = 1 - \dfrac{1}{3} = \dfrac{2}{3}$

$$P(X = 2) = \binom{5}{2}\left[\frac{1}{3}\right]^2 \left[\frac{2}{3}\right]^{5-2} = \frac{80}{243}$$

(b) $$\text{mean} = np = 5 \times \frac{1}{3} = 1.667$$

$$\text{Standard deviation} = \sqrt{\text{variance}} = \sqrt{npq} = \sqrt{5 \times \frac{1}{3} \times \frac{2}{3}} = 1.054$$

### 3.10.2 Poisson Distribution

Let $X$ be a discrete random variable that can assume values 0, 1, 2... Then the probability function of $X$ is given by Poisson distribution:

$$f(x) = P(X = x) = \frac{\lambda^x e^{-\lambda}}{x!} \qquad x = 0, 1, 2, \cdots \qquad (3.10.2.1)$$

where $\lambda$ is a positive constant.
The properties of Poisson distribution are

$$\text{mean} = \lambda \quad \text{variance} = \lambda$$

### 3.10.3 Poisson Approximation to Binomial Distribution

In Binomial distribution if $n$ is large and $p$ is close to zero, then it can be approximated by Poisson distribution with $\mu = np$. In practice, $n \geq 50$ and $np \leq 5$ give satisfactory approximation.

It can be seen from Eq. 3.10.1.1 that if $n$ is large, then calculation of a desired probability considering Binomial distribution is tedious. On the other hand, calculation of a desired probability considering Poisson distribution is fairly simple as seen from Eq. 3.10.2.1.

### Example 3.10.1.2

If 4% of the total items made by a factory are defective, find the probability that less than 2 items are defective in a sample of 50 items.
*Solution*
Let us solve the problem using binomial distribution

$$n = 50, p = 0.04, q = 1 - p = 1.0.04 = 0.96$$

$$P(X = 0) = \binom{50}{0}(0.04)^0 (0.96)^{50}$$

$$P(X = 1) = \binom{50}{10}(0.04)^1 (0.96)^{49}$$

$$P(X < 2) = P(X = 0) + P(X = 1)$$

It is obvious that the calculation of a desired probability is tedious.
Now let us solve the problem by using Poisson approximation.
Since $n = 50$ and $np = 50 \times 0.04 = 2\ (< 5)$, the Poisson approximation is valid, with $\lambda = np = 2$.

Hence,

$$P(X = 0) = \frac{(2)^0 e^{-2}}{0!} = 0.13534$$

$$P(X = 1) = \frac{(2)^1 e^{-2}}{1!} = 0.27068$$

$$P(X < 2) = P(X = 0) + P(X = 1) = 0.13534 + 0.27068 = 0.40602$$

### 3.10.4 Normal (or Gaussian) Distribution

This is the most important continuous probability distribution as most of the natural phenomenon are characterized by random variables with normal distribution. The importance of normal distribution is further enhanced because of the *Central Limit Theorem* (discussed in Sec. 3.10.7). The density function for Normal (or Gaussian) distribution is given by

$$f(x) = \frac{1}{\sigma\sqrt{2\pi}} e^{-(x-\mu)^2/(2\sigma^2)} \quad -\infty < x < \infty \qquad (3.10.4.1)$$

where $\mu$ and $\sigma$ are mean and standard deviation, respectively

The properties of normal distribution are :

$$\text{mean} = \mu \qquad \text{variance} = \sigma^2$$

The corresponding distribution function is

$$F(x) = P(X \le x) = \frac{1}{\sigma\sqrt{2\pi}} \int_{-\infty}^{x} e^{-(v-\mu)^2/2\sigma^2} dv$$

$$= \frac{1}{2} + \frac{1}{\sigma\sqrt{2\pi}} \int_{0}^{x} e^{-(v-\mu)^2/2\sigma^2} dv \qquad (3.10.4.2)$$

Let $Z$ be the *standardized random variable* corresponding to $X$. Thus if $Z = \dfrac{x-\mu}{\sigma}$, then the mean of $Z$ is zero and its variance is 1. Hence,

$$f(z) = \frac{1}{\sqrt{2\pi}} e^{-z^2/2} \qquad (3.10.4.3)$$

$f(z)$ is known as *standard normal density function*. The corresponding distribution function is

$$F(z) = P(Z \le z) = \frac{1}{\sqrt{2\pi}} \int_{-\infty}^{z} e^{-u^2/2} du = \frac{1}{2} + \frac{1}{\sqrt{2\pi}} \int_{0}^{z} e^{-u^2/2} du \qquad (3.10.4.4)$$

The integral of Eq. 3.10.4.4 is not easily evaluated. However, it is related to the error function, whose tabulated values are available in mathematical tables. The error function of $z$ is defined as

$$\text{erf } z = \frac{2}{\sqrt{\pi}} \int_{0}^{z} e^{-u^2} du \qquad (3.10.4.5)$$

The error function has values between 0 and 1.

$$erf(0) = 0 \text{ and } erf(\infty) = 1$$

The complementary error function of $Z$ is defined as

$$erfc(Z) = 1 - erf(z) = \frac{2}{\sqrt{\pi}} \int_z^\infty e^{-u^2} du \qquad (3.10.4.6)$$

The relationship between $F(z)$ $erf(z)$ and $erfc(z)$ is as follows

$$F(z) = \frac{1}{2}\left[1 + erf\left(\frac{z}{\sqrt{2}}\right)\right] = 1 - \frac{1}{2} erfc\left(\frac{z}{\sqrt{2}}\right) \qquad (3.10.4.7)$$

A graph of $f(z)$ is known as standard normal curve and is shown in Fig. 3.10.4.1 Here the areas within 1,2,

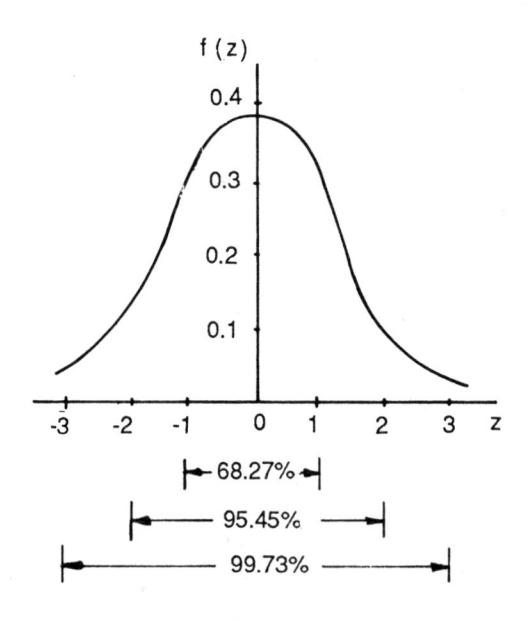

Fig. 3.10.4.1 Standard Normal Curve

and 3 standard deviations of mean (i.e., $-1 < z < 1$, $-2 < z < 2$ and $-3 < z < 3$, respectively) are 0.6827, 0.9545, and 0.9973 respectively. The total area under the curve $f(z)$ is 1. The table giving the areas under $f(z)$ for positive values of $z$ is available. From this table the areas between any two values of $Z$ can be found using symmetry about $Z = 0$.

### 3.10.5 Normal Approximation to Binomial Distribution

If $n$ is large and if neither $p$ nor $q$ is too close to zero, then the Binomial distribution can be closely approximated by a normal distribution with standardized random variable given by

$$Z = \frac{X - np}{\sqrt{npq}}$$

In practice $np \geq 5$ and $nq \geq 5$ give satisfactory approximation.

## Example 3.10.5.1

Find the probability of 4 to 7 heads inclusive in 10 tosses of a fair coin.

*Solution*

First method: (using Binomial distribution)

$$n = 10, p = q = \frac{1}{2}$$

$$P(X = 4) = \binom{10}{4}\left(\frac{1}{2}\right)^4\left(\frac{1}{2}\right)^6 = \frac{105}{512}, P(X = 5) = \binom{10}{5}\left(\frac{1}{2}\right)^5\left(\frac{1}{2}\right)^5 = \frac{63}{256}$$

$$P(X = 4) = \binom{10}{6}\left(\frac{1}{2}\right)^6\left(\frac{1}{2}\right)^4 = \frac{105}{512}, P(X = 7) = \binom{10}{7}\left(\frac{1}{2}\right)^7\left(\frac{1}{2}\right)^3 = \frac{15}{128}$$

Thus,

$$P(4 \leq X \leq 7) = \frac{105}{512} + \frac{63}{256} + \frac{105}{512} + \frac{15}{128} = 0.7734$$

Second method: (using Normal approximation)

$np = nq = 5$. Hence Normal approximation is permissible.

$$\mu = np = 5, \sigma = \sqrt{npq} = \sqrt{10 \times \frac{1}{2} \times \frac{1}{2}} = 1.58$$

Treating the data as continuous, 4 to 7 heads can be considered as 3.5 to 7.5.

3.5 in standard units $= \dfrac{3.5 - 5}{1.58} = -0.95$

7.5 in standard units $= \dfrac{7.5 - 5}{1.58} = 1.58$

The desired probability P is then given by the shaded area in Fig. 3.10.5.1. Hence,

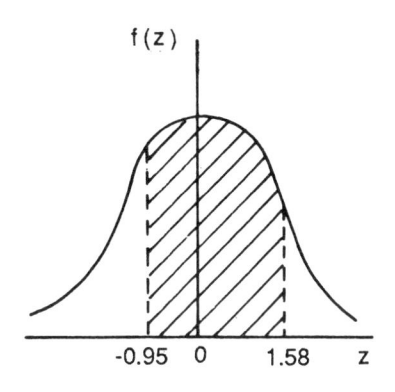

Fig. 3.10.5.1 Standard Normal Curve for Example 3.10.5.1

$$P = \text{area between} -0.95 \text{ and } 1.58$$
$$= (\text{area between} -0.95 \text{ and } 0) + (\text{area between } 0 \text{ and } 1.58)$$
$$= 0.3289 + 0.4429 = 0.7718$$

The percentage error using Normal approximation is just 0.21%.

### 3.10.6 Normal Approximation to Poisson Distribution

As Binomial distribution has relationship with both Poisson and Normal distributions, one would expect that there should be some relationship between Poisson and Normal distributions. In fact it is found to be so. It has been seen that the Poisson distribution approaches Normal distribution as $\lambda \to \infty$.

### 3.10.7 Central Limit Theorem

We have seen that both Binomial and Poisson distributions approach Normal distribution as the limiting case. The Central Limit Theorem indicates that the probability density of a sum of $N$ independent random variables tends to approach a normal density as the number $N$ increases. The mean and variance of this normal density are the sums of mean and variance of $N$ independent random variables. For example, the electrical noise in communication system is due to a large number of randomly moving charged particles. Hence, according to the Central Limit Theorem, the instantaneous value of noise will have a normal distribution, a fact that has been established experimentally.

## 3.11 UNIFORM AND OTHER DISTRIBUTIONS

A random variable $X$ is said to have a *uniform distribution* in the region $a \leq x \leq b$ if its density function is

$$f(x) = \begin{cases} \dfrac{1}{b-a} & a \leq x \leq b \\ 0 & \text{otherwise} \end{cases} \tag{3.11.1}$$

A uniform distribution is shown in Fig. 3.11.1.

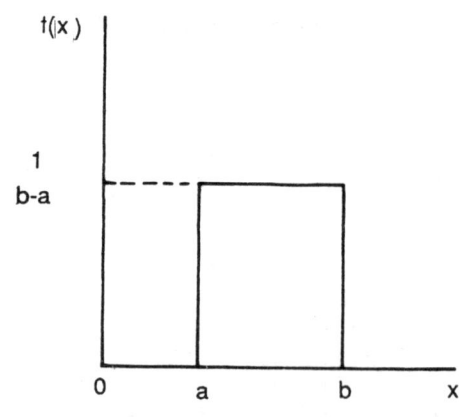

Fig. 3.11.1 Uniform Distribution

The properties are

$$\text{mean} = \frac{1}{2}(a+b) \qquad \text{variance} = \frac{1}{12}(b-a)^2$$

There are several other distributions, some of which are given in Table 3.11.1

Table 3.11.1

| Distribution | Density function | Properties |
|---|---|---|
| Cauchy | $f(x) = \dfrac{a}{\pi(x^2 + a^2)}$, $a > 0$, $-\infty < x < \infty$ | $\mu = 0$, variance does not exist |
| Rayleigh | $f(x) = \begin{cases} \dfrac{x}{a^2} e^{-r^2/(2a^2)} & 0 \le r \le \infty \\ 0 & a > 0 \end{cases}$ | Attains maximum at $r = a$ <br><br> $\mu = a\sqrt{\dfrac{\pi}{2}}, \; \sigma^2 = 2a^2$ |
| Gamma | $f(x) = \begin{cases} \dfrac{x^{\alpha-1} e^{x/\beta}}{\beta^\alpha \, \gamma(\alpha)} & x > 0 \; (\alpha, \beta > 0) \\ 0 & x \le 0 \end{cases}$ <br> $\gamma(\alpha)$ is gamma function | $\mu = \alpha\beta, \; \sigma^2 = \alpha\beta^2$ |
| Beta | $f(x) = \begin{cases} \dfrac{x^{\alpha-1}(1-x)^{\beta-1}}{B(\alpha, \beta)} & 0 < x < 1 \; (\alpha, \beta > 0) \\ 0 & \text{otherwise} \end{cases}$ <br> $B(\alpha, \beta)$ is beta function | $\mu = \dfrac{\alpha}{\alpha + \beta}$ <br><br> $\sigma^2 = \dfrac{\alpha\beta}{(\alpha + \beta)^2 (\alpha + \beta + 1)}$ |

## 3.12 RANDOM PROCESSES

Let us consider an experiment of measuring the temperature of a room. Let there be a collection of thermometers. Each thermometer reading is a random variable which can take on any value from the sample space $S$. Also, at different times the readings of thermometers may be different. Thus the room temperature is a function of both the sample space and the time. In this example, we have extended the concept of random variable by taking into consideration the time dimension. Here we assign a time function $x(t,s)$ to every outcome $S$. There will be a family of all such functions. This family of functions $X(t,S)$ is known as *random process* or *stochastic process*. A random process $X(t,S)$ represents an ensemble or a set or a family of time functions where $t$ and $S$ are variables. In place of $x(t,s)$ and $X(t,S)$, the short notations $x(t)$ and $X(t)$ are often used.

Figure 3.12.1 shows a few members of the ensemble. [$x_1(t)$ is the reading of first thermometer, $x_2(t)$ is the reading of second thermometer and so on.] Each member is also known as *sample function* or *ensemble member* or *realization of the process*. A random process represents single time functions when $t$ is variable and $s$ is fixed. $x_1(t)$ and $x_2(t)$ are examples of single time functions.

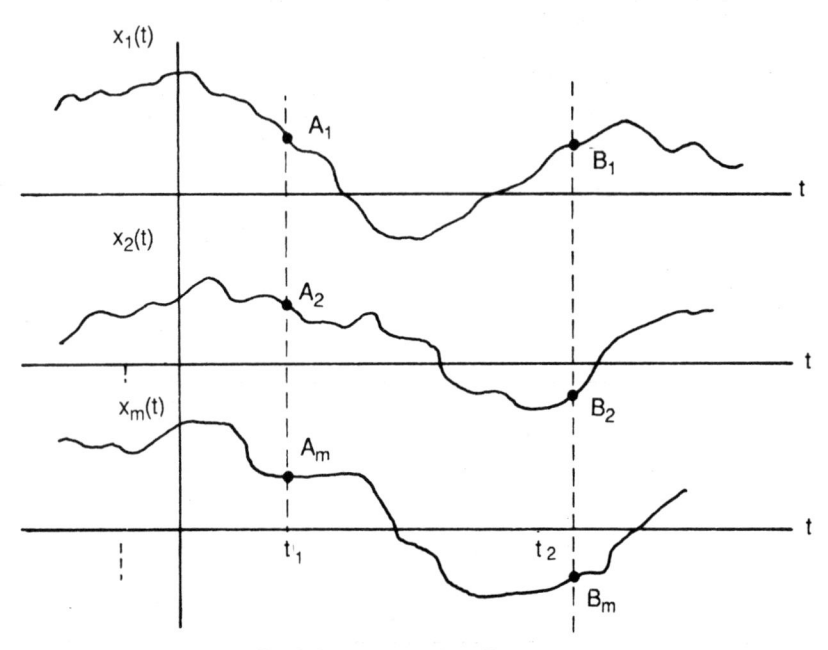

Fig.3.12.1 A Random Process

To determine the statistics of the room temperature, say mean value, we may follow one of the following two procedures:

1. We may fix t to some value, say $t_1$. The result is a random variable $X(t_1, S) = X(t_1) = [A_1 \; A_2 \cdots A_m]$. The mean value of $X(t_1), E[X(t_1)]$, can now be calculated. It is known as *ensemble average*. It may be noted that ensemble average is a function of time. There is an ensemble average corresponding to each time. Thus at time $t_2$, we have

$$X(t_2, s) = X(t_2) = [B_1 \; B_2 \cdots B_m]$$

The ensemble average corresponding to time $t_2 = E[X(t_2)]$, can also be found out. Similarly, ensemble average corresponding to any time can be found out.

2. We may consider a sample function, say $x_1(t)$ over the entire time scale. Then the mean value of $x_1(t)$ is defined as

$$< x_1(t) > \; = \; \lim_{T \to \infty} \frac{1}{2T} \int_{-T}^{T} x_1(t) dt$$

Similarly, we can find mean values of other sample functions. The expected value of all mean values is known as *time average* and is given as

$$<X(t)> = E[<x(t)>]$$

A random process for which mean values of all sample functions are same is known as a *regular random process*. In this case,

$$<x_1(t)> = <x_2(t)> = \cdots = <x(t)> = <X(t)>$$

For some processes, ensembles average is independent to time; i.e.,

$$E[X(t_1)] = E[X(t_2)] = \cdots = E[X(t)]$$

Such processes are known as *stationary processes in restricted sense*. (Here it is restricted to mean. It may also be restricted to second moment, third moment etc.) If all the statistical properties of a random process are independent of time, then it is known as *stationary process in strict sense*. When we say *stationary process*, then it is meant that the process is stationary in strict sense.

When an ensemble average is equal to the time average, then the process is known as *ergodic process in restricted sense*. When all statistical ensemble properties are equal to statistical time properties, then the process is known as *ergodic process in strict sense*. When we say *ergodic process*, then it is meant that the process is ergodic in strict sense.

It may be noted that ergodic process is a subset of a stationary process, i.e., if a process is ergodic, then it is also stationary, but the reverse is not necessarily true. The process which is not stationery is known as *non-stationery process*. It is obvious that the random process of Fig. 3.11.1 (room temperature) is non-stationary, because the ensemble averages at different times are not always the same.

## 3.13 MARKOV PROCESSES

Many times a given random variable $x_n$ is statistically dependent upon some finite number of previously occurring random variables. Thus, if

$$fc(x_n / x_{n-1}, x_{n-2} \cdots) = fck(x_n / x_{n-1}, x_{n-2}, \cdots, x_{n-k}) \tag{3.13.1}$$

then we say that $\{x_n\}$ is a $k^{\text{th}}$ order *Markov Process*. Here the occurrence of $x_n$ is conditioned on $K$ previously occurring random variables $x_{n-1}, x_{n-2}, \cdots, x_{n-k}$.

By putting $K = 1, 2,...$, we get first order Markov process, second-order Markov process,..., respectively. e.g. the first order Markov process is given by

$$f_c(x_n / x_{n-1}, x_{n-2} \cdots) = f_{c1}(x_n / x_{n-1}) \tag{3.13.2}$$

and the second order Markov process is given by

$$f_c(x_n / x_{n-1}, x_{n-2} \cdots) = f_{c2}(x_n / x_{n-1}, x_{n-2}) \tag{3.13.3}$$

A particular class of Markov processes is discrete time, discrete valued Markov processes, where the random variables are discrete and they can assume only a finite set of possible values. Such Markov processes are also known as *Markov chains*. The order of Markov chain depends upon the number of previously occurring random variables which condition the random variable $x_n$.

Let us consider the first order Markov chain where the outcome at any time $t_n$ depends only on the outcome at the immediately proceeding time $t_{n-1}$. Thus, for each pair $(x_{n-1}, x_n)$, the transition probability can be defined as the conditional probability

$$P_{ij} = P(x_n = \psi_j / x_{n-1} = \psi_i) = P(\psi_j / \psi_i) \qquad (3.13.4)$$

where $\psi_j$ and $\psi_i$ are the outcomes of the random variables $x_n$ and $x_{n-1}$ respectively. The transition matrix $T$ is defined as a $q \times q$ square matrix

$$T = [P_{ij}] = \begin{bmatrix} P_{11} & P_{12} & \cdots & P_{1q} \\ P_{21} & P_{22} & \cdots & P_{2q} \\ \cdots & \cdots & \cdots & \cdots \\ P_{q1} & P_{q2} & \cdots & P_{qq} \end{bmatrix} \qquad (3.13.5)$$

where $q$ is the number of possible outcomes of $x_i$. The conditions to be satisfied for the matrix given in Eq. 3.13.5 are

(i) $$P_{ij} \geq 0 \quad i, j = 1, 2, \cdots, q \qquad (3.13.6)$$

(ii) $$\sum_{j=1}^{q} P_{ij} = 1 \quad i = 1, 2 \cdots, q \qquad (3.13.7)$$

Matrices whose elements satisfy Eqs 3.13.6 and 3.13.7 are also known as *Stochastic Matrices* or *Markov Matrices*. In addition if the following condition is also satisfied

(iii) $$\sum_{i=1}^{q} P_{ij} = 1 \quad j = 1, 2, \cdots, q \qquad (3.13.8)$$

then the matrix is known as *doubly stochastic matrix*.

## Interpretations of the above Conditions

The stochastic matrix must be a square matrix. All the elements in it must lie between 0 and 1. The summation of all rows must be unity. In addition, if summation of all columns is zero, then it is a doubly stochastic matrix.

Let there be a column vector known as state distribution vector $\pi^{(n)}$, where

$$\pi^{(n)} = \begin{bmatrix} P_1^{(n)} & P_2^{(n)} & \cdots & P_q^{(n)} \end{bmatrix} \qquad (3.13.9)$$

which gives the probabilities of all possible outcomes at time $t_n$. The initial state distribution vector is given by

$$\pi^{(0)} = \pi = \begin{bmatrix} P_1^{(0)} & P_2^{(0)} & \cdots & P_q^{(0)} \end{bmatrix} \qquad (3.13.10)$$

A Markov chain is completely defined by the transition matrix and the state distribution vectors.

## PROBLEMS

3.1 Prove both forms of De Morgan's theorem by using Venn diagrams.
3.2 A box contains 2 White, 3 Black and 4 Red balls. Three balls are drawn in succession. What is the probability that they are of

(i) different colours?
(ii) same colour?
3.3 Three cards are drawn from an ordinary deck of 52 playing cards. What is the probability that
(i) all three cards are even numbered of black colour?
(ii) one of them is an ace, and other two are of same colour as the ace?
3.4 Two fair dice are thrown. $X$ denotes the total of the numbers on the two dice. Find the probability function
and distribution function of $X$.
3.5 A function is given by

$$f(x) = \begin{cases} C(x^2 + 1) & 1 < x < 3 \\ 0 & \text{otherwise} \end{cases}$$

(a) Find the value of $C$ for $f(x)$ to be a density function,
(b) Find the probability that $x$ lies between 1 and 2.

3.6 The joint probability function of two discrete random variables $X$ and $Y$ is given by

$$f(x, y) = Cx^2y \quad \begin{aligned} x &= 1, 2 \\ y &= 0, 1, 2 \end{aligned}$$
$$= 0 \quad \text{otherwise}$$

Find
(a) the value of $C$,
(b) $P(X > 1, Y \leq 1)$ and
(c) Marginal probability functions of $X$ and $Y$.

3.7 The joint probability function of two continuous random variables is given by

$$f(x, y) = \begin{cases} C(2x + 3y) & 1 < x < 3 \quad 0 < y < 2 \\ 0 & \text{otherwise} \end{cases}$$

Find
(a) the value of $C$,
(b) $P(X < 2, Y > 1)$ and,
(c) $P[(X + Y) > 3]$

3.8 For the random variable of Prob. 3.6, find
(a) $E(X^2), E(Y)$
(b) $E(3X + 2Y^2)$

3.9 For the random variable of Prob. 3.7, find
(a) $E(X^2), E(Y)$
(b) $E(3X + 2Y^2)$

3.10 Prove that $E(cX) = cE(X)$ and $E(X + Y) = E(X) + E(Y)$.

3.11 A fair coin is tossed five times. What is the probability of getting two heads? Also find the mean and standard deviation for the number of heads.

3.12 Only 5% of the total students of a college take part in sports. Special marks are awarded to these students. What is the probability that more than two students will get these special marks out of 10 students chosen at random?

3.13 A fair coin is tossed fifteen times. Find the probability of a head appearing more than six times, and less than, or exactly ten times.

# FOUR
# NOISE

## INTRODUCTION

Undesired electrical signals which are introduced with a message signal during the transmission, or processing of the latter, are called *noise*. Noise, thus, is an unwanted signal that corrupts a desired message signal. In general, noise may be predictable, or unpredictable (random), in nature. The predictable noise can be estimated, and eliminated, by proper engineering design. Some examples of such noise are: power supply hum, spurious oscillation in feedback amplifiers, ignition radiation pick-up, radiation pickup generated by electrical appliances, and fluorescent lighting. The predictable noise, generally is man-made and can be reduced or eliminated.

Unpredictable noise varies randomly with time and, as such, we have no control over this noise. Identification of the message signal at the receiver depends upon the amount of noise accompanied by the message during the process of communication. In the absence of noise, identification of the message signal at the receiver is perfect. The presence of noise complicates the systems of communication. The amount of noise power present in the received signal decides the minimum power level of the desired message signal at the transmitter. The term *noise* is normally used to refer the unpredictable, or random noise.

## 4.1 SOURCES OF NOISE

There are various sources of random noise. They are broadly classified as:
 (a) External Noise, and
 (b) Internal Noise.

The external noise is created outside the circuit, and includes:

(i) Erratic Natural Disturbances: This type of noise does not occur regularly. It is caused by lightning, electrical storms, and intergalactic or other atmospheric disturbances. This noise is unpredictable in nature, and is known as *atmospheric* or *static* noise. Besides this, the extraterrestrial noise is also created by erratic natural disturbances. The atmospheric noise is less severe above 30 MHz.

(ii) Man-made Noise: This noise is because of the undesired pick-ups from electrical appliances, such as motors, switch gears, automobile and aircraft ignitions, etc. This type of noise is under human control, and can be eliminated by removing the source of the noise. This noise is effective in frequency range of 1MHz-500 MHz.

The *internal noise* is created by the active and passive components present within the communication circuit itself. This type of noise is also known as *fluctuation noise*. It is caused by spontaneous fluctuations in the physical system. Examples of such fluctuations are; (a) thermal motion of the free electrons inside a resistor, known as *Brownian motion*, which is random in nature; (b) the random emission of electrons in vacuum tubes; and (c) the random diffusion of electrons and holes in a semiconductor.

The fluctuation noise is very significant, and will be treated in greater detail. The two important types of fluctuation noise are (i) *shot noise*, and (ii) *thermal noise*.

## 4.2 SHOT NOISE

Shot noise appears in active devices due to the random behaviour of charge carriers (electrons and holes). In electron tubes, shot noise is generated due to the random emission of electrons from cathodes; in semiconductor devices, it is caused due to the random diffusion of minority carriers, or random generation and recombination, of electron-hole pairs.

Current in electron devices (tubes or solid state) flows in the form of discrete pulses, every time a charge carrier moves from one point to the other (e.g., cathode to plate). Therefore, although the current appears to be continuous, it is still a discrete phenomena. The nature of current variation with time is shown in Fig. 4.2.1.

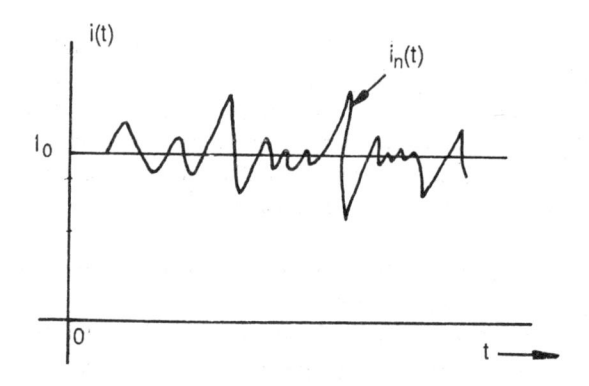

Fig. 4.2.1

The current fluctuates about a mean value $I_o$. This current $i_n(t)$ which wiggles around the mean value is known as *shot noise*. The wiggling nature of the current is not visualized by normal instruments, and normally we think that the current is a constant equal to $I_o$. The wiggling nature of the current can be observed in a fast sweep oscilloscope.

The total current $i(t)$ may be expressed as

$$i(t) = I_o + i_n(t) \qquad (4.2.1)$$

where $I_o$ is the constant (mean), and $i_n(t)$ is the fluctuating (noise) current.

### Power Density Spectrum of Shot Noise in Diodes

The time varying component $i_n(t)$ of the current $i(t)$ specified by Eq. 4.2.1 is random in nature, and it cannot be expressed as a function of time, i.e., it is an *indeterministic* function. However, this indeterministic stationary random function $i_n(t)$ can be specified by its power density spectrum.

The number of electrons contributing to the random stationary current $i_n(t)$ are large. Assuming that the electrons do not interact with each other during their movement, or emission (e.g., temperature limited diode current), the process may be considered statistically independent. According to central limit theorem, such a process has a Gaussian distribution. Hence, shot-noise is Gaussian-distributed with a zero mean.

The total diode current may be taken as the sum of the current pulses, each pulse being formed by the transit of an electron from cathode to anode. It can be seen that for all practical purposes the power density spectrum of the statistically independent non-interacting random noise current $i_n(t)$ is given by (Prob. 4.1a)

$$S_i(\omega) = qI_o \qquad (4.2.2)$$

where $q$ is the electronic charge ($q = 1.59 \times 10^{-19}$ coulombs), and $I_o$ is the mean value of the current in amperes. The power density spectrum in Eq. 4.2.2 is frequency independent. This type of frequency independence is only up to a frequency range decided by the transit time of an electron to reach from anode to cathode. Beyond this frequency range, the power density varies with frequency as shown in Fig. 4.2.2a. The transit time of an electron, in a diode, depends on anode voltage V and is given as

$$\tau = 3.36 \times \frac{d}{\sqrt{V}} \mu \sec.$$

where $d$ is spacing between anode and cathode.

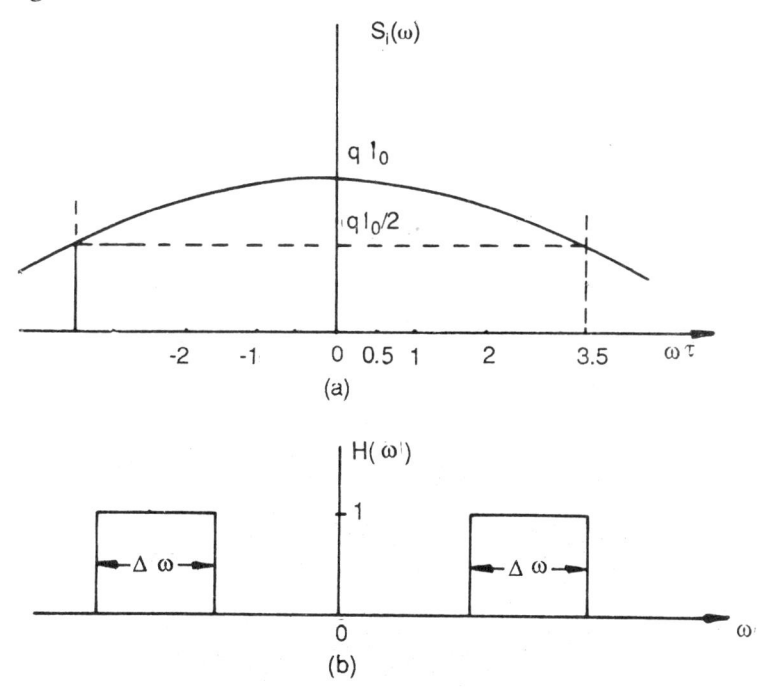

Fig. 4.2.2 Shot Noise (a) Power Density Spectrum (b) Bandwidth of Measuring System

For instance, in a diode with $d = 2$mm and $V = 40$ volts, we have $\tau \cong 10^{-3} \mu$ sec. In Fig. 4.2.2 a, the power density curve may be considered flat close to the origin, i.e., $|\omega\tau| \leq 0.5$. Therefore $S_i(\omega)$ can be considered constant up to $|\omega\tau| = 0.5$. For $\tau = 10^{-3} \mu$ sec, the maximum frequency up to which power density $S_i(\omega)$ remains constant is given by

$$\omega = 0.5 \times 10^9 = 5 \times 10^8 \text{ rad}/s$$

This is equivalent to a linear frequency $f = \dfrac{\omega}{2\pi} \cong 80$ MHz. Therefore for all practical purposes, the $S_i(\omega)$ may be considered to be frequency-independent below 100 MHz. This frequency limit covers the frequency range of most of the practical communication systems, except UHF and microwaves.

## Schottky Formula

Power density spectrum of shot noise current $i_n(t)$ for statistically independent process is given by Eq. 4.2.2 The mean square value (average power) of the randomly fluctuating noise current will be

$$\overline{i_n^2} = 2qI_o(\Delta f) \tag{4.2.3}$$

where $2\Delta f$ is the bandwidth (including negative frequency) of the measuring system involved, as shown in Fig. 4.2.2b (of course below 100 MHz). Equation 4.2.3 is known as Schottky formula.

Equation 4.2.2 has been developed assuming that electrons contributing the diode current do not interact with each other, as in the case of a temperature limited region of a thermionic diode. There may be cases where electrons contributing the diode current interact with each other as in a space-charge limited region of a thermionic diode. In such cases, power density spectrum is given by

$$S_i(\omega) = \alpha q I_o \tag{4.2.4}$$

where $\alpha$ is a smoothing constant, and ranges between 0.01 to 1. The space-charge has a smoothing effect and $\alpha$ depends on the tendency of interacting electrons to smooth out and yield a constant current. The more is the smoothing effect, the greater is the value of $\alpha$. It is expressed as

$$\alpha = 1.288\, kT_c g_d / qI_o$$

where $T_c$ is cathode temperature in degrees Kelvin; $k$ is the Boltzmann constant ($k = 1.38 \times 10^{-23}$ Joules per degree Kelvin), and $g_d$ is the dynamic conductance of the diode. Substituting this value of $\alpha$, Eq. 4.2.4 becomes

$$S_i(\omega) = 1.288\, kT_c g_d \tag{4.2.5}$$

An equivalent circuit of a noisy diode in terms of a noiseless diode is shown in Fig. 4.2.3.

Noisy diode      Noiseless diode

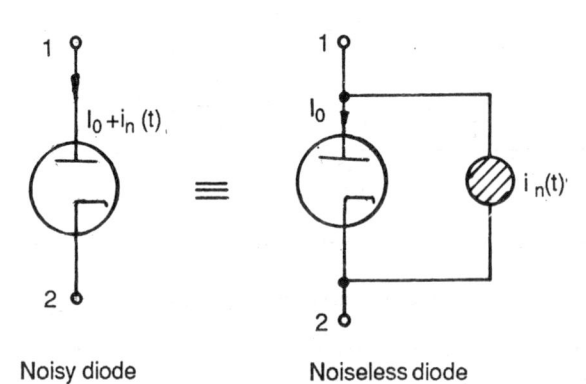

Fig. 4.2.3

## Shot Noise in Triodes

The power density spectrum of the noise current in triodes is similar to the diode and is given as

$$S_i(\omega) = 2kT(2.5 g_m) \tag{4.2.6}$$

where $g_m$ is the dynamic transconductance of the triode. A noisy triode is represented by a noiseless triode and a noise current source, as shown in Fig. 4.2.4 (a). The noise current in plate circuit is caused by a noise

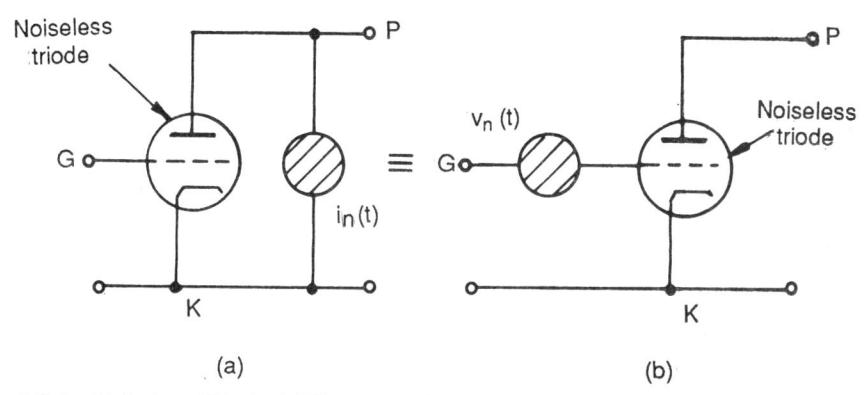

Fig. 4.2.4 Noiseless Triode (a) With Noise Current Source; (b) With Noise Voltage Source

voltage $v_n(t) = i_n(t) / g_m$ in grid circuit as in Fig. 4.2.4 (b). Since power is proportional to square of the voltage, or current, the relation between power density spectrum of voltage source, and that of current source is given by

$$S_v(\omega) = \frac{1}{g_m^2} S_i(\omega) = \frac{1}{g_m^2} 2kT (2.5 g_m) = 2kT \left( \frac{2.5}{g_m} \right)$$
$$= 2kT R_{eq} \tag{4.2.7}$$

where

$$R_{eq} = \frac{2.5}{g_m} \tag{4.2.8}$$

The expression for power density spectrum of noise voltage source in Eq. 4.2.7 has been developed in terms of an equivalent resistance $R_{eq}$. This has been done intentionally to establish an analogy between a shot noise and a resistor noise to be discussed in Sec. 4.3

## Additional Sources of Noise in Electron Tubes and Solid State Devices

### (i) Partition Noise

The shot-noise in multigrid tubes (tetrode, pentode etc.) is the same as in a triode, and the expression for $S_v(\omega)$ given by Eq. 4.2.7 holds true. But the multigrid tubes contain more than one grid and the partition of electrons emitted from cathode among the various grids is random in nature. This gives rise to another source of noise in multigrid tubes, called *partition noise*. This noise may also may be included in shot-noise expression by modifying the equivalent resistance as given below for pentodes:

$$R_{eq}^1 = 1 + \frac{7.7 I_s}{g_m^1} R_{eq}$$

where $g_m^1$ is the transconductance of a Pentode tube, $I_s$ the screen grid current, and $R_{eq}$ is the equivalent resistance of a pentode operated as a triode. Similarly, partition noise arises in multielectrode semiconductor devices like transistors, FET etc., due to random partition of charge carriers.

(ii) *Flicker Noise*

Noise of this type arises due to imperfections in cathode surface of electron tubes and surfaces around the junctions of semiconductor devices. This noise can be reduced by proper processing of the surfaces to avoid slow varying conditions at the surfaces. The power density spectrum of flicker noise is inversely proportional to frequency i.e., $S(\omega) \alpha \dfrac{1}{f}$.

Hence, this noise becomes significant at very low frequencies, generally below a few kHz.

## 4.3 RESISTOR NOISE

The noise arising due to random motion of free charged particles (usually electrons) in a conducting media, such as a resistor, is called resistor noise. This noise is also known as *Johnson noise*, after J.B. Johnson who, investigated this type of noise in conductors. The random agitation is a universal phenomenon at atomic levels and is caused by the energy supplied through flow of heat. The intensity of random motion is proportional to thermal (heat) energy supplied (i.e., temperature), and is zero at a temperature of absolute zero. This noise is also known as *thermal noise*. The path of the electron motion is random because of their collisions with lattice structure. The net motion of all the electrons gives rise to an electric current to flow through the resistor, causing the noise.

### Power Density Spectrum of Resistor Noise

The free electrons contributing to resistor noise are large in number. If their random motion is assumed to be statistically independent, then the central limit theorem predicts the resistor noise to be, Gaussian, distributed with a zero mean. It can be shown that the power density spectrum of the current contributing the thermal noise is given by

$$S_i(\omega) = \dfrac{2kTG}{1 + \left(\dfrac{\omega}{\alpha}\right)^2} \qquad (4.3.1)$$

where $T$ is ambient temperature in degree Kelvin, $G$ is the conductance of the resistor in *mhos*, $k$ is the Boltzmann constant, and $\alpha$ is the average number of collisions per second per electron.

The variation of power density spectrum with frequency is shown in Fig. 4.3.1

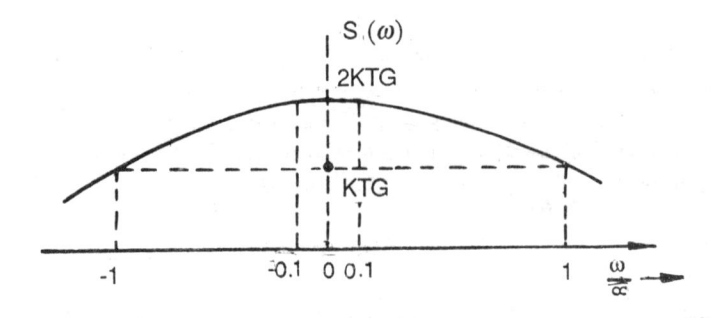

Fig. 4.3.1 Power Density Spectrum of the Resistor Noise Current

It is obvious from the figure that the spectrum may be considered to be flat for $\frac{\omega}{\alpha} \leq 0.1$. The power density spectrum $S_i(\omega)$ for this range of frequency is nearly constant and is given by

$$S_i(\omega) = 2kTG \qquad (4.3.2)$$

The value of $\alpha$ is of the order of $10^{14}$ and hence the frequency range corresponding to $\frac{\omega}{\alpha} < 0.1$ is of the order of $10^{13}$ Hz. Therefore, the frequency independent expression of $S_i(\omega)$ given by Eq. 4.3.2 holds up to a frequency range of $10^{13}$ Hz. This range covers almost all the practical applications in communication systems. Hence, for all practical purposes, the power density spectrum $S_i(\omega)$ is considered to be independent of frequency, and is given by Eq. 4.3.2.

## Equivalent Circuit of a Noise Resistor

A noisy resistor $R$ can be represented by a noiseless conductance $G$ in parallel with a thermal noise current source $i_n(t)$ as shown in Fig. 4.3.2 (a). The Thevenin equivalent of Fig. 4.3.2a is shown in Fig. 4.3.2b, which

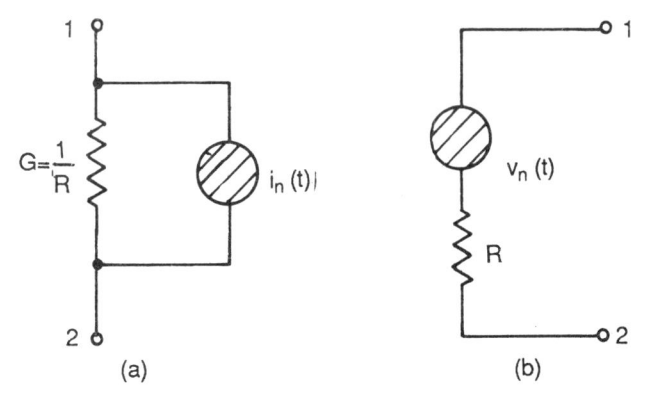

Fig. 4.3.2 Noiseless Resistor (a) With Noise Current Source (b) With Noise Voltage Source

represents the noiseless resistor $R$ in series with a thermal noise voltage source $v_n(t)$. Current $i_n(t)$ and voltage $v_n(t)$ are related as

$$v_n(t) = R i_n(t) \qquad (4.3.3)$$

Now, since the power density spectrum is a function of the square of voltage or current, the relation between the power density spectrum $S_i(\omega)$ of the current source $i_n(t)$ and $S_v(\omega)$ of the voltage source $v_n(t)$, is given as

$$S_v(\omega) = R^2 S_i(\omega) = R^2(2kTG) = R^2\left(2kT\frac{1}{R}\right)$$

$$= 2kTR \qquad (4.3.4)$$

## Power of Thermal Noise Voltage

The power density spectrum $S_v(\omega)$ of a thermal noise voltage $v_n(t)$ is independent of frequency. Since the power density spectrum is the power per unit bandwidth, noise power increases with an increase in bandwidth, and becomes infinite as the bandwidth tends to infinity. This is obvious from the relation for noise power $P_n$ given by

$$P_n = \frac{1}{2\pi} \int_{-\infty}^{\infty} S_v(\omega) d\omega = \frac{1}{2\pi} \int_{-\infty}^{\infty} 2kTR \, d\omega$$

The integral becomes infinite when integrated over an infinite bandwidth.

However, for a finite bandwidth of $2\Delta f (-\Delta f$ to $\Delta f)$ the noise power (m.s. value) is given by

$$P_n = S_v(\omega).2\Delta f = 4kTR(\Delta f) \qquad (4.3.5a)$$

Since, the power of a signal is the same as its mean square value,

$$P_n = \overline{v_n^2} = 4kTR(\Delta f)$$

The corresponding rms value is given by

$$v_n = \sqrt{4kTR(\Delta f)} \qquad (4.3.5b)$$

Note that $\Delta f$ is one sided (positive) bandwidth. The thermal noise power contribution is limited only by the *bandwidth* of the circuit.

## White Noise

White light contains all colour frequencies. In the same way, white noise, too, contains all frequencies in equal amount. The power density spectrum of a white noise is independent of frequency; which means it contains all the frequency components in equal amount. When the probability of occurrence of a white noise *level* is specified by a Gaussian distribution function, it is known as *White Gaussian noise*. Equation 4.2.7 already shows that the power density spectrum of that noise is independent of the operating frequencies. Hence, shot noise and thermal noise may be considered as white Gaussian noise for all practical purposes. The power density of white noise is

$$S_w(\omega) = N_0/2$$

A function containing all the frequency components in equal amount is known as Delta function. Can a white noise be specified by a Delta function? We cannot, because the noise signal is not specified as a function of time, as it is a non-deterministic waveform. Actually, the main difference between a white noise and the Delta function is that the white noise is represented by power density spectrum in which phase spectrum has no significance, whereas the Delta function is specified by spectral density function which has an amplitude spectrum associated with a phase spectrum. In other words, white noise contains all frequency components, but the phase relationship of the components is random, whereas the Delta function has all frequency components with equal magnitude, and the *same relative phase*. The inverse Fourier transform of white noise is specified by autocorrelation function wherein phase relationship has no significance. Thus, the auto-correlation function of a white noise is a Delta function. This is shown in Fig. 4.3.3. where it is obvious that Delta function and the power density spectrum of white noise are a Fourier transform pair. White noise has infinite power and is not physically realizable. But, its concept is helpful in convenient mathematical analysis of systems. The autocorrelation is zero for $\tau \neq 0$; i.e., any two samples of white noise are uncorrelated, and also if white noise is Gaussian, the two samples are statistically independent.

## Generalized Nyquist Theorem

By this theorem the power density spectrum of a single resistor is extended to a generalized linear bilateral passive network, usually containing resistance $R$, capacitance $C$, and inductance $L$. According to this

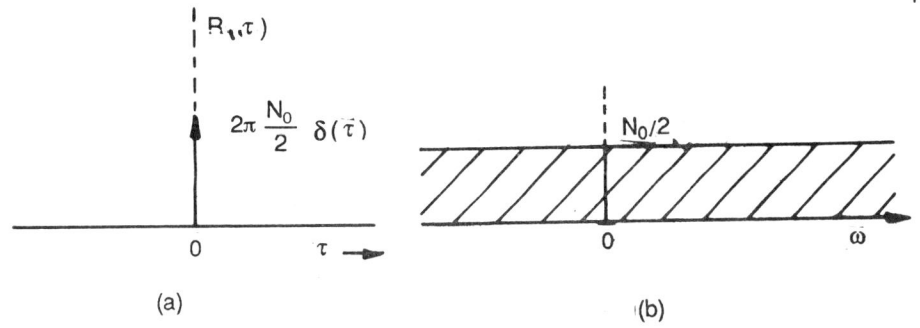

Fig. 4.3.3 (a) Autocorrelation; (b) Power Spectrum of White Noise

theorem, a linear bilateral passive network (*RLC* network) is replaced by its Thevenin, or Nortan equivalent. The resistor noise of the network is evaluated by considering the equivalent resistance (in Thevenin's equivalent), or equivalent conductance (in Norton's equivalent) as a source of thermal noise.

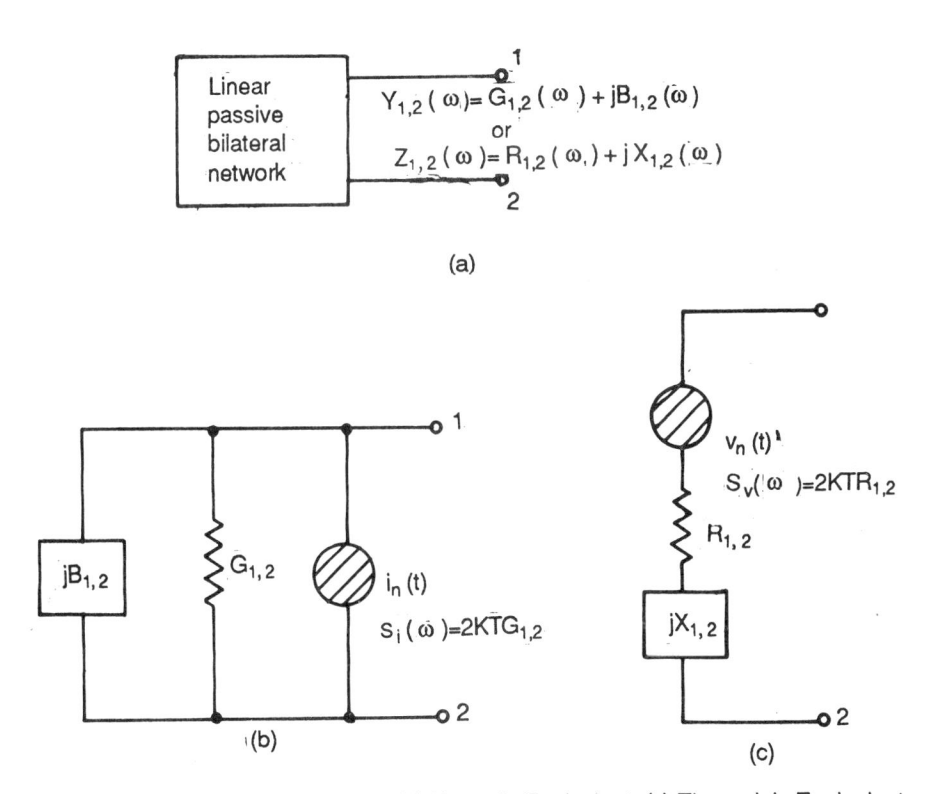

Fig. 4.3.4 (a) An RLC Network (b) Norton's Equivalent; (c) Thevenin's Equivalent

Consider a linear passive bilateral network shown in Fig. 4.3.4(a). The Norton's and Thevenin's equivalent across the terminals 1 and 2 are shown in Figs 4.3.4 (b) and (c), respectively. The Norton's equivalent in Fig. 4.3.4(b) is representing the network in terms of conductance $G_{1,2}(\omega)$ and susceptance $B_{1,2}(\omega)$ forming an equivalent admittance of the network given by

$$Y_{1,2}(\omega) = G_{1,2}(\omega) + jB_{1,2}(\omega)$$

The source of the thermal noise is conductance $G_{1,2}(\omega)$ and hence the network can be represented by a noiseless admittance $Y_{1,2}(\omega)$ in parallel with a noise current source with power density spectrum given by

$$S_i(\omega) = 2kT\, G_{1,2}(\omega) \qquad (4.3.6)$$

Similarly, the Thevenin's equivalent shown in Fig. 4.3.4c is represented by a noiseless impedance $Z_{1,2}(\omega) = R_{1,2}(\omega) + jX_{1,2}(\omega)$ in series with a noise voltage source $v_n(t)$, with the power density spectrum given by

$$S_v(\omega) = 2kT\, R_{1,2}(\omega) \qquad (4.3.7)$$

Here, the source of resistor noise is $R_{1,2}(\omega)$.

*Note*:

(1) The theorem is applicable only to passive networks, and not the networks containing active elements, such as tubes and transistors.

(ii) In general $R_{1,2}$ and $G_{1,2}$ are frequency dependent and hence the power density spectrum is also frequency-dependent, unlike the white noise.

### Example 4.3.1

Find the power density spectrum of the thermal noise voltage across terminals 1,2 of a passive network shown in Fig. 4.3.5(a)

*Solution*

The admittance of the network across terminals 1,2 is given by

$$Y_{1,2} = 1 + j2\omega + \frac{1}{1+j\omega} = \frac{2 - 2\omega^2 + 3j\omega}{1 + j\omega}$$

Therefore, the impedance across terminal 1,2 is

$$Z_{1,2} = \frac{1}{Y_{1,2}} = \frac{1 + j\omega}{2(1 - \omega^2) + 3j\omega}$$

By separating the real and the imaginary parts, we get

$$Z_{1,2} = \frac{(2 + \omega^2) - j\omega(1 + 2\omega^2)}{4 + 4\omega^4 + \omega^2}$$

The real part $R_{1,2}$ is

$$R_{1,2} = \mathrm{Re}[Z_{1,2}] = \frac{2 + \omega^2}{4 + \omega^2 + 4\omega^4}$$

This resistance $R_{1,2}$ is the source of noise as shown in Fig. 4.3.5b. The noise density spectrum of $v_n(t)$ is given by

$$S_v(\omega) = 2kT R_{1,2} = \frac{2kT(2 + \omega^2)}{4 + \omega^2 + 4\omega^4}$$

### Example 4.3.2

Evaluate the thermal noise voltage developed across a resistor of 700 ohms. The bandwidth of the measuring

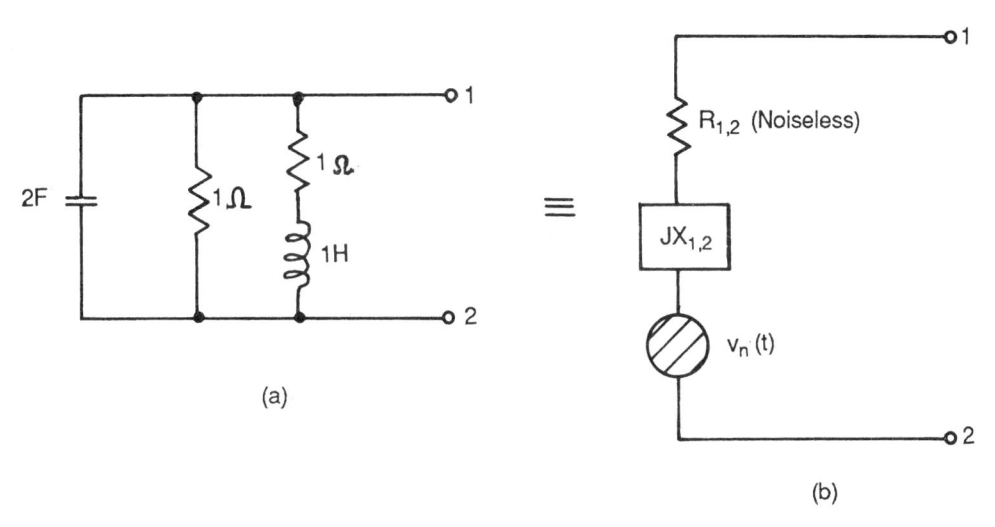

Fig. 4.3.5

instruments is 7 MHz, and the ambient temperature is 27° C.

*Solution*
The mean square value of the noise voltage is given by

$$\overline{v^2_n} = 4kTR(\Delta f)$$

Therefore the rms value of the noise voltage is given by

$$V_{rms} = \sqrt{\overline{v^2_n}} = \sqrt{4kT(\Delta f).R}$$

$$= \sqrt{4 \times 1.38 \times 10^{-23} \times 300 \times 7 \times 10^6 \times 700}$$

$$= 9\mu \text{ volts}$$

## 4.4 CALCULATION OF NOISE IN LINEAR SYSTEMS

We will now discuss the calculation of noise power at the output of a linear system. We will consider first the cases with a single noise source, and then study multiple noise sources.

### (i) Single Noise Source

Consider a linear system shown in Fig. 4.4.1. The system is excited by a single noise voltage source $v_{ni}(t)$ producing an output $v_{no}(t)$. Let us assume that the system has noiseless elements.

From Eq. 2.4.14 in Chapter 2, it is obvious that the power density spectrum of noise voltage at the input, and output, is related as

$$S_{no}(\omega) = S_{ni}(\omega)|H(\omega)|^2 \qquad (4.4.1)$$

The mean square value of output noise signal (i.e., output noise power $P_o$) can be evaluated by integrating $S_{no}(\omega)$ over the bandwidth under consideration i.e.,

**204** COMMUNICATION SYSTEMS: Analog and Digital

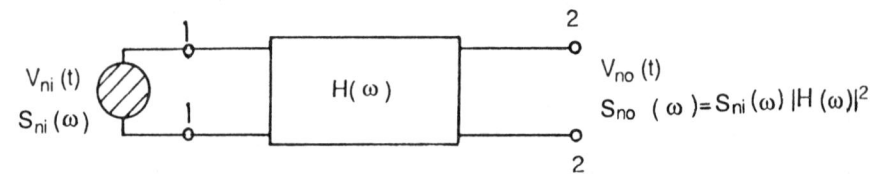

Fig. 4.4.1 Input and Output Noise

$$\overline{v^2_{no}} = P_o = \frac{1}{2\pi}\int_{-\infty}^{\infty} S_{ni}(\omega)|H(\omega)|^2 d\omega \qquad (4.4.2)$$

The rms noise voltage is obtained by taking square root of Eq. 4.4.2.

$$V_{rms} = \sqrt{\overline{v^2_{no}}} = \left[\frac{1}{2\pi}\int_{-\infty}^{\infty} S_{ni}(\omega)|H(\omega)|^2 d\omega\right]^{\frac{1}{2}} \qquad (4.4.3)$$

### Example 4.4.1

Find the root mean square (rms) value of the noise voltage at 27°C developed across the capacitor terminal of the circuit in Fig. 4.4.2a.

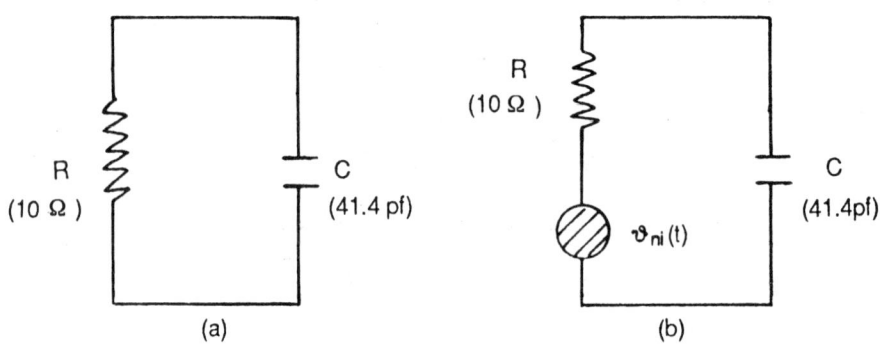

Fig. 4.4.2

*Solution*
The only source of noise in the circuit is resistor $R$, and hence it can be replaced by an input noise voltage source $v_{ni}(t)$ and a noiseless resistor $R$, as shown in Fig. 4.4.2b. The power density spectrum of the input noise voltage source is given by

$$S_{ni}(\omega) = 2kTR$$

The transfer function of the network is

$$H(\omega) = \frac{(1/j\omega C)}{R + \frac{1}{j\omega C}} = \frac{1}{j\omega CR + 1} = \frac{1 - j\omega CR}{1 + \omega^2 C^2 R^2}$$

The magnitude of $H(\omega)$ is

$$|H(\omega)| = \frac{\sqrt{1+\omega^2 C^2 R^2}}{1+\omega^2 C^2 R^2} = \frac{1}{\sqrt{1+\omega^2 C^2 R^2}}$$

The power density spectrum of the output noise voltage $v_{no}(t)$ can be evaluated by using Eq. 4.4.1,

$$S_{no}(\omega) = S_{ni}(\omega)|(\omega)|^2 = \frac{2kTR}{1+\omega^2 C^2 R^2} \qquad (4.4.4)$$

The mean square value of the noise voltage $v_{no}(t)$ is given as

$$\overline{v_{no}^2} = P_o = \frac{2kTR}{\pi}\int_0^\infty \frac{1}{1+\omega^2 C^2 R^2}d\omega = \frac{2kTR}{\pi CR}\tan^{-1}\omega CR \Big|_0^\infty$$

$$= \frac{kT}{C} \qquad (4.4.5)$$

The rms noise voltage is given by

$$V_{rms} = \sqrt{\overline{v_{no}^2}} = \sqrt{\frac{kT}{C}} \qquad (4.4.6)$$

Numerical:- $k = 1.38 \times 10^{-23}$ Joules/°K

$$T = 273 + 27 = 300°K, C = 41.4 pf$$

Therefore,

$$V_{rms} = \sqrt{\frac{1.38 \times 10^{-23} \times 300}{41.4 \times 10^{-12}}} = 10^{-5} \text{ volts} = 10\mu v$$

## (ii) Multiple Noise Source

In communication systems, we may involve the systems with multiple sources of noise, such as resistors and active devices. When all the sources are independent of each other, the mean square value of the response can be obtained by using the principle of superposition.

### *Principle of Superposition of Power Spectra*

This theorem states that for a system involving multiple independent sources of noise, the mean square value of the response is equal to the sum of the mean square values of the responses evaluated by individual sources at a time. Similarly, the power density spectrum of the response can also be evaluated by using the superposition theorem.

Consider a system with two independent input sources $f_1(t)$, and $f_2(t)$, at two different terminals producing a response $y(t)$ as shown in Fig. 4.4.3. Let $f_1(t)$ alone produces a response $y_1(t)$ and $f_2(t)$ alone produces a response $y_2(t)$ to the linear system. Now, if $f_1(t)$ and $f_2(t)$ are independent sources with at least one of them having zero mean value, then according to the principle of superposition of spectra.

$$\overline{y^2(t)} = \overline{y_1^2(t)} + \overline{y_2^2(t)} \qquad (4.4.7)$$

**206** COMMUNICATION SYSTEMS: Analog and Digital

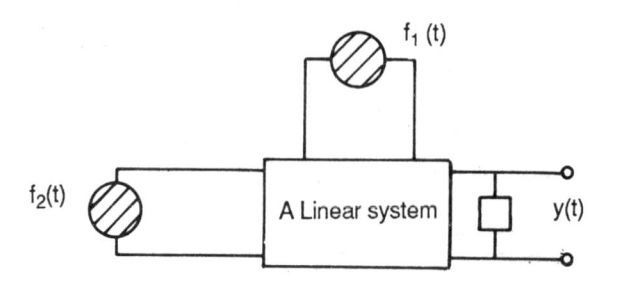

Fig. 4.4.3 Superposition of Spectra

and
$$S_y(\omega) = S_{y_1}(\omega) + S_{y_2}(\omega) \qquad (4.4.8)$$

The result can be extended to more than two sources

$$\overline{y_1^2(t)} = \overline{y_1^2(t)} + \overline{y_2^2(t)} + \overline{y_3^2(t)} \cdots \overline{y_n^2(t)} \qquad (4.4.9a)$$

and

$$S_y(\omega) = S_{y_1}(\omega) + S_{y_2}(\omega) + S_{y_3}(\omega) + \cdots + S_{y_n}(\omega) \qquad (4.4.9b)$$

The theorem can be proved by solving the integrals defining the mean square value and the power density spectrum. (Prob. 4.1b).

Thus, the problem of linear systems can be solved by using two methods; (a) generalized Nyquist theorem (Sec. 4.3) and (b) principle of superposition of spectra. However, in general, the Nyquist theorem proves to be the better solution.

### Example 4.4.2

A single stage amplifier is shown in Fig. 4.4.4(a). Find the number of sources of noise; (b) Develop an expression for the power density spectrum, and the mean square value of the output in terms of the power density spectrum of individual input sources and the corresponding transfer functions.

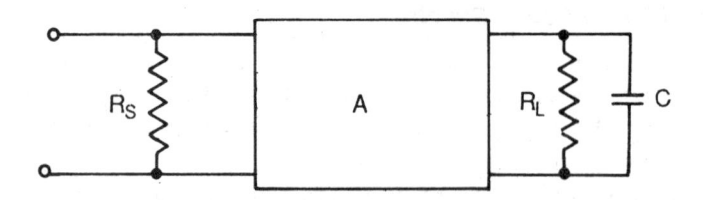

Fig.4.4.4 (a) Amplifier

*Solution*
(a) There are two sources of resistor noise viz., $R_s$, and $R_L$. Besides these, there is an active device (single stage amplifier) producing shot noise. This shot noise can be represented by an analogous resistor noise with equivalent resistance given by

$$R_{eq} = \frac{2.5}{g_m}$$

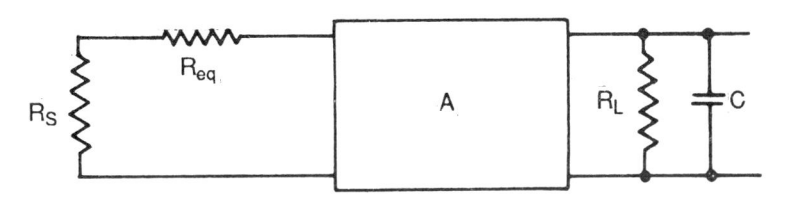

Fig. 4.4.4 (b) Amplifier with Shot Noise Equivalent Resistance, $R_{eq} = \dfrac{2.5}{g_m}$

where $g_m$ is the mutual transconductance of the device used in the amplifier. This is shown in Fig. 4.4.4b. Thus, we have three source of noise

(i) Resistor $R_s$ producing voltage $v_{n1}$ with power density spectrum $S_i(\omega) = 2kTR_s$
(ii) Resistor $R_L$ producing voltage $v_{n2}$ with $S_i(\omega) = 2kTR_L$ and
(iii) Shot noise voltage $v_{n3}$ represented by $R_{eq}$ with power density spectrum $S_i(\omega) = 2kTR_{eq}$.

Let the output voltage produced by the three sources be $v_{no}$ with a power density spectrum $S_{no}(\omega)$. Let $H_1(\omega), H_2(\omega),$ and $H_3(\omega)$ represent the transfer functions relating to output $v_{no}$ with corresponding input noise sources. The power density spectrum at the output considering only $v_{n1}$ as the input is equal to $S_1(\omega)|H_1(\omega)|^2$. Similarly, the power density spectrum at the output, considering $v_{n2}$ and $v_{n3}$ alone at the input respectively are given as $S_2(\omega)|H_2(\omega)|^2$ and $S_3(\omega)|H_3(\omega)|^2$. Therefore, by superposition theorem the power density spectrum of the output voltage $v_{no}$ is the sum of the three spectra i.e.,

$$S_{no}(\omega) = |H_1(\omega)|^2 S_1(\omega) + |H_2(\omega)|^2 S_2(\omega) + |H_3(\omega)|^2 S_3(\omega) \qquad (4.4.10a)$$

The mean square value of the output voltage is given by

$$\overline{v_{no}^2} = \frac{1}{\pi} \int_0^\infty \left[ |H_1(\omega)|^2 S_1(\omega) + |H_2(\omega)|^2 S_2(\omega) + |H_3(\omega)|^2 S_3(\omega) \right] d\omega \qquad (4.4.10b)$$

## Noise in Reactive Circuits

Let us consider the noise generated in a parallel resonant circuit shown in Fig. 4.4.5(a). The inductance $L$ is assumed to be ideal, and any resistance present is included in resistance $R_s$. The only source of noise voltage is resistance $R_s$ with the mean square value given by

$$\overline{v_{ni}^2} = 4kTR_s(\Delta f)$$

The equivalent circuit assuming that $R_s$ is noiseless in series with a noise voltage sources $v_{ni}$ is shown in Fig. 4.4.5 (b). We have to find the noise voltage $v_{no}$ developed at the output.

The noise current generated due to $v_{ni}$ is given by

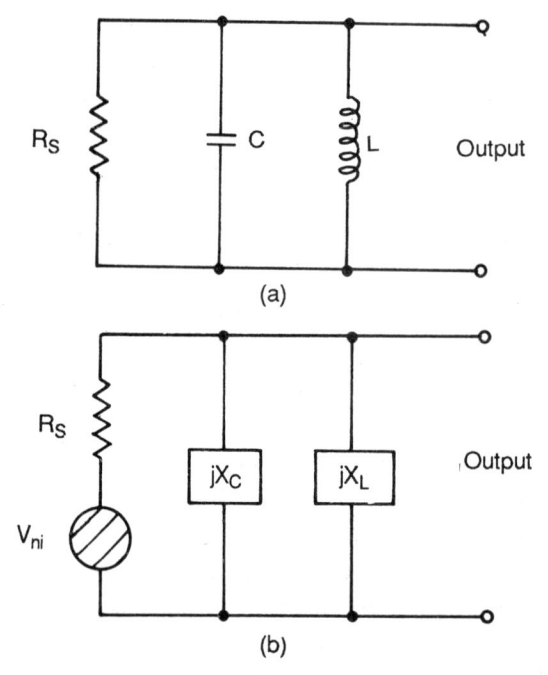

**Fig. 4.4.5** (a) A Resonant Circuit (b) Noiseless $R_s$

$$I_{rms} = \frac{V_{ni}}{Z}$$

Where $Z$ is impedance of the circuit, and $v_{ni}$ is the rms value of $v_{no}$. At resonance $Z = R_s$, hence

$$I_{rms} = \frac{V_{ni}}{R_s}$$

The rms value of the output voltage $v_{no}$ is given by

$$V_{no} = I_{rms} \cdot X_c = \frac{V_{ni}}{R_s} \cdot X_c$$

where $X_c$ is reactance of the capacitor $C$

Since $X_c = Q \cdot R_s$ at resonance,

$$V_{no} = \frac{V_{ni}}{R_s} \times Q \cdot R_s = Q \cdot V_{ni}$$

The m.s. value of the output voltage is,

$$\overline{v_{no}^2} = \overline{QV_{ni}^2}$$

$$= Q^2 \cdot 4kTR_s(\Delta f) = 4kT(\Delta f)(Q^2 R_s) \qquad (4.4.11a)$$

Let $R_p = Q^2 R_s$, where $R_p$ is known as equivalent parallel resistance, then

$$\overline{v_{no}^2} = 4kTR_p(\Delta f) \qquad (4.4.11b)$$

## Example 4.4.3

A parallel tuned circuit is resonated at 200 MHz with a Q of 10, and a capacitance of 10 pf. The temperature of the circuit is 17°C. What noise voltage will be observed across the circuit by a wide band voltmeter?

*Solution*

$$Q = \frac{f}{\Delta f}$$

where $f$ is resonant frequency and $\Delta f$ is bandwidth of the tuned circuit. Hence,

$$\Delta f = \frac{f}{Q} = \frac{200}{10} = 20 \text{ MHZ}$$

$$Q = \frac{1}{\omega C R_s}$$

or

$$R_s = \frac{1}{Q \omega C} = \frac{1}{10 \times 2\pi \times 200 \times 10^6 \times 10 \times 10^{-12}} \cong 8 \text{ ohms}$$

$$v_{no} = \left(\overline{v_{no}^2}\right)^{\frac{1}{2}} = \sqrt{4kT(\Delta f)(Q^2 \cdot R_s)}$$

$$= \sqrt{4 \times 1.38 \times 10^{-23} \times (273 + 17) \times 20 \times 10^6 \times 10^2 \times 8} = 16 \mu v$$

## 4.5 NOISE BANDWIDTH

The *noise bandwidth* is an important parameter for specifying the noise power at the output of a *bandpass linear system*. Consider a linear bandpass system shown in Fig. 4.5.1a. The square of transfer function, i.e. $|H(\omega)|^2$ is plotted in Fig. 4.5.1b. Only the positive half of the plot $|H(\omega)|^2$ is shown assuming that the negative half is symmetrical about vertical axis.

The noise power (m.s. value) at the output of the system as given by Eq. 4.4.2 is

$$P_o = \overline{v_{no}^2} = \frac{1}{2\pi} \int_{-\infty}^{\infty} S_{ni}(\omega) |H(\omega)|^2 d\omega = \frac{1}{\pi} \int_{0}^{\infty} S_{ni}(\omega) |H(\omega)|^2 d\omega$$

For all practical purposes, the input noise power density is assumed to be constant with frequency. Let us take this constant value as $C$. Thus

$$S_{ni}(\omega) = C$$

Then, the output noise power is given by

$$P_o = \frac{C}{\pi} \int_{0}^{\infty} |H(\omega)|^2 d\omega$$

The integral in the right hand side is the area under the curve *square of transfer function* i.e., $|H(\omega)|^2$ shown in Fig. 4.5.1b. We can say that power $P_o$ is proportional to the area under the $|H(\omega)|^2$ curve i.e.,

# 210 COMMUNICATION SYSTEMS: Analog and Digital

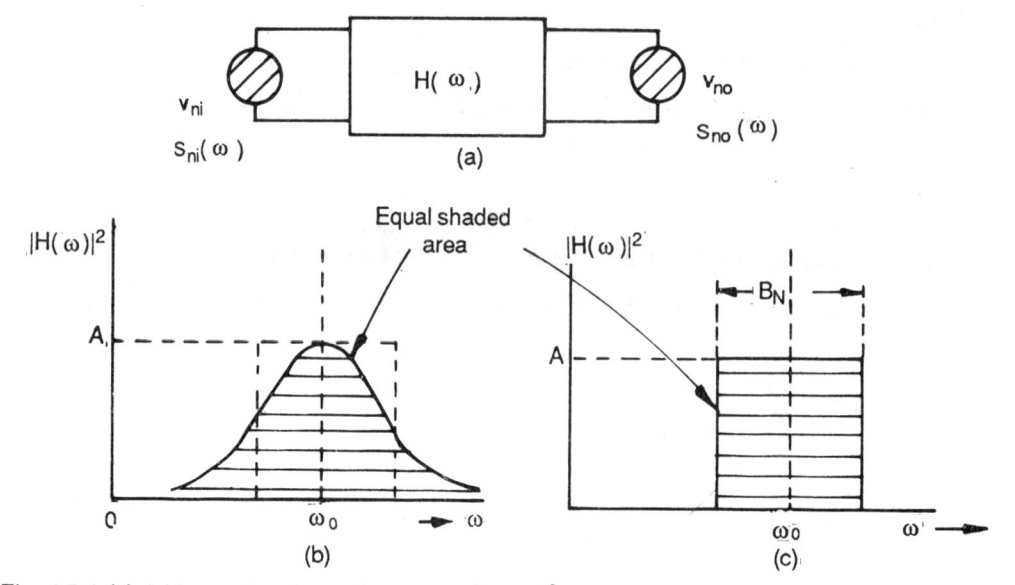

Fig. 4.5.1 (a) A Linear Bandpass System (b) $|H(\omega)|^2$ of the Actual System; (c) $|H(\omega)|^2$ of the Ideal System

$$P_o = \frac{C}{\pi} \times \text{area under the curve } |H(\omega)|^2$$

Let us consider an ideal bandbass system with rectangular characteristic of square transfer function such that the area under this curve is same as that of an actual system, and the height of the ideal curve is equal to A, which is the maximum value of $|H(\omega)|^2$ in actual system. Then bandwidth of the ideal system is called as *equivalent noise bandwidth* denoted by $B_N$.

The characteristic of the ideal system is shown in Fig. 4.5.1c. The area under this ideal rectangular characteristic specifying the power of the signal is given by

$$A \times B_N$$

where

$$A = |H(\omega_o)|^2$$

Equating the areas of actual and ideal systems,

$$A \times B_N = \int_0^\infty |H(\omega)|^2 \, d\omega$$

which gives the expression for Noise Bandwidth $B_N$ as

$$B_N = \frac{1}{A} \int_0^\infty |H(\omega)|^2 \, d\omega \tag{4.5.1}$$

$$= \frac{\text{Area under the } |H(\omega)|^2 \text{ curve}}{\text{Maximum value of the } |H(\omega)|^2 \text{ curve}}$$

The noise power, in terms of is obtained as follows,

$$P_o = \overline{v_{no}^2} = \frac{C}{\pi} \times [\text{area under the curve } |H(\omega)|^2 \text{ of actual system}]$$

$$= \frac{C}{\pi} \times [\text{area under the curve } |H(\omega)|^2 \text{ of ideal system}]$$

$$= \frac{C}{\pi} \times [AB_n]$$

Hence,

$$P_o = \overline{v_{no}^2} = \frac{C\, AB_N}{\pi} \tag{4.5.2}$$

Here, C is a constant equal to the power density spectrum of noise voltage at the input of the system. Thus, we can define *equivalent noise bandwidth* by using Eq. 4.5.1 as follows:
"*Equivalent noise bandwidth* is the bandwidth of that ideal bandpass system which produces the same noise power as the actual system."

## Relation Between Noise Bandwidth and 3-dB Bandwidth

We know that the 3-dB bandwidth of a system is defined as the range of frequencies for which power does not fall below half (3-dB) of the maximum power. In actual systems a 3-dB bandwidth denoted by $B_{3\text{-dB}}$ is somewhat less than the equivalent noise bandwidth,

$$B_N > B_{3\text{-dB}}$$

However, for ideal systems both are same.
It can be verified that for a $RC$ low pass filter, (Prob. 4.3)

$$B_N = \frac{\pi}{2} \cdot B_{3\text{-dB}} \tag{4.5.3}$$

## 4.6 AVAILABLE POWER

The available power of a source is defined as the maximum power that can be drawn from the source. The well-known maximum power transfer theorem states that maximum power is drawn from a source when it drives a matched load. The impedance of a matched load is a complex conjugate of the source impedance.

Consider a noise voltage source $v_n(t)$ with a source impedance $Z_s = R_s + j X_s$ shown in Fig. 4.6.1. The noise source is driving a load impedance $Z_L$ which is matched with source impedance, i.e.,

$$Z_L = Z^*_s = R_s - j X_s$$

The rms current flowing through the load is given by

$$= \frac{\sqrt{\overline{v_n^2}}}{Z} = \frac{\sqrt{\overline{v_n^2}}}{2R_s} \qquad (\text{since } Z = 2R_s)$$

**212** COMMUNICATION SYSTEMS: Analog and Digital

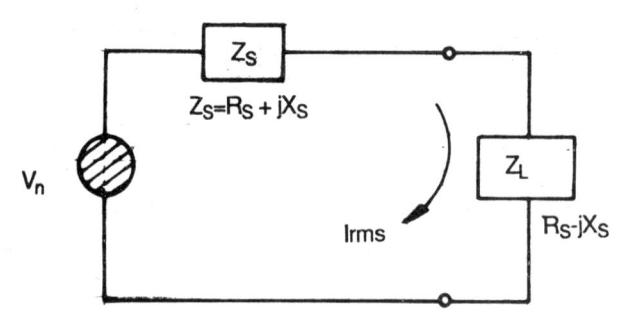

Fig. 4.6.1

The available power $P_a$ is given by

$$P_a = I^2_{rms} \times R_s = \frac{\overline{v_n^2}}{4R_s^2} \cdot R_s$$

$$P_a = \frac{\overline{v_n^2}}{4R_s} = \frac{P_n}{4R_s} \tag{4.6.1}$$

where $P_n$ is the normalized power, and it gives the power developed by $v_n$ across 1 ohm resistor. The power developed across the total circuit resistance $2R_s$ will be

$$P_{no} = \frac{P_n}{2R_s} \tag{4.6.2}$$

Therefore, the noise source $v_n$ supplies a noise power equal to $\frac{P_n}{2R_s}$ out of which half of the power i.e., $P_n/4R_s$ is available across load and, the remaining half is dissipated across the source resistance $R_s$.

Similarly, the available power density is given by

$$S_a(\omega) = \frac{S_v(\omega)}{4R_s} \tag{4.6.3}$$

where $S_v(\omega)$ is the power density of the noise voltage $v_n(t)$. It is obvious from Eq. 4.6.1 that for a noise source $v_n$ the available power is $\frac{1}{4R_s}$ times the normalized power. Similarly, the available power density is $\frac{1}{4R_s}$ times the normalized power density spectrum.

It can be shown that (Prob. 4.4) for a current source $i_n(t)$ with a power density spectrum $S_i(\omega)$ and internal admittance $Y_s = G_s + jB_s$, the available power density is given by

$$S_a(\omega) = \frac{S_i(\omega)}{4G_s} \tag{4.6.4}$$

### Available Power of an R-L-C Network

A passive network containing R-L-C elements can be represented by Thevenin's equivalent as shown in Fig. 4.6.2.

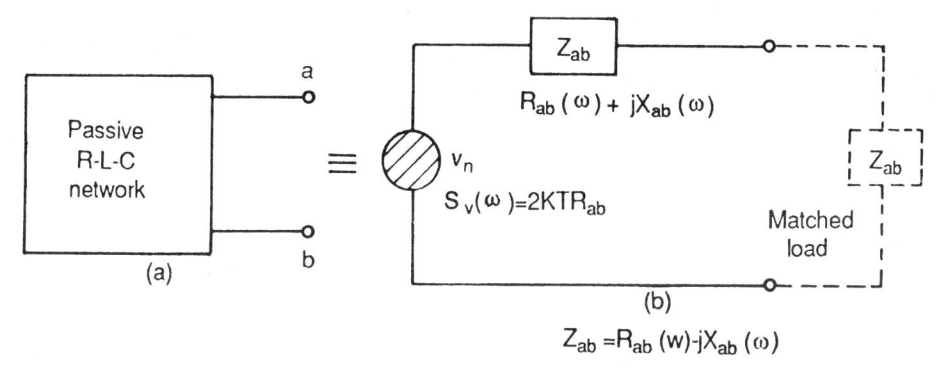

NOISE 213

[Fig. 4.6.2 (a) A Passive Network; (b) Thevenin's Equivalent]

Fig. 4.6.2 (a) A Passive Network; (b) Thevenin's Equivalent

The equivalent impedance across the terminal a,b may be put in the form,

$$Z_{ab} = R_{ab}(\omega) + jX_{ab}(\omega)$$

In general, resistance $R_{ab}(\omega)$ may be frequency-dependent. The only noise voltage source $v_n(t)$ produced is due to resistance $R_{ab}$ which has a power density spectrum given by

$$S_v(\omega) = 2kTR_{ab}$$

The available power density spectrum $S_a(\omega)$ may be obtained using Eq. 4.6.3

$$S_a(\omega) = \frac{S_v(\omega)}{4R_{ab}} = \frac{2kTR_{ab}}{4R_{ab}}$$

or

$$S_a(\omega) = \frac{kT}{2} \qquad (4.6.5)$$

which is independent of $R_{ab}$.

The available noise power may be obtained by using Eq. 4.6.1,

$$P_a = \frac{\overline{v_n^2}}{4R_{ab}}$$

The mean square value of the thermal noise generated by the resistance $R_{ab}$ over a bandwidth $(\Delta f)$ (one-sided bandwidth) may be obtained with the help of Eq. 4.3.5,

$$\overline{v_n^2} = 4kTR_{ab}(\Delta f)$$

Hence, the available noise power $P_a$ becomes

$$P_a = \frac{\overline{v_n^2}}{4R_{ab}} = \frac{4kTR_{ab}(\Delta f)}{4R_{ab}} = kT(\Delta f)$$

or

$$P_a = kT(\Delta f) \qquad (4.6.6)$$

Thus the available noise power is independent of the source resistance $R_{ab}$.

## 4.7 NOISE TEMPERATURE

Reorienting Eqs 4.6.5, and 4.6.6, respectively, we get

$$T = \frac{2S_a(\omega)}{k} \tag{4.7.1}$$

and

$$T = \frac{P_a}{k(\Delta f)} \tag{4.7.2}$$

The temperature given by the above equations, represents the actual physical temperature of R-L-C network (all components are assumed to be on the same common temperature).

The above expressions of temperature $T$ do not hold true for the following cases:
(i) For networks containing active devices which may amplify the signals.
(ii) For networks that contain noise source other than resistor noise, such as shot noise or atmospheric noise etc.

However, it is possible to extend the above expressions (given by Eqs 4.7.1 and 4.7.2) of noise temperature to account for the cases involving two-port networks with active elements and other source of noise. This can be done by defining a generalized temperature, called *effective noise temperature* or, *noise temperature* denoted by $T_n$. Let $P_a$ denote the available noise power of a two-port network containing active elements and other sources of noise alongwith resistors noise. Let $S_a(\omega)$ be the corresponding generalized noise power density. The equivalent noise temperature $T_n$ is thus given by

$$T_n = \frac{P_a}{k(\Delta f)} \tag{4.7.3}$$

and

$$T_n = \frac{2S_a(\omega)}{k} \tag{4.7.4}$$

The noise temperature $T_n$ need not be the actual physical temperature of any part of the network. For a network containing only passive components (R-L-C), the noise temperature $T_n$ is the same as physical temperature $T$ (ambient temperature). Thus we can define the noise temperature as follows:

Equivalent noise temperature of a system is defined as the temperature at which a noisy resistor, has to be maintained such that by connecting this resistor to the input of the noiseless version of the system, it produces the same available noise power at the output of the system, as is produced by all the sources of noise in the actual system.

### Examples of Equivalent Noise Temperature

#### (i) Two-port Network with Active Elements

Consider an amplifier system shown in Fig. 4.7.1. Here, the resistor R is the source of the thermal noise

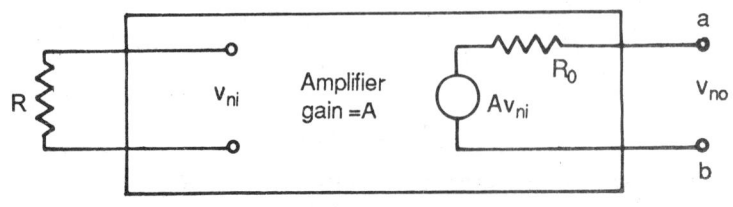

Fig. 4.7.1

at a temperature $T$. This noise source is connected to the input of the amplifier of gain $A$. We will make the following assumptions for simplicity:
- (a) Input impedance of the amplifier is infinite.
- (b) Amplifier output resistance is a noiseless resistor $R_o$.
- (c) Resistor noise is white, i.e., it has frequency-independent power density.

The rms noise voltage at the input due to the resistor $R$ over a bandwidth $(\Delta f)$ is given by;

$$V_{ni} = \sqrt{4kTR(\Delta f)}$$

Hence, rms value of amplifier output voltage

$$V_{no} = A\sqrt{4kTR(\Delta f)}$$

So, the mean square value of the output voltage is

$$\overline{v_{no}^2} = A^2 4kTR(\Delta f)$$

The available noise power at the output is (see Eq. 4.6.1).

$$P_a = \frac{\overline{v_{no}^2}}{4R_o} = \frac{A^2 4kTR(\Delta f)}{4R_o} = \frac{kTRA^2(\Delta f)}{R_o}$$

The equivalent noise temperature can be obtained by using Eq. 4.7.3,

$$T_n = \frac{P_a}{k(\Delta f)} = A^2 \left(\frac{R}{R_o}\right) T \qquad (4.7.5)$$

Thus, $T_n$ depends on gain A (active element), and $(R/R_o)$. This noise temperature may assume any value including $T_n = 0$. This is not necessarily the physical temperature of any part of the network.

### (ii) Other Sources of Noise

Let us consider the concept of noise temperature in an antenna made of a loop of wire. If we assume that the antenna has a negligibly small resistance, then the noise generated by the antenna itself will be zero. However, noise may be induced to the antenna by other sources of noise such as,

(a) lightning, automobile ignition, fluorescent lights, etc. This type of noise dominates the AM systems (lower radio frequencies), and has little effect on FM systems (higher radio frequencies). Practically the power density spectrum of such a noise falls off appreciably above 50 MHz.

(b) thermal radiation from physical bodies at a temperature higher than 0°K, like the sun, the stars, and other cosmic bodies.

If an antenna is completely shielded from the above sources of noise, then the antenna noise temperature would be zero, but in that case, the antenna will not receive any desired signals as well. On the other hand, if the antenna is not properly shielded, and noise is induced from the above source, it will have a non-zero antenna temperature $T_n$. For the frequency range of operation $(\Delta f)$ in which power density spectrum may be assumed frequency-independent, the noise temperature is as given by Eq. 4.7.2.

## 4.8 NOISE IN TWO-PORT NETWORKS

In communication systems, signals are processed in a number of ways. For example, signals may be amplified by an amplifier, or a signal may be subjected to a frequency mixer for conversion to an intermediate frequency, etc. The processing stages (like amplifiers, mixers, etc.) have two set of terminals: one for input and the other for output. Such systems are called two-port networks (i.e., input-port and output-port). Each port of the network may have resistors, as well as active devices as sources of noise within it. The message signal at the input of the port is accompanied by noise sources, and while passing through the two-port network, additional noise is added to the signal. Thus, at the output of a two-port, the total noise is due to contribution of two noise sources:

(i) internal noise generated within the network and (ii) noise that may be amplified or attenuated by the network. The quantitative analysis of these noise sources is characterized by two parameters: (i) equivalent noise temperature and (ii) noise figure.

We will develop the expressions for these parameters later on. Let us first define a more fundamental parameter associated with the two-port network, known as signal to noise ratio.

## Signal to Noise Ratio

The ratio of a signal power to the accompanying noise power is referred as signal to noise ratio and is denoted by $S/N$. If a signal voltage $v_s(t)$ is accompanied by a noise voltage source $v_n(t)$, the ratio of signal to noise power (m.s. value) is

$$\frac{S}{N} = \frac{\overline{v_s^2}}{\overline{v_n^2}} \qquad (4.8.1)$$

Since the power density spectrum is power per unit bandwidth, Eq. 4.8.1 can be expressed in terms of power density as follows:

$$\frac{S}{N} = \frac{S_s(\omega)}{S_n(\omega)} = \frac{\text{power density spectrum of signal voltage}}{\text{power density spectrum of noise voltge}} \qquad (4.8.2)$$

The $S/N$ ratio is always referred to the power ratio, unless stated otherwise. Sometimes, the ratio of a signal voltage (rms) to a noise voltage (rms) is also specified, but in that case, it is specifically stated that $S/N$ *voltage ratio* is being considered.

The $S/N$ ratio at the input, and output of a *noise free* two-port network is the same, because both signal and noise are processed identically. However, practical two-port networks introduce additional noise, hence the $S/N$ ratio at its output deteriorates, i.e., the $S/N$ ratio is decreased. The comparison of $S/N$ ratio at the input and output of a two-port network provides the noisiness indication of the network. This noisiness is expressed in terms of noise-figure, or *equivalent noise temperature*.

## Available Power Gain

Let us define a term *available power gain* which may be useful in the study of noise figure, and noise temperature of two-port networks. The arrangement shown in Fig. 4.8.1 represents a signal $v_s(t)$ having source impedance $Z_s = R_s + jX_s$. The signal $v_s(t)$ is accompanied by a noise source $v_n(t)$ caused by $R_s$. The impedance may be considered to be noiseless. The $S/N$ ratio across the terminal $a,b$ in terms of power density spectra is given by

$$\frac{S}{N} = \frac{S_s(\omega)}{S_n \omega}$$

The ratio of the available signal power to the available noise power denote by $(S/N)_a$ can be obtained by considering $v_s(t)$ and $v_n(t)$, independent of each other. Thus

NOISE 217

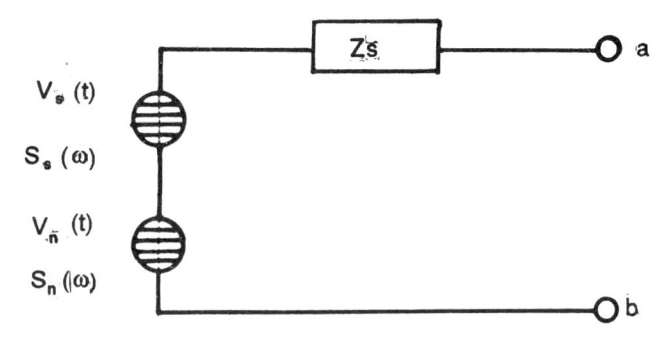

**Fig. 4.8.1 A Signal Accompanied with Noise**

$$\left(\frac{S}{N}\right)_a = \frac{\text{available power density of signal}}{\text{available power density of noise}} = \frac{S_s(\omega)/4R_s}{S_n(\omega)/4R_s} \quad (4.8.3)$$

Equation 4.8.3 states a significant result: a signal to noise power density ratio across any terminals is the same as the available signal to power density ratio across the terminals.

The ratio of *available* power density at the input and the output is referred to as *available power gain* and is denoted by $G_a$,

$$G_a = \frac{\text{Available power density at the output of the two - port network} (S_o)_a}{\text{Available power density at the input of the two - port network } (S_i)_a}$$

In general, available power gain is a function of frequency, and is denote by $G_a(\omega)$. Thus

$$G_a(\omega) = \frac{(S_o)_a}{(S_i)_a} \quad (4.8.4)$$

The available power gain is related with signal transfer through the network, and depends on both driving signal and parameters of network. This does not depend on load impedance.

Some useful results can be derived from Eq. 4.8.4. The available power density spectrum at the output is $G_a(\omega)$ times available power density spectrum at the input of a two-port network, i.e.,

$$(S_o)_a = G_a(\omega)(S_i)_a$$

Note that power density may belong to a message signal, or a noise signal. Available output power will be

$$P_{ao} = \frac{1}{2\pi} \int_{-\infty}^{\infty} G_a(\omega)(S_i)_a \, d\omega \quad (4.8.5)$$

If the input is a white noise voltage source with a power density spectrum $S_{ni} = \frac{kT}{2}$, then Eq. 4.8.5 becomes

$$(P_{ao})_{noise} = \frac{1}{2\pi} \int_{-\infty}^{\infty} \frac{kT}{2} G_a(\omega) \, d\omega \quad (4.8.6)$$

It can be verified that the available power gain of $n$-cascaded stages with available gains $G_{a1}, G_{a2}, G_{a3}, \ldots$

$G_{an}$ is given by the product (Prob. 4.6)

$$G_a = G_{a1} G_{a2} G_{a3} \ldots G_{an} \qquad (4.8.7)$$

This is analogous to cascaded amplifiers.

## Effective Input Noise Temperature

The noise temperature is referred at the input of a two-port network which accounts for the internal noise produced by the network, and thereafter the network is considered to be noise free. Consider a two-port network shown in Fig. 4.8.2. The input is white noise source $v_{ni}$ with the available power density $\frac{kT}{2}$. Let $G_a(\omega)$ be the available power gain of the network. If network is *noise-free*, then the available power density at the output will be (see Eq. 4.8.4)

$$(S'_{no})_a = G_a(\omega)\frac{kT}{2} \qquad (4.8.8)$$

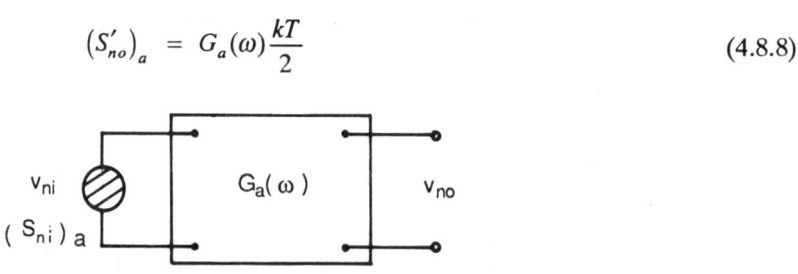

Fig. 4.8.2

However, a practical two-port network introduces its own noise and hence the available power density $(S_{no})_a$ at the output of a noisy network is higher than $(S'_{no})_a$ the one given by Eq. 4.8.8 for a noise-free network. This increase in noise power density may be accounted in terms of noise-temperature at the input, and then the network may be considered noise-free. The input noise temperature of the network which accounts for the internal noise generated by the network is known as *effective input noise temperature*, denoted by $T_e$. The output power density is now given as

$$(S_{no})_a = G_a(\omega)\frac{kT}{2} + G_a(\omega)\frac{kT_e}{2}$$

$$= (T + T_e)\frac{k}{2}G_a(\omega) \qquad (4.8.9)$$

$T$ is the noise temperature of the source, and $T_e$ is the noise temperature accounting for the network-generated noise. $T_e$ depends on the source as well as the network.

### Example 4.8.1

An antenna having a noise temperature 30 °K is connected at the input of a receiver which has an equivalent input noise temperature of 270 °K. The midband available gain of the receiver is $10^{10}$ and the corresponding noise bandwidth is 1.5 MHz. Find the available output noise power.
Solution
Power density at the input of the receiver is

$$(S_{ni})_a = \frac{kT}{2} = k\left(\frac{T_{ant} + T_e}{2}\right)$$

NOISE  219

The available noise power at the output of the receiver is given by

$$(P_{ao})_{\text{noise}} = (S_{no})_a \times \text{noise bandwith } B_N$$
$$= (S_{ni})_a \cdot G_a(\omega) \cdot B_N$$

Here, $T_{ant} = 30° K$, $T_e = 270° K$
Therefore,

$$(P_{ao})_{\text{noise}} = G_a(\omega) k (T_{ant} + T_e) B_N$$
$$= 10^{10} \times 1.38 \times 10^{-23} \times 300 \times 1.5 \times 10^6 = 62 \mu \text{ watt}$$

## 4.9 NOISE-FIGURE

The amount of noise power contributed by a two-port network is also characterized by its *noise figure*. It is defined as the ratio of the total noise power spectral density $S_{no}$ at the output of the two-port network to the noise power spectral density $S'_{no}$ at the output, assuming the network to be entirely noiseless. This ratio is the same as the ratio of corresponding available power densities (refer Eq. 4.8.3). Thus the noise figure $F$ of a two-port network is

$$F = \frac{(S_{no})_a}{(S'_{no})_a} \qquad (4.9.1)$$

When a two-port network is assumed to be noiseless, the power density at the output is solely due to the noise source at the input. Hence, alternately, we can define,

$$F = \frac{\text{power density of the total noise at the output of network}}{\text{power density at the output solely due to the source}}$$

The total noise power density at the output $S_{no}$ is equal to the sum of the power density solely due to the source $S'_{no}$ and the power density contributed by network itself denoted by $S''_{no}$ i.e.,

$$S_{no} = S'_{no} + S''_{no}$$

and noise figure in Eq. 4.9.1 may be written as

$$F = \frac{S'_{no} + S''_{no}}{S'_{no}} \quad 1 + \frac{S''_{no}}{S'_{no}} \qquad (4.9.2)$$

### Noise-Figure in Terms of Available Power Gain $G_a(\omega)$

Consider a two-port network (e.g., an amplifier) with available power gain $G_a(\omega)$. Let us assume that input noise source is represented by a resistor noise with a power density spectrum $(S_{ni})_a = \frac{kT}{2}$.

Now, if the two-port network is noiseless then output available power density spectrum $(S'_{no})_a$ is obtained by using by Eq. 4.8.4

$$(S'_{no})_a = \frac{kT}{2} G_a(\omega)$$

But, the network itself introduces additional noise and hence actual noise power density spectrum is higher and is denoted by $(S_{no})_a$. Hence the noise figure is given by, (see Eq. 4.9.1).

$$F = \frac{(S_{no})_a}{G_a(\omega)\frac{kT}{2}} \qquad (4.9.2)$$

It is obvious that

(a) For a noiseless network, $(S_{no})_a = (S'_{no})_a$ hence $F = 1$ (i.e., 0 dB).

(b) For a noisy network $(S_{no})_a > (S'_{no})_a$ hence $F > 1$

Thus, the minimum value of noise figure is unity which is for an ideal (noiseless) network.

### Noise-figure in Terms of Network Transfer Function $H(\omega)$

Let $S_{ni}$ be the power density spectrum of the noise voltage at the input of a network. If the network transfer function is $H(\omega)$, the power density spectrum at the output of the noiselss network will be

$$S'_{no} = S_{ni}\,|H(\omega)|^2$$

and the corresponding Noise-Figure will be

$$F = \frac{S_{no}}{S_{ni}\,|H(\omega)|^2} \qquad (4.9.3)$$

### Noise-figure in Terms of Equivalent Input Noise Temperature

Reorienting Eq. 4.8.9 we may write:

$$\frac{(S_{no})_a}{G_a(\omega)\frac{k}{2}} = T + T_e$$

Substituting this value in Eq. 4.9.2,

$$F = \frac{T + T_e}{T} = 1 + \frac{T_e}{T} \qquad (4.9.4)$$

or

$$T_e = T(F - 1) \qquad (4.9.5)$$

where $T$ and $T_e$ are noise temperature of the source, and the network, respectively.

### Spot Noise-figure and Average Noise-figure

The noise-figure defined by Eq. 4.9.2 refers to a particular *spot* in the frequency spectrum and is known as *spot* (or *spectral*) noise-figure. We may define the average noise-figure $F$ over a frequency range from $\omega_1$ to $\omega_2$ as follows:

$$\overline{F} = \frac{\int_{\omega_1}^{\omega_2} G_a(\omega)\,F(\omega)\,d\omega}{\int_{\omega_1}^{\omega_2} G_a(\omega)\,d\omega} \qquad (4.9.6)$$

# NOISE

## Noise-figure in Terms of S/N Ratio

In practical systems, we are not only interested in noise but also the desired signal accompanying noise. In this case, signal to noise ratio is a more useful parameter. Such a situation is shown in Fig. 4.9.1. The input of the two-port network consists of a message signal $v_s$ and a noise signal $v_n$. The noise signal is due to resistor

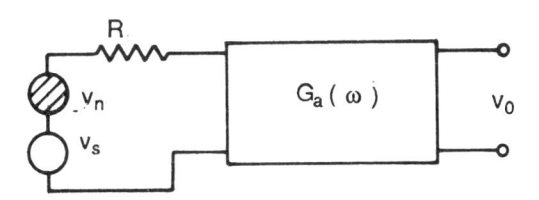

Fig. 4.9.1

$R$, and has available power density $(S_{ni})_a = \dfrac{kT}{2}$. Let it be that signal $v_s$ has available power density denoted by $(S_{si})_a$. The network has an available power gain given by $G_a(\omega)$. Hence, at the output, the available spectral power density of message signal is given by

$$(S_{so})_a = G_a(\omega)(S_{si})_a \tag{4.9.7}$$

The available power density of the output noise voltage is given by (Eq. 4.9.1.)

$$(S_{no})_a = F(S'_{no})_a \tag{4.9.8}$$

Here, $(S'_{no})_a$ is power density at the output of a noiseless network when no noise is introduced by networks, hence, we can write $(S'_{no})_a$ in terms of input noise power density $(S_{ni})_a$ as given below:

$$(S'_{no})_a = G_a(\omega)(S_{ni})_a \tag{4.9.9}$$

From Eqs. 4.9.8 and 4.9.9 we get

$$(S_{no})_a = G_a(\omega) F(S_{ni})_a \tag{4.9.10}$$

Substituting $G_a(\omega)$ from Eq. 4.9.7 in Eq. 4.9.10, we get

$$(S_{no})_a = \dfrac{(S_{so})_a}{(S_{si})_a} F(S_{ni})_a$$

The reorientation of the above equation yields an expression for spot noise-figure as given below:

$$F = \dfrac{(S_{si})_a / (S_{ni})_a}{(S_{so})_a / (S_{no})_a} \tag{4.9.11a}$$

Here $F$ is defined by ratio of ratios. The numerator is a ratio of the input signal power density spectrum to the input noise power density spectrum. The denominator is a ratio of the output signal power density spectrum to output noise power density spectrum.

Equation 4.9.11a can be extended to define the average noise-figure $F$ over a frequency range $f_1$ to $f_2$. Assuming that spectral power densities of the signal and noise are uniform over the bandwidth under consideration, Eq. 4.9.11a may be written as

$$\overline{F} = \frac{S_i/N_i}{S_o/N_o} \qquad (4.9.11b)$$

where $S_i/N_i$ is the input signal to noise ratio, and $S_o/N_o$ is the output signal to noise ratio.

A relation between signal to noise ratio at the input of a network and signal to noise ratio at the output of the network can be expressed in terms of noise temperatures. From Eq. 4.9.4 the noise figure is given by (assuming $\overline{F} = F$)

$$\overline{F} = 1 + \frac{T_e}{T}$$

Comparing this $F$ with the one given by Eq. 4.9.11b we get

$$S_i/N_i = \left(1 + \frac{T_e}{T}\right)\frac{S_o}{N_o} \qquad (4.9.12)$$

The noise-figure can be expressed in various other forms. If the available gain of the network is a constant $(G_a)$ over the frequency range of interest, then, $S_o = G_a S_i$ and $\overline{F} = F$. In this case Eq. 4.9.11b may be expressed as

$$F = \frac{1}{G_a}\frac{N_o}{N_i} \qquad (4.9.13)$$

Further, the output noise power $N_o$ can be expressed as the sum of the output noise power due to input source $(G_a N_i)$ and noise power contributed by two-port network $(N_{tp})$ i.e.,

$$N_o = G_a N_i + N_{tp} \qquad (4.9.14)$$

Substituting this value of $N_o$ in Eq. 4.9.13 we get

$$F = 1 + \frac{N_{tp}}{G_a N_i} \qquad (4.9.15)$$

The power contributed by the two-port network may be written as

$$N_{tp} = G_a(F-1)N_i \qquad (4.9.16)$$

### Example 4.9.1

Determine the noise figure of a common base transistor amplifier. The equivalent circuit of the transistor is shown in Fig. 4.9.2.

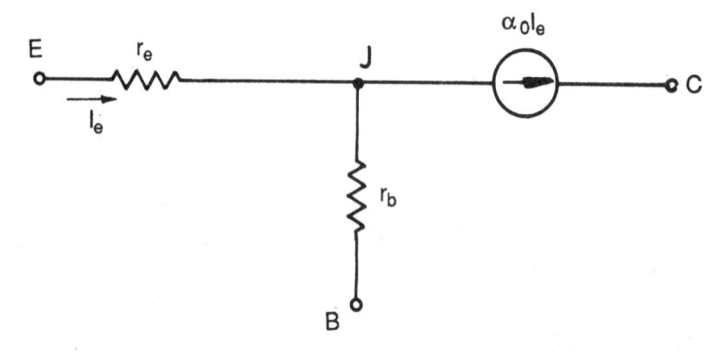

Fig. 4.9.2 Equivalent Circuit of Transistor (Z-parameter)

## Solution
The transistor has three main noise sources, viz., (a) shot noise (b) partition noise and (c) thermal noise. These are shown in Fig. 4.9.3.

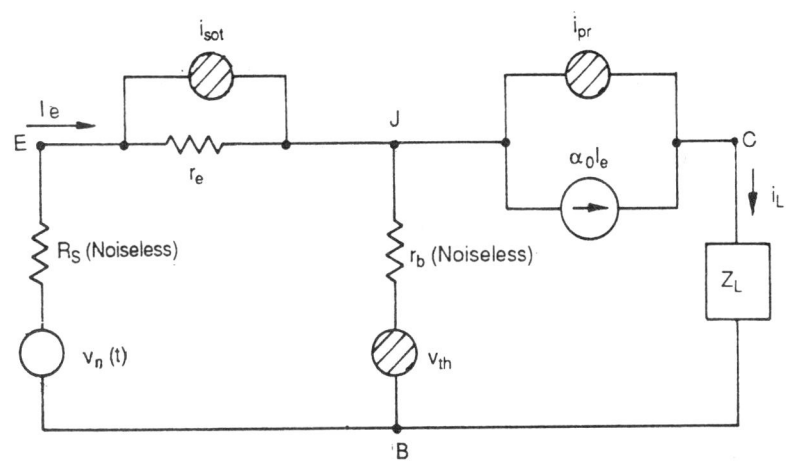

Fig. 4.9.3 Various Sources of Noise in a Transistor

### Shot Noise
This noise is represented by a current source in the emitter arm *EJ*. The power density spectrum (assuming it is uniform) of the noise current source is given by

$$S'_{sot} = qI_e \tag{4.9.17}$$

where $I_e$ is the average d-c emitter current.

### Partition Noise
This noise is represented by a current source $i_{pr}(t)$ in the collector arm *JC*. If $\alpha_o$ is the d.c current gain in a common base transistor, the power density spectrum of the current source $i_{pr}(t)$ at low frequencies is given by

$$S_{pr}(\omega) = qI_e \alpha_o / (1 - \alpha_o) \tag{4.9.18}$$

### Thermal Noise
The base of the transistor has a physical resistance, known as *base spreading* resistance, denoted by $r_b$. This resistance is the source of thermal noise and is represented by a voltage source $v_{th}$ in the base arm *JB*. The power density spectrum of this noise voltage source is

$$S_{th} = 2kT r_b \tag{4.9.19}$$

The transistor is driven by a source and deliver the power to a load. The source resistance $R_s$ also produces a thermal noise. This is shown in Fig. 4.9.3 by a noise voltage source $v_n(t)$. The power density spectrum of this noise voltage source is given by

$$S_{R_s} = 2kTR_s \tag{4.9.20}$$

## Transfer Function

The transfer function relating the load current $i_L$ to the noise current $i_{so}$ may be shown to be

$$H_{sot}(\omega) = \frac{\alpha_o r_e}{R_s + r_b(1-\alpha_o) + r_e} \tag{4.9.21a}$$

Similarly, the transfer function relating the current $i_L$ to the partition noise current $i_{pr}$ is

$$H_{pr}(\omega) = \frac{r_b + r_e + R_s}{R_s + r_b(1-\alpha_o) + r_e} \tag{4.9.21b}$$

and the transfer function relating the current $i_L$ to the thermal noise voltage $v_{th}(t)$ is given by

$$H_{th}(\omega) = \frac{\alpha_o}{R_s + (1-\alpha_0) + r_e} \tag{4.9.21c}$$

The transfer function relating the current $i_L$ to the noise voltage source $v_n(t)$ is given by

$$H_s(\omega) = \frac{\alpha_o}{R_s + r_b(1-\alpha_o) + r_e} \tag{4.9.21d}$$

The noise-figure $F$ as defined by Eq. 4.9.2 may be represented in terms of the available power density spectrum as follows:

$$F = 1 + \frac{(S''_{no})_a}{(S'_{no})_a} \tag{4.9.22}$$

Here $(S''_{no})_a$ is the available power density contributed by the transistor alone at the load terminal, and is given as

$$(S''_{no})_a = S_{sot}|H_{sot}|^2 + S_{pr}(\omega)|H_{pr}|^2 + S_{th}|H_{th}|^2 \tag{4.9.23}$$

Similarly, $(S'_{no})_a$ is the available power density contributed at the load by the source alone, and is given by

$$(S_{no})_a = S_{R_s}(\omega)|H_s(\omega)|^2 \tag{4.9.24}$$

Therefore, the noise-figure $F$ is given by

$$F = 1 + \frac{r_b}{R_s} + \frac{qI_e}{2kTR_s}\left[r_e^2 + \frac{1-\alpha_o}{\alpha_o}(r_e + r_b + R_s)^2\right] \tag{4.9.25}$$

The emitter resistance $r_e$ of a transistor is given in terms of the emitter current $I_e$ as follows:

$$r_e = \frac{kT}{qI_e} \tag{4.9.26}$$

Hence, the noise-figure may be written as

$$F = 1 + \frac{r_b + r_e/2}{R_s} + \frac{(r_b + r_e + R_s)^2(1-\alpha_o)}{2\alpha_o r_e R_s} \tag{4.9.27}$$

This expression is valid for a frequency range $\omega < \omega_a \sqrt{1-\alpha_o}$, where $\omega_a$ is the $\alpha$ cut-off frequency of the transistor. At higher frequencies partition noise increases, and Eq. 4.9.27 does not hold. Also at lower frequencies (< 1 kHz), the flicker noise is large and above expression does not hold. In other words, Eq. 4.9.27 holds for intermediate frequencies only. The source resistance $R_s$ can be optimized to yield minimum noise-figure. The variation of $F$ with frequency is shown in Fig. 4.9.4.

Fig. 4.9.4 Noise Figure Variation in a Transistor

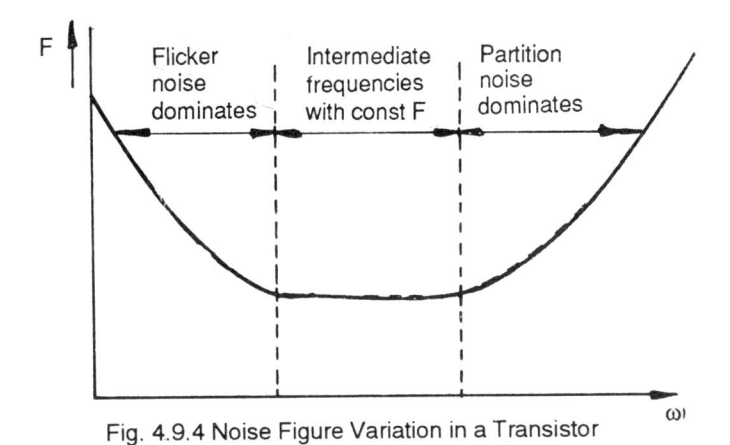

### Noise-figure in a Common Emitter Amplifier

A procedure similar to the common base amplifier may be adopted to find the noise figure in a common emitter amplifier. The noise-figure can be shown to be the same as that in a common base amplifier (Prob. 4.5)

## 4.10 CASCADED STAGES

Let us consider the two-port networks involving more than one stage in a cascade. We will evaluate the noise figure, and equivalent noise temperature, of such cascaded amplifiers.

### Noise-Figure of Cascaded Amplifiers

A cascade of two stages is shown in Fig. 4.10.1

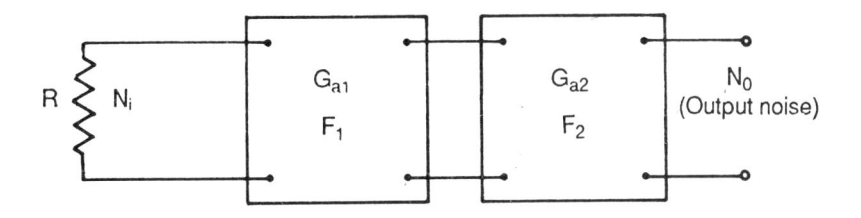

Fig. 4.10.1 Cascaded Two-Port Network

The available gain and the noise-figure of the first stage is $G_{a1}$, and $F_1$, respectively. Similarly, the second stage has $G_{a2}$ and $F_2$ as the available gain and the noise-figure respectively.

Let $N_i$ be the noise power generated by resistor $R$ at the input of the first stage. The noise power at the final output due to $N_i$ is given by (assuming that the noise contributed by the stages is zero).

$$N_{o1} = G_{a1} G_{a2} N_i$$

The first stage introduces its own internal noise. From Eq. 4.9.16, the contribution of this noise at the output of first stage is equal to $G_{a1}(F_1 - 1)N_i$. This noise is amplified by second stage and appears at its output as given by

$$N_{o2} = G_{a1}.G_{a2}(F_1 - 1)N_i$$

Similarly, the noise contributed by the second stage to the final output noise power can be evaluated to be

$$N_{o3} = G_{a2}(F_2 - 1)N_i$$

The total noise power at the output thus is,

$$\begin{aligned} N_o &= N_{o1} + N_{o2} + N_{o3} \\ &= G_{a1} G_{a2} N_i + G_{a1} G_{a2} (F_1 - 1) N_i + G_{a2}(F_2 - 1)N_i \end{aligned} \qquad (4.10.1)$$

Equation 4.10.1 can be rearranged as

$$\frac{1}{G_{a1} G_{a2}} \frac{N_o}{N_i} = 1 + F_1 - 1 + \frac{F_2 - 1}{G_{a1}} = F_1 + \frac{F_2 - 1}{G_{a1}} \qquad (4.10.2)$$

The overall available gain $G_a$ of the cascaded stages is given by

$$G_a = G_{a1} G_{a2}$$

Substituting this value in Eq. 4.10.2, we get

$$\frac{1}{G_a} \frac{N_o}{N_i} = F_1 + \frac{F_2 - 1}{G_{a1}}$$

The left-hand side of the above equation represents the overall noise-figure $F$ of the cascaded amplifier (refer Eq. 4.9.13). Hence, the overall noise-figure $F$ of the two cascaded amplifier is

$$F = F_1 + \frac{F_2 - 1}{G_{a1}} \qquad (4.10.3)$$

The expression can be extended to a multistage amplifier as furnished below:

$$F = F_1 + \frac{F_2 - 1}{G_{a1}} + \frac{F_3 - 1}{G_{a1}.G_{a2}} + \frac{F_4 - 1}{G_{a1}.G_{a2}.G_{a3}} + \cdots + \frac{F_n}{G_{a1}.G_{a2}\ldots G_{a(n-1)}} \qquad (4.10.4)$$

Equation 4.10.4 shows that the contribution to overall noise-figure is primarily by the first stage, and contribution by succeeding stages becomes smaller and smaller. This is known as the Friss formula.

## Equivalent Noise-temperature of Cascaded Stages

Let us assume that the first and second stages of Fig. 4.10.1 are characterized by equivalent noise temperatures $T_{e1}$ and $T_{e2}$ respectively. Reproducing Eq. 4.9.4, we have

$$F - 1 = \frac{T_e}{T}$$

where $T_e$ is the overall noise temperature of the two cascaded stages. Substituting noise-figures in terms of the corresponding equivalent noise temperature in Eq. 4.10.3, we get

$$1 + \frac{T_e}{T} = 1 + \frac{T_{e1}}{T} + \frac{T_{e2}}{G_{a1} T}$$

or

$$T_e = T_{e1} + \frac{T_{e2}}{G_{a1}} \qquad (4.10.5)$$

## NOISE

The expression can be extended to multistage amplifiers. The equivalent noise temperature of a multistage networks is, thus, given as

$$T_e = T_e + \frac{T_{e2}}{G_{a1}} + \frac{T_{e3}}{G_{a1}G_{a2}} + \frac{T_{e4}}{G_{a1}G_{a2}G_{a3}} + \cdots + \frac{T_{en}}{G_{a1}G_{a2}\cdots G_{a(n-1)}} \quad (4.10.6)$$

### The Cascode Amplifier

It is obvious from Eqs 4.10.4 and 4.10.6 that contribution to the net output noise by succeeding stages in the cascade system becomes progressively smaller, because the denominator increases by product of gains. The main contribution of noise is by the first and second term in Eq. 4.10.6. The contribution by the following stages reduces rapidly (since $G_{a1}G_{a2}$ or $G_{a1}G_{a2}G_{a3}$ are large denominators). This is also obvious from Ex. 4.10.2. Thus a low noise amplifier can be obtained if, (i) the first stage amplifier has low noise temperature $T_{e1}$ (hence a low $F_1$), and, (ii) the first stage amplifier has a high power gain, so that $G_{a1}$ is large, and the term $\frac{T_{e2}}{G_{a1}}$ is small. The other terms will obviously be much smaller.

It should be noted that the power gain can be increased by increasing the voltage gain, but this can lead to instability. Hence, the first stage should be so designed that it has a high available power gain, but a low voltage gain to avoid instability. An amplifier having a large available power gain but a low voltage gain is called a cascode amplifier. An available power gain of 30 dB and $T_e = 4\,°K$, can be obtained by a properly designed cascode amplifier, and by cooling it with liquid nitrogen.

### Example 4.10.1

Figure 4.10.2 shows a typical microwave receiver used in satellite communication. Evaluate (i) the overall noise-figure of the receiver and (ii) the overall equivalent temperature of the receiver. Assume that ambient temperature $T = 17°\,C$.

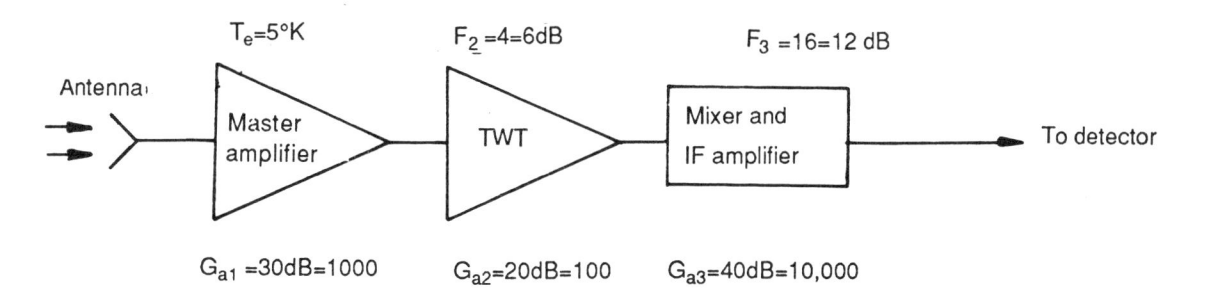

Fig. 4.10. 2 A Microwave Receiver

(i) The noise-figure of the first stage (master amplifier) is

$$F_1 = 1 + \frac{T_e}{T} = 1 + \frac{5}{273 + 17} = 1 + \frac{5}{290}$$

Similarly, noise-figure contributed by the first, second and third stage yield a net Noise-figure as

228  COMMUNICATION SYSTEMS: Analog and Digital

$$F = \left(1 + \frac{5}{290}\right) + \frac{4-1}{1000} + \frac{16-1}{100000} = 1.02$$

The main contribution is made by the first term.

(ii) The overall noise temperature $T_e = T(F - 1) = 290(0.02) = 5.8° K$

### Example 4.10.2

The noise-figure of the individual stages of a two-stage amplifier is 2.03, and 1.54 respectively. The available power gain of the first stage is 62. Evaluate the overall noise-figure.

*Solution*

$$F = F_1 + \frac{F_2 - 1}{G_a}$$

$$F_1 = 2.03, F_2 = 1.54, \text{ and } G_a = 62$$

$$F = 2.03 + \frac{1.54}{62} = 2.03 + 0.0247 = 2.0547 \simeq F_1$$

Thus, the noise-figure of the first stage is dominant, and the following stages contribute little towards the overall noise-figure.

### Example 4.10.3

The amplifier shown in Fig. 4.10.3a has an effective bandwidth of 4 MHz, and a voltage gain of 100. Determine the rms noise voltage at the output. The operating temperature is 27°C. The noise is contributed only by the imput resistance.

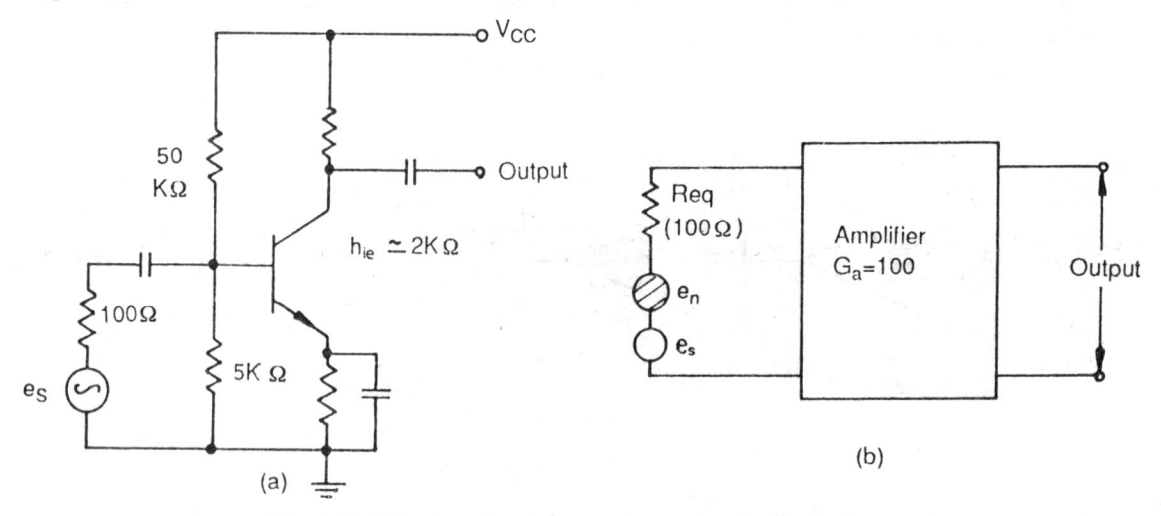

Fig. 4.10.3 (a) Amplifier Circuit, (b) Noise Equivalent Circuit

*Solution*

The equivalent input resistance of the amplifier is

$$R_{eq} = 100\Omega \,||\, 50K\Omega \,||\, 5K\Omega \,||\, h_{ie}$$

$h_{ie} = 2K\Omega$; hence $R_{eq} \cong 100\,\Omega$ (Fig. 4.10.3b.)

$$T = 273 + 27 = 300°K$$

Hence, the rms noise voltage is the input is

$$V_{ni} = \sqrt{4kTR(\Delta f)} = \sqrt{4 \times 1.38 \times 10^{-23} \times 300 \times 100 \times 4 \times 10^6} = 6.62\,\mu V$$

After amplification, the noise voltage is

$$V_{no} = 6.62\,\mu V \times 100 = 0.662\,mV$$

## Equivalent Noise Resistance

Let us define the term *equivalent noise resistance* or simply, *noise resistance* denoted by $R_{eq}$ for cascaded amplifier. The equivalent noise resistance of an amplifier, or receiver is due to the noise contribution of (a) input resistance of the first stage, (b) equivalent noise resistance of the first stage; and, (c) equivalent noise resistance of the subsequent stages referred to the input of the first stage.

The equivalent noise resistance is the input resistance that will produce the same amount of noise at the output, as does the actual amplifier or receiver. Thus, the amplifier, or receiver may be replaced by a noiseless amplifier or receiver with an equivalent noise resistance at its input. This resistance will produce the same amount of noise at the output as does the actual receiver or amplifier.

Consider a multistage system shown in Fig. 4.10.4, $A_n$s, and $R_n'$s represent voltage gain, and input resistance of the individual stages, respectively. Similarly $E_n'$s represents the rms noise voltage at the input of corresponding stages.

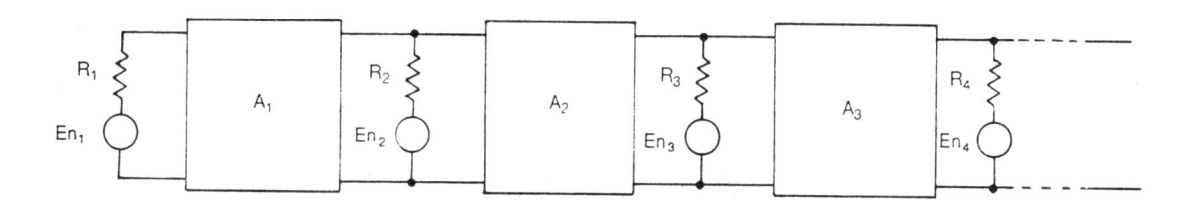

Fig. 4.10.4 Cascaded Stages

It is obvious from Fig. 4.10.4 that $E_{n3}$ is produced by $R_3$. Hence from Eq. 4.3.5b, we have

$$E_{n3} = [4kTR_3(\Delta f)]^{\frac{1}{2}} \quad (4.10.7)$$

This is the noise voltage at the output of the second stage. If this noise is transferred to the input of the second stage, it becomes,

$$E'_{n3} = \frac{E_{n3}}{A_2} = \sqrt{\frac{4kTR_3(\Delta f)}{A_2^2}}$$

$$= \sqrt{4kTR'_3(\Delta f)}$$

where

$$R'_3 = R_3 / A_2^2$$

This is equivalent noise resistance referred to at the input of second stage, and is obtained by dividing the output resistance with the square of the gain of the stage. The resistance $R_2$ is already present at the input of the seconds stage, and hence, the total noise resistance at the input of the second stage is,

$$R_{2t} = R_2 + R_3' = R_2 + R_3/A_2^2$$

Similarly, this resistance $R_{2t}$ may be transferred at the input of the first stage to yield

$$R_2' = \frac{R_{2t}}{A_1^2} = \frac{R_2}{A_1^2} + \frac{R_3}{A_1^2 A_2^2}$$

and the net noise resistance referred to, at the input of the first stage is

$$R_{1t} = R_1 + R_2' = R_1 + \frac{R_2}{A_1^2} + \frac{R_3}{A_1^2 A_2^2}$$

This result may be extended to $n$-stages cascaded networks, as

$$R_{1t} = R_1 + \frac{R_2}{A_1^2} + \frac{R_3}{A_1^2 A_2^2} + \frac{R_4}{A_1^2 A_2^2 A_3^2} + \cdots + \frac{R_n}{A_1^2 A_2^2 \cdots A_{n-1}^2} \cdots \quad (4.10.8)$$

## Noise-Figure in Terms of Noise Resistance

A multistage amplifier with an overall gain $A$ and an input resistance $R_1$, is represented by a single block in Fig. 4.10.5. The amplifier is excited by a generator having a resistance $R_a$.

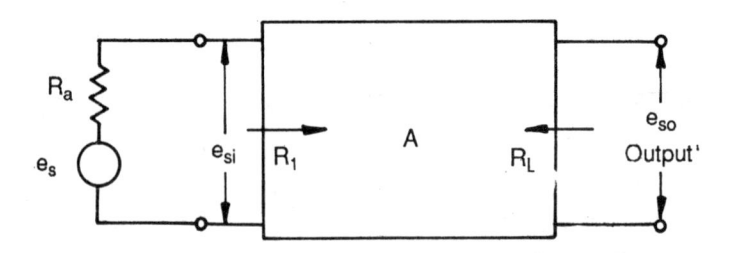

Fig. 4.10.5

For evaluating the noise-figure, we require both the input and output noise power. We will assume an equivalent noise resistance $R_{eq}$ which excludes $R_1$, the input resistance, then

$$R_{eq} = R_{1t} - R_1$$

where $R_{1t}$ is total noise resistance of the amplifier system. It can be shown that the noise-figure $F$, in terms of equivalent noise resistance $R_{eq}$ is given by (Prob. 4.7)

$$F = 1 + R_{eq} \cdot \frac{(R_a + R_1)}{R_a R_1} \quad (4.10.9)$$

For $R_1 \gg R_a$, we have

$$F = 1 + \frac{R_{eq}}{R_a} \quad (4.10.10)$$

## Example 4.10.4

The first stage of a two-stage r.f amplifier has an output resistance of 20 KΩ and a voltage gain of 10. The input resistance, and the noise resistance is 500Ω and 2 KΩ, respectively. The second stage has an output resistance of 400 KΩ, a voltage gain of 20, an input resistance of 80 KΩ and noise resistance of 10 KΩ. Compute equivalent noise resistance of the two-stage amplifier and its noise-figure. The amplifier is driven by a generator whose output impedance is 40Ω.

*Solution*

The net noise resistance $R_{it}$ can be obtained, using Eq. 4.10.8

$$R_3 = 400 \text{ K}\Omega$$

$$R_2 = \frac{20 \times 80}{20 + 80} + 10 = 26 \text{ K}\Omega$$

$$R_1 = 500 + 2000 = 2500 \text{ }\Omega$$

$$R_{1t} = R_1 + \frac{R_2}{A_1^2} + \frac{R_3}{A_1^2 A_2^2} = 2500 + \frac{26000}{10^2} + \frac{400,000}{(10 \times 20)^2}$$

$$= 2770 \text{ }\Omega$$

$$R_{eq} = R_{1t} - R_1$$

$$= 2770 - 500 = 2270 \text{ }\Omega$$

$$F = 1 + \frac{2770(40 + 500)}{40 \times 500} = 75.79$$

## 4.11 MEASUREMENT OF NOISE-FIGURE

In order to know the noise-figure of a two-port network, it is advisable to measure it under the exact conditions of usage. This is because, even when the noise-figure has been specified by the manufacturer, the data may not be useful under the desired conditions of usage. The noise-figure depends on the following main factors:

(i) source impedance; (ii) temperature, (iii) d.c. current level, and (iv) frequency range in use. We will discuss a method for the measurement of the noise-figure in which a calibrated noise source is used. The set up is shown in Fig. 4.11.1. A vacuum diode is used as noise source. The amplifier has a source resistance $R_s$. This method uses the following definition of noise figure.

$$F = \frac{\text{total noise power at the output} (N_{to})}{\text{noise power at the output due solely to the source noise} (N_{so})}$$

The Following steps are adopted in this method

(i) The diode is operated in a temperature-limited region so that it offers an infinite dynamic resistance, i.e., behaves like an open circuit. This may be adjusted by $E_d$.

(ii) Diode circuit (noise source) is disconnected, and the mean square value (power) of the output noise voltage is measured. The reading represents total noise power at the output, $N_o$.

(iii) Now, the diode (noise source) is connected. The m.s. value of the output noise voltage is enhanced due to the shot-noise contribution of the diode. The power density spectrum of the shot noise is $qI_d$, where $I_d$ is the d.c. current through the diode, and $q$ is the electronic charge. The amount of noise power given by the diode source can be controlled by adjusting voltage $E_d$.

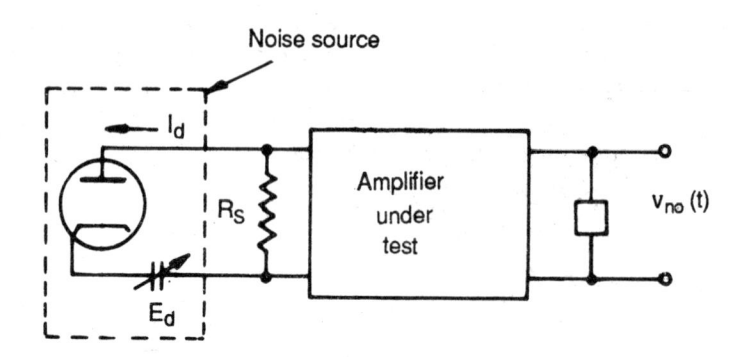

[Figure 4.11.1: Noise source circuit with diode (I_d, E_d), R_s, Amplifier under test, and output v_no(t)]

Fig. 4.11.1

(iv) Adjust $E_d$ so that the m.s. value of the output noise voltage becomes $2N_o$ i.e., twice the reading in step (ii). Read the diode current $I_d$.

The average noise-figure is evaluated by the relation

$$\overline{F} = \frac{qI_dR_s}{2kT} \qquad (4.11.1)$$

The Eq. 4.11.1 can be proved as follows.

*Proof*

The diode represents a noise source with a power density spectrum $qI_d$. By addition of this noise power in the circuit, the output noise power increases by $N_o$. Therefore, the noise power at the output solely due to the input noise source is $N_o$.

Now, let us find the $N_{so}$, the noise power at the output solely due to the source resistance $R_s$. This resistance generates, at the input of an amplifier, a noise source with a power density spectrum $2kTG_s$ where $G_s = (1/R_s)$. Since diode noise source with a power density $qI_d$ produces at the output a noise power $N_o$, the resistance noise source with a power density spectrum $2kTG_s$ will produce a noise power at the output given by

$$N_{so} = 2kTG_s\left(\frac{N_o}{qI_d}\right) = \frac{2kTG_sN_o}{qI_d}$$

Hence the noise figure is given by

$$\overline{F} = \frac{N_o}{N_{so}} = \frac{qI_d}{2kTG_s} = \frac{qI_dR_s}{2kT}$$

This proves the Eq. 4.11.1.

This method provides an average noise figure. In order to measure a spot noise figure $F$, measurement over the entire frequency range has to be performed, considering the narrow band each time. This can be done by using tunable filters at the output.

## 4.12 SIGNAL IN PRESENCE OF NOISE

A message signal is corrupted by noise added during the different processes of communication system. It is desirable to detect the signal in presence of large noise. We will see in the following paragraph that a signal component can be identified by determining the autocorrelation of a noise corrupted signal.

Consider a random process consisting of a sinusoidal message signal (a message signal is assumed to be sinusoidal because any message signal can be resolved in terms of sinusoidal signals), intermixed with white noise component $\omega(t)$. The intermixed signal is given by

$$s(t) = A\cos(\omega_c t + \theta) + \omega(t) \tag{4.12.1}$$

where $\theta$ is a uniformly distributed random variable. The white noise signal is assumed to have a zero mean, and a spectral density function equal to $N/2$. We will find the autocorrelation function of $s(t)$. The two components of $s(t)$ are uncorrelated, hence the autocorrelation function can be obtained by finding the autocorrelation functions of the two components separately and then adding them. The autocorrelation of white noise is an impulse function $\frac{N}{2}\delta(\tau)$ (refer Fig. 4.3.3). Autocorrelation function of a sinusoidal component is $(A^2/2)\cos\omega_c\tau$. Therefore, autocorrelation function of $s(t)$ is given by

$$R_s(\tau) = \frac{A^2}{2}\cos\omega_c\tau + \frac{N}{2}\delta(\tau) \tag{4.12.2}$$

Figure 4.12.1 shows the plot of Eq. 4.12.2. Obviously, the autocorrelation function is same as the sinusoidal message signal, except at $\tau=0$. Thus, the sinusoidal wave can be easily recognized by observing $R_s(\tau)$ beyond $|\tau|>0$. Autocorrelation receivers are used in radars as matched filter to maximize the output $S/N$ ratio.

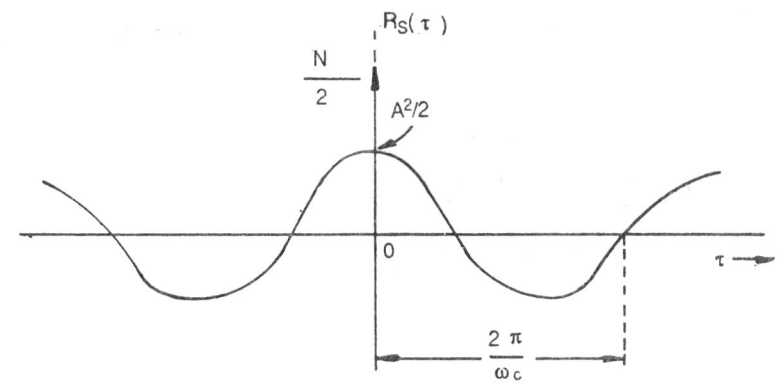

Fig. 4.12.1 Autocorrelation of a Sine Function Corrupted with Noise

## 4.13 NARROW BAND NOISE

In communication systems, message signals intermixed with noise are usually passed through bandpass filters (e.g., tuned circuit in receivers). The bandpass filters have narrow bandwidths in the sense that the bandwidth is small as compared to centre frequency. The wideband noise accompanied with the desired signal is also passed through this bandpass filter. Hence, in general, we have to deal with bandpass noise for evaluating the noise performance of a communication system. We will develop a mathematical model of this narrow band noise, and study the related properties.

**234 COMMUNICATION SYSTEMS: Analog and Digital**

Consider a narrow band Gaussian noise $n(t)$ with the spectral density function $S_n(\omega)$ shown in Fig. 4.13.1. The spectrum is centered about $\omega_c$ and has a bandwidth of $2\omega_m$ radians.

The power density curve may be approximated by a set of delta functions as shown in Fig. 4.13.1. This is due to the fact that an element area given by $S_n(\omega_c + K_1 \Delta\omega).\Delta\omega$ becomes a delta function under the limit

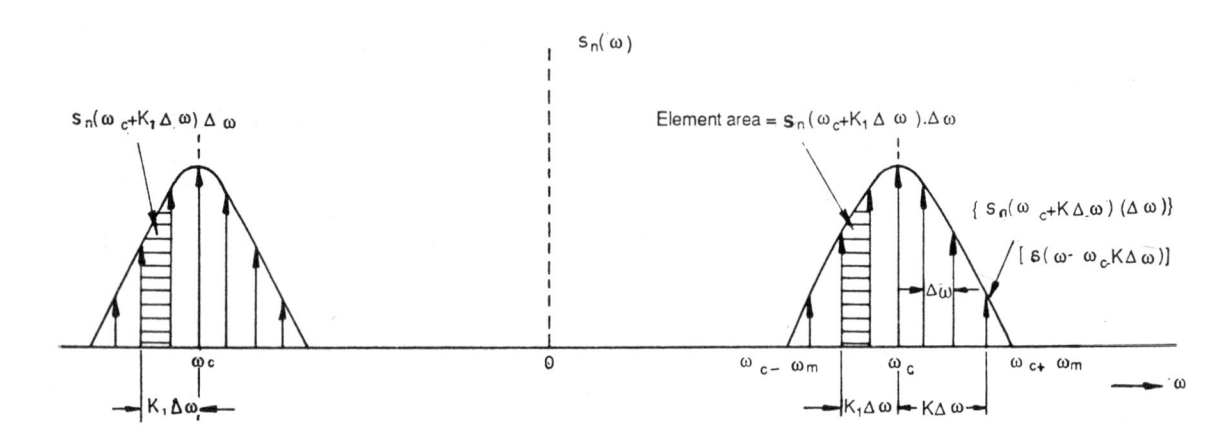

Fig. 4.13.1

$\Delta\omega \to 0$. The pair of delta functions corresponding to this element area will be located at $\omega = \pm(\omega_c - K_1\Delta\omega)$ and have a strength equal to the area $S_n(\omega_c + K_1\Delta\omega).\Delta\omega$. The corresponding pair of delta function is denoted by

$$\{S_n(\omega_c + K_1\Delta\omega)\Delta\omega\}\,\delta[\omega \pm (\omega_c - K_1\Delta\omega)]$$

Similarly, a pair of delta functions located at $\pm(\omega_c + K\Delta\omega)$ is given by

$$\{S_n(\omega_c + K\Delta\omega).\Delta\omega\}\,\delta[\omega \pm (\omega_c + K\Delta\omega)]$$

The characteristic curve may be approximated by summation of similar delta functions located at a uniform spacing of $\Delta\omega$. The entire curve can be represented in terms of delta function as follows:

$$S_n(\omega) = \lim_{\Delta\omega \to 0}\sum_K \{S_n(\omega_c + K\Delta\omega)\Delta\omega\}[\delta(\omega - \omega_c - K\Delta\omega) + \delta(\omega + \omega_c + K\Delta\omega)] \quad (4.13.1)$$

We have seen in Chapter 1 that the spectrum of $\cos\omega_c t$ consists of a pair of impulses located at $\pm\omega_c$. Eq. 4.13.1 belongs to such spectrum, and can be represented in time domain as the sum of cosine functions given below:

$$n(t) = \lim_{\Delta\omega \to 0}\sum_K A_K \cos[(\omega_c + K\Delta\omega)t + \theta_K] \quad (4.13.2)$$

where $A_K$ is defined by

$$A_K = \sqrt{4S_n(\omega_c + K\Delta\omega)\Delta\omega} \quad (4.13.3)$$

and $\theta_K$ belongs to a set of values of an independent random variable $\{\theta_K\}$ each with a uniform distribution defined by

$$p(\theta_k) = \begin{cases} \dfrac{1}{2\pi}, & 0 \leq \theta_K \leq 2\pi \\ 0, & \text{elsewhere} \end{cases} \quad (4.13.4)$$

As $\Delta\omega$ goes to zero, the number of sine functions in Eq. 4.13.2 tends to infinity. According to central limit theorems, it can be defined by a Gaussian process. Equation 4.13.2 is a mathematical description of a *sine wave model* of narrow band noise $n(t)$. This noise is Gaussian distributed with a zero mean, and a variance defined by

$$\sigma^2 = \sum_K \dfrac{A_K^2}{2} \quad (4.13.5)$$

## Narrow Band Noise in Terms of Quadrature Components

The expansion of the sinusoidal function in Eq. 4.13.2 yields

$$\cos\left[(\omega_c + K\Delta\omega)t + \theta_K\right] = \cos\omega_c t \cdot \cos(K\Delta\omega t + \theta_K) - \sin\omega_c t \cdot \sin(K\Delta\omega t + \theta_K)$$

Substituting this expanded form in Eq. 4.13.2, we get

$$n(t) = n_c(t)\cos\omega_c t - n_s(t)\sin\omega_c t \quad (4.13.6)$$

where

$$n_c(t) = \lim_{\Delta\omega \to 0} \sum_K A_K \cos(K\Delta\omega t + \theta_K) \quad (4.13.7)$$

and

$$n_s(t) = \lim_{\Delta\omega \to 0} \sum_K A_K \sin(K\Delta\omega t + \theta_K) \quad (4.13.8)$$

It is obvious that term $n_c(t)$ is in phase with the carrier wave $\cos\omega_c t$; whereas $n_s(t)$ is 90° out of phase with the carrier wave term. Hence $n_c(t)$ is referred to as *in-phase component* and $n_s(t)$ as *quadrature component* of the narrow-band noise $n(t)$.

## Properties of the Noise Components

(1) The quadrature components $n_c(t)$ and $n_s(t)$ are low frequency (slowly varying) signals. In Eqs 4.13.7 and 4.13.8, the frequency $\Delta\omega$ is small, and hence $n_c(t)$ and $n_s(t)$ represent low frequency sinusoids. If $n(t)$ represents a bandpass signal limited to $\omega_c + \omega_m$, then $n_c(t)$ and $n_s(t)$ will be limited to $\omega_m$. Therefore, components $n_c(t)$ and $n_s(t)$ can be derived from $n(t)$ by using a scheme shown in Fig. 4.13.2. The low pass filters have a bandwidth $\omega_m$ much less than carrier frequency $\omega_c$. Hence, $n_s(t)$ and $n_c(t)$ have slow time variations as compared to $\omega_c$.

(2) Both the components $n_c(t)$, and $n_s(t)$, have the same spectral power density given by

$$S_{n_c}(\omega) = S_{n_s}(\omega) = \begin{cases} S_n(\omega + \omega_c) + S_n(\omega - \omega_c), & -\omega_m \leq \omega \leq \omega_m \\ 0, & \text{otherwise} \end{cases} \quad (4.13.9)$$

[Fig. 4.13.2]

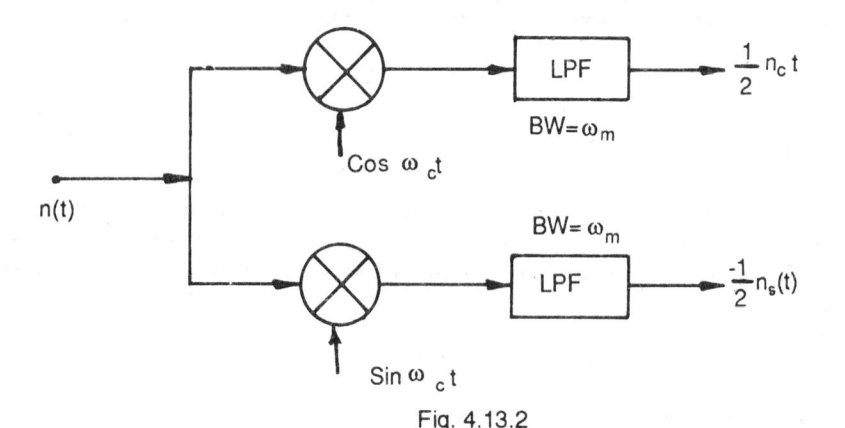

## Proof

From Eqs 4.13.7 and 4.13.8, the power density spectra of $n_c(t)$ and $n_s(t)$ is given by

$$S_{n_c}(\omega) = S_{n_s}(\omega) = \lim_{\Delta\omega \to 0} \sum_K \frac{A_K^2}{4}[\delta(\omega - K\Delta\omega) + \delta(\omega + K\Delta\omega)] \quad (4.13.10)$$

Similarly from Eq. 4.13.2, the power density of $n(t)$ is given as

$$S_n(\omega) = \lim_{\Delta\omega \to 0} \sum_K \frac{A_K^2}{4}[\delta(\omega - \omega_c - K\Delta\omega) + \delta(\omega + \omega_c + K\Delta\omega)] \quad (4.13.11)$$

Now, using the frequency shifting theorem of power density spectrum on 4.13.11, get

$$S_n(\omega + \omega_c) = \lim_{\Delta\omega \to 0} \sum_K \frac{A_K^2}{4}[\delta(\omega - K\Delta\omega) + \delta(\omega + 2\omega_c + K\Delta\omega)] \quad (4.13.12)$$

and

$$S_n(\omega - \omega_c) = \lim_{\Delta\omega \to 0} \sum_K \frac{A_K^2}{4}[\delta(\omega - 2\omega_c + K\Delta\omega) + \delta(\omega + K\Delta\omega)] \quad (4.13.13)$$

Adding Eqs 4.13.12 and 4.13.13, and eliminating the terms lying outside $-\omega_m \leq \omega \leq \omega_m$, we will get Eq. 4.13.9 which proves the theorem.

When $n(t)$ has a symmetrical spectral density curve about $\pm \omega_c$ Eq. 4.13.9 becomes

$$S_{n_c}(\omega) = S_{n_s}(\omega) = \begin{cases} 2\delta_n(\omega + \omega_c), & -\omega_m \leq \omega \leq \omega_m \\ 0, & \text{elsewhere} \end{cases} \quad (4.13.14)$$

(3) The powers (m.s. values) of $n(t), n_c(t)$ and $n_s(t)$, are identical,

$$\overline{n^2(t)} = \overline{n_c^2(t)} = \overline{n_s^2(t)} \quad (4.13.15)$$

The power is obtained by finding the area under the curve of the power density spectrum. It can be seen that the area under the power density curves of $n(t), n_c(t)$ and $n_s(t)$, are identical (Prob. 4.16).

(4) If $n(t)$ is a Gaussian with a zero mean, then

(i) $n_c(t)$ and $n_s(t)$ are statistically independent,
(ii) $n_c(t)$ and $n_s(t)$ are also Gaussian with zero mean, and
(iii) the variance of $n(t)$, $n_c(t)$, and $n_s(t)$ is identical
(5) Power of $n(t)$ is equally divided into its quadrature components.
From Eq. 4.13.15, we have

$$\frac{\overline{n_s^2}}{2} + \frac{\overline{n_c^2}}{2} = \frac{\overline{n^2(t)}}{2} + \frac{\overline{n^2(t)}}{2} = \overline{n^2(t)} \qquad (4.13.16)$$

It is obvious from Eq. 4.13.16 that the power of $n(t)$ is equally divided into its quadrature components $n_c(t)$ and $n_s(t)$.

### Envelope of Narrow Band Noise $n(t)$

The noise signal $n(t)$ given by Eq. 4.13.6 can be written in an alternate form as given below:

$$n(t) = R(t) \cos\left[\omega_c t + \theta(t)\right] \qquad (4.13.17)$$

where

$$R(t) = \sqrt{n_c^2(t) + n_s^2(t)} \qquad (4.13.18a)$$

and

$$\theta(t) = \tan^{-1}\left[\frac{n_s(t)}{n_c(t)}\right] \qquad (4.13.18b)$$

The function $R(t)$ and $\theta(t)$ are called envelope and phase of $n(t)$ respectively.

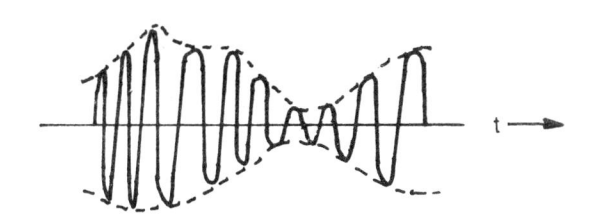

Fig. 4.13.3  Wave-form of Narrow Band Noise $n(t)$

Equation 4.13.7 represents a waveform shown in Fig. 4.13.3. The waveform has a sinusoidal appearance having both amplitude and angle modulation.

## PROBLEMS

4.1 (a) Prove Eq. 4.2.2.
  (b) State, explain and prove the principle of superposition of spectra for noise signals.

4.2  A single stage vacuum tube amplifier is shown in Fig. Prob 4.2. Show that the power density spectrum of the output noise voltage is given by $S_{no}(\omega) = \dfrac{2.52 \times 10^{-5}}{10^{10} + \omega^2}$ and the rms value of the noise voltage is 11.2 μV.

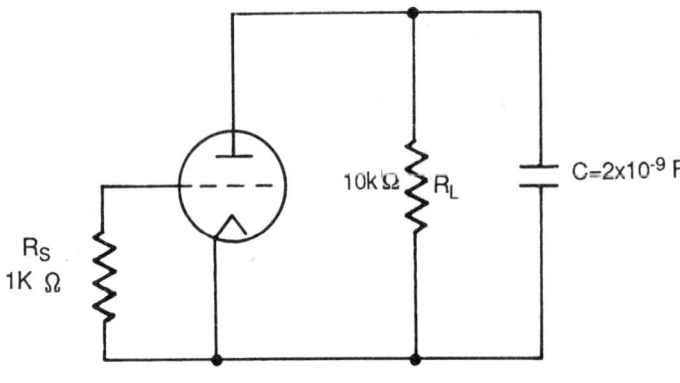

**Fig. Prob. 4.2**

$g_m = 2.5 \times 10^{-3}$ ℧
$r_p = 10^4$ Ω
$\mu = g_m r_p$

$R_S = 1K\ \Omega$, $R_L = 10k\ \Omega$, $C = 2 \times 10^{-9}$ F

4.3 Show that, for a R.C lowpass filter the noise bandwidth $B_N$ and system bandwidth $B_{3dB}$ are related as

$$B_N = \frac{\pi}{2} B_{3dB}$$

4.4 Show that for a current source $i_n(t)$ with the power density spectrum $S_i(\omega)$ and internal admittance $Y = G_s + jB_s$, the available power density $S_a(\omega)$ is given by

$$S_a(\omega) = \frac{S_i(\omega)}{4G_S}$$

4.5 Find the noise figure of a common emitter transistor amplifier, and show that the noise figure is same as that of a common base amplifier given by Eq. 9.27.

4.6 Show that the available power gain $G_a$ of an amplifier with $n$-cascaded stages is given by the product of the available gains of the individual stage $G_{a1}, G_{a2}, G_{a3}, \cdots, G_{an}$.

4.7 Prove the relation given in Eq. 4.10.9.

4.8 Determine the noise figure of a common base transistor amplifier if

$r_b = 200$ ohms, $r_e = 20$ ohms, $\alpha_o = 0.96$, $R_s = 1K$ ohms.

Derive the formula used, if any.

4.9 Two resistors each of 2K ohms are at temperature 300° K and 400° K respectively. Find the power density spectrum of noise voltage at the terminals formed by (a) series combination; (b) parallel combination of these resistors.

4.10 An antenna has a resistance of 100 ohms. The receiver connected to it has an equivalent noise resistance of 60 ohms. Find the receiver noise-figure.

4.11 The equivalent noise temperature of a parametric amplifier is 40° K. What is its noise figure if the ambient temperature is 27°C?

4.12 Determine equivalent noise bandwidth of circuits with the following transfer functions. Take d.c as reference frequency.

(a) $$H(\omega) = Ke^{-a|\omega|}$$

(b) $$H(\omega)=G_{2\omega_m}(\omega)$$

4.13 Determine the equivalent noise bandwidth referred to $\omega_c$ if
$$H(\omega) = S_a[(\omega - \omega_c)t_o] + S_a[(\omega + \omega_c)t_o]$$

4.14 Give reasons why a coscode amplifier has improved noise performance.

4.15 Determine the power density spectrum of the noise current flowing through a series R-L-C network. Also, determine the r.m.s value of the noise current in an R-L circuit.

4.16 Plot the power density spectrum of $n_c(t), n_s(t)$ and $n(t)$ assuming that $n_c(t)$ and $n_s(t)$ have identical spectra limited to $|\omega| \leq \omega_m$. Show that areas under the three curves are equal.

# Five
# AMPLITUDE MODULATION SYSTEMS

## INTRODUCTION

Modulation is a fundamental requisite of communication systems. It is defined as the process by which some characteristic of a signal called *carrier* is varied in accordance with instantaneous value of another signal called *modulating* signal. Signals containing information or intelligence to be transmitted are referred to as modulating signals. This information bearing signal is also called *baseband signal*. The term baseband designates the band of frequencies representing the signal supplied by the source of information. The carrier frequency is greater than the modulating frequencies. The signal resulting from the process of modulation is called *modulated signal*.

### Continuous Wave or Analog Modulation

When the carrier wave is continuous in nature the modulation process is known as continuous wave ($CW$) modulation or analog modulation. The carrier wave is usually a sinusoidal signal. Amplitude modulation, and angle modulation are examples of this type of modulation.

### Pulse Modulation Systems

When the carrier wave is a pulse-type waveform, the modulation process is known as pulse modulation. *PAM, PWM, PCM, DM* etc., are examples of this type of modulation.

### Need for Modulation

Modulation is needed in a communication system to achieve the following basic needs.

### (1) Multiplexing

Simultaneous transmission of multiple messages (more than one message) over a channel is known as multiplexing. The channel is referred to here, as the media of transmission. The channel may be a pair of wires (called transmission lines) or free space. If transmitted without modulation, the different message signals over a single channel will interfere with one another. This is because their baseband (spectrum) is identical. However, different message signals can be transmitted over a single channel without interference using multiplexing techniques. There are two types of multiplexing techniques, viz. (i) frequency division multiplexing, and (ii) time division multiplexing. The frequency division multiplexing uses analog modulation systems, whereas, the time division multiplexing uses pulse modulation systems.

Multiplexing helps in transmitting a number of messages simultaneously over a single channel and therefore the number of channels needed will be less. This reduces the cost of installation and maintenance of more channels.

## (2) *Practicability of Antennas*

When free space is used as a communication media, messages are transmitted and received with the help of antennas. The message signal is radiated by an antenna at the transmitter. The antenna radiates effectively when its height is of the order of the wavelength of the signal being transmitted. In broadcast systems, the maximum audio frequency transmitted from a radio station is of the order of 5 KHz. If this signal were to be transmitted without modulation, the height of antenna needed for an effective radiation would be half of the wavelength given as

$$\frac{\lambda}{2} = \frac{c}{2f} = \frac{3 \times 10^8}{2 \times 5 \times 10^3} = 30,000 \text{ meters} = 30 \text{ Km}$$

It would be impracticable to construct and install such an antenna.

However, the height of the antenna can be reduced by analog modulation technique and yet achieve effective radiation. As modulation provides frequency shifting or frequency translation, audio frequency signals at radio stations (3Hz-5kHz) are translated to higher frequency spectrum (radio frequency ranges). The higher radio frequencies with smaller wavelength act as carrier of the audio frequencies (modulating signal). Therefore, the height of the antenna required is much reduced and becomes practical. For example if an audio frequency is translated to a radio frequency carrier of frequency 1 MHz, the antenna height required will be

$$\frac{\lambda}{2} = \frac{3 \times 10^8}{2 \times 10^6} = 150 \text{ meters}$$

This antenna height can be practically achieved.

## Narrowbanding

Assuming that the practicability of an antenna height is not a problem, there is yet another problem caused by direct radiation of baseband signal. Let us assume that baseband signal in a broadcast system is radiated directly with the frequency range extending from 50 Hz to 10 kHz, the ratio of highest to lowest wavelength is 200. If an antenna is designed for 50 Hz, it will be too long for 10 kHz and vice versa, and we may require a wide-band antenna which can operate for band-edge ratio of 200, which is practically impossible. However, suppose the audio signal is translated to radio range frequency say 1 MHz then ratio of lowest to highest frequency will be $\frac{10^6 + 50}{10^6 + 10^4} \cong \frac{1}{1.01}$, which is approximately unity; and the same antenna will be suitable for the entire band extending from $(10^6 + 50)$ Hz to $(10^6 + 10^4)$ Hz. Thus frequency translation converts a wideband signal to a narrowband. Here bandwidth is referred to as the ratio of the edge frequencies of the band. This is called *narrowbanding*.

## Types of Analog Modulation

Usually a carrier wave in analog modulation is sinusoidal, given by
$e_c = A \cos(\omega_c t + \theta)$
This sine wave is expressed by three parameters
(i) amplitude $A$
(ii) frequency $\omega_c$ (angular frequency)
(iii) phase angle $\theta$

Any of the three parameters can be varied in accordance with the baseband signal. Accordingly, the modulation process is termed as *amplitude modulation, frequency modulation,* or *phase modulation.* The frequency modulation and phase modulation both effectively, vary the total angle $\phi = \omega_c t + \theta$ of the carrier wave, and hence are combined in a single heading—*Angle modulation.* Thus, we can say that there are two types of analog (CW)modulation as under:

1. Amplitude modulation: A modulation process in which amplitude of the carrier is varied in accordance with the instaneous value of the modulating signal is known as amplitude modulation.

2. Angle-Modulation: A modulation process in which total phase angle of the carrier is varied in accordance with the modulating signal is known as angle modulation. The total angle can be varied either by *frequency modulation* or *phase modulation.*

The main aim of CW modulation (amplitude or angle) is frequency translation, to achieve frequency division multiplexing and practical antenna designs.

## 5.1 SUPPRESSED CARRIER SYSTEMS (DSB-SC)

The modulation theorem of Fourier transform states that the spectrum of $f(t) \cos\omega_c t$ is same as that of $f(t)$, except that it is translated by $\pm \omega_c$. Mathematically,
if
$$f(t) \leftrightarrow F(\omega)$$
then
$$f(t) \cos \omega_c t \leftrightarrow \frac{1}{2}[F(\omega + \omega_c) + F(\omega - \omega_c)] \qquad (5.1.1)$$

Therefore, a method for achieving frequency translation is to multiply the modulating signal $f(t)$ with a

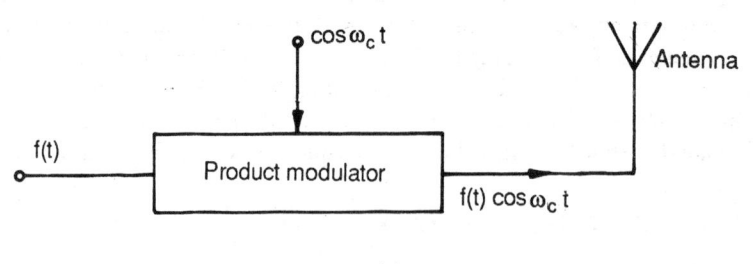

(a)

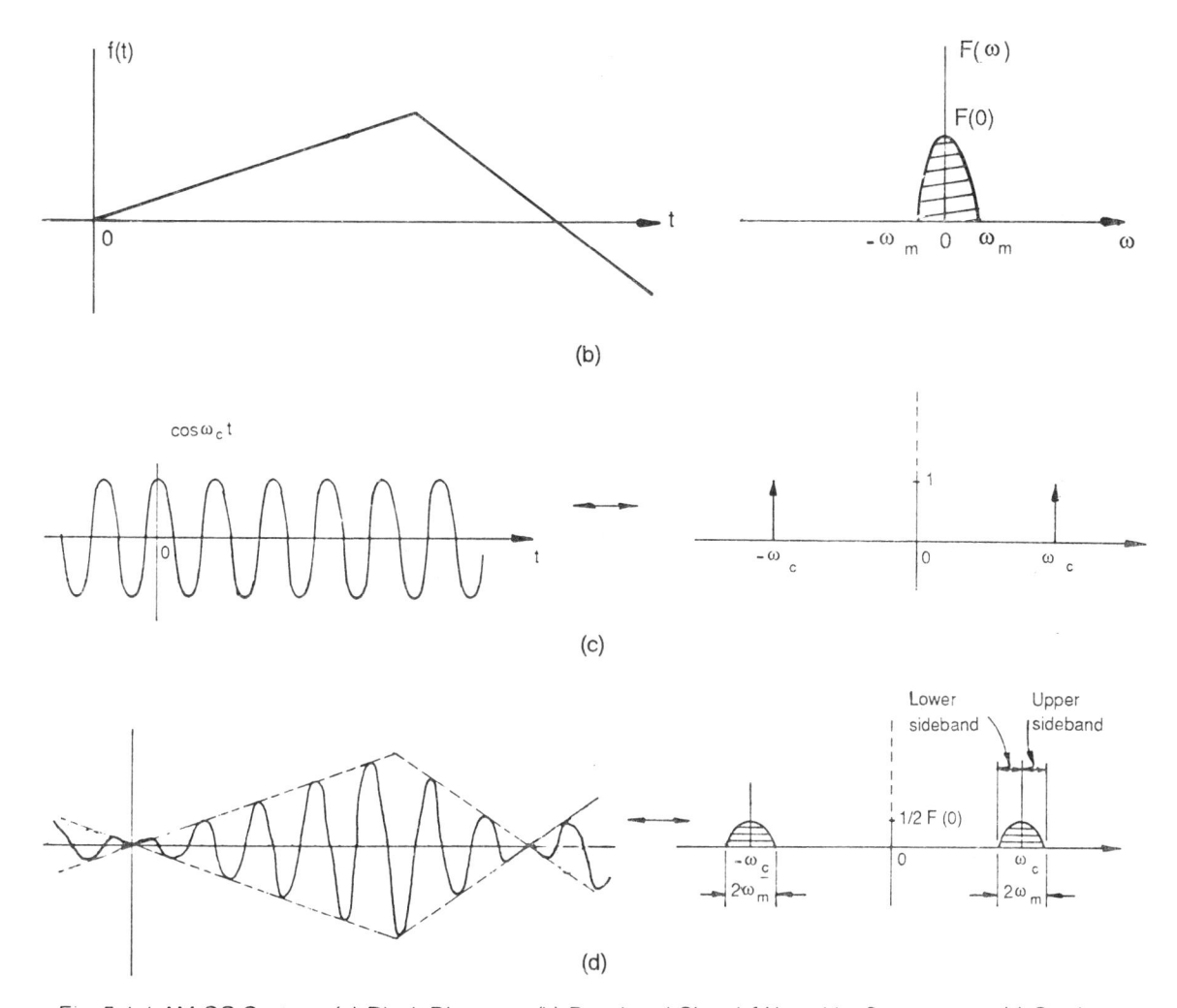

Fig. 5.1.1 AM-SC System (a) Block Diagram; (b) Baseband Signal f (t) and its Spectrum; (c) Carrier and its Specturm; (d) Modulated Signal and its Specturm

sinusoidal carrier signal $\cos \omega_c t$. The schematic diagram of such a product modulator is shown in Fig. 5.1a. This is used at transmitter end of a communication system. The multiplication yields amplitude variation in the carrier wave in accordance with modulating signal $f(t)$ as shown in Fig. 5.1.1d. Therefore $f(t) \cos \omega_c t$ is an amplitude modulated signal, known as *Amplitude Modulated Suppressed Carrier (AM-SC)* signal. *The AM-SC signal exhibits phase-reversal at zero crossings* which is obvious from the waveform of Fig. 5.1.1d. From the spectrum in Fig. 5.1.1d, it is obvious that the impulses at $\pm \omega_c$ are missing which means

(a) the carrier term $\omega_c$ is suppressed in the spectrum. Hence, it is called suppressed carrier system (AM-SC).

(b) the baseband (0 to $\omega_m$) is present twice in the modulated spectrum. The modulated signal consists of $\pm \omega_c + \omega_m$ and $\pm \omega_c - \omega_m$ frequency terms. Positive and associated negative frequency terms are necessary for a real signal. The two terms mentioned above are called sidebands. The term $\pm \omega_c + \omega_m$ is called upper sideband and $\pm \omega_c - \omega_m$ is called lower sideband. Thus, this system produces two sidebands corresponding

**244** COMMUNICATION SYSTEMS: Analog and Digital

to each frequency component in modulating signal. The system is therefore, termed as a *double sideband suppressed carrier* (DSB-SC) amplitude modulation system.

## Recovery of Baseband Signal: Synchronous Detection

The AM-SC system is used at the transmitter for shifting the baseband signal (with maximum frequency $\omega_m$) to a higher carrier frequency $\pm \omega_c$. This modulated signal is transmitted from the transmitter and it reaches the receiver via a propagating media. At the receiver, the original baseband signal $f(t)$ is desired to be recovered from the modulated signal. This is achieved by retranslating the baseband signal from a higher spectrum (centered at $\pm \omega_c$) to the original spectrum. This process of retranslation is known as *demodulation* or *detection*. The original baseband is recovered from the modulated signal by the detection process.

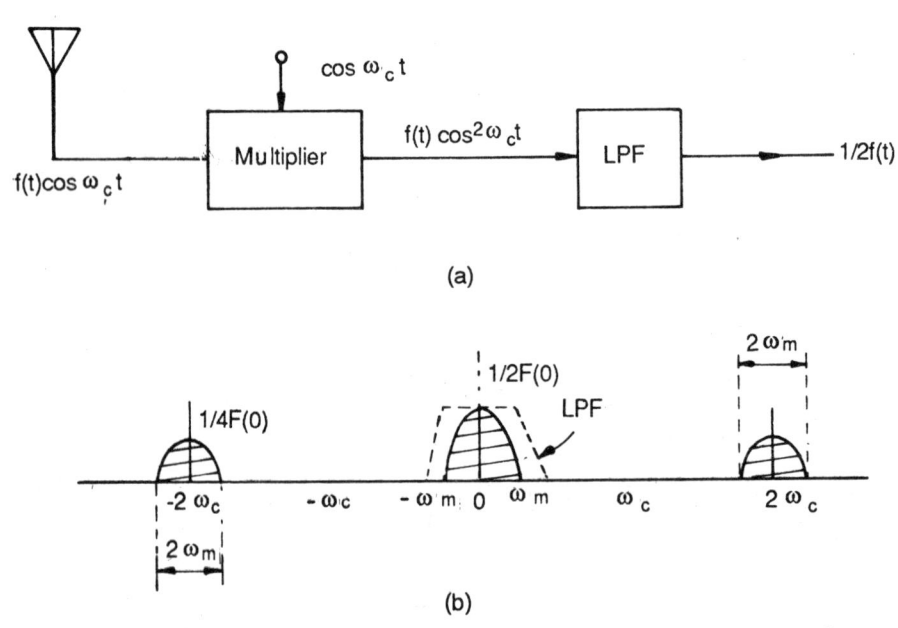

Fig. 5.1.2 Synchronous Detection (a) Block Diagram; (b) Spectrum of $f(t) \cos^2 \omega_c t$

A method for detecting the AM-SC signal shown in Fig. 5.1.2a, is used at the receiver end for recovery of the message signal. The method of retranslation is similar to that of translation. Here modulated signal $f(t) \cos \omega_c t$ is multiplied with $\cos \omega_c t$ (locally generated carrier), and passed through a low pass filter. The signal $f(t) \cos \omega_c t$ when multiplied with $\cos \omega_c t$, yields

$$f(t) \cos^2 \omega_c t = \frac{1}{2} f(t)[1 + \cos 2\omega_c t]$$

$$= \frac{1}{2} f(t) + \frac{1}{2} f(t) \cos 2\omega_c t \qquad (5.1.2)$$

It is clear from right-hand side of the Eq. 5.1.2 that the term $\frac{1}{2} f(t) \cos 2\omega_c t$ centered near $\pm 2\omega_c$ can be

AMPLITUDE MODULATION SYSTEMS 245

bypassed by a low pass filter, and at the output of the low pass filter original baseband signal $\frac{1}{2}f(t)$ is recovered. The spectrum of $f(t)\cos^2\omega_c t$ is obtained by taking Fourier transform of *RHS* of Eq. 5.1.2,

$$f(t)\cos^2\omega_c t \leftrightarrow \frac{1}{2}F(\omega) + \frac{1}{4}[F(\omega + 2\omega_c) + F(\omega - 2\omega_c)] \tag{5.1.3}$$

The spectrum is shown in Fig. 5.1.2b. The spectrum reveals that the original baseband signal (0 to $\omega_m$) is present alongwith another spectrum centered around $\pm 2\omega_c$. When passed through a lowpass filter (with a cut-off frequency $\omega_m$), the original baseband signal appears at the output of the filter. The spectrum centered near $\pm 2\omega_c$ is not allowed to pass through the lowpass filter. Note that $\omega_c >> \omega_m$ and $2\omega_c$ is still greater than $\omega_m$, and is easily filtered out. Thus, the original message signal $f(t)$ is recovered from the AM-SC signal.

## Synchronous, or Coherent, or Homodyne Detection

The detection process in Fig. 5.1.2a needs a local oscillator. The frequency and phase of the locally generated signal and the carrier signal at the transmitter must be identical. In other words, local oscillator signal must be exactly coherent or synchronized with the carrier wave at the transmitter, both in frequency and phase, otherwise the detected signal will be distorted. Hence, this method of recovery is known as synchronous detection, or coherent detection, also known as homodyne detection. The demerits of the synchronous detection is that it requires an additional system at the receiver to ensure that locally generated carrier is synchronized with the transmitter carrier, making the receiver complex and costly.

## Effect of Phase and Frequency Errors in Synchronous Detection

The frequency and phase of the local oscillator signal in synchronous detection must be identical to the transmitted carrier. Any discrepancy in frequency and/or phase causes a distortion in the detected output at the receiver. Let us examine the nature of distortion caused by phase or frequency discrepancy.

Let a modulated signal reaching the receiver signal be $f(t)\cos\omega_c t$. Assuming, a locally generated signal with frequency and phase error equal to $\Delta\omega$ and $\phi$ respectively, the product of the two signals in the synchronous detector yields,

$$e_d(t) = f(t)\cos\omega_c t \cdot \cos[(\omega_c + \Delta\omega)t + \phi]$$

$$= \frac{1}{2}f(t)\{\cos[(\Delta\omega)t + \phi] + \cos[(2\omega_c + \Delta\omega)t + \phi]\}$$

When this signal is passed through a low pass filter with a cut off frequency $\omega_m$, the terms centered around $\pm 2\omega_c$ are filtered out, and the filter output is given by

$$e_o(t) = \frac{1}{2}f(t)\cos[(\Delta\omega)t + \phi] \tag{5.1.4}$$

The baseband signal $f(t)$ is multiplied by a slow-time varying function $\cos[(\Delta\omega)t + \phi]$ that distorts the message signal $f(t)$. Let us consider the following special cases,
(i) when the frequency error $\Delta\omega$, and phase error $\phi$ are both zero, then Eq. 5.1.4 yields,

$$e_o(t) = \frac{1}{2}f(t)$$

i.e., there is no distortion in the detected output.

(ii) when there is only the phase error, i.e.,

$$\Delta\omega = 0, \text{ but } \phi \neq 0$$

then Eq. 5.1.4 yields,

$$e_o(t) = \frac{1}{2} f(t) \cos \phi$$

which shows that the output is multiplied by $\cos\phi$. When $\phi$ is time independent, *there is no distortion; rather, there is only attenuation*. The output is maximum when $\phi = 0$, and minimum when $\phi = 90°$. However, in general, $\phi$ randomly varies with time due to random variation of propagation media (ionosphere). This causes undesirable distortion in the detected output.

**Quadrature Null Effect**

The detected output is zero when $\phi = 90°$. This is called a quadrature null effect, because the signal is zero when the local carrier is in phase quadrature with the transmitted carrier.

(iii) when there is only the frequency error, i.e.,

$$\Delta\omega \neq 0, \text{ and } \phi = 0$$

then Eq. 5.1.4 yields

$$e_o(t) = \frac{1}{2} f(t) \cos(\Delta\omega) t$$

Here, the multiplying factor $\cos(\Delta\omega)t$ is time-dependent, and causes distortion in the detected output. The error $\Delta\omega$ is usually small, and hence a message $f(t)$ is multiplied by a slow varying sinusoidal signal. This is a more serious distortion. Hence, frequency error should be avoided.

(iv) when both errors are non-zero, i.e., $\Delta\omega \neq 0$ and $\phi \neq 0$, Eq. 5.1.4 itself provides the detected output. In this case, the constant phase error provides attenuation, and the frequency error causes distortion in the detected output. Thus, we get an attenuated and distorted output in the receiver.

**Synchronization Techniques**

The phase and frequency of the locally generated carrier in synchronous detector is extremely critical. Precision phase and frequency control of the local carrier needs an expensive and a complicated circuitry at the receiver. Some synchronization techniques are given as follows:

*1. Pilot Carrier*

A small amount of carrier signal known as pilot carrier is transmitted alongwith the modulated signal from the transmitter. This small amount of carrier signal is known as pilot carrier. This pilot carrier, seperated at the receiver by an appropriate filter, is amplified, and is used to phase lock the locally generated carrier at the receiver. The phase locking provides synchronization. This system, where a weak carrier is transmitted alongwith AM-SC signal, is also refferred to as *partially suppressed carrier* system, as the carrier is not totally suppressed. The process in which a larged carrier is transmitted alongwith AM-SC signal is called **amplitude modulation**. This is dealt in the next section. The large carrier simplifies the reception system. The AM-SC with partially suppressed carrier is equivalent to an overmodulated AM signal.

## 2. Costa's Receiver

This system, used for synchronous detection of AM-SC signal, is shown in Fig. 5.1.3.

The system has two synchronous detectors—one detector is fed with a locally generated carrier which is in phase with the transmitted carrier. This detector is known as *inphase coherent detector* or *I-channel*. The other synchronous detector employs a local carrier which is in phase quadrature with the transmitted carrier, and is known as *Quadrature phase coherent detector* or *Q-channel*. Combined, the two detectors constitute a negative feedback system which synchronizes the local carrier with the transmitted carrier.

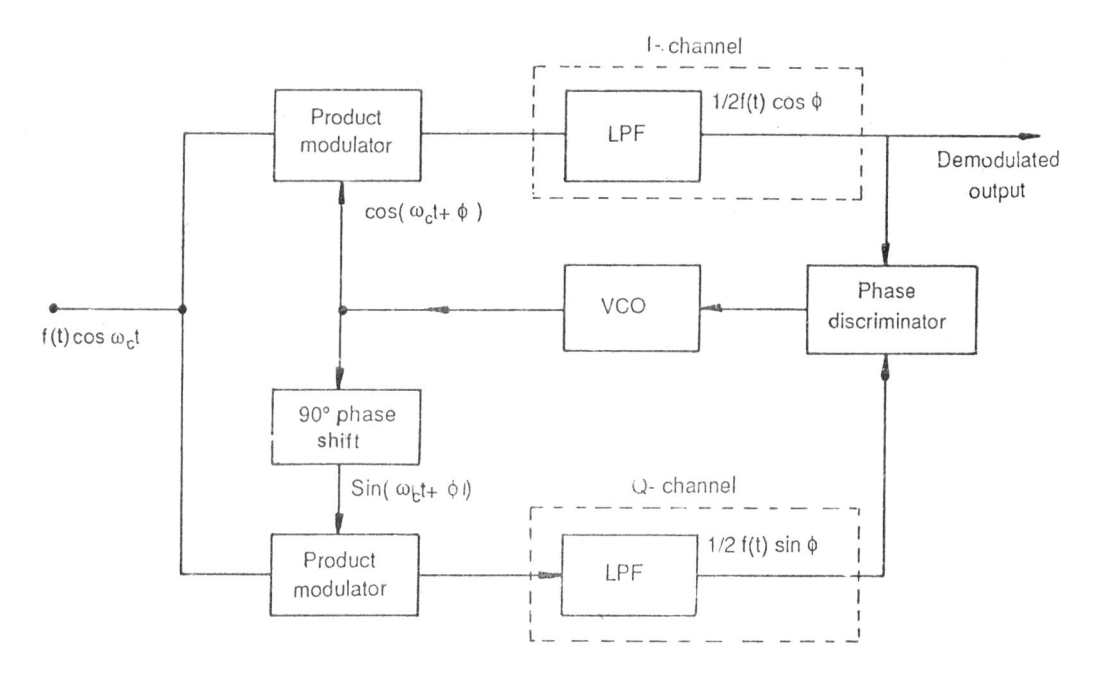

Fig. 5.1.3 Costa's Receiver

### Principle of Operation

To start with, assume that the local carrier signal is synchronized with the transmitted carrier, and $\phi \neq 0$. As shown in Fig. 5.1.3, the output of the I-channel is the desired modulating signal (as $\cos \phi = 1$), but the output of Q-channel is zero (as $\sin \phi = 0$) due to the quadrature null effect. Now, assuming that the local oscillor-frequency drifts slightly, i.e., $\phi$ is a small non-zero quantity, I-channel output is almost unchanged, but Q-channel output now is not a zero, rather, some signal will appear at its output, proportional to $\sin \phi$. The output of the Q-channel,

(i) is proportional to $\phi$ (as $\sin \phi = \phi$ for small $\phi$)

(ii) will have a polarity same as the I-channel for one direction of phase shift in local oscillator, whereas, the polarity will be opposite to I-channel for the other direction of phase shift.

The phase discriminator provides a dc control signal which may be used to correct local oscillator phase error. The local oscillator is a voltage controlled oscillator (VCO). Its frequency can be adjusted by an error control dc signal.

*Limitation:* The Costa's receiver ceases phase control when there is no modulation, i.e., $f(t) = 0$. The phase control re-establishes itself on the reappearance of modulation. The re-establishment is so rapid that distortion is not perceptible in voice communication.

**248** COMMUNICATION SYSTEMS: Analog and Digital

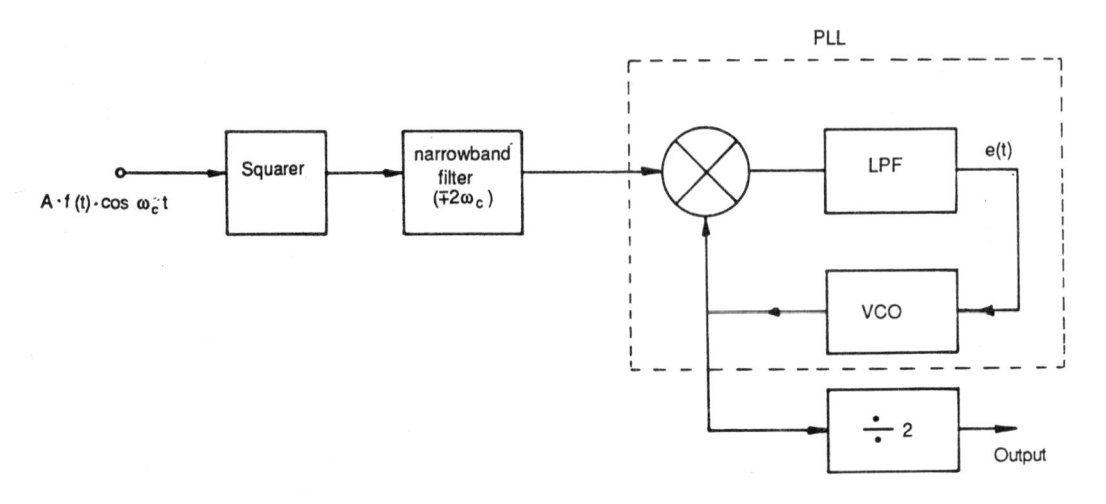

*Fig. 5.1.4 Squaring Circuit for Synchronization*

### 3. Squaring Loop

In this method, the received signal is squared by a squaring circuit as shown in Fig. 5.1.4. The output of the squarer is given as,

$$[A \cdot f(t) \cdot \cos \omega_c t]^2 = A^2 f^2(t) \cos^2 \omega_c t$$

For convenience, let us assume that $f(t)$ is a single tone sinusoid $\cos \omega_m t$, then the output of squarer becomes,

$$[A \cos \omega_c t \cdot \cos \omega_m t]^2 = A^2 \cos^2 \omega_m t \cos^2 \omega_c t$$

$$= \frac{A^2}{4} (1 + \cos 2\omega_m)(1 + \cos 2\omega_c t)$$

$$= \frac{A^2}{4}[1 + \cos 2\omega_m t + \cos 2\omega_c t + \cos 2\omega_c t \cos 2\omega_m t] \quad (5.1.5)$$

The term $\cos 2\omega_c t$ may be obtained by using a narrowband filter centered at $\pm 2\omega_c$. This frequency $\pm 2\omega_c$ is kept constant by tracking through a phase locked loop (PLL). The PLL uses negative feedback technique for providing a constant frequency signal, $\cos 2\omega_c t$. Any drift in frequency is corrected by an error signal $e(t)$, generated at the output of the lowpass filter of PLL, as shown in Fig. 5.1.4. PLL has been discussed in detail in the next chapter. The VCO output is frequency divided by 2, to yield a synchronized local carrier of frequency $\omega_c$. This local carrier is used in synchronous detector. The frequency division may be accomplished by using *bistable multivibrator*.

### Example 5.1.1

In an AM-SC system, the modulating signal is a single-tone sinusoid $E_m \cos \omega_m t$ which modulates a carrier signal $E_c \cos \omega_c t$. Plot the spectrum of the modulated wave.

## Solution

The AM-SC system is a product modulator, and its output yields,

$$e(t) = E_m \cos \omega_m t \cdot E_c \cos \omega_c t$$

$$= \frac{1}{2} E_m E_c [\cos(\omega_c + \omega_m)t + \cos(\omega_c - \omega_m)t]$$

The Fourier transform is given as,

$$E(\omega) = \frac{\pi}{2} E_m E_c [\delta\{\omega + (\omega_c + \omega_m)\} + \delta\{\omega - (\omega_c + \omega_m)\} + \delta\{\omega + (\omega_c - \omega_m)\} + \delta\{\omega - (\omega_c - \omega_m)\}]$$

The spectrum is shown in Fig. 5.1.5. The two sidebands are $\omega_c - \omega_m$ and $\omega_c + \omega_m$.

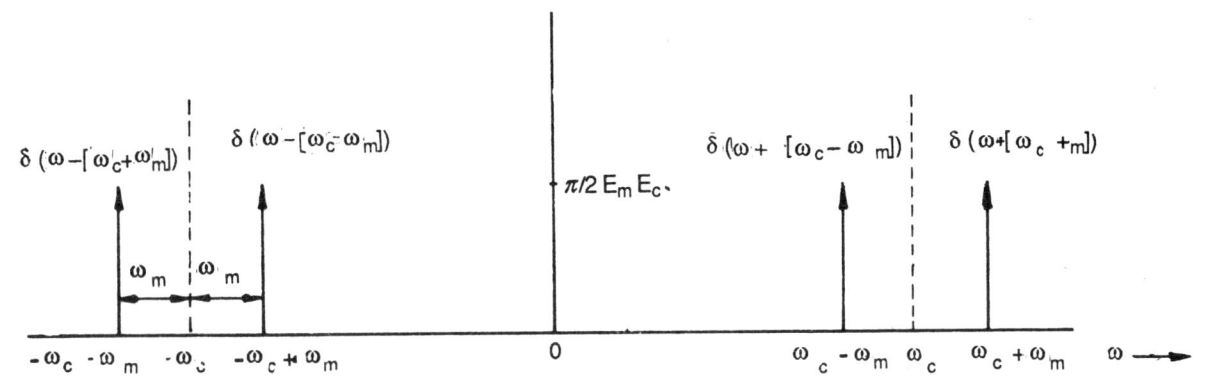

Fig. 5.1.5 Single-Tone AM-SC

## Example 5.1.2

The modulating signal $f(t)$ in an AM-SC system is a multiple-tone signal given by

$$f(t) = E_1 \cos \omega_1 t + E_2 \cos \omega_2 t + E_3 \cos \omega_3 t$$

The signal $f(t)$ modulates a carrier $E_c \cos \omega_c t$. Plot the single-sided trigonometric spectrum and, find the bandwidth of the modulated signal. Assume that,

$$\omega_3 > \omega_2 > \omega_1, \text{ and}$$

$$E_3 > E_2 > E_1$$

## Solution

The AM-SC system yields the product of modulating and carrier signals. Hence, the modulated signal is given as

$$e(t) = f(t) \cdot E_c \cos \omega_c t$$

$$= (E_1 \cos \omega_1 t + E_2 \cos \omega_2 t + E_3 \cos \omega_3 t) \cdot E_c \cos \omega_c t$$

$$= E_1 E_c \cos \omega_c t \cdot \cos \omega_1 t + E_2 E_c \cos \omega_c t \cdot \cos \omega_2 t + E_3 E_c \cos \omega_c t \cdot \cos \omega_3 t$$

$$= \frac{1}{2} E_1 E_c \left[ \cos(\omega_c + \omega_1)t + \cos(\omega_c - \omega_1)t \right]$$

$$+ \frac{1}{2} E_2 E_c \left[ \cos(\omega_c + \omega_2)t + \cos(\omega_c - \omega_2)t \right]$$

$$+ \frac{1}{2} E_3 E_c \left[ \cos(\omega_c + \omega_3)t + \cos(\omega_c - \omega_3)t \right]$$

Thus, the modulated signal consists of two sidebands (upper and lower) corresponding to each of the three tones present in $f(t)$. The spectrum is shown in Fig. 5.1.6.

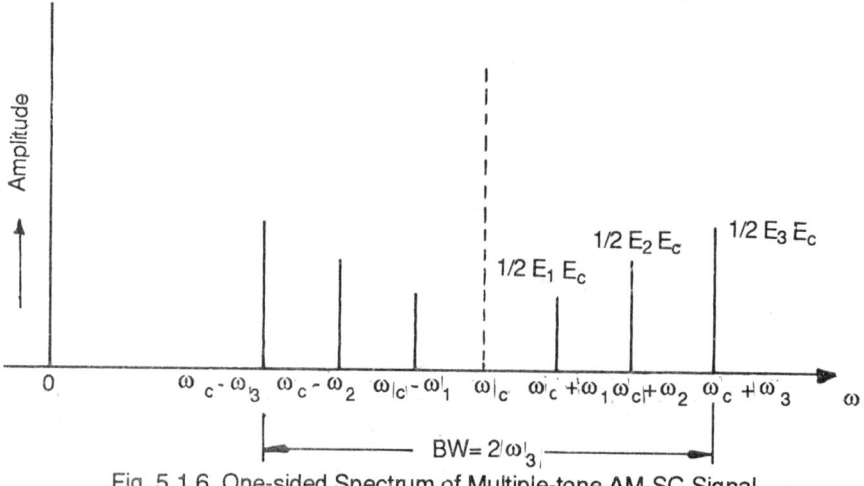

Fig. 5.1.6 One-sided Spectrum of Multiple-tone AM-SC Signal

The bandwidth of the modulated signal is $2\omega_3$. This is obvious from Fig. 5.1.6. The bandwidth is twice the highest frequency-tone present in the modulating signal $f(t)$.

## Phasor Representation of AM-SC Signals

The amplitude variation in an AM-SC system can be explained with the help of a phasor diagram. Let us consider that a single-tone modulating signal $E_m \cos \omega_m t$ modulates a carrier $E_c \cos \omega_c t$. The resulting modulated signal is given as

$$e(t) = E_m E_c \cos \omega_m t \cos a\omega_c t$$

$$= \frac{1}{2} E_m E_c \left[ \cos(\omega_c + \omega_m)t + \cos(\omega_c - \omega_m)t \right] \qquad (5.1.6)$$

Assume that a coordinate system rotates anticlockwise at an angular frequency $\omega_c$. The phasor for the carrier frequency term is fixed and oriented in horizontal direction, as shown by the dotted line in Fig 5.1.7. The phasor for the upper-sideband term $\frac{1}{2} E_m E_c \cos(\omega_c + \omega_m)t$ rotates anticlockwise at an angular frequency of $\omega_m$. Similarly, the phasor for the lower-sideband term $\frac{1}{2} E_m E_c \cos(\omega_c - \omega_m)t$ rotates

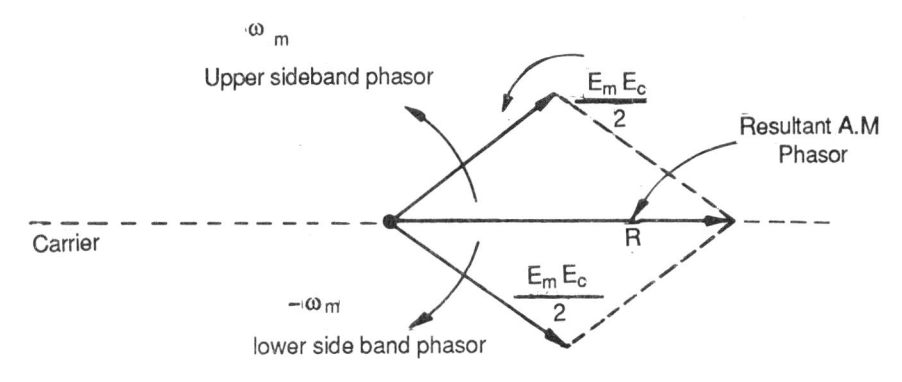

Fig. 5.1.7 Phasor Diagram of AM-SC

clockwise at the same angular frequency $\omega_m$. The resultant amplitude of the modulated wave at any instant is the vector sum of the two sideband phasors. The carrier phasor is absent in Eq. 5.1.6 and does not contribute to amplitude variation. Therefore, the carrier phasor is shown by a dotted line in Fig. 5.1.7.

The resultant amplitude $R$ of the modulated signal varies as the phasors corresponding to the upper, and lower sidebands rotate with time. Let us consider two specific cases:

(i) At $t = 0$, both the phasors have a maximum value of $\frac{1}{2}E_m E_c$, lying in the same direction, yielding a maximum resulstant amplitude, equal to $E_m E_c$ in the axial direction.

(ii) At $t = \dfrac{\pi}{2(\omega_c + \omega_m)}$ i.e., $(\omega_c + \omega_m)t = \dfrac{\pi}{2}$, both the phasors have zero amplitude and axial resultant is zero. Similarly, amplitude of the resultant wave can be obtained at any other instant, and the amplitude variation in modulated signals can be interpreted.

The amplitude variations in AM-SC signal can also be interpreted by considering the resultant wave as a sum of the waves corresponding to each sideband. For example, in Eq. 5.1.6, the modulated wave $e(t)$ is constituted by superposing the wave forms of the two sidebands. Although the two side bands have constant amplitudes, their sum will show an amplitude variation because of the unequal frequencies of the two sidebands. The amplitude variation in the AM-SC signal is illustrated in the Fig. 5.1.8. Note the *phase reversal at zero crossings of the AM-SC waveform.*

A similar interpretation has been obtained by the phasor diagram. It is worth noting how the two wave forms of constant amplitude construct an amplitude varying waveform. The envelope of AM-SC wave crosses zero axis, and hence an envelope detractor cannot be used for the recovery of the baseband signal.

## GENERATION OF DSB-SC SIGNALS

### Chopper-type (switching) modulator

Figure 5.1.1 shows that DSB-SC signals can be generated by a product modulator where the baseband signal $f(t)$ and the carrier signal $\cos \omega_c t$, are multiplied. However, multiplication of $f(t)$ by a sinusoidal carrier is not the only way. The same effect is achieved by multiplying $f(t)$ with any periodic wave having the fundamental frequency $\omega_c$. The reason is that a periodic wave contains sinusoidal components of frequency $0, \pm \omega_c, \pm 2\omega_c,$

252  COMMUNICATION SYSTEMS: Analog and Digital

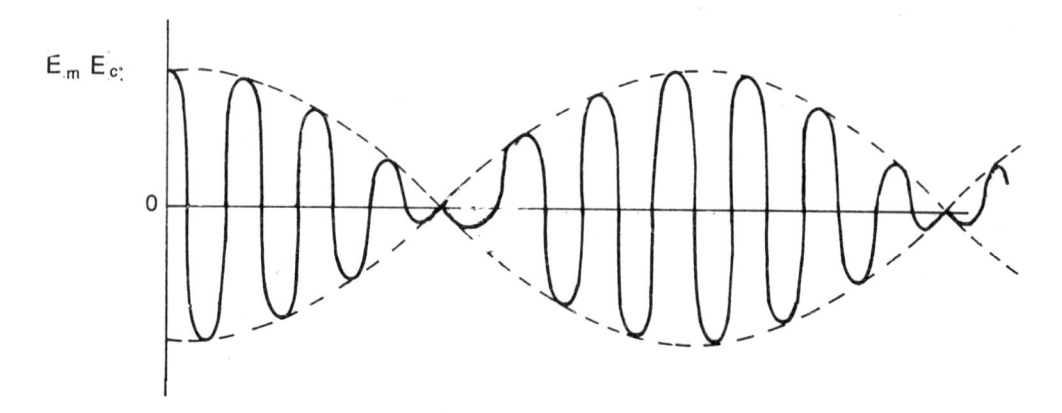

Fig. 5.1.8 Amplitude Variation in AM-SC Signal

etc. and multiplication shifts the spectrum $F(\omega)$ to $0, \pm \omega_c, \pm 2\omega_c$ etc. By using a bandpass filter the spectrum centered around $\pm \omega_c$ can be separated. This provides the DSB-SC signal. A convenient method of generating DSB-SC (AM-SC) signal is by multiplying $f(t)$ with a square wave $p(t)$, as shown in Fig.5.1.9.

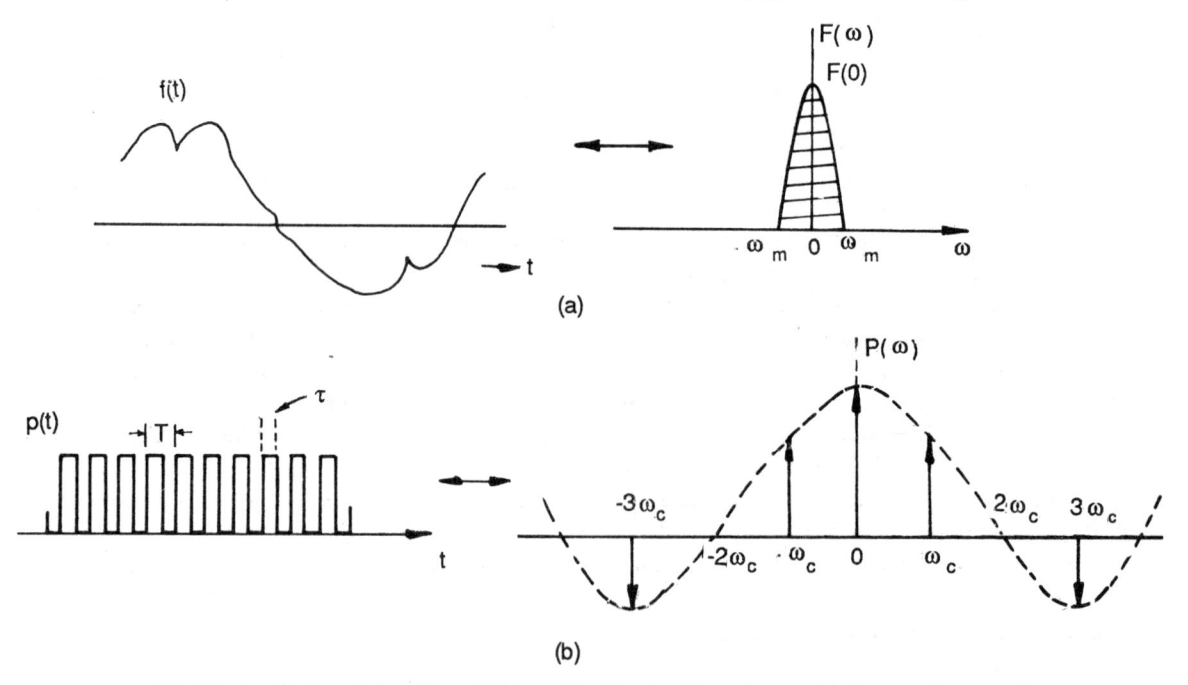

Fig 5.1.9 (a) Baseband Signal $f(t)$ and its Fourier Transform; (b) Square Signal $p(t)$ and its Fourier Transform

Figure 5.1.9a shows a baseband signal and its Fourier transform. The square wave with its Fourier transform is shown in Fig. 5.1.9 b. The fundamental frequency of the square wave $p(t)$ is $\omega_c \left( \text{i.e., period } T = \frac{2\pi}{\omega_c} \right)$,

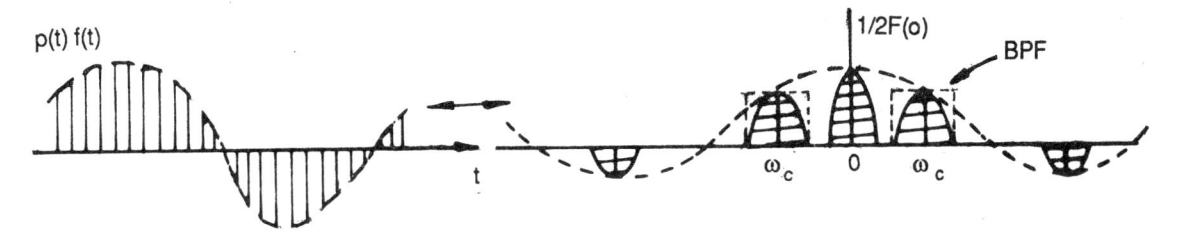

(c) Chopped Signal and its Fourier Transform

Fig. 5.1.9 Chopping of Baseband

and its amplitude is assumed to be unity. The square wave $p(t)$ has a duty cycle $(\tau/T) = \dfrac{1}{2}$, and hence its Fourier transform is,

$$P(\omega) = \pi \sum_{n=-\infty}^{\infty} Sa\left(\frac{n\pi}{2}\right) \delta(\omega - n\omega_c) \tag{5.1.7}$$

Figure 5.1.9c shows the product $f(t)p(t)$ and its spectrum. The product signal is a chopped version of $f(t)$ at a chopping rate of $\omega_c$. The Fourier transform of this chopped signal can be obtained by taking convolution of $P(\omega)$ and $F(\omega)$, i.e.,

$$f(t)\, p(t) \leftrightarrow \frac{1}{2\pi}[F(\omega)\, P(\omega)]$$

substituting $P(\omega)$ from Eq. 5.1.7, we get

$$f(t)\, p(t) \leftrightarrow \frac{1}{2}\left[F(\omega) * \sum_{n=-\infty}^{\infty} Sa\left(\frac{n\pi}{2}\right)\delta(\omega - n\omega_c)\right]$$

$$\leftrightarrow \frac{1}{2} \sum_{n=-\infty}^{\infty} Sa\left(\frac{n\pi}{2}\right) F(\omega - n\omega_c) \tag{5.1.8}$$

The spectrum in Fig. 5.1.9c represents Eq. 5.1.8, where $F(\omega)$ repeats with its amplitude decaying according to sampling function $Sa\left(\dfrac{n\pi}{2}\right)$. Since we are only interested in the spectrum centered around $\pm \omega_c$, this can be achieved by passing this product signal through a bandpass filter that allows to pass only signals centered near $\pm \omega_c$, and rejects other frequency components. Only those sidebands centered near $\pm \omega_c$ are present in the filtered output, which is a DSB-SC signal. Thus, we get AM-SC (DSB-SC) signal by passing a chopped signal $f(t)$ through a bandpass filter, as shown Fig. 5.1.10.

## Frequency Conversion or Mixing

Frequency translation is the process of converting one frequency space into another, and hence it is also known as frequency conversion. Similarly, frequency translation provides the sum and the difference of frequencies

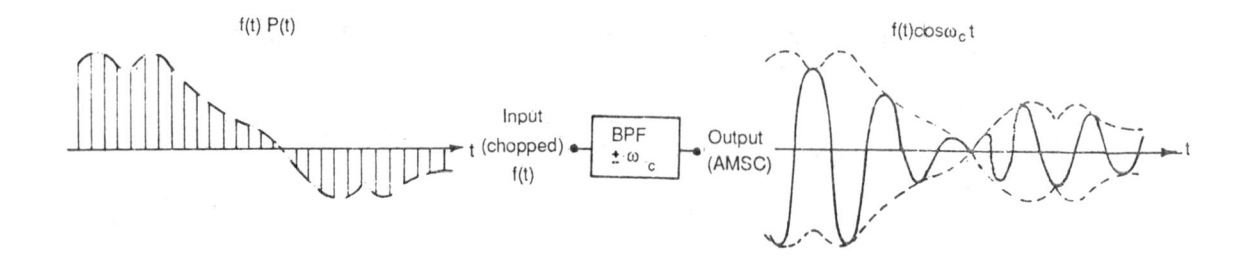

**Fig. 5.1.10 Bandpass Filtering of Chopped Signal**

($\omega_c + \omega_m$ and $\omega_c - \omega_m$), and therefore, known as frequency mixing. A modulator and a demodulator, both perform frequency translation, and hence can provide frequency conversion or mixing. That is why they are also termed as Frequency convertors, or frequency mixers.

## Switching Modulator Circuits

The baseband signal $f(t)$ can be chopped by a switching or chopper circuit. One such chopper circuit is shown in Fig. 5.1.11

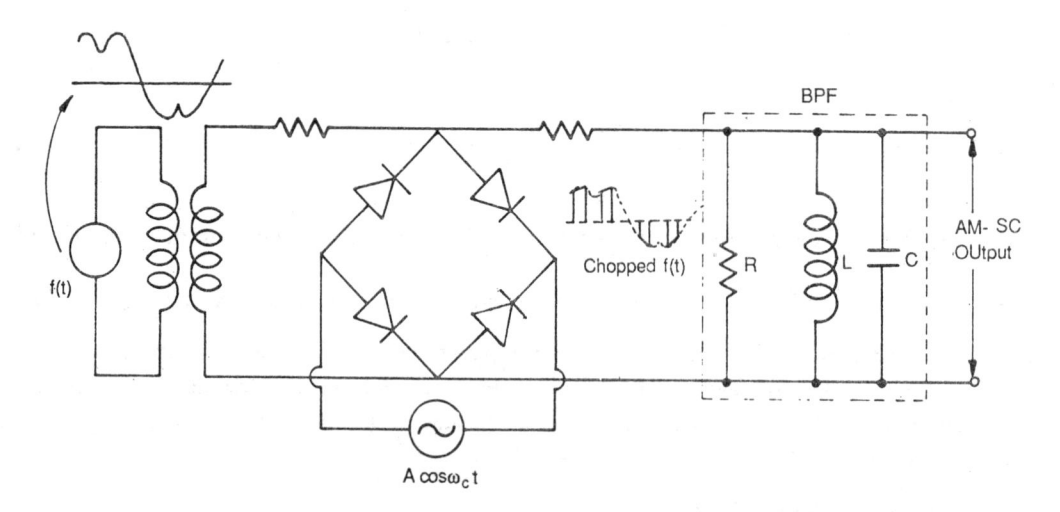

**Fig. 5.1.11 A Chopper-type Balanced Modulator (Ring Modulator)**

AMPLITUDE MODULATION SYSTEMS 255

The diodes are used as a switching device. The carrier signal is such that its amplitude $A > |f(t)|_{max}$ and, $\omega_c > \omega_m$, where $\omega_m$ is the maximum frequency component present in the baseband signal $f(t)$. For a positive cycle of the carrier signal, all the four diodes are forward biased (note that the carrier amplitude A is greater than the highest magnitude of $f(t)$, and hence the bias of the diodes is decided by the polarity of the carrier amplitude). The four conducting diodes provide a short circuit to the baseband signal $f(t)$ and the output is zero for the period of positive half cycles. On the other hand, for negative cycles of the carrier, all the four diodes are reverse biased, and provide a path for the signal $f(t)$ to reach at the output. Hence, for a positive half cycle of the carrier, no output appears, whereas, for the negative half cycle, $f(t)$ appears at the output, and thus, a chopped version of $f(t)$ is obtained at the output of the chopper circuit. The parallel R-L-C network tuned to $\omega_c$ acts as a bandpass filter and at the output of this filter the DSB-SC signal is achieved. The circuit consists of four diodes connected in a ring form; hence the circuit is known as a ring modulator. The circuit is referred to as a *double-balanced modulator* as it is balanced with respect to the baseband signal as well as the carrier.

### Time Varying Linear Systems

The ring modulator behaves like a linear time varying circuit. It is linear because multiplication of $f(t)$ by a constant will increase the output by the same constant. It is time varying because the circuit parameter changes with time periodically. Thus, any time varying linear system can provide modulation. A time-invariant linear system cannot generate modulation because such a system cannot provide frequency change in the output. Modulation can be achieved by two methods:
 (i) linear time-variant circuits
 (ii) non-linear circuits

### (i) Linear Time-variant Circuits

Circuits whose gain or transfer function can be varied with time, come under this category. The block schematic of such system is shown in Fig. 5.1.12. In Fig. 5.1.12a, the gain varies linearly with $f(t)$, and in Fig. 5.1.12b the gain varies linearly with $\cos \omega_c t$. Both the systems are time varying having a linear input output relation. In the ring modulator system, the gain is varied by a carrier signal $\cos \omega_c t$ between *zero* and *unity* at the rate of carrier frequency; hence it is an example of linear time-variant system.

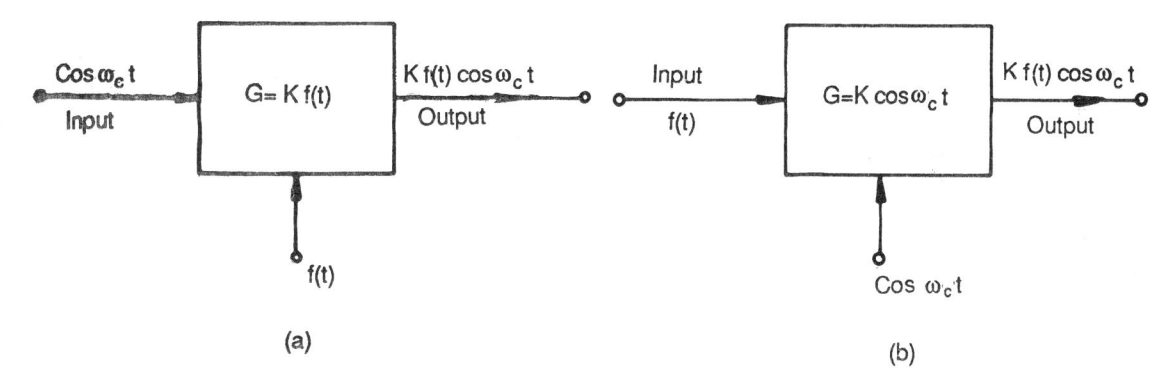

Fig. 5.1.12 Time-variant Linear System as a Modulator

*Amplifying devices*: The diode circuit discussed above does not provide any amplification; However, we have devices like transistors, FET, vacuum tubes etc., which can provide amplification alongwith modulation. The gain parameter of such devices ($h_{fe}$, $g_m$, $\mu$ etc.) can be varied with time, and hence, such devices can provide modulation. Some circuits using amplifying devices are discussed in the next section.

### (II) Non-linear Circuits

Devices, or circuits, having non linear characteristics can also provide modulation. The characteristics of a non-linear device/circuit is shown in Fig. 5.1.13. A non-linear characteristic is mathematically represented by a power series. Neglecting the terms higher than the second order, the output '$i$' is related with the input '$e$' as follows,

$$i = ae + be^2 \qquad (5.1.9)$$

where $a$ and $b$ are the coefficients of the series. Devices like diodes, transistors, FET etc., exhibit such characteristic and can be used to provide modulation.

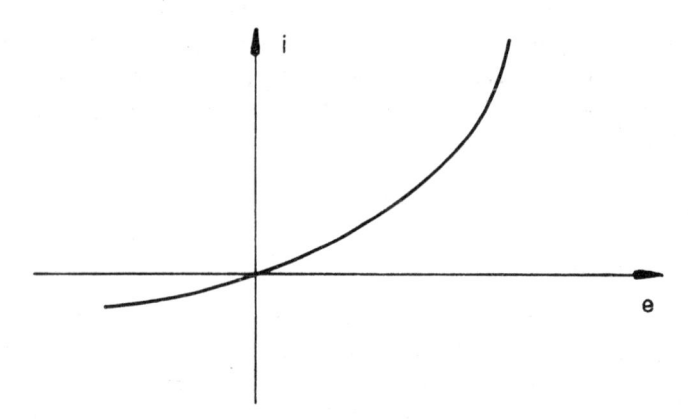

Fig. 5.1.13 Non-linear Characteristic

### Balanced Modulators

In a balanced modulator, two non-linear devices are connected in the balanced mode, so as to suppress the carrier wave. Some of these circuits are discussed below :

**(a) Using diodes:** A balanced modulator using two diodes as non linear elements is shown in Fig. 5.1.14. In this figure voltages $e_1$ and $e_2$ are expressed as,

$$e_1 = \cos \omega_c t + f(t)$$

and

$$e_2 = \cos \omega_c t - f(t)$$

AMPLITUDE MODULATION SYSTEMS 257

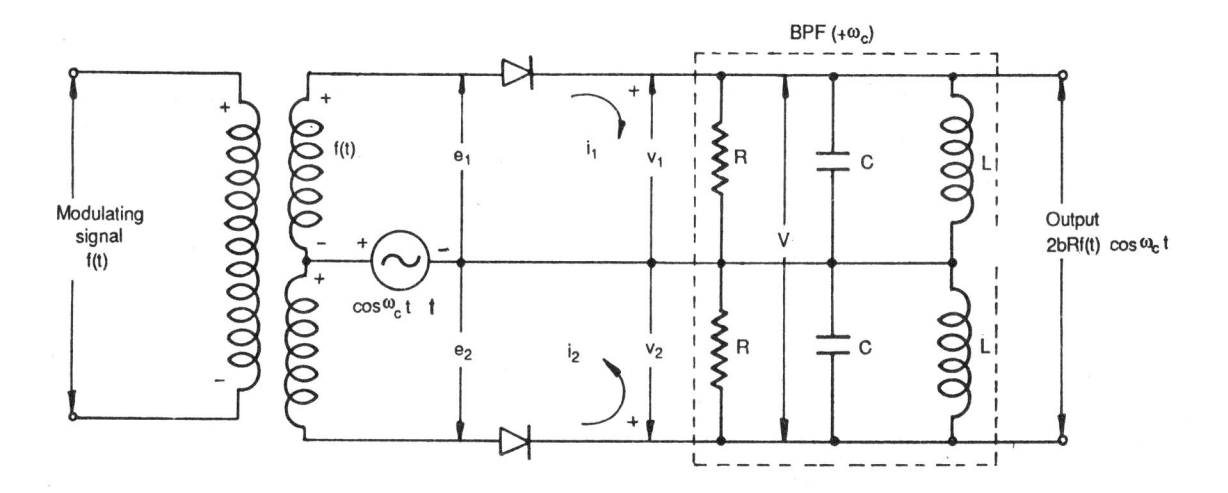

Fig. 5.1.14 Balanced Modulator Using Diodes

Currents $i_1$ and $i_2$ are obtained using Eq. 5.1.9,

$$i_1 = ae_1 + be_1^2$$
$$= a[\cos \omega_c t + f(t)] + b[\cos \omega_c t + f(t)]^2 \quad (5.1.10)$$

and

$$i_2 = ae_2 + be_2^2$$
$$= a[\cos \omega_c t - f(t)] + b[\cos \omega_c t - f(t)]^2 \quad (5.1.11)$$

The voltage $V_o$, at the input of the bandpass-filter is given by,

$$V_o = V_1 - V_2 = i_1 R - i_2 R$$
$$V_o = 2R[af(t) + 2b f(t) \cos \omega_c t]$$

The output of BPF centered around $\pm \omega_c$ is given by

$$\text{output} = 2bRf(t) \cos \omega_c t$$
$$= Kf(t) \cos \omega_c t \quad (5.1.12)$$

Equation 5.1.12 represents the desired DSB-SC signal.

**(b) Using Transistors:** Figure 5.1.15 shows a balanced modulator circuit for generating DSB-SC signal using transistors. The principle of operation of the circuit is identical to the diode circuit. The transistors are operated in the non-linear region. The voltage and currents shown in Fig. 5.1.15 are analogous to Fig. 5.5.14. The equations relating to voltages and currents are identical to the ones of diode balance modulator circuit.

The circuit yields DSB-SC signal represented by Eq. 5.1.12. Transistors are amplifying devices. The circuit can be fabricated using other amplifying devices like FET or electron tubes. The diode circuit is cheaper, whereas amplifying devices provide power gain.

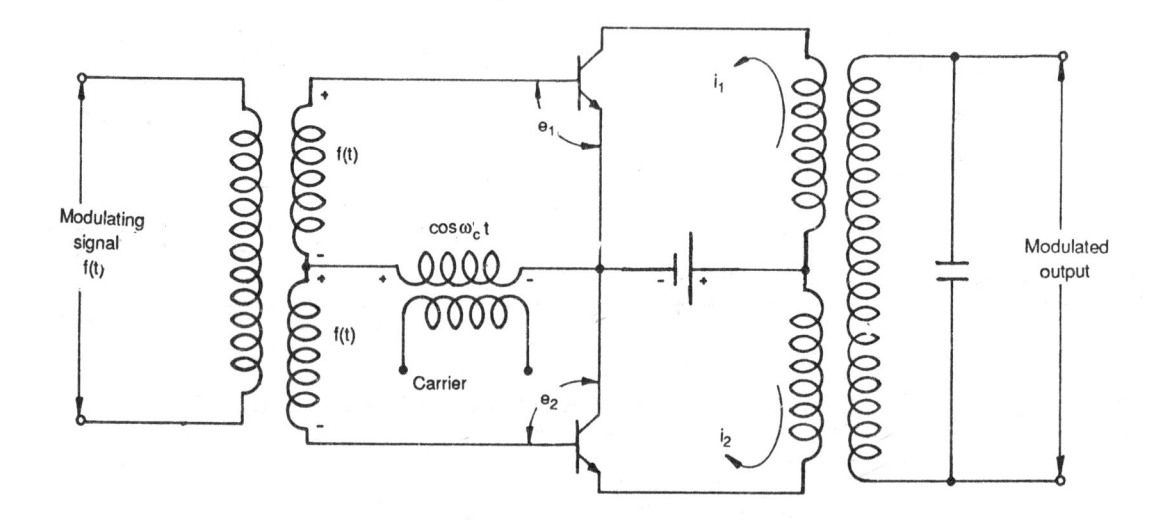

Fig. 5.1.15 Balanced Modulator Using Transistors

## 5.2 SINGLE SIDEBAND MODULATION (SSB)

The double sideband suppressed carrier system (DSB-SC), discussed in Section 5.1 doubles the bandwidth of the modulated signal as compared to the baseband. This is because the baseband appears twice in the modulated (DSB-SC) signal, as shown in Fig. 5.2.1 b. The baseband ranges between 0 and $\omega_m$ having bandwidth of $\omega_m$ (considering only positive frequency) as in Fig. 5.2.1 a. This bandwidth becomes $2\omega_m$ (i.e. twice) after modulation as in Figure 5.2.1b.

The message signal (baseband) appears twice in the DSB-SC signal, and it unnecessarily increases the bandwidth. Lower the bandwidth of the modulated signal, more is the number of channels that can be accommodated in a given frequency space. It is, therefore, desirable to transmit only one sideband, as this contains the entire information content in the message signal (once), and at the same time it reduces the bandwidth by half. This means, we can accommodate twice the number of channels in a given frequency space by using a single sideband in place of both the sidebands. Modulation of this type which provides a single sideband with suppressed carrier is known as single sideband suppressed carrier system (SSB-SC). The spectrum of SSB-SC is shown in Fig. 5.2.1c (utilizing lower sidebands) and Fig. 5.2.1d (utilizing upper sideband). Both the spectra of SSB-SC signals are symmetrical about the vertical axis, so that they represent real signals. The bandwidth of SSB-SC signal is $\omega_m$ (again considering only positive frequencies) same as the bandwidth of the baseband signal.

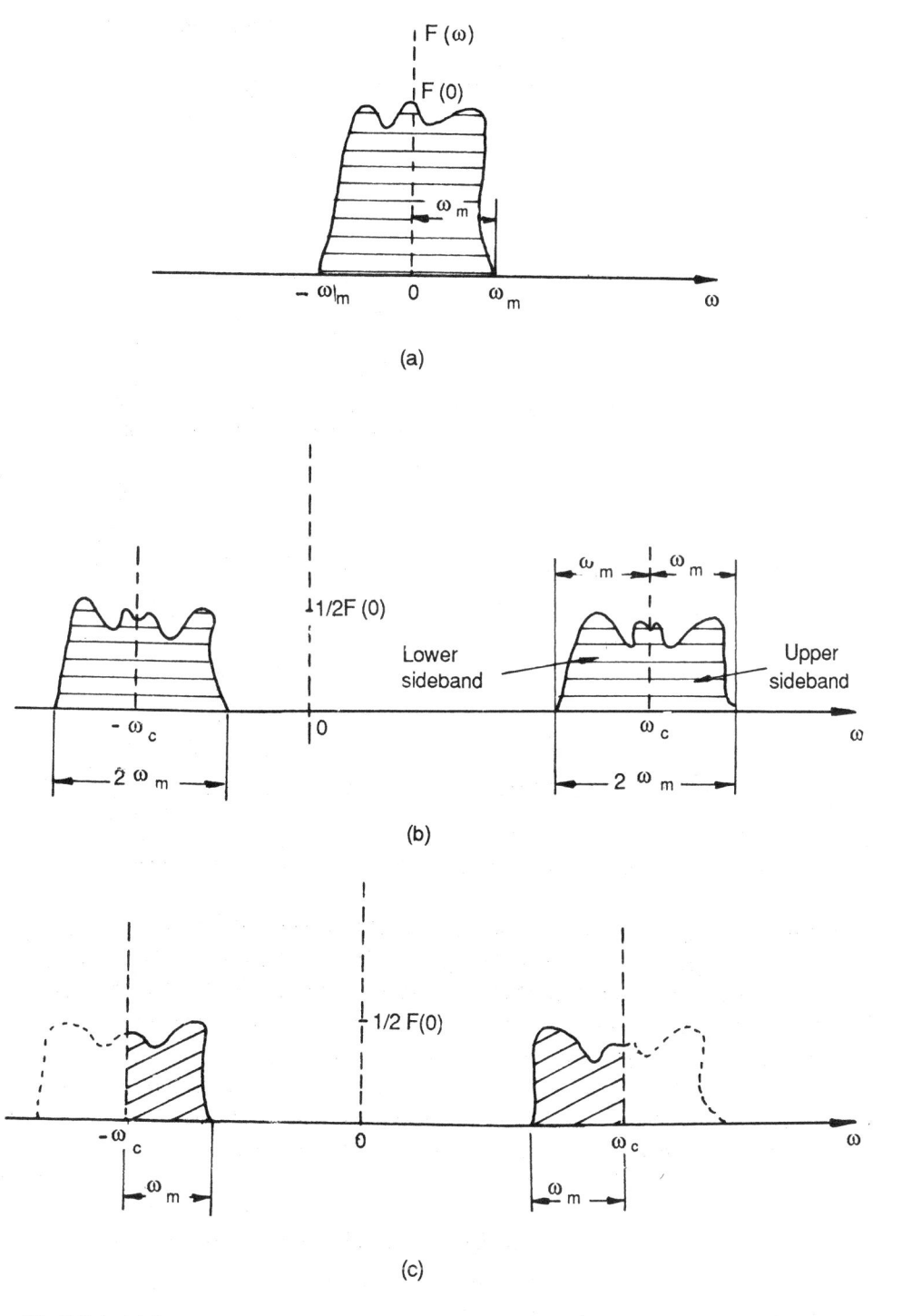

Fig 5.2.1 (a) Baseband Signal; (b) DSB-SC Signal; (c) SBC-SC (lower sidebands)

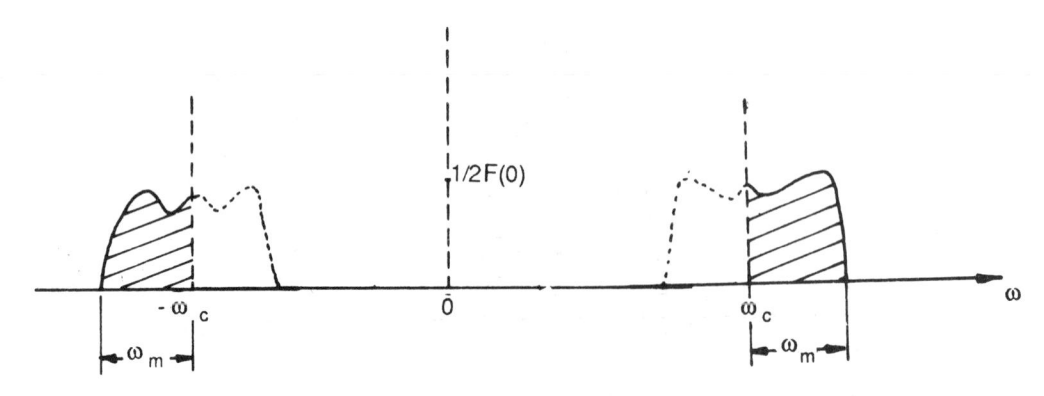

Fig. 5.2.1 (d) SSB-SC (Upper sidebands)

## Time Domain Description of the SSB-SC Wave

An expression can be conveniently derived to represent the SSB wave in time domain, considering a specific case of single-tone modulating signal. The general expression can be derived by using the concept of a pre-envelope. We will first consider the simple case of a single-tone modulating signal $f(t)$.

## SSB-SC with Single-Tone Modulating Signal

Consider a single-tone modulating signal given by

$$f(t) = \cos \omega_m t$$

The spectrum of this modulating signal is a pair of impulses at $\omega = \pm \omega_m$ as shown in Fig. 5.2.2(a). Let this signal modulate a carrier $\cos \omega_c t$, the resulting AM-SC (DSB-SC) spectrum is shown in Fig. 5.2.2b. In order to get the SSB-SC waveform, one of the sidebands is eliminated. Figure 5.2.2c represents the spectrum of SSB-SC with lower sidebands. It is obvious from the spectrum that this corresponds to a time domain signal $\cos(\omega_c - \omega_m)t$, since the spectrum of cosine function consists of two impulses in its frequency domain. Thus, the SSB-SC (lower sideband) signal may be expressed as

$$\cos(\omega_c - \omega_m)t = \cos \omega_m t \cos \omega_c t + \sin \omega_m t \sin \omega_c t \tag{5.2.1a}$$

Similarly, the expression for the single-tone SSB-SC with upper sidebands (Fig. 5.2.2d) may be given as

$$\cos(\omega_c + \omega_m)t = \cos \omega_m t \cos \omega_c t - \sin \omega_m t \sin \omega_c t \tag{5.2.1b}$$

Equation 5.2.1a and 5.2.1b may be combined to be written as

$$\varphi_{SSB}(t) = \cos \omega_m t \cos \omega_c t \pm \sin \omega_m t \sin \omega_c t \tag{5.2.2}$$

where (+) sign indicates the lower sideband and (−) sign indicates the upper sideband. The terms $\sin \omega_m t$, and $\sin \omega_c t$ may be written as

$$\sin \omega_m t = \cos\left(\omega_m t - \frac{\pi}{2}\right) \tag{5.2.3a}$$

$$\sin \omega_c t = \cos\left(\omega_c t - \frac{\pi}{2}\right) \tag{5.2.3b}$$

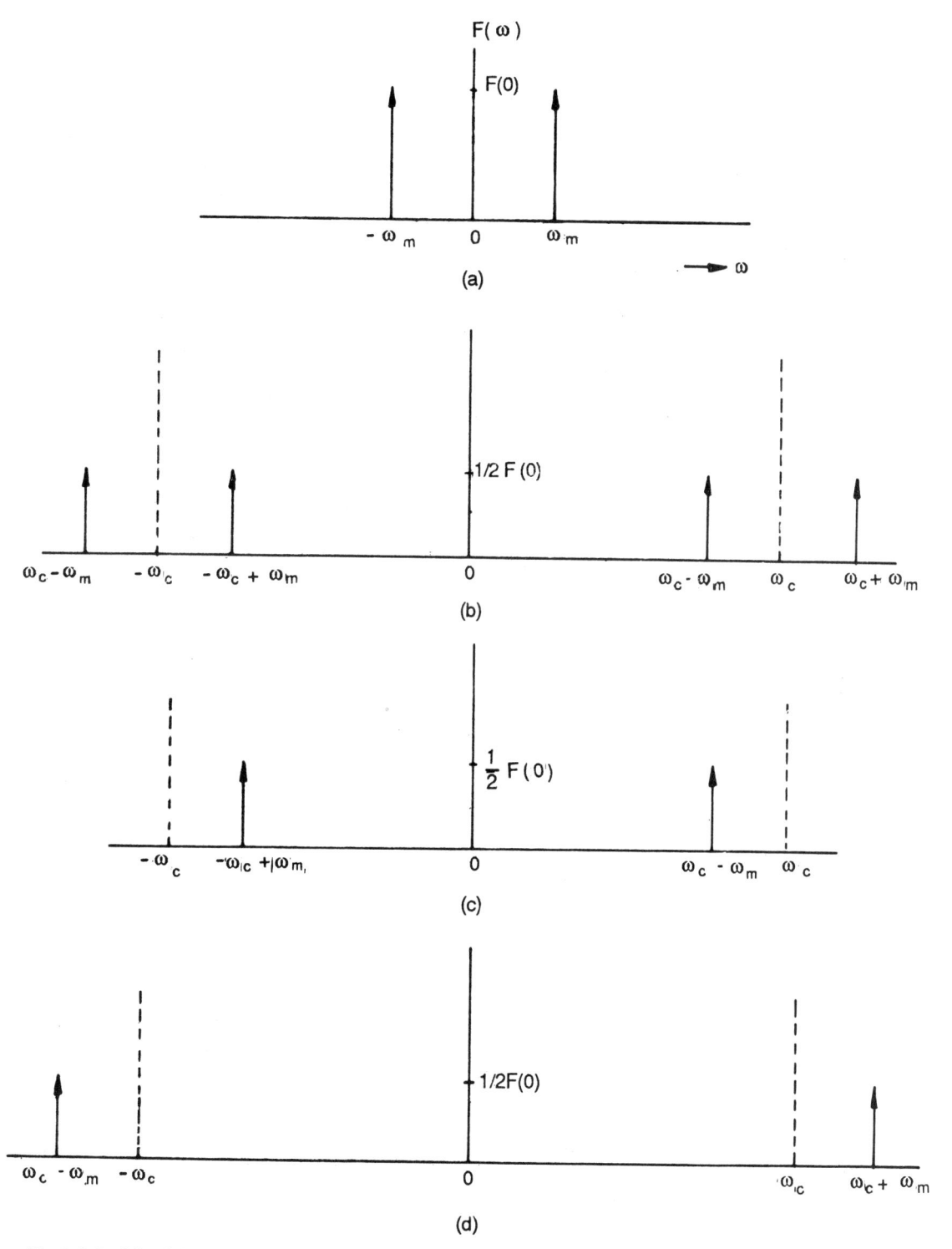

Fig.5.2.2 SSB-SC With Single-Tone Modulation (a) Spectrum of $\cos \omega_m t$; (b) Spectrum of AM-SC; (c) Spectrum of SSB-SC with Lower Sideband; (d) Spectrum of SSB-SC with Upper Sideband

Thus, the sine terms can be obtained from the corresponding cosine terms, by giving a phase shift of $\left(-\frac{\pi}{2}\right)$.

Equation 5.2.2 is an expression for SSB-SC for a specific case of single-tone modulation, but this paves the path for reaching a general expression for SSB-SC. In Eq. 5.2.2, the term $\sin \omega_m t$ is obtained by giving a phase shift of $\left(-\frac{\pi}{2}\right)$ to the modulating frequency $\cos \omega_m t$. Similarly, in a general modulating signal $f(t)$, if all the frequency components are shifted by $\left(-\frac{\pi}{2}\right)$, it may lead to a SSB-SC signal. Note that $f(t)$ can be expressed as a continuous sum of sinusoidal signals. Thus, Eq. 5.2.2 may be extended for a SSB-SC signal modulated by a general modulating signal $f(t)$ as given below:

$$\varphi_{SSB}(t) = f(t) \cos \omega_c t \pm f_h(t) \sin \omega_c t \qquad (5.2.4)$$

where $f_h(t)$ is a signal obtained by shifting the phase of every component present in $f(t)$ by $\left(-\frac{\pi}{2}\right)$. Similar to Eq. 5.2.2, the (+) sign corresponds to the lower sideband and (−) sign corresponds to the upper sideband.

### Hilbert Transform

It can be seen that $f_h(t)$ obtained by giving $\left(-\frac{\pi}{2}\right)$ phase shift to each frequency component present in $f(t)$, actually represents the *Hilbert transform* of $f(t)$, (Example 5.2.1), i.e., $f_h(t)$ is Hilbert transform of $f(t)$ defined as

$$f_h(t) = \frac{1}{\pi} f(t) * \frac{1}{t}$$

$$= \frac{1}{\pi} \int_{-\infty}^{\infty} \frac{f(\tau)}{t-\tau} d\tau \qquad (5.2.5a)$$

The *inverse Hilbert transform* is defined as

$$f(t) = -\frac{1}{\pi} \int_{-\infty}^{\infty} \frac{f_h(\tau)}{t'-\tau} d\tau \qquad (5.2.5b)$$

### Example 5.2.1

Show that if every frequency component of a signal $f(t)$ is shifted by $\frac{\pi}{2}$, the resultant signal $f_h(t)$ is the Hilbert transform of $f(t)$.

### Solution

The situation may be regarded as though the signal $f(t)$ is passed through a phase shifting system $H(\omega)$, and the output is $f_h(t)$ as shown in Fig. 5.2.3.

AMPLITUDE MODULATION SYSTEMS 263

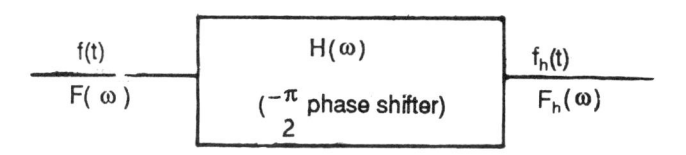

```
   f(t)      ┌─────────────────┐   f_h(t)
  ─────────→ │      H(ω)       │ ─────────→
   F(ω)      │ (-π/2 phase     │   F_h(ω)
             │      shifter)   │
             └─────────────────┘
```

Fig. 5.2.3 A Phase Shifting System

The characteristics of the system are specified as follows:

(i) The magnitude of the frequency components present in $f(t)$ remains unchanged when it is passed through the system i.e., $H(\omega) = 1$, and

(ii) The phase of the positive frequency components is shifted by $-\frac{\pi}{2}$. Since the phase spectrum $\theta(\omega)$ has an odd symmetry, the phase of the negative frequency components is shifted by $+\frac{\pi}{2}$. $H(\omega)$ and $\theta(\omega)$ are plotted in Fig. 5.2.4.

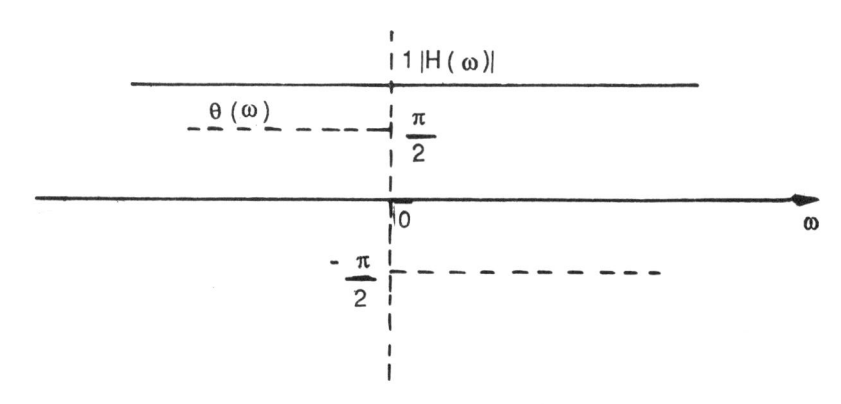

Fig. 5.2.4 Transfer Function of $-\frac{\pi}{2}$ Phase Shifter

The transfer function is given as

$$H(\omega) = |H(\omega)| e^{j\theta(\omega)} = 1 \cdot e^{j\theta(\omega)} \qquad (5.2.6)$$

It is evident from Fig. 5.2.4 that

$$\theta(\omega) = \begin{cases} +\frac{\pi}{2}, & \omega < 0 \text{ (i.e., negative frequencies)} \\ -\frac{\pi}{2}, & \omega > 0 \text{ (i.e., positive frequencies)} \end{cases}$$

Therefore, Eq. 5.2.6 may be written as

$$H(\omega) = \begin{cases} e^{j\frac{\pi}{2}} & \omega < 0 \\ e^{-j\frac{\pi}{2}} & \omega > 0 \end{cases}$$

## 264 COMMUNICATION SYSTEMS: Analog and Digital

It is known that

$$e^{j\frac{\pi}{2}} = \cos\frac{\pi}{2} + j\sin\frac{\pi}{2} = j$$

and

$$e^{-j\frac{\pi}{2}} = \cos\left(-\frac{\pi}{2}\right) + j\sin\left(-\frac{\pi}{2}\right) = -j$$

Hence, $H(\omega)$ becomes

$$\frac{H(\omega)}{j} = \begin{cases} 1 & \omega < 0 \\ -1 & \omega > 0 \end{cases}$$

$$= -\text{sgn}(\omega) \text{ (Example 1.5.4)}$$

or

$$H(\omega) = -j\,\text{sgn}(\omega) \tag{5.2.7}$$

The response $F_h(\omega)$ of the system is related to input $F(\omega)$, as

$$F_h(\omega) = F(\omega)H(\omega)$$

where

$$f(t) \leftrightarrow F(\omega) \text{ and } f_h(t) \leftrightarrow F_h(\omega)$$

Substituting $H(\omega)$ from Eq. 5.2.7

$$F_h(\omega) = -jF(\omega)\,\text{sgn}(\omega) \tag{5.2.8}$$

Taking the inverse Fourier transform of both sides of Eq. 5.2.8

$$f_h(t) = F^{-1}[-jF(\omega)\,\text{sgn}(\omega)]$$

The time domain of sgn $(\omega)$ is given by

$$\frac{1}{\pi t} \leftrightarrow \text{sgn}(\omega)$$

Using time convolution theorem,

$$f_h(t) = \frac{1}{\pi}\left[f(t) * \frac{1}{t}\right]$$

$$= \frac{1}{\pi}\int_{-\infty}^{\infty}\frac{f(\tau)}{t-\tau}d\tau$$

which is the Hilbert transform of $f(t)$. Some applications of the Hilbert transform are:
(i) generation of SSB signals
(ii) design of minimum phase type filters
(iii) representation of bandpass signals

## AMPLITUDE MODULATION SYSTEMS

### Properties of the Hilbert Transform

The properties of a Hilbert transform are as under:
(1) A signal $f(t)$ and its Hilbert transform $f_h(t)$ have the same energy density spectrum.
(2) A signal $f(t)$ and its Hilbert transform $f_h(t)$ have the same autocorrelation function.
(3) A signal $f(t)$ and its Hilbert transform $f_h(t)$ are mutually orthogonal, i.e.

$$\int_{-\infty}^{\infty} f(t) f_h(t) dt = 0$$

(4) If $f_h(t)$ is a Hilbert transform of $f(t)$, then the Hilbert transform of $f_h(t)$ is $-f(t)$, that is, if

$$H[f(t)] = f_h(t)$$

then

$$H[f_h(t)] = -f(t)$$

where $H$ denotes the Hilbert transform.

### Pre-envelop or Analytic Signal

The concept of pre-envelop, also known as the *analytic function* is useful in deriving the general expression of the SSB-SC signal. The pre-envelope of a real valued signal $f(t)$ is defined as

$$f_p(t) = f(t) + j f_h(t) \qquad (5.2.9)$$

where $f_h(t)$ is a Hilbert transform of $f(t)$. Obviously, the pre-envelop $f_p(t)$ is a complex-valued signal. The real part of $f_p(t)$ is $f(t)$, and the imaginary part is its Hilbert transform $f_h(t)$. The complex conjugate of the pre-envelop denoted by $f_p^*(t)$ is given as

$$f_p^* = f(t) - j f_h(t) \qquad (5.2.10)$$

### SSB-SC for a General Modulating Signal

We have already derived an expression for a generalized SSB-SC signal by extending the concept of SSB-SC for a single-tone modulation. We will now derive the same expression (Eq. 5.2.4), by using the concept of pre-envelope, or analytic signal.

Pre-envelope of a function $f(t)$ as defined by Eq. 5.2.9 is reproduced as follows

$$f_p(t) = f(t) + j f_h(t)$$

The Fourier transform of $f_p(t)$ is the sum of the Fourier transforms of $f(t)$ and $f_h(t)$,

$$F[f_p(t)] = F[f(t)] + j F[f_h(t)]$$

or

$$F_p(\omega) = F(\omega) + j[-jF(\omega) \operatorname{sgn}(\omega)] \qquad \text{(refer Eq. 5.2.8)}$$

$$= F(\omega) + F(\omega) \operatorname{sgn}(\omega) \qquad (5.2.11a)$$

**266** COMMUNICATION SYSTEMS: Analog and Digital

It is known that,
$$\text{sgn}(\omega) = \begin{cases} 1 & \text{for } \omega > 0 \\ -1 & \text{for } \omega < 0 \end{cases}$$

Hence,
$$F_p(\omega) = \begin{cases} 2F(\omega), & \omega > 0 \\ 0, & \omega < 0 \end{cases} \tag{5.2.11b}$$

Figures 5.2.5 a and b show $F(\omega)$ and $F_p(\omega)$ respectively. It is evident from Fig. 5.2.5b that $F_p(\omega)$ vanishes for negative frequencies.

Similarly, we can find the Fourier transform of $f_p^*(t)$ defined by Eq. 5.2.10,
$$F_p^*(\omega) = F(\omega) - j[-jF(\omega)\,\text{sgn}(\omega)]$$
$$= F(\omega) - F(\omega)\,\text{sgn}(\omega) \tag{5.2.12a}$$

which may be written as,
$$F_p^*(\omega) = \begin{cases} 0, & \omega > 0 \\ 2F(\omega), & \omega < 0 \end{cases} \tag{5.2.12b}$$

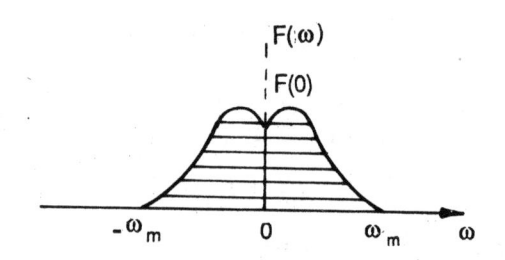

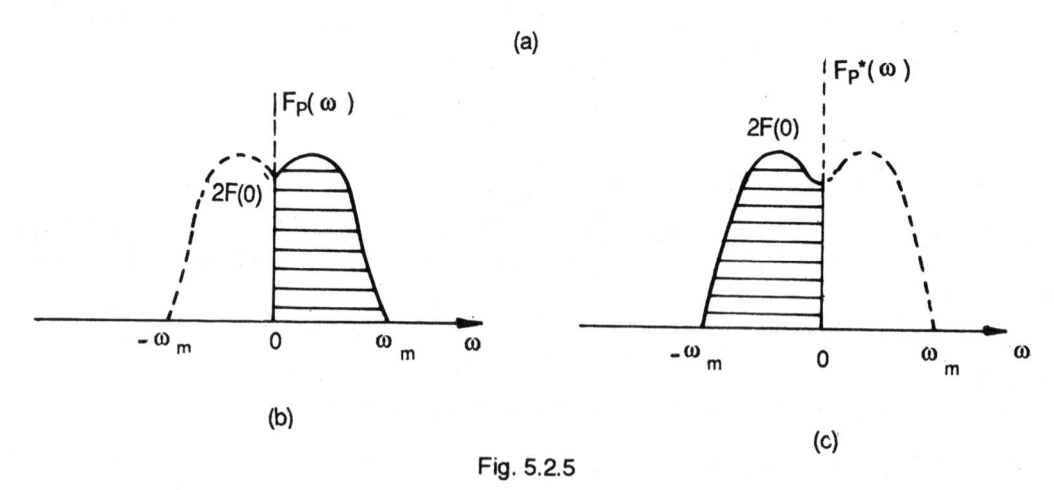

Fig. 5.2.5

This is plotted in Fig. 5.2.5c. Evidently, $F_p(\omega)$ vanishes for positive frequencies.

Now, take an SSB-SC wave consisting of only the lower sidebands of a general modulating signal $f(t)$ as shown in Fig. 5.2.6.
We observe the following points in Fig. 5.2.6.

AMPLITUDE MODULATION SYSTEMS 267

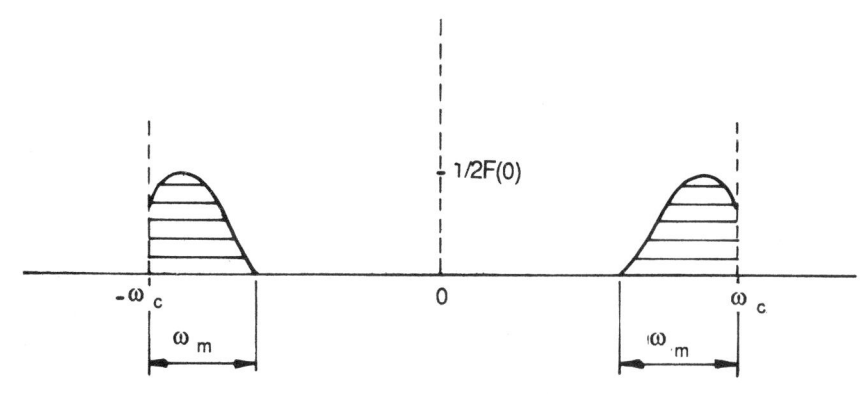

Fig. 5.2.6 SSB-SC Spectrum Consisting of Lower Sidebands

(i) The right-hand portion of the figure represents a spectrum of $\frac{1}{4} f_p^*(t) e^{j\omega_c t}$. This is equivalent to shifting the spectrum $F_p^*(\omega)$ towards right by $\omega_c$.

(ii) Similarly, left-hand portion of Fig. 5.2.6 represents the spectrum of

$$\frac{1}{4} f_p(t) e^{-j\omega_c t}$$

Therefore, Fig. 5.2.6 represents the spectrum of a combined signal $f_p(t) e^{+j\omega_c t} + f_p(t) e^{-j\omega_c t}$. Hence, the time domain representation of the SSB-SC spectrum shown in Fig. 5.2.6 is given as

$$\varphi_{SSB}(t) = \frac{1}{4}\left[f_p^*(t) e^{j\omega_c t} + f_p(t) e^{-j\omega_c t}\right] \tag{5.2.13}$$

Substituting in terms of Hilbert transform from Eqs 5.2.9 and 5.2.10, we get

$$\varphi_{SSB}(t) = \frac{1}{4}[f(t) - jf_h(t)] e^{j\omega_c t} + \frac{1}{2}[f(t) + jf_h(t)] e^{-j\omega_c t}$$

$$= \frac{1}{2} f(t) \left[\frac{e^{j\omega_c t} + e^{-j\omega_c t}}{2}\right] + \frac{j}{2} f_h(t) \left[\frac{e^{-j\omega_c t} - e^{j\omega_c t}}{2}\right]$$

$$= \frac{1}{2}[f(t) \cos \omega_c t + f_h(t) \sin \omega_c t]$$

which is the time domain description of the SSB-SC signal consisting of only the lower sidebands. Similarly, we can derive an expression for an SSB-SC signal consisting of the upper sidebands,

$$\varphi_{SSB}(t) = \frac{1}{2}[f(t) \cos \omega_c t - f_h(t) \sin \omega_c t]$$

Therefore, the time domain description of the SSB-SC wave is represented by the expression

$$\varphi_{SSB}(t) = f(t) \cos \omega_c t \pm f_h(t) \sin \omega_c t$$

where + and − signs corresponds to the lower sidebands and upper sidebands, respectively.

## Generation of SSB-SC Signals

SSB-SC waves can be generated by two methods:
(i) frequency discrimination method, or filter method;
(ii) phase discrimination method, or phase-shift-method or phasing method

### Frequency-Discriminator Method

First a DSB-SC signal is generated by using a simple product modulator, or a balanced modulator, and then one of the sidebands is filtered out by an appropriate bandpass filter. The schematic diagram is shown in Fig. 5.2.7. The design of bandpass filter is very critical and puts some limitations on the modulating and carrier frequencies.

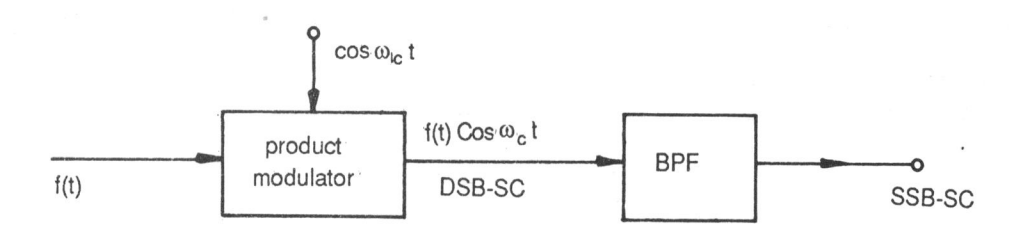

Fig. 5.2.7 Filter Method for Generating SSB-SC

*Limitations*:
1. The frequency discrimination method is useful only when the baseband is restricted at its lower edge, so that the upper and lower sidebands are non-overlapping as shown in Fig. 5.2.8b. Here, the baseband starts from 300 Hz and therefore, the upper and lower sidebands are separated by 600 Hz, i.e., *twice of the lowest frequency component of the baseband*. The desired (upper) sideband can be easily obtained by a band pass filter. The filter need not have a sharp cutoff, and hence, simple in design and construction. However, the filter should be highly selective. Such filters are realized by a crystal that acts as a resonant circuit with a $Q$ factor lying between 1000 to 2000. The filter must reject the unwanted sideband by at least 40 dB as shown in Fig. 5.2.8b. When the baseband is not restricted at its lower edge and contains very low frequencies, the upper and lower sidebands may overlap each other, as shown in Fig. 5.2.8a. The baseband starts from d.c., and hence there is no separation between the two sidebands; rather they touch each other at $\omega_c$. Therefore, a sharp cutoff bandpass filter is needed to separate the desired (upper) sideband from the lower one. It is impossible to realize such an ideal filter.

The filter method is used for speech communication where lowest spectral component is 70 Hz, and it can be taken as 300 Hz without affecting the intelligibility of the speech. However, the system is not useful for video communication where the baseband starts from as low as d.c.

2. The frequency discrimination method has another restriction that the baseband should be appropriately related to the carrier frequency. The design of the bandpass filter becomes very difficult if the carrier frequency is much higher than the bandwidth of the baseband signal. Let us consider an example in which a voice signal with a baseband ranging from 300 Hz to 300 kHz modulates a carrier wave of 10 MHz. The bandpass filter requires a selectivity that can provide an attenuation of 40 dB within 600 Hz at a carrier frequency of 10 MHz, a percentage frequency change of only 0.006 per cent. The design and construction of such a sharp selective

AMPLITUDE MODULATION SYSTEMS 269

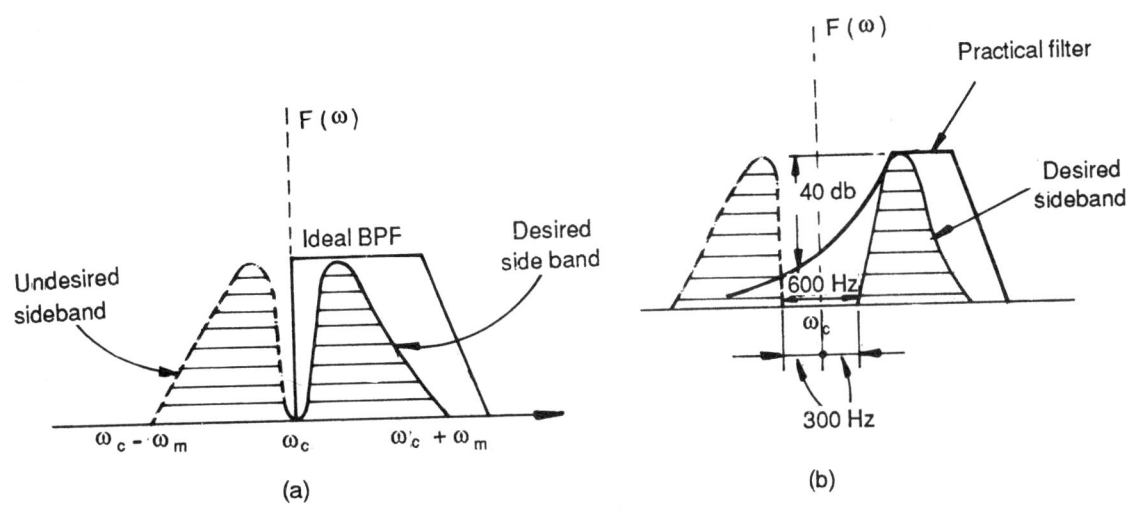

Fig. 5.2.8 Filter Requirement for SSB-SC Generation (a) Overlapping Sidebands, (b) Non-overlapping Sidebands

filter is very complex and difficult. This problem is solved by performing the translation of the baseband signal to final carrier frequency in *several stages*. This approach is illustrated in Fig. 5.2.9 which involves two translation stages.

In Fig. 5.2.9, the first carrier frequency $f_{c1}$ is selected to be of frequency 100 kHz. The upper sideband of the output of the first balanced modulator ranges from 100.3 kHz to 103 kHz. The filter $H_1(\omega)$ following this

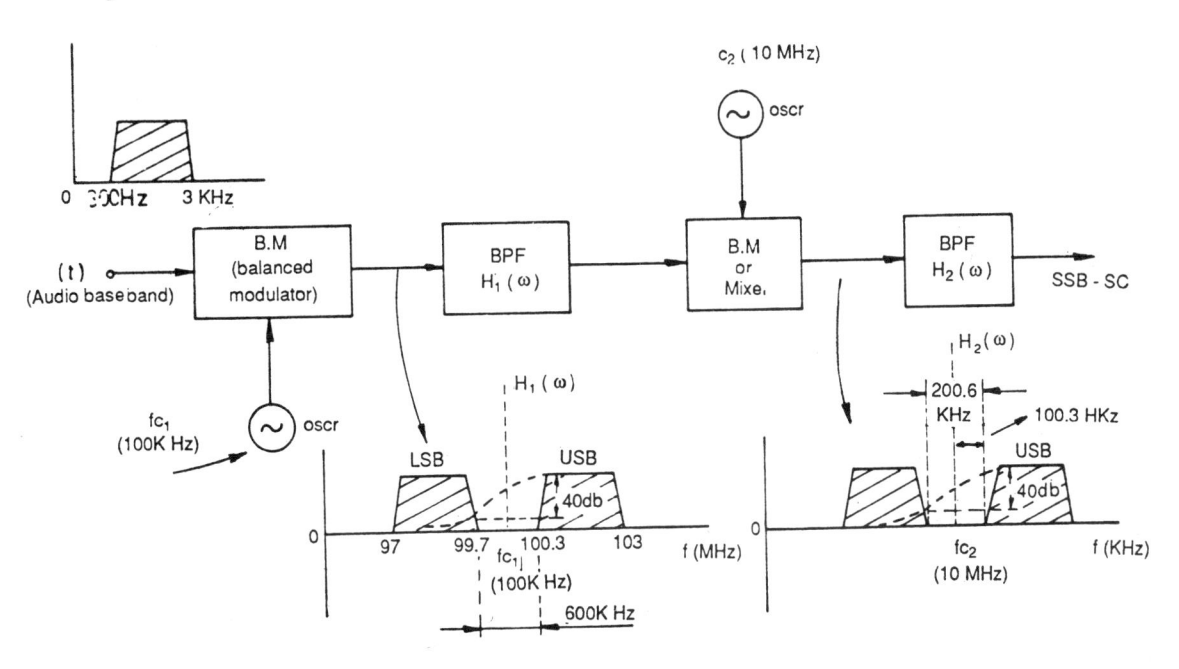

Fig. 5.2.9 SSB-SC with Two Translation Stages

balanced modulator needs a selectivity which can provide an attenuation of 40 dB within 600 Hz at a carrier frequency of 100 kHz, a percentage change of 0.6 per cent. This selectivity requirement is only a hundredth of the selectivity required in the case of the carrier frequency of 10 MHz. Therefore, the design of the bandpass filter to select desired sideband (e.g. upper sideband) becomes simpler. The upper sideband (100.3 kHz to 103 kHz) selected by this filter is applied to a second balanced modulator using a carrier frequency $f_{c2}$ of 10 MHz. Then the second bandpass filter $H_2(\omega)$ needs to provide an attenuation of 40 dB within 200.6 kHz at a carrier frequency of 10 MHz, a percentage change of about 2 per cent. The design of this filter becomes simpler than the first filter $H_1(\omega)$.

The second balanced modulator in Fig. 5.2.9 can be replaced by a mixer circuit which is a simpler frequency translation device. A mixer, however, has the disadvantage that it produces not only the desired sum and difference frequencies, but also the undesired input frequencies. But this problem is not so serious because the input frequencies (including carrier frequency) are well out of the passband of the second filter $H_2(\omega)$. Thus, while generating SSB, if the carrier is of the MHz order, or more, the frequency translation should be achieved in more than one stage for convenience in appropriate filter design.

## Phase Discrimination or Phasing Method

In this method, the time domain description of the SSB-SC wave given in Eq. 5.2.4 is used. A lower sideband SSB-SC signal is given by the expression

$$\varphi_{SSB}(t) = f(t) \cos \omega_c t + f_h(t) \sin \omega_c t$$

The block diagram of phasing method is shown in Fig. 5.2.10.

The product term $f(t) \cos \omega_c t$ is an AM-SC signal which can be generated by a simple product modulator, or a balanced modulator. The product term $f_h(t) \sin \omega_c t$ is generated by passing $f_h(t)$ and $\sin \omega_c t$ through another product, or balanced modulator. The function $f_h(t)$ is generated by passing $f(t)$ through a wideband $-\frac{\pi}{2}$ phase-shifter. A wideband phase-shifter is needed so that all the frequency component present in the baseband signal are shifted by 90°, keeping their amplitudes unchanged. Signal $\sin \omega_c t$ is obtained by passing $\cos \omega_c t$ through a simple $(-90°)$ phase-shifter. The product terms $f(t)\cos \omega_c t$ and $f_h(t)\sin \omega_c t$, are added in an adder to generate the SSB-SC signal.

The merits of this phasing method are that it does not require any sharp cut off filter, and it is possible to generate the desired sideband in a single frequency translation step, regardless of how large the carrier frequency maybe. Inspite of these merits, the phasing method is less popular than the filter method. The reason is that the following constraints should be precisely met in order to suppress the carrier and undesired side band.

(i) Each balanced modulator need to be carefully balanced in order to suppress the carrier.

(ii) In practice, it is difficult to design a wideband phase shifting network, particularly where the baseband extends over many octaves. This network is needed to generate Hilbert transform of $f(t)$.

(iii) Each modulator should have equal sensitivity to the baseband signal.

(iv) The carrier phase-shifting network must provide an exact 90° phase shift at the carrier frequency.

The most difficult problem arises while designing a wideband phase shifting network to generate Hilbert transform. The wideband constant phase shift is obtained over a given frequency range by including a phase-shifting network in each path of modulation, i.e., before each of the two balanced modulators. This type of arrangement is shown in Fig. 5.2.11. The two phase shifting networks are denoted by $\alpha$ and $\beta$. The phase-shifts $\alpha$ and $\beta$ are related by

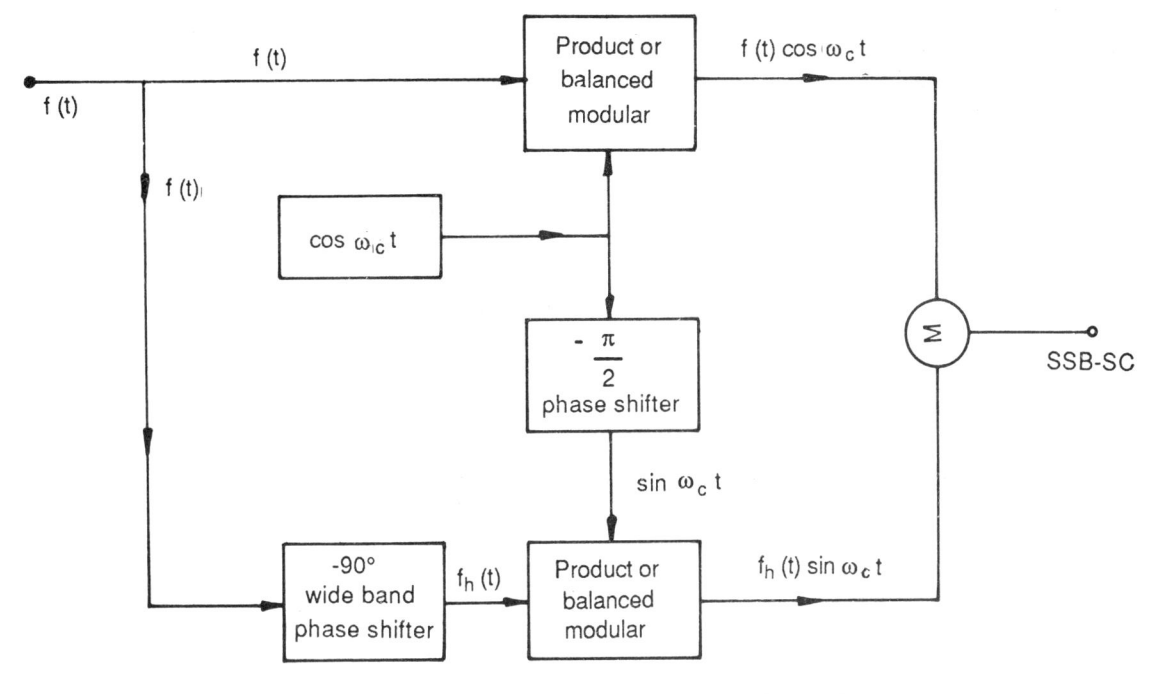

Fig. 5.2.10 Phase Shift Method for SSB-SC Generation

$$\beta - \alpha = \frac{\pi}{2}$$

The phase-shifts $\alpha$ and $\beta$ are adjusted to provide a constant phase-shift over a given frequency range within the specified tolerance.

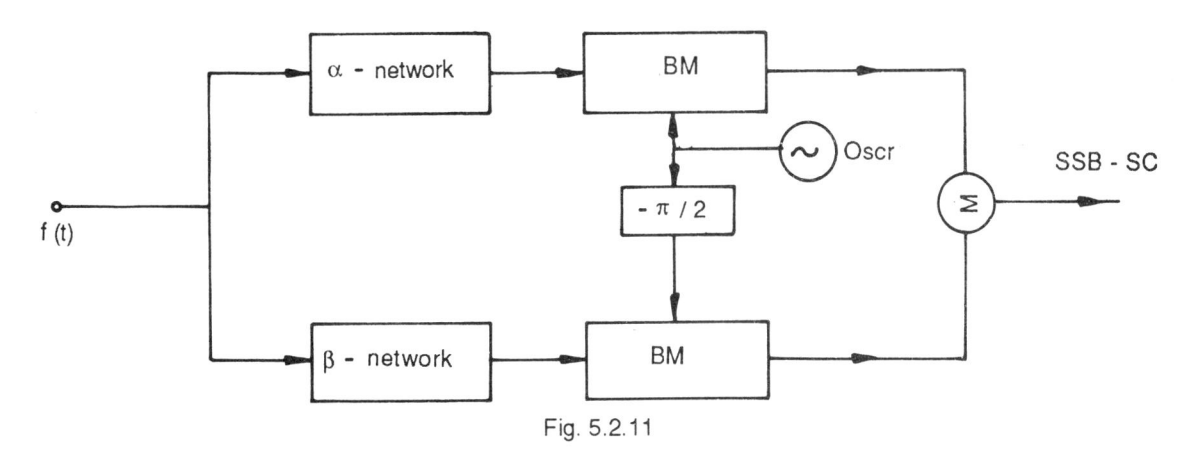

Fig. 5.2.11

## Detection of SSB-SC Signals

The baseband signal $f(t)$ can be recovered from the SSB-SC signal by using the synchronous detection technique, as discussed for DSB-SC signals. By synchronous detection, the spectrum of an SSB-SC signal, centered about $\omega = \pm \omega_c$, is retranslated to the baseband spectrum, centered about $\omega = 0$. The process of synchronous detection involves multiplication of the received SSB-SC singal with a locally generated carrier as shown in Fig. 5.2.12 (a)

The output of the multiplier is,

$$e_d(t) = \varphi_{SSB}(t) \cos \omega_c(t) = [f(t) \cos \omega_c t \pm f_h(t) \sin \omega_c t] \cos \omega_c t$$

$$= f(t)\cos^2 \omega_c t \pm f_h(t)\sin \omega_c t \cos \omega_c t$$

$$= \frac{1}{2} f(t)\{1 + \cos 2\omega_c t\} \pm \frac{1}{2} f_h(t) \sin 2\omega_c t$$

$$= \frac{1}{2} f(t) + \frac{1}{2}[f(t) \cos 2\omega_c t \pm f_h(t) \sin 2\omega_c t] \quad (5.2.15a)$$

(a)

(b)

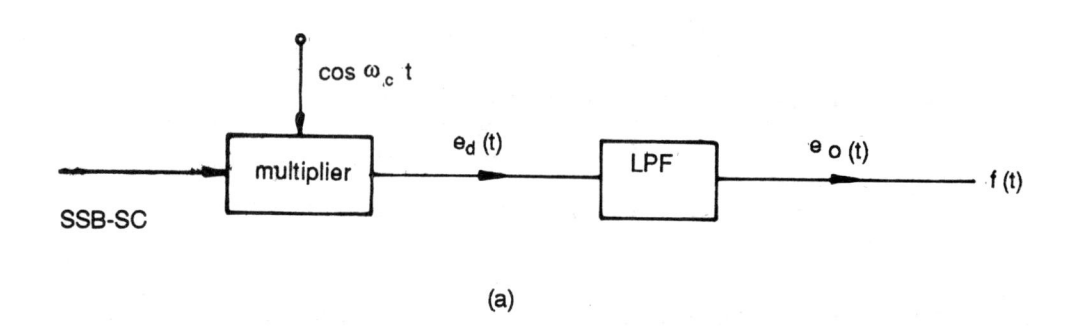

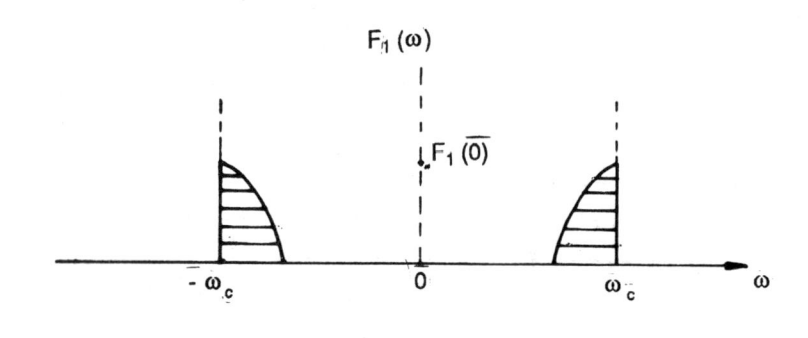

AMPLITUDE MODULATION SYSTEMS 273

(c)

(d)

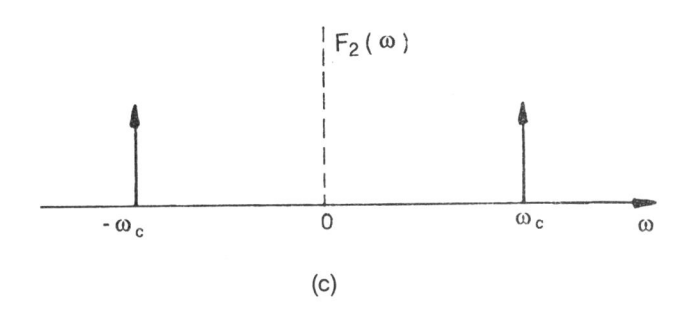

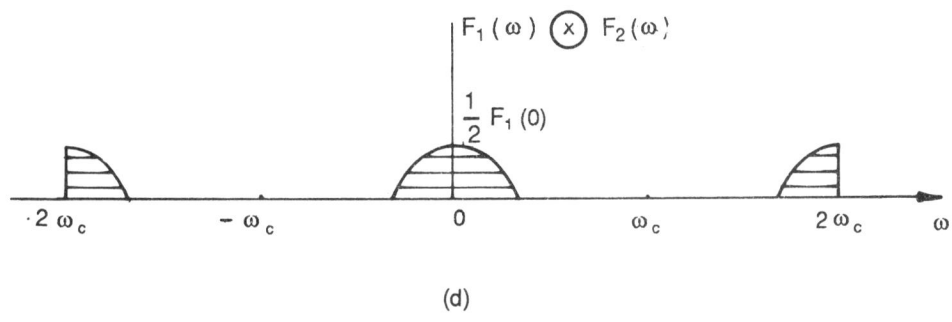

Fig. 5.2.12 Detection of SSB-SC Signal (a) Synchronous Detector; (b) SSB-SC (lower sideband); (c) Spectrum of $\cos \omega_c t$; (d) Convolution of Figures (b) and (c)

when $e_d(t)$ is passed through a low pass filter, the terms centered about $\pm 2\omega_c$ are filtered out and we get, at the output of detector, $e_o$ given by

$$e_o(t) = \frac{1}{2} f(t) \tag{5.2.15b}$$

The frequency domain explanation is evident from Fig. 5.2.12. Multiplication of the SSB signal with $\cos \omega_c t$ in the time domain is equivalent to convolution of their spectra, as shown in Fig. 5.2.12 d. The component centered around $\pm 2\omega_c$ is filtered out; whereas, the messaged signal centered at $\omega = 0$ appears at the output. The synchronous detection can be achieved by a ring demodulator circuit, or a modulator circuit using non-linear devices.

## Effect of Phase and Frequency Error

As in DSB-SC signals, the synchronous detection of SSB-SC signals also needs phase and frequency synchronisation. A detected output will be distorted when there is a discrepancy in phase and frequency of the locally generated carrier and transmitted carrier. Distortion due to discrepancy in frequency is somewhat same in both DSB-SC and SSB-SC detection; however, distortion because of phase discrepancy is more serious in the latter.

Let us presume that the locally generated carrier has a frequency error of $\Delta\omega$, and a phase-error of $\phi$, then, the output of multiplier in Fig. 5.2.12a is given as

$$e_d(t) = [f(t)\cos \omega_c t + f_h(t)\sin \omega_c t][\cos\{(\omega_c + \Delta\omega)t + \phi\}]$$

$$= \frac{1}{2}f(t)\cos(\Delta\omega t + \phi) + \frac{1}{2}f(t)\cos[(2\omega_c + \Delta\omega)t + \phi]$$

$$- \frac{1}{2}f_h(t)\sin(\Delta\omega t + \phi) + \frac{1}{2}f_h(t)\sin[(2\omega_c + \Delta\omega)t + \phi]$$

The terms centered around $\pm 2\omega_c$ are filtered out by a low pass filter, and the signal appearing at the output of the filter is

$$e_o(t) = \frac{1}{2}f(t)\cos(\Delta\omega t + \phi) - \frac{1}{2}f_h(t)\sin(\Delta\omega t + \phi) \tag{5.2.16}$$

Consider the following cases,
(i) When there is no phase and frequency error, ($\Delta\omega = 0, \phi = 0$) then

$$e_o(t) = \frac{1}{2}f(t) \text{ (without any distortion)}$$

(ii) When phase error is zero, and frequency error, $\Delta\omega$, then from 5.2.16,

$$e_o(t) = \frac{1}{2}f(t)\cos(\Delta\omega)t - \frac{1}{2}f_h(t)\sin(\Delta\omega)t$$

The signal is slow variant with time causing a distortion similar to that observed in DSB-SC detection. The amplitude fluctuation at a rate of $\Delta\omega$ is known as *Warble* and is tolerable upto $\Delta\omega = 30$ HZ.

(iii) When the frequency error is zero and phase error is equal to $\phi$, then

$$e_o(t) = \frac{1}{2}[f(t)\cos\phi - f_h(t)\sin\phi] \tag{5.2.17}$$

since $f(t)$ and $f_h(t)$ contain the same frequency spectrum they cannot be separated. We will see now that the undesired term $f_h(t)\sin\phi$ introduces a phase distortion, shifting every component of $f(t)$ in phase by $\phi$ radians. The Fourier transform of Eq. 5.2.17 yields

$$E_o(\omega) = \frac{1}{2}[F(\omega)\cos\phi - F_h(\omega)\sin\phi] \tag{5.2.18}$$

Substituting this value of $F_h(\omega)$ from Eq. 5.2.8 in Eq. 5.2.18, and recalling the definition of sgn ($\omega$) we can write,

$$E_o(\omega) = \begin{cases} \dfrac{1}{2}F(\omega)\,e^{j\phi}, & \omega > 0 \\ \dfrac{1}{2}F(\omega)\,e^{-j\phi}, & \omega < 0 \end{cases} \tag{5.2.19}$$

Multiplication by $e^{j\phi}$ in frequency domain is equivalent to phase shift of $\phi$ in time domain. Hence Eq. 5.2.19 shows that every component of $f(t)$ is shifted in phase by $\phi$ radians, causing phase distortion in the detected output. In DSB-SC, the phase error introduced only attenuation, and not distortion. The phase distortion does not seriously affect voice communication, as it may change only the quality of speech and not its intelligibility. This is because the human ear is less sensitive to phase distortion (distortion being known as Donald Duck Voice effect). However, phase distortion is intolerable in video signal and music transmission.

## Phasor Diagram and Waveform of SSB Signals

Let us consider a single-tone modulating signal $E_1 \cos \omega_1 t$ that modulates a carrier $E_c \cos \omega_c t$ to generate an SSB-SC signal. The resulting signal is given as (considering lower sideband)

$$\varphi_{SSB}(t) = \frac{1}{2} E_1 E_c \cos(\omega_c - \omega_1)t \qquad (5.2.20)$$

The signal represents a single sinusoid with an amplitude $\frac{1}{2} E_1 E_c$ and a frequency $(\omega_c - \omega_1)$. Therefore, the phasor diagram consists of a single phasor, and there will be no amplitude fluctuation in the modulated wave. This is obvious from the phasor diagram in Fig. 5.2.13a, and the waveform drawn in Fig. 5.2.13b. It is also from the waveform is that both amplitude and frequency of the wave are constant, .i.e., there is no amplitude modulation.

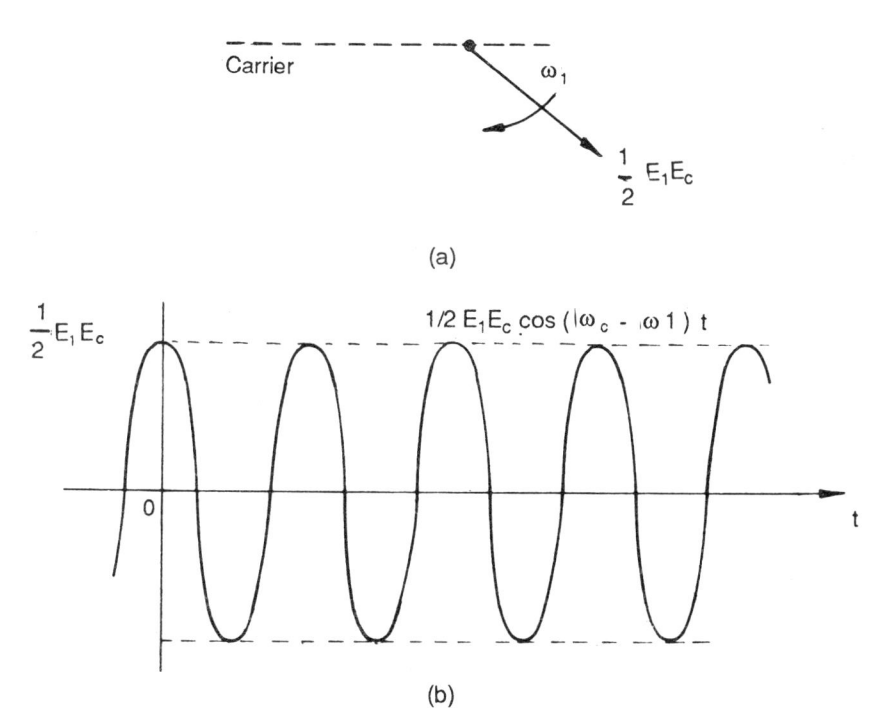

Fig. 5.2.13 (a) Phasor Diagram of a Single-Tone SSB-SC Signal; (b) Waveform of a Single-Tone SSB-SC Signal

Therefore, an SSB-SC signal modulated by a signal tone modulating signal does not carry any useful information. However, the baseband signals appearing in actual systems are multiple-tone rather than single tone. Consider a two tone baseband given by

$$f(t) = E_1 \cos \omega_1 t + E_2 \cos \omega_2 t$$

Figure 5.2.14a shows the phasor diagram of an SSB-SC signal modulated by the baseband signal $f(t)$. The resultant phasor varies with time as the two phasors rotate. The amplitude variation in the resulting waveform is shown in Fig. 5.2.14b. Thus the multiple tone SSB-SC signal shows amplitude variation.

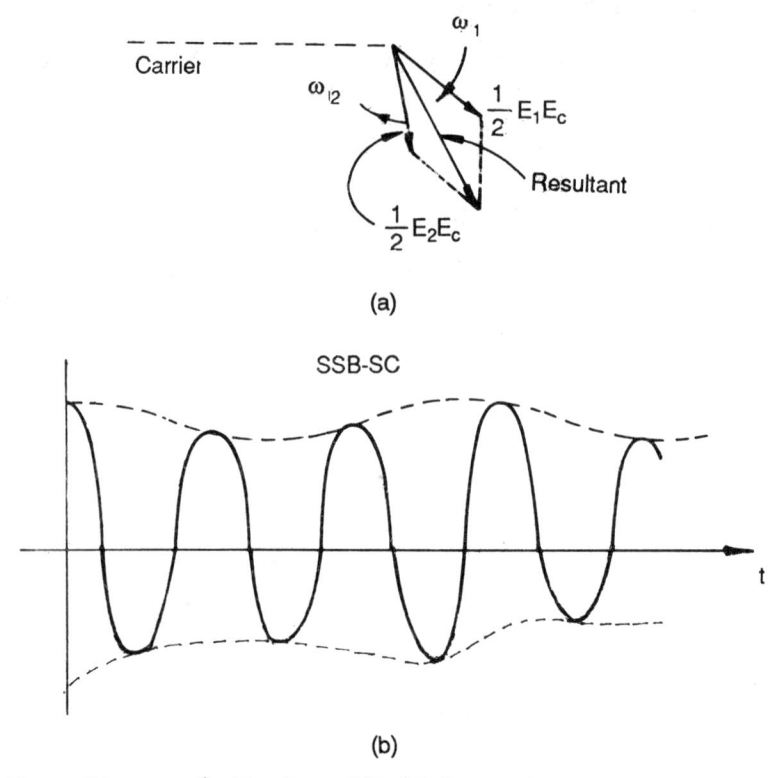

Fig.5.2.14 (a) Phasor Diagram of a Two-Tone SSB-SC Signal; (b) Waveform of a Two-Tone SSB-SC Signal

## Waveform of SSB with Large Carrier

Now, let us consider an SSB signal in which a large carrier is also present. This carrier may be introduced at the transmitter as in AM, or may be added at the receiver. When a large carrier is added to the SSB-SC signal the detection becomes easy. When a large carrier is present, the SSB waveform shows amplitude variations, even for a single-tone modulating frequency. The expression for such a wave is obtained by adding a large carrier terms in the expression of SSB-SC, given in Eq. 5.2.20 i.e.,

$$\varphi_{SSB}(t) = \frac{1}{2} E_1 E_c \cos(\omega_c - \omega_1)t + E_c \cos \omega_c t; \quad E_c > E_1 \tag{5.2.21}$$

The phasor diagram is shown in Fig. 5.2.15. The SSB-SC waveform is formed by superposition of two waveforms (lower sideband and large carrier term).

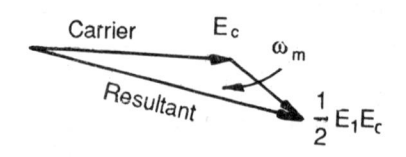

Fig. 5.2.15 Phasor Diagram of an SSB With Large Carrier

## Demodulation of the SSB Signal with Large Carrier

When a large carrier is introduced, a synchronous detector is not necessary for recovering the baseband signal. An envelope detector provides the approximate baseband signal. An envelope detector is simpler and cheaper than synchronous detector. An expression for a SSB-signal with a large carrier $A \cos \omega_c t$ can be written as

$$\varphi_{SSB}(t) = f(t) \cos \omega_c t + f_h(t) \sin \omega_c t + A \cos \omega_c t \qquad (5.2.22)$$

Here it should be kept in mind that the carrier is added to an SSB-SC signal (and not multiplied with it as in synchronous detection). This technique is called *carrier reinsertion technique*.
Equation 5.2.22 can be written as

$$\varphi_{SSB}(t) = [A + f(t)] \cos \omega_c t + f_h(t) \sin \omega_c t$$

$$= e(t) \cos (\omega_c t + \theta)$$

where $e(t) = \sqrt{[A + f(t)]^2 + f_h^2(t)}$ is the envelope of the wave, and $\theta = -\tan^{-1}\left[\dfrac{f_h(t)}{A + f(t)}\right]$ is the phase of the wave. When this SSB signal is applied to an envelop detector, the output of the detector will be the envelope $e(t)$:

$$e(t) = \left[\{A + f(t)\}^2 + f_h^2(t)\right]^{\frac{1}{2}}$$

$$= A\left[1 + \frac{2f(t)}{A} + \frac{f^2(t)}{A^2} + \frac{f_h^2(t)}{A^2}\right]^{\frac{1}{2}}$$

which can be expanded in the form a Binomial series, given by

$$e(t) = A\left[1 + \frac{2f(t)}{2A} + \text{higher terms}\right]$$

since $A >> |f(t)|$ and $|f_h(t)|$, higher terms may be negleted, then

$$e(t) \simeq A + f(t)$$

Thus the output of the envelope detector is close to the desired baseband signal $f(t)$. The carrier re-insertion technique is also useful for the detection of DSB-SC signals.

The large carrier can be added either at the transmitter or at the receiver. When the carrier is added at the receiver, the synchronization problem remains as it is, and this technique has no advantage. When carrier is added at the transmitter, this mode has the advantage of both SSB and AM systems. It needs only half the bandwidth, at the same time, detection becomes simple.

## Compatible Single Sideband

An SSB signal can be generated in which the carrier is suppressed, even then it can be detected with an envelope detector (simple diode detector). Such a signal is compatible for reception using a commercial AM radio receiver. The signal with this characteristic is termed as *compatible single sideband* (CSSB). However, such a signal involves complex signal processing, and the system is presently impractical for commercial applications.

## 5.3 VESTIGIAL-SIDEBAND MODULATION (VSB)

The major advantage of the SSB-SC modulation is that it reduces the bandwidth requirement to half as compared to DSB-SC modulation. But SSB-SC signals are relatively difficult to generate due to the difficulty

278 COMMUNICATION SYSTEMS: Analog and Digital

in isolating the desired sideband. The required filter must have a very sharp cut off characteristic, particularly when the baseband signal contains extremely low frequencies (e.g., television and telegraphic signals). This difficulty is overcome by a scheme known as *vestigial sideband* modulation, which is a compromise between SSB-SC and DSB-SC modulation.

In vestigial sideband modulation (VSB) the desired sideband is allowed to pass completely. Whereas just a small portion (called trace or vestige) of the undersired sideband is also allowed, as shown in Fig. 5.3.1. The transmitted vestige of the undesired sideband compensates for the loss of the wanted (upper) sideband.

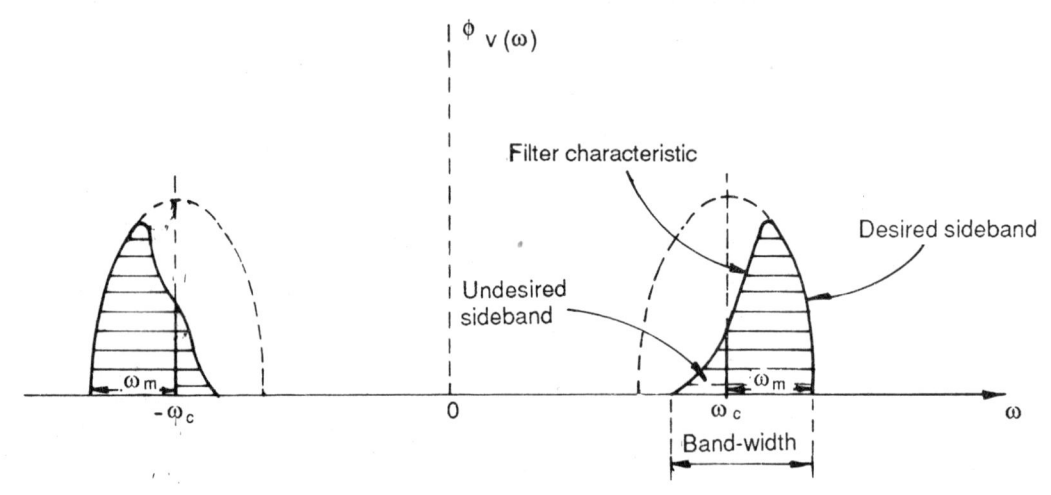

Fig. 5.3.1 Spectrum of a Vestigial Sideband Signal

It is evident from figure 5.3.1 that the filter required need not have a sharp cut-off, which is an advantage of the VSB system. However, as compared to SSB-SC, the bandwidth of the VSB becomes larger; although it remains much smaller than the DSB-SC signal. Thus, VSB is a compromise between DSB-SC and SSB-SC signals. VSB is used in television for transmission of picture signals.

## Generation and Detection of VSB Signal

(i) *Filter method:* A VSB-SC signal can be generated by passing a DSB-SC signal through an appropriate filter as in Fig. 5.3.2a. The system is similar to DSB-SC, except that the filter characteristic should be appropriate. The filter characteristic desired for VSB can be derived by analysing the demodulation technique of the VSB-SC signal which is nothing but the synchronous detection shown in Fig. 5.3.2(b)

*Filter characteristics:* Let us find the characteristics of the filter with a transfer function $H(\omega)$ shown in Fig. 5.3.3a that may generate a VSB-SC signal from a DSB-SC signal. A DSB-SC signal has a spectrum given by

$$\varphi_{DSB}(\omega) = \frac{1}{2}F(\omega + \omega_c) + \frac{1}{2}F(\omega - \omega_c)$$

The output of the filter $H(\omega)$ will be given by

$$\varphi_v(\omega) = \frac{1}{2}H(\omega)[F(\omega + \omega_c) + F(\omega - \omega_c)] \tag{5.3.1}$$

where

$$\varphi_v(t) \leftrightarrow \varphi_v(\omega)$$

## AMPLITUDE MODULATION SYSTEMS

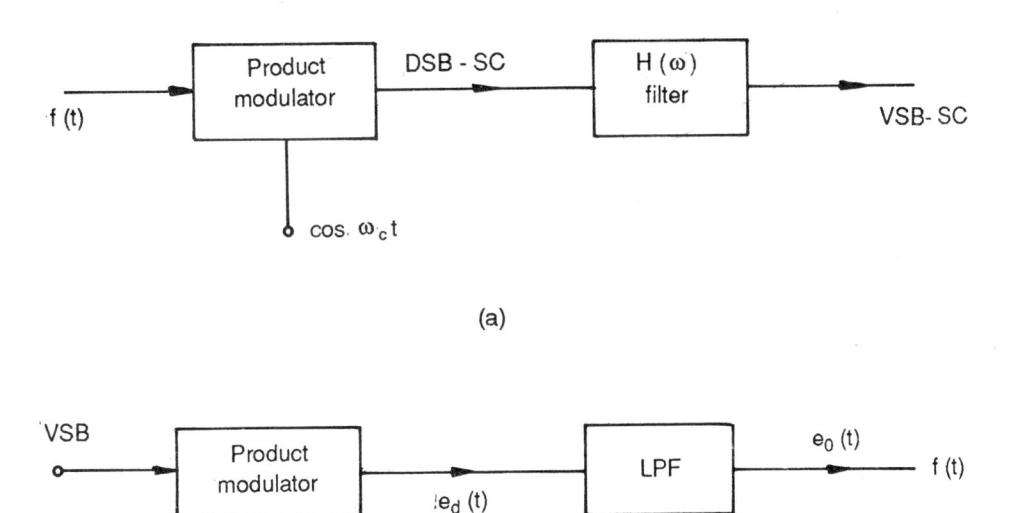

Fig. 5.3.2 (a) VSB-SC Modulator (filter method); (b) VSB-SC Synchronous Detector

This signal $\varphi_v(t)$ is passed through a product modulator for synchronous detection as shown in Fig. 5.3.2b. The output of this product modulator denoted by $e_d(t)$ will be given as

$$e_d(t) = \varphi_v(t) \cos \omega_c t$$

or

$$e_d(t) \leftrightarrow \frac{1}{2}[\varphi_v(\omega + \omega_c) + \varphi_v(\omega - \omega_c)]$$

Substituting the value of $\varphi_v(\omega)$ from Eq. 5.3.1, we get

$$e_d(t) \leftrightarrow \frac{1}{4}[\{F(\omega + 2\omega_c) + F(\omega)\} H(\omega + \omega_c)$$
$$+ \{F(\omega) + F(\omega - 2\omega_c)\} H(\omega - \omega_c)]$$

The signal $e_d(t)$ is passed through a low pass filter and, hence, the components centered around $\pm 2\omega_c$ are filtered out. Thus, the output of L P F of the synchronous detector will be given as

$$e_o(t) = \frac{1}{4} F(\omega) \{H(\omega + \omega_c) + H(\omega - \omega_c)\} \qquad (5.3.2)$$

For getting a distortionless reception, the output $e_o(t)$ should be given as

$$e_o(t) = C_1 F(\omega) \qquad (5.3.3)$$

where $C_1$ is a constant.

## 280 COMMUNICATION SYSTEMS: Analog and Digital

When comparing Eq. 5.3.2 with Eq. 5.3.3, we get the following condition for a frequency range $|\omega|<\omega_m$ (since the baseband is limited to $\omega_m$)

$$H(\omega + \omega_c) + H(\omega - \omega_c) = C \text{ (constant)}, \qquad |\omega|<\omega_m \qquad (5.3.4)$$

The terms $H(\omega + \omega_c)$, and $H(\omega - \omega_c)$, represent $H(\omega)$ shifted by $-\omega_c$ and $+\omega_c$, respectively, as shown in Fig. 5.3.3b and c. The sum of these shifted spectrum should be constant over the frequency range $|\omega| < \omega_m$ as shown in Fig. 5.3.3d. It is obvious from Fig. 5.3.3e that $H(\omega + \omega_c) + H(\omega - \omega_c)$ will be constant over $|\omega| < \omega_m$ only if the cut-off characteristic of filter $H(\omega)$ has a complementary symmetry (odd symmetry) around the carrier frequency.

(ii) *Phase-Discrimination Method for Generation of VSB-SC Wave:* Like SSB-SC signals, VSB-SC signals also can be generated by using the phase discrimination method. The principle of this method will become clear once we determine the time domain description of the VSB-SC wave.

## Time Domain Description of VSB Wave

The frequency domain description of the VSB signal is given by Eq. 5.3.1 as follows:

$$\varphi_v(\omega) = \frac{1}{2}[F(\omega + \omega_c) + F(\omega - \omega_c)]H(\omega)$$

The above frequency domain description can be considered to have two parts: (i) $\frac{1}{2}[F(\omega + \omega_c) + F(\omega - \omega_c)]$ with its equivalent time domain as $f(t)\cos \omega_c t$ and (ii) $H(\omega)$ with its time domain equivalent unit $h(t)$. Therefore, time domain of $\varphi_v(t)$ can be written by using time convolution theorem,

$$\varphi_v(t) = h(t) \circledast f(t) \cos \omega_c t \qquad (5.3.5)$$

$$= \int_{-\infty}^{\infty} h(\tau) f(t - \tau) \cos \omega_c (t - \tau) d\tau$$

$$= \int_{-\infty}^{\infty} h(\tau) f(t - \tau)[\cos \omega_c t \cos \omega_c \tau + \sin \omega_c t \sin \omega_c \tau]$$

The above equation may be split into two parts:

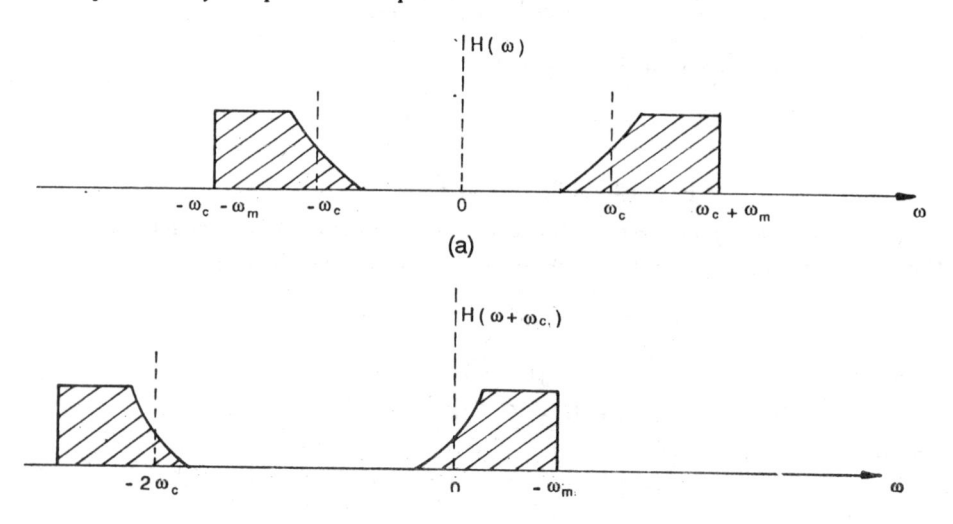

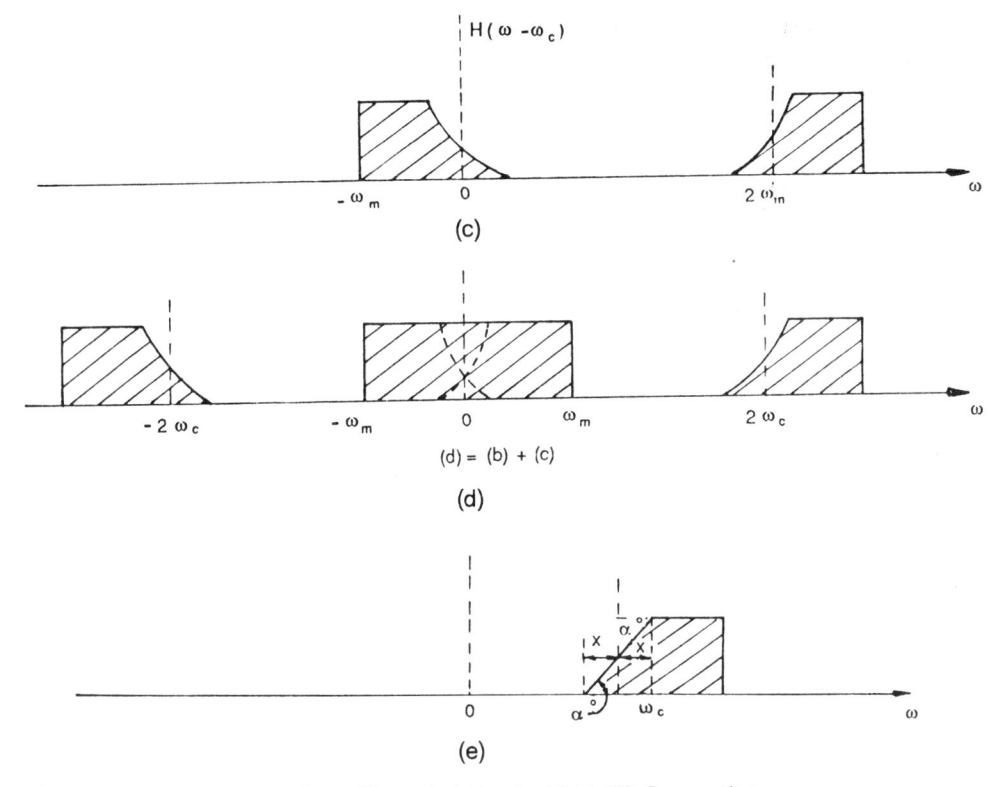

Fig. 5.3.3 Filter Characteristics for VSB-SC Generation

$$\varphi_v(t) = \cos \omega_c t \int_{-\infty}^{\infty} h(\tau) \cos \omega_c \tau \, f(t-\tau) d\tau$$

$$+ \sin \omega_c t \int_{-\infty}^{\infty} h(\tau) \sin \omega_c \tau \, f(t-\tau) d\tau$$

The equation can be written in a simplified form by defining the in-phase and the quadrature-phase components denoted by $f_c(t)$, and $f_s(t)$, respectively. The simplified version is similar to the expression of the SSB-SC wave in time domain. Thus, the time domain description of VSB-SC signal is,

$$\varphi_v(t) = \frac{1}{2} f_c(t) \cos \omega_c t + \frac{1}{2} f_s(t) \sin \omega_c t \tag{5.3.6}$$

where

$$f_c(t) = 2 \int_{-\infty}^{\infty} h(\tau) \cos \omega_c \tau \, f(t-\tau) d\tau \tag{5.3.7}$$

and

$$f_s(t) = 2 \int_{-\infty}^{\infty} h(\tau) \sin \omega_c \tau \, f(t-\tau) d\tau \tag{5.3.8}$$

Note that the factor 1/2 in Eq. 5.3.6 is introduced for convenience.

## In Phase component $f_c(t)$

It is evident from Eq. 5.3.7 that $f_c(t)$ can be obtained by passing $f(t)$, through a filter with an impulse response $h_c(t)$ given by

$$h_c(t) = 2h(t) \cos \omega_c t \qquad (5.3.9)$$

The frequency domain of $h_c(t)$ provides the transfer function of the in-phase filter,

$$H_c(\omega) = H(\omega - \omega_c) + H(\omega + \omega_c) \qquad (5.3.10)$$

From Eqs. 5.3.4 and 5.3.10,

$$H_c(\omega) = C = 1 \quad \text{(say)}$$

or

$$h_c(t) = \delta(t)$$

Obviously, when a function $f(t)$ is passed through a filter with $H_c(\omega)=1$ the output is function $f(t)$ itself. hence

$$f_c(t) = f(t) \qquad (5.3.11)$$

Thus, the in-phase component $f_c(t)$ is same as the baseband signal $f(t)$.

## Quadrature-phase Component $f_s(t)$

A reasoning similar to the one used for in-phase component reveals that $f_s(t)$ is obtained by passing $f(t)$ through a filter having unit impulse response given by

$$h_s(t) = 2h(t) \sin \omega_c t$$

or

$$H_s(\omega) = \frac{1}{j}[H(\omega - \omega_c) - H(\omega + \omega_c)] \qquad (5.3.12)$$

Thus, we need a filter with the transfer function $H_s(\omega)$ given by Eq. 5.3.12 to get the quadrature component $f_s(t)$.

The time domain description of the VSB-SC signal is given as follows: (using Eqs 5.3.6 and 5.3.11)

$$\varphi_v(t) = \frac{1}{2}f(t) \cos \omega_c t + \frac{1}{2}f_s(t) \sin \omega_c t \qquad (5.3.13)$$

This time domain description given in Eq. 5.3.13 provides a basis for VSB signal generation as shown in Fig. 5.3.4. Note that DSB-SC and SSB-SC are two special cases of VSB-SC wave defined by Eq. 5.3.13. It becomes DSB-SC when $f_s(t)$ equals zero, and SSB-SC when $f_s(t) = f_h(t)$.

The block diagram and Eq. 5.3.13 correspond to the VSB signal with a full upper-sideband and a partial lower sideband. Likewise, Eq. 5.3.13 will represent a VSB signal having a full lower-sideband and a partial upper-sideband if the plus sign, is replaced by a minus sign. Accordingly, the lower path signal in the block diagram of Fig. 5.3.4 will be subtracted instead of added.

## Detection of VSB Signal with Large Carrier

Such signals can be detected by using an envelope detector. However, distortion is introduced due to the

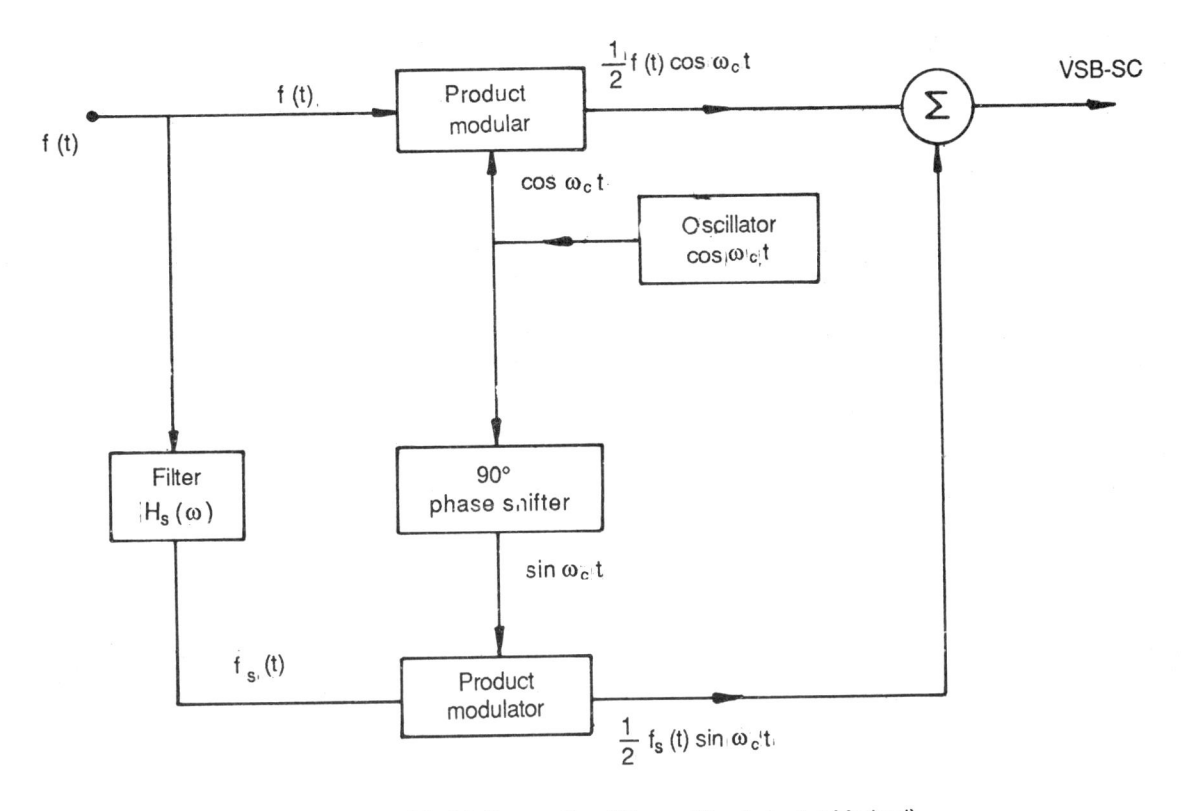

Fig. 5.3.4 VSB-SC Generation (Phase-Discriminator Method)

quadrature component $f_s(t)$. This can be seen by the simple analysis given below:
When a large carrier $A \cos \omega_c t$ is added to the Eq. 5.3.13 scaled by a factor $(m_a A)$, the modified VSB at the input of the envelope detector is,

$$\varphi_v(t) = \frac{1}{2} A m_a f(t) \cos \omega_c t + \frac{1}{2} A m_a f_s(t) \sin \omega_c t + A \cos \omega_c t$$

The output of the envelope detector is the magnitude of the signal $\varphi_v(t)$,

$$|\varphi_v(t)| = A \left[1 + \frac{1}{2} m_a f(t)\right] \left[1 + \left\{\frac{\frac{1}{2} m_a f_s(t)}{1 + \frac{1}{2} m_a f(t)}\right\}\right]^{\frac{1}{2}} \qquad (5.3.14)$$

It is obvious from Eq. 5.3.14 that quadrature component $f_s(t)$ produces distortion. This distortion can be minimized by: (i) reducing $m_a$, (ii) increasing the bandwidth of VSB to reduce $f_s(t)$. In commercial television broadcasting, the vestigial sideband occupies a width of about one-sixth of a full sideband which keeps the distortion within tolerable limit for $m_a = 1$.

## 5.4 AMPLITUDE MODULATION WITH LARGE CARRIER (AM)

The conventional amplitude modulation system used in broadcast systems consists of a large carrier power alongwith upper and lower sidebands, and is referred to as AM. The main advantage of this system is that the receiver is made simple and less expensive. In a *broadcast system*, a transmitter has to transmit power for a large number of receivers. In other words, the transmitter is associated with a large number of receivers, and hence a less costly receiver is needed. However, a single large power transmitter can be installed to feed power for a multitude of receivers. The AM system is preferred for this type of application.

The AM signal is obtained by adding a large carrier power $A \cos \omega_c t$ to the DSB-SC signal

$$\varphi_{AM}(t) = f(t) \cos \omega_c t + A \cos \omega_c t \tag{5.4.1}$$

where $A > |f(t)|_{max}$ i.e., carrier amplitude is larger than the maximum value of a modulating signal. The spectrum of the AM signal can be obtained by evaluating the Fourier transform of Eq. 5.4.1.

$$\varphi_{AM}(\omega) = \frac{1}{2}[F(\omega + \omega_c) + F(\omega - \omega_c)] + \pi A [\delta(\omega + \omega_c) + \delta(\omega - \omega_c)] \tag{5.4.2}$$

The function $f(t)$ and its spectrum is shown in Figs 5.4.1a and b, respectively.

Note that $\omega_c > \omega_m$ where $\omega_m$ is the maximum frequency component in $f(t)$. The AM wave and its spectrum is shown in Figs 5.4.1c and d respectively.

The spectrum of AM is identical to that of the DSB-SC signal except that there are two additional impulses located at $\omega = \pm \omega_c$. These impulses represent the presence of a carrier of frequency $\omega_c$ and amplitude A. The waveform in Fig. 5.4.1c can be interpreted as follows:
The AM wave in Eq. 5.4.1 can be written as

$$\varphi_{AM}(t) = [A + f(t)] \cos \omega_c t \tag{5.4.3}$$

Equation 5.4.3 represents a wave of frequency and aplitude $A + f(t)$. Thus, the amplitude of the wave is changing around $A$ in accordance with the value of the modulating signal $f(t)$. The frequency of the AM signal is unchanged and is equal to $\omega_c$. The variation in the aplitude about $A$ becomes obvious when comparing the carrier $A \cos \omega_c t$ with the AM wave shown in Fig. 5.4.1c.

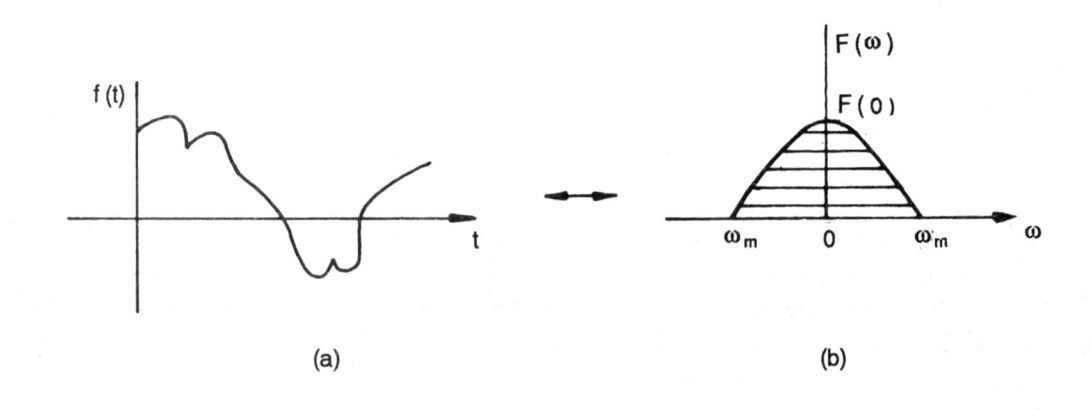

(a)

(b)

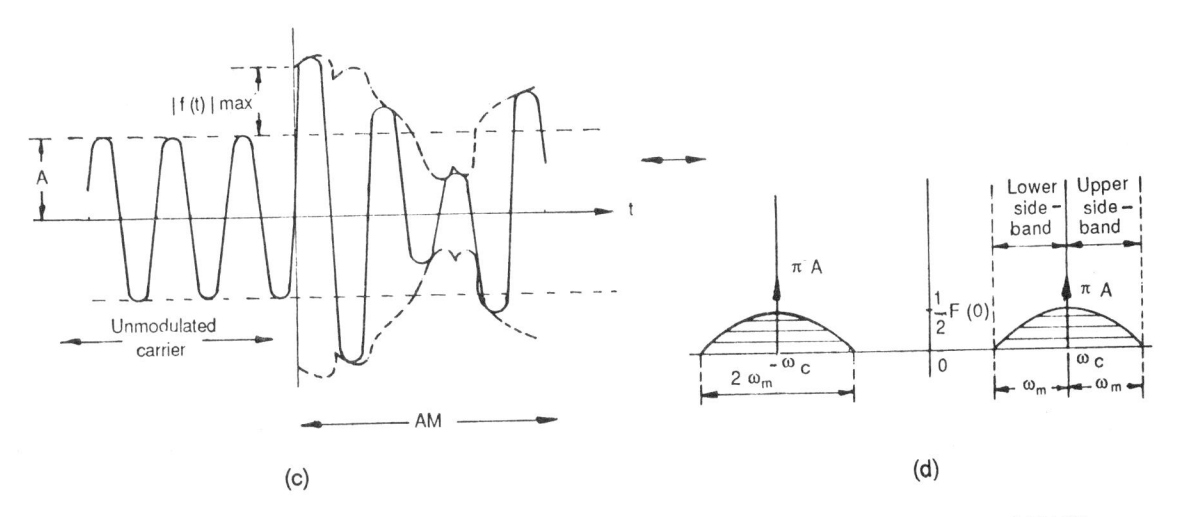

Fig. 5.4.1 (a) Baseband Signal $f(t)$; (b) Spectrm $F(\omega)$; (c) AM Wave; (d) Spectrum of AM Wave

**Envelope of the AM Wave**

The AM wave has a time varying amplitude called as the envelope of the AM wave. Figure 5.4.1c shows that the envelope of the wave consists of the baseband signal. The expression for the AM wave is given by Eq. 5.4.3.

$$\varphi_{AM} = [A + f(t)]\cos \omega_c t = E(t) \cos \omega_c t$$

where $E(t)$ is the envelope of the AM wave. This envelope consists of the baseband signal $f(t)$. Therefore, the baseband signal can be recovered from an AM wave by detecting the envelope. By using an envelope detector, the detection is less costly and simple. This is an important feature of the AM wave. This will be discussed later in Section 5.6.

An envelope is defined as the magnitude of the pre-envelope of a function. The pre-envelope is defined as in Eq. 5.2.9.

**Example 5.4.1**

Find the pre-envelope, and the envelope, of a function $f(t)$ given by $f(t) = A \cos \omega_c t$
*Solution*
By using Eq. 5.2.9, the pre-envelope is given by

$$f_p(t) = f(t) + j f_h(t)$$

$$= A\left[\cos \omega_c t + j \cos\left(\omega_c t - \frac{\pi}{2}\right)\right] = A \cos \omega_c t + j \sin \omega_c t \quad (5.4.4)$$

$$= A e^{j\omega_c t}$$

The envelope is the magnitude of $f_p(t)$,

$$|f_p(t)| = |Ae^{j\omega_c t}| = A$$

Thus, the envelope of $A \cos \omega_c t$ is a constant A which is the amplitude of the sinusoidal wave.

## Example 5.4.2

A baseband signal $f(t) = Sa(t)$ is used to amplitude modulate a carrier cos 13t. Draw the waveform and spectrum of AM signal.

*Solution*

The baseband modulating signal is a sample function given by

$$f(t) = Sa(t) = \frac{\sin(t)}{t}$$

The sampling function and its spectrum is shown in Fig. 5.4.2a and b respectively. The maximum frequency term in the spectrum is 1 rad/sec. The waveform and spectrum of AM is shown in Figs 5.4.2c and d respectively. The AM spectrum is obtained by using frequency shifting property of the Fourier transform. The two impulses at $\pm \omega_c$ indicate the presence of a large carrier.

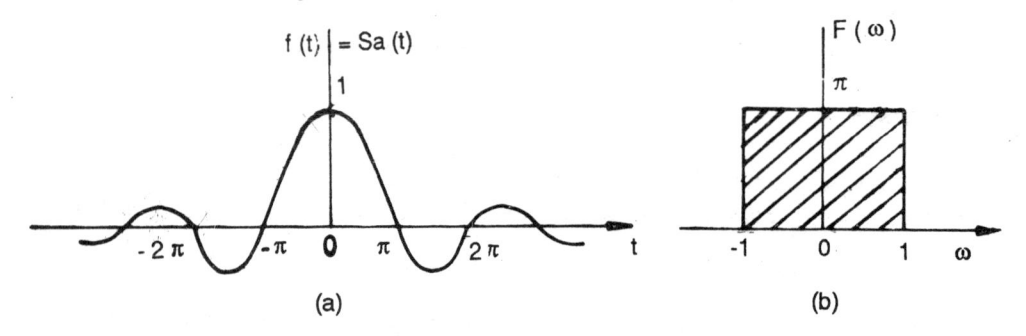

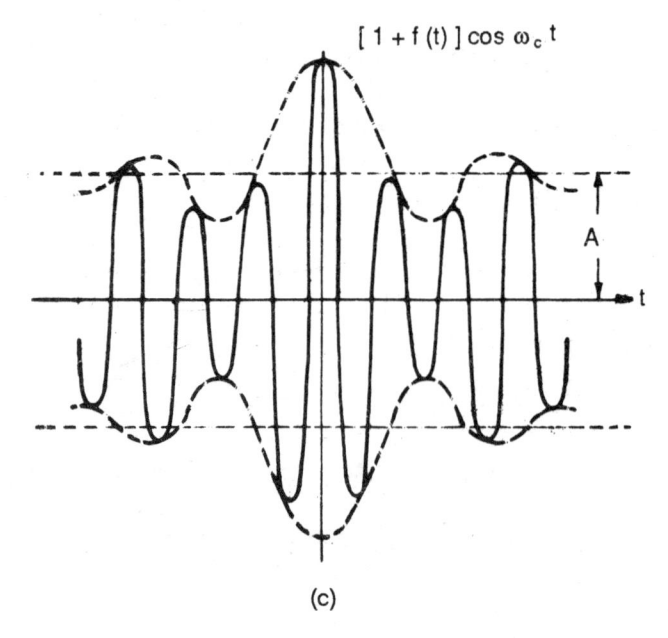

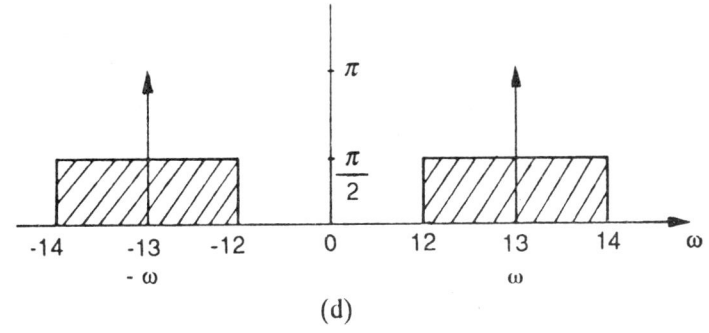

Fig. 5.4.2 (a) Sa(t); (b) F(ω); (c) AM Waveform; (d) Spectrum of AM

## BANDWIDTH

The bandwidth of the AM wave is twice the bandwidth of the baseband signal. This is obvious from Fig. 5.4.1b and d. The bandwidth is the same as that of DSB-SC signal. The AM wave also produces two sidebands (upper and lower) corresponding to each frequency component present in the baseband signal. Hence the transmission bandwidth of AM signal is

$$B = 2\omega_m \tag{5.4.5}$$

where $\omega_m$ is the maximum frequency component of the baseband.

## MODULATION INDEX

The extent of amplitude variation in AM about an unmodulated carrier amplitude is measured in terms of a factor called *modulation index* defined as

$$m_a = \frac{|f(t)|_{\max}}{\text{carrier amplitude } A}$$

$$= \frac{\text{maximum excursion in AM amplitude about } A}{\text{Carrier amplitude } A} \tag{5.4.6}$$

This factor is also known as *depth of modulation, degree of modulation and modulation factor.*

## OVERMODULATION AND ENVELOPE DISTORTION

The baseband signal is preserved in the envelope of the AM signal only if

$$|f(t)|_{\max} \leq A$$

i.e., $m_a = \dfrac{|f(t)|_{\max}}{A}$ is less than or equal to unity. The absolute value of $m_a$ multiplied by 100 is called percentage modulation. Therefore, the envelope preserves the baseband signal only if the percentage modulation is less than, or equal to, 100 percent. However, if $m_a > 1$ or the percentage modulation is greater than 100, the envelope does not preserve the baseband signal, rather, the baseband signal recovered from the envelope is distorted. This type of distortion is known as *envelope distortion*. An AM signal with $m_a > 100\%$ is known as *overmodulated signal*. Consider a baseband shown in Fig. 5.4.3a modulating a carrier $A \cos \omega_c t$.

The following three cases arise depending on the amplitude of the baseband signal relative to carrier amplitude A.

### (i) AM with $m_a < 1$

The waveform of this type of signal is shown in Fig. 5.4.3b. Here, the amplitude of the baseband signal is less than A.

$$|f(t)|_{max} < A$$

As is evident from the waveform, the envelope is not reaching the zero-amplitude axis of the AM wave, and the baseband signal is fully preserved in the envelope of the AM wave. An envelope detector can recover the baseband signal without distortion.

### (ii) AM with $m_a = 1$

The waveform is shown in Fig. 5.4.3c. Here, the amplitude of the baseband is equal to A,

$$|f(t)|_{max} = A$$

and hence, the waveform envelope just touches the zero-amplitude axis. The baseband signal remains preserved in the envelope and can be recovered by using an envelope detector. However, this is a limiting factor as even a small variation in the magnitude of $f(t)$ will cause envelope distortion.

### (iii) Overmodulated Signal with $m_a > 1$

An overmodulated signal is shown in Fig. 5.4.3d. Where, the modulation index is more than 100%. Hence,

$$|f(t)|_{max} < A$$

i.e., the amplitude of a baseband signal exceeds carrier amplitude. Therefore, the portion of the envelope crosses the zero-amplitude axis, for both positive and negative excursions cancelling each other. If the resulting waveform is observed on a cathode ray oscilloscope, a trace will appear as shown in Fig. 5.4.3e. The envelope and the baseband signal are not same. This is called *envelope distortion*. An envelope detector provides distorted baseband signal. A portion of the detected baseband signal is clipped.

The above discussion point out clearly that a baseband signal can be recovered from an AM signal by using a less costly and simple envelope detector, provided the percentage modulation does not exceed 100% i.e., the signal is not overmodulated. The limiting value of $m_a = 100\%$ is also avoided to be on safer side. However, an overmodulated signal can be recovered using a costly and as well as complex technique; synchronous detection. This is illustrated in Ex. 5.4.3.

### Example 5.4.3

Show that an AM signal can be recovered, irrespective of the value of percentage modulation by using synchronous detection technique.

*Solution*

Consider the general expression for AM signal given by Eq. 5.4.3, in which baseband signal $f(t)$ modulates a carrier $A \cos \omega_c t$. In synchronous detection $\varphi_{AM}(t)$ given by Eq. 5.4.3 is multiplied by $\cos \omega_c t$ and then passed through low filter. Multiplication with $\cos \omega_c t$ yields

$$\varphi_{AM}(t) \cos \omega_c t = [A + f(t)] \cos^2 \omega_c t = [A + f(t)]\left[\frac{1}{2}(1 + \cos 2\omega_c t)\right]$$

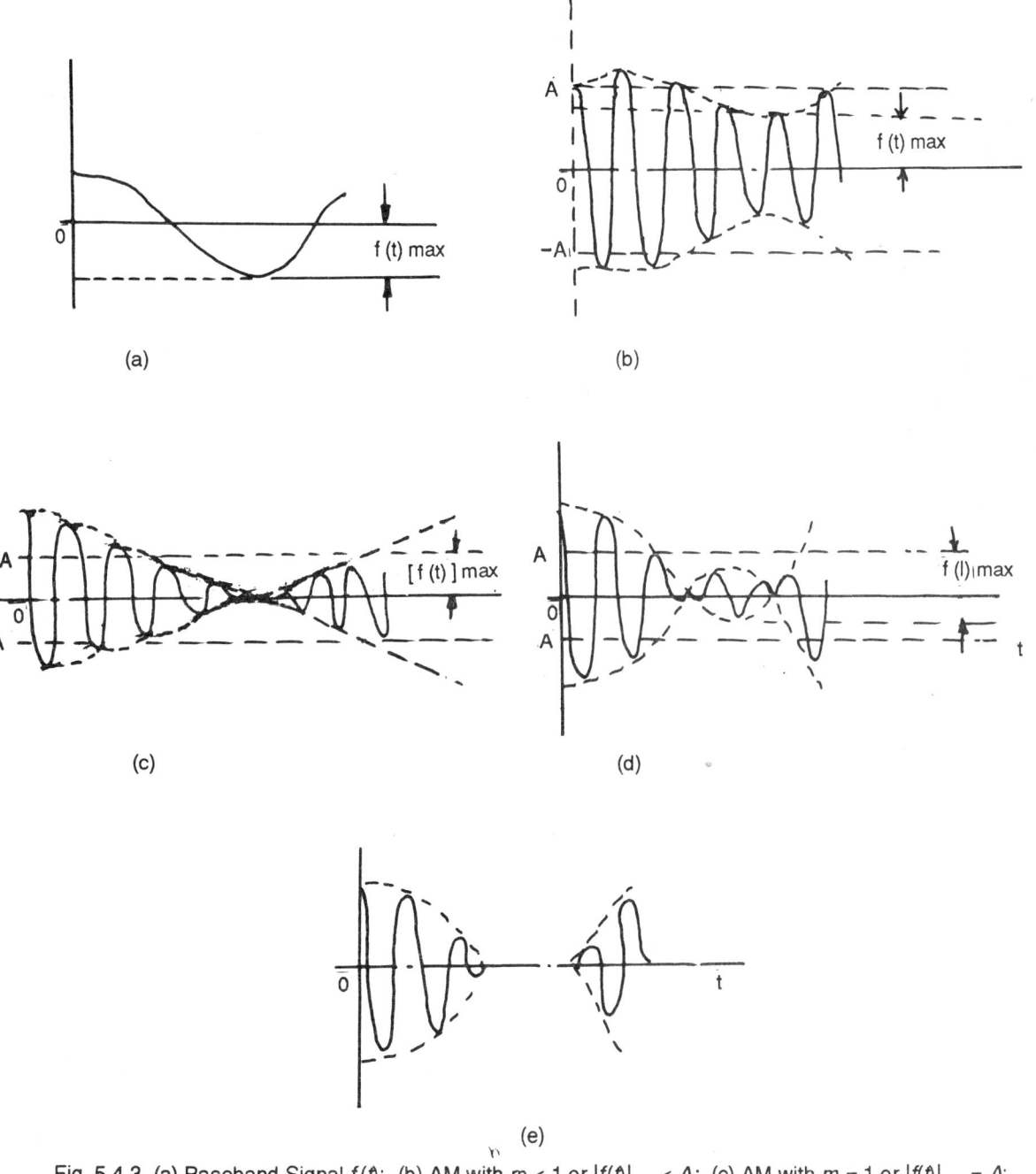

Fig. 5.4.3 (a) Baseband Signal $f(t)$; (b) AM with $m_a < 1$ or $|f(t)|_{max} < A$; (c) AM with $m_a = 1$ or $|f(t)|_{max} = A$; (d) AM with $m_a > 1$ or $|f(t)|_{max} > A$; (e) Overmodulated AM as Observed on CRO

$$= \frac{A}{2} + \frac{1}{2} f(t) + \frac{A}{2} \cos 2\omega_c t + \frac{1}{2} f(t) \cos 2\omega_c t$$

When this multiplied signal is passed through a low pass filter, the terms centered around at $2\omega_c t$ are filtered out, and we get at the output of synchronous detector.

$$e_o(t) = \frac{1}{2}[A + f(t)]$$

The constant dc term $\frac{1}{2} A$ can be removed by a capacitor and hence we are able to recover the baseband signal $f(t)$. Note than in the above process, we have applied no where any condition on a magnitude of $f(t)$ relative to the carrier amplitude. Thus, a baseband signal is recovered irrespective of percentage modulation. Therefore, even overmodulated signal can be recovered using synchronous detection. However synchronous detection is a costly and complex phenomenon and is not desirable in broadcast systems, where a large number of receivers are associated with a transmitter. When overmodulated, the receiver tends to be complex as well as expensive, and the main advantage of AM is lost.

### Partially Suppressed Carried AM

An overmodulated AM signal where the carrier amplitude A is less than the maximum value of a baseband signal and thus has a percentage modulation greater than 100%, is equivalent to a partially suppressed carrier AM. The carrier is not totally suppressed, rather it is present in a smaller quantity. This type of AM signal is analogous to an AM-SC system with a pilot carrier.

### Example 5.4.4

*Single Tone AM*

A single tone modulating signal $e_m = E_m \cos \omega_m t$ amplitude modulates a carrier $e_c = E_c \cos \omega_c t$.
(i) Derive an expression for the AM wave $e(t)$.
(ii) Derive an expression for modulation index.
(iii) Draw the AM waveform and its spectrum.
(iv) Draw the single sided spectrum of the AM wave, and find the bandwidth.
*Solution*
The general expression for the AM wave is given by

$$\varphi_{AM}(t) = [A + f(t)] \cos \omega_c t$$
$$= A \cos \omega_c t + f(t) \cos \omega_c t$$

Here,
$$f(t) = E_m \cos \omega_m t \quad \text{and} \quad A \cos \omega_c t = E_c \cos \omega_c t$$

Hence,
$$e(t) = E_c \cos \omega_c t + E_m \cos \omega_c t \cos \omega_m t \quad (5.4.7a)$$
$$= E_c \left[1 + \frac{E_m}{E_c} \cos \omega_m t\right] \cos \omega_c t$$

$$= E_c \left[1 + m_a \cos \omega_m t\right] \cos \omega_c t \qquad (5.4.7b)$$

where, $m_a = \dfrac{E_m}{E_c}$ is the modulation index.

The desired AM expression is given by Eq. 5.4.7b.

(ii) The modulation index $m_a = \dfrac{E_m}{E_c}$ is given above. This can also be derived using definition of modulation index (El. 5.4.6)

$$m_a = \frac{|f(t)|_{max}}{A}$$

Here, $|f(t)|_{max}$ the maximum value of the modulating signal is $E_m$ and carrier amplitude $A = E_c$. Hence,

$$m_a = \frac{E_m}{E_c}$$

(iii) The spectrum of AM signal is obtained using frequency shifting property of Fourier transform. The AM waveform and the spectrum are shown in Fig. 5.4.4.

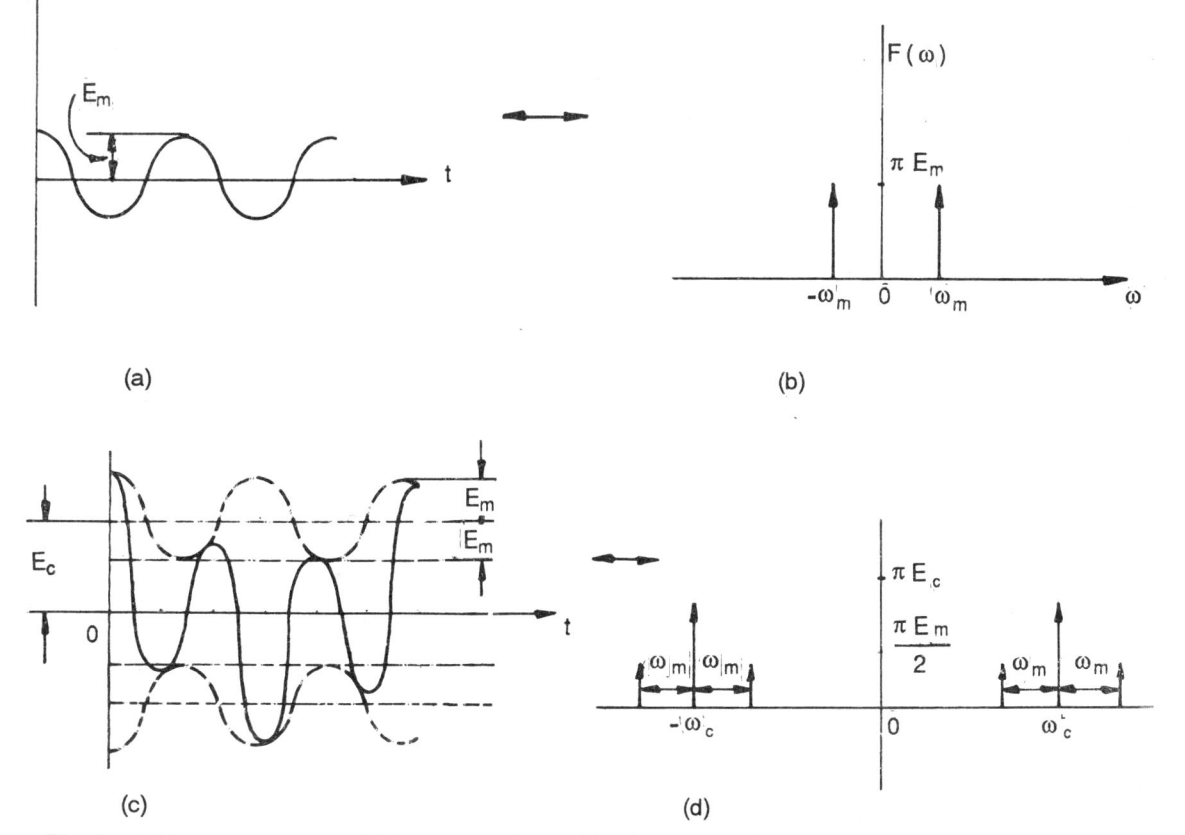

Fig. 5.4.4 (a) $e_m = \cos \omega_m t$; (b) Spectrum of $e_m$; (c) AM Wave; (d) Spectrum (Double-sided) of an AM wave

Figure 5.4.4a and b show the modulating signal and its spectrum, whereas c and d show AM signal and its spectrum.

(iv) A single-sided spectrum is the plot of trigonometric series. The trigonometric expression can be obtained by expanding Eq. 5.4.7b.

$$e(t) = E_c \left[1 + m_a \cos \omega_m t\right] \cos \omega_c t$$

$$= E_c \cos \omega_c t + m_a \cos \omega_m t \cos \omega_c t \qquad (5.4.8a)$$

After trigonometric manipulations, we get

$$e(t) = E_c \cos \omega_c t + \frac{m_a E_c}{2} (2 \cos \omega_m t \cos \omega_c t)$$

$$= E_c \cos \omega_c t + \frac{m_a E_c}{2} \left[\cos (\omega_c + \omega_m) t + \cos (\omega_c - \omega_m) t\right]$$

$$= E_c \cos \omega_c t + \frac{m_a E_c}{2} \cos (\omega_c + \omega_m) t + \frac{m_a E_c}{2} \cos (\omega_c - \omega_m) t \qquad (5.4.8b)$$

Equation 5.4.8b reveals that the AM signal has three frequency components:
(a) carrier frequency $\omega_c$ with amplitude $E_c$
(b) upper sideband ($\omega_c + \omega_m$) with amplitude $\frac{m_a E_c}{2}$
(c) lower sideband ($\omega_c - \omega_m$) with amplitude $\frac{m_a E_c}{2}$

Figure 5.4.5 shows the plot of these three frequency components in a single-sided spectrum of the AM wave. Note the presence of the large carrier term in the spectrum.

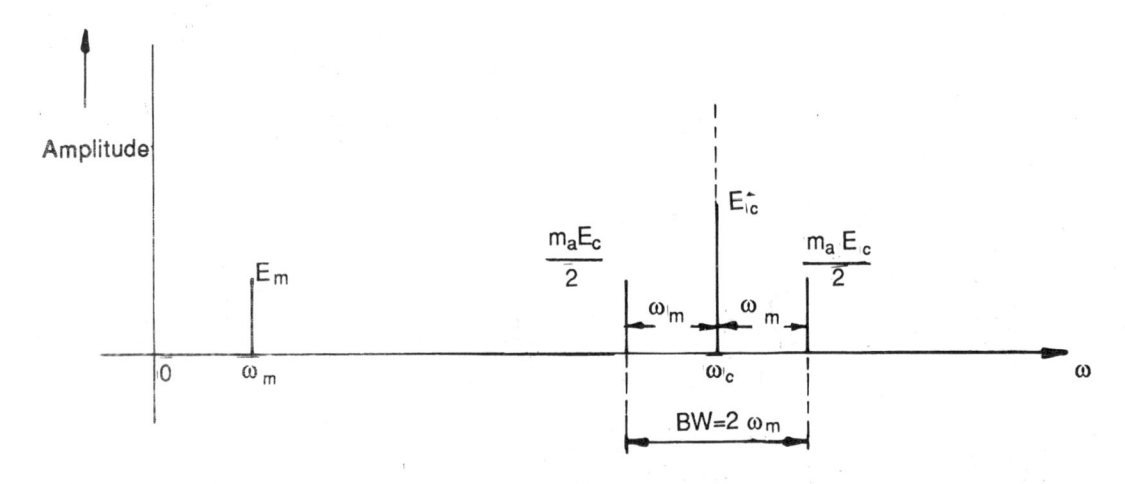

Fig. 5.4.5 Spectrum (Single-sided) of AM Wave

It is apparent from the above figure that the bandwidth of the signal is $2\omega_m$.

## PHASOR REPRESENTATION OF AN AM SIGNAL

The relative amplitude and phase of various frequency components present in AM signal can be better understood by its phasor diagram. For convenience, we will consider a single-tone modulating signal.

The three frequency components present in AM signal given by Eq. 5.4.8b are represented by the three phasors shown in Fig. 5.4.6. $E_c$ represents the carrier phasor. The two sidebands $(\omega_c + \omega_m)$ and $(\omega_c - \omega_m)$ are represented by two phasors rotating in opposite directions with angular frequency $\omega_m$. At any given time,

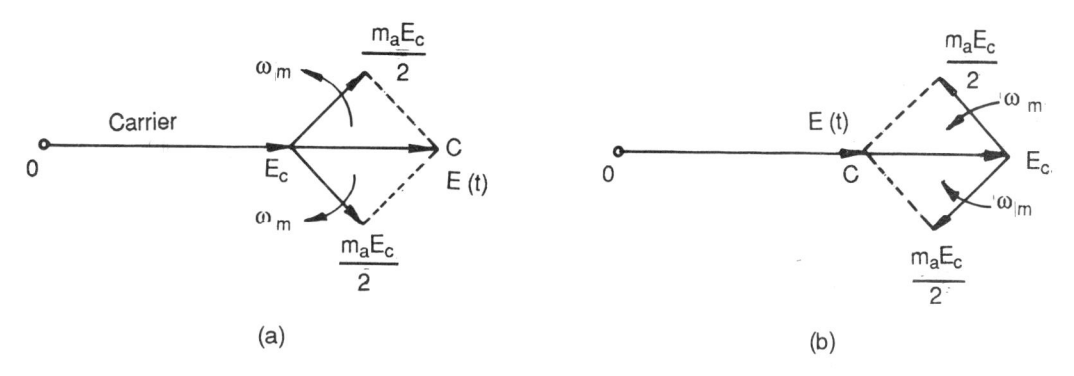

Fig. 5.4.6 Phasor Diagram of AM Wave (a) Maximum Amplitude of the AM Wave
(b) Minimum Amplitude of the AM Wave

the net amplitude, $E(t)$, *represented by phasor OC*, depends on the position of the sideband phasors, with respect to the carrier phasor. Figures 5.4.6 a and b illustrate two such situations causing maximum and minimum resultants respectively. The resultant of the two sideband phasors (*OC*) is always parallel to the carrier phasor.

The waveform of the AM signal can be interpreted as superposition of three waveforms: (i) carrier wave $E_c \cos\omega_c t$; (ii) upper-sideband $\frac{m_a E_c}{2}\cos(\omega_c + \omega_m)t$; and (iii) lower-side band $\frac{m_a E_c}{2}\cos(\omega_c - \omega_m)t$. Note that each of three waveforms has a constant amplitude but the superposed waveform shows amplitude variation.

### Experimental Determination of Modulation Index

The modulation index can be determined experimentally by observing the AM waveform on a cathode ray oscilloscope (CRO). A single-tone modulating signal $E_m \cos\omega_m t$ modulating a carrier $E_c \cos\omega_c t$ results in an AM wave as shown in Fig. 5.4.7. This waveform is applied to Y-deflecting plates of a CRO, as shown in Fig. 5.4.8. The trace is observed on the CRO screen. The maximum and minimum value of the envelope of the AM signal is measured with the help of the CRO. The modulation index is given by

$$m_a = \frac{E_{\max} - E_{\min}}{E_{\max} + E_{\min}} \qquad (5.4.9)$$

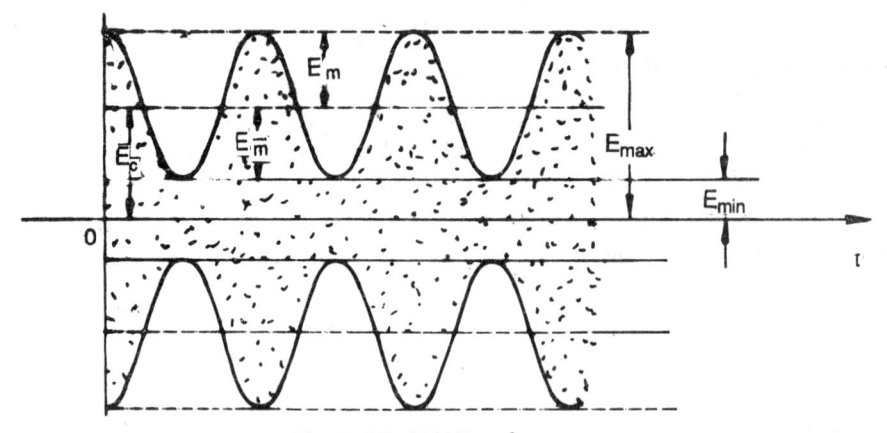

Fig. 5.4.7 AM Waveform

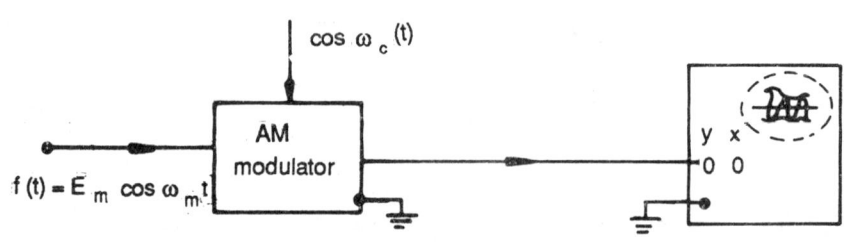

Fig. 5.4.8 Measurement of $m_a$ by CRO

## Derivation

Equation 5.4.9 is derived with the help of the waveform shown in Fig. 5.4.7. It is obvious from the figure that

$$E_m = E_{max} - E_c = E_c - E_{min}$$

By definition,
$$m_a = \frac{E_m}{E_c}$$

This $m_a$ may be written in two equivalent ways

$$m_a = \frac{E_{max} - E_c}{E_c} \qquad (5.4.10)$$

and,

$$m_a = \frac{E_c - E_{min}}{E_c} \qquad (5.4.11)$$

By adding Eqs 5.4.10 and 5.4.11, we get

$$2 m_a = \frac{E_{max} - E_{min}}{E_c}$$

or
$$m_a = \frac{E_{max} - E_{min}}{2E_c} \qquad (5.4.12)$$

Subtracting Eq. 5.4.11 from Eq. 5.4.10, we get
$$0 = E_{max} + E_{min} - 2E_c \qquad (5.4.13)$$
or
$$2E_c = E_{max} + E_{min} \qquad (5.4.14)$$

Substituting $2E_c$ from Eq. 5.4.14 into Eq. 5.4.12, we get
$$m_a = \frac{E_{max} - E_{min}}{E_{max} + E_{min}}$$

## Example 5.4.5

A multiple-tone modulating signal $f(t)$, consisting of three frequency components, is given by
$$f(t) = E_1 \cos \omega_1 t + E_2 \cos \omega_2 t + E_3 \cos \omega_3 t$$
where
$$\omega_3 > \omega_2 > \omega_1 \text{ and } E_1 > E_2 > E_3 \qquad (5.4.15)$$

This signal $f(t)$ modulates a carrier $e_c = E_c \cos \omega_c t$.
(a) Derive an expression for AM wave.
(b) Draw a single-sided spectrum, and find the bandwidth of the AM wave.
*Solution*
(i) The expression for an AM signal can be derived from the general Eq. 5.4.3
$$\varphi_{AM}(t) = [A + f(t)] \cos \omega_c t = A \cos \omega_c t + f(t) \cos \omega_c t$$
Here
$$A = E_c \text{ and, } f(t) \text{ is given by Eq. 5.4.15,}$$
Therefore,
$$\varphi_{AM}(t) = E_c \cos \omega_c t + (E_1 \cos \omega_1 t + E_2 \cos \omega_2 t + E_3 \cos \omega_3 t) \cos \omega_c t$$
$$= E_c [1 + m_1 \cos \omega_1 t + m_2 \cos \omega_2 t + m_3 \cos \omega_3 t] \cos \omega_c t \qquad (5.4.16)$$
where
$$m_1 = \frac{E_1}{E_c}; m_2 = \frac{E_2}{E_c}; \text{ and } m_3 = \frac{E_3}{E_c} \qquad (5.4.17)$$

Equation 5.4.16 represents the desired AM wave.
It is obvious that $m_1$ is the modulation index corresponding to the tone $E_1 \cos \omega_1 t$ of the modulating signal $f(t)$. Similarly, $m_2$, and $m_3$, correspond to second, and third tone present in $f(t)$. Equation 5.4.16 can be generalized for $n$ tones as follows:
$$\varphi_{AM}(t) = E_c[1 + m_1 \cos \omega_1 t + m_2 \cos \omega_2 t + ... + m_n \cos \omega_n t] \cos \omega_c t \qquad (5.4.18)$$

(ii) The spectrum can be obtained by expanding Eq. 5.4.16, as follows

$$\varphi_{AM}(t) = E_c \cos \omega_c t + m_1 E_c \cos \omega_c t \cos \omega_1 t + m_2 E_c \cos \omega_c t \cos \omega_2 t + m_3 E_c \cos \omega_c t \cos \omega_3 t$$

$$= E_c \cos \omega_c t + \frac{m_1 E_c}{2}[\cos(\omega_c + \omega_1)t + \cos(\omega_c - \omega_1)t]$$

$$+ \frac{m_2 E_c}{2}[\cos(\omega_c + \omega_2)t + \cos(\omega_c - \omega_2)t]$$

$$+ \frac{m_3 E_c}{2}[\cos(\omega_c + \omega_3)t + \cos(\omega_c - \omega_3)t] \quad (5.4.19)$$

The expression represents the presence of the following frequency components,
(i) carrier wave of frequency $\omega_c$ with amplitude $E_c$.
(ii) upper and lower sidebands corresponding to the tone $\omega_1$ having frequencies $(\omega_c + \omega_1)$, and $(\omega_c - \omega_1)$, respectively. Both the sidebands have same amplitudes equal to $\frac{m_1 E_c}{2}$.
(iii) Similar upper and lower sidebands for the tones $\omega_2$ and $\omega_3$.
The single-sided spectrum is drawn in Fig 5.4.9 with the help of Eq. 5.4.19. The spectrum can be extended to $n$ tones, by using Eq. 5.4.18.

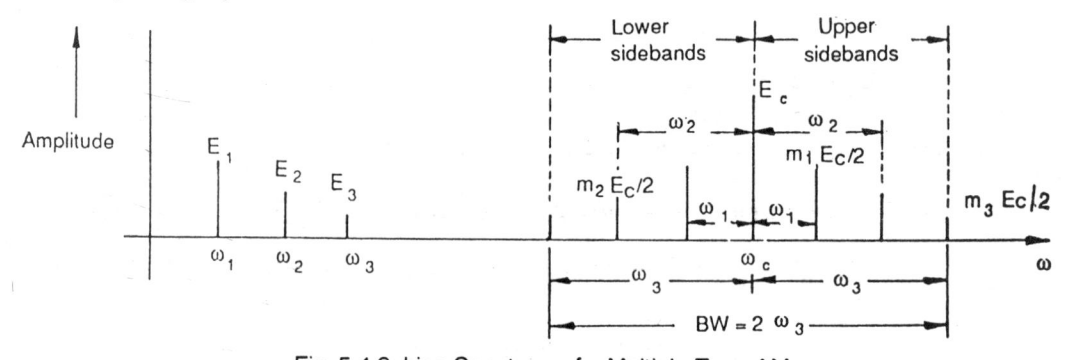

Fig. 5.4.9 Line Spectrum of a Multiple-Tone AM

*Bandwidth* $= 2\omega_3$, twice the highest frequency component present in $f(t)$.

## Superposition Theorem of Spectra

We can observe from the spectrum shown in Fig. 5.4.9 that the AM spectrum follows the **superposition theorem**. The theorem states that the sideband spectrum of a multiple-tone AM signal is equal to the sum of the sideband-spectrum of the individual tone modulation, i.e., the multiple-tone **AM spectrum in** can be obtained by steps given below:
(i) Find the spectrum of single tone modulation considering $\omega_1$ only,
(ii) Find the spectrum of single tone modulation considering $\omega_2$ only,
(iii) Find the spectrum of single tone modulation considering $\omega_3$ only,
(iv) Add the spectra of (i), (ii) and (iii) excluding the carrier term of each spectrum,
(v) Add the carrier term only once.

This is shown graphically in Fig. 5.4.10. Figure (a) shows the AM spectrum corresponding to $\omega_1$ only. Similarly, Figs (b) and (c) correspond to AM spectrum for $\omega_2$ and $\omega_3$ respectively. The sum of the spectrums in Figs (a), (b), and (c), and carrier terms provides the resulting spectrum shown in Fig. 5.4.9. Therefore, we can say that AM spectrum follows the superposition theorem of spectra.

*Linear Modulation*: The modulation system following the superposition theorem of spectra is known as linear modulation system. Thus, *AM is a linear modulation system*. We will see in the next chapter that frequency modulation violates the law of superposition of spectra, and hence *FM is a non-linear modulation system*.

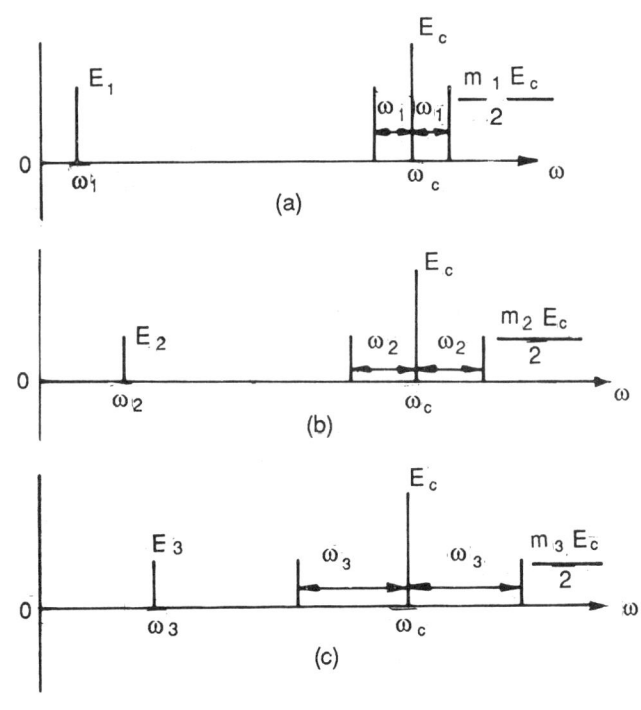

Fig. 5.4.10

## Example 5.4.6

An amplitude modulated signal is given by

$$\varphi_{AM}(t) = 10\cos(2\pi \cdot 10^6 t) + 5\cos(2\pi \cdot 10^6 t)\cos(2\pi \cdot 10^3 t) + 2\cos(2\pi \cdot 10^6 t)\cos(4\pi \cdot 10^3 t) \text{ volts}$$

Find the various frequency components present and the corresponding modulation indices. Draw the line spectrum and find bandwidth.

*Solution*

The expression can be written in the form of Eq. 5.4.18 as follows:

$$\varphi_{AM}(t) = 10\left[1 + \frac{5}{10}\cos(2\pi \cdot 10^3 t) + \frac{2}{10}\cos(4\pi \cdot 10^3 t)\right]\cos(2\pi \cdot 10^6 t)$$

$$= 10[1 + 0.5\cos(2\pi \cdot 10^3 t) + 0.2\cos(4\pi 10^3 t)]\cos(2\pi \cdot 10^6 t)$$

comparing this Equation with Eq. 5.4.18, we get

$$E_c = 10, m_1 = 0.5, m_2 = 0.2, \omega_1 = 2\pi.10^3,$$

$$\omega_2 = 4\pi\,10^3, \text{ and } \omega_c = 2\pi\,10^6$$

The expression corresponds to an AM with a two-tone ($\omega_1$ and $\omega_2$) modulating signal with modulation indices of 0.5 and 0.2, respectively. Therefore, the terms present in AM are,

(i) carrier term with frequency $\omega_c = 2\pi\,10^6$ rad/s.
or

$$f_c = \frac{\omega_c}{2\pi} = 10^6 \text{ Hz} = 1 \text{ MHz}$$

(ii) Upper sidebands

(a) 
$$\omega_c + \omega_1 = 2\pi\,10^6 + 2\pi\,10^3 = 2\pi(10^6 + 10^3) \text{ rad/s}$$

or

$$f_c + f_1 = \frac{\omega_c + \omega_1}{2\pi} = (10^6 + 10^3) \text{ Hz} = 1.001 \text{ MHz}$$

(b) Similarly,

$$f_c + f_2 = (10^6 + 2\times 10^3) \text{ Hz} = 1.002 \text{ MHz}$$

(iii) Lower sidebands,

(a)
$$f_c - f_1 = (10^6 - 10^3) \text{ Hz} = 0.999 \text{ MHz}$$

(b)
$$f_c - f_2 = (10^6 - 2\times 10^3) \text{ Hz} = 0.998 \text{ MHz}$$

Figure 5.4.11 shows the line spectrum having the above frequency components. The amplitude of the sidebands are:

(a) Each sideband of $\omega_1$ $\quad \frac{m_2 E_c}{2} = \frac{0.5 \times 10}{2} = 2.5$ volts

(b) Each sidebands of $\omega_2$ $\quad \frac{m_2 E_c}{2} = \frac{0.2 \times 10}{2} = 1.0$ volts

Bandwidth = $2\omega_m$, where $\omega_m$ is the maximum frequency component present in the modulating signal. Here $\omega_m = 2$ kHz
So bandwidth = $2 \times 2$ kHz = 4 kHz

**Power Content in AM Signal**

The AM signal is represented by Eq. 5.4.1 as reproduced below

$$\varphi_{AM}(t) = A \cos \omega_c t + f(t) \cos \omega_c t$$

The total power P of the AM wave is the sum of the carrier power $P_c$, and sideband power $P_s$. Remember that we are considering the normalized power, i.e., power dissipated in 1 ohms resistance.

AMPLITUDE MODULATION SYSTEMS 299

Fig. 5.4.11

### Carrier power

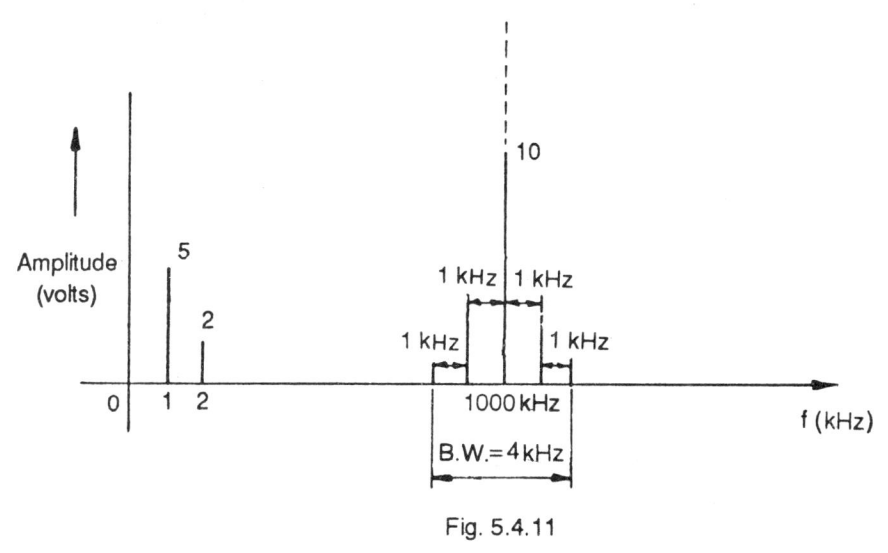

The carrier power $P_c$ is the mean square value of the carrier term $A\cos\omega_c t$

$$P_c = \overline{[A \cos \omega_c t]^2} = \frac{A^2}{2} \qquad (5.4.20)$$

### Sideband Power

The sideband power $P_s$ is the mean square value of the sideband term $f(t) \cos \omega_c t$,

$$P_s = \overline{[f(t)\cos \omega_c t]^2} = \frac{1}{2\pi} \int_0^{2\pi} f^2(t) \cos^2 \omega_c t \, d(\omega t)$$

$$= \frac{1}{2\pi} \int_0^{2\pi} \frac{1}{2} f^2(t)[1 + \cos 2\omega_c t] \, d(\omega t)$$

$$= \frac{1}{2\pi} \int_0^{2\pi} \frac{1}{2} f^2(t) \, d(\omega t) + \frac{1}{2\pi} \int_0^{2\pi} \frac{1}{2} f^2(t) \cos 2\omega_c t \, d(\omega t)$$

The second integral-term is filtered out by the BPF centered around $\omega_c$. Hence the, sideband power is obtained by solving the first integral.

$$P_s = \frac{1}{2} \overline{f^2(t)} \qquad (5.4.21)$$

This is half the baseband signal-power. The total sideband power $P_s$ is due to the equal contributions of the upper and lower sidebands. Hence, half of the $P_s$ is contributed by the lower sidebands and the remaining half by the upper sideband. Therefore, power carried by the upper and the lower sidebands is,

$$(P_s)_{LSB} = (P_s)_{USB} = \frac{P_s}{2} = \frac{1}{4}\overline{f^2(t)} \qquad (5.4.22)$$

The total power of the AM signal $P$ is sum of the carrier power $P_c$ defined by Eq. 5.4.20, and the sideband power $P_s$ given by Eq. 5.4.21. Therefore, total AM power is,

$$P = P_c + P_s = \frac{1}{2}A^2 + \frac{1}{2}\overline{f^2(t)}$$

$$= \frac{1}{2}[A^2 + \overline{f^2(t)}] \qquad (5.4.23)$$

## Transmission Efficiency of AM signal

The total modulated power of an AM signal is given by 5.4.23. The useful message (baseband) power out of this total power is the power carried by the sidebands, i.e., $P_s$. The large carrier power $P_c$ does not carry any message, rather, this power is a waste from the transmission point of view. However, this large carrier power $P_c$ is transmitted alongwith the sideband power for convenient and cheap detection. Therefore, $P_s$ is the only useful message power present in the AM wave. The amount of useful message power $P_s$ present in AM wave is expressed by a term called *transmission efficiency* denoted by *Eff*. It is defined as *percentage of total power contributed by the sidebands*. Mathematically, the transmission efficiency is given as

$$(Eff)_{AM} = \frac{P_s}{P} \times 100 = \frac{\frac{1}{2}\overline{f^2(t)}}{\frac{1}{2}A^2 + \frac{1}{2}\overline{f^2(t)}} \times 100 = \frac{100\overline{f^2(t)}}{A^2 + \overline{f^2(t)}} \qquad (5.4.24)$$

It will be seen in Ex. 5.4.7 that maximum transmission efficiency of the AM is 33.33%. This means, that only one-third of the total power is carried by the sidebands and the rest two third is a waste, and is transmitted only for a low cost reception system.

## Transmission Efficiency in Suppressed Carrier Systems

In suppressed Carrier systems, the carrier power is totally suppressed, and only the sidebands are transmitted. Therefore, the total AM power is carried by the sidebands and no power is contributed by the carrier i.e., $P_c = 0$. Hence, the efficiency for a suppressed carrier (SC) system is

$$(Eff)_{sc} = \frac{100\overline{f^2(t)}}{0 + \overline{f^2(t)}} = 100 \text{ per cent}$$

Thus, the entire power in suppressed carrier systems is carried by sidebands, of course, at the cost of expensive and complex receiver.

## AMPLITUDE MODULATION SYSTEMS

**Example 5.4.7** Power of a Single-tone AM signal

A carrier $A\cos\omega_c t$ is modulated by a single-tone modulating signal $f(t) = E_m \cos\omega_m t$. Find
(a) total modulated power,
(b) rms (root mean square) value of the modulated signal, and
(c) transmission efficiency for a 100 percent modulation

*Solution*

(a) Carrier power (unmodulated) $P_c = \overline{(A\cos\omega_c t)^2} = \dfrac{A^2}{2}$

Sideband power $P_s = \dfrac{1}{2}\overline{f^2(t)} = \dfrac{1}{2}\overline{(E_m \cos\omega_m t)^2} = \dfrac{1}{2}\dfrac{E_m^2}{2} = \dfrac{1}{4}E_m^2$

The total modulated power, $P$ is the sum of $P_c$ and $P_s$

$$P = \frac{A^2}{2} + \frac{1}{4}E_m^2 = \frac{A^2}{2}\left[1 + \frac{1}{2}\left(\frac{E_m}{A}\right)^2\right]$$

Putting $\dfrac{E_m}{A} = m_a$, and $\dfrac{A^2}{2} = P_c$

$$P = P_c\left(1 + \frac{m_a^2}{2}\right) \tag{5.4.25}$$

where $m_a$ is the modulation index.

The total sideband power in Eq. 5.4.25 is $P_c\left(\dfrac{m_a^2}{2}\right)$, out of which one half is contributed by the lower sideband and the other half by the upper sideband. Therefore, the power carried by each sideband is $P_c(m_a^2/4)$.

(b) The rms value is under root of the ms value (power) of the AM signal

$$V_{rms} = \sqrt{P} = \sqrt{P_c\left(1 + \frac{m_a^2}{2}\right)}$$

Now, by putting $\sqrt{P_c} = V_{c_{rms}}$ (rms value of unmodulated carrier),

$$V_{rms} = V_{c_{rms}}\sqrt{1 + \frac{m_a^2}{2}} \tag{5.4.26}$$

(c) The transmission efficiency is given by Eq. 5.4.24

$$Eff = \frac{\overline{f^2(t)}}{A^2 + \overline{f^2(t)}} \times 100$$

Here

$$\overline{f^2(t)} = \frac{E_m^2}{2},$$

# 302 COMMUNICATION SYSTEMS: Analog and Digital

Hence,

$$Eff = \frac{(E_m^2/2)}{A^2 + (E_m^2/2)} \times 100$$

Dividing the numerator and denominator by $(A^2/2)$ we get,

$$Eff = \frac{m_a^2}{2 + m_a^2}$$

where
$$m_a = \frac{E_m}{A}, \qquad (5.4.27)$$

For a 100 per cent modulation, $m_a = 1$, and

$$Eff = \frac{1}{3} \times 100 = 33.3\% \qquad (5.4.28)$$

i.e. for a 100 per cent modulation the sidebands carry only one-third of the total AM power. This is the maximum power which can be carried by the sidebands in AM. If $m_a$ is less than 100 per cent the sideband power will be further reduced.

### Example 5.4.8

Power in Multiple-tone Modulation

A carrier $A \cos \omega_c t$ is modulated by a modulating signal $f(t) = E_1 \cos \omega_1 t + E_2 \cos \omega_2 t + E_3 \cos \omega_3 t$. Derive expressions for (a) total modulated power (b) net modulation index of the AM wave.

**Solution**

The AM wave is given by

$$\begin{aligned}
\varphi_{AM}(t) &= A \cos \omega_c t + f(t) \cos \omega_c t \\
&= A \cos \omega_c t + [E_1 \cos \omega_1 t + E_2 \cos \omega_2 t + E_3 \cos \omega_3 t] \cos \omega_c t \\
&= A \left[ 1 + \frac{E_1}{A} \cos \omega_1 t + \frac{E_2}{A} \cos \omega_2 t + \frac{E_3}{A} \cos \omega_3 t \right] \cos \omega_c t \\
&= A [1 + m_1 \cos \omega_1 t + m_2 \cos \omega_2 t + m_3 \cos \omega_3 t] \cos \omega_c t
\end{aligned}$$

where $m_1 = \frac{E_1}{A}$, $m_2 = \frac{E_2}{A}$, and $m_3 = \frac{E_3}{A}$, are modulation indices of the corresponding frequency components. The expression can be further expanded as

$$\varphi_{AM}(t) = A \cos \omega_c t + \underbrace{m_1 A \cos \omega_1 t \cos \omega_c t}_{\text{sidebands of } \omega_1} + \underbrace{m_2 A \cos \omega_2 t \cos \omega_c t}_{\text{sidebands of } \omega_2} + \underbrace{m_3 A \cos \omega_3 t \cos \omega_c t}_{\text{sidebands of } \omega_3}$$

Therefore, the total sideband power is the sum of the powers contributed by the sidebands of $\omega_1, \omega_2,$ and $\omega_3$.

$$\begin{aligned}
P_s &= \frac{1}{2} \left[ \overline{(m_1 A \cos \omega_1 t)^2} + \overline{(m_2 A \cos \omega_2 t)^2} + \overline{(m_3 A \cos \omega_3 t)^2} \right] \\
&= \frac{1}{2} \left[ \frac{m_1^2 A^2}{2} + \frac{m_2^2 A^2}{2} + \frac{m_3^2 A^2}{2} \right] = \frac{1}{4} A^2 [m_1^2 + m_2^2 + m_3^2]
\end{aligned}$$

The desired expression for the total modulated power of the AM signal is

$$P = P_c + P_s = \frac{A^2}{2} + \frac{1}{4}A^2[m_1^2 + m_2^2 + m_3^2]$$

$$= P_c\left[1 + \frac{m_1^2 + m_2^2 + m_3^2}{2}\right] \quad (5.4.29)$$

where,

$$P_c = \frac{A^2}{2}$$

The expression can be extended to $n$-modulating terms,

$$P = P_c\left[1 + \frac{m_1^2}{2} + \frac{m_2^2}{2} + \cdots + \frac{m_n^2}{2}\right] \quad (5.4.30)$$

(b) Let us assume that $m$ is the net modulation index of an AM wave. The power of the AM wave can be expressed as Eq. 5.4.25.

$$P = P_c\left(1 + \frac{m^2}{2}\right)$$

Comparing this expression with Eq. 5.4.30, we get

$$m^2 = m_1^2 + m_2^2 + m_3^2 + \cdots + m_n^2 \quad (5.4.31)$$

or

$$m = \sqrt{m_1^2 + m_2^2 + m_3^2 + \cdots + m_n^2}$$

which is the desired expression for the net modulation index.

### Example 5.4.9

Find total modulated power, sideband power, and net modulation index for the AM signal given in Ex. 5.4.6 expressed in volts.

*Solution*

The AM signal in Ex. 5.4.6 is given as

$$\varphi_{AM} = 10 \cos(2\pi\, 10^6 t) + 5 \cos(2\pi\, 10^6 t) \cos(2\pi\, 10^3 t) + 2 \cos(2\pi\, 10^6 t) \cos(4\pi\, 10^3 t)$$

The expression can be written as

$$\varphi_{AM} = 10[1 + 0.5 \cos 2\pi.10^3 t + 0.2 \cos 4\pi\, 10^3 t] \cos 2\pi.10^6 t$$

The signal is equivalent to the expression

$$\varphi_{AM} = A[1 + m_1 \cos \omega_1 t + m_2 \cos \omega_2 t] \cos \omega_c t$$

Here

$$A = 10\,V, m_1 = 0.5, \text{ and } m_2 = 0.2$$

Unmodulated carrier power $P_c = \dfrac{A^2}{2} = \dfrac{10^2}{2} = 50$ watt.

**304** COMMUNICATION SYSTEMS: Analog and Digital

The power of the modulated signal is given by (Eq. 5.4.29),

$$P = P_c(1 + m_1^2 + m_2^2) = 50\left(1 + \frac{.25 + 0.04}{2}\right) = 57.25 \text{ watts}$$

Sideband power $P_s = P - P_c = 57.25 - 50 = 7.25$ watts.
Net modulation index $m$ is given by Eq. 5.4.31,

$$m = \sqrt{m_1^2 + m_2^2} = \sqrt{0.5^2 + 0.2^2} = 0.539$$

## 5.5 GENERATION OF AM WAVES

The methods employed for generating AM-SC signals (sec.5.1) can also be used to generate AM signals (Here AM refers to signals with large carriers). The methods for generation of AM waves are broadly divided into two parts. These are:
 (i) Using linear time invariant circuits, and (ii) Using non-linear circuits.
 Circuits like switching or chopper modulator, collector modulator, drain modulator etc. are examples of the first category. Non-linear characteristics of devices, such as transistors, diodes, field effect transistors etc., are used in the second category. We will discuss some of the methods from both the categories.

### Chopper-type (switching) Modulator

The principle of this method is similar to that of AM-SC signal, except that chopping action is done on a combined carrier plus message signal, i.e., $[A \cos \omega_c t + f(t)]$. Note that in the AM-SC generation only $f(t)$ was chopped.

The chopping is done at the rate of carrier frequency, and the chopping action is equivalent to multiplication of $A \cos \omega_c t + f(t)$ with a periodic square signal $p(t)$ of frequency $\omega_c$. The spectrum of the chopped signal is, therefore obtained by convolving the spectrum of $[A + f(t)]$ with the spectrum of the square pulse, which is a sampling function. This is shown in Fig. 5.5.1.

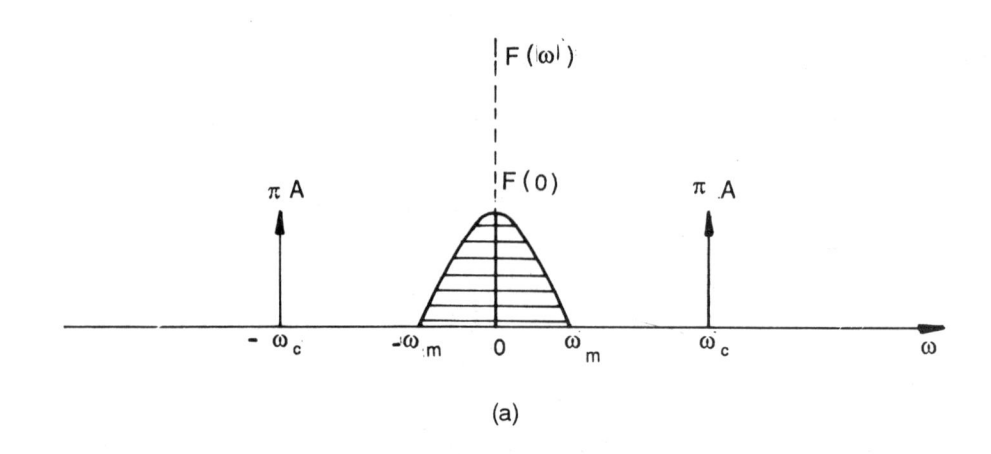

(a)

(b)

(c)

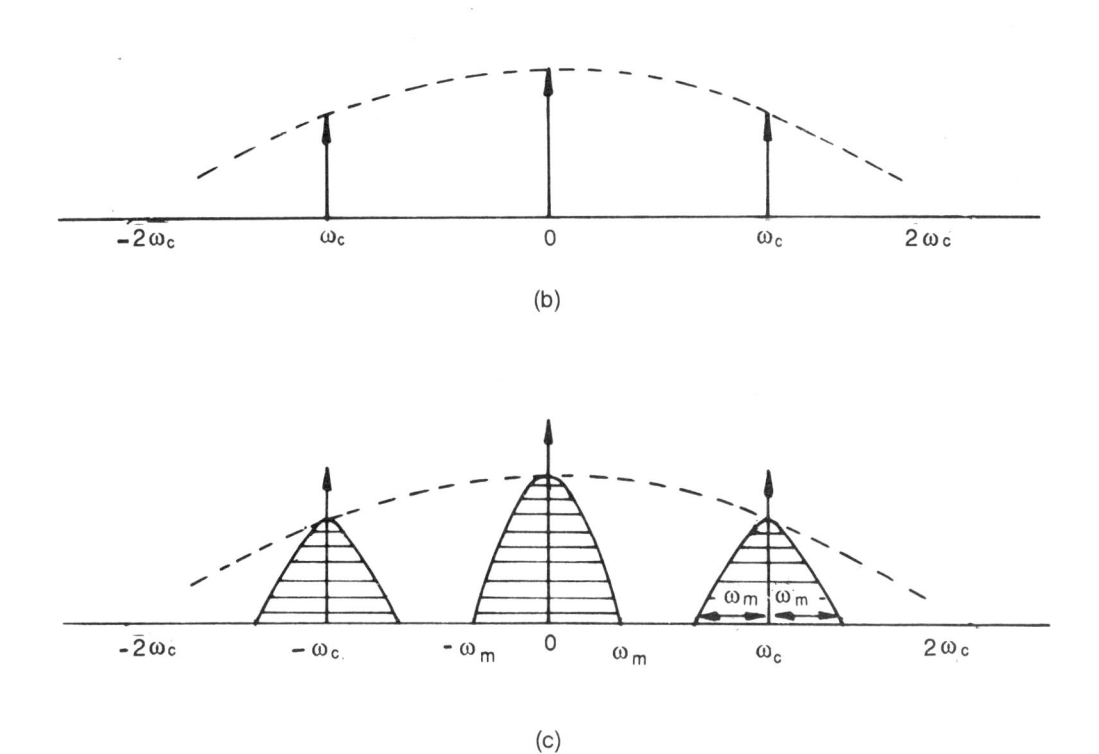

Fig. 5.5.1 (a) Spectrum of A cos $\omega_c t + f(t)$; (b) Spectrum of Square Signal; (c) Spectrum of Chopped Signal

Figure 5.5.1a shows the spectrum of $[A \cos \omega_c t + f(t)]$. The impulses at $\pm \omega_c$ correspond to the carrier $A \cos \omega_c t$. Figure 5.5.1b shows the spectrum of square signal $p(t)$ which is a sampling function. Figure 5.5.1c represents the spectrum of the chopped signal, which is obtained by convolving Figs (a) and (b). When passed through a bandpass filter. The spectrum in Fig. 5.5.1c yields spectrum $F(\omega)$ shifted by $\pm \omega_c$, along with impulses at $\pm \omega_c$, which is the spectrum of an AM signal.

## Circuit Diagram

The switching action can be obtained by a ring modulator used for AM-SC generation. A simple diode can perform the job of switching, provided $A \gg |f(t)|_{max}$. A simple diode chopper circuit used for AM generation is shown in Fig. 5.5.2.

The diode is forward biased for every positive half cycle of the carrier, and behaves like a short circuited switch. The signal appears at the input of the bandpass filter. For a negative half cycle of the carrier, the diode is reverse biased and behaves like an open switch. The signal does not reach the filter, and no output is obtained. Thus, the signal is chopped at the rate of the carrier frequency. This circuit is a sort of time-invariant system.

## Modulators Using Amplifying Devices

The diode modulator circuit does not provide amplification, and hence, it can be used for low power applications. However, amplifying devices like transistors, FET, electron tubes etc. can provide

**306** COMMUNICATION SYSTEMS: Analog and Digital

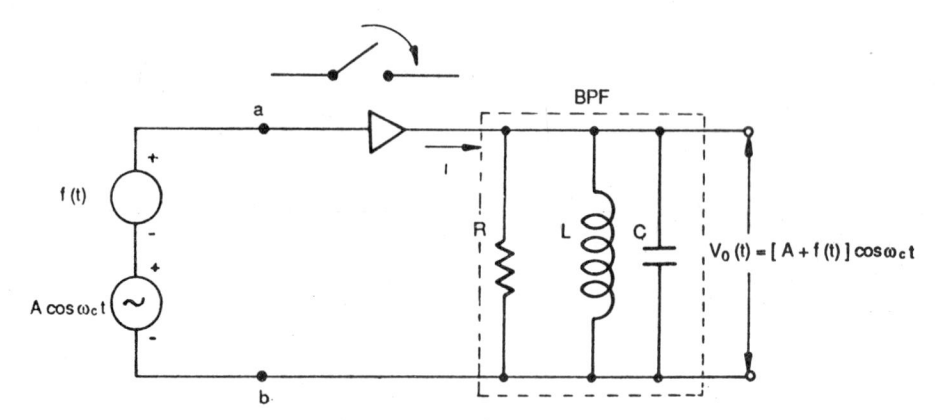

Fig. 5.5.2 A Simple Diode Switching Modulator

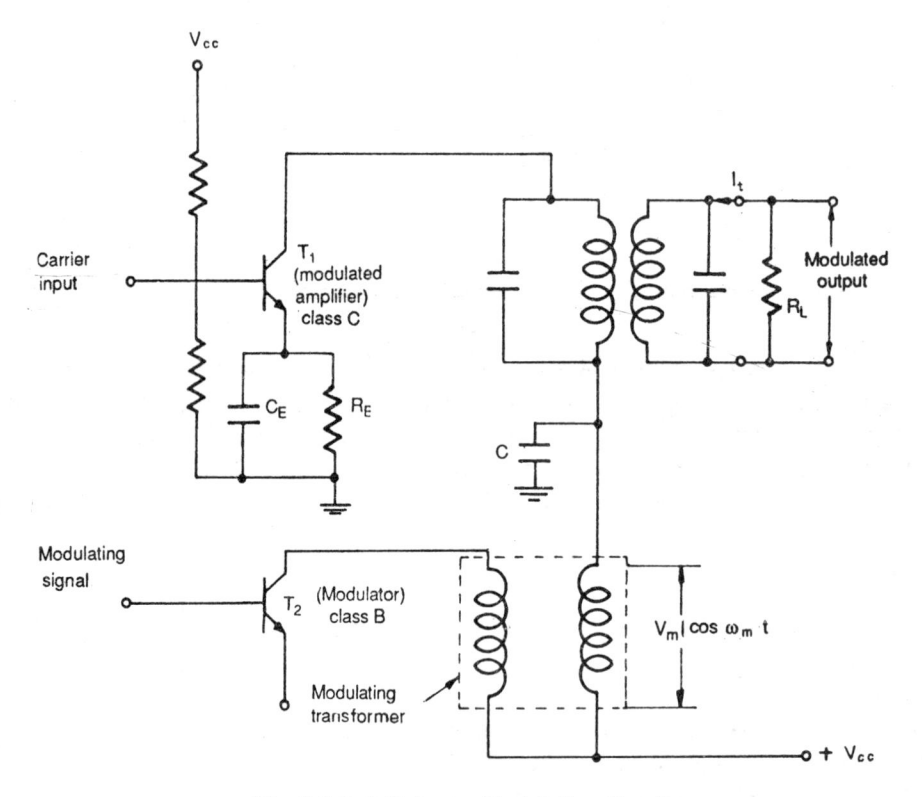

Fig. 5.5.3. A Collector Modulation Circuit

amplification. Each one of them can be used for generating amplitude modulation by varying their gain parameters ($h_{fe}, g_m, \mu$ etc.) in accordance with the modulating signal. A very popular circuit used for this purpose is known as *modulated class C amplifier*. If the device used is a transistor, the method is known as *collector modulation.* Similarly, if FET, or electron tube device is used, the method is known as *drain modulation,* or *plate modulation* respectively. Figure 5.5.3 shows a collector modulation circuit.

Transistor $T_1$ forms a radio frequency amplifier in class C mode for higher efficiency. The carrier signal is applied at the base of $T_1$. $V_{cc}$ is the collector supply used for biasing. The transistor $T_2$ forms a class B amplifier used to amplify the baseband (modulating) signal. After amplification, the modulating signal appears across the modulating transformer. This modulating signal exists in series with the collector supply, $V_{cc}$. Capacitor C offers low path for the carrier signal, and as such, the carrrier is prevented from flowing through the modulation transformer.

## *Principle of Operation*

In a class C amplifier, the amplitude of the output voltage is a definite fraction of (or at the most equal to) the supply voltage $V_{cc}$. A linear relation exists between the output tank current $I_t$ and the supply variable voltage $V_c$ (assume $V_c$ to be a variable supply). The output voltage is a replica of input waveform. But the amplitude of the output voltage is approximately equal to the supply voltage $V_{cc}$, as shown in Fig. 5.5.4a. If R is the resistance at resonance of the output resonant circuit, then the amplitude of output voltage will be $RI_t \simeq V_{cc}$.

Thus, the unmodulated carrier is amplified by the class C modulated amplifier, and its amplitude remains constant equal to $V_{cc}$, as no voltage appears across the modulation transformer in the absence of modulating voltage. However, when a modulating voltage $V_m \cos \omega_m t$ appears across the modulation transformer, it is added with the supply voltage. The net effect is a *slow variation in supply voltage* $V_{cc}$: slow because as compared to the carrier frequency, the modulating frequency is small. This slow variation in supply voltage changes the amplitude of the carrier voltage at the output of the modulated class C amplifier, as shown in Fig. 5.5.4b. The envelope of the output voltage is identical with the modulating voltage, and thus an AM wave is generated. The slowly varying supply voltage $V_c$ is given by

$$V_c = V_{cc} + V_m \cos \omega_m t \qquad (5.5.1)$$

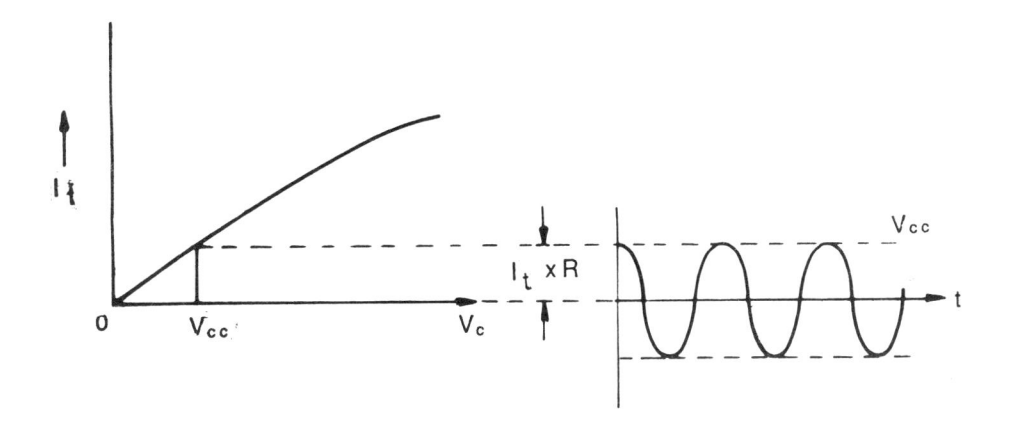

(a)

**Fig. 5.5.4**

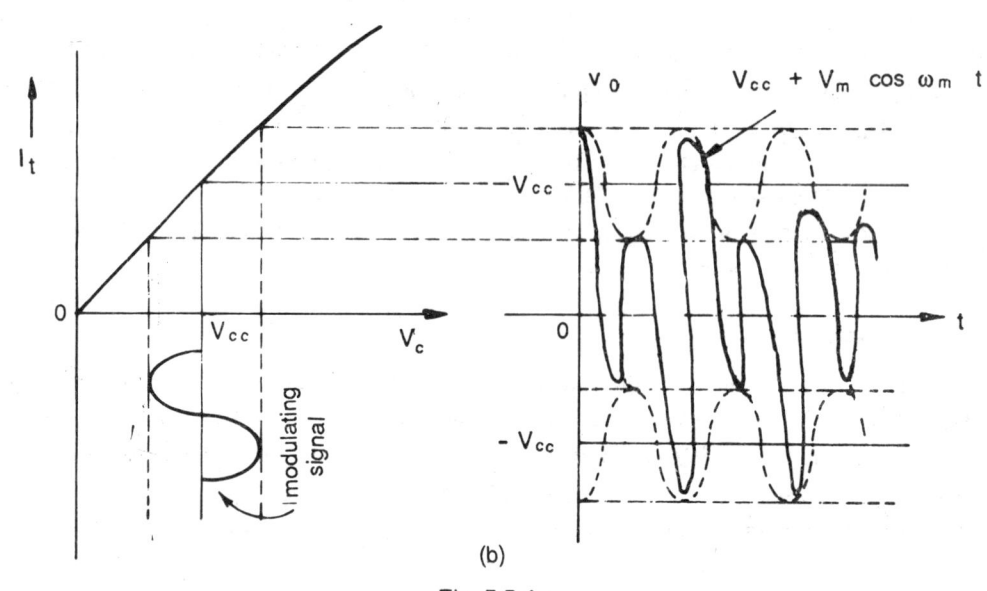

$$= V_{cc}[1 + m_a \cos \omega_m t] \quad (5.5.2)$$

and the modulated output voltage $V_o$ is given by

$$V_o = V_{cc}[1 + m_a \cos \omega_m t]\cos \omega_c t$$

where $m_a$ the modulation index given by

$$m_a = \frac{V_m}{V_{cc}}$$

## Overmodulation

When $V_m > V_{cc}$, the envelope crosses the zero axis and an envelope distortion occurs due to overmodulation as shown in Fig.5.4.3d.

## Collector Circuit Efficiency

The modulated power delivered to the output load depends on the input supplied by supply voltage and power dissipation in the collector circuit. Out of the total power input in the collector circuit, only a part of it reaches the output load, the remaining power is lost in the collector circuit. This loss is due to the collector resistance, and other dissipating components. Let us assume that

$P_{in}$ = input power in the collector circuit

$P_{out}$ = output power delivered to a load $R_L$ and

$P_d$ = Power dissipation in the collector circuit

then

$$P_d = P_{in} - P_{out}$$

## AMPLITUDE MODULATION SYSTEMS

$$= P_{in}\left(1 - \frac{P_{out}}{P_{in}}\right) \tag{5.5.3}$$

The ratio of output power and input power, in the collector circuit is defined as *collector circuit efficiency* denoted by $\eta_c$

$$\eta_c = \frac{P_{out}}{P_{in}} \tag{5.5.4}$$

Hence,

$$P_d = P_{in}(1 - \eta_c) \tag{5.5.5}$$

The power input in the collector circuit is the multiplication of a slow varying supply voltage $V_{cc}(1 + m_a \cos \omega_m t)$, and the resulting collector current $I_c(1 + m_a \cos \omega_m t)$, where $I_c$ is the average current in the absence of modulation. This multiplication provides instantaneous input power at any moment. The average input power is obtained by taking the average of this instantaneous power over a cycle. Thus,

$$P_{in} = \int_0^{2\pi} V_{cc}(1 + m_a \cos \omega_m t) I_c(1 + m_a \cos \omega_m t) d\omega t$$

$$= V_{cc} I_c + \frac{m_a^2}{2} V_{cc} I_c = V_{cc} I_c \left(1 + \frac{m_a^2}{2}\right) \tag{5.5.6}$$

$$= P_{cc}\left(1 + \frac{m_a^2}{2}\right) \tag{5.5.7}$$

where $P_{cc} = V_{cc} I_c$, is the power supplied by $V_{cc}$.

The total input power is composed of two parts: $P_{cc}$ and $\frac{m_a^2}{2} P_{cc}$. The second part represents the sideband power introduced by the modulating amplifier in the collector circuit. Let us denote this power by $P_m$, i.e., $P_m = \frac{m_a^2}{2} P_{cc}$. This modulating power is introduced in the form of a varying power supply. When this modulating signal does not appear across the modulation transformer in the collector circuit, then $m_a = 0$, and the total input power is only $P_{cc}$. Thus $P_{cc}$ the unmodulated carrier power. substituting the value of $P_{in}$ from Eq. 5.5.6 into Eq. 5.5.5, we get

$$P_d = V_{cc} I_c\left(1 + \frac{m_a^2}{2}\right)(1 - \eta_c) \tag{5.5.8}$$

Thus, dissipation *in the collector circuit* increases with an increase in modulation index. This dissipation generates heat. In large power transmitters, if the modulation index is large, the necessary arrangement for cooling is made.

When the modulation index $m_a = 0$, the power dissipation, $P_{do}$ is,

$$P_{do} = V_{cc} I_c(1 - \eta_c)$$

Equation 5.5.8 can be written in terms of $P_{do}$ as follows:

$$P_d = P_{do}\left(1 + \frac{m_a^2}{2}\right) \qquad (5.5.9)$$

From Eq. 5.5.4 and 5.5.6 we get

$$P_{out} = \eta_c P_{in}' = \eta_c V_{cc} I_c \left(1 + \frac{m_a^2}{2}\right) \qquad (5.5.10)$$

### Example 5.5.1

An amplitude modulated amplifier has a radio frequency output of 50 watts at 100% modulation. The internal loss in the modulator is 10 watts:
(a) What is the unmodulated carrier power?
(b) What power output is required from the modulator (baseband signal)?
(c) If the percentage modulation is reduced to 75%, how much output is needed from the modulator (baseband signal)?

*Solution*

$$P_{out} = 50 \text{ watts}$$
$$P_{in} = P_{out} + P_d$$
$$= 50 + 10 = 60 \text{ watts}$$

(a) unmodulated carrier power $P_{cc}$ is related with $P_{in}$ as,

$$P_{in} = P_{cc}\left(1 + \frac{m_a^2}{2}\right)$$

or

$$60 = P_{cc}\left(1 + \frac{1}{2}\right) = \frac{3}{2} P_{cc}$$

Hence

$$P_{cc} = 60 \times \frac{2}{3} = 40 \text{ watt}$$

(b) Power required from the modulator (baseband amplifier) is obtained by subtracting unmodulated power from the total input power:

$$P_m = P_{in} - P_{cc} = 60 - 40 = 20 \text{ watts}$$

(c) Power needed from the modulator ($P_m$) is given by

$$P_m = P_{cc}\left(\frac{m_a^2}{2}\right) = \frac{40 \times (.75)^2}{2} = 11.25 \text{ watts}$$

### AM Generation Using Non-linear Circuits

A simple diode can be used as a non-linear modulator by restricting its operation to non-linear region of its characteristic. The undesired frequency terms are filtered out by a bandpass filter. The non-linear circuit is exactly similar to the circuit shown earlier in Fig. 5.5.2 for AM generation using switching modulator. However, in this case, the total input to the diode $[f(t) + A \cos \omega_c t]$ is kept so small that the diode operates in a non-linear region of its *V-I* characteristic. Thus, this method is useful only for small signal amplitude modulation.

The input voltage to the diode is given by

$$e = f(t) + A\cos\omega_c t$$

The resulting current *i* will be given by (using the non-linear relation in Eq. 5.1.9)

$$i = ae + be^2 = a[f(t) + A\cos\omega_c t] + b[f(t) + A\cos\omega_c t]^2$$

The expansion of this equation provides the presence of a number of frequency components, such as $\omega_c, 2\omega_c, \omega_m, 2\omega_m, (\omega_c \pm \omega_m), (\omega_c \pm 2\omega_m)$ and so on. All these components, except the desired ones i.e., $\omega_c$ and $\omega_o \pm \omega_m$ are filtered out by a bandpass filter tuned to $\omega_c$.

The diode modulator does not provide amplification. Also, a single diode is unable to balance out the undesired frequency completely. These limitations can be eliminated by using amplifying devices like transistor, FET, or an electron tube in a balanced mode.

## Balanced Modulator

A transistor balanced modulator circuit used for AM generation is shown in Fig. 5.5.5.

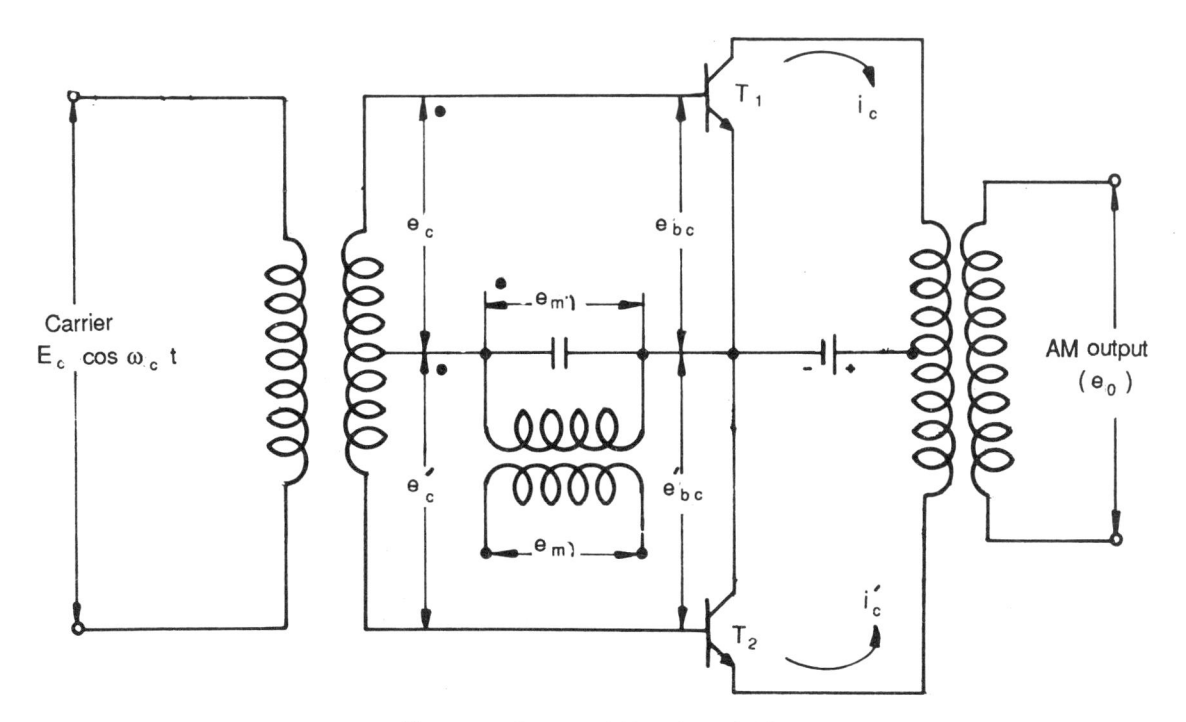

Fig. 5.5.5 Balanced Modulator (AM)

The circuit is similar to the AM-SC generator shown earlier in Fig. 5.1.15 except that feeding points of the carrier and the modulating signal are interchanged. The carrier voltage across the two windings of a center-tap transformer are equal, and opposite in phase i.e., $e_c = -e'_c$.

The input voltage to transistor $T_1$ is given by
$$e_{bc} = e_c + e_m = E_c \cos \omega_c t + E_m \cos \omega_m t \qquad (5.5.11)$$
Since, both $e_c$ and $e_m$ are in phase.
Similarly, the input voltage to transistor $T_2$ is given by
$$e'_{bc} = e'_c + e_m = - E_c \cos \omega_c t + E_m \cos \omega_m t \qquad (5.5.12)$$
This is because $e'_c$ and $e_c$ are in phase opposition i.e., $e_c = - e'_c$
By the non-linearly relationship,
$$i_c = a_1 e_{bc} + a_2 e_{bc}^2$$
and
$$i'_c = a_1 e'_{bc} + a_2 e'^{2}_{bc}$$
Substituting the values of $e_{bc}$ and $e'_{bc}$ from Eqs 5.5.11 and 5.5.12, we get
$$i_c = a_1 (E_c \cos \omega_c t + E_m \cos \omega_m t) + a_2 [E_c \cos \omega_c t + E_m \cos \omega_m t]^2$$

$$= a_1 [E_c \cos \omega_c t + E_m \cos \omega_m t] + a_2 [E_c^2 \cos^2 \omega_c t + E_m^2 \cos^2 \omega_m t + 2 E_m E_c \cos \omega_c t \cos \omega_m t] \qquad (5.5.13)$$
Similarly,
$$i'_c = a_1 [- E_c \cos \omega_c t + E_m \cos \omega_m t] + a_2 [E_c^2 \cos^2 \omega_c t + E_m^2 \cos^2 \omega_m t - 2 E_c E_m \cos \omega_c t \cos \omega_m t] \qquad (5.5.14)$$

The output AM voltage $e_o$ is given as
$$e_o = K(i_c - i'_c) \qquad (5.5.15)$$
This is because the currents $i_c$ and $i'_c$ flow in opposite directions in a tuned circuit. $K$ is a constant depending on impedance and other circuit parameters.
Substituting Eqs 5.5.13 and 5.5.14 in Eq. 5.5.15 we get
$$e_o = 2 K a_1 E_c \cos \omega_c t + 4 K a_2 E_c E_m \cos \omega_c t \cos \omega_m t$$
The other terms are balanced out. We can write,
$$e_o = 2 K E_c a_1 \left[ 1 + \frac{2 a_2 E_m}{a_1} \cos \omega_m t \right] \cos \omega_c t$$
$$= 2 K a_1 E_c [1 + m_a \cos \omega_m t] \cos \omega_c t \qquad (5.5.16)$$
where $m_a = \dfrac{2 a_2 E_m}{a_1}$ is the modulation index.

## Advantage of a Balanced Modulator Over a Simple Non-linear Circuit

In simple non-linear circuits, the undesired non-linear terms (harmonics) are eliminated by a bandpass filter. Hence, the bandpass filter must be carefully designed. But in a balanced modulator, the undesired non-linear terms are automatically balanced out and, at the output we get only the desired terms, so filter design is not so stringent.

## 5.6 DEMODULATION OF AM WAVES

The process of extracting a baseband (modulating) signal from the modulated signal is known as *demodulation*. As stated earlier, AM signals with large carrier are detected by using the *envelope detector*. The envelope

# AMPLITUDE MODULATION SYSTEMS

detector employs circuits that extract the envelope of the AM wave. Evidently, the envelope of the AM wave is the baseband signal. However, a low-level modulated signal can be detected by using square law detectors in which a device operating in the non-linear region is used to detect the baseband signal. Thus, we can categorize the detectors into two parts: (i) Square Law detectors, and (ii) Envelope detectors.

## Square Law Detectors

Square law detectors are used for detecting low level modulated signals (say below 1 volt) so that operating region of device-characteristic is restricted to non-linear region. The circuit is very similar to a square law (non-linear) modulator. The only difference is in filter circuits. In a detector, the filter is a low-pass filter instead of a bandpass filter. A simple square law detector is shown in Fig. 5.6.1a. The non-linear characteristic, modulated input voltage and resulting diode current waveforms are shown in Fig. 5.6.1b. The dc source $V_d$ is used to adjust the operating point. The operation is limited to the non-linear region due to which the lower half portion of the current waveform is compressed. This causes *envelope distortion*. The average value of the diode current does not remain constant, rather it varies with time as shown in Fig. 5.6.1b.

The distorted diode current is given by the non-linear (square law) relation

$$i = a_1 e + a_2 e^2$$

where $e$ is input modulted voltage,

$$e = E_c (1 + m_a \cos \omega_m t) \cos \omega_c t$$

Hence,

$$i = a_1 [E_c (1 + m_a \cos \omega_m t) \cos \omega_c t] + a_2 [E_c (1 + m_a \cos \omega_m t) \cos \omega_c t]^2 \quad (5.6.1)$$

The expansion of Eq. 5.6.1 reveals the presence of the components $2\omega_c, 2(\omega_c \pm \omega_m), \omega_m$ and $2\omega_m$ besides the *input frequency terms*. This current is passed through a low pass filter which allows to pass the frequencies

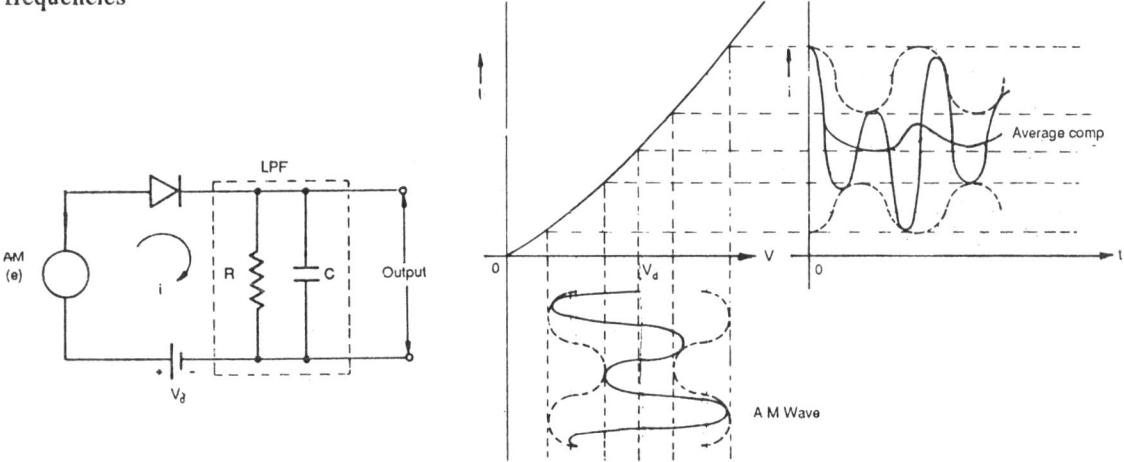

Fig. 5.6.1 Square Law Detector (a) Circuit Diagram (b) Characteristics

below $\omega_m$ and suppresses the other higher components. Thus, the baseband signal with frequency $\omega_m$ is recovered.

## Distortion

The non-linear characteristics of the diode produces additional frequency components. Frequencies centered about $\omega_c$ and $2\omega_c$ are easily suppressed by using a low pass filter, as they are far away from $\omega_m$. But $2\omega_m$, is very close to $\omega_m$ and hence it cannot be totally suppressed by the low pass filter. Therefore, component $2\omega_m$ introduces distortion. This distortion term $2\omega_m$ cannot be completely eliminated and is always present as distortion. Thus, a square law detector cannot provide a distortionless AM detection. However, the degree of distortion can be reduced by keeping the magnitude of the component $2\omega_m$ much less than the amplitude of the desired component $\omega_m$.

## Linear Diode Detector

A diode operating in a linear region of its characteristic can extract the envelope of an AM wave. Such a detector is called *envelope detector*. This detector is extremely popular in commercial receiver circuits because it is very simple and less expensive, and at the same time provides satisfactory performance for the reception of broadcast programmes. The circuit diagram of a linear diode detector is shown in Fig. 5.6.2a. The tuned transformer provides perfect tuning at the desired carrier frequency. R-C forms the time constant network. When the modulated carrier at the input of detector is 1 volt, or more, the operation takes place in the linear region of the diode characteristic. The idealized linear characteristic of the diode along with input voltage and output current waveforms are shown in Fig. 5.6.2b.

### *Operation*

Let us first assume that the capacitor is absent. The output waveform will be a half-rectified carrier wave as shown in Fig. 5.6.2b. The diode will present a low resistance $r_d$ (forward biased). The total resistance of the circuit becomes $r_d + R$. A current flowing in the diode circuit through this total resistance, will result in a half-rectified carrier voltage across resistance $R$.

Now, consider that the capacitance has been introduced in the circuit. For the positive half-cyle of a rectified wave, the capacitor is charged to the peak value of the carrier voltage through the resistance $(r_d + R)$. But, for a negative half-cycle, the diode is reverse biased, and the carrier voltage is disconnected from the R-C circuit. So the capacitor starts discharging through the resistance $R$ with a time constant $\tau = RC$. If the time constant is properly chosen, the voltage acrross the capacitor does not fall appreciably during the small pepriod of a negative half-cycle, and by that time the next positive cycle appears. This positive cycle further charges the capacitor to the peak value of the carrier voltage and the process continues. Thus, the voltage across the capacitor is just like a spiky baseband signal, i.e., voltage across the capacitor is same as envelope of the

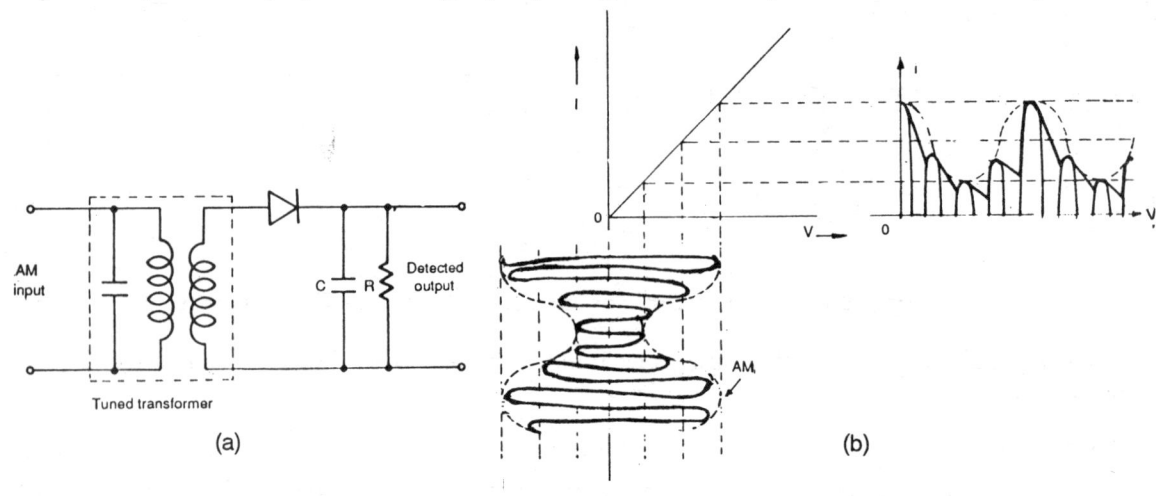

AMPLITUDE MODULATION SYSTEMS 315

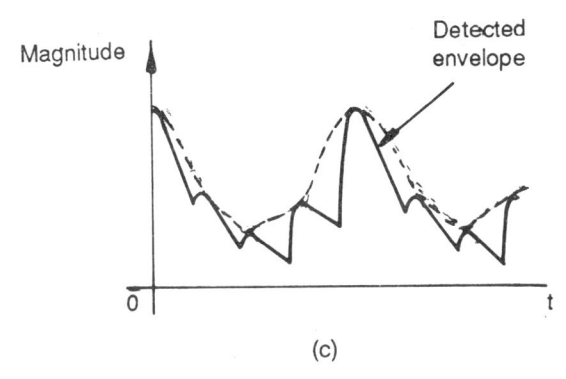

Fig. 5.6.2 Linear Diode Detector  (a) Circuit  (b) Characteristics
(c) Detected output

modulated carrier, but spikes are introduced due to charging and discharging of the capacitor. The resulting detected baseband signal is shown in Fig. 5.6.2c. The spikes can be reduced to a neglible amount by keeping the time constant and R-C large, so that the capacitor discharges negligibly small. But the large value of R-C creates another problem called *diagonal clipping,* and hence, we cannot increase it beyond a certain limit. The choice of time constant is an important point for consideration.

## *Choice of Time Constant R-C*

It is desired to keep the time constant R-C very high as compared to time period of the carrier wave in order to minimize spikes or fluctuation in the detected envelope. On the other hand, if it is kept too high, the discharge curve becomes approximately horizontal. In that case, negative peaks of the detected envelope may be completely, or partially missing. The recovered baseband signal is distorted at negative peaks as shown in Fig. 5.6.3. This type of distortion is known as *diagonal clipping*. An optimum value of the time constant has to be chosen which provides a compromise between following two facts,
 (i) The spike, or fluctuation in a detected envelope should be minimum.
 (ii) Negative peaks of detected envelope should not be missed even partially, (diagonal clipping).

The above two factors can be compromised if we make R-C large enough so that it follows the entire envelope of rectified carrier wave and at the same time no portion of the envelope is missed. The negative peak will be totally maintained in the detected output and no portion will be clipped, if the discharge curve follows the modulation envelope. For this, the rate of discharge of capacitor $\geq$ rate of decrease of modulation envelope. Let us derive a relation which satisfies this condition. Consider a single-tone AM signal given by

$$\varphi_{AM}(t) = A[1 + m_a \cos \omega_m t] \cos \omega_c t$$

## *Rate of Decay of Envelope*

The envelope of the AM voltage is, given as

$$e = A[1 + m_a \cos \omega_m t] \tag{5.6.2}$$

**316** COMMUNICATION SYSTEMS: Analog and Digital

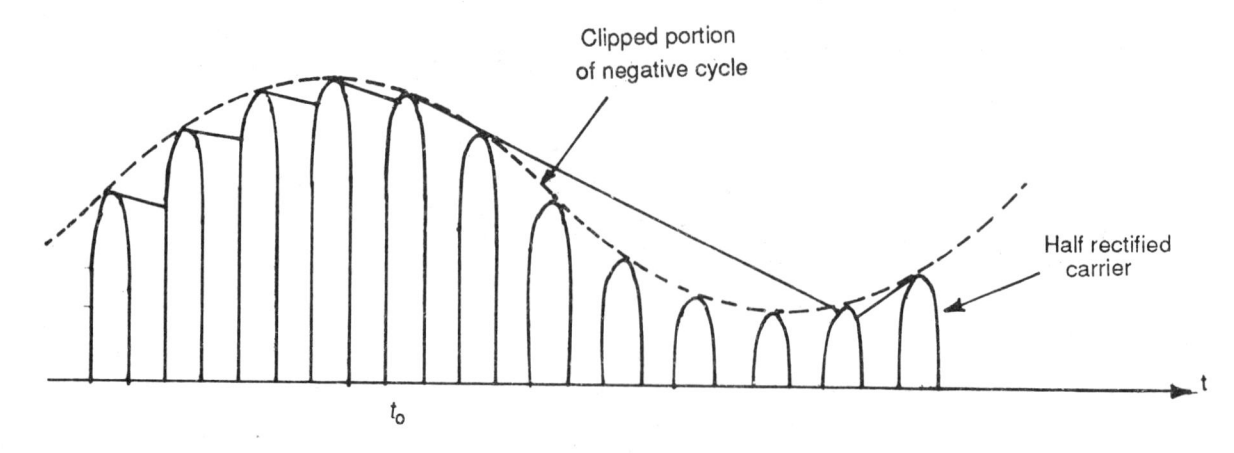

Fig. 5.6.3 Diagonal Clipping

The rate of change (slope) of this envelope is

$$-\frac{de}{dt} = A m_a \omega_m \sin \omega_m t \qquad (5.6.3)$$

The negative slope indicates the *decay of the voltage.*
The slope at an instant $t_o$ will be

$$-\left(\frac{de}{dt}\right)_{t=t_o} = A m_a \omega_m \sin \omega_m t_o \qquad (5.6.4)$$

This is the rate of decay of the envelope at an instant $t = t_o$.

### Rate of Discharge of Capacitor

The capacitor $C$ discharges exponentially. Let us assume that the capacitor starts discharging at $t = t_o$. Before the capacitor starts discharging at this instant, it has been charged to a voltage equal to the value of the envelope at $t = t_o$. This initial voltage across the capacitor is equal to the envelope voltage $e$ at $t = t_o$, given by

$$e_o = (e)_{t=t_o} = A[1 + m_a \cos \omega_m t_o] \qquad (5.6.5)$$

The capacitor discharges exponentially from the initial voltage $e_o$. The capacitor voltage at any instant $t$ is given as

$$e_{cap} = e_o \, e^{-\frac{t-t_o}{RC}} \qquad (5.6.6)$$

The rate of change of capacitor voltage (slope of the discharge curve) is,

$$-\frac{de_{cap}}{dt} = -\frac{d}{dt}\left[e_o \, e^{-\frac{(t-t_o)}{RC}}\right]$$

AMPLITUDE MODULATION SYSTEMS 317

$$= \frac{e_o}{RC} e^{-\frac{(t-t_o)}{RC}} \tag{5.6.7}$$

This rate of change at $t = t_o$ is given as

$$-\left(\frac{de_{cap}}{dt}\right)_{t=t_o} = \frac{e_o}{RC}$$

Substituting $e_o$ from Eq. 5.6.5,

$$-\left(\frac{de_{cap}}{dt}\right)_{t=t_o} = \frac{A}{RC}[1 + m_a \cos \omega_m t_o] \tag{5.6.8}$$

To avoid diagonal clipping the slope of the discharge-curve at $t=t_o$ given by Eq. 5.6.8 must be equal to, or greater than, the envelope-decay rate given in Eq. 5.6.4. Thus,

$$\frac{A}{RC}[1 + m_a \cos \omega_m t_o] \geq A\, m_a\, \omega_m \sin \omega_m t_o$$

This inequality can be written as

$$\frac{1}{RC} \geq \frac{\omega_m\, m_a \sin \omega_m t_o}{1 + m_a \cos \omega_m t_o} \tag{5.6.9a}$$

At an instant $t$, Eq. 5.6.9a can be given as

$$\frac{1}{RC} \geq \frac{\omega_m\, m_a \sin \omega_m t}{1 + m_a \cos \omega_m t} \tag{5.6.9b}$$

It is desired that $(1/RC)$ is always greater than, or equal to, the right handside (RHS) of Eq. 5.6.2b. This can be achieved by maximizing the RHS. The condition for maximizing the RHS is obtained by equating the derivative of RHS of Eq. 5.6.9b to zero.

$$\frac{d}{dt}\left[\frac{\omega_m\, m_a \sin \omega_m t}{1 + m_a \cos \omega_m t}\right] = 0$$

Solution of the above equation yields,

$$\cos \omega_m t = -m_a \tag{5.6.10a}$$

or

$$\sin \omega_m t = \sqrt{1 - m_a^2} \tag{5.6.10b}$$

substituting these values of $\sin \omega_m t$ and $\cos \omega_m t$ in Eq. 5.6.9b, we get

$$\frac{1}{RC} \geq \omega_m\, \frac{m_a \sqrt{1 - m_a^2}}{1 - m_a^2}$$

or

$$\frac{1}{RC} \geq \frac{\omega_m\, m_a}{\sqrt{1 - m_a^2}} \tag{5.6.11}$$

If $m_a << 1$, (small modulation index) then Eq. 5.6.11 reduces to

$$\frac{1}{RC} \geq \omega_m m_a \qquad (5.6.12)$$

Thus, Eq. 5.6.11 provides the desired relation for obtaining optimum value of time constant $RC$ in terms of modulation index, and modulating frequency.

### Example 5.6.1

An amplitude modulated wave $10\left[1 + 0.6 \cos 2\pi\, 10^3 t\right] \cos 2\pi\, 10^6 t$ is to be detected by a linear diode detector. Find (a) the time constant $\tau$ (b) the value of resistance $R$ if the capacitor used is $100\, pF$.

*Solution*
From Eq. 5.6.11,

(a)
$$\frac{1}{RC} \geq \frac{\omega_m m_a}{\sqrt{1 - m_a^2}}$$

Here
$$\omega_m = 2\pi\, 10^3,\ m_a = 0.6$$

Hence
$$\frac{1}{RC} \geq \frac{2\pi\, 10^3 \times 0.6}{\sqrt{1 - 0.6^2}} = 5.886 \times 10^3$$

In the limiting case
$$\tau = RC = \frac{10^{-3}}{5.886} = 0.17\ \text{ms}$$

(b) $R = \dfrac{\tau}{C} = 0.17 \times 10^3/100 \times 10^{-12} = 0.17 \times 10^{13}\ \Omega$

## 5.7 AM TRANSMITTERS AND RECEIVERS

A transmitter not only performs the modulation process, but also raises the power level of a modulated signal to the desired extent for effective radiation. The AM transmitters are divided into two categories:
(i) Low-level modulation; (ii) High-level modulation.

### Block Diagram of the AM Transmitter

(i) *Low power level modulation*: The block diagram for this type of transmitter is shown in Fig. 5.7.1. This system is used in low power transmitters. In this method, the modulation process takes place at a low power level, and the signal is then amplified by a class B power amplifier to raise to the desired power level. The limitation of the system is lower efficiency of the class B r.f. amplifier as compared to class C, the latter cannot be used as it does not faithfully reproduce the modulated signal consisting of carrier and sideband terms. The amplifier should possess sufficient bandwidth to amplify these terms equally. The class C amplifier have narrow bandwidth, which may cause sideband cuttings, and thereby introduce distortions. The remedy is to increase the power level of the carrier before modulation to the desired extent, and then perform the modulation. This is done in another type of transmitter known as *high-level transmitter*.

AMPLITUDE MODULATION SYSTEMS 319

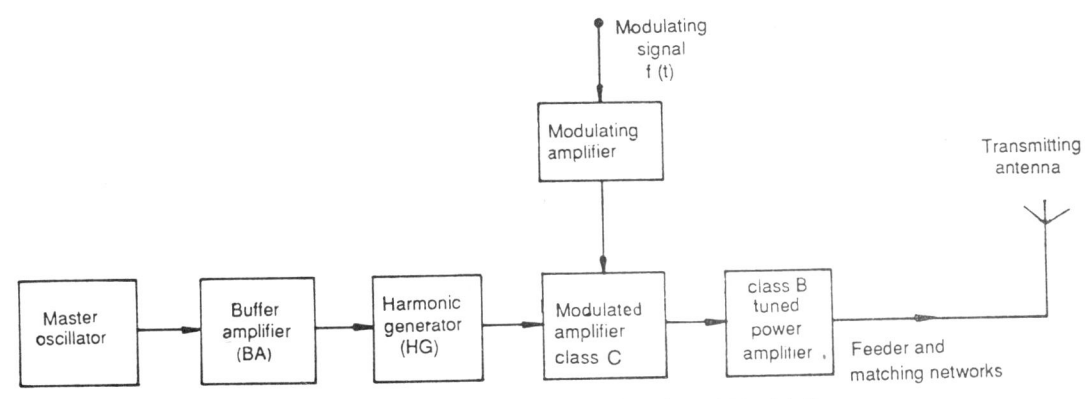

Fig.5.7.1 AM Transmitter Using Low Level Modulation

(ii) *Higher power level modulation*: The block diagram for this type of transmitter is shown in Fig. 5.7.2. In high level power modulation, the carrier is first amplified by a class C amplifier, and then modulation is done at the high power level. Here, the class C amplifier have to amplify only the carrier frequency, and hence, can have high gain and efficiency. This type of transmitter is complex, and more expensive than the low level transmitters. It is, however suitable for high power transmission. Each constituent stage of the block diagrams is described briefly as follows:

## 1. Master Oscillator (MO)

The Master Oscillator generates a stable sub-harmonic carrier frequency (i.e., the fraction of a desired carrier frequency). This stable sub-harmonic oscillation is generated by using a crystal oscillator and then frequency is raised to the desired value by harmonic generators. The sub-harmonic frequency is used because stable crystal oscillators are easy to make at lower carrier frequencies. For example, if one wants 1 MHz of carrier frequency, it is better to make a crystal oscillator of 250 kHz which is the fourth sub-harmonic of 1 MHz. Note that a 250 kHz crystal oscillator will be much more stable than a 1 MHz crystal oscillator. So, the master oscillator generates a sub-harmonic of 250 kHz, and then it is increased four times by a harmonic generator to get the desired carrier frequency of 1 MHz. The limitation of the crystal oscillator is that it can provide only single frequency, and this frequency cannot be altered. This is not a problem in a broadcast transmitter which normally transmits only one carrier frequency.

### *Stability of MO Frequency*

The carrier frequency generated by an MO ought to be very stable. Any change in the MO frequency will cause interference with other transmitting stations and the receiver will accept programmes from more than one transmitter. There are several reasons for the change in the frequency of a master oscillator. They are:

    i) *Frequency drift*: This means a slow variation in frequency with time. The frequency drift permitted in medium wave transmitters is $\pm$ 20 Hz, whereas, for shortwave and UHF transmitters the drift can be tolerated up to $\pm 0.002\%$ of carrier frequency. The frequency drift occurs due to variation of circuit parameters with temperature variation or aging.

    ii) *Frequency Scintillation*: This means an abrupt change in the MO frequency due to abrupt changes in load on the master oscillator. The carrier frequency scintillation is avoided by using a buffer amplifier between the master oscillator and the harmonic generator. Any variation of load current is handled by the buffer amplifier, and the master oscillator is not affected. The following measures are adopted in order to keep the MO frequency stable:

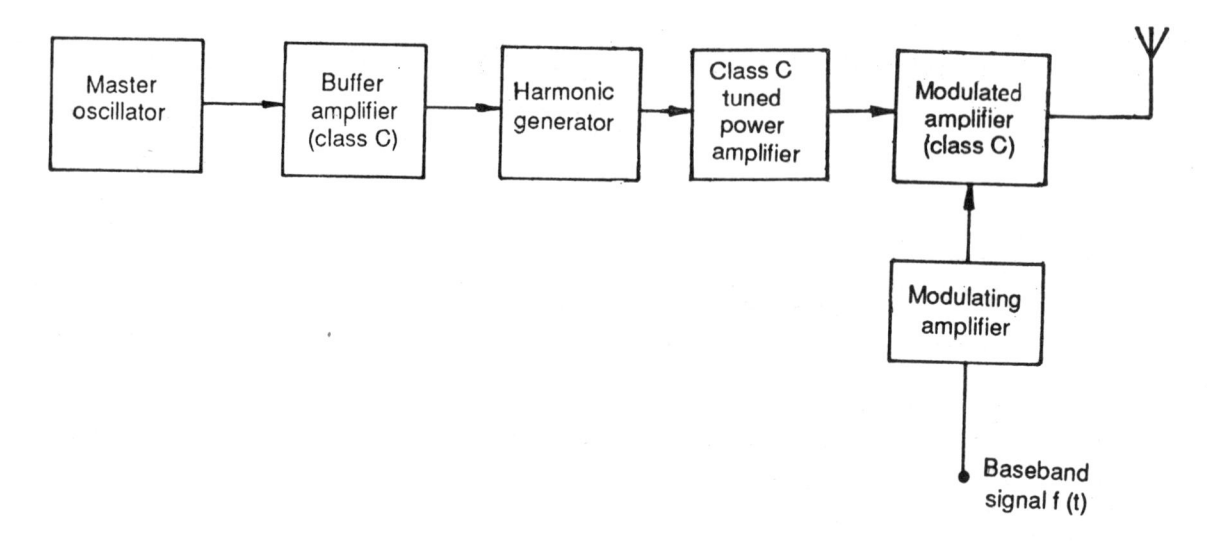

Fig. 5.7.2 AM Transmitter Using High Level Modulation

(a) Oscillator must be kept in a constant-temperature chamber so that the circuit parameter does not change with a change in temperature.
(b) Stabilized power supply should be used in the oscillator circuit.
(c) Effective Q of the tuned circuits should be very high.
(d) The active devices used in the oscillator circuit should have a high value of the ratio $(g_m/C)$ where $g_m$ is the mutual transconductance, and $C$, is the device capacitance.
(e) Master oscillator should generate a sub-harmonic of the carrier frequency.
(f) Buffer amplifier must be used between the master oscillator and other stages.

## 2. Buffer Amplifier

This is a tuned amplifier providing a high input impedance at the master oscillator frequency. Any variation in load current does not affect the master oscillator due to this high input impedance of buffer amplifier at the operating frequency of the master oscillator. Thus, buffer amplifier *isolates the MO from the succeeding stages*, so that the loading effect may not change the frequency of the MO. A harmonic generator stage can work as a buffer, because its input and output tank circuits are tuned to different frequencies. If the current flow is large at a frequency in the output circuit, the current flow at input circuit will be small, as it is tuned to a different frequency, and thereby offers a high input impedance.

## 3. Harmonic Generator

It is an electronic circuit that generates harmonics of its input frequency. The principle of harmonic generation is the same as that of a non-linear modulator. When a signal is given to a non-linear circuit, it generates harmonics of input frequency. The desired harmonic is selected by a properly tuned circuit. One of the various possible circuit arrangements is shown in Fig. 5.7.3. The circuit uses a class C tuned amplifier. It is well know that in the class C operation the current flows in the form of pulses recurring with the same frequency as the input signal. These pulses contain harmonics of input frequency $\omega_c$. This can be seen by Fourier analysis of the pulse train. If this pulse train is passed through a resonant circuit tuned to $n\omega_c$, the $n^{th}$ harmonic will be selected, and the others will be rejected. Thus, the output of the harmonic generator will have a frequency $n\omega_c$. The class C operation is obtained by biasing the device much below cut-off.

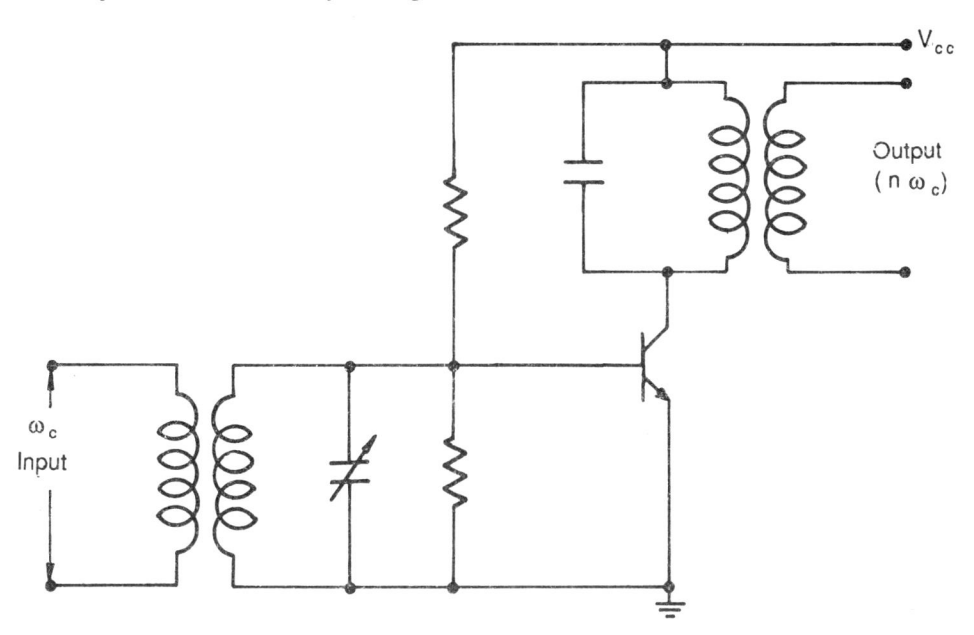

Fig. 5.7.3 A Class C Harmonic Generator

## 4. Driver Amplifier or Intermediate Power Amplifier

One or more stages of a class C tuned amplifier is used to increase the power level of a carrier signal to provide a large drive to the modulated class C amplifier. The output of the harmonic generator provides a low power carrier signal. This power is amplified to raise the power to desired level to drive the final amplifier stage. The number of class C stages depends on the desired power level, and also the gain of each stage. This stage is termed as Driver amplifier or intermediate power amplifier (IPA).

## 5. Modulation System

The collector modulation circuit is used for modulation in high power transmitters. The modulating amplifier is a class A, or class B, amplifier amplifying the baseband signal.

## 6. Feeder and Antenna

The transmitter power is fed to a transmitting antenna for effective radiation. The length of the antenna (a conductor) should be of the order of the wavelength for effective radiation. The antenna is normally located at a distance from the transmitter and hence power from the transmitter, is fed to the antenna through a properly designed transmission line called feeder. The impedance of the feeder line must be properly matched with the transmitter impedance at one end, and with the antenna impedance at the other end.

### *Neutralization*

Neutralization circuits are used in transmitters to eliminate the spurious oscillation caused by undesired feedback of energy through interelectrode capacitances.

## AM Receiver

A radio receiver is an electronic circuit that picks up a desired modulated radio frequency signal, and recovers the baseband signal from the modulated signal. A receiver performs the following functions which are explained with the help of the block diagram of a simple receiver known as *tuned radio frequency* (TRF) receiver shown in Fig. 5.7.4.

### (i) Interception

This function is performed by a receiving antenna (small conductor). The radio waves coming from various transmitting stations arrive at this antenna. These radio waves contain electrical energy in the form of an electromagnetic wave. When this electromagnetic wave is intercepted by the receiving antenna, a voltage is induced in it. A single antenna intercepts radio waves of all frequencies. The desired frequency is selected by the selection process.

### (ii) Selection

This is the process by which the receiver selects a particular carrier frequency, and rejects others, so that at any time a signal from only one transmitter is received. This is done by a tank circuit. The resonant frequency of this circuit can be changed with the help of a variable capacitor, or inductor. Thus, although a large number of modulated carriers coming from various transmitters is intercepted by the antenna, only one gets selected, which is the one for which the receiver is tuned.

### (iii) R.F. Amplification

The selected carrier (radio signal) is amplified by a class C tuned amplifier. This is necessary to raise the carrier voltage level so that the linear diode detector following this stage may operate in linear region. The increased carrier voltage also suppresses the effect of noise.

### (iv) Detection

Detection is the process of recovering a baseband signal from a modulated carrier. It has been discussed in detail in Sec. 5.6

### (v) Audio Amplification

The detected audio signal is further amplified so that it can drive the speaker. This stage consists of a R-C coupled amplifier followed by a class B pushpull power amplifier.

### (vi) Reproduction

Reproduction is the process by which an electrical signal is converted into a desired physical message. In a commercial broadcast receiver, the output of the audio amplifier is fed into a speaker that produces sound according to the input audio signal.

## TRF Receiver

A TRF receiver performs all the above functions and works satisfactorily at medium wave frequencies. At higher radio frequencies the performance of a TRF becomes poor. The performance of receiver is improved by a technique known as *heterodyning*. A receiver based on this technique is known as

AMPLITUDE MODULATION SYSTEMS 323

*superheterodyne receiver.* The performance of a receiver is judged from its various features such as selectivity, sensitivity and fidelity.

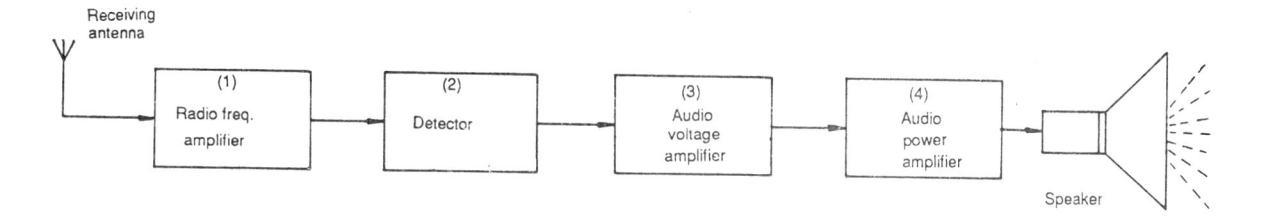

Fig. 5.7.4 TRF Receiver

## Features of a Receiver

The various types of receivers can be compared in their performances by the following features.

### (i) Selectivity

Selectivity is receivers' ability to distinguish between two adjacent carrier frequencies. By this feature it is decided how perfectly the receiver is able to select the desired carrier frequency and reject the others. Selectivity depends on the sharpness of the resonance curve of a tuned circuit involved in the receiver. The sharper the resonance curve, the better the selectivity. Better selectivity means that the receiver has greater capability to reject undesired signals.

The sharpness of the resonance curve depends on $Q$ of the resonant circuit. The higher is the $Q$, more is the selectivity. In Fig. 5.7.5, curve 1 is sharper than curve 2, and hence has better selectivity.

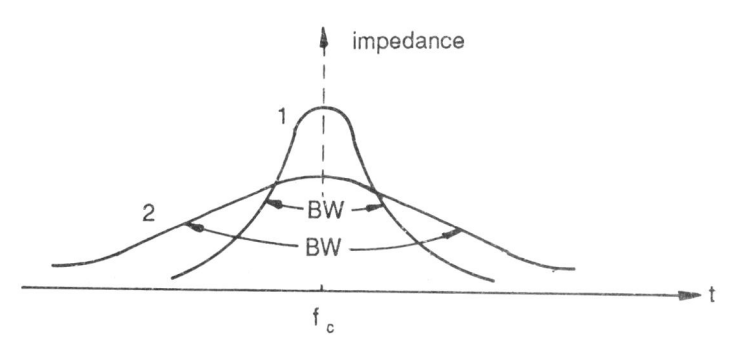

Fig. 5.7.5 Resonance Curve of a Tuning Circuit

### (ii) Sensitivity

The ability of a receiver to detect the weakest possible signal is known as sensitivity. A receiver with a good

## (iii) Fidelity

The ability of a receiver to reproduce faithfully all frequency components present in the baseband signal is called fidelity. If any component is missed, or attenuated considerably, fidelity suffers and the reproduced signal is distorted. This feature is mainly decided by the audio amplifier which amplifies the baseband signal.

### Superheterodyne Receiver

This receiver is popularly used. The hetrodyning gives a far better performance than the TRF receiver. The block diagram of a superheterodyne receiver is given in Fig. 5.7.6. Significant feature of the heterodyne receiver is that all incoming radio frequencies are converted into a single *intermediate frequency* $f_i$ by the heterodyning (mixing) process. The incoming carrier ($f_c$) and a locally generated signal ($f_l$) are mixed in mixer, also referred to as first detector. The mixer generates the sum and difference frequencies at the output. The difference frequency ($f_l - f_c$) is selected by a properly tuned circuit. The local oscillator frequency is higher than the carrier i.e., $f_l > f_c$. That is why the receiver is called *superheterodyne* receiver. The intermediate frequency (IF) in commercial radio receivers is fixed to 455 kHz.

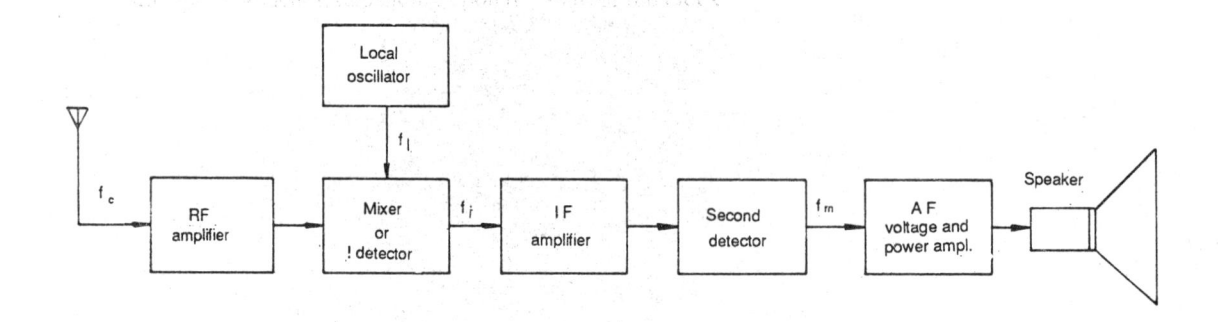

Fig. 5.7.6 Superheterodyne Receiver

### Advantages of a Superheterodyne Receiver

All the incoming carriers frequencies (from medium waves to short waves) are converted into a fixed IF frequency of 455 kHz. Therefore, all the succeeding stages have to operate on a fixed frequency. Hence, the circuit design becomes simple with improved performance. Let us now see how the performance is improved.

**Sensitivity** : The IF signal is amplified in an intermediate frequency amplifier. This is to be designed for a fixed frequency band centered near 455 kHz. Naturally, it is easy to design and develop a high gain amplifier at a fixed carrier frequency. The sensitivity of a superheterodyne receiver is primarily decided by the IF amplifier stage. A high gain IF amplifier provides a better sensitivity than the TRF receiver. In the TRF receiver, the sensitivity is decided by an RF amplifier, which amplifies the entire band of medium and short waves rather than a fixed carrier frequency. It is difficult to design a RF stage that provides high gain for such a wide range of carrier frequencies (medium waves to short waves). Also, high gain AF amplifiers suffer from instability problem.

**Selectivity:** The selectivity of a superheterodyne receiver is also decided mainly by the IF amplifier. The tuned circuit associated with the IF amplifier operates at a fixed centre frequency of 455 kHz. Since the band width of the baseband signal is about 10 kHz, the $Q$ required by a tuned circuit is given by

$$Q = \frac{f_c}{\text{bandwidth}} \tag{5.7.1}$$

where $f_c$ is the centre frequency
Therefore

$$Q = \frac{455 \text{ kHz}}{10 \text{ kHz}} = 45.5$$

As the tuning circuit for this fixed and practical value of $Q$ is easy to design, a better selectivity than that of a TRF receiver can be achieved.

The selectivity of a TRF receiver is poor because of the following reasons:
(i) The $Q$ factor requirement of the tuned circuits employed in RF amplifier is not fixed. Consider a tuned circuit required to have a bandwidth of 10 kHz at the lower edge of the broadcasting medium wave (550 kHz). The $Q$ required by the tuned circuit is 55. At the higher edge of the medium wave (1650 kHz), the $Q$ requirement is 165. Thus, the $Q$ required varies with carrier frequency. This variation in $Q$ is difficult to achieve in practice.
(ii) The $Q$ required by the tuned circuits for upper side of the short wave (say 20 MHz) is 2000. It is impossible to achieve such a high value of $Q$ with ordinary tuned circuits.

**Detector Design** The detector stage of a superheterodyne receiver operates on a fixed frequency, IF, and hence it can be designed to detect the entire baseband without introducing much distortion, as the time constant RC of the detector remains fixed. In a TRF, the detector has to detect signals of various carrier frequencies. The single detector cannot give a good performance for a large range of radio frequencies, as the desired time constant does not remain fixed.

## Advantages of TRF Receiver

TRF receivers are cheaper and simpler than superheterodyne receivers. Also, they do not have problems such as image signal rejection or tracking and alignment, that arise in a superheterodyne receiver.

## Constituent Stages of a Superheterodyne Receiver

**RF Amplifier** This is a class C tuned voltage amplifier. The main functions of this stage are:
 (a) Amplification of the received radio signal to provide better sensitivity and improved signal to noise ratio.
 (b) Rejection of the unwanted signals and an improved adjacent channel selectivity.
 (c) Rejection of the image signal.

## Image Signal

This is a signal whose frequency is above the local oscillator frequency by the same amount, as the desired signal frequency is below the local oscillator frequency. This is illustrated by the spectrum shown in Fig. 5.7.7. It is obvious from the figure that

$$f_\ell - f_c = f_i$$

or

$$f_c = f_\ell - f_i \qquad (5.7.2)$$

The desired signal frequency $f_c$ is below local oscillator frequency $f_\ell$ by an amount $f_i$, the IF. By definition, the image signal $f_c'$ is above $f_\ell$ by the same amount $f_i$ i.e. image signal $f_c'$ is given by,

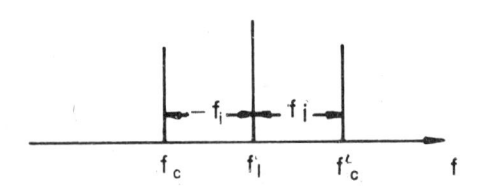

$$f_c' = f_\ell + f_i \qquad (5.7.3)$$

Fig. 5.7.7 Image Signal

which is obvious from Fig. 5.7.7. The difference between $f_c'$ and $f_c$ is $2f_i$. From Eqs 5.7.2 and 5.7.3

$$f_c' = f_\ell + f_i = f_c + f_i + f_i$$
$$= f_c + 2f_i \qquad (5.7.4)$$

Thus, the image signal is $2f_i$ more than the desired carrier signal. If somehow, this image signal is intercepted by an antenna and reaches the mixer, it produces the same IF as produced by the desired signal $f_c$. The desired signal produces $f_\ell - f_c = f_i$ whereas the image signal produces $f_c' - f_\ell = f_i$. The spurious IF signal produced by image signal is also amplified by the IF amplifier and produces interference in the receiver output. Therefore, the image signal should be rejected. The RF amplifier adds one more tuned circuit per stage, and thereby increases the image frequency attenuation.

**Mixer Stage**

A mixer (Fig. 5.7.8) is a non-linear device that mixes the incoming signal of frequency $(f_c)$ with a local oscillator voltage of frequency $(f_l)$ and generates an output voltage of an intermediate frequency $(f_\ell - f_c)$ as shown in Fig. 5.7.7. The non-linear mixer circuit produces the sum and difference frequency components $(f_\ell \pm f_c)$ alongwith the input frequencies and their harmonics. The desired intermediate frequency $f_i = (f_\ell - f_c)$ is selected by a tuned circuit known as *input intermediate frequency transformer (IFT)*. The IFT is tuned by adjusting the core of the transformer. The process is known as *inductive tuning*.

The mixer is also known as *first detector*. Some mixer circuits use separate devices for mixing and generating local oscillator voltage. Such circuits are referred as separately excited mixers. Another type of mixer circuit, known as *self excited mixer* utilises a single device as mixer and local oscillator. They are also referred to as *frequency convertors*. The frequency convertors and mixers perform the job of *frequency-changers*.

AMPLITUDE MODULATION SYSTEMS 327

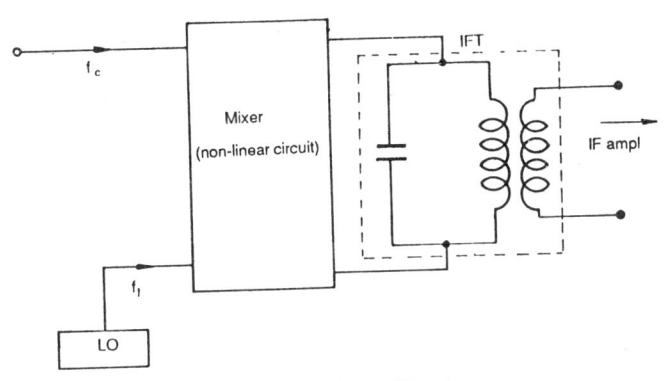

Fig. 5.7.8 A Mixer-Circuit

Figure 5.7.9 shows a frequency convertor circuit used in broadcast receivers. In this case, the receiver circuit does not contain RF amplifier stage. The incoming signal is applied at the base of a transistor. The local oscillator (LO) voltage develops across the emitter of the same transistor. Tuning capacitors of antenna and local oscillator circuits are mounted on the same rotating shaft, i.e., they are ganged together to provide single dial tuning. The ganged capacitor is shown by a dotted line in Fig. 5.7.9.

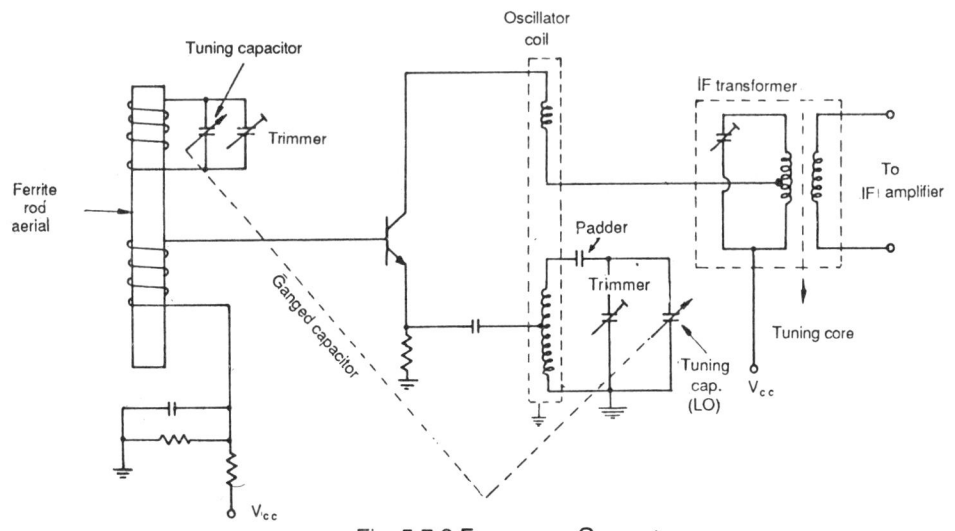

Fig. 5.7.9 Frequency Convertor

***Tracking and Alignment:*** For convenient tuning, it is common practice to gang the tuning capacitors of local oscillator, and front-end tuning circuit including the RF amplifier (if present). The ganged capacitor forms a mechanically coupled system for *simultaneous Tuning* of a number of resonant circuits by a single knob. The *front-end circuits* including RF circuit and mixer input tuned circuit are tuned to incoming carrier *signal frequency* $f_c$. The local oscillator has to be *simultaneously tuned* to a frequency exactly higher than the signal frequency $f_c$ by an amount equal to the intermediate frequency $f_i$. The simultaneous tuning is obtained by using a ganged capacitor in which different sections are made as nearly identical as possible. The oscillator coil has less inductance than the coil of the RF section to achieve the *tracking* i.e., to keep the difference of local oscillator frequencies and incoming signal frequency exactly equal to the intermediate frequency.

It is not feasible practically to keep a constant difference between the local oscillator and incoming carrier frequencies for the *entire band*. The best that can be done is to adjust the difference exactly equal to IF at two points along the dial, allowing some error along the rest of the dial. This error is known as *tracking error*. The precise *alignment* of tuned circuits to achieve a zero tracking error at two points along the dial is known as two point tracking. However, it is possible to precisely align the tuned circuits at three points along the dial by including a series capacitance known as *padder* and a shunt capacitor known as *trimmer* as shown in Fig. 5.7.9. This process is called as *three point tracking*. The process of alignment of the receiver for achieving three point tracking is as follows:

(a) select a proper inductance to achieve exact IF in the middle of the band when tuning capacitor is positioned in the centre.

(b) the tuning capacitor is then moved to high frequency end, and the trimmer is adjusted to obtain exact IF again.

(c) finally, the tuning capacitor is moved to low frequency end, and padder is adjusted to obtain exact IF.

The process is repeated two or three times for exact alignment at each tracking point. Three point tracking in the broadcast medium wave band is done at 600 kHz, 950 kHz and 1500 kHz. A tracking error as low as 3 kHz can be achieved for the entire band, which is considered to be negligible. In short wave bands, three point tracking is not required, and only a trimmer is sufficient to achieve two point tracking.

## Local Oscillator

The type of local oscillator circuit used in a receiver depends on factors such as operating frequency, tuning range and stability. The superheterodyne receivers up to 36 MHz use mostly the *Armstrong* or the *Hartley* oscillator.

In superheterodyne receivers, the *local oscillator frequency is always kept higher than the signal frequency* by an amount equal to the intermediate frequency, for the following reasons.

(a) The maximum to minimum capacitance ratio of the two sections (signal and local oscillator sections) of the ganged capacitor is quite close. For medium wave band (550 kHz-1650 kHz), the maximum to minimum capacitance ratio required by the signal section of the ganged capacitor is

$$\frac{C_{max}}{C_{min}} = \left(\frac{1650}{550}\right)^2 = 9:1$$

Now consider the local oscillator section. When local oscillator frequency is kept higher, the maximum to minimum capacitance ratio required is

$$\frac{C_{max}}{C_{min}} = \left(\frac{1650 + 455}{550 + 455}\right)^2 = \left(\frac{2105}{1005}\right)^2 = 4.4:1$$

which is quite close to that of signal section. The usual ganged capacitors available have a capacitance ratio of 10:1, which is well within the limit imposed by the tuning capacitors of both the sections. Therefore, the ganged capacitor with identical sections serves the purpose of tuning the front-end as well as local oscillator circuit.

If the local oscillator frequency is kept lower, the maximum to minimum capacitance ratio required is,

$$\frac{C_{max}}{C_{min}} = \left(\frac{1650 - 455}{550 - 455}\right)^2 = \left(\frac{1195}{95}\right)^2 = 156:1$$

Thus ratio is beyond the limit imposed by tuning capacitor of signal section and cannot be covered by the oscillator in one sweep. The ganged capacitor with identical sections will not serve the purpose. Note that ganged capacitors with identical sections are *easy to fabricate*.

(b) The tracking errors can be reduced to a great extent by keeping constant ratio of local oscillator frequency to signal frequency for the entire band. This ratio varies between 1005/550 = 1.83 to 2105/1650 = 1.28 for medium wave band when *local oscillator frequency* is kept *above* the incoming carrier *signal frequency*. This variation from 1.83 to 1.28 is quite small as compare to the situation when local oscillator operates *below* the signal frequency, giving a frequency ratio variation between 550/95 = 5.79 to 1650/1195 = 1.38. The variation of ratio from 5.79 to 1.38 is quite large and can result in severe tracking problems.

## IF Amplifier

The intermediate frequency (IF) amplifiers are tuned voltage amplifiers. Most of the receiver gain is provided by this stage. More than one stage of IF amplifier is used to get a good sensitivity. The output of the IF amplifier appears across a tuned transformer circuit. For $n$ stages of IF amplifier, the total number of IF transformers needed is $(n + 1)$.

## Choice of Intermediate Frequency

The intermediate frequency in a commercial AM receiver has a fixed value of 455 kHz. This value is chosen as a compromise between two conflicting factors.

(i) adjacent channel selectivity and easy *tracking* for which IF should be low, and
(ii) image signal rejection for which IF should be high.

For a proper channel selectivity the intermediate frequency $f_i$ should be low. A lower $f_i$ needs a lower $Q$, and the proper tank circuit can be easily designed; whereas if $f_i$ is large, it will require a large $Q$ and the tank circuit design will be complex. For example, if $f_i$ = 455 KHz and the baseband is 10 kHz, the $Q$ value desired by the tank circuit is,

$$Q = \frac{455}{10} = 45.5$$

The tuned circuit for this $Q$ can be easily designed. But, if $f_i$ = 10 MHz (say), the $Q$ desired is

$$Q = \frac{10 \times 10^6}{10 \times 10^3} = 1000$$

The design of a tuned circuit with such a high $Q$ is impossible. Hence, for a better selectivity, $f_i$ should be lower. Also, a low value of IF makes the difference between signal and local oscillator frequency small, and as a result *tracking* becomes easy.

The value of $f_i$ should be large for image signal rejection. From Eq. 5.7.4, we have

$$f_c' = f_c + 2f_i$$

or

$$\frac{f_c'}{f_c} = 1 + (2f_i / f_c)$$

Therefore, if $f_i$ is kept large, the image signal $f_c'$ will be far apart from desired signal $f_c$ and can be easily rejected. In commercial AM receivers, the intermediate frequency is kept fixed at 455 kHz in order to achieve a compromise between the conflicting factors. This value of IF provides good image signal rejection at broadcast band even if no RF amplifier is used, as the ratio $f_c'/f_c$ is large. At short waves, however, the image signal rejection becomes poor due to a phenomenon known as *double spotting*. In this phenomenon, signals from the same short waves station is picked at two nearby points on the receiver tuning dial. Double spotting is harmful because it can mask a weak station by a nearby strong station at the spurious point on the dial. The spurious point corresponds to an image signal. Double spotting can be avoided by a good front end selectivity to avoid image signal. The RF amplifier add more tuning circuits and improve the front and selectivity of the receiver at short waves.

### Second Detector

The linear diode detector discussed in the previous section is popularly used in radio receivers because of its simple circuit and low cost. The automatic volume control (AVC) bias is obtained from this stage in order to keep the receiver output substantially constant with time for any variations in receiver input voltage. The magnitude of the receiver input voltage varies with time due to fading, or when the receiver is tuned from one station to another having a different signal strength. The AVC eliminates the effect of these variations. Figure 5.7.10a provides a "simple AVC" bias if diode D in AVC circuit is removed. The AVC circuit samples a fraction of the detector output and converts it to AVC bias voltage. The AVC bias is applied to RF and IF stages to provide them a negative bias. As the input of the receiver signal increases, the AVC bias voltage also increases, and in turn, the negative bias to RF and IF amplifiers is increased, thereby reducing their gain. The output of the receiver is, thus, maintained constant. The simple AVC circuit has a disadvantage in that it is operative even for low magnitude signals at the receiver input. This deteriorates the sensitivity of the receiver. Therefore, it is desirable that AVC should be operative only when the input signal is strong. This arrangement is made by a circuit called *delayed AVC*, which is obtained by incorporating a forward biased diode D in the AVC filter circuit as shown in Fig. 5.7.10a. The AVC is operative only if the signal is more than the diode bias voltage. The AVC characteristic can be further improved by amplifying the delayed AVC bias with the help of a dc amplifier. This is known as *amplified and delayed* AVC. The characteristics of the various types of AVC circuits are shown in Fig. 5.7.10b. Obviously amplified and delayed AVC is very much near to the idea AVC. An ideal AVC remains inoperative for signal strength between certain limits, and provides a constant output for signal strength exceeding this limit.

### Audio Amplifier

This stage consists of an RC coupled voltage amplifier followed by a push-pull power amplifier. The fidelity of the receiver is determined by the frequency response characteristic of the stage. The more the bandwidth of the stage, the better the fidelity.

## 5.8  NOISE IN AMPLITUDE MODULATED SYSTEMS

In communication systems message signal travels from the transmitter to the receiver via a channel (radio, or line). This is shown in Fig. 5.8.1. The channel introduces additive noise in the message and, hence, the message reaching the receiver becomes corrupted. As the receiver detects both noise and message signals, it reproduces a noisy message at the output. The noise characteristics of a modulation system is evaluated by a parameter known as *figure of merit* denoted by $\gamma$. It is defined as the ratio of output signal to noise ratio to input signal to noise ratio of a receiver.

AMPLITUDE MODULATION SYSTEMS 331

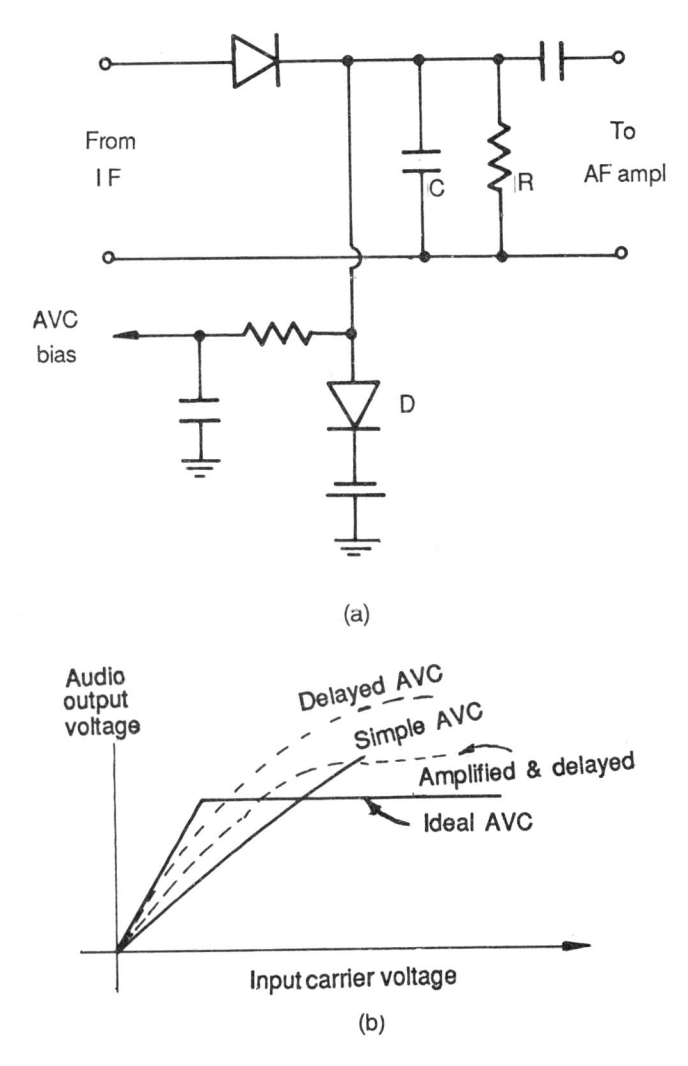

Fig. 5.7.10 (a) A Detector Circuit with AVC; (b) AVC Characteristics

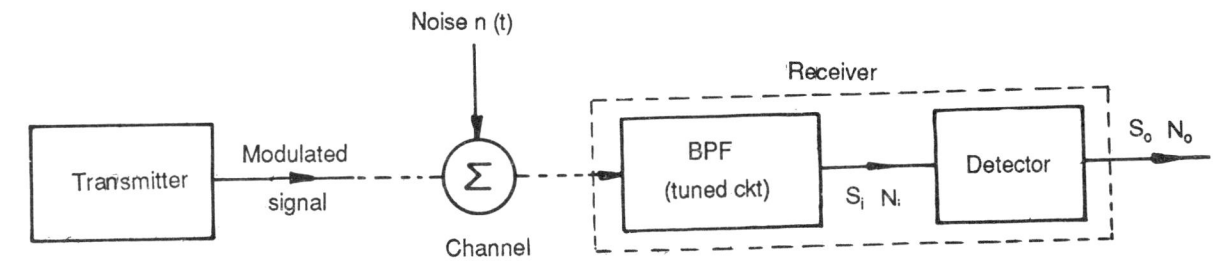

Fig. 5.8.1 Channel Noise in Communication System

$$\gamma = \frac{S_o/N_o}{S_i/N_i} \qquad (5.8.1)$$

where $N_i$ is the input noise power in a frequency range equal to the message bandwidth $2f_m$ including negative frequencies. A modulation system with higher $\gamma$ has a better noise performance and adverse effect of noise is less.

## Calculation of Figure of Merit for Various Communication Systems

The noise performance of various communication systems can be compared by evaluating their figure of merit $\gamma$. For convenience in analysing the following assumptions are made;

(i) *Channel Noise is Additive.* It is assumed that interference effect of noise is obtained by simple addition of signal $f(t)$ and noise $n(t)$.

(ii) *Channel Noise is White and Gaussin.* Channel noise $n(t)$ is white in nature, i.e., it is uniformly distributed over the entire band of frequencies under consideration. Therefore, the power density spectrum of the noise is uniform over the frequency band under consideration. Hence, the total noise power $N$ is obtained simply by multiplying the *noise power density spectrum* $\eta/2$ with the bandwidth, i.e.

$$N = (\eta/2) \times \text{bandwidth} \qquad (5.8.2)$$

The noise amplitude has a Gaussian distribution.

(iii) *Band Pass Noise at the Input of Detector*: The first stage of every receiver is a tuned circuit that acts as a band pass filter. This filter allows only the narrowband signal centered about carrier frequency $\pm f_c$ and rejects the other frequencies. Therefore, the noise signal lying outside this band is also rejected, and the bandwidth of the noise at the demodulator input is same as that of the modulated signal. The white noise at the input of the detector has a constant power density spectrum over the passband as shown in Fig. 5.8.2. The power density is given by,

$$S_{ni} = \frac{\eta}{2} \qquad (5.8.3)$$

(iv) *The Input Noise Power $N_i$*: For evaluating the figure of merit (Eq. 5.8.1) *the input noise power $N_i$ is determined for baseband* i.e., $2f_m$ Hz (including negative frequencies). From Eq. 5.8.2.

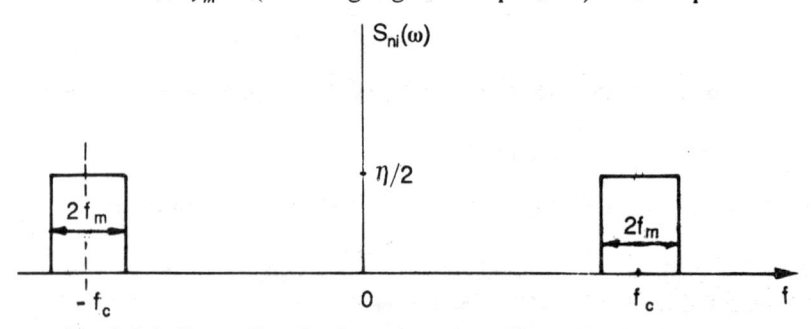

Fig. 5.8.2 Power Density Spectrum of the Bandpass White Noise

$$N_i = (\eta/2)(2f_m) = \eta f_m \qquad (5.8.4)$$

Equation 5.8.4 will be taken as *reference input noise power* for evaluating $\gamma$ of various AM and FM receivers. However, $N_i$ in Eq. 5.8.4 is the true input noise power only for the case of SSB systems.

## Noise Calculation for Various AM Systems

The performance of various amplitude modulated systems can be studied by evaluating their individual figure of merit.

**1. DSB-SC:** The DSB-SC system uses synchronous detection at the receiver. A schematic diagram of the transmitter and receiver is shown in Fig. 5.8.3.

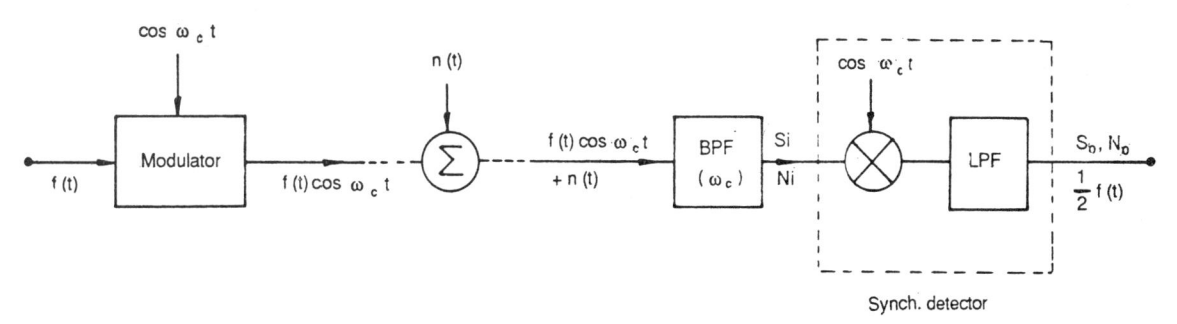

Fig. 5.8.3 A DSB-SC System

**Noise Power:** The input noise power is given by Eq. 5.8.4. Now we will evaluate the noise power $N_o$ at the output of synchronous detector. The noise signal at the input of this detector is a bandpass noise given by,

$$n(t) = n_c(t) \cos \omega_c t - n_s(t) \sin \omega_c t$$

This signal is multiplied by $\cos \omega_c t$ in the synchronous detector. The multiplied signal $n_d$ is given by

$$\begin{aligned} n_d(t) &= n(t) \cos \omega_c t = [n_c(t) \cos \omega_c t - n_s(t) \sin \omega_c t] \cos \omega_c t \\ &= n_c(t) \cos^2 \omega_c t - n_s(t) \cos \omega_c t \sin \omega_c t \\ &= \frac{1}{2} n_c(t)[1 + \cos 2\omega_c t] - \frac{1}{2} n_s(t) \sin 2\omega_c t \\ &= \frac{1}{2} n_c(t) + \frac{1}{2} n_c(t) \cos 2\omega_c t - \frac{1}{2} n_s(t) \sin 2\omega_c t \end{aligned}$$

The noise signal $n_d(t)$ is passed through a low pass filter. The terms $\frac{1}{2} n_c \cos 2\omega_c t$ and $\frac{1}{2} n_s(t) \sin 2\omega_c t$ centered near $\pm 2\omega_c$ are filtered out by the low pass filter, and hence, the noise power $n_o(t)$ appearing at the output of the synchronous detector is only $\frac{1}{2} n_c(t)$ i.e.,

$$n_o(t) = \frac{1}{2} n_c(t) \tag{5.8.5}$$

Since the power density spectrum is proportional to the mean square value of the signal, the power density spectrum of $n_o(t)$ is related to the power density spectrum of $n_c(t)$ as follows:

$$S_{n_o}(\omega) = \frac{1}{4} S_{n_c}(\omega) \tag{5.8.6}$$

It will be seen later in Ex. 5.8.1a that for a DSB-SC system,

$$S_{n_c}(\omega) = \eta, \quad |f| \le f_m \tag{5.8.7}$$

Substituting this value of $S_{n_c}(\omega)$ in Eq. 5.8.6, we get

$$S_{n_o}(\omega) = \frac{\eta}{4}, \quad |f| \le f_m \tag{5.8.8}$$

The bandwidth of the baseband signal at the output of the detector is $2 f_m$. Hence, the output noise power $N_o$ is given by,

$$N_o = S_{n_o}(\omega) \times (\text{Bandwidth}) = \frac{\eta}{4} \times 2 f_m = \frac{\eta f_m}{2} \tag{5.8.9}$$

From Eq. 5.8.4 we have

$$N_i = \eta f_m$$

Therefore,

$$\frac{N_o}{N_i} = \frac{1}{2} \tag{5.8.10}$$

**Signal Power**: The modulated signal at the input of the detector is $f(t) \cos\omega_c t$. Therefore, the input signal power $S_i$ is the ms value of $f(t) \cos \omega_c t$

$$S_i = \overline{[f(t) \cos \omega_c t]^2} = \frac{1}{2}\overline{f^2(t)} \tag{5.8.11}$$

The signal voltage at the output of the detector is $\frac{1}{2}f(t)$ (refer Eq. 5.1.2). Therefore, the output signal power is the m.s. value of $\frac{1}{2}f(t)$,

$$S_o = \overline{\left[\frac{1}{2}f(t)\right]^2} = \frac{1}{4}\overline{f^2(t)} \tag{5.8.12}$$

Comparing with 5.8.11

$$S_o = \frac{1}{2} S_i \tag{5.8.13a}$$

or

$$\frac{S_o}{S_i} = \frac{1}{2} \tag{5.8.13b}$$

Figure of merit $\gamma$ is given by,

$$\gamma = \frac{S_o/N_o}{S_i/N_i} = \frac{S_o/S_i}{N_o/N_i}$$

Substituting $(S_o/S_i)$ from Eq. 5.8.13b and $(N_o/N_i)$ from Eq. 5.8.10, we get

$$\gamma = 1 \tag{5.8.13c}$$

Thus the figure of merit for a DSB-SC system is unity. Therefore, the $S/N$ ratios at the input and output of the detector are identical, and there is no improvement in $S/N$ ratio.

## Example 5.8.1

Show that for a DSB-SC system, the power densities of various components of bandpass noise are related as

$$S_{n_c}(\omega) = S_{n_s}(\omega) = 2S_n(\omega) = \eta, \quad |\omega| \leq \omega_m \quad (5.8.14)$$

Also show that for an SSB-SC system,

$$S_{n_c}(\omega) = S_{n_s}(\omega) = S_n(\omega) = \frac{\eta}{2}, \quad |\omega| \leq \omega_m \quad (5.8.15)$$

The power densities in Eq. 5.8.14, and 5.8.15 are zero elsewhere except for $|\omega| \leq \omega_m$

**Solution**

The bandpass noise signal is represented as

$$n(t) = n_c(t) \cos \omega_c t - n_s(t) \sin \omega_c t$$

The power density spectrum of $n(t)$ is related to the power density spectrums of $n_c(t)$ and $n_s(t)$ as given below (Eq. 4.13.9)

$$S_{n_c}(\omega) = S_{n_s}(\omega) = \begin{cases} S_n(\omega + \omega_c) + S_n(\omega - \omega_c), & |\omega| \leq \omega_m \\ 0 & \text{elsewhere} \end{cases}$$

Therefore, $S_{n_c}(\omega)$ and $S_{n_s}(\omega)$ can be obtained simply by shifting the spectrum $S_n(\omega)$ by an amount $-\omega_c$ and $+\omega_c$ and then, superposing the two shifted spectra, as shown in Fig. 5.8.4. It is obvious from the superposed spectrum shown in Fig. 5.8.4d that,

$$S_n(\omega + \omega_c) + S_n(\omega - \omega_c) = 2S_n(\omega) = \eta, \quad |\omega| \leq \omega_m$$

Therefore, using the relations defined by Eq. 4.13.9, we get

$$S_{n_c}(\omega) = S_{n_s}(\omega) = 2S_n(\omega) = \eta, \text{ for } |\omega| \leq \omega_m$$

The power densities are zero elsewhere. This proves Eq. 5.8.14.

In a similar way, we can proceed for SSB-SC system. The spectrum $S_n(\omega)$ for a SSB-SC system is shown in Fig. 5.8.5a in which the spectrum exists only for upper sidebands. The spectrum has been shifted by $-\omega_c$ in Fig. 5.8.5b, and by $+\omega_c$ in Fig. 5.8.5c. The superposition of the shifted spectra is shown in Fig. 5.8.5d. It is obvious from the Fig. 5.8.5d and Eq. 4.13.9 that,

$$S_{n_c}(\omega) = S_{n_s}(\omega) = S_n(\omega) = \frac{\eta}{2}, \quad |\omega| \leq \omega_m$$

The power densities are zero elsewhere. This proves Eq. 5.8.15

## 2. SSB-SC

Like DSB-SC systems, the SSB-SC system too uses synchronous detection at the receiver, and hence, the procedure for evaluating the figure of merit is similar.

**Noise-Power:** The input noise power $N_i$ is given by Eq. 5.8.4. The noise power density at the output of the synchronous detector is same as given by Eq. 5.8.6 for the DSB-SC case.

$$S_{n_o}(\omega) = \frac{1}{4} S_{n_c}(\omega)$$

The power density spectrum $S_{n_c}(\omega)$ for SSB-SC signal is equal to $\eta/2$ (refer Eq. 5.8.15). Hence,

**336** COMMUNICATION SYSTEMS: Analog and Digital

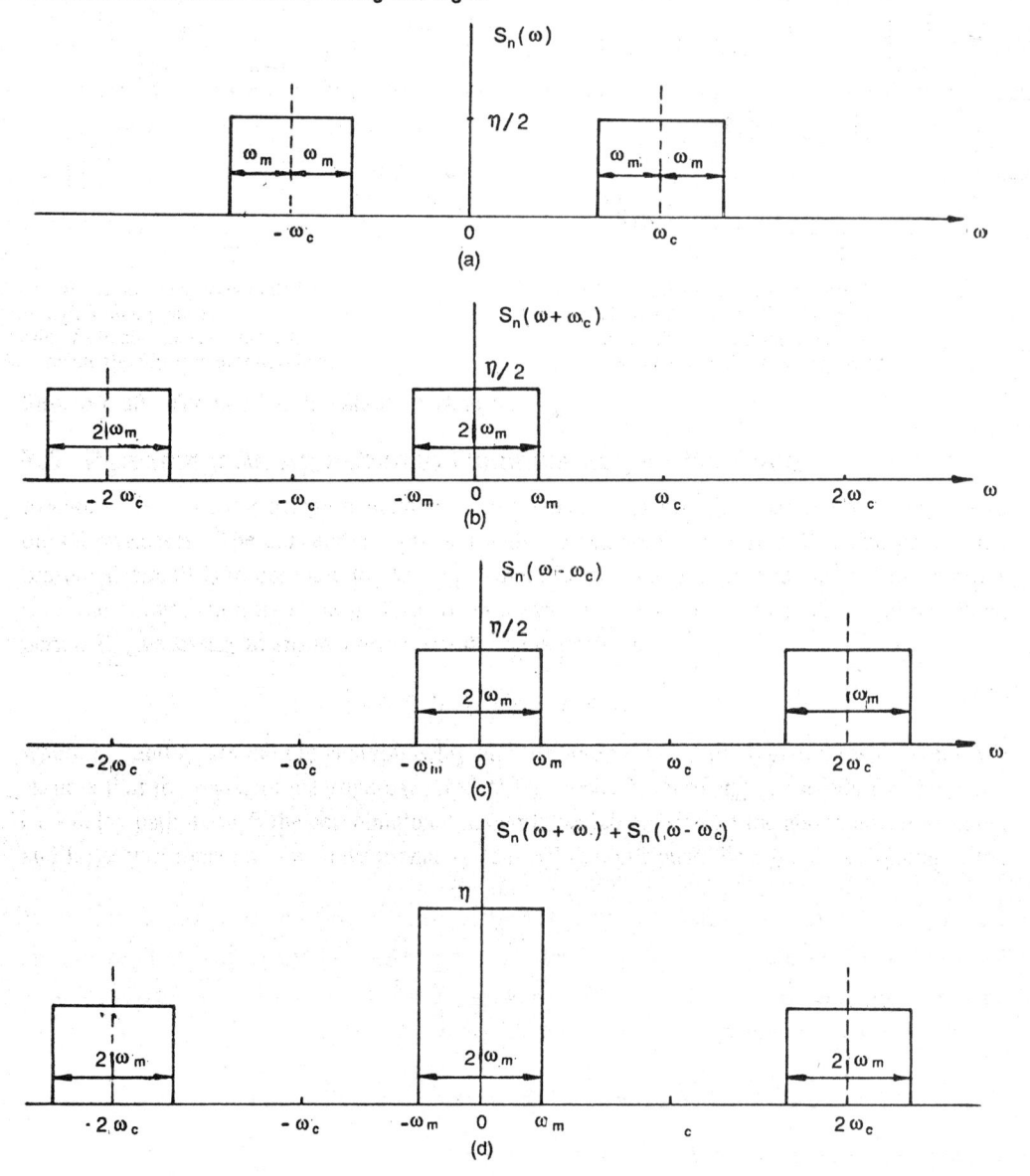

**Fig. 5.8.4** Noise Power Density in SSB-SC System (a) Spectrum of $n(t)$ (b) Spectrum (a) Shifted by $-\omega_c$ (c) Spectrum (a) shifted by $+\omega_c$ (d) Super position of (b) and (c)

$$S_{n_o}(\omega) = \frac{1}{4} \times \frac{\eta}{2} = \frac{\eta}{8} \qquad (5.8.16)$$

The bandwidth of the detected signal at the output of the detector is $2 f_m$. It should be kept in mind that both SSB-SC and DSB-SC produce identical spectra at the synchronous detector output. This is obvious from Figs 5.1.2 and 5.2.12. In both the cases, the detected output is $1/2 f(t)$, with identical spectra. Therefore, the noise power at the output of the detector is

$$N_o = S_{n_o}(\omega) \times (\text{bandwidth}) \qquad (5.8.17)$$

$$= \frac{\eta}{8} \times 2 f_m = \frac{\eta f_m}{4} \tag{5.8.17}$$

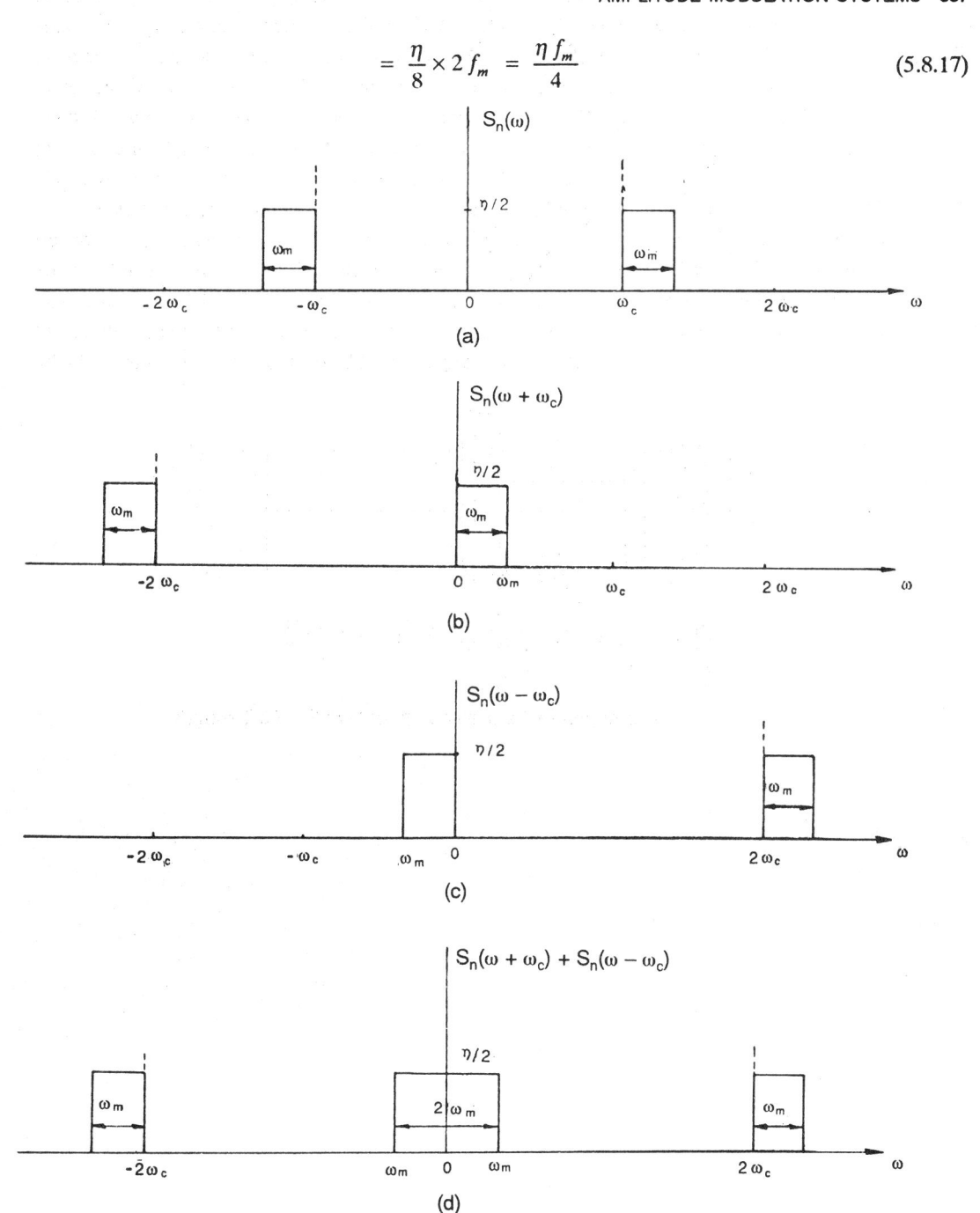

Fig. 5.8.5 Noise Power Density in SSB-SC System (a) Spectrum $n(t)$ (b) Spectrum (a) shifted by $-\omega_c$ (c) Spectrum of (a) shifted by $+\omega_c$ (d) Superposition of (b) and (c)

**338** COMMUNICATION SYSTEMS: Analog and Digital

From Eq. 5.8.4
$$N_o = \eta f_m$$
Hence,
$$N_o = \frac{N_i}{4} \tag{5.8.18}$$

**Signal Power**: The signal power at the input and output of the synchronous detector in a SSB-SC receiver can be evaluated by observing the spectrum shown in Fig. 5.2.12. The signal power is proportional to the area under the power density spectrum (Parseval's theorem). The power density spectrum of the signal is proportional to square of the Fourier transform. It is obvious from Fig. 5.2.12b that the area under the power density spectrum of a SSB-SC signal is same as the area under the power density spectrum $F(\omega)$. Hence, the power of the SSB-SC signal is same as the power (m.s. value) of the baseband signal $f(t)$, i.e.,

$$S_i = \overline{\varphi^2_{SSB}(t)} = \overline{f^2(t)} \tag{5.8.19}$$

The detector output in the SSB-SC system is $1/2 f(t)$. Hence, the output of the detector has the signal power

$$S_o = \overline{\left[\frac{1}{2} f(t)\right]^2} = \frac{1}{4} \overline{f^2(t)} = \frac{1}{4} S_i$$

or
$$\frac{S_o}{S_i} = \frac{1}{4} \tag{5.8.20}$$

The figure of merit is given by

$$\gamma = \frac{S_o/N_o}{S_i/N_i} = \frac{S_o/S_i}{N_o/N_i} \tag{5.8.21a}$$

Substituting $S_o/S_i$ from 5.8.20, and $N_o/N_i$ from Eq. 5.8.18,
$$\gamma = 1 \tag{5.8.21b}$$

This is same as the figure of merit of the DSB-SC system, which means both are identical from noise performance point of view inspite of the fact that SSB-SC has half bandwidth. This is because, the large bandwidth in the DSB-SC system increases the signal as well as noise power.

### 3. Amplitude-Modulation System (envelope detector)

In amplitude modulation system, a large carrier is accompanied with the two sidebands. If $n_i(t)$ is the additive noise signal, the input to the detector is given by

$$\varphi_{AM}(t) = [A + f(t)] \cos \omega_c t + n_i(t) \tag{5.8.22}$$

*The input signal power $S_i$ is given as,*

$$S_i = \text{ms value of carrier} + \text{ms value of sidebands}$$
$$= \frac{1}{2}\left[A^2 + \overline{f^2(t)}\right] \tag{5.8.23}$$

**Output Power**: The output of the envelope detector (popularly used in AM receivers) will be the envelope of the AM signal $\varphi_{AM}(t)$ given in Eq. 5.8.22. Therefore, in order to find the detected output, we need to find the envelope of $\varphi_{AM}(t)$. Substituting quadrature representation for $n_i(t)$, Eq. 5.8.22 becomes

$$\varphi_{AM}(t) = [A + f(t)] \cos \omega_c t + n_c(t) \cos \omega_c t - n_s(t) \sin \omega_c t$$
$$= [A + f(t) + n_c(t)] \cos \omega_c t - n_s(t) \sin \omega_c t \qquad (5.8.24)$$

After trigonometric manipulation Eq. 5.8.24 can be written as

$$\varphi_{AM}(t) = A(t) \cos \{\omega_c t + \psi(t)\} \qquad (5.8.25)$$

where $A(t)$, and $\psi(t)$, respectively, are randomly time varying amplitude and phase angle of $\varphi_{AM}(t)$. They are given by

$$A(t) = \sqrt{[A + f(t) + n_c(t)]^2 + n_s^2(t)} \qquad (5.8.26a)$$

$$\psi(t) = \tan^{-1}\left[\frac{n_s(t)}{A + f(t) + n_c(t)}\right] \qquad (5.8.26b)$$

The time varying amplitude $A(t)$ is the envelope of $\varphi_{AM}(t)$ and hence, the output of the envelope detector will be the envelope $A(t)$. The envelope $A(t)$ has both signal and noise components. The noise performance depends on the relative magnitudes of the signal and noise. The analysis can be carried out for two cases, viz.
(i) small noise case and (ii) large noise case.

### (i) Small-Noise Case:

In this case, noise is taken to be much smaller than the signal,

$$n_i(t) << [A + f(t)]$$

A phasor-representation of envelope $A(t)$ is shown in Fig. 5.8.6. The noise component $n_c(t)$ is shown to be in phase with signal $[A + f(t)]$, whereas, $n_s(t)$ is in phase quadrature. Since $n_i(t) << [A + f(t)]$, the noise component $n_s(t)$ is also much smaller than $[A + f(t)]$.

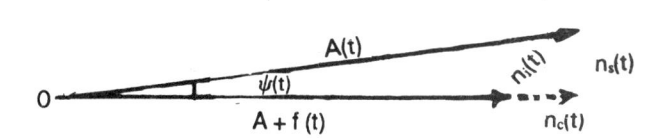

Fig. 5.8.6 Phasor Diagram of the Envelope $A(t)$

It is evident from the phasor diagram that if $n_s(t)$ is much small, then $\psi(t)$ is also much small and may be taken as zero. When $\psi(t) = 0$, the envelope $A(t)$ becomes,

$$A(t) = A + f(t) + n_c(t) \qquad (5.8.27)$$

Therefore the output of the envelope detector contains a useful signal $f(t)$ and a noise component $n_c(t)$. The carrier amplitude $A$ carries no useful information. The signal power $S_o$ and the noise power $N_o$, at the output of the detector, are evaluated as follows:

**Output Signal Power $S_o$:** It is the m.s. value of useful signal $f(t)$, i.e.,

$$S_o = \overline{f^2(t)} \qquad (5.8.28)$$

**340** COMMUNICATION SYSTEMS: Analog and Digital

***Output Noise Power:*** The noise signal $n_o(t)$ at the output of the detector is $n_c(t)$ with the power density given by

$$S_{n_o}(\omega) = S_{n_c}(\omega) \qquad (5.8.29a)$$

As the AM system contains both sidebands, $S_{n_c}(\omega)$ will be identical to that of DSB-SC system, given by Eq. 5.8.7, i.e., $S_{n_o}(\omega) = S_{n_c}(\omega) = \eta$. The output noise power is given as,

$$N_o = S_{n_o}(\omega)(BW) = \eta \cdot (2 f_m) = 2\eta f_m \qquad (5.8.29b)$$

The figure of merit $\gamma$ is obtained with the help of Eqs 5.8.23, 5.8.4, 5.8.28, and 5.8.29b:

$$\gamma = \frac{S_o/N_o}{S_i/N_i} = \frac{\overline{f^2(t)}/(2\eta f_m)}{\frac{1}{2}\left[A^2 \overline{f^2(t)}\right]/\eta f_m}$$

$$= \frac{\overline{f^2(t)}}{A^2 + \overline{f^2(t)}} \qquad (5.8.30)$$

From Eq. 5.8.30, it is clear that the noise performance improves with a reduction in carrier amplitude $A$ and it is maximum when $A = 0$, which is equivalent to the suppressed carrier AM system. Thus, AM system remains always inferior to the suppressed system from the point of view of 'noise performance'. The greatest value of $\gamma$ that can be achieved will depend on the minimum possible value of carrier amplitude $A$. The minimum value of $A$ is equal to the maximum value of $f(t)$ for avoiding the overmodulation in the envelope detector. Therefore, the best noise performance is achieved when carrier amplitude A is equal to the maximum value of $f(t)$ -i.e., for 100% modulation. This is verified by Ex. 5.8.2.

### Example 5.8.2

### Single-tone modulation

Derive an expression for the figure of merit $\gamma$ when the modulating signal $f(t)$ is a single-tone sinusoid given by $f(t) = m_a A \cos \omega_m t$, where $m_a$ is the modulation index, and $A$ is the carrier amplitude. Find the value of $\gamma$ when the depth of modulation is: (a) 100%; (b) 50%; and (c) 30%.

**Solution**

$$f(t) = m_a A \cos \omega_m t \qquad (5.8.31)$$

The power content in $f(t)$ is given as

$$\overline{f^2(t)} = \frac{m_a^2 A^2}{2}$$

From Eq. 5.8.30, the figure of merit is given as

$$\gamma = \frac{m_a^2 A^2/2}{A^2 + \frac{m_a^2 A^2}{2}}$$

or

$$\gamma = \frac{m_a^2}{2 + m_a^2} \qquad (5.8.32)$$

(a) When $m_a = 100\%$, i.e., $m_a = 1$

$$\gamma = \frac{1}{2+1} = \frac{1}{3} = 0.33$$

Thus, the AM system must transmit three times more power as compared to the suppressed carrier AM system in order to achieve the same noise performance.

(b) $\qquad m_a = 50\% = \frac{1}{2}$

$$\gamma = \frac{1/2}{2+1/2} = \frac{1}{5} = 0.2$$

which is less than the one in case (a)

(c) $\qquad m_a = 30\% = 0.3$

$$\gamma = \frac{0.3}{2+0.3} = \frac{3}{23} = 0.13$$

which is less than even case (b).

Thus, *the maximum value of $\gamma$ in envelope detector is 0.33 for 100% modulation.*

**Synchronous Detection of AM System :** Even if the synchronous detection is used for AM demodulation, the noise performance is identical to the envelope detector (Prob. 5.12). Moreover, the synchronous detection is both complex and costly; hence, it is never used for AM detection.

**(ii) Large Noise Case :**

In this case, $n_i(t) >> [A + f(t)]$

Therefore quadrature components of $n_i(t)$ are also much larger than $[A + f(t)]$. Here, the noise terms dominate, and the performance of the envelope detector is completely different from the low noise case. The envelope of the modulated signal at the input of the detector is given by Eq. 5.8.26a,

$$A(t) = \sqrt{[A + f(t) + n_c(t)]^2 + n_s(t)^2}$$

$$= \sqrt{[A + f(t)]^2 + 2n_c(t)[A + f(t)] + n_c^2(t) + n_s^2(t)}$$

Since the noise dominates over the signal, the independent signal term $[A + f(t)]^2$ is very small as compared to the other terms. Therefore, $[A + f(t)]^2$ may be ignored, and the expression for $A(t)$ becomes

$$A(t) = \sqrt{n_c^2(t) + n_s^2(t) + 2n_c(t)[A + f(t)]}$$

$$= \sqrt{[n_c^2(t) + n_s^2(t)]\left[1 + \frac{2n_c(t)[A + f(t)]}{n_c^2(t) + n_s^2(t)}\right]}$$

Putting

$$R(t) = \sqrt{n_c^2(t) + n_s^2(t)}; \text{ and; } \theta(t) = \tan^{-1}\left(\frac{n_s(t)}{n_c(t)}\right)$$

we get

$$A(t) = \sqrt{R^2(t)\left[1 + 2\{A + f(t)\}\cdot\frac{n_c(t)}{R^2(t)}\right]}$$

$$= \sqrt{R^2(t)\left[1 + \frac{2\{A + f(t)\}}{R(t)}\cos\theta(t)\right]}$$

$$= R(t)\left[1 + \frac{2\{A + f(t)\}}{R(t)}\cos\theta(t)\right]^{\frac{1}{2}} \tag{5.8.33}$$

Since the noise component $R(t) \gg 2\{A + f(t)\}$, Eq. 5.8.33 can be written in approximate form as

$$A(t) = R(t)\left[1 + \frac{1}{2}\frac{2\{A + f(t)\}}{R(t)}\cos\theta(t)\right]$$

$$= R(t) + A\cos\theta(t) + f(t)\cos\theta(t) \tag{5.8.34}$$

It is evident from Eq. 5.8.34, that envelope $A(t)$ which appears at the output of the envelope detector has no exclusive term proportional to the baseband (modulating) signal $f(t)$. The term $f(t)$ is multiplied with a large noise term $\cos\theta(t)$ and hence, $f(t)$ cannot be separated from the noise. As the modulating term $f(t)$ is completely mingled with noise, it carries no useful information. The loss of the message $f(t)$ in an envelope detector due to the presence of the large noise is referred to as the *threshold effect*.

### Threshold Effect in an Envelope Detector

We have seen that when a noise is large as compared to the signal at the input of the envelope detector, the detected output has a message signal completely mingled with the noise. It means that if the input signal-to-noise ratio $(S_i / N_i)$ is below a certain level, called *threshold level*, the noise dominates over the message signal. *Threshold* is defined as value of the input *signal to noise ratio $(S_i / N_i)$ below which the output signal to noise ratio $(S_o / N_o)$ deteriorates much more rapidly than the input* signal to noise ratio. The threshold effect starts in an envelope detector whenever the carrier power to noise power ratio approaches unity or less. It is important to stress that the threshold effect is the property of an envelope detector; on the other hand, *such an effect is not observed in a coherent detector* (prob 5.13).

## 5.9 COMPARISON OF VARIOUS AM SYSTEMS

So far we have studied the various forms of amplitude modulation, accordingly we can divide them into two broad categories, viz., suppressed carrier AM system; and AM systems with a large carrier amplitude. The suppressed carrier system can be further subdivided into three parts: (a) DSB-SC; (b) SSB-SC; and (c) VSB-SC. It is worthwhile to compare the various AM systems based on the following points:

### (i) Receiver End

For the receiver, an AM is advantageous, because the envelope detector used in such a receiver is very simple and inexpensive. Therefore, the AM system is preferably used in public communication systems where a transmitter is associated with a large number of receivers. The receivers cost is important in such cases, as, for example in a radio broadcast system, a receiver is needed almost in every house, and can be treated as an essential commodity. Therefore, the cost of the receiver must be low. AM systems with a large carrier are advantageous in this case. The receivers of a suppressed carrier system are complex and costly as they, need additional synchronizing circuits. Therefore, suppressed carrier system is undesirable for public communication systems.

## (ii) Transmitter End

The suppressed carrier systems need low power transmitters as they transmit only sideband power and no carrier power is transmitted. The low power transmitters used in suppressed carrier systems are less expensive than the large power AM transmitters. Suppressed carrier systems are useful in point to point communication where we need many transmitters but only few receivers per transmitter. The expensive receivers may be justified in this case for the gain achieved by less expensive transmitters.

## (iii) Generation of Modulated Signals

AM is easier to generate, even for large power level. On the other hand, balanced modulators that are required to generate suppressed carrier signals are more difficult to design. Further, SSB-SC is more difficult to generate than DSB-SC. VSB-SC is easier to generate than SSB-SC, but at the cost of higher channel-width.

## (iv) Bandwidth

SSB is most advantageous from bandwidth point of view. This system has a minimum bandwidth, although its generation is most difficult. SSB-SC is used in long-range high-frequency communications, especially in audio-communication, where phase-distortion is not significant. Amateur operated radio uses SSB-SC systems. VSB-SC system is next to the SSB-SC. It has bandwidth between DSB and SSB and has less problems in its generation. Picture signal transmission in televisions uses vestigial sideband transmission that cuts down the bandwidth of a channel as compared to the DSB-SC system. The DSB system has a maximum bandwidth. It should be recalled that the higher bandwidth of a channel *reduces the number of available channels in a given frequency space.*

## (v) Selective Fading

The random variation of a signal strength at the receiver due to multipath propagation is called fading. The signals from a transmitter arrive at the receiver by multiple paths, each path having a different length. The path length varies randomly due to random changes in the ionospheric conditions. Signals arriving at the receiver by different path lengths randomly differ in phase. The strength of a resultant received signal varies randomly with time and this produces a fading effect. The fading is frequency sensitive and hence carrier and sidebands may suffer by different amounts of fading. This is called *selective fading*. This type of fading disturbs the magnitude ratio of carrier and sidebands. In AM, carrier amplitude may be considerably reduced as compared to sidebands due to more sever fading in the former. This effect produces over modulation in AM signal, and when detected by an envelope detector, it will be heavily distored. Also, the selective fading becomes severe at high frequencies. Therefore, at high frequencies, AM is inferior to suppressed carrier systems.

Selective fading produces more distortion in the DSB-SC system as against in the SSB-SC system, because it disturbs the phase relationship of two sidebands of the DSB-SC signal. Since SSB-SC has only one sideband this type of distortion is minimum.

## (vi) Transmission Efficiency

Efficiency of a suppressed carrier system is 100%, whereas, in an AM system the maximum efficiency is only, 33.3% corresponding to $m_a = 1$.

## (vii) Noise Performance

SSB-SC and DSB-SC systems have $\gamma = 1$, hence they are equivalent from the noise performance point of view.

The AM system (with a large carrier) using the envelope detector has maximum $\gamma = \frac{1}{3}$, and also it suffers from the threshold effect. Therefore, the AM system is always inferior to suppressed carrier systems from the noise performance point of view.

### (viii) Non-Linear-Distortion

Non-linear distortion occurs due to non-linear characteristics of the systems. This type of distortion causes two undesired effects: (a) harmonic generation; and (b) generation of intermodulation frequencies. In the harmonic distortion, harmonics of the input frequencies are generated. In intermodulation distortion, the sum and difference of input frequencies, and their harmonics are generated. Obviously, non-linear distortions will be more severe in the systems involving signals with more number of frequency terms, because this causes more number of harmonics and intermodulation frequency terms. Therefore, non-linear distortion is expected to be maximum in AM and minimum in SSB-SC systems, whereas in DSB-SC it is in between the AM and SSB-SC. Non-linear distortion not only modifies the quality of a signal, but also it may produce interchannel cross-talk in frequency division multiplexing. This problem is controlled by suitable design of filters.

## 5.10 FREQUENCY DIVISION MULTIPLEXING (FDM)

As discussed in the beginning of this chapter multiplexing is the process of simultaneous transmission of several messages over a common channel without interference. Let us suppose, that we want to transmit $n$ messages each one band limited to $\omega_m$, e.g., telephone signals from $n$ subscribers. Each message is centered near zero frequency. The spectra of individual messages are shown in Fig. 5.10.1a. The bandwidth of each signal depends on the type of message. For example, a telephonic message has a bandwidth of 3 kHz and a broadcast signal has a bandwidth of 5 kHz. If these $n$ signals are transmitted simultaneously over a common channel (line or radio) without multiplexing, they will interfere with each other, and no useful information will be produced at the receiver. However, they can be transmitted without interference if they are multiplexed. In frequency division multiplexing, each baseband signal is translated by analog modulation (amplitude or angle) to different carrier frequencies. The corresponding spectrum of multiplexed signals using AM is shown in Fig. 5.10.1b. Each carrier is separated from the neighbouring one by at least $2\omega_m$. The multiplexed signals can be transmitted over a common channel without interference.

Figure 5.10.2 shows the schematic diagram of a frequency division multiplexing system. At the transmitter, each message is superimposed on a separate carrier wave using AM. At the receiver, the various carrier frequencies are selected using bandpass filters tuned to appropriate carrier frequencies, and demodulated by separate detectors.

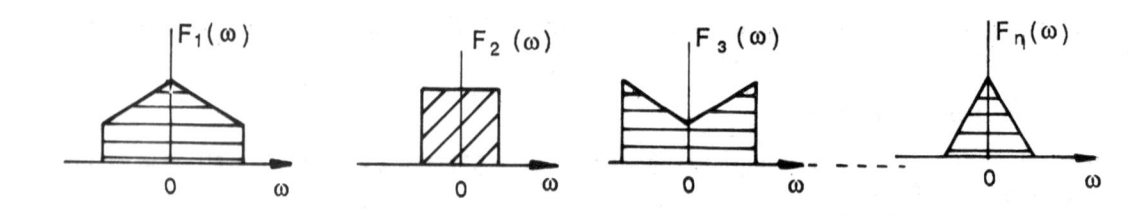

(a)

AMPLITUDE MODULATION SYSTEMS 345

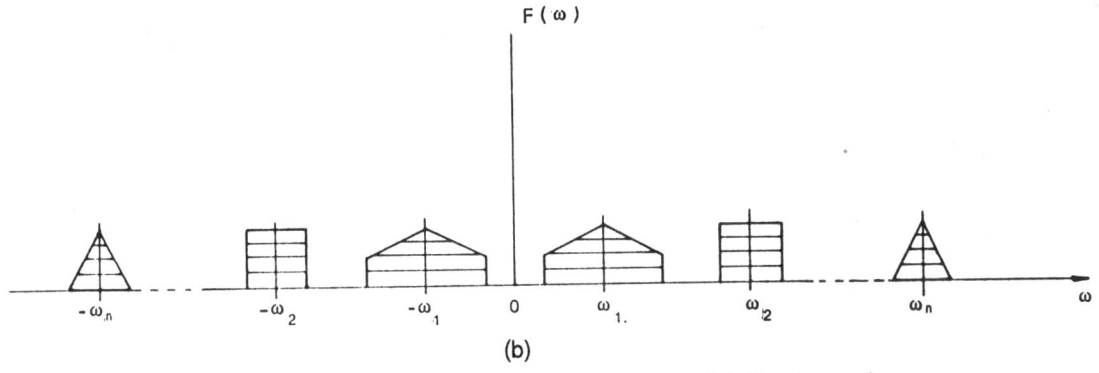

Fig. 5.10.1 (a) Spectrum of n-messages  (b) Spectrum of Frequency Division Multiplexed Signal

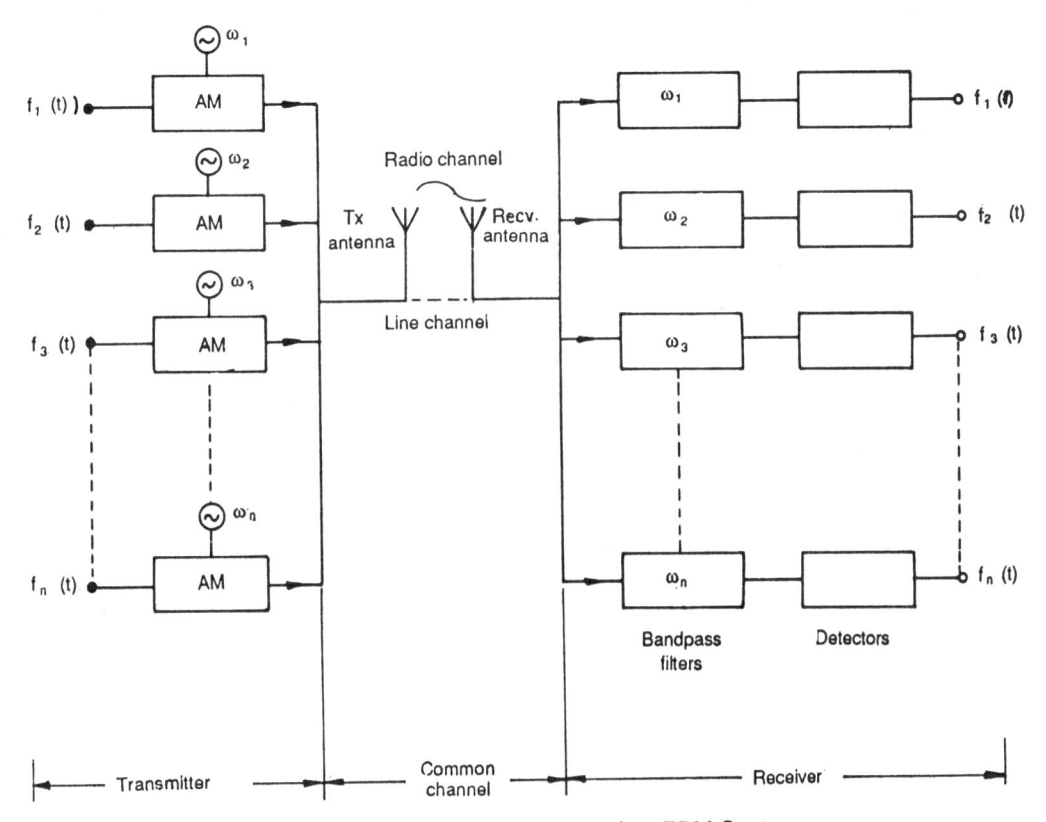

Fig.5.10.2 Schematic Diagram of an FDM System

## PROBLEMS

5.1  A voltage $v = 200(1 + 0.4 \sin \omega_m t) \sin \omega_c t$ is applied to a resistor of 100 ohms. Find the power dissipated by each of the frequency components present in the voltage v.

5.2  An amplitude-modulated voltage is given by $v = 50(1 + 0.2 \cos 100 t + 0.01 \cos 3500 t) \cos 10^6 t$. State all frequency components (in Hz) present in the voltage, and find modulation index for each modulating voltage term. What is the effective modulation index of v.

5.3 An amplitude-modulated amplifier provides an output of 106 watts at 100% modulation. The internal loss is 20 watt.
(a) What is the unmodulated carrier power?
(b) What is the sideband power?

5.4. Find the various frequency components and their amplitudes in the voltage given below:

$e = 50(1+0.7\cos 5000t - 0.3\cos 1000t) \sin 5 \times 10^6 t$. Draw the single sided-spectrum. Also evaluate the modulated and sideband powers.

5.5 Define amplitude modulation

The rms value of a radio frequency voltage is 200 volts before modulation. When it is modulated by a sinusoidal audio frequency voltage, its rms voltage becomes 242 volts. Calculate the modulation index.

5.6 An AM transmitter has an unmodulated carrier power of 10 kW. It can be modulated by a sinusoidal modulating voltage to a maximum depth of 40%, without overloading. If the maximum modulation index is reduced to 30% what is the extent upto which the unmodulated carrier power can be increased to avoid overloading?

5.7 A sinusoidal carrier $e_o = 100 \cos(2\pi.15^5 t)$ is amplitude modulated by a sinusoidal voltage $e_m = 50 \cos(2\pi.10^3 t)$ up to a modulation depth of 50%. Calculate the frequency and amplitude of each sideland and rms voltage of the modulated carrier.

5.8 The rms voltage of a carrier wave is 5 volts before modulation, and 5.9 volts after modulation. What is the percentage of modulation? Calculate the modulated power if the unmodulated power is 2 kW.

5.9 In a collector-modulated class C amplifier, the collector dissipation is 60 watts. The amplifier efficiency is 80%. Find the unmodulated carrier power if the depth of modulation is 70%.

5.10 In a linear collector modulated amplifier, the amplitude of the modulating voltage is 60 volts, and the collector supply voltage is 150 volts. The dc collector current under an unmodulated condition is 45 amperes. The unmodulated carrier power is 1 kW: Calculate
(i) the modulation index: (ii) the circuit efficiency: (iii) the plate dissipation; and (iv) the modulated carrier power.

5.11 In a linear collector-modulated amplifier, the plate supply is 300 volts, and the dc collector current under an unmodulated condition is 20 amperes. The sinusoidal modulating voltage appearing in the collector circuit of the modulated amplifier has an amplitude of 150 volts. The unmodulated carrier power is 5 kW. Calculate (i) the modulation index; (ii) the carrier power after modulation (iii) the collector circuit efficiency, and (iv) the collector dissipation under modulated as well as unmodulated condition.

5.12 Show that the figure of merit $\gamma$ for an AM system using synchronous detection is identical to the envelope detector.

5.13 Show that an AM system using synchronous detection does not suffer from the threshold effect.

5.14 A channel has a uniform noise power density spectrum $S_n(\omega) = 0.25 \times 10^{-3}$. A DSB-SC signal with carrier frequency of 200 kHz is transmitted over this channel. The modulating signal $f(t)$ is band limited to 10 kHz. The power of the sideband signal is 5 kW. The incoming signal at the receiver is filtered through an ideal bandpass filter before it is fed to the demodulator.
(a) what is the transfer function of this filter at the receiver?
(b) Find the S/N ratio at the demodulator input and output.
(c) Find and sketch the noise power density spectrum at the demodulator output.

AMPLITUDE MODULATION SYSTEMS 347

5.15 A carrier wave of a frequency of 20 kHz is amplitude-modulated by a modulating signal $f(t) = \cos 2\pi 10^3 t + \cos 4\pi 10^3 t$. Find the expression for the corresponding SSB-SC signal.
5.16 Show that the collector circuit efficiency in a collector modulation remains unchanged after modulation.
5.17 Repeat Prob. 5.14 for SSB-SC, (a) with upper side-bands and, (ii) with lower sidebands.
5.18 Show that the squaring circuit shown in Fig. 5.1.4 cannot be used for synchronization in SSB-SC system.

# SIX
# ANGLE MODULATION SYSTEMS

## INTRODUCTION

Angle modulation is another significant process of modulation used for message transmission. The frequency modulation has an important advantage over amplitude modulation i.e. interference due to noise is considerably reduced in the former. However, this advantage of noise-immunity is at the cost of increased bandwidth and, hence comparatively a less number of channels can be accommodated in a given frequency space.

## 6.1 DEFINITION

Angle modulation is the process of varying the total phase angle of a carrier wave in accordance with the instantaneous value of the modulating (baseband) signal, keeping amplitude of the carrier constant.

Let us consider an unmodulated carrier wave given by

$$\varphi(t) = A\cos(\omega_c t + \theta_0) \qquad (6.1.1)$$

or

$$\varphi(t) = A\cos\psi \qquad (6.1.2)$$

where $\psi = (\omega_c t + \theta_0)$ is the total phase angle of the carrier wave.

Equation 6.1.2 may be considered as the real part of a rotating phasor $A e^{j\psi}$. Let it be denoted by $\tilde{\varphi}$ i.e.,

$$\tilde{\varphi} = A e^{j\psi} \qquad (6.1.3)$$

and

$$\varphi(t) = Re[A e^{j\psi}] = A Re[\cos\psi + j\sin\psi] = A\cos\psi \qquad (6.1.4)$$

The phasor $\tilde{\varphi}$ rotates at a constant angular velocity $\omega_c$ provided $\theta_0$ is independent of time. Obviously $\theta_0$ is the phase angle of the unmodulated carrier at, $t = 0$. The constant angular velocity $\omega_c$ of the phasor $\tilde{\varphi}$ is related to its total phase angle $\psi(t)$ as derived below

$$\psi = \omega_c t + \theta_0$$

differentiating the above equation, we get

$$\frac{d\psi}{dt} = \omega_c \qquad (6.1.5)$$

This derivative $\frac{d\psi}{dt}$ is constant with time for an unmodulated carrier. But, in general, this derivative may not be constant with time, rather it may vary with time. Accordingly, angular velocity of the phasor $\tilde{\varphi}$ will also change with time. This time dependent angular velocity is called *instantaneous angular velocity*, and is denoted by $\omega_i$. Thus, Eq. 6.1.5 for this case becomes

$$\frac{d\psi}{dt} = \omega_i \qquad (6.1.6)$$

or

$$\psi = \int \omega_i \, dt \qquad (6.1.7)$$

Note that $\omega_i$ is time dependent.

## Concept of Instantaneous Frequency

The time dependent angular velocity $\omega_i$ of the phasor $\varphi$ provides a time varying instaneous frequency $f_i$ of the carrier wave $\varphi(t)$. This implies that the frequency of the carrier wave changes from one cycle to another. The instantaneous frequency $f_i$ may vary with some function of time. Figure 6.1.1a shows a variation in the frequency of a carrier. The variation is linear with time from a value $\omega_0$ and $2\omega_0$. Figure 6.1.1b shows the corresponding waveform of the carrier with varying frequencies.

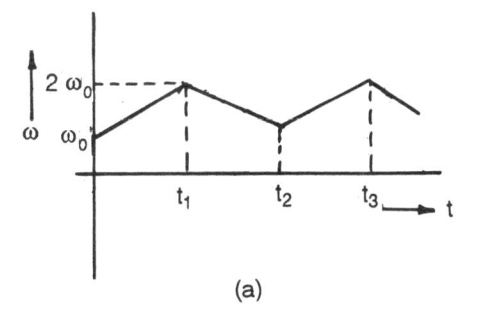

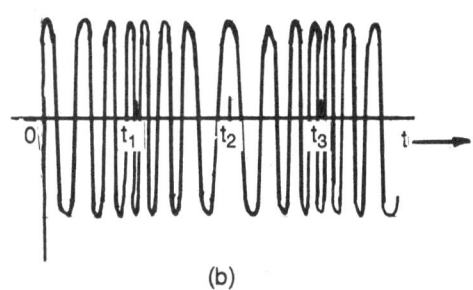

Fig. 6.1.1. Concept of Instantaneous Frequency (a) Variation of $\omega$ with Time
(b) Wave-form of a Carrier Wave with Varying Frequency

If this variation in frequency is made according to the baseband signal, the carrier wave is called the frequency modulated wave.

## Example 6.1.1

Determine and draw the instantaneous frequency of a wave having a total phase angle given by,

$$\psi(t) = 2000t + \sin 10t$$

*Solution*

The instantaneous angular velocity of the wave is the derivative of the total phase angle, i.e.,

$$\omega_i = \frac{d\psi}{dt} = \frac{d}{dt}[2000t + \sin 10t] = 2000 + 10 \cos 10t$$

This is plotted in Fig. 6.1.2

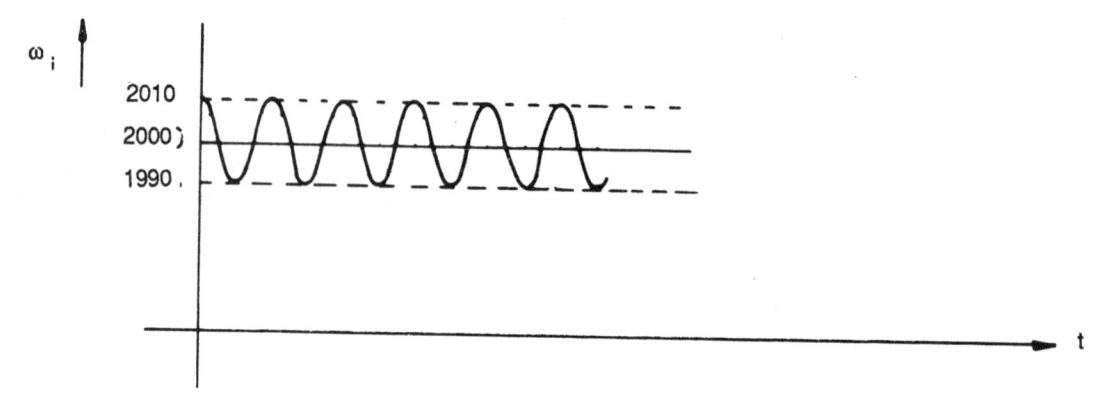

Fig. 6.1.2

The frequency $f_i = \frac{\omega_i}{2\pi}$, so the variation in $f_i$ is identical to the variation in $\omega_i$.

## Example 6.1.2

Plot the instantaneous frequency and the corresponding frequency modulated wave, modulated by a rectangular pulse, shown in Fig. 6.1.3a.

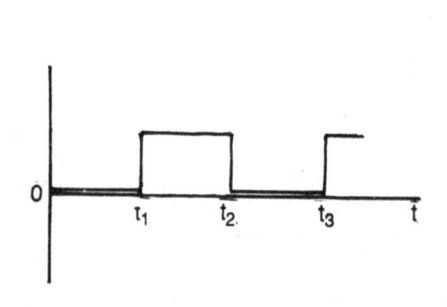

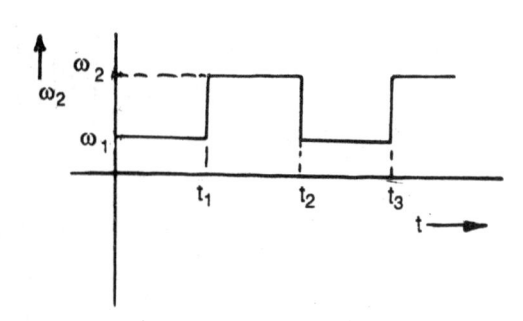

(a)

(b)

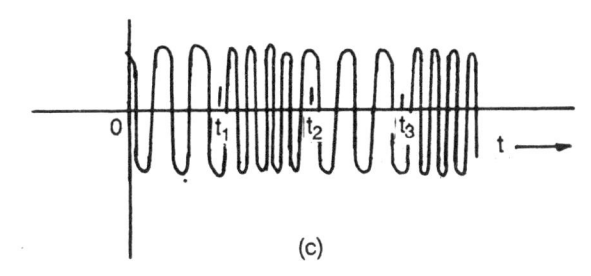

Fig. 6.1.3 (a) Rectangular Modulating Wave (b) Angular Frequency Variation (c) FM Wave

*Solution*

The variation of frequency with time is shown in Fig. 6.1.3b, and the corresponding FM wave is shown in Fig. 6.1.3c.

### Example 6.1.3

Draw the waveform of an FM wave when the modulating signal is a sawtooth wave as shown in Fig. 6.1.4a, and it modulates a carrier wave $\cos \omega_c t$.

*Solution*
FM wave is shown in Fig 6.1.4b

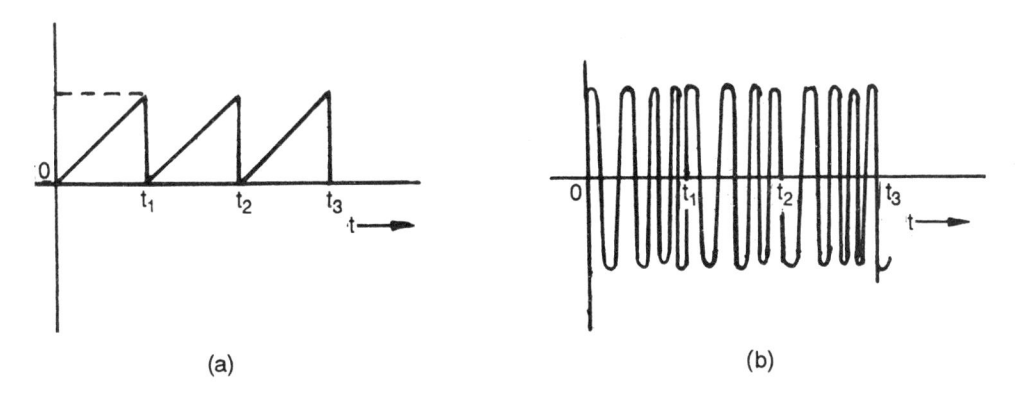

Fig. 6.1.4 (a) Saw-tooth Modulationg Signal (b) FM Wave

The frequency of the FM wave in Fig. 6.1.4b is changing linearly with time over a cycle of modulating signal.

### Types of Angle Modulation

Consider an unmodulated carrier wave given by

$$\varphi(t) = A\cos(\omega_c t + \theta_0) = A\cos\psi$$

where $\psi$ is the total phase angle of the carrier given as

$$\psi = (\omega_c t + \theta_0)$$

If this angle $\psi$ is made to vary in accordance with the instantaneous value of the modulating signal, the carrier is said to be angle modulated. There are two ways of varying the phase angle $\psi$, and, accordingly, we have two types of angle modulation.

## 1. Phase Modulation (PM)

In this type of angle modulation, the phase angle $\psi(t)$ is varied linearly with a modulating signal $f(t)$ about an unmodulated phase angle $\omega_c t$. That is to say, the instantaneous value of a phase angle $\psi_i$ is equal to the phase of an unmodulated carrier $(\omega_c t)$ plus a time varying component proportional to $f(t)$. Mathematically,

$$\psi_i(t) = \omega_c t + K_p f(t) \tag{6.1.8}$$

Note that $\theta_0$ is time independent and hence, has been ignored.

The proportionality constant $K_p$ is known as *phase sensitivity* of the modulator, expressed in radians/volts. The carrier wave, after phase modulation has the phase angle given by Eq. 6.1.8, and is represented as

$$\varphi_{PM} = A \cos \psi_i(t)$$

$$= A \cos[\omega_c t + K_p f(t)] \tag{6.1.9}$$

## 2. Frequency Modulation (FM)

In this type of angle modulation, the instantaneous frequency $\omega_i$ is varied linearly with a modulating signal $f(t)$ about an unmodulated frequency $\omega_c$. In other words the instantaneous value of the angular frequency $\omega_i$ is equal to the frequency $\omega_c$ of the unmodulated carrier plus a time varying component proportional to $f(t)$. Mathematically,

$$\omega_i = \omega_c + K_f f(t) \tag{6.1.10}$$

The term $K_f$ represents the *frequency sensitivity* of the modulator expressed in Hz/volt. Let us compare this frequency $\omega_i$ with the corresponding instantaneous frequency in a phase modulated wave which may be obtained by differentiating Eq. 6.1.8,

$$\omega_i = \frac{d\psi_i}{dt} = \omega_c + K_p \frac{df(t)}{dt} \tag{6.1.11}$$

Comparing this with Eq. 6.1.10, we find that in PM the instantaneous frequency varies linearly with the time derivative of $f(t)$ whereas in FM, it varies linearly with $f(t)$.

The total phase angle of the FM wave can be obtained by integrating Eq. 6.1.10, i.e.,

$$\psi_i = \int \omega_i \, dt = \int [\omega_c + K_f f(t)] dt$$

$$= \omega_c t + K_f \int f(t) \, dt \tag{6.1.12}$$

The corresponding FM wave is given by

$$\varphi_{FM}(t) = A \cos \psi_i$$

$$= A \cos \omega_c t + K_f \int f(t) \, dt \tag{6.1.13}$$

ANGLE MODULATION SYSTEMS 353

If we assume, for convenience, that the phase angle of the carrier at $t = 0$ is zero, then the limit of integration will be 0 to $t$. FM is then, represented by

$$\varphi_{FM}(t) = A\cos\omega_c t + K_f \int_0^t f(t)\,dt \qquad (6.1.14)$$

## Relationship between PM and FM

PM and FM are closely related in the sense that the net effect of both is variation in total phase angle. In PM, phase angle varies linearly with $f(t)$ whereas in FM phase angle varies linearly with the integral of $f(t)$. In other words, we can get FM by using PM, provided that at first the modulating signal is integrated, and then applied to the phase modulator. This is shown in the block-diagram of Fig. 6.1.5.

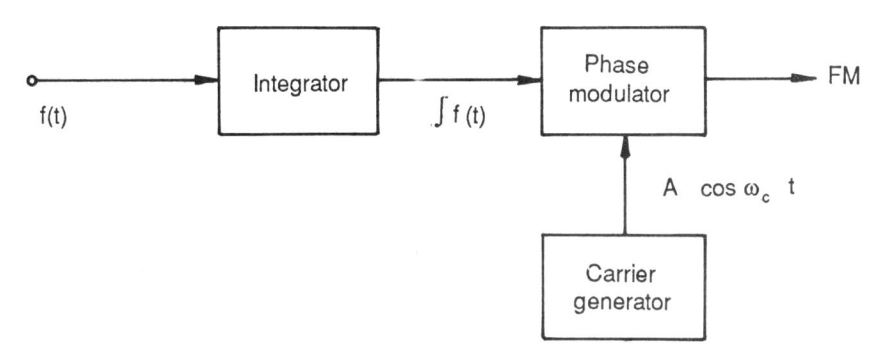

Fig. 6.1.5 FM Generation Using Phase Modulator

Equation 6.1.11 reveals that the converse is also true i.e., we can generate a PM wave using frequency modulator provided that $f(t)$ is first differentiated and then applied to the frequency modulator as shown in Fig. 6.1.6. *The phase modulated waveform can be obtained by drawing an FM wave corresponding to the derivative of $f(t)$* (Prob. 6.1c and 6.3b).

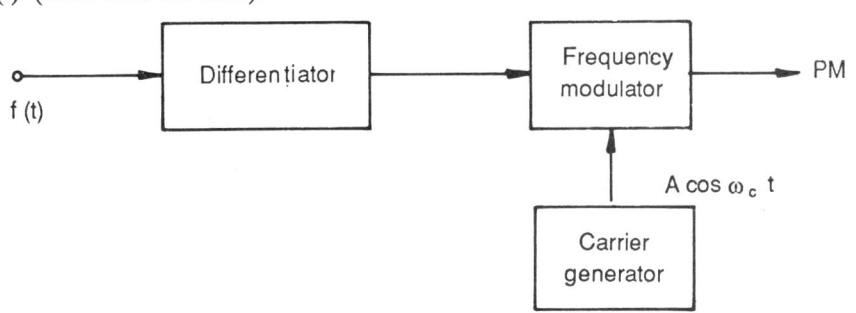

Fig. 6.1.6 PM Generation Using Frequency Modulator

## Phasor Representation of PM and FM Waves

The signals $\varphi_{PM}$ and $\varphi_{FM}$ can be considered as real part of phasors $\tilde{\varphi}_{PM}$, and $\tilde{\varphi}_{FM}$ respectively, given by (refer Eq. 6.1.3)

$$\tilde{\varphi}_{PM} = A e^{j[\omega_c t + K_p f(t)]} \qquad (6.1.15a)$$

and

$$\tilde{\varphi}_{FM} = A e^{j\left[\left(\omega_c t + K_f \int f(t) dt\right)\right]} \qquad (6.1.15b)$$

Let us assume that

$$\int f(t) dt = g(t)$$

then

$$\tilde{\varphi}_{FM} = A e^{j\left[\omega_c t + K_f g(t)\right]} \qquad (6.1.16)$$

The two categories of angle modulation, viz, FM and PM are closely related. Hence, their analysis is similar except for a slight difference. So, we will analyse FM in detail and then point out the distinguishing features of the two.

## Frequency Deviation

The instantaneous frequency of FM signal varies with time. The maximum change in instantaneous frequency from the average, i.e., $\omega_c$, is known as frequency deviation. The frequency deviation can be determined with the help of Eq. 6.1.10

$$\omega_i = \omega_c + K_f f(t)$$

The maximum change in $\omega_i$ from the average value $\omega_c$ will depend on the magnitude and sign of $K_f f(t)$. The frequency deviation denoted by $\Delta\omega$ will be either positive or negative depending on the sign of $K_f f(t)$. But the amount of deviation in both cases is decided by the maximum magnitude, $|K_f f(t)|_{max}$. This is shown in Fig. 6.1.7. Thus,

$$\Delta\omega = |K_f f(t)|_{max} \qquad (6.1.17)$$

The frequency deviation is a useful parameter for determining the bandwidth of the FM signals.

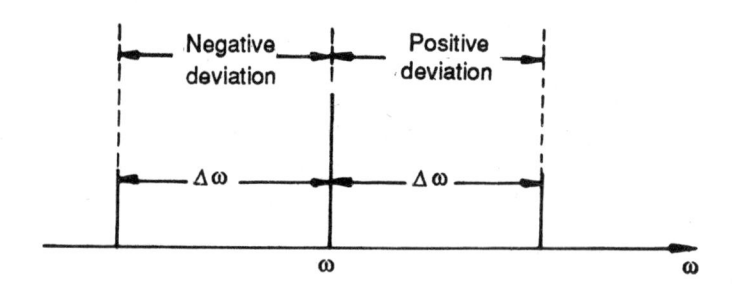

Fig. 6.1.7 Frequency Deviation

## Example 6.1.4

A single tone modulating signal $f(t) = E_m \cos \omega_m t$, frequency modulates a carrier $A \cos \omega_c t$. Find the frequency deviation.

*Solution*

The deviation $\Delta\omega$ is given by Eq. 6.1.17

$$\Delta\omega = |K_f\, f(t)|_{max}$$
$$= |K_f\, E_m \cos \omega_m t|_{max}$$

Since the maximum magnitude of $\cos \omega_m t = 1$, the deviation is,

$$\Delta\omega = K_f E_m \quad (6.1.18a)$$

and the frequency sensitivity is

$$K_f = \frac{\Delta\omega}{E_m} \quad \text{radian per volt} \quad (6.1.18b)$$

## Types of Frequency Modulation

The bandwidth of an FM signal depends on the deviation $K_f\, f(t)$. When the deviation is high, the bandwidth will be large, and vice-versa. According to Eq. 6.1.17, the deviation is controlled by $K_f\, f(t)$. Thus, for a given $f(t)$, the deviation, and hence, bandwidth will depend on frequency sensitivity $K_f$. If $K_f$ is too small then the bandwidth will be narrow and vice-versa. Thus, depending on the value of $K_f$, we can divide FM into two categories, viz.

(i) *Narrowband FM*: When $K_f$ is small, the bandwidth of FM is narrow. We will see in the next section that the bandwidth of a narrow band FM is the same as that of AM, which is twice the baseband.

(ii) *Wideband FM*: When $K_f$ has an appreciable value, then the FM signal has a wide bandwidth. We will see later that unlike AM, the bandwidth of FM is too large; ideally infinite.

## 6.2 NARROWBAND FM

The general expression for FM in the phasor form is given by Eq. 6.1.16

$$\tilde{\varphi}_{FM}(t) = A e^{j[\omega_c t + K_f\, g(t)]}$$

For a narrowband FM, $K_f\, g(t) \ll 1$ for all values of $t$.
Hence

$$e^{jK_f\, g(t)} \simeq 1 + jK_f\, g(t)$$

and FM phasor expression becomes

$$\tilde{\varphi}_{FM}(t) \simeq A[1 + jK_f\, g(t)] e^{j\omega_c t}$$

The FM signal is the real part of its phasor representation,

$$\varphi_{FM}(t) \simeq Re[\tilde{\varphi}_{FM}(t)] = A\cos \omega_c t - AK_f\, g(t) \sin \omega_c t \quad (6.2.1)$$

Similarly, the narrowband PM is given by

$$\varphi_{PM}(t) \simeq A\cos \omega_c t - AK_p\, f(t) \sin \omega_c t \quad (6.2.2)$$

The expressions for FM and PM are very much similar to the expression of the AM signal, with only a slight modification. Hence, the bandwidth of a narrowband FM is same as that of the AM. It can be seen that $g(t)$ and $f(t)$ have the same bandwidth.

$$g(t) = \int_0^t f(t)\,dt$$

so

$$F[g(t)] = \frac{1}{j\omega}[F(\omega)]$$

or

$$G(\omega) = \frac{1}{j\omega}F(\omega)$$

Thus, if $F(\omega)$ is band limited to $\omega_m$, then, $G(\omega)$ also is restricted to $\omega_m$. Both FM and PM expressions have carrier and sideband terms, these are similar to the AM expression except with *differing phase relations between carrier and sideband terms*. Actually, this is the difference that makes the amplitude of FM constant. This is in spite of the fact that it has the same frequency components as AM where amplitude variation is observed. This has been discussed in Ex. 6.2.2 with the help of phasor diagrams.

### Generation of Narrowband FM

Equations 6.2.1 and 6.2.2 suggest methods for generating the narrowband FM, and PM, respectively. The sideband terms are obtained by a balanced modulator, as in DSB-SC amplitude modulation systems and then the carrier term is added to sideband terms. The methods for generating narrowband FM and PM are shown in Figs 6.2.1 and 6.2.2, respectively. The block diagrams satisfy the corresponding expression for PM and FM.

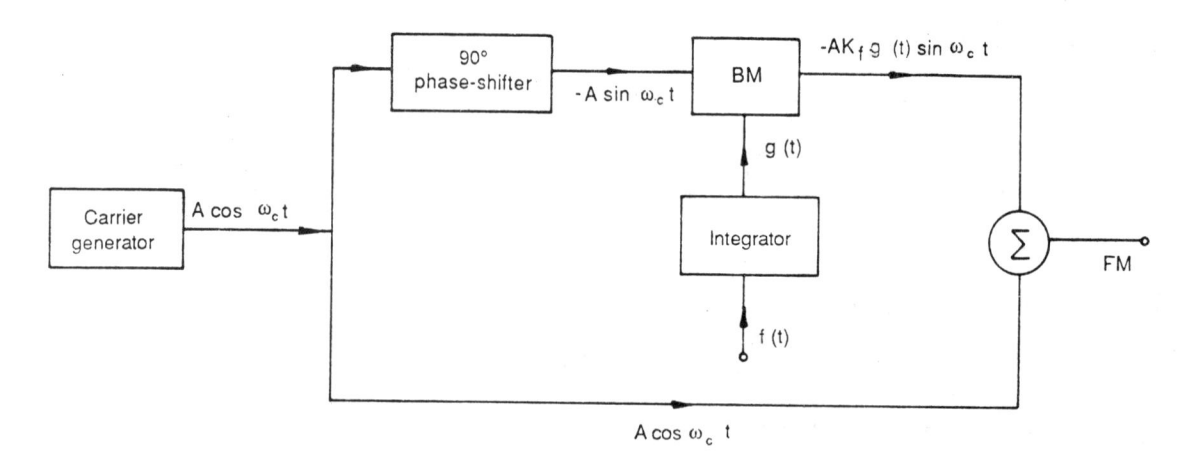

Fig. 6.2.1 Narrowband FM Generation

### Example 6.2.1

### Single Tone Frequency Modulation

A carrier $A\cos\omega_c t$ is frequency modulated by a single tone modulating signal, $f(t) = E_m \cos\omega_m t$. Find
(a) an expression for the FM wave, and
(b) an expression for a narrowband FM.

## ANGLE MODULATION SYSTEMS

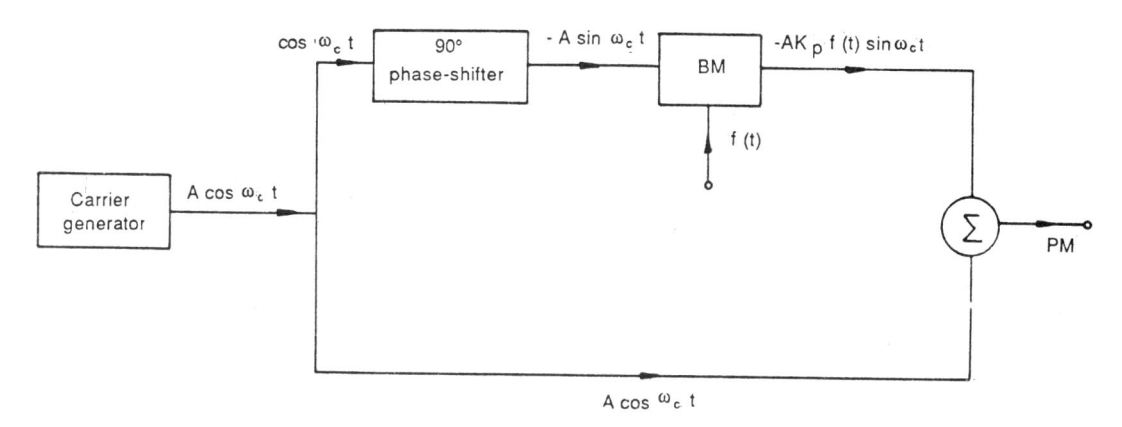

Fig. 6.2.2 Narrow band PM Generation

*Solution*
(a) The modulating signal is given by
$$f(t) = E_m \cos \omega_m t$$
The instantaneous frequency of the resulting modulated signal can be obtained by using Eq. 6.1.10.
$$\omega_i = \omega_c + K_f f(t) = \omega_c + K_f E_m \cos \omega_m t \quad (6.2.3)$$
From Eq. 6.1.17 we find that $K_f E_m = \Delta\omega$ the frequency deviation. Hence, Eq. 6.2.3 can be written as
$$\omega_i = \omega_c + \Delta\omega \cos \omega_m t \quad (6.2.4)$$
The phase angle of the modulated wave is obtained by integrating Eq. 6.2.4
$$\psi_i = \int \omega_i \, dt = \int [\omega_c + \Delta\omega \cos \omega_m t] dt$$
$$= \omega_c t + \frac{\Delta\omega}{\omega_m} \sin \omega_m t$$
$$= \omega_c t + m_f \sin \omega_m t \quad (6.2.5)$$
where
$$m_f = \frac{\Delta\omega}{\omega_m} \quad (6.2.6)$$
$$= \frac{K_f E_m}{\omega_m} \quad (6.2.7)$$

The term $m_f$ is known as the *modulation index* of the FM wave. It is defined as the "ratio of frequency deviation to the modulating frequency". Equation 6.2.7 reveals that the modulation index is directly proportional to the amplitude of the modulating signal as in AM. However, the denominator in the frequency modulation index is $\omega_m$ whereas in amplitude modulation index the denominator was $E_c$. The modulation index plays an important role in deciding the bandwidth of the FM system. It was stated earlier that the bandwidth of FM depends on the frequency deviation $\Delta\omega$, which, in turn, depends on $m_f$.

The FM signal is given by

$$\varphi_{FM}(t) = A\cos\psi_i$$

Substituting $\psi_i$ from Eq. 6.2.5

$$\varphi_{FM}(t) = A\cos[\omega_c t + m_f \sin\omega_m t] \qquad (6.2.8a)$$

This is the expression for a single-tone FM. The maximum change in total phase angle from the centre phase $\omega_c t$ is known as *phase deviation*, in FM, denoted by $\Delta\theta$. From Eq. 6.2.8a,

$$\Delta\theta = m_f \text{ radians} \qquad (6.2.8b)$$

Thus, a phase deviation equal to $m_f$ is produced in FM. Substituting $m_f$ from Eq. 6.2.6, we get a relationship between *phase deviation and frequency deviation in FM*, given by,

$$\Delta\theta = \frac{\Delta\omega}{\omega_m} = m_f \qquad (6.2.8c)$$

The modulation index $m_f$ decides whether an FM wave is a narrowband or a wideband because it is directly proportional to frequency deviation $\Delta\omega$. It will be seen later in Sec. 6.8 that $m_f = 0.5$ is the transition point between a narrowband and a wideband FM. If $m_f < 0.5$, then FM is a narrowband, otherwise, it is a wideband.

(b) The narrowband FM is defined by Eq. 6.2.1,

$$\varphi_{FM}(t) = A\cos\omega_c t - AK_f g(t)\sin\omega_c t$$

where

$$g(t) = \int f(t)\,dt$$

Here

$$f(t) = E_m \cos\omega_m t$$

Hence

$$g(t) = \int E_m \cos\omega_m t = \frac{E_m}{\omega_m}\sin\omega_m t$$

Substituting this value of $g(t)$ in the expression of FM, we get

$$\varphi_{FM}(t) = A\cos\omega_c t - AK_f \frac{E_m}{\omega_m}\sin\omega_m t \sin\omega_c t$$

Since

$$\frac{K_f E_m}{\omega_m} = m_f$$

$$\varphi_{FM}(t) = A\cos\omega_c t - Am_f \sin\omega_m t \sin\omega_c t \qquad (6.2.9)$$

This is the desired expression for a narrowband single-tone FM.

## Example 6.2.2

### Phasor Diagram of Narrowband FM

The narrowband FM has no amplitude variation, in spite of the fact that it has same frequency components as AM. Justify the statement with the help of phasor diagrams.

## ANGLE MODULATION SYSTEMS

*Solution*
Let us write the expressions for single-tone AM and single-tone FM.

$$\varphi_{AM}(t) = A\cos\omega_c t + A m_a \cos\omega_m t \cos\omega_c t$$

and

$$\varphi_{FM}(t) = A\cos\omega_c t - A m_f \sin\omega_m t \sin\omega_c t$$

The sidebands present in AM and narrowband FM can be observed by a trigonometric manipulation of the above expression, i.e.

$$\varphi_{AM}(t) = A\cos\omega_c t + \frac{1}{2} m_a A \left[\cos(\omega_c + \omega_m)t + \cos(\omega_c - \omega_m)t\right] \quad (6.2.10)$$

$$\varphi_{FM}(t) = A\cos\omega_c t + \frac{1}{2} m_f A \left[\cos(\omega_c + \omega_m)t - \cos(\omega_c - \omega_m)t\right] \quad (6.2.11)$$

Note the difference between the expressions of AM and FM. The carrier and both the sideband terms in AM have identical phase relations. But in FM the lower sideband is negative, i.e., 180° out of phase as compared to the carrier. This difference makes the amplitude of FM approximately constant. This approximation is because the narrowband FM defined by Eq. 6.2.1 is only approximate. We have taken an approximation that deviation is very small. Hence only two sidebands are present. Strictly speaking, higher sidebands are also present however, in a negligible amount. If they are also considered, the amplitude of FM is *exactly* the same as that of the unmodulated carrier.

Let us draw the phasor diagrams of AM and FM waves from Eqs 6.2.10 and 6.2.11. Consider a co-ordinate system rotating counter-clockwise at an angular frequency $\omega_c$. The carrier phasor is fixed, and is aligned in a horizontal direction. The sideband phasors rotate at an angular velocity $\omega_m$ relative to the carrier and in opposite directions to each other. The phasor diagram of AM has already been discussed in Chapter 5, and is reproduced in Fig. 6.2.3a for reference. In AM, the resultant amplitude of the carrier varies as the sideband vector rotates. The amplitude varies from maximum value of $A + m_a A$ (when the resultant of sidebands are in phase with the carrier, as shown in line $OA$) to a minimum value of $A - m_a A$ (when resultant of the sidebands are in phase opposition with the carrier as shown in line $OB$).

The phasor diagram of the narrowband FM is shown in Fig. 6.2.3b. The only difference is that the lower sideband phasor is reversed (opposite) as compared to the lower sideband phasor of AM. The net resultant yields approximately the same amplitude as the unmodulated carrier, i.e., phasor $OA' = OB'$. This justifies the statement that there is no amplitude variation in FM, in spite of the fact that both AM and FM have the same frequency components. Also, one more point worth noting is that the resultant of the two sidebands $(A'B')$ in narrowband FM is always perpendicular to the carrier phasor $(OA')$, while in AM the resultant $(CA)$ is always parallel to the carrier phasor $(OC)$. In Fig. 6.2.3b, by resolving the two sideband phasors in the direction of $A'B'$, we get

$$A'B' = m_f E_c \sin\omega_m t$$

*Spectrum of Narrowband FM*: The narrowband FM has one carrier term, and two sideband terms. The fact is obvious from the time domain expression of FM given in Eq. 6.2.11. This can also be seen from the frequency domain representation of the narrowband FM. The spectrum can be obtained by taking Fourier transform of Eq. 6.2.1, i.e.,

$$\Phi_{FM}(\omega) = F\left[A\cos\omega_c t - AK_f g(t)\sin\omega_c t\right]$$

Let,

$$g(t) \leftrightarrow G(\omega)$$

then,

$$\Phi_{FM}(\omega) = \pi A \left[\delta(\omega - \omega_c) + \delta(\omega + \omega_c)\right]$$
$$+ \frac{jAK_f}{2}\left[G(\omega + \omega_c) - G(\omega - \omega_c)\right] \quad (6.2.12)$$

The frequency components present in narrowband FM are same as in AM, and hence its bandwidth is also identical to AM.

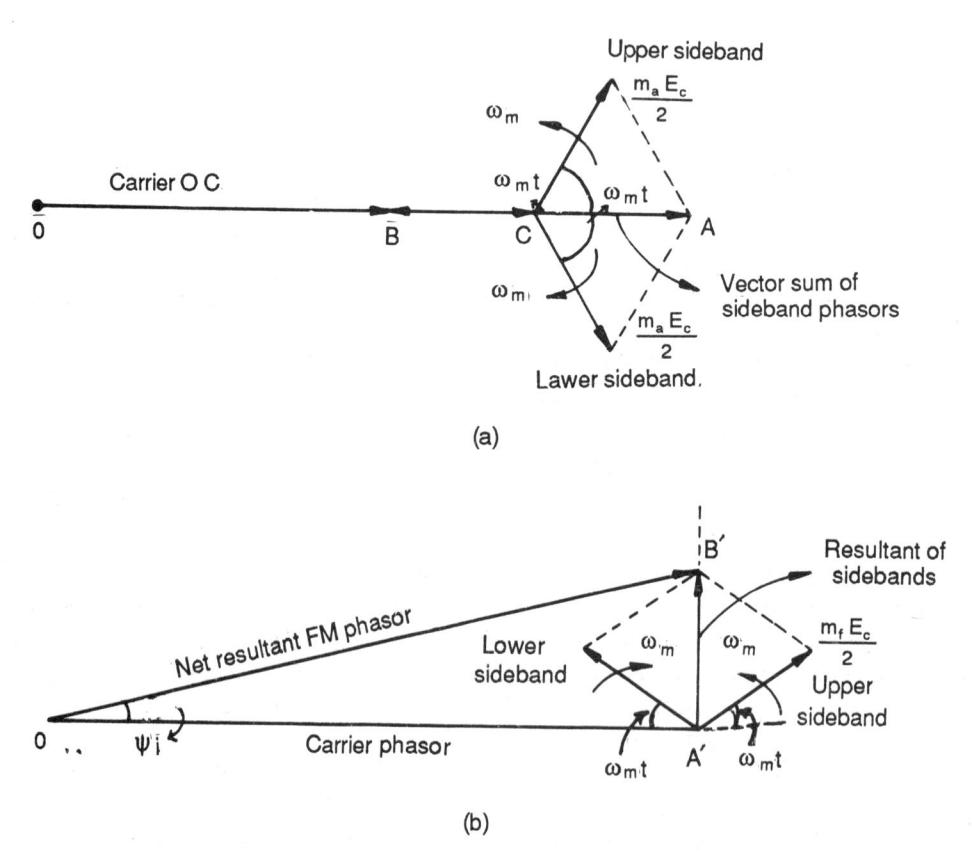

Fig. 6.2.3. Phasor Diagrams (a) AM (b) FM

## Example 6.2.3

A single-tone modulating signal $\cos(15\pi 10^3 t)$ frequency modulates a carrier of 10 MHz and produces a frequency deviation of 75 kHz. Find (i) the modulation index and (ii) phase deviation produced in the FM wave. (iii) If another modulating signal produces a modulation index of 100 while maintaining the same deviation, find the frequency and amplitude of the modulating signal, assuming $K_f = 15$ kHz per volt.

*Solution*

(i) The FM modulation index is given by

$$m_f = \frac{\Delta\omega}{\omega_m} = \frac{\Delta f}{f_m} = \frac{75 \times 10^3}{7.5 \times 10^3} = 10$$

(ii) The FM wave is given as
$$\varphi_{FM} = A\cos(\omega_c t + m_f \sin\omega_m t)$$
The instantaneous total phase is given by
$$\psi_i = \omega_c t + m_f \sin\omega_m t$$
The phase deviation from $\omega_c t$ is given by (refer 6.2.8b)
$$\Delta\theta = m_f$$
Therefore, the phase deviation is given as, $\Delta\theta = 10$ rad

(iii)
$$\Delta\theta = \frac{\Delta f}{f_m} = m_f$$

Now, $\quad m_f = 100, \Delta f = 75$ kHz

$$f_m = \frac{\Delta f}{m_f} = \frac{75 \times 10^3}{100}$$

or $\quad f_m = 750$ Hz

From Eq. 6.1.18a the amplitude of the modulating signal $E_m$ is given by

$$E_m = \frac{\Delta f}{K_f} = \frac{75}{15} = 5 \text{ Volt}$$

## 6.3 WIDEBAND FM

As discussed in the last section, when $m_f$ is large, the FM produces a large number of sidebands and the bandwidth of FM is quite large. Such systems are called wideband FM. In this section, we will analyse the wideband FM signals, restricting our analysis to the single-tone FM system. The analysis will be extended for multiple-tone FM systems in the next section.

### Sideband Terms Produced in Frequency Modulation

Let us consider the expression of a single tone FM signal given by Eq. 6.2.8a. This expression is a real part of the exponential phasor given by

$$\tilde{\varphi}_{FM}(t) = Ae^{j(\omega_c t + m_f \sin\omega_m t)}$$
$$= Ae^{j\omega_c t} e^{jm_f \sin\omega_m t} \tag{6.3.1}$$

The second exponential in Eq. 6.3.1 is a periodic function of period $1/f_m$ and can be expanded in the form of a complex Fourier series:

$$e^{jm_f \sin\omega_m t} = \sum_{n=-\infty}^{\infty} F_n e^{jn\omega_m t}, \quad -\frac{1}{2f_m} \leq t \leq \frac{1}{2f_m} \tag{6.3.2}$$

where the Fourier coefficient $F_n$ is defined by

$$F_n = f_m \int_{-\pi/\omega_m}^{\pi/\omega_m} e^{j(m_f \sin\omega_m t)} e^{-jn\omega_m t} dt \tag{6.3.3}$$

Putting $\omega_m t = x$, we get

$$F_n = \frac{1}{2\pi} \int_{-\pi}^{\pi} e^{j(m_f \sin x - nx)} dx$$

The integral on the righthand side is recognized as the $n^{th}$ order Bessel function of the first kind and argument $m_f$. This function is denoted by $J_n(m_f)$

Hence
$$F_n = J_n(m_f) \quad (6.3.4)$$

Substituting Eq. 6.3.4 in Eq. 6.3.2, we get

$$e^{jm_f \sin \omega_m t} = \sum_{n=-\infty}^{\infty} J_n(m_f) e^{jn\omega_m t} \quad (6.3.5)$$

Next, substituting Eq. 6.3.5 in Eq. 6.3.1, we get

$$\tilde{\varphi}_{FM}(t) = A e^{j\omega_c t} \sum_{n=-\infty}^{\infty} J_n(m_f) e^{jn\omega_m t} = A \sum_{n=-\infty}^{\infty} J_n(m_f) e^{j(\omega_c + n\omega_m)t}$$

The real part of the right hand side gives the expression for FM signal,

$$\varphi_{FM}(t) = A \sum_{n=-\infty}^{\infty} J_n(m_f) \cos(\omega_c + n\omega_m)t \quad (6.3.6)$$

## Bessel Function

The Bessel function $J_n(m_f)$ can be expanded in a power series given by

$$J_n(m_f) = \sum_{m=0}^{\infty} \frac{(-1)^m \left(\frac{1}{2} m_f\right)^{n+2m}}{m!(n+m)!} \quad (6.3.7)$$

The Bessel function $J_n(m_f)$ versus $m_f$ has been plotted in Fig. 6.3.1 for different positive integer values of $n$. Some important properties of the Bessel functions are as follows:

1.  $$\left. \begin{array}{l} J_n(m_f) = J_{-n}(m_f), \quad n \text{ even} \\ J_n(m_f) = -J_{-n}(m_f), \quad n \text{ odd} \end{array} \right] \quad (6.3.8)$$

2. For a small value of $m_f$:

$$\left. \begin{array}{l} J_o(m_f) \simeq 1; \quad J_1(m_f) \simeq m_f/2 \\ J_n(m_f) \simeq 0, \quad n > 1 \end{array} \right\} \quad (6.3.9)$$

3.
$$\sum_{n=-\infty}^{\infty} J_n^2(m_f) = 1 \quad (6.3.10)$$

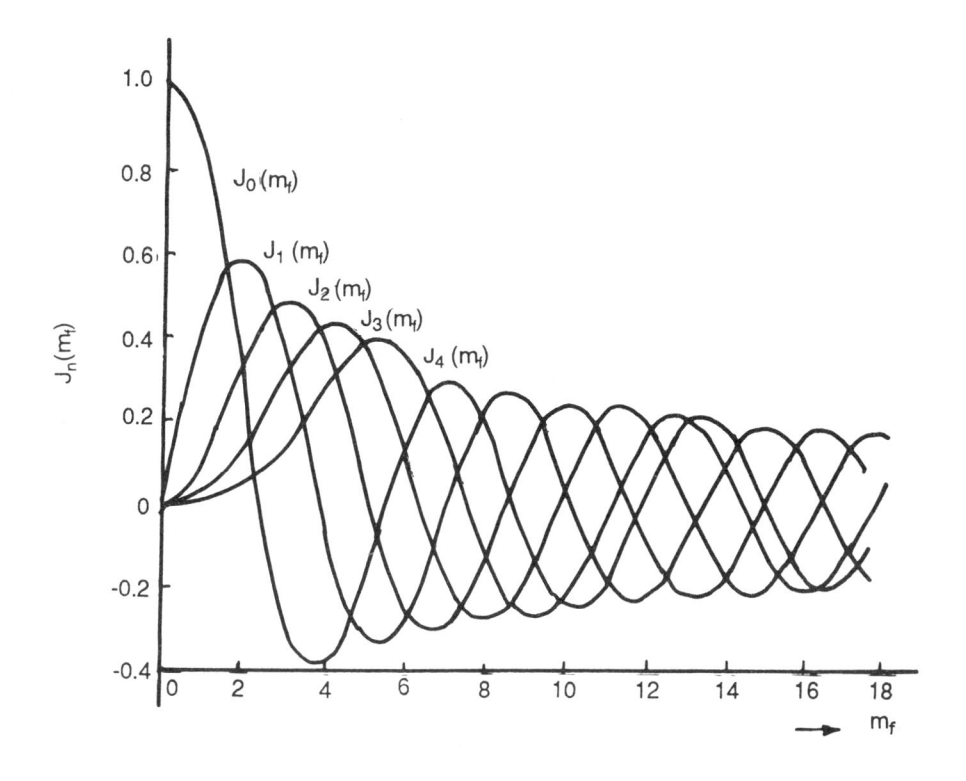

Fig. 6.3.1 Bessel Function

By using the first property (i.e., Eq. 6.3.8), we can express Eq. 6.3.6 as

$$\varphi_{FM}(t) = AJ_o(m_f)\cos\omega_c t$$
$$+ AJ_1(m_f)\left[\cos(\omega_c + \omega_m)t - \cos(\omega_c - \omega_m)t\right]$$
$$+ AJ_2(m_f)\left[\cos(\omega_c + 2\omega_m)t + \cos(\omega_c - 2\omega_m)t\right]$$
$$+ AJ_3(m_f)\left[\cos(\omega_c + 3\omega_m)t - \cos(\omega_c - 3\omega_m)t\right]$$
$$+ \cdots + \cdots \qquad (6.3.11)$$

From Eq. 6.3.11, we may make the following observations:
1. *Frequency Components*: The FM signal has the following frequency components.
(a) Carrier term $\cos\omega_c t$ with magnitude $AJ_o(m_f)$ i.e., the magnitude of the carrier term is reduced by a factor $J_o(m_f)$. It is evident from Fig. 6.3.1 that the maximum value of $J_o(m_f)$ is unity when $m_f = 0$, which is equivalent to no modulation. Further, the carrier term is absent at $m_f = 2.4, 5.52$, and so on for which $J_o(m_f)$ is zero.

(b) Theoretically infinite number of sidebands are produced, and the amplitude of each sideband is decided by the corresponding Bessel function $J_n(m_f)$. The presence of infinite number of sidebands makes the ideal bandwidth of the FM signal infinite. However, the sidebands with small amplitudes are ignored. The sidebands having considerable amplitudes are known as *significant sidebands*. They are finite in numbers.

2. *Narrowband FM* : For a small value of $m_f$ (less than 0.6), only $J_o(m_f)$ and $J_1(m_f)$ are significant, and the higher terms are negligible. Thus, FM has a carrier term and only one pair of sidebands. This is equivalent to a narrowband FM.

3. *Power Content in FM Signal* : Since the amplitude of FM remains unchanged, the power of the FM signal is same as that of unmodulated carrier. This can also be proved from the Parseval's theorem which states that the total power of a signal is equal to the sum of the power of individual components present in it. From Eq. 6.3.11, we have

$$\overline{\varphi_{FM}^2(t)} = \frac{A^2}{2} \sum_{n=-\infty}^{\infty} J_n^2(m_f) = \frac{A^2}{2}$$

because the summation in above equation yields unity (refer Eq. 6.3.10). Thus, we find that the FM power is the same as that of the unmodulated carrier (i.e., $A^2/2$).

4. *Transmission Efficiency:* We have seen that the total power of FM signal is same as that of the unmodulated carrier, which is unlike AM where the total power depends on the modulation index. In FM, out of the total power $A^2/2$, the power caried by the carrier term depends on the value of $J_o(m_f)$, and the power carried by a sideband depends on the value of corresponding $J_o(m_f)$. If we adjust $m_f = 2.4, 5.52,$, or any such value so that, $J_o(m_f) = 0$, then the power carried by the carrier term in the FM signal will be zero. All the power being carried by the sidebands provides a 100 per cent transmission efficiency. For other values of $m_f$, some power is carried by the carrier also, and efficiency is less than 100 per cent. Therefore, by adjusting the value of $m_f$ we can get FM efficiency much more than AM is 33%, and may approach equal to the efficiency of AM-SC (i.e., 100%). Thus, FM has the efficiency between AM and AM-SC.

5. *Spectrum of FM Signal:* The number of significant sidebands produced in an FM signal depends on the value of $m_f$ which, in turn, depends on maximum frequency deviation ($\Delta f$) and modulating frequency $f_m$.

6. *International Regulation for Frequency Modulation:* The following values are prescribed by CCIR (Consultative Committee for International Radio) which must be followed by commercial FM broadcast stations in order to avoid interference. These are as under :

(a) Maximum frequency deviation ± 75 kHz
(b) Frequency stability of the carrier ± 2 kHz
(c) Allowable bandwidth per channel = 200 kHz.

The significant sidebands determine the bandwidth of an FM signal. They are further explained in Ex. 6.3.1.

### Example 6.3.1

Consider an FM broadcast signal which has been modulated by a single-tone modulating signal of frequency $f_m$ = 15 kHz. The frequency deviation is the same as allowed by the international regulation. Find the significant sidebands and the bandwidth of the FM signal as a result of these sidebands.

*Solution*

$$m_f = \frac{\Delta f}{f_m} = \frac{75 \text{ kHz}}{15 \text{ kHz}} = 5$$

We have to find the value of $J_n(m_f)$ for $m_f = 5$. It can be obtained from Fig. 6.3.1. The amplitude of different frequency components present in FM signal in terms of the unmodulated carrier amplitude $A$ are as under:

(i) Carrier term has an amplitude of $0.18\,A$

(ii) First sideband pair (lower and upper) has amplitude of $0.33\,A$

(iii) Similarly, second, third and fourth sideband pairs have amplitudes of 0.05 A, 0.36 A, and 0.39 A, respectively.

(iv) The higher sidebands terms have very small amplitudes. Beyond the 8th sideband pair, the amplitudes are less than $0.01\,A$

*Significant Sidebands:* The sidebands having amplitudes more than or equal to 1% of the carrier amplitude (i.e., 01 A) are known as *significant sidebands*, and need to be considered for evaluation of FM bandwidth. Sidebands with amplitudes less than 1% of the carrier amplitude are considered negligible.

The significant sidebands are plotted in the FM spectrum shown in Fig. 6.3.2.

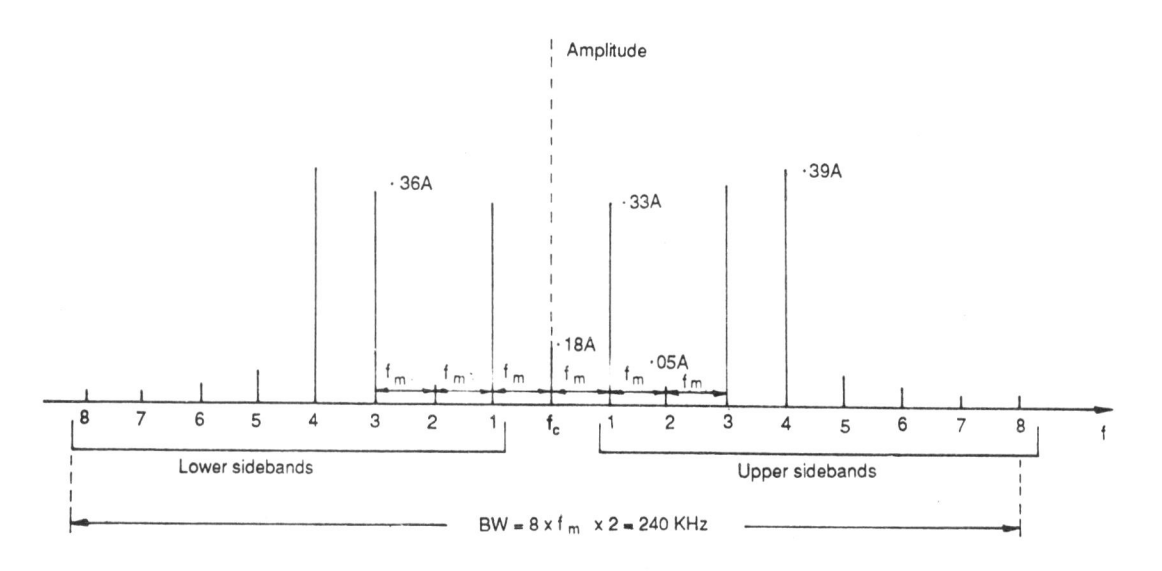

Fig. 6.3.2  FM Spectrum

The spectrum has eight pairs of significant sidebands separated from each other by $f_m = 15$ kHz. Hence, the bandwidth of the FM signal is given by

$$\text{Bandwidth} = 2n f_m = 2 \times 8 \times 15 = 240 \text{ kHz}$$

**6. Effect of Variation in $m_f$ on the Spectrum of FM Wave**: The number of sidebands produced in FM increases with increase in $m_f$. The modulation index $m_f$ is given by

$$m_f = \frac{K_f E_m}{\omega_m} = \frac{\Delta \omega}{\omega_m}$$

Therefore, $m_f$ can be increased either, (i) by reducing $f_m$ but keeping amplitude of modulating signal $E_m$ fixed, or (ii) by increasing $E_m$, but keeping $f_m$ fixed. The FM spectra for the first case are shown in Fig. 6.3.3 for $m_f = 1$, and 2. In this case $E_m$ is fixed i.e., $\Delta f$ is maintained constant. The increase in $m_f$ produces more number of sidebands crowding into the fixed frequency interval of $2\Delta f$. The FM spectra for the second case is plotted in Fig. 6.3.4 again for $m_f = 1$, and 2. In this case $f_m$ is fixed and increase in $m_f$ is caused due to increase in $\Delta f$. Note that in either case the bandwidth is approximately $2\Delta f$.

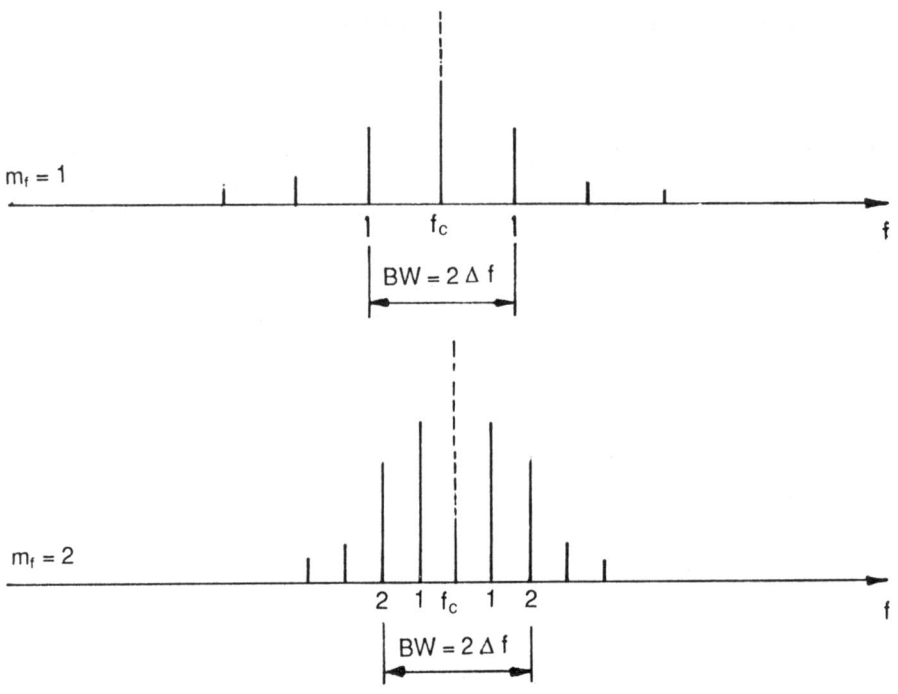

Fig. 6.3.3  FM Spectra for the Case of Fixed $E_m$ and Varying $f_m$

*Transmission Bandwidth of FM Waves*: The number of significant sidebands $n$ produced in an FM wave can be obtained from the plot of the function $J_n(m_f)$. For $n > m_f$, the values of $J_n(m_f)$ are negligible, particularly when $m_f \gg 1$. Therefore, the number of significant sidebands produced in wideband FM may be considered to be an integer approximately equal to $m_f$, i.e.,

$$n \simeq m_f; \quad m_f \gg 1$$

The upper sidebands are separated by $\omega_m$ radians, and form a frequency span of $n\omega_m$. Similar span is produced by the lower sidebands. Therefore, the bandwidth of FM wave is given by (assuming $n$ sidebands)

$$BW = 2n.\omega_m \text{ radians/s} \tag{6.3.12}$$

Since

$$n \simeq m_f$$

$$BW = 2m_f \omega_m$$

Also,

$$m_f = \frac{\Delta\omega}{\omega_m}$$

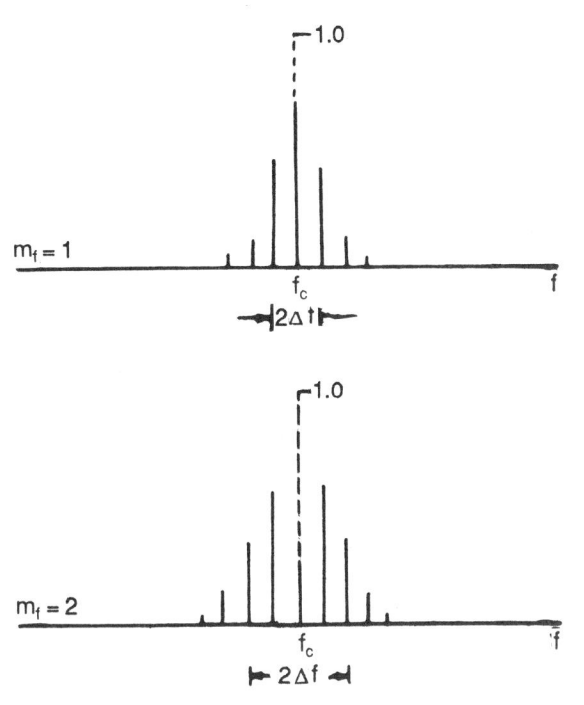

[FM spectra figure showing $m_f = 1$ with $2\Delta f$ indicated, and $m_f = 2$ with $2\Delta f$ indicated, both centered at $f_c$]

Fig. 6.3.4 FM Spectra for the Case of Fixed $f_m$ and Varying $E_m$

Hence,

$$BW = 2\frac{\Delta\omega}{\omega_m}\omega_m \simeq 2(\Delta\omega) \text{ radians}$$

or

$$BW = 2(\Delta f) \text{ Hz} \tag{6.3.13}$$

Thus, the approximate bandwidth of a wideband FM system is given as twice the frequency deviation. This approximation holds true for $m_f \gg 1$. For a smaller value of $m_f$, the bandwidth may be more than $2(\Delta\omega)$. This is evident from the graph shown in Fig. 6.3.5.

## Carson's Rule

An empirical formula for the bandwidth of a single-tone wideband FM is given by Carson's rule. According to this rule, the FM bandwidth is given by

$$BW = 2(\Delta\omega + \omega_m) = 2(\Delta\omega) + 2\omega_m \tag{6.3.14}$$

Since,

$$\omega_m = \frac{\Delta\omega}{m_f},$$

$$BW = 2(\Delta\omega) + \frac{2\Delta\omega}{m_f}$$

$$= 2\Delta\omega\left(1 + \frac{1}{m_f}\right) \text{ radians} \tag{6.3.15a}$$

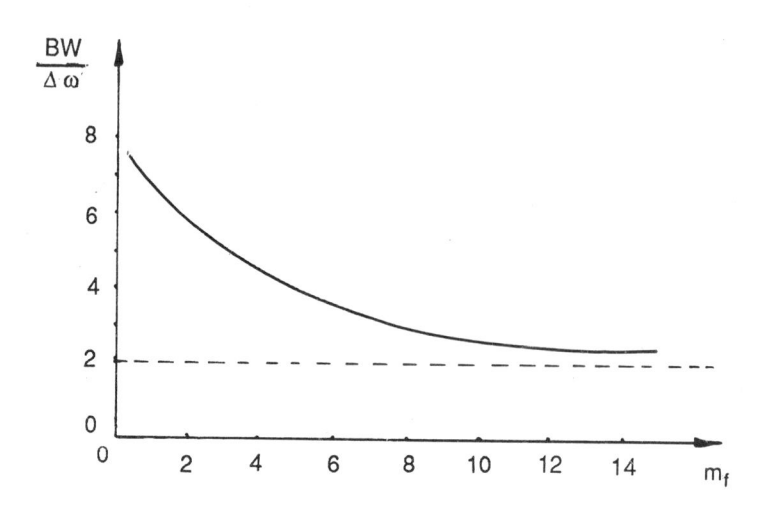

Fig. 6.3.5 Variation of Bandwidth with $m_f$

or

$$BW = 2\Delta f \left(1 + \frac{1}{m_f}\right) \text{ Hz} \qquad (6.3.15b)$$

Special cases:
(i) when

$$\Delta\omega \ll \omega_m \text{ (narrowband FM), i.e., } m_f \ll 1$$

then

$$BW = 2\omega_m$$

which is equivalent to AM
(ii) When

$$\Delta\omega \gg \omega_m \text{ (wideband FM), i.e., } m_f \gg 1 \qquad (6.3.16)$$

then

$$BW = 2(\Delta\omega)$$

For larger values of $m_f$, the bandwidth relation in Eq. 6.3.16 has a very small error and can be assumed to be true bandwidth for all practical purposes.

### Example 6.3.2

The maximum deviation allowed in an FM broadcast system is 75 kHz. If the modulating signal is a single-tone sinusoid of 10 kHz, find the bandwidth of the FM signal. What will be the change in the bandwidth, if modulating frequency is doubled? Determine the bandwidth when modulating signals' amplitude is also doubled.

*Solution*

$$\Delta f = 75 \text{ kHz}, f_m = 10 \text{ kHz}$$

From Eq. 6.3.14,
$$BW = 2(\Delta f) + 2f_m = 75 \times 2 + 10 \times 2 = 170 \text{ kHz}$$
When modulating frequency is doubled, the bandwidth changes slightly,
$$BW = 75 \times 2 + 2 \times 20 = 190 \text{ kHz}$$
When the modulating signal amplitude is doubled, the frequency deviation becomes $2 \times 75 = 150$ kHz. Now, the bandwidth becomes
$$BW = 2 \times 150 + 2 \times 20 = 340 \text{ kHz}$$
Thus, the bandwidth is almost doubled.

### Bandwidth of FM signal with Arbitrary Modulating Signal f(t)

So far, we have discussed the bandwidth for a single-tone modulating signal. Let us now consider that the modulating signal is an arbitrary function $f(t)$. For estimating the working bandwidth in this case, we first determine the value of *deviation ratio D* defined as,

$$D = \frac{\text{Peak frequency deviation corresponding to the maximum possible amplitude of } f(t)}{\text{The maximum frequency component present in the modulating signal } f(t)}$$

$D$ plays the same role for evaluating the bandwidth as $m_f$ does in the single-tone modulation system. Assuming $D$ as equivalent to $m_f$, the transmission bandwidth of FM is obtained by using Carson's rule.

### Example 6.3.3

A carrier $A \cos \omega_c t$ is modulated by a signal $f(t) = 2\cos 10^4 \cdot 2\pi t + 5 \cos 10^3 \cdot 2\pi t + 3 \cos 10^4 \cdot 4\pi t$. Find the bandwidth of the FM signal by using Carson's rule. Assume $K_f = 15 \times 10^3$ Hz per volt. Also find modulation index $m_f$.

*Solution*

The maximum frequency component in $f(t)$ is 20 kHz. The second tone in $f(t)$ has the maximum amplitude, i.e., $E_m = 5$ volt. Therefore, the frequency deviation $\Delta f$ is given by,

$$\Delta f = 15 \times 10^3 \times 5 = 75 \text{ kHz}$$

And deviation ratio is given by

$$D = \frac{75 \text{ kHz}}{20 \text{ kHz}} = \frac{7.5}{2}$$

The bandwidth is given as

$$BW = 2(\Delta f)\left(1 + \frac{1}{m_f}\right)$$

Here $m_f$ is equivalent to $D$, i.e., $D = m_f = \frac{7.5}{2}$, hence,

$$BW = 2 \times 75\left(1 + \frac{2}{7.5}\right) = 190 \text{ kHz}$$

## 6.4 SOME REMARKS ABOUT PHASE MODULATION

As stated earlier, a phase-modulated signal can be analysed in a similar way as the FM signal. However,

# 370 COMMUNICATION SYSTEMS: Analog and Digital

there are some distinct features of phase modulation as compared to FM. We will discuss the distinguishing features of PM in this section.

The main distinguishing feature of PM is that the deviation in the carrier frequency $\omega_c$ in the phase modulated signal is linearly proportional to the modulating frequency $\omega_m$. Recall that in FM the deviation is independent of the modulating frequency. The expression for frequency deviation in PM can be derived as follows:

The total phase angle of the PM signal is given by Eq. 6.1.8.

$$\psi_i(t) = \omega_c t + K_p f(t)$$

For a single-tone modulating signal $\quad f(t) = E_m \cos \omega_m t$
Hence,

$$\psi_i(t) = \omega_c t + K_p E_m \cos \omega_m t$$

The maximum departure in the phase is $K_p E_m$. This is known as phase deviation denoted by $\theta_d$. Thus $\theta_d = K_p E_m$, and the expression for the PM wave is

$$\varphi_{PM} = A \cos \psi_i(t) = A \cos [\omega_c t + \theta_d \cos \omega_m t]$$

The instantaneous frequency corresponding to $\psi_i$ is given by

$$\omega_i = \frac{d\psi}{dt} = \omega_c - K_p E_m \omega_m \sin \omega_m t \qquad (6.4.1)$$

The maximum departure in the frequency from $\omega_c$ is $K_p E_m \omega_m$. Therefore, a frequency deviation produced in PM is given as

$$(\Delta\omega)_{PM} = K_p E_m \omega_m \qquad (6.4.2a)$$

which is dependent on the modulating frequency $\omega_m$.
The frequency deviation in FM is, $(\Delta\omega)_{FM} = K_f E_m$. Therefore, for an equivalent bandwidth in PM and FM, we have

$$K_p \omega_m = K_f \qquad (6.4.2b)$$

## Bandwidth of PM

The PM bandwidth is given by Carson's rule,

$$(BW)_{PM} \simeq 2(\Delta\omega) = 2K_p E_m \omega_m$$

Thus, the bandwidth of the PM signal varies tremendously with a change in the modulating frequency $\omega_m$. This is illustrated in Ex. 6.4.1. Note that the FM bandwidth varies slightly with $\omega_m$. This fact is also obvious from Ex. 6.4.1. The phase modulation index $m_p$ is the same, as the deviation $\theta_d$, and is given by

$$m_p = K_p E_m = \theta_d \qquad (6.4.3)$$

### Example 6.4.1

A modulating signal $5 \cos 2\pi\, 15 \times 10^3 t$, angle modulates a carrier $A \cos \omega_c t$.
(i) Find the modulation index and the bandwidth for (a) the FM systems, (b) the PM system.
(ii) Determine the change in the bandwidth and the modulation index for both FM and PM, if $f_m$ is reduced to 5 kHz.

Assume $K_p = K_f = 15$ kHz/volt
*Solution*
(i) Here $E_m = 5V$, $f_m = 15$ kHz
(a) *FM System*
Frequency deviation $\Delta f = K_f E_m = 15 \times 10^3 \times 5 = 75$ kHz
Therefore
$$m_f = \frac{75}{15} = 5$$
$$\text{Bandwidth} = 2(\Delta f + f_m) = 2(75 + 15) = 2 \times 90 = 180 \text{ kHz}$$
b) *PM System*
Frequency deviation,
$$(\Delta f) = K_p E_m f_m = 15 \times 10^3 \times 5 \times 15 \times 10^3 = 1125 \text{ MHz}$$
The corresponding bandwidth is
$$BW = 2(\Delta f + f_m) \simeq 2\Delta f = 2(1125 \text{ MHz}) = 2250 \text{ MHz}$$
The bandwidth is quite large as compared to FM. This is because the modulation index in PM is quite large,
$$m_p = K_p E_m = 5 \times 15 \times 10^3 = 75,000$$
This tremendous value of the modulation index $m_p$ produce a large number of sidebands, and this is the reason for a large bandwidth.
(ii) now $f_m = 5$ kHz.
For FM: the deviation $(\Delta f)$ is independent of $f_m$, and remains 75 kHz.
Hence,
$$m_f = \frac{75}{5} = 15$$
The Bandwidth $= 2(\Delta f + f_m) = 2(75 + 5) = 160$ kHz. Thus, in FM the modulation index changes considerably with a change in the modulating frequency, but the *bandwidth changes only slightly*.
For PM: deviation $(\Delta f)$ is dependent on $f_m$, and is given as
$$\Delta f = K_p E_m f_m = 15 \times 10^3 \times 5 \times 5 \times 10^3 = 375 \text{ MHz}$$
Hence,
$$BW = 2(\Delta f + f_m) \simeq 2(\Delta f) = 750 \text{ MHz}$$
Thus, the bandwidth has changed considerably as compared to case (b).
The modulation index is independent of $f_m$,
$$m_p = K_p E_m = 5 \times 10^3 \times 5 = 75 \text{ kHz}$$
which is same as in case (b).

Thus, in PM, the bandwidth changes considerably with a change in the modulation frequency but the modulation index $m_p$ remains unchanged. On the other hand, in FM, the bandwidth changes only slightly, but the modulation index changes considerably. Actually, the change in $f_m$ adjusts the modulation index, and hence the sidebands produced in FM in such a way that the bandwidth remains almost unchanged. But, in PM, the modulation index $m_p$ is unaffected by the change in $f_m$ and hence the number of sidebands generated remains the same. Since the separation $f_m$ between the sidebands $(f_m)$ is changed, the bandwidth varies tremendously.

## 6.5 MULTIPLE-TONE WIDEBAND FM (NON-LINEAR MODULATION)

In Sec. 6.3 we dealt with the simplest case of the modulating signal $f(t)$ consisting a single-tone sinusoid. In general, the modulating signal $f(t)$ is multi-tone in nature and comprises of a band of sinusoidal signals. For convenience, let us take a modulating signal $f(t)$ consisting only two frequency tones given by

$$f(t) = E_1 \cos \omega_1 t + E_2 \cos \omega_2 t \tag{6.5.1}$$

This modulating signal, frequency modulates a carrier $A \cos \omega_c t$. The resulting modulated signal is given by

$$\varphi_{FM}(t) = A \cos\left[\omega_c t + K_f \int f(t) dt\right]$$

The integration of $f(t)$ is obtained by integrating Eq. 6.5.1.

$$\int f(t) dt = \frac{E_1}{\omega_1} \sin \omega_1 t + \frac{E_2}{\omega_2} \sin \omega_2 t$$

Therefore, the FM signal is given by

$$\varphi_{FM}(t) = A \cos\left[\omega_c t + \frac{K_f E_1}{\omega_1} \sin \omega_1 t + \frac{K_f E_2}{\omega_1} \sin \omega_2 t\right] \tag{6.5.2}$$

Let $m_1$ and $m_2$ denote the frequency modulation indices of the first and second tone, respectively. Then, by Eq. 6.2.7, we can write

$$m_1 = \frac{K_f E_1}{\omega_1}, \text{ and } m_2 = \frac{K_f E_2}{\omega_2} \tag{6.5.3}$$

Equation 6.5.2 can be expressed in terms of modulation indices as,

$$\varphi_{FM}(t) = A \cos[\omega_c t + m_1 \sin \omega_1 t + m_2 \sin \omega_2 t] \tag{6.5.4}$$

Equation 6.5.4 represents the expression for the FM wave with a two-tone modulating signal and can be expanded as (Prob. 6.17)

$$\varphi_{FM}(t) = A \sum_{l=-\infty}^{\infty} \sum_{n=-\infty}^{\infty} J_l(m_1) J_n(m_2) \cos[\omega_c + l\omega_1 + n\omega_2] \tag{6.5.5}$$

The expression comprises of the following four types of frequency terms:

(i) A carrier frequency component $\omega_c$ with an amplitude

$$J_o(m_1) J_o(m_2) A$$

(ii) A set of sidebands corresponding to first tone $\omega_1$. These sidebands have amplitudes $J_l(m_1) J_o(m_2)$ and frequencies $(\omega_c \pm l\omega_1)$ where $l = 1, 2, 3,...$

(iii) A set of sidebands corresponding to second tone $\omega_2$. These sidebands have amplitudes $J_o(m_1) J_n(m_2)$ and frequencies $(\omega_c \pm n\omega_2)$ where $n = 1, 2, 3,...$

(iv) A set of *cross-modulation terms* $(\omega_c \pm l\omega_1 \pm n\omega_2)$ with amplitudes $J_l(m_1) J_n(m_2)$ where $l = 1, 2, 3,...$ and $n = 1, 2, 3,...$

Let us compare the above results with the results of the multiple tone AM signals. In amplitude modulation, each modulating tone produced its own sidebands, and no cross-modulation terms were produced. This modulation system follows the law of superposition of spectra, and is called as *linear modulation*. The FM spectrum, on the other hand, does not follow the superposition of spectra due to the presence of additional cross-modulation frequency terms. Such a modulation system is called *non-linear modulation*. In non-linear modulation systems like FM, or PM, the spectrum of a multiple-tone modulated signal cannot be obtained

ANGLE MODULATION SYSTEMS    373

simply by adding the spectra corresponding to individual tones. The additional cross-modulation terms has also to be associated. The bandwidth of a multiple tone wideband FM can be obtained using Carson's rule (Ex. 6.3.3).

### Linearization of FM Signals

FM behaves like a linear modulation when the modulation index is small. We have seen in Sec. 6.2 that for $m_f \ll 1$, FM behaves like AM. Thus, if $m_1$ and $m_2$ are much less than 1, in the two-tone modulating signal case, FM behaves like an amplitude modulation, producing only two sidebands corresponding to each modulating tone. Thus, it follows the law of superposition of spectra and exhibits the property of linear modulation.

## 6.6  FM MODULATORS AND TRANSMITTERS

The FM modulator circuits used for generating FM signals can be grouped into two categories:
   (1) The Parameter Variation Method (Direct Method)
   (2) The Armstrong Method (Indirect Method)

### (1)  Parameter Variation Method (Direct Method)

In this method, the modulating signal directly modulates the carrier. The carrier signal is generated by an electronic oscillator circuit. The oscillator circuit involves a parallel resonant (tuned LC) circuit. The frequency of oscillation of the carrier generator is given by,

$$\omega_c = \frac{1}{\sqrt{LC}} \qquad (6.6.1)$$

The carrier frequency $\omega_c$ can be made to vary according to the modulating signal $f(t)$, if L or C is varied according to $f(t)$. An oscillator whose frequency is controlled by a modulating voltage is known as *voltage controlled oscillator* (VCO). The frequency of VCO is varied in accordance with the modulating signal by shunting a voltage variable capacitor with its tuned circuit. This voltage variable capacitor, is known as *varicap* or *varactor*. Reverse biased semiconductor diodes exhibit such a property. The capacitance of bipolar transistors (BJT) and field-effect transistors (FET) also varies by the Miller-effect. This Miller capacitance can be utilized for frequency modulation. The electron tubes may also provide variable reactance (inductive or capacitive), proportional to modulating signal. Such tubes are known as *reactance tubes* and can be used for FM generation.

The inductance L of the tuned circuit can also be varied according to the modulating signal $f(t)$. The inductors are wound on a magnetic cores (ferrite-cores). The permeability of the magnetic material (ferrite) is a function of the external magnetic field. This field is created by another coil, wound on the same core, through which a current proportional to $f(t)$ is passed. The inductance L, of the inductor depends on its permeability, $\mu$, which in turn, is controlled by the modulating signal $f(t)$. Thus, the inductance L is controlled by $f(t)$. The ferrite core has low loss at high frequency providing a high Q (quality factor), which is important for the stability of high-frequency oscillators. The FM circuit involving such inductors is known as *saturable reactor modulator*. Frequency modulation may also be achieved from voltage controlled devices, such as PIN diode, klystron oscillators and multivibrators.

### *FM Modulation Circuit Using Varactor Diode*

The varactor diode is a semiconductor diode whose junction capacitance changes appreciably with dc bias voltage. This diode is shunted with the tuned circuit (tank circuit) of the carrier oscillator as shown in Fig. 6.6.1.

**374** COMMUNICATION SYSTEMS: Analog and Digital

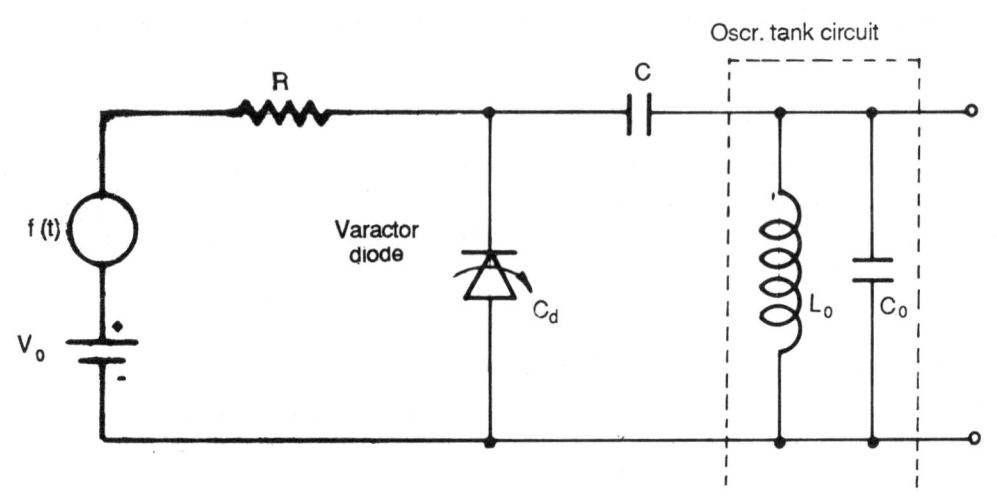

**Fig. 6.6.1** Varactor Diode FM Modulator

The capacitor $C$ is kept much smaller than the diode capacitance $(C_d)$, in order to keep the radio frequency (r.f) voltage from oscillator across the diode small as compared to $V_o$. $V_o$ is the polarizing voltage to maintain a reverse bias across the varactor diode. Also, reactance of $C$ at the highest modulating frequency is kept large as compared to R, so that the shunting of the modulating signal through the tank circuit is avoided.

*Working Principle:* The capacitance $C_d$ of the diode is given by the relation

$$C_d = K(v_D)^{-1/2} \tag{6.6.2}$$

where $v_D$ is total instantaneous voltage across the diode given by

$$v_D = V_o + f(t) \tag{6.6.3}$$

and $K$ is a constant of proportionality.

The total capacitance of the oscillator tank circuit is $(C_o + C_d)$ and, hence, the instantaneous frequency of oscillation $\omega_i$ is given as, (using Eq. 6.6.1.).

$$\omega_i = \frac{1}{\sqrt{L_o(C_o + C_d)}} \tag{6.6.4a}$$

Substituting $C_d$ from Eq. 6.6.2, we get

$$\omega_i = \left[L_o\left(C_o + Kv_D^{-1/2}\right)\right]^{-1/2} \tag{6.6.4b}$$

This frequency $\omega_i$ is dependent on $v_D$ which in turn depends on the modulating signal $f(t)$. Therefore, the oscillator frequency $\omega_i$ is dependent on the modulating signal $f(t)$ and thus frequency modulation is generated.

## Distortion due to Non-Linearity

It is evident from Eq. 6.6.4b that the frequency $\omega_i$ is not changing linearly with the diode voltage $v_D$. This non-linearity produces distortion because of the frequency variations caused by the higher harmonics of the modulating frequency. Let us evaluate the order of distortion. Assume that the oscillator tank circuit comprises of only the diode capacitance $C_d$, and $C_o$ is absent. Equation 6.6.4b then becomes

… ANGLE MODULATION SYSTEMS

$$\omega_i = \left[L_o K v_D^{-1/2}\right]^{-1/2} = (L_o K)^{-1/2} (v_D)^{1/4}$$

$$= \frac{v_D^{1/4}}{(L_o K)^{1/2}} \tag{6.6.5}$$

The right-hand side of Eq. 6.6.5 can be represented by a Taylor series about polarising voltage $V_o$ as given below:

$$\frac{v_D^{1/4}}{(L_o K)^{1/2}} = \frac{V_o^{1/4}}{(L_o K)^{1/2}} + \frac{(v_D - V_o)}{4(L_o K V_o^{3/2})^{1/2}} - \frac{(3 v_D - V_o)^2}{16(L_o K V_o^{7/2})^{1/2}} \tag{6.6.6}$$

The higher order terms can be neglected if $(v_D - V_o)$ is small.

In this case $v_D - V_o = \Delta V = f(t)$, (refer Eq. 6.6.3). For a single-tone modulating voltage $f(t) = V_m \sin \omega_m t$,

$$v_D - V_o = \Delta V = V_m \sin \omega_m t \tag{6.6.7a}$$

and

$$(v_D - V_o)^2 = V_m^2 \sin^2 \omega_m t = \frac{V_m^2}{2} (1 - \cos 2\omega_m t) \tag{6.6.7b}$$

From Eqs 6.6.5, 6.6.6 and 6.6.7, we have

$$\omega_i = \frac{v_D^{1/4}}{(L_o K)^{1/2}} = \frac{V_o^{1/4}}{(L_o K)^{1/2}} + \frac{V_m \sin \omega_m t}{4(L_o K V_o^{3/2})^{1/2}} - \frac{3V_m^2}{(32 L_o K V_o^{7/2})^{1/2}} + \frac{3V_m^2 \cos 2\omega_m t}{32(L_o K V_o^{7/2})^{1/2}} \tag{6.6.8}$$

The frequency variation corresponding to the fundamental term is due to $\sin \omega_m t$. Similarly, the frequency variation corresponding to the second harmonic is due to term $\cos 2\omega_m t$. This variation belonging to the undesired second harmonic is the cause of distortion. The percentage second harmonic distortion is defined as the ratio of amplitude of the term $\cos 2\omega_m t$ and the fundamental term $\sin \omega_m t$.

$$\text{Percentage second harmonic distortion} = \frac{3V_m^2 / 32(L_o K V_o^{7/2})^{1/2}}{V_m / 4(L_o K V_o^{3/2})^{1/2}} \times 100$$

$$= \frac{3V_m}{8V_o} \times 100 \tag{6.6.9}$$

Therefore, by adjusting the proper ratio of $V_m$ and $V_o$, the second harmonic distortion may be considerably reduced. Ignoring the effect of the second harmonic of the modulating signal, Eq. 6.6.8 becomes

$$\omega_i = \frac{V_o^{1/4}}{(L_o K)^{1/2}} + \frac{V_m \sin \omega_m t}{4(L_o K V_o^{3/2})^{1/2}} \tag{6.6.10}$$

$$= \omega_c + (\Delta \omega) \sin \omega_m t \tag{6.6.11}$$

where $\omega_c$ is the carrier frequency in the absence of the modulating signal, and $(\Delta \omega)$ is the frequency deviation due to the modulating frequency term. From Eqs 6.6.10 and 6.6.11, we see that the carrier frequency $\omega_c$ corresponding to $V_m = 0$ is

$$\omega_c = \frac{V_o^{1/4}}{(L_o K)^{1/2}} \qquad (6.6.12a)$$

and frequency deviation due to the modulating signal $V_m \sin \omega_m t$ is

$$\Delta\omega = \frac{V_m}{4\left[L_o K V_o^{3/2}\right]^{1/2}} \qquad (6.6.12b)$$

The frequency modulation index is given by

$$m_f = \frac{\Delta\omega}{\omega_m} = \frac{V_m}{4\omega_m\left[L_o K V_o^{3/2}\right]^{1/2}} \qquad (6.6.13)$$

Thus, the modulation index not only depends on the modulating voltage $V_m$, but also on the polarizing voltage $V_o$.

### Example 6.6.1

A semiconductor junction diode is used to modulate the frequency of an oscillator. The junction capacitance is the total tuning capacitance of the oscillator tank circuit. When a d.c bias voltage of 15 volts is applied to the diode, the oscillator frequency generated is 5 MHz. If a single-tone modulating voltage 4 sin 12560 $t$ modulates the carrier: find (a) the percentage second harmonic distortion; and (b) the frequency modulation index.

*Solution*

The polarizing voltage $V_o = 15$ volt
The oscillator frequency before modulation is 5 MHz i.e., $= 10\pi \, 10^6$ rad/s
From Eq. 6.6.12a

$$(L_o K)^{1/2} = \frac{V_o^{1/4}}{\omega_c} = \frac{15^{1/4}}{10\pi \, 10^6}$$

The modulating frequency $\omega_m = 12560$ rad/s and its amplitude $V_m$ is 4V.

(a) The percentage second harmonic distortion is given by Eq. 6.6.9.

Percentage distortion $= \dfrac{3V_m}{8V_o} \times 100 = \dfrac{3 \times 4}{8 \times 15} \times 100 = 10\%$

(b) The modulation index is given by Eq. 6.6.13.

$$m_f = \frac{V_m}{4\omega_m (L_o K V_o^{3/2})^{1/2}} = \frac{V_m}{4\omega_m (L_o K)^{1/2} V_o^{3/4}}$$

Substituting the values, we get

$$m_f = \frac{4 \times 10\pi \times 10^6}{4 \times 12560 \times 15^{1/4} \times 15^{3/4}} = \frac{\pi \, 10^6}{1256 \times 15} = 166.66$$

### Miller Capacitance Frequency Modulator

The Miller capacitance of amplifying devices (electron tube, BJT or FET) can be used for FM generation. A modulator circuit utilizing the Miller capacitance of an FET is shown in Fig. 6.6.2. The input capacitance $C_i$ of the FET is controlled by its transconductance. This capacitance, known as Miller Capacitance is given by

$$C_i = C_{gs} + C_{gd}(1 + g_m R_L) \tag{6.6.14}$$

where $C_{gs}$ is gate to the source capacitance; and $C_{gd}$ is gate to the drain capacitance. The transconductance $g_m$ of the FET is given as

$$g_m = g_{mo}\left(1 - \frac{V_{gs}}{V_p}\right) \tag{6.6.15}$$

where $V_{gs}$ in the gate to the source voltage, $V_p$ is the pinch-off voltage, and $g_{mo}$ is the transconductance at $V_{gs} = 0$. From Eqs 6.6.14 and 6.6.15, we get

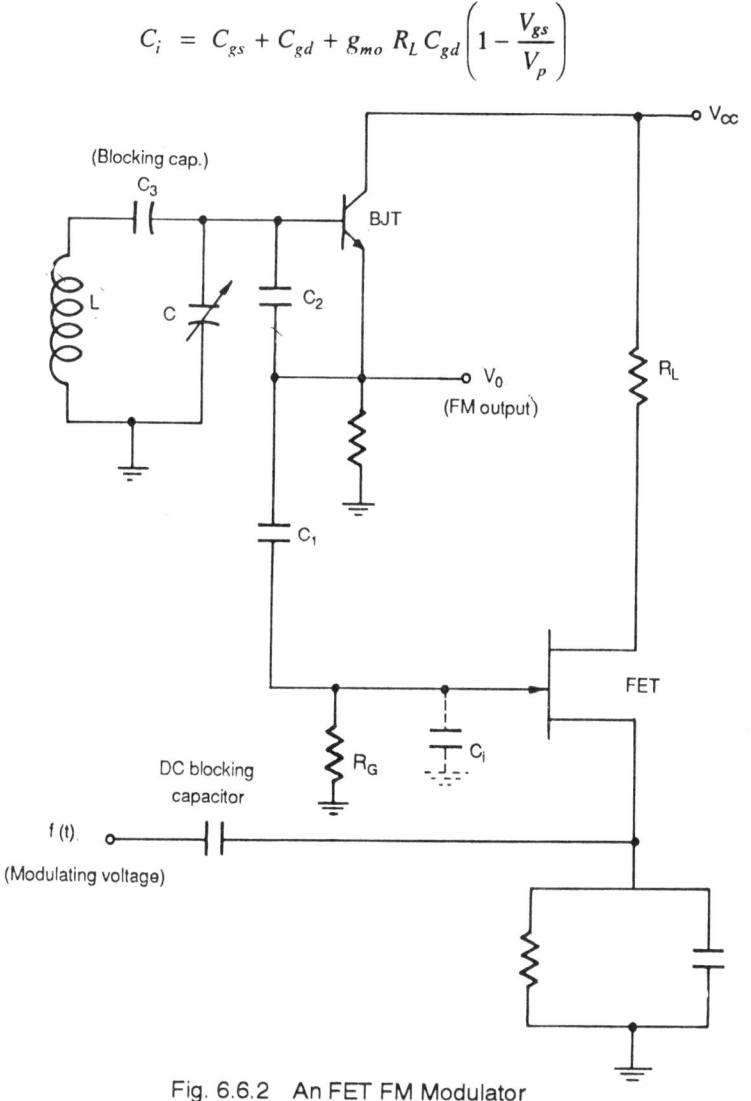

$$C_i = C_{gs} + C_{gd} + g_{mo} R_L C_{gd}\left(1 - \frac{V_{gs}}{V_p}\right) \tag{6.6.16}$$

Fig. 6.6.2 An FET FM Modulator

The capacitance $C_i$ in series with $C_1$ forms the feedback path needed for oscillation. $C_1 > C_i$, so that the reactance of $C_1$ is smaller than that of $C_i$. The instantaneous gate to source voltage is given by, $V_{gs}$ = constant bias voltage $V_{gs}$ + time varying voltage $v_{gs}$. However, only the time varying part $v_{gs}$, which is proportional to the modulating voltage $f(t)$ takes part in frequency modulation. Therefore, Eq. 6.6.16 becomes

$$C_i = C_{gs} + C_{gd} + g_{mo} R_L C_{gd} \left(1 - \frac{v_{gs}}{V_p}\right) \qquad (6.6.17)$$

The capacitance $C_2 > C_i$, in order to get large voltage gain of transistor amplier, as the gain is given by

$$X_{C_1} / X_{C_2}$$

where $X$ represent the capacitive reactance.

Now, since $C_1$ and $C_2$ both are large as compared to $C_i$ they do not play any significant role in the frequency determining circuit (as their reactance is small). The frequency of oscillation is, thus, primarily decided by an adjustable capacitor $C$, shunted by the Miller capacitor $C_i$. Therefore, the total capacitance contributing to the frequency of oscillation is given by

$$C_T = C + C_i$$

Substituting $C_i$ from Eq. 6.6.17, we get

$$C_T = C + C_{gs} + C_{gd} + g_{mo} R_L C_{gd} - \frac{g_{mo} R_L C_{gd} v_{gs}}{V_p}$$

or

$$C_T = C_o + B \qquad (6.6.18)$$

where $C_o$ and $B$ are constants, given as

$$C_o = C + C_{gs} + C_{gd} + g_{mo} R_L C_{gd} \qquad (6.6.19a)$$

and

$$B = -\frac{g_{mo} R_L C_{gd}}{V_p} \qquad (6.6.19b)$$

Thus, the tuning capacitance $C_T$ is controlled by $v_{gs}$ which is proportional to the modulating signal $f(t)$. The frequency of oscillation is given by

$$\omega_i = \frac{1}{\sqrt{L(C_o + B v_{gs})}} \qquad (6.6.20)$$

where $L$ is the inductance of the tuning coil.

Obviously the frequency $\omega_i$ is not linearly varying with $v_{gs}$ which causes harmonic distortion.

*Harmonic Distortion*: The non-linearity in Eq. 6.6.20 causes frequency modulation due to the harmonic terms of the modulating signal which creates distortion. For evaluating the amount of distortion, Eq. 6.6.20 can be expanded into McClauren's series given as

$$\omega_i = [L(C_o + B v_{gs})]^{-\frac{1}{2}} = A_o + A_1 v_{gs} + A_2 v_{gs}^2 + \cdots \qquad (6.6.21)$$

The coefficient $A_o$ is obtained by letting $v_{gs} = 0$ on both sides of the Eq. 6.6.21. Thus,

ANGLE MODULATION SYSTEMS

$$A_o = (1/LC_o)^{1/2} \tag{6.6.22a}$$

The coefficient $A_1$ is obtained by taking the first derivative of both sides of Eq. 6.6.21 with respect to $v_{gs}$ and then letting $v_{gs} = 0$. Thus,

$$A_1 = \frac{B}{2(LC_o^3)^{1/2}} \tag{6.6.22b}$$

Similarly, other coefficients are also obtained by taking the higher derivatives and letting $v_{gs} = 0$. Thus,

$$A_2 = \frac{3B^2}{8(LC_o^5)^{1/2}} \tag{6.6.22c}$$

and so on.
Substituting the values of the coefficients, Eq. 6.6.21 becomes

$$\omega_i = \frac{1}{(LC_o)^{1/2}} - \frac{B}{2(LC_o^3)^{1/2}} v_{gs} + \frac{3B^2}{8(LC_o^5)^{1/2}} v_{gs}^2 + \cdots \tag{6.6.23}$$

Let us consider the case of a single-tone modulating signal. In this case,

$$v_{gs} = f(t) = V_m \cos \omega_m t \tag{6.6.24a}$$

and

$$v_{gs}^2 = V_m^2 \cos^2 \omega_m t = \frac{V_m^2}{2}[1 + \cos 2\omega_m t] \tag{6.6.24b}$$

Substituting the value of $v_{gs}$ in Eq. 6.6.23 and neglecting the terms higher than the second order, we get

$$\omega_i = \frac{1}{(LC_o)^{1/2}} - \frac{BV_m \cos \omega_m t}{2(LC_o^3)^{1/2}} + \frac{3B^2}{8(LC_o^5)^{1/2}} \frac{V_m^2}{2}(1 + \cos 2\omega_m t)$$

$$= \frac{1}{(LC_o)^{1/2}} - \frac{BV_m \cos \omega_m t}{2(LC_o^3)^{1/2}} + \frac{3B^2 V_m^2}{16(LC_o^5)^{1/2}} + \frac{3B^2 V_m^2}{16(LC_o^5)^{1/2}} \cos 2\omega_m t + \cdots \tag{6.6.25}$$

The percentage second harmonic distortion is the ratio of amplitude of the harmonic term $\cos 2\omega_m t$ and that of the fundamental signal term $\cos \omega_m t$. Therefore,

$$\text{Percentage second harmonic distortion} = \frac{3B^2 V_m^2/16(LC_o^5)^{1/2}}{BV_m/2(LC_o^3)^{1/2}} \times 100$$

Putting the value of $B$ from Eq. 6.6.19b, we get,

$$\text{Percentage second harmonic distortion} = \frac{3g_{mo} R_L C_{gd} V_m}{8 C_o V_P} \times 100 \tag{6.6.26}$$

**380** COMMUNICATION SYSTEMS: Analog and Digital

If the second harmonic term is adjusted to be negligible, Eq. 6.6.25 can be written as

$$\omega_i = \frac{1}{(LC_o)^{1/2}} - \frac{BV_m}{2(LC_o^3)^{1/2}} \cos\omega_m t$$

$$\omega_c + (\Delta\omega)\cos\omega_m t \qquad (6.6.27)$$

where $\omega_c$ is the oscillator frequency in the absence of the modulating signal, and $(\Delta\omega)$ is the frequency deviation, i.e.,

$$\omega_c = \frac{1}{(LC_o)^{1/2}} \qquad (6.6.28a)$$

$$(\Delta\omega) = \frac{BV_m}{2(LC_o^3)^{1/2}} \qquad (6.6.28b)$$

The modulation index is given as

$$m_f = \frac{\Delta\omega}{\omega_m}$$

$$= \frac{BV_m}{2\omega_m(LC_o^3)^{1/2}} \qquad (6.6.29a)$$

Substituting $B$ from Eq. 6.6.19b, we get

$$m_f = \frac{g_{mo} R_L C_{gd} V_m}{2\omega_m V_p (LC_o^3)^{1/2}} \qquad (6.6.29b)$$

### Example 6.6.2

A Miller capacitance FM modulator utilizes an FET with the following parameters,

$$g_{mo} = 3 \text{ m mho}, \ V_p = 2.5 \text{V}, \ C_{gs} = 2\text{pF}, \ C_{gd} = 1.5\text{pF}$$

and $C_o = 10\text{pF}$. The oscillator frequency without modulation is 10 MHz. The modulating signal is, $0.5 \cos 2\pi 10^3 t$. Determine (i) The percentage second harmonic distortion if $R_L = 2K$ ohms. (ii) The frequency modulation index.

*Solution*
(i) The percentage harmonic distortion is obtained by using Eq. 6.6.26. Substituting the values, we get

$$\text{percentage harmonic distortion} = \frac{3 \times 3 \times 10^{-3} \times 2 \times 10^3 \times 1.5 \times 10^{-12} \times 0.5}{8 \times 10 \times 10^{-12} \times 2.5} \times 100 = 6.75\%$$

(ii) $\qquad V_m = 0.5 \text{ volts}, \ \omega_m = 1000 \times 2\pi, \ \omega_c = 10 \text{ MHz}$

From Eq. 6.6.28a, we get

$$2\pi.10 \times 10^6 = \frac{1}{(LC_o)^{1/2}}$$

or

$$L^{1/2} = \frac{1}{2\pi.10^7 \times C_o^{1/2}}$$

Therefore

$$L^{1/2} C_o^{3/2} = \frac{1}{2\pi.10^7 C_o^{1/2}} C_o^{3/2}$$

or

$$(LC_o^3)^{1/2} = \frac{C_o}{2\pi.10^7} = \frac{10 \times 10^{-12}}{2\pi.10^7} = \frac{10^{-18}}{2\pi}$$

$$m_f = \frac{3 \times 10^{-3} \times 2 \times 10^3 \times 1.5 \times 10^{-12} \times 0.5}{2 \times 2\pi \times 10^3 \times 2.5 \times (10^{-18}/2\pi)} = 900$$

## Reactance Tube Modulator

A vacuum tube circuit shown in Fig. 6.6.3 exhibits inductive, or capacitive reactance at its output terminal $t_1, t_2$. This reactance is proportional to the modulating signal $f(t)$ applied at its grid. This reactance shunts the tank-circuit of the oscillator. The oscillator frequency is controlled by the modulating signal $f(t)$ and an FM signal is generated. The equivalent circuit is shown in Fig. 6.6.4. The plate resistance $r_P$ is so large that it behaves like an open circuit, and as such may be ignored. The plate current $I$ is given as

$$I = g_m E_g + \frac{V_p}{Z_1 + Z_2} \tag{6.6.30}$$

where $g_m$ is the transconductance of the tube.
The grid voltage $E_g$ is given by

$$E_g = \frac{Z_2}{Z_1 + Z_2} V_p \tag{6.6.31}$$

From Eqs 6.6.30 and 6.6.31, we get

$$I = \frac{(1 + g_m Z_2) V_p}{(Z_1 + Z_2)}$$

The admittance $Y_{t_1 t_2}$, across the terminal $t_1, t_2$ is given as

$$Y_{t_1, t_2} = \frac{I}{V_p} = \frac{(1 + g_m Z_2)}{(Z_1 + Z_2)}$$

The impedance $Z_1$ and $Z_2$ can be designed in such a way that $Z_1 \gg Z_2$ and $|g_m Z_2| \gg 1$.

**382** COMMUNICATION SYSTEMS: Analog and Digital

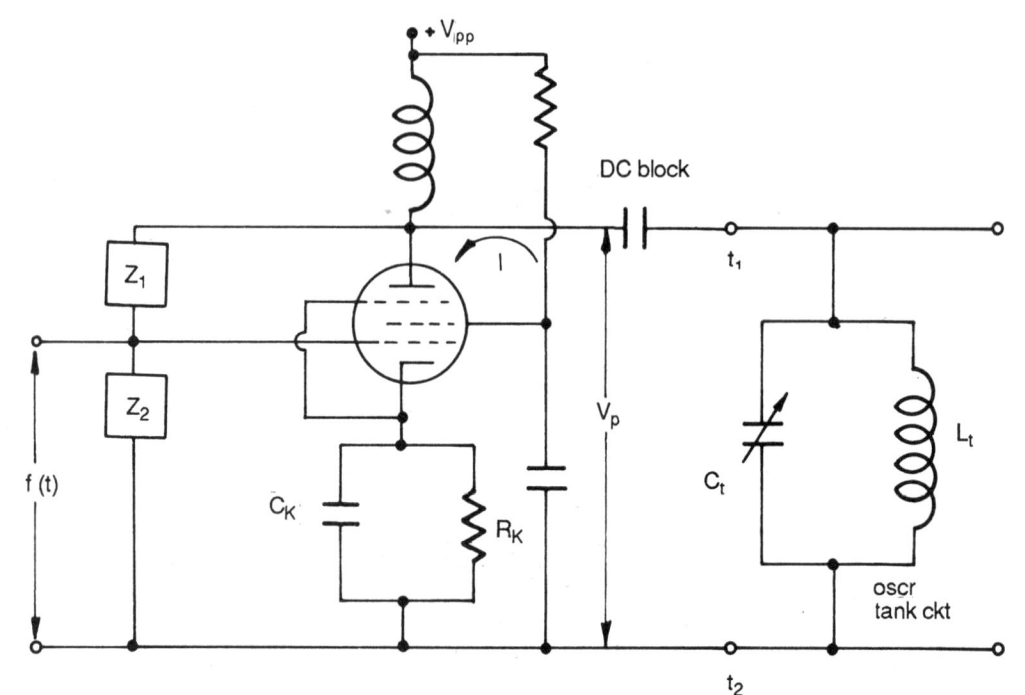

Fig. 6.6.3 Reactance Tube Modulator

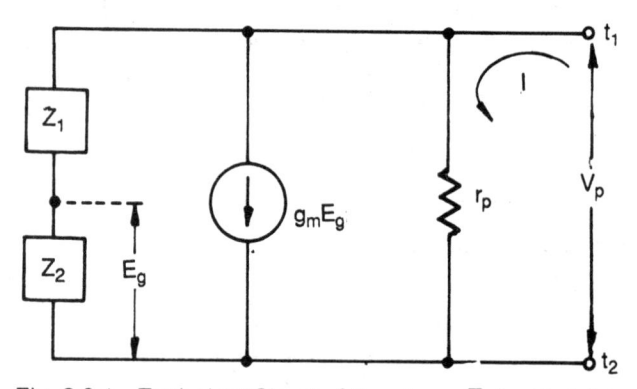

Fig. 6.6.4 Equivalent Circuit of Reactance Tube Modulator

Hence,

$$Y_{t_1,t_2} = \frac{g_m Z_2}{Z_1} \tag{6.6.32}$$

The transconductance $g_m$ is controlled by its grid voltage. Let a signal proportional to $f(t)$ be applied at the grid, then

$$g_m = g_{mo} + b f(t)$$

where $g_{mo}$ is the transconductance when the grid voltage is zero, and $b$ is a constant of proportionality. Putting this value of $g_m$ in Eq. 6.6.32, we get,

$$Y_{t_1,t_2} = \frac{Z_2}{Z_1}[g_{mo} + bf(t)] \tag{6.6.33}$$

Thus, the admittance across $t_1$, $t_2$ is controlled by the modulating signal. The admittance $Y_{t_1,t_2}$ may represent a *capacitive* or an *inductive* reactance depending upon the choice of $Z_1$ and $Z_2$.

## Capacitive Reactance Tube

The capacitive reactance is observed at the terminal $t_1$, $t_2$ when $Z_1$ is chosen to be a capacitance $C_1$, and $Z_2$ a resistance $R$. In this case,

$$Z_1 = 1/j\omega C_1, \text{ and } Z_2 = R \tag{6.6.34}$$

From Eqs 6.6.33 and 6.6.34, we get

$$Y_{t_1,t_2} = j\omega C_1 R [g_{mo} + bf(t)]$$

which is equivalent to a capacitance, given by

$$C = C_1 R [g_{mo} + bf(t)] = C_o + Bf(t) \tag{6.6.35}$$

where $C_o = g_{mo} R C_1$, and $B = b R C_1$, are the constants. The above expression for $C$ is similar to the Miller capacitance of FET given by Eq. 6.6.18 and is controlled by the modulating signal $f(t)$. This capacitance shunts the tank capacitance $C_t$ of the oscillator tank circuit and generates an FM signal.

If we assume that the tank capacitance of an oscillator is totally contributed by the reactance tube, (i.e., $C_t = 0$), the frequency of oscillation is given by

$$\omega_i = \frac{1}{\sqrt{L_t [C_o + Bf(t)]}} \tag{6.6.36}$$

Just like the FET modulator, the frequency $\omega_i$ is not linearly varying with the modulating signal $f(t)$ causing harmonic distortion. The distortion can be evaluated by using the same procedure as discussed earlier for the FET modulator.

## Inductive Reactance Tube

The inductive reactance is observed by interchanging $Z_1$ and $Z_2$ of the capacitive reactance tube, i.e.,

$$Z_1 = R \text{ and } Z_2 = \frac{1}{j\omega C_1} \tag{6.6.37}$$

Here $Z_1$ is a resistance $R$, and $Z_2$ is a capacitance $C_1$. From Eqs 6.6.33 and 6.6.37, we get

$$Z_{t_1 t_2} = \frac{1}{Y_{t_1 t_2}} = j\omega C_1 R / [g_{mo} + bf(t)] \tag{6.6.38}$$

This is equivalent to an inductance $L$, given as

$$L = \frac{C_1 R}{g_{mo} + bf(t)} \tag{6.6.39}$$

# 384 COMMUNICATION SYSTEMS: Analog and Digital

Thus, the effective inductance $L$ offered by the reactance tube is controlled by the modulating signal $f(t)$. Therefore, the frequency of the oscillator may be varied according to the modulating signal by incorporating this inductance $L$ in the tank circuit of the oscillator. If we assume that this inductance $L$ is the only inductance in the tank circuit of Fig. 6.6.3 (i.e., $L_t = 0$), the frequency of oscillation is given by

$$\omega_i = \frac{1}{\sqrt{L C_t}} \tag{6.6.40}$$

where $C_t$ is the tank capacitance. The harmonic distortion due to non linearty exists in this case also.

## Demerits of the Direct Method

(i) In the direct method it is difficult to obtain a high order stability in carrier frequency. This is because the carrier generation is directly affected by the modulating signal. The modulating signal directly controls the tank circuit of the carrier generator and, hence a stable oscillator circuit like the crystal oscillator, cannot be used (the crystal oscillator provides a stable but fixed frequency). Thus carrier generation cannot be of high stability which is an essential requirement.

A remedy to this problem is the indirect method (Armstrong method) of FM generation. In this method, the carrier oscillator is not required to respond to the modulating signal directly; rather, the carrier generation is isolated from other parts of the circuit. Hence, stable crystal oscillators can be used for generating carrier signal.

(ii) The non-linearity produces a frequency variation due to harmonics of the modulating signal and, hence the FM signal is distorted. Proper care has to be taken for keeping this distortion minimum.

In spite of the above demerits, the direct method is used for high power FM generation in many applications.

## FM Transmitter Using Direct Method of Frequency Modulation

Figure 6.6.5 shows the block diagram of an FM transmitter. The modulator circuit uses parameter-variation method. A pre-emphasis circuit is used to reduce the effect of noise at higher audio frequencies for threshold improvement. The details about this circuit are discussed in Sec. 6.10. Like the AM transmitter, the carrier oscillator generates sub-harmonic of final carrier frequency to achieve frequency stability. A stable oscillation frequency at a lower radio frequency (say 4 MHz) is generated by an oscillator and then it is raised to the final carrier (say 96 MHz) by frequency multipliers. It should be kept in mind that more frequency stability can be obtained if the carrier oscillator operates at low radio frequency. The multiplying circuit not only increases the carrier frequency, but also the frequency deviation by the same factor. This fact is proved in Ex. 6.6.3.

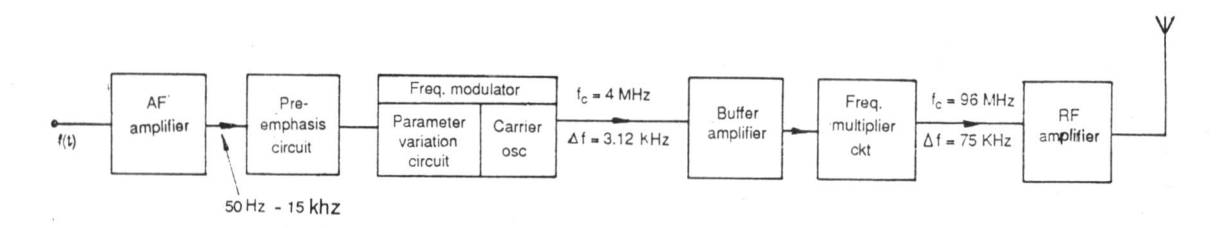

Fig. 6.6.5 Block Diagram of an FM Transmitter using Direct Modulation

## Example 6.6.3

Show that a non-linear square-law device used for frequency multiplication of an FM signal doubles the carrier frequency as well as the frequency deviation.

*Solution*

A square law frequency multiplier system squares the signal applied at its input. Let us assume that the FM signal at the input of a square law multiplier is given by

$$\varphi_{FM}(t) = \cos\left[\omega_c t + K_f g(t)\right]$$

The output of the square law device is

$$\left[\varphi_{FM}(t)\right]^2 = \cos^2\left[\omega_c t + K_f g(t)\right] = \frac{1}{2}\left[1 + \cos\{2\omega_c t + 2K_f g(t)\}\right]$$

Thus, the carrier frequency is doubled to $2\omega_c$. The sensitivity $K_f$ is also doubled to $2K_f$. The frequency deviation is proportional to $K_f$ hence, the frequency deviation is also doubled. This result can be generalized to any order of multiplication. The multiplier circuit not only multiplies the carrier frequency, but also the frequency deviation (and hence, the modulation index) by same factor. This fact is used in Armstrong method of FM generation.

### Frequency Stabilization System in FM Transmitters

We have seen that the direct method has the disadvantages of frequency unstability. Variations in environmental conditions (temperature, humidity), supply voltages, aging of active devices etc. can also cause frequency unstability. Therefore, FM transmitter must incorporate some auxiliary means for frequency stabilization. A stabilization scheme utilizing feedback principle is shown in Fig. 6.6.6. A stable crystal oscillator provides the reference frequency. The output of the crystal oscillator and the frequency modulator is fed into a mixer, and the difference frequency term is extracted at the output of the mixer. The mixer output then, is fed into a frequency discriminator. The frequency discriminator circuit provides an error voltage whose instantaneous value is proportional to the instantaneous frequency of its input. When the FM wave has frequency exactly equal to the assigned carrier frequency (no drift), the error signal is zero. But, whenever the transmitter carrier frequency drifts from the assigned value, a dc error signal of proper polarity is generated. The amplified error voltage of proper polarity is applied to a *VCO* in order to correct the transmitter frequency to the assigned value.

### Armstrong (Indirect) Method of FM Generation

In the Armstrong method, frequency stability of a higher order can be obtained because the crystal oscillator can be used as a carrier generator. The basic principle of this method is to generate a narrowband FM (NBFM) indirectly by using the phase-modulation technique, and then converting this NBFM to a wideband FM (WBFM), as shown in Fig. 6.6.7. The distortion is low in NBFM as the modulation index is small. The phase modulation is preferred because of its easy generation schemes. The generation of NBFM is illustrated in Fig. 6.2.1. The multiplier circuit, apart from multiplying the carrier frequency, also increases the frequency deviation (refer Ex. 6.6.3), and thus the NBFM (small deviation) is converted into WBFM (large deviation).

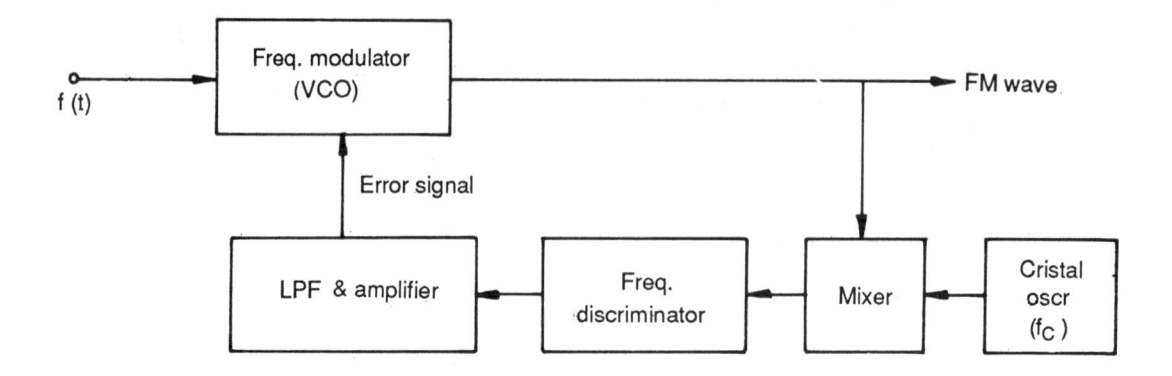

Fig. 6.6.6 A Frequency Stabilization Scheme

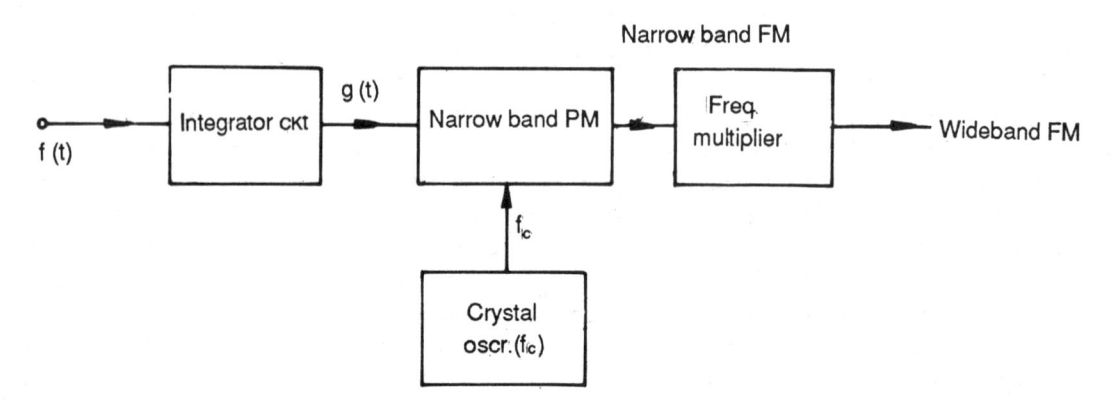

Fig. 6.6.7 Armstrong Method for FM Generation

## Armstrong FM Transmitter

The block diagram of an Armstrong transmitter (indirect-method) is shown in Fig. 6.6.8. The multiplication process is performed in several stages in order to increase the carrier frequency as well as frequency deviation to the assigned value. The desired multiplication factors for carrier frequency and frequency deviation are not the same. If the multiplication is done in one shot, the carrier frequency as well as the deviation is multiplied by the same factor (Ex. 6.6.3) which is not desired. Hence, multiplication is done in stages. For example, consider the typical values mentioned in Fig. 6.6.8. Initially, the carrier frequency $f_{c1}$ and the deviation $\Delta f$ generated by NBFM, has the following typical values,

$$f_{c1} = 200 \text{ kHz}, (\Delta f) \simeq 25 \text{ Hz}$$

which correspond to $m_f = 0.5$ and $f_m = 50$ Hz. The multiplication factor is determined by the lower frequency limit of the baseband spectrum which is (25 Hz - 15 kHz)

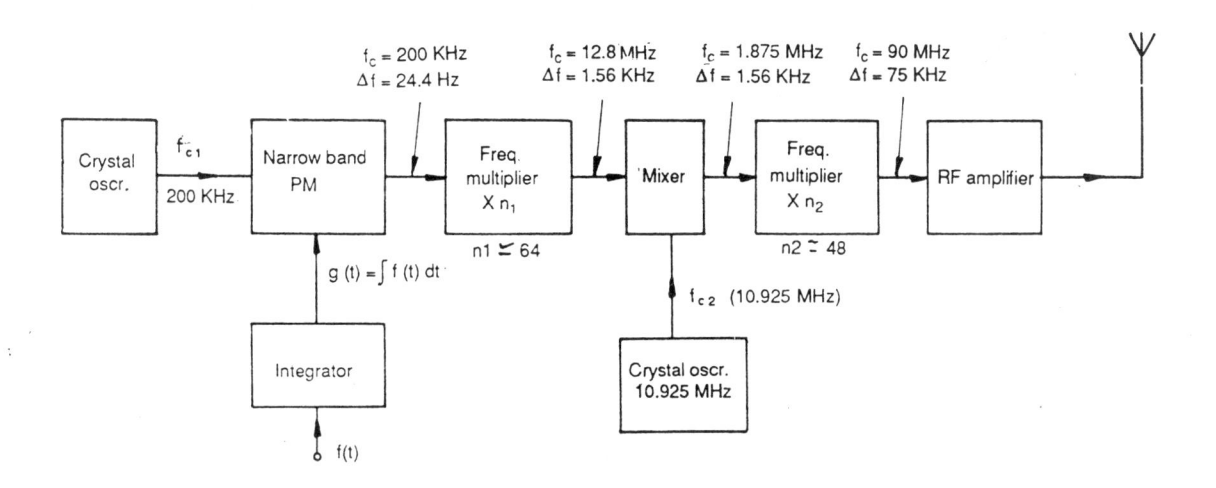

Fig. 6.6.8 Block Diagram of an Armstrong FM Transmitter

The final carrier frequency and the deviation desired at the transmitter output is as follows

$$f_c = 90 \text{ MHz}, (\Delta f) = 75 \text{ kHz}$$

Therefore, the multiplication factor (one shot) needed for a desired deviation is $\dfrac{75 \text{ kHz}}{25 \text{ Hz}} = 3000$; whereas, the multiplication factor needed to achieve the desired carrier frequency is $\dfrac{90 \text{ MHz}}{200 \text{ kHz}} = 450$ only. Thus, if we adjust the multiplication factor of 3000 for achieving the desired deviation, the final carrier frequency generated will be 600 MHz, which is much higher than the assigned value of 90 MHz. However, by performing multiplication in two stages, we can achieve both, the carrier frequency $f_c$ and deviation $(\Delta f)$ as assigned. The design procedure is as follows:

(i) Choose the multiplication factors of multiplier stages so that the net multiplication achieves the desired deviation $(\Delta f)$

$$n_1, n_2 = 3000 \qquad (6.6.41a)$$

(ii) Before second multiplication, the carrier frequency at the output of the first multiplier is shifted downward to $(n f_1 - f_2)$ by mixing it with a sinusoidal signal of frequency $f_{c2}$. The shifted frequency is increased $n_2$ times by second multiplier to get the assigned carrier frequency $f_c = 90$ MHz. Therefore,

$$n_2 (n_1 f_{c1} - f_{c2}) = f_c$$

Substituting the values, we get

$$n_2(0.2n_1 - 10.925) = 90 \qquad (6.6.41b)$$

Solving Eqs 6.6.41a and b, we get $n_1 = 64.3$ and $n_2 = 46.7$. For convenience in the design of the multiplier circuits, the values are rounded off, so that the multiplication may be done by factors of 2 and 3, i.e., $n_1 = 64 = 2^6$ and $n_2 = 48 = 3 \times 2^4$. Using these multiplication factors, we get the set of values indicated in Fig. 6.6.3 for initial value of $\Delta f = 24.4$ Hz.

## 6.7 FM DEMODULATORS AND RECEIVERS

The process of extracting a modulating signal from a frequency modulated carrier is known as demodulation. Electronic circuits that perform the demodulation process are called FM detectors. The FM detector performs the extraction in two steps:

1. It converts the frequency modulated (FM) signal into a corresponding amplitude modulated (AM) signal by using frequency dependent circuits, i.e., circuits whose output voltage depends on input frequency. Such circuits are called as *frequency discriminators*.

2. The original modulating signal $f(t)$ is recovered from this AM signal by using a linear diode envelope detector.

A simple R-L circuit can be used as a discriminator, but this circuit has a poor sensitivity as compared to a tuned LC circuit. The various types of LC discriminators will be discussed in this section. The FM discriminators suffer from the threshold effect in the presence of excessive noise. The threshold can be improved by pre-emphasis, and de-emphasis circuits, and various other techniques discussed later in this chapter.

### Types of FM Discriminators

FM discriminators can be divided into two types:

#### (1) Slope Detectors

The principle of operation depends on the slope of the frequency response characteristic of a frequency selective network. The two main FM discriminators, which use detuned resonant circuits come under this category, and are as follows:
 (i) Single-tuned discriminator circuit, or simple slope detector.
 (ii) Stagger-tuned discriminator circuit, or balanced slope detector.

#### (2) Phase difference Discriminators

The following two circuits come under this category:
 (a) Foster-Seely discriminator.
 (b) Ratio detector.

*Simple Slope Detector (single-tuned circuit)* The circuit diagram of this detector is shown in Fig. 6.7.1a. The circuit consists of a tuned circuit which is slightly detuned from the carrier frequency $\omega_c$. This circuit converts the FM signal into an AM signal. The AM signal is then detected by a diode detector. The frequency response characteristic of this detuned (slightly off-tuned at $\omega_c$) is shown in Fig. 6.7.1b. The slope of the characteristic curve is given as

$$\alpha = \frac{de_{AM}}{d\omega} \tag{6.7.1}$$

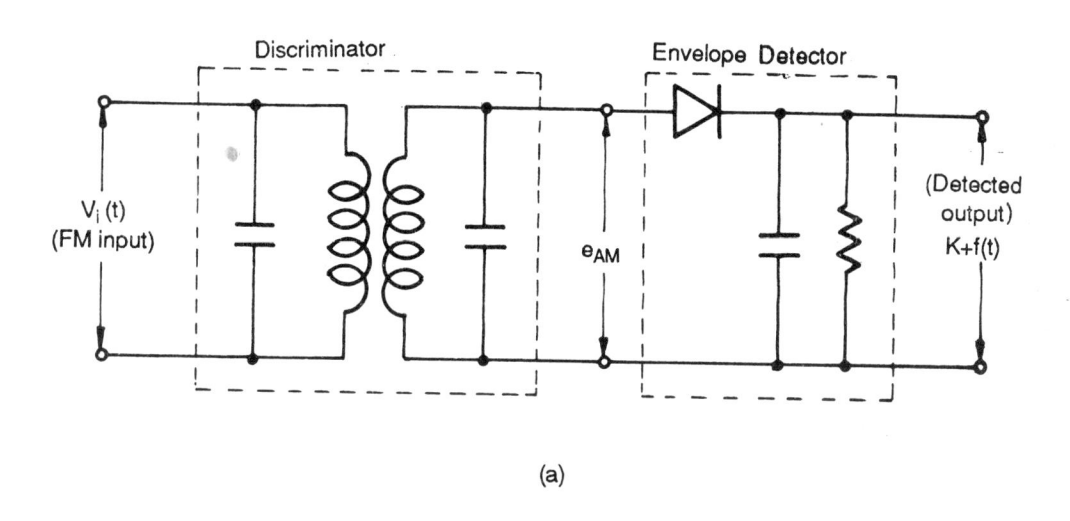

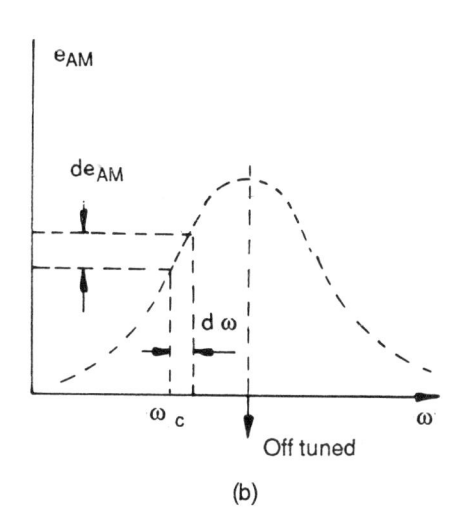

Fig. 6.7.1 (a) Circuit diagram of a Simple Slope Detector
(b) Simple Slope Detector — Characteristic

A small variation in the frequency ($\Delta\omega$) of the input signal will produce a change in the amplitude of $e_{AM}$ by an amount $e_{AM} = \alpha(\Delta\omega)$. Thus, frequency variations at the input of the discriminator produces amplitude variations at its output. In this way, the FM signal is converted to a AM signal, which is detected by an envelope detector to recover the modulating signal $f(t)$.

Although the circuit is simple and inexpensive, it suffers from the following demerits:

(i) The circuit's non-linear characteristic causes a harmonic distortion. The non-linearity is obvious from the fact that the slope is not the same at every point of the characteristics.

(ii) It does not eliminate the amplitude variations and the output is sensitive to any amplitude variations in the input FM signal, which is not a desirable feature. A good discriminator circuit should respond only to frequency variations, and not to amplitude variations.

The first limitation (non-linearity) is removed by using a stagger-tuned LC circuits (balanced slope detector).

**Balanced Slope Discriminator:** The circuit diagram consists of two LC circuits as shown in Fig. 6.7.2a. The frequency response is shown in Fig. 6.7.2b.

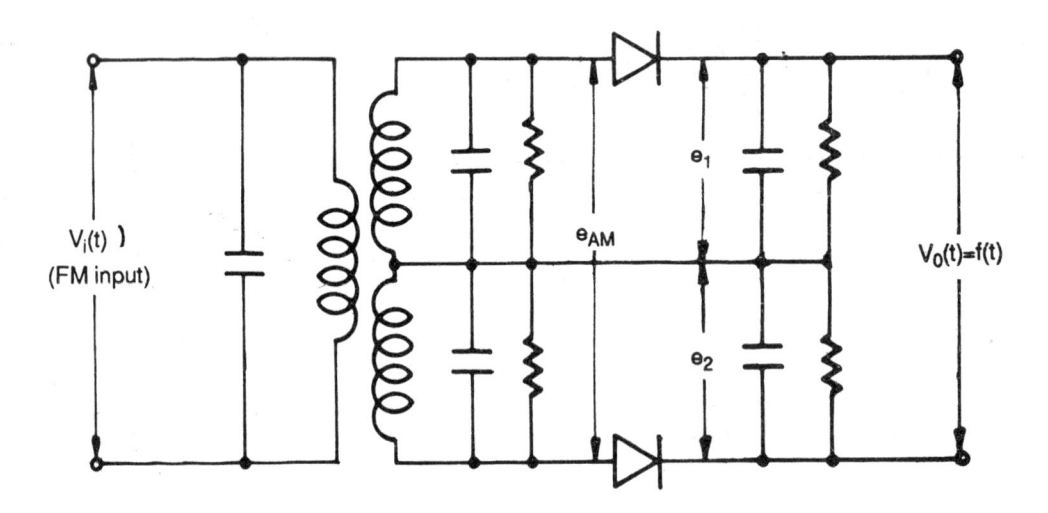

Fig.6.7.2(a) Circuit Diagram of a Balanced Slope Detector

The two tuned circuits are in the stagger-tuned mode, i.e., one is tuned above the carrier frequency $\omega_c$ (curve $e_1$) and the other is tuned below $\omega_c$ (curve $e_2$). The resultant curve ($e_1 + e_2$) is *linear* as shown by the dotted line. The slope = $\dfrac{de_{AM}}{d\omega}$ is identical at all the points on the linear portion of the characteristic curve. Thus, no harmonic distortion is caused when the operation is restricted to the linear region.

*Disadvantages of the Balanced Slope-Discriminator:* (i) The linear characteristic is limited to a small frequency deviation ($\Delta\omega$). The frequency deviation ($\Delta\omega$) for which the resultant characteristic is linear depends on the 3dB bandwidth of each tuned circuit which is equal to 2W. As shown in Fig. 6.7.2b, the linearity is limited to a range of 3W. Therefore, the deviation ($\Delta\omega$) for which the curve is linear and gives a satisfactory output, is given by

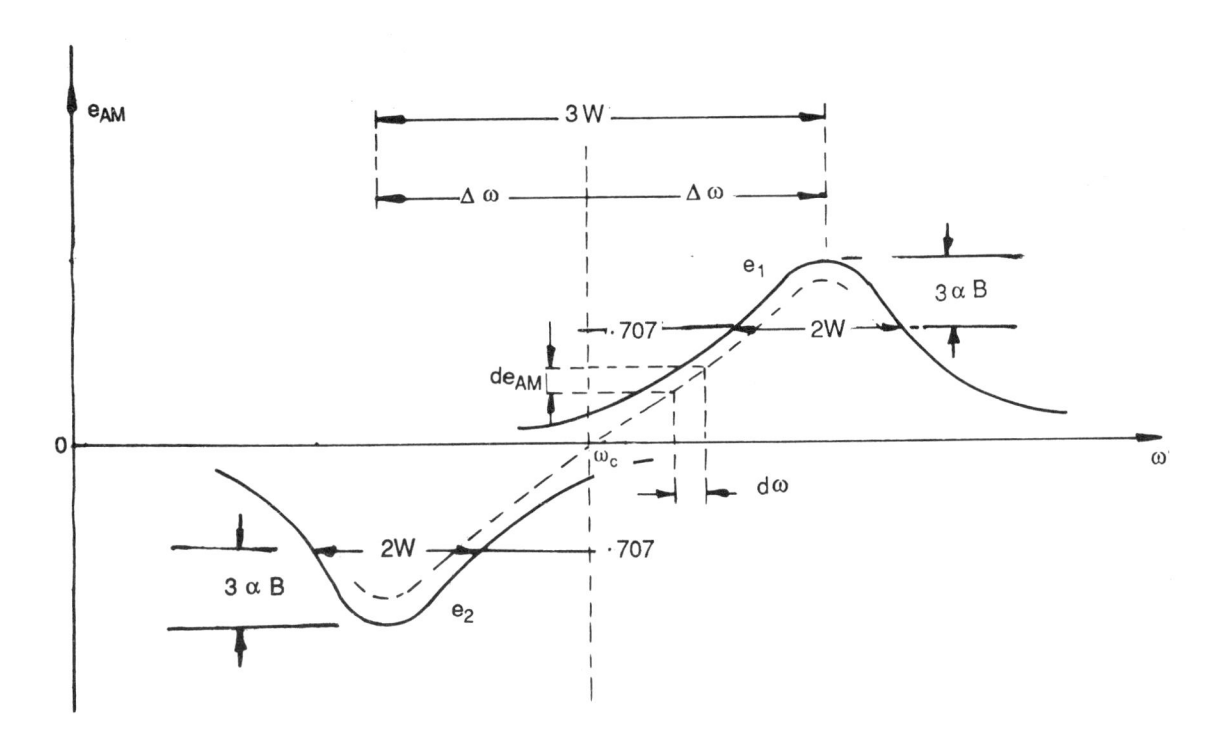

Fig. 6.7.2 (b) Balanced Slope Detector — Characteristic

$$2(\Delta\omega) = 3W$$

or
$$\Delta\omega = 1.5\,W \qquad (6.7.2)$$

If the frequency deviation of an input FM wave is more than the one given by relation 6.7.2 distortion will occur due to non-linearity of the characteristic. Therefore, the operation of this detector is limited to small deviation only.

(ii) The discriminator characteristics depends critically on the amount of detuning of the resonant circuits.
(iii) The tuned circuit output is not purely bandlimited and, hence, the low pass RC filter of the envelope detector introduces distortion.

In spite of the above disadvantages, this circuit can be designed to keep the distortions within tolerable limits and provide a satisfactory operation.

## Phase Difference Discriminators

*Foster-Seeley (Center-Tuned) Discriminator*: This type of discriminator is very widely used. The circuit arrangement is shown in Fig. 6.7.3a. The circuit comprises of an inductively coupled double-tuned circuit in which both primary and secondary are tuned to same frequency (intermediate frequency $f_{if}$). The centre of the secondary is connected to top of the primary (collector end) through a capacitor C. This capacitor C performs the following functions

(i) It blocks the dc from primary to secondary. (ii) It couples the signal frequency from primary to center-tapping of the secondary. The primary voltage $V_3$ (signal voltage) thus appears across the inductance L. Almost, entire $V_3$ appears across L, except a small drop across C. By proper choice of C and L the drop across C may be kept negligible.

The center tapping of the secondary has an equal and opposite voltage across each half winding. Therefore, $V_1$ and $V_2$ are equal in magnitude but opposite in phase. The radio frequency voltages $V_{a1}$ and $V_{a2}$ applied to the diodes $D_1$ and $D_2$ are given as

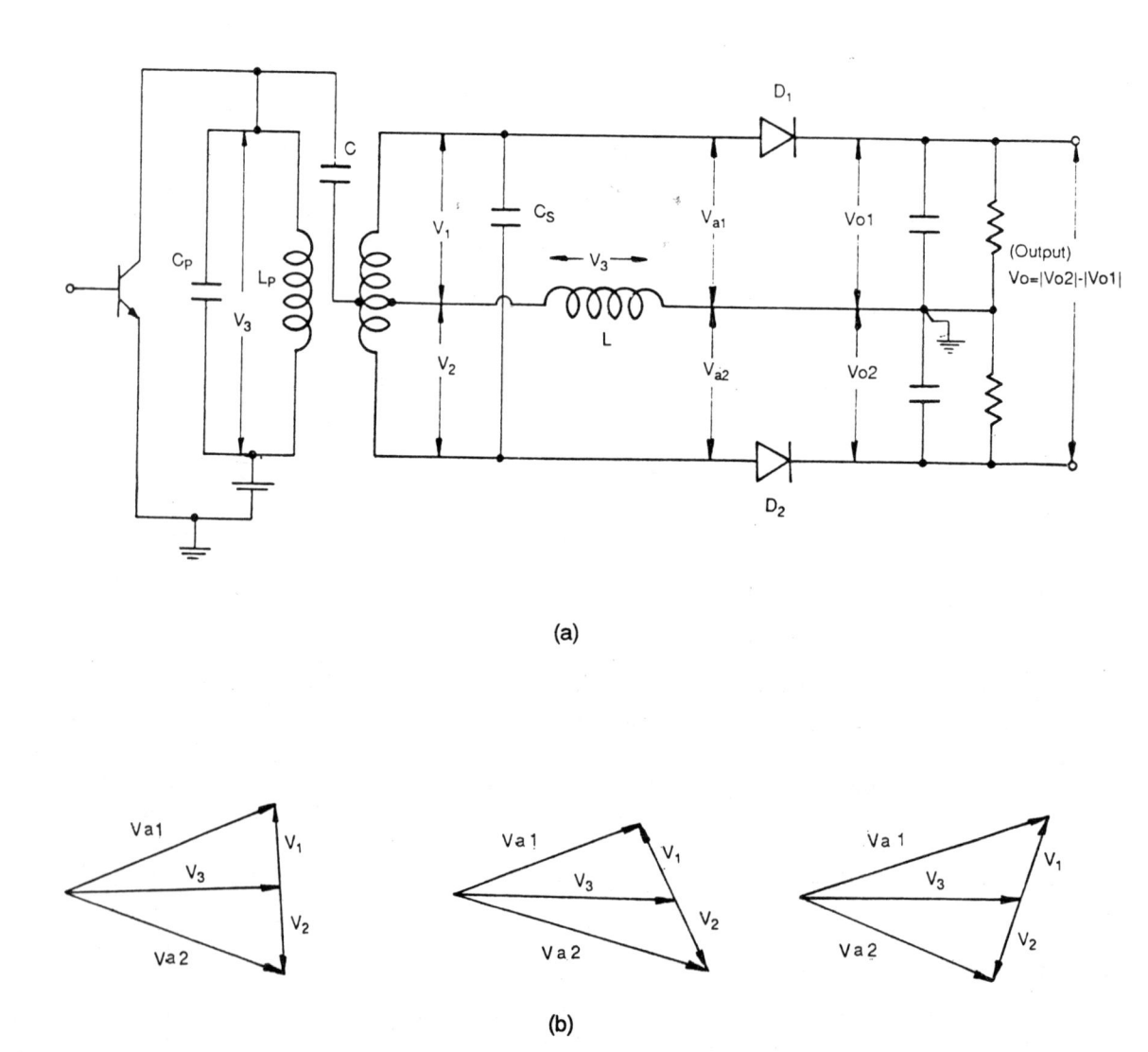

Fig. 6.7.3. (a) Circuit diagram of a Foster-Seely Discriminator  (b) Phasor Diagram

Voltage

+ve

Peak separation

← Δf → ← Δf →

$V_{a2}$ $V_{a1}$

$|V_{a2}|-|V_{a1}|$

0   fif   f

(c)

Relative amplitude of output

Increased amplitude
Reduced amplitude

Instantaneous freq

(d)

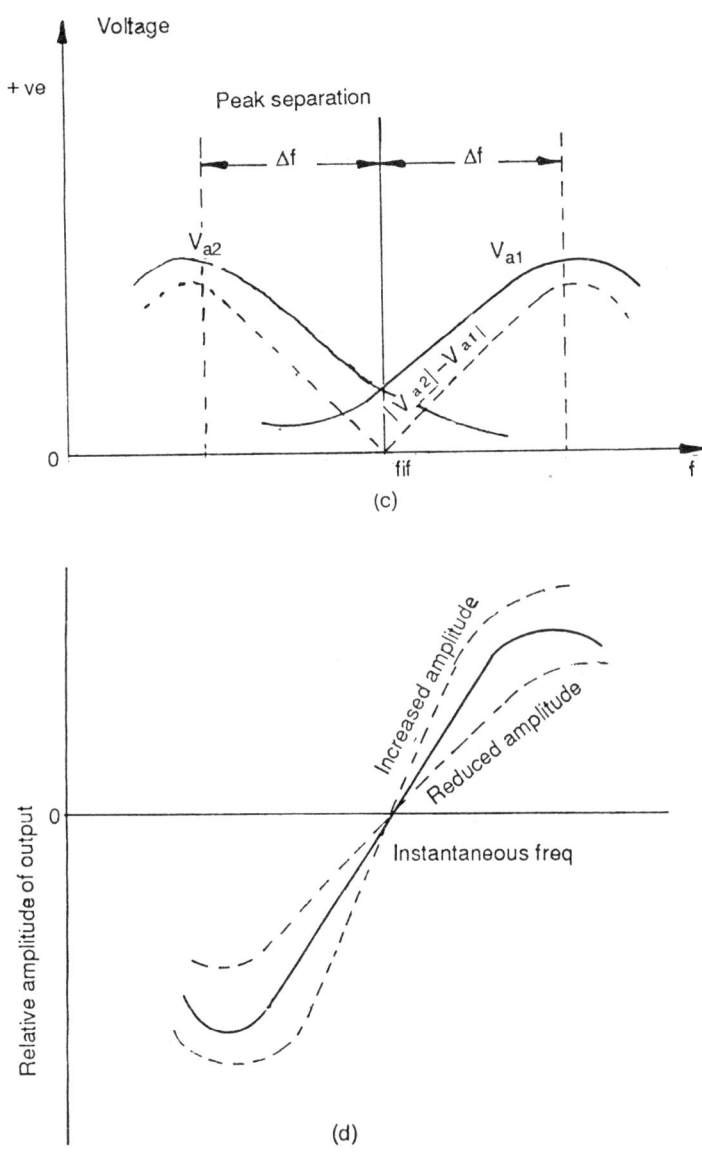

Fig. 6.7.3  Discriminator Characteristics

$$V_{a1} = V_3 + V_1; V_{a2} = V_3 - V_2 \qquad (6.7.3)$$

Voltage $V_{a1}$ and $V_{a2}$ depend on the phasor relations between $V_1$, $V_2$ and $V_3$. The phasor diagram for different frequencies are shown in Fig. 6.7.3b. The phasor diagrams $V_1$ and $V_2$ are always equal and are in phase opposition. However the phase position of $V_1$ and $V_2$ relative to $V_3$ will depend on the tuned secondary at the resonance, or off the resonance, as discussed as follows.

1. *At resonance*: When an input voltage has a frequency equal to the resonant frequency $f_{if}$ of the tuned secondary, $V_3$ is inphase quadrature (90 out phase) with $V_1$ and $V_2$. This is shown in the first phasor diagram. The resultant voltages $V_{a1}$ and $V_{a2}$ are equal in magnitude.

2. *Off resonance*: When the input voltage is off the resonant frequency $f_{if}$ of the tuned secondary, the phase position of $V_1$ and $V_2$ relative to $V_3$ will be different from 90°. Let $Q_s$ be the quality factor of the tuned secondary. When an input signal frequency is above the resonant frequency $f_{if}$ by an amount ($f_{if}/2Q_s$) the phase different between $V_3$ and $V_1$ is 45°, as shown in the second phasor diagram. Since $V_2$ is in phase opposition of $V_1$, the phase difference between $V_3$ and $V_2$ is 135°. The phasor diagram reveals that $V_{a1}$ is reduced whereas $V_{a2}$ is increased. The situation is reversed when the input voltage has a frequency below $f_{if}$ which is evident from the third phasor diagram. Thus, the amplitude of the voltages $V_{a1}$ and $V_{a2}$ will vary with the instantaneous frequency $f$ in the manner shown in Fig. 6.7.3c. The RF voltage $V_{a1}$ and $V_{a2}$ are separately rectified by the diodes $D_1$ and $D_2$ respectively to produce voltages $V_{o1}$ and $V_{o2}$. The RF components are by-passed by the capacitors leaving only modulating frequency component and a d.c. term. The voltage $V_{o1}$ and $V_{o2}$ then represent the amplitude variations of $V_{a1}$ and $V_{a2}$ respectively. The diodes are so arranged that the output voltage $V_o$ is equal to the arithmetic difference $|V_{o2}|-|V_{o1}|$ i.e.,

$$V_o = |V_{o2}| - |V_{o1}| \qquad (6.7.4)$$

Therefore, the voltage $V_o$ will vary with instantaneous frequency in accordance with the difference $|V_{a2}|-|V_{a1}|$, as shown by the dotted curve in Fig. 6.7.3c. This dotted curve is called discriminator characteristic. It is zero at resonance, positive above resonance and negative below resonance. The discriminator characteristic is linear for the region between the peaks of $V_{a1}$ and $V_{a2}$. This range is known as *peak-separation range*.

The discriminator works satisfactorily over a range of instantaneous frequency only slightly less than the peak separation range. The peak separation range, therefore, must be more than the twice the frequency deviation ($\Delta f$).

*Disadvantage*: The Foster-Seeley discriminator has the disadvantage that any variation in amplitude of the input FM signal due to noise modifies the discriminator characteristic as shown by the dotted lines in the Fig. 6.7.3d. The undesired frequency components corresponding to amplitude variations are produced in the detected output and the output gets distorted. The distortion is reduced using a limiter circuit in the FM receiver.

### Ratio Detector

Ratio detector is an improvement over the Foster-Seeley discriminator and is widely used. As it does not respond to amplitude variations, a limiter is not needed. The circuit diagram of a ratio detector is shown in Fig. 6.7.4. The circuit is similar to the circuit of Foster-Seeley discriminator (Fig. 6.7.3a), except the following:

(i) The polarity of diode $D_2$ has been reversed

(ii) The output $V_o$ is taken from the center tap of a resistor R that shunts the load impedance of the two diodes. The output voltage varies with the input signal frequency (FM) exactly in the same way as it does in the Foster-Seeley discriminator, but its magnitude is reduced to half. This can be shown as follows:

For the time being, we will ignore the two resistances (both denoted by $\dot{R}$) and the capacitance $C$. The voltage $V_{o1}$ and $V_{o2}$ have the same magnitude as in the case of a Foster-Seeley discriminator, but $V_{o2}$ is now reversed in polarity. Therefore the voltage $V_R$ is now sum of $V_{o1}$ and $V_{o2}$ unlike in the Foster-Seeley case, where it was the difference of $V_{o2}$ and $V_{o1}$. Thus,

$$V_R = |V_{o1}| + |V_{o2}| \qquad (6.7.5)$$

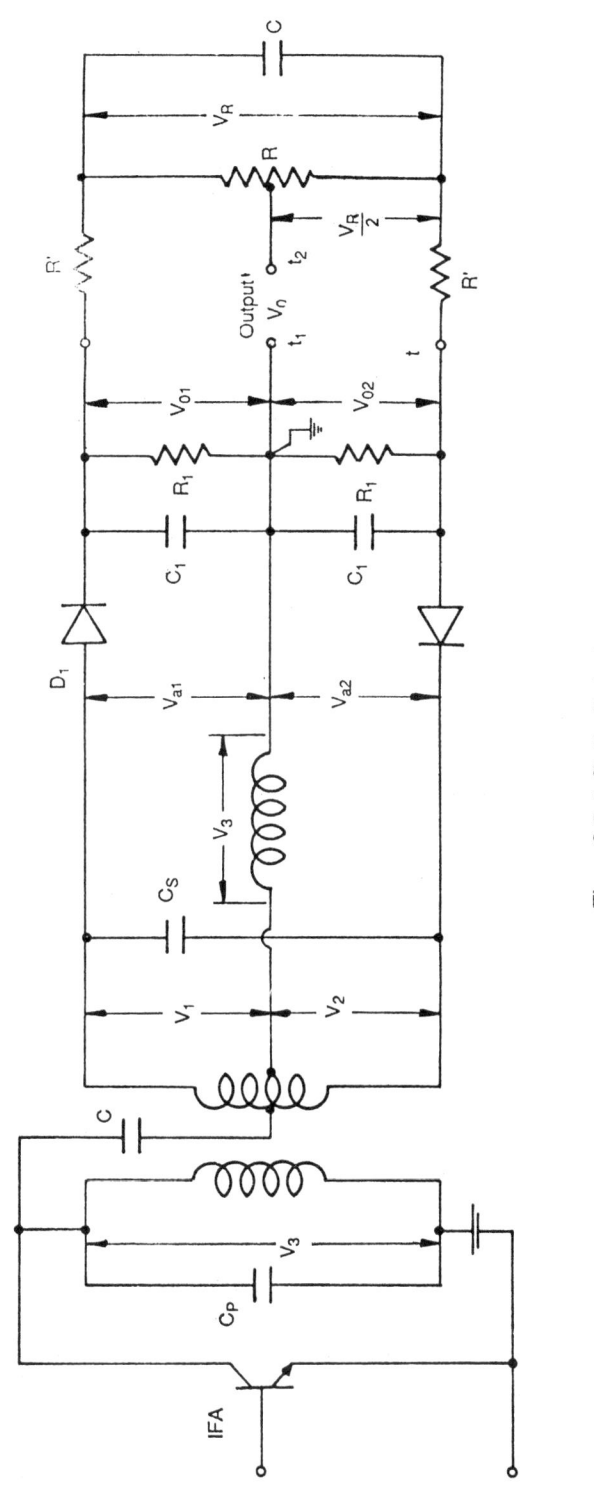

Fig. 6.7.4 Ratio Detector

The output voltage $V_o$ is taken across the terminal $t_1, t_2$. From the circuit diagram

$$V_o = V_{t_1,t_2} = |V_{t,t_1}| - |V_{t,t_2}| = |V_{o2}| - \left|\frac{V_R}{2}\right|$$

But

$$V_R = |V_{o1}| + |V_{o2}|$$

Hence,

$$V_o = V_{o2} - \frac{V_{o1} + V_{o2}}{2}$$

or

$$V_o = \frac{V_{o2} - V_{o1}}{2} \qquad (6.7.6)$$

This output is only half of the output given by Eq. 6.7.4 for the Foster-Seeley discriminator. Thus, the ratio detector has exactly same behaviour except that its output is reduced.

### Merits of the Ratio Detector

The circuit has an advantage over the Foster-Seeley discriminator that it does not respond to the amplitude variations simultaneously present in the FM input signal. The effect of amplitude fluctuations (AM) is suppressed by shunting the capacitor $C$ with $R$ and introducing the two resistances $R_1$ of a suitable value in the circuit. The reactance of $C$ at the lowest modulation frequency must be small as compared to $R_2$. Then the capacitance $C$ bypasses the undesired amplitude fluctuations but not the amplitude variations that are generated due to FM.

### Demerits of the Ratio Detector

The ratio detector may not tolerate the long-period variation in signal strength. This requires an AGC signal. The dc voltage across $C$ may be used as an AGC control voltage

### FM Receiver

The function of the FM receiver is to intercept the FM signal incoming from an FM transmitter, and then recover the original modulating signal. Figure 6.7.5 shows the block diagram of an FM broadcast receiver which is similar to an AM superheterodyne receiver.

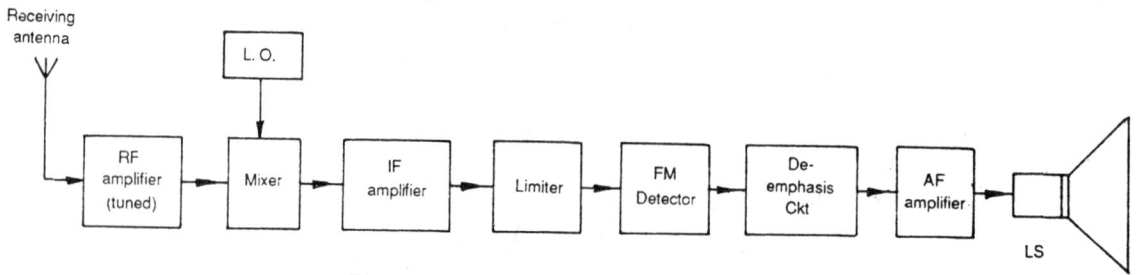

Fig. 6.7.5 Block Diagram of FM Receiver

The stages of the block diagram are described as follows:

## RF Amplifier

This stage amplifies the received radio signal. FM uses radio frequency ranging from 40 MHz to 1GHz for various application such as FM broadcasting, television sound transmission, police radio, military systems etc. The bandwidth needed for an RF amplifier may also be large (150 kHz). The RF amplifier also reject the image signal, as in AM receivers.

## Frequency Mixer and Local Oscillator

Separate active devices are used for the mixer and local oscillator as the frequency involved is in VHF and UHF range. The IF frequency in an FM receiver is much higher than AM. The FM broadcast receivers use 10.7 MHz as intermediate frequency.

## IF Amplifier

This stage amplifies the intermediate frequency signals. It comprises of multistage double-tuned or stagger-tuned amplifiers to provide a high gain and a high overall bandwidth (150 kHz). This stage is responsible for sensitivity and selectivity of the receiver.

## Limiter

The limiter keeps the IF amplifier output voltage constant to a pre-determined value and removes all amplitude fluctuations due to noise, or any other interference. This is essential because the FM detector needs a constant amplitude FM voltage at its input for a satisfactory operation. The limiter may be ignored if the ratio detector has been used. Simple diodes and amplifying devices are used in making a limiter circuit. Figure 6.7.6 shows an FET limiter circuit. The leaky type bias provided at the gate by $R_G - C_G$ combination gives the limiting action.

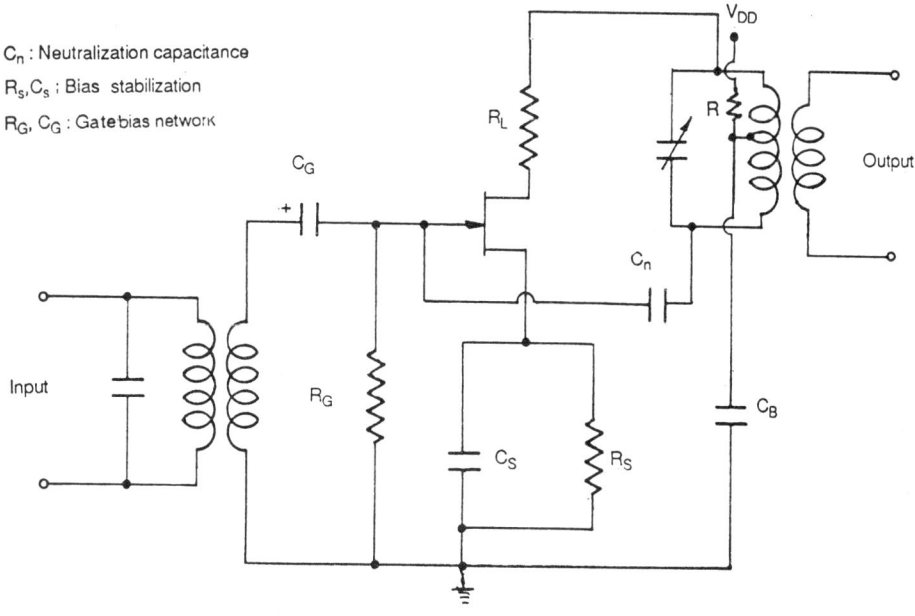

Fig. 6.7.6 A Limiter Circuit

As the input voltage increases, the bias at CS increases in such a way that the gain of the amplifier is reduced and the output voltage remains constant irrespective of variations in input voltage amplitude. Thus, the limiting action is achieved.

### FM Detector

It recovers the modulating signal from the IF signal. Various types of FM detector circuit have already been discussed. De-emphasis circuit does the inverse job of the pre-emphasis circuit. The high modulating frequency terms boosted by pre-emphasis are brought back to original amplitude level by de-emphasis circuit. This circuit will be discussed in Sec. 6.10.

### AF Amplifier and Speaker

This stage amplifies the audio frequency (AF) modulating signal recovered by the FM detector. The amplifier has wider bandwidth than that of AM receiver. The loud speaker converts the electrical signal into sound signal.

## 6.8 NOISE IN ANGLE MODULATED SYSTEMS

### Frequency Modulation

The noise performance of amplitude modulated systems has been discussed in Sec. 5.8 On the same lines we will discuss the noise performance of angle modulated (frequency and phase) signal. The block diagram of FM system is shown in Fig. 6.8.1. The bandpass filter selects the carrier alongwith sidebands.

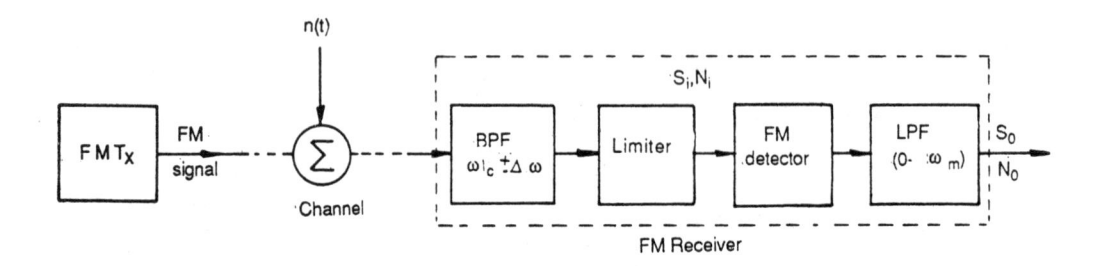

Fig. 6.8.1 FM System

The FM detector comprises a frequency discriminator followed by the envelope detector. The input of the FM detector is maintained constant by a limiter. The output of the receiver is the baseband signal $f(t)$. Let us evaluate the input and output signal powers $(S_i, S_o)$ and noise powers $(N_i, N_o)$ for evaluating the noise performance.

### Signal Power

It has been observed in Sec. 6.3 that the power of the FM signal is same as that of an unmodulated carrier. This is because the amplitude of the signal remains unchanged after modulation. Therefore

ANGLE MODULATION SYSTEMS 399

$$S_i = \overline{(A\cos\omega_c t)^2} = \frac{A^2}{2} \tag{6.8.1}$$

The output power of the signal and noise can be calculated independently i.e. for evaluating the signal output power, the noise is assumed to be absent, and vice-versa. This assumption can be justified (Prob 6.4). The output $S_o(t)$ of the detector is proportional to instantaneous frequency $\omega_i$ of the FM signal applied at its input,

$$S_o(t) \, \alpha \, \omega_i$$
$$= K\omega_i \tag{6.8.2}$$

where $K$ is the constant of proportionality. The angular frequency depends on the modulating signal $f(t)$ and is given as

$$\omega_i = \omega_c + K_f f(t)$$

Hence,

$S_o(t) = K\omega_c + KK_f f(t)$ in which the useful message signal is $KK_f f(t)$. The signal power due to this message term is

$$S_o = K^2 K_f^2 \overline{f^2(t)} \tag{6.8.3}$$

## Noise Power

The input noise power is evaluated for the baseband width i.e. bandwidth of the message signal $f(t)$ which is $2\omega_m$. The input noise power is same as given by Eq. 5.8.4.

$$N_i = \eta f_m \tag{6.8.4}$$

The output noise power is computed assuming the message signal term $f(t)$ to be zero. Therefore, input $g_i(t)$ to the detector will have a carrier term plus noise term.

$$g_i(t) = A\cos\omega_c t + n_i(t)$$

Substituting quadrature components of $n_i(t)$ we get

$$g_i(t) = A\cos\omega_c t + n_c(t)\cos\omega_c t - n_s(t)\sin\omega_c t$$
$$= [A + n_c(t)]\cos\omega_c t - n_s(t)\sin\omega_c(t)$$

After trigonometric manipulations, $g_i(t)$ can be written as

$$g_i(t) = A(t)\cos\{\omega_c t + \psi(t)\} \tag{6.8.5a}$$
$$= A(t)\cos\theta(t) \tag{6.8.5b}$$

where

$$A(t) = \sqrt{[A + n_c(t)]^2 + n_s^2(t)} \tag{6.8.6a}$$

and

$$\psi(t) = \tan^{-1}\left\{\frac{n_s(t)}{A + n_c(t)}\right\} \tag{6.8.6b}$$

Obviously,

$$\theta(t) = \omega_c t + \psi(t) \tag{6.8.7}$$

**400** COMMUNICATION SYSTEMS: Analog and Digital

Since, the output of the FM detector depends on the instantaneous frequency $\omega_i$ which is $\frac{d}{dt}\{\theta(t)\}$, hence $A(t)$ is not of our interest. We are interested only in $\theta(t)$ and $\psi(t)$ which are responsible for the FM detector output. Again the analysis, depends on the fact whether the noise signal is small or large in comparison to the carrier amplitude. We will take up two cases separately, viz.

(i) $A \gg n_c(t)$ and $n_s(t)$ (small noise case)
(ii) $A \ll n_c(t)$ and $n_s(t)$ (large noise case)

*Small Noise Case:* In this case $A \gg n_c(t)$ and $n_s(t)$ and hence Eq. 6.8.6b becomes

$$\psi(t) \tan \simeq \tan^{-1}\left[\frac{n_s(t)}{A}\right] \tag{6.8.8a}$$

It is well known that for a small angle $\phi$, $\tan \phi = \phi$, hence,

$$\psi(t) \simeq \frac{n_s(t)}{A} \tag{6.8.8b}$$

The corresponding value of $\theta(t)$ is given as, (from Eq. 6.8.7)

$$\theta(t) = \omega_c t + \frac{n_s(t)}{A} \tag{6.8.9}$$

The instantaneous frequency is given as

$$\omega_i = \frac{d}{dt}[\theta(t)] = \frac{d}{dt}\left[\omega_c t + \frac{n_s(t)}{A}\right]$$

$$= \omega_c + \frac{\dot{n}_s(t)}{A} \tag{6.8.10a}$$

Where

$$\dot{n}_s(t) = \frac{d}{dt}\{n_s(t)\} \tag{6.8.10b}$$

The detected output $e_d(t)$ is proportional to $\omega_i$, hence

$$e_d(t) = K\left[\omega_c + \frac{\dot{n}_s(t)}{A}\right] = K\omega_c + \frac{K\dot{n}_s(t)}{A}$$

The noise part of this detected output is given as

$$n_o(t) = \frac{K\dot{n}_s(t)}{A} \tag{6.8.11}$$

The power density spectrum of $n_o(t)$ is given by

$$S_{n_o}(\omega) = \frac{K^2}{A^2} S_{\dot{n}_s}(\omega) \tag{6.8.12}$$

where $S_{\dot{n}_s}(\omega)$ is the power density spectrum of $\dot{n}_s(t)$. Assume that $n_s(t)$ has the power density spectrum $S_{n_s}(\omega)$, then its derivative $\dot{n}_s(t)$ will have its power density spectrum given by,

$$S_{\dot{n}_s}(\omega) = \omega^2 S_{n_s}(\omega)$$

ANGLE MODULATION SYSTEMS 401

By putting this value in Eq. 6.8.12, we get

$$S_{n_o}(\omega) = \frac{K^2 \omega^2}{A^2} S_{n_s}(\omega) \qquad (6.8.13)$$

The detected output is bandlimited to $\omega_m$ and hence Eq. 6.8.13 holds only for $|\omega| \leq \omega_m$, and is zero elsewhere. The value of $S_{n_s}(\omega)$ for a white noise is similar to the DSB-SC case and is given as follows (refer Eq. 5.8.7)

$$S_{n_s}(\omega) = \eta, \qquad |\omega| \leq \omega_m$$

Substituting this value in Eq. 6.8.13, we get

$$S_{n_o}(\omega) = \begin{cases} \dfrac{K^2 \omega^2 \eta}{A^2}, & |\omega| \leq \omega_m \\ 0, & \text{elsewhere} \end{cases} \qquad (6.8.14)$$

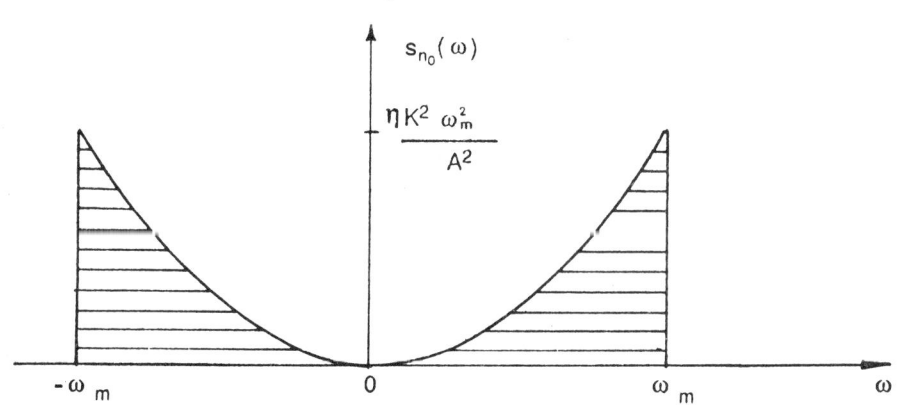

Fig. 6.8.2  Noise Power Density Spectrum at the Discriminator Output

The spectrum $S_{n_o}(\omega)$ is shown in Fig. 6.8.2. It is evident that the spectrum is a parabolic function of frequency $\omega$.

The output noise power $N_o$ is given by

$$N_o = \frac{1}{\pi} \int_0^{\omega_m} S_{n_o}(\omega) d\omega = \frac{K^2 \eta}{\pi A^2} \int_0^{\omega_m} \omega^2 d\omega$$

$$= \frac{\eta K^2 \omega_m^3}{3 \pi A^2} \qquad (6.8.15)$$

*Noise-quieting effect*: Equation 6.8.15 reveals that the output noise power of the FM detector varies inversely with carrier power $(A^2/2)$. This decrease in the noise power with an increase in carrier power is known as *noise-quieting effect* of the carrier.

The output signal to noise ratio $(S_o / N_o)$ is obtained by Eqs 6.8.3 and 6.8.15

$$\frac{S_o}{N_o} = \frac{3 \pi K_f^2 \overline{f^2(t)} A^2}{\eta \omega_m^3} \qquad (6.8.16)$$

## Figure of Merit $\gamma$

The figure of merit $\gamma$ for the FM system may be obtained by using Eq. 6.8.1, 6.8.4, and 6.8.16,

$$\gamma_{FM} = \frac{(S_o/N_o)}{(S_i/N_i)} = \frac{3K_f^2 \overline{f^2(t)} (2\pi f_m)}{\omega_m^3}$$

$$= \frac{3K_f^2 \overline{f^2(t)}}{\omega_m^2} \qquad (6.8.17)$$

## Example 6.8.1

### Single tone Modulation

A single-tone modulating signal $E_m \cos\omega_m t$, frequency modulates a carrier $A\cos\omega_c t$. show that
  (a) The detector output signal to noise ratio is proportional to the square of bandwidth of the FM signal
  (b) The figure of merit $\gamma$ is given by

$$\gamma_{FM} = (3/2) m_f^2 \qquad (6.8.18)$$

*Solution*

$$f(t) = E_m \cos\omega_m t$$
$$K_f f(t) = K_f E_m \cos\omega_m t$$

The mean square value of both the sides yields

$$K_f^2 \overline{f^2(t)} = K_f^2 \frac{E_m^2}{2} \qquad (6.8.19a)$$

From Eq. 6.1.18 a

$$K_f^2 E_m^2 = (\Delta\omega)^2 \qquad (6.8.19b)$$

and hence

$$K_f^2 \overline{f^2(t)} = \frac{(\Delta\omega)^2}{2} \qquad (6.8.20)$$

Substituting this value in Eq. 6.8.16, we get

$$\frac{S_o}{N_o} = \frac{3\pi(\Delta\omega)^2 A^2}{2\eta \omega_m^3} = 3\pi \left(\frac{\Delta\omega}{\omega_m}\right)^2 \frac{(A^2/2)}{2\pi f_m \eta} \qquad (6.8.21)$$

This shows that $(S_o/N_o)$ is proportional to $(\Delta\omega)^2$. Now the bandwidth of FM is twice the deviation $(\Delta\omega)$, and hence, the $(S_o/N_o)$ ratio is proportional to the square of bandwidth of the FM signal.

(b) By definition $\frac{\Delta\omega}{\omega_m} = m_f$, hence Eq. 6.8.21 becomes

$$\frac{S_o}{N_o} = \frac{3}{2} m_f^2 \frac{(A^2/2)}{\eta f_m} = \frac{3}{2} m_f^2 \left(\frac{S_i}{N_i}\right)$$

Therefore,

$$\gamma_{FM} = \frac{(S_o/N_o)}{(S_i/N_i)} = \frac{3}{2} m_f^2 \qquad (6.8.22)$$

## Noise Performance of FM as Compared to AM

It is worthwhile to compare the noise performance of FM with that of AM. The comparison may be conveniently made by considering the single-tone modulation. From Eq. 6.8.22.

$$\gamma_{FM} = \frac{3}{2} m_f^2$$

For AM with a large carrier, we have from Eq. 5.8.32

$$\gamma_{AM} = \frac{m_a^2}{2 + m_a^2}$$

The best noise performance is obtained for $m_a = 1$,

$$\gamma_{AM} = \frac{1}{3}$$

Therefore

$$\frac{\gamma_{FM}}{\gamma_{AM}} = \frac{1}{2} m_f^2 \qquad (6.8.23a)$$

The performance of FM and AM will be same if the right-hand side is unity,

$$\frac{1}{2} m_f^2 = 1$$

or

$$m_f = \frac{\sqrt{2}}{3} \simeq 0.5 \qquad (6.8.23b)$$

This shows that FM has no improvement over AM when $m_f \leq 0.5$. *Thus $m_f = 0.5$ may be considered as a transition point between the wideband and narrowband FM*. It should be recalled that narrowband FM and AM are identical in bandwidth. This implies that the noise improvement depends on the bandwidth. Since narrowband FM and AM have the same bandwidth, there is no improvement in the noise performance of narrowband FM over AM. The improvement in noise performance is at the expanse of the bandwidth as shown below,

Bandwidth of FM signal, $W_{FM} = 2(\Delta\omega) = 2 m_f \omega_m$
and
Bandwidth of AM signal, $W_{AM} = 2\omega_m$
Hence

$$\frac{W_{FM}}{W_{AM}} = m_f$$

Substituting this value of $m_f$ in Eq. 6.8.23a, we get

$$\frac{\gamma_{FM}}{\gamma_{AM}} = \frac{1}{2}\left(\frac{W_{FM}}{W_{AM}}\right)^2 \qquad (6.8.25)$$

This reveals that if a bandwidth ratio is increased by a factor 2, the $(\gamma_{FM}/\gamma_{AM})$ increases by a factor 4 i.e. 6 dB, which verifies the fact that noise performance is improved by increasing the bandwidth. This exchange of bandwidth and noise performance studied in detail by Hartley and Shannon, is feasible in FM and not in AM, because AM has a fixed bandwidth.

## Phasor Representation of FM Noise

A qualitative study of FM noise can be done by its phasor diagram representation. We will assume that the message signal is absent i.e. $f(t) = 0$, and only $n_i(t)$ is associating the carrier at the input of the detector. Then the input of the detector is given as

$$g_i(t) = [A + n_c(t)] \cos \omega_c t - n_s(t) \sin \omega_c t \qquad (6.8.26)$$

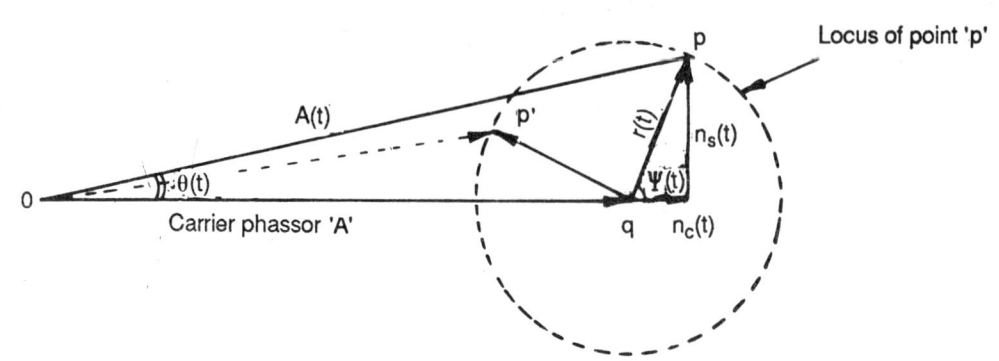

Fig. 6.8.3 Phasor Diagram of FM Noise

Obviously, the term $n_c(t)$ is in phase with carrier term A and the term $n_s(t)$, in phase quadrature with $n_c(t)$. The phasor diagram is shown in Fig. 6.8.3. The coordinate system is assumed to be rotating counter clockwise with a frequency $\omega_c$. The resultant of quadrature components of the noise is $r(t)$.
From Eqs 6.8.5a and b,

$$g_i(t) = A(t) \cos[\omega_c t + \psi(t)] = A(t) \cos \theta(t) \qquad (6.8.27)$$

where $A(t)$ is the resultant of the carrier vector A and the noise vector $r(t)$. $\psi(t)$ is the angle that $A(t)$ makes with the carrier vector A. Noise parameters $r(t)$ and $\psi(t)$ vary randomly with time, accordingly the noise vector $r(t)$ rotates around the point $q$. The rotation path is locus of point $p$ and is random in nature. For convenience if we assume $r(t)$ constant, the path will be circular as shown by the dotted curve. As this point $p$ rotates around the point $q$, the resultant vector $A(t)$ and its phase $\theta(t)$ also varies. One such case is shown by point $p'$ for which both $A(t)$ and $\theta(t)$ is less as compared to point $p$. For a low-noise case, $r(t) \ll A$ and hence, locus of $p$ is restricted near the point $q$. But as the noise signal $r(t)$ increases, the locus reaches near the origin and even may sweep the origin 'O' if noise vector $r(t)$ exceeds the carrier amplitude A (large noise case).

The interfering noise vector $r(t)$ modulates the carrier both in amplitude and in angle. The amplitude variation in $A(t)$ does not produce any adverse effect in the FM receiver, as it is not responded to by the FM receiver due to the features of limiter, or ratio detector. However, the angle fluctuation produces a phase deviation of $\pm \psi(t)$ about the carrier phasor. This phase deviation is responded by FM receiver. As the noise signal $r(t)$ increases in comparison to carrier, the phase deviation $\psi(t)$ also increases. This increases the distortion produced by noise. This type of distortion may be caused by any type of interfering signal accompanying the FM signal.

### Example 6.8.2

A single-tone modulating signal with frequency $f_m = 25$ Hz modulates a carrier $A \cos \omega_c t$ and produces a deviation of 75 kHz. This FM signal reaches the receiver alongwith an interfering noise signal $n_i(t)$.

(i) Find the phase deviation produced if the noise voltage magnitude $r(t)$ is, (a) one fourth, (b) one third; and (c) half the carrier amplitude A, (ii) What is the maximum possible phase deviation produced by noise in a low-noise cases; (iii) Find the phase deviation produced by the modulating signal and hence show that the phase deviation produced by noise is much smaller than that is produced by low noise.

*Solution*

(i) The phasor diagram for case (a) and (c) is shown in Fig. 6.8.4. The maximum value of phase deviation about the carrier phasor in each case is obtained when $A(t)$ is tangenial to the circular locus.

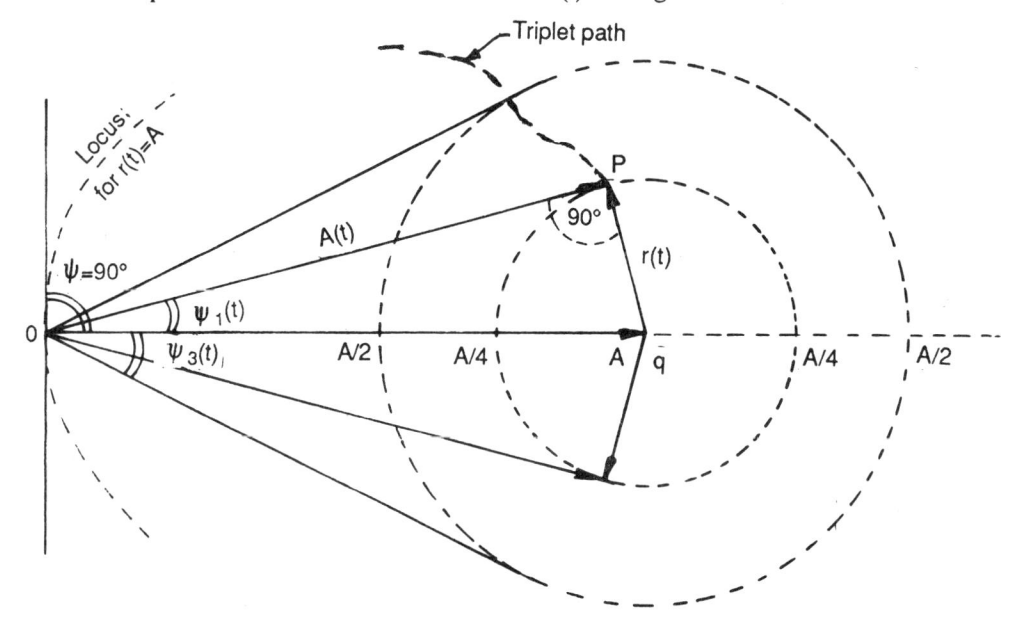

Fig. 6.8.4 Phase Deviation in Low Noise Case

It is obvious from the right angle triangle $opq$ that the phase deviation $\psi(t)$ is given as

$$\psi(t) = \pm \sin^{-1}\left[\frac{r(t)}{A}\right] \tag{6.8.28a}$$

(a) when the noise signal $r(t) = \dfrac{A}{4}$, the corresponding phase deviation is

$$\psi_1(t) = \sin^{-1}\left(\frac{1}{4}\right) = 14.5°$$

(b) similarly, we can draw the locus for $r(t) = \dfrac{A}{3}$, hence,

$$\psi_2(t) = \sin^{-1}\left(\frac{1}{3}\right) = 19.47° \tag{6.8.28b}$$

(c) Here $r(t) = A/2$

$$\psi_3(t) = \sin^{-1}\left(\frac{1}{2}\right) = 30° = \frac{\pi}{6} \text{ radians}$$

(ii) In the low-noise case, the noise vector $r(t)$ is always less than the carrier phasor $A$. The maximum possible phase deviation, therefore, is obtained when $r(t)$ approaches the carrier magnitude $A$. In this case, $r(t) = A$ and the maximum possible phase deviation in the low-noise case is given by

$$\psi = \sin^{-1}(1) = \pm \frac{\pi}{2} \text{ radians} \qquad (6.8.28c)$$

Note that as long as the noise input $r(t)$ is less than the carrier amplitude $A$, the locus of $A(t)$ never goes beyond the origin, and the maximum possible deviation is $\pm \frac{\pi}{2}$.

(iii) The modulating signal has a frequency deviation $\Delta f$ = 75 kHz, and the modulating frequency $f_m$ = 5 kHz.

Hence, the frequency modulation index is given by

$$m_f = \frac{\Delta f}{f_m} = \frac{75 \text{ kHz}}{5 \text{ kHz}} = 15$$

The corresponding phase deviation is given by Eq. 6.2.8b

$$\Delta \theta = m_f = 15 \text{ radians}$$

This is much greater than the maximum possible phase deviation produced by noise which is $\frac{\pi}{2}(= 1.57)$ radians (refer Eq. 6.8.28c).

## Capture Effect

It is obvious from Ex 6.8.2, that the phase deviation produced by the noise signal in FM is much smaller than the phase deviation produced by the modulating signal, provided that noise is smaller than the carrier. Thus, in the low-noise case, the distortion produced by noise at the output of the FM detector is negligible in comparison to the desired modulating signal. In other words, noise is almost suppressed by the signal. The noise suppression characteristic of FM can also be applied to the common channel interference. This phenomena, known as *capture effect*, is defined as follows:

"When FM signals from two transmitters operated on the same or nearly same carrier frequency reach the receiver simultaneously, the signal of a weak magnitude is suppressed by a strong signal, and the FM receiver reproduces only the strong signal". This effect is similar to noise-suppressing action, with the weaker signal playing the role of the noise. The weak signals due to common channel and adjacent channel interferences are suppressed. This is a very useful feature of FM system. AM lacks this particular feature.

In AM, the interfering signal causes appreciable variation in the amplitude of the resultant vector $A(t)$ which appears at the AM detector output as distortion.

## Large Noise Case

It is evident from Eq. 6.8.22 that under a low-noise case, the noise performance in FM goes on improving as the modulation index and bandwidth of FM is increased. But this improvement does not continue for an infinitely large increase in the bandwidth. This is because the input noise power introduced by the circuit components and devices increases with bandwidth ($\Delta f$), and it reaches a point when the noise signal $r(t)$ becomes higher than the carrier amplitude A. Then S/N ratio at the receiver input becomes less than unity. It is observed experimentally that when a carrier to noise ratio becomes even slightly less than unity, an impulse of noise is generated. This *noise impulse* appears at the output of the FM discriminator in the form a 'click' sound. The discriminator output noise is shown in Fig. 6.8.5. For low noise case, the instantaneous circuit noise $n(t)$ has Gaussian probability density. This type of noise is referred to as *smooth noise* and is always suppressed

ANGLE MODULATION SYSTEMS 407

by the message signal. But when noise is large and the carrier to noise ratio becomes slightly less than unity, even momentarily, a noise impulse is superimposed on the smooth noise as shown in Fig. 6.8.5b. which is referred to as *spike noise* or *impulse-noise*.

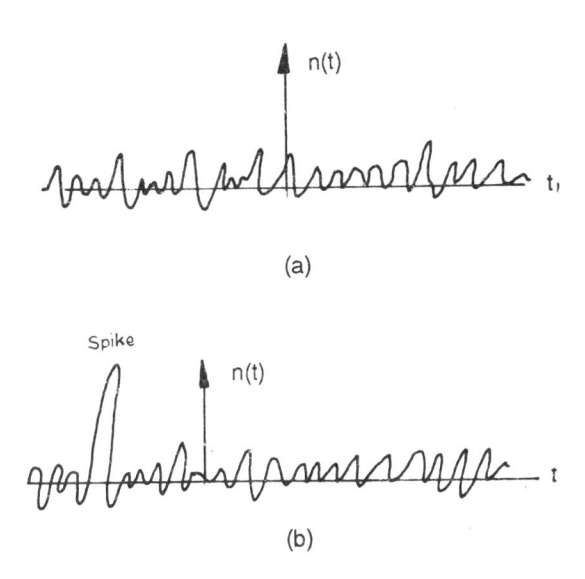

Fig. 6.8.5 Circuit Noise at Discriminator Output (a) Low-Noise case, (b) Large Noise case

## Threshold Effect in FM

When the carrier to noise ratio is slightly less than unity, the frequency of spike generation is small, and each spike produces individual clicking sound in the receiver. But, when the carrier to noise ratio is further decreased so that the ratio is moderately less than unity, the spikes are generated rapidly and the clicks merge into a *sputtering sound*. This phenomena is known as *threshold effect* in FM. The minimum carrier to noise ratio for which the FM noise improvement is not deteriorated significantly as compared to the small-noise case is defined as *threshold*. If the carrier to noise ratio is less than the threshold value, the sputter effect causes distortion in the FM receiver and the desired message signal cannot be resolved from the noise. *The threshold effect is more severe in FM than AM.*

## Generation of Spikes and Triplet Pulses

A positive or negative impulse (spike) is generated at the discriminator output under large noise condition; whereas a *pulse-triplet* (combination of three pulses) is generated under low-noise condition. A pulse-triplet contributes very small noise energy at the detector output as compared to a spike.

### Pulse Triplet Generation

The phasor diagram representation for low noise case, and a triplet path is shown earlier in Fig. 6.8.4. The end point of the resultant vector $r(t)$ rotates near the point $q$, and as the noise increases, the end point moves away from the point $q$. The locus of the end point of $A(t)$ may reach near the origin '$O$', but will never encircle the origin. This locus is known as *triplet path*, because this type of path causes a triplet pulse at the FM discriminator output. When the noise vector $r(t)$ is slightly less than the carrier vector $A$, $\psi(t)$ varies from $\pi$ to $-\pi$ (or vice-versa), as shown in Fig 6.8.6a. The waveform $\psi(t)$ from a pulse doublet for each phase-reversal over a small interval, e.g., $t_1$ to $t_2$; and $t_3$ to $t_4$. The output of the discriminator is proportional to the instantaneous frequency $\omega_i$ ($= d\psi/dt$) and is obtained by differentiating the waveform of Fig. 6.8.6a. This differentiated waveform ($d\psi/dt$) generates a pulse triplet for each pulse doublet, as shown in Fig.6.8.6b.

(a)

(b)

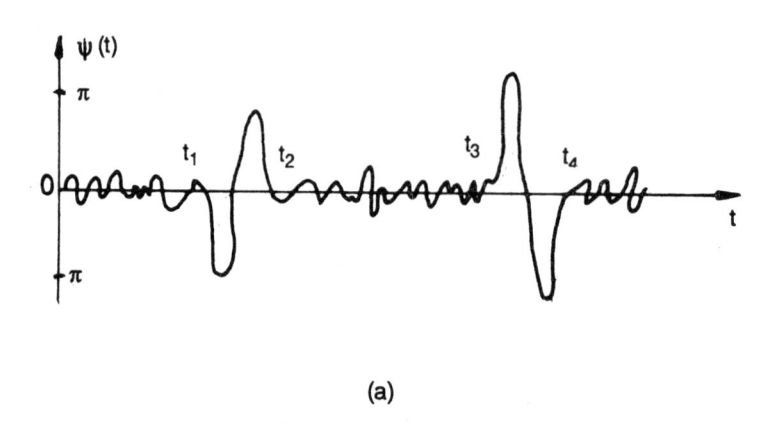

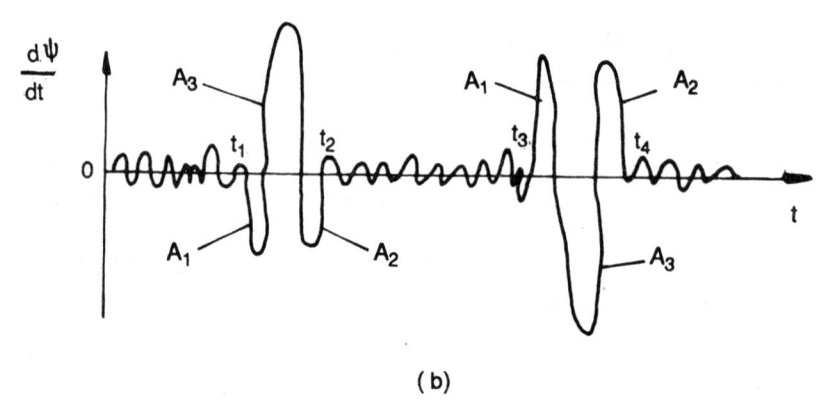

Fig. 6.8.6 Pulse Triplet Generation (a) Phasor Variation; (b) Noise Output of Discriminator (triplet)

The important point to be noted is that the total algebraic area under each triplet is zero because total negative area is equal to the positive area (or vice-versa), i.e.,

$$A_1 + A_2 = A_3$$

Thus, the total change in $\psi(t)$ over the interval of a triplet is zero and no noise is produced at the discriminator output.

### Spike generation

Figure 6.8.7a shows the phasor diagram representation for a large-noise case. When the noise phasor $r(t)$ is large as compared to the carrier vector $A$, the locus of the end point $P$ of the resultant $A(t)$ moves away from point $q$ and may even rotate about the origin. The locus encircles the origin and is referred to as *spike path*, since it generates a noise-spike. Phase angle $\psi(t)$ changes by a multiple of $2\pi$ during a small time interval as shown in Fig. 6.8.7b. The phase angle is increasing by $2\pi$ during intervals $(t_1, t_2)$ and $(t_3, t_4)$; whereas it is decreasing by $2\pi$ during the interval $(t_5, t_6)$.

Note that the phase angle $\psi(t)$ changes in any one direction over an interval (either increasing or decreasing), and not positive to negative (or vice versa) as in the low-noise case. Hence, an undirectional impulse appears over an interval at the FM discriminator output which is obvious from the wave form shown in Fig. 6.8.7c. These noise impulses shown in Fig. 6.8.7c are called noise-spikes. When $A(t)$ rotates counter-clockwise, the phase $\psi(t)$ changes by $-2\pi$ and generates a positive spike; and if $A(t)$ rotates clockwise the phase $\psi(t)$ changes by $-2\pi$ generating a positive spike.

The area under each spike in Fig 6.8.7c is $2\pi$. This can be seen by integrating the first impulse occuring for the interval $(t_1, t_2)$.

$$A = \int_{t_1}^{t_2} \frac{d\psi}{dt} dt = [\psi]_{t_1}^{t_2} = 2\pi \tag{6.8.29}$$

Each spike has equal area of $2\pi$ and, hence, smaller the duration, greater is its height. These spikes behave like shot noise.

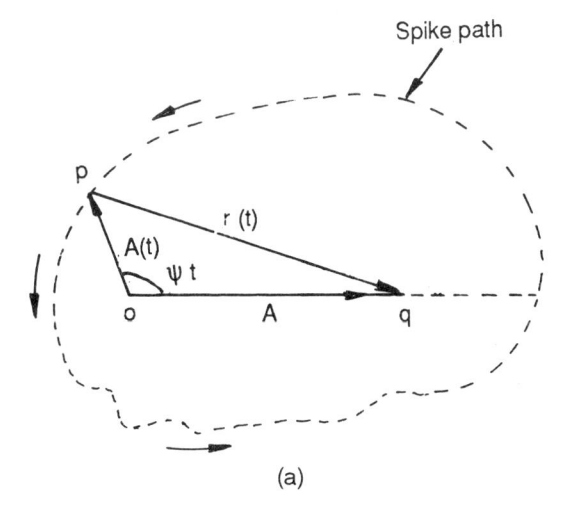

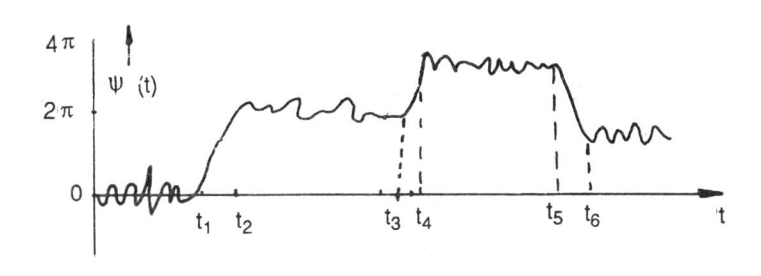

```
           dψ
           ──
           dt

                          A            A          t₅  t₆
            0 ─────────┬──┬────────┬──┬──────────┬────┬──────→
                      t₁ t₂       t₃ t₄          A    t
                                                 
                   A = 2π for all spikes
```

(c)

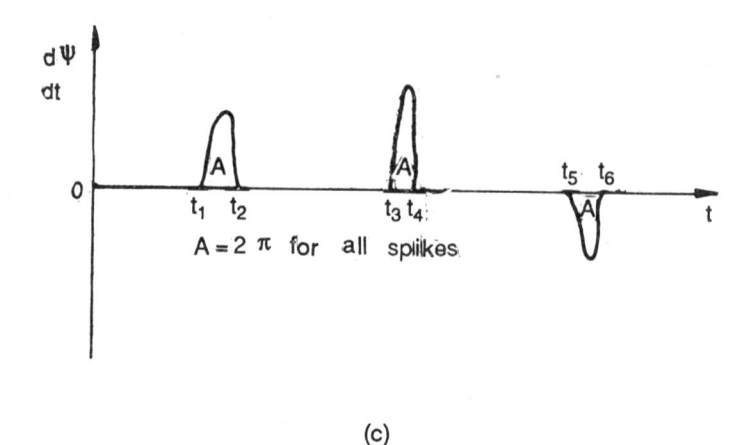

Fig. 6.8.7 Spike Generation (a) Phasor Diagram; (b) Phase Angle Variation; (c) Noise-output of discriminator

## Noise Energy at the output of Baseband Filter

The FM detector consists of a low pass filter (0 to $\pm f_m$) to allow only the baseband modulating signal. We will now illustrate that noise spike (large-noise case) generates a large noise energy at the output of the lowpass filter, and this causes the threshold effect. But the pulse-triplet (low-noise case) generates almost negligible noise energy at the output of a baseband filter; so that noise is completely suppressed by the modulating signal and no threshold is observed. This is obvious from the Fourier transforms of noise-spike and noise-triplet shown in Fig. 6.8.8.

According to Parseval's theorem, energy of a signal is proportional to the area under its $|F(\omega)|^2$ curve. At the output of the low pass filter, the area centered in the range 0 to $\pm \omega_m$ is of interest. It is obvious from Fig. 6.8.8 that in the vicinity of $\omega = 0$, the area under $|F(\omega)|$ is much more for a spike and much less for a triplet. This can also be proved from a property of Fourier transform, according to which Fourier transform of a function $f(t)$ at $\omega = 0$ is equal to the area under the function $f(t)$. We have already seen that the area under a triplet curve is almost zero, and hence its Fourier transform near $\omega = 0$ is negligibly small (Fig 6.8.8b). But the area under the impulse curve is $2\pi$ and, hence, its Fourier transform near $\omega = 0$ is quite large (Fig 6.8.8a).

Thus, we arrive at the conclusion that a spike produces large energy at the output of an FM detector causing threshold, whereas, a triplet produces negligible energy, avoiding threshold effect.

## Spike-duration

Since the spectrum in Fig 6.8.8a extends over a 0 to $\pm \Delta f$ the spike duration will be given by the reciprocal of $(\Delta f)$, i.e., $(1/\Delta f)$

## Frequency of spike occurrence (unmodulated carrier)

If a carrier frequency is in the centre of a band, the noise spectral components are symmetrically placed on both sides of carrier frequency, and the average number of positive spikes per second $f_+$ is same as the average number of negative going spikes per second $f_-$. For an ideal IF filter of bandwidth $B$, centered at a carrier frequency $f_c$, the frequency of spike generation, $f$ is given by (S.O.Rice),

ANGLE MODULATION SYSTEMS 411

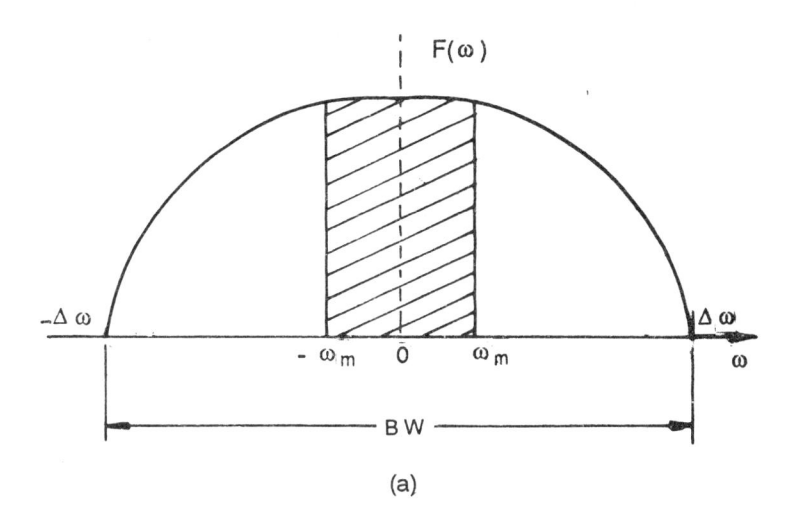

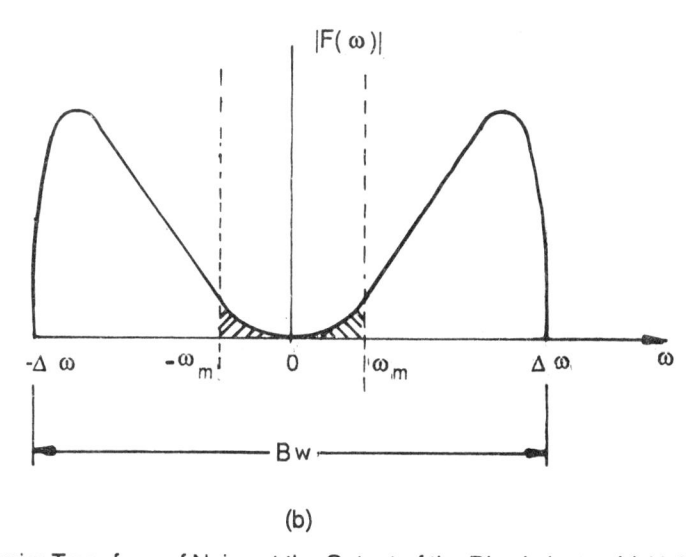

**Fig. 6.8.8** Fourier Transform of Noise at the Output of the Discriminator (a) Noise Spike Case; (b) Noise Triplet Case

$$f = f_+ = f_- = \frac{B}{2\sqrt{3}} \, erfc\,(\sqrt{\rho}) \tag{6.8.30}$$

where $\rho$ is the carrier power to noise power ratio given by

$$\rho = \frac{A^2}{2B\eta} \tag{6.8.31}$$

and *erfc* denotes the complementary error function. $\eta/2$ is the noise spectral density. Equation 6.8.31 reveals that the frequency of spike occurrence is proportional to the bandwidth $B$.

## Output Signal to Noise Ratio $(S_o/N_o)$

The approximate value of output signal to noise ratio $(S_o/N_o)_{FM}$. For a large-noise case is given as (S.O.Rice).

$$(S_o/N_o)_{FM} = \frac{3\rho(B/2f_m)^3}{1 + 4\sqrt{3}\,\rho\,(B/2f_m)^2\,erfc\,(\sqrt{\rho})} \qquad (6.8.32)$$

where $f_m$ is the bandwidth of the modulating signal. The continuous curve 'a' in Fig. 6.8.9 shows a typical plot of Eq. 6.8.32 for $(B/2f_m) = 5$. It is obvious from the curve that when carrier to noise ratio is too large (above 10 dB), the curve becomes linear. The linearity can be explained from Eq. 6.8.32. When $\rho$ is tending to infinity (large), the denominator approaches unity, and Eq. 6.8.32 becomes

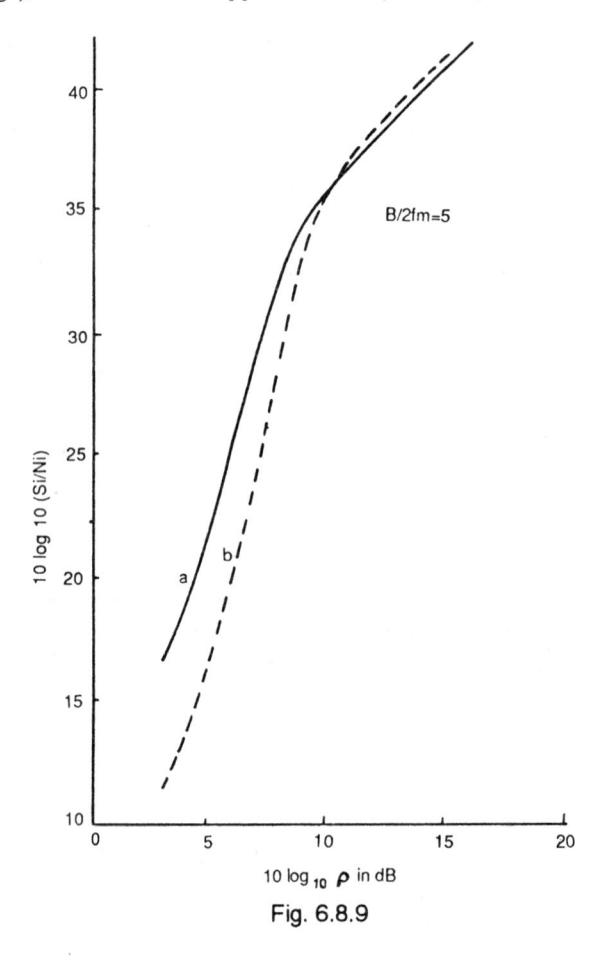

Fig. 6.8.9

$$\left(\frac{S_o}{N_o}\right)_{FM} = 3\rho(B/2f_m)^3 \qquad (6.8.33)$$

Accordingly, $(S_o/N_o)$ has a linear variation with $\rho$ and threshold effect does not occur. When $\rho$ is less than 10 dB (i.e. carrier to noise ratio is less than 10), non-linearity in the curve 'a' is observed.

## Effect of Modulation

Curve 'b' in Fig. 6.8.9 shows the effect of the single-tone modulation. In this case, non-liner variation is observed for $\rho$ less than 11 dB which is very close to experimentally observed threshold value of 13 dB. Below this threshold value of $\rho$, $(S_o/N_o)$ falls more rapidly than the unmodulated curve, and the *threshold effect is more severe*. But above this threshold value of $\rho$, the noise performance is better than the unmodulated case.

In practice, the threshold effect occurs when input carrier to noise ratio is below 13 dB (i.e., 20). If $\rho$ is kept above 20, the average number of spikes generated per second is very small and the threshold effect may be avoided. Hence, for avoiding the FM threshold, $\rho \geq 20$. From Eq. 6.8.31, we have

$$\frac{A^2}{2\eta B} \geq 20$$

or

$$\frac{A^2}{2} \geq 20\eta B \tag{6.8.34}$$

Hence, the FM threshold effect may be avoided by keeping the carrier power $\left(\dfrac{A^2}{2}\right)$ greater than $20\,\eta\,B$.

When a carrier is modulated by a single tone modulating frequency, and $\rho < 20$, then the baseband signal appearing at the discriminator output is accompanied by noise-spikes. The rate of spike generation is increased after modulation. Negative spikes appear most commonly in positive extremity, whereas, positive spikes appear in negative extremity as shown in Fig. 6.8.10.

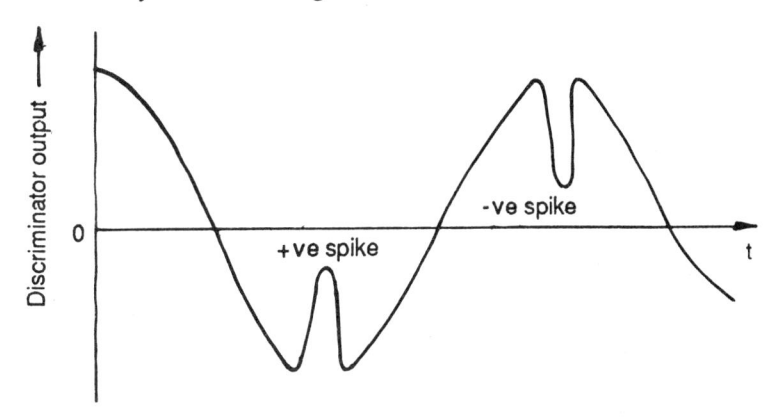

Fig.6.8.10 Spikes in Single-tone Modulating Signal

## Comparison of Threshold in Analog Modulation Systems

It is worth comparing the noise performance and threshold in various analog modulation systems. The comparison has been shown in Fig. 6.8.11. The figure shows the plots $(S_o/N_o)$ vs $(S_i/N_i)$ based on the expressions of the figure of merit $\gamma$ derived earlier in this chapter for different modulation systems. It should be kept in mind that the curves showing threshold at lower $(S_i/N_i)$ are better because in this case the threshold can be avoided with comparatively weaker signals.

The following conclusions can be drawn from the plots shown in Fig. 6.8.11.

*(i) Comparison between AM-Systems :* The SSB-SC and the AM-SC systems have no threshold effect, which is obvious from the complete linearity of the curves; whereas, AM (envelope detector) exhibits a threshold effect for low $(S_i/N_i)$. Hence noise performance of the suppressed AM system is better than AM using envelope detector, since for the same $(S_i/N_i)$, the former provides a higher $(S_o/N_o)$. Thus, suppressed carrier AM systems using coherent detection is superior to AM that uses envelope detection from the noise performance point of view; SSB-SC having the lowest bandwidth is optimum.

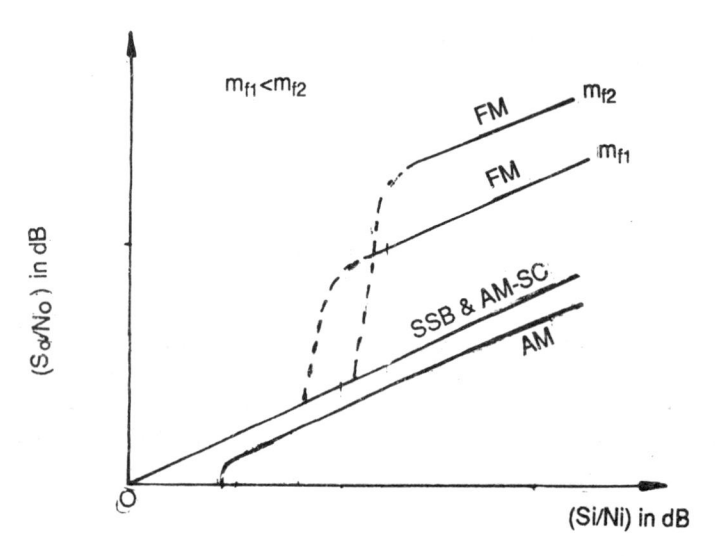

Fig.6.8.11 Threashold Comparison in Analog Modulation System

*(ii) Comparison Between AM and FM :* FM is superior to AM above threshold since the former provides more $(S_o/N_o)$ for an equivalent $(S_i/N_i)$. But below threshold, FM is inferior because the threshold occurs at higher $(S_i/N_i)$. Also, $(S_o/N_o)$ falls more rapidly than AM. Hence, in large-noise cases, suppressed carrier AM, using coherent detection, is most suitable.

*(iii) Comparison Between FM Systems of Different Bandwidths :* It is obvious from the two plots of FM that, above threshold, (low-noise) system with a larger modulation index $m_{f2}$ (large bandwidth) is superior than the one with a smaller modulation index $m_{f1}$ (lower bandwidth). However, below threshold, the system with a larger bandwidth exhibits faster decay in $(S_o/N_o)$ and threshold occurs at higher $(S_i/N_i)$. Hence it is inferior to the lower bandwidth FM.

From the above discussion we arrive at the ultimate conclusion that, for large-noise systems (below threshold), coherent detection is most suitable. On the contrary, for small-noise systems (above threshold), the wideband FM is most appropriate. Narrowband FM is equivalent to AM.

### Noise in Phase Modulation

The phase modulation is equivalent to FM, except that the modulating signal $f(t)$ is first differentiated, and then it frequency modulates the carrier. If we use an FM detector to demodulate a phase-modulated signal, the modulated signal will be a derivative of $f(t)$, i.e., $\dfrac{df}{dt}$. The original modulating signal can be obtained by integrating the output of the FM demodulator. Figure 6.8.12 shows the schematic diagram of phase modulation, and demodulation using a frequency modulator.

ANGLE MODULATION SYSTEMS 415

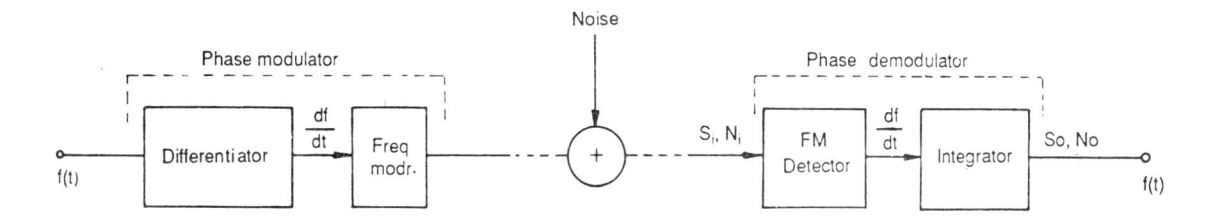

Fig. 6.8.12 Phase Modulation and Demodulation

The differentiator circuit preceding the frequency modulator is equivalent to a pre-emphasis circuit, whereas, the integrator is equivalent to a de-emphasis circuit. It will be discussed in the next section that, these circuits are used to improve the noise performance in FM receiver. Thus, *a frequency modulation system equipped with the facility of pre-emphasis* and *de-emphasis*, is equivalent to a phase-modulation system. Obviously, FM has better noise performance as compared to phase modulation. This can be seen by evaluating the figure of merit for phase modulation system.

### Figure of Merit ($\gamma_{pm}$)

The figure of merit $\gamma_{pm}$ for phase modulation can be derived by adopting the procedure used for frequency modulation. The expressions for the input signal and noise power is identical to the FM case, and is given as (Eqs 6.8.1 and 6.8.4).

$$S_i = \frac{A^2}{2} \text{ and, } N_i = \eta f_m \qquad (6.8.35)$$

The signal power at the FM detector output is proportional to $\dfrac{df}{dt}$ and, after integrator circuit, the output becomes $f(t)$. Hence, the output signal power is obtained by using Eq. 6.8.3,

$$S_o = K^2 K_p^2 \overline{f^2(t)} \qquad (6.8.36)$$

Similarly, the noise signal at the output of FM detector is similar to Eq. 6.8.11 and is given as

$$n_o(t) = \frac{K \dot{n}_s(t)}{A}$$

After integration (assumed ideal), the noise power at the output of the PM detector is

$$n_o(t) = \frac{K n_s(t)}{A} \qquad (6.8.37a)$$

The power density spectrum of $n_o(t)$ is given as

$$S_{n_o}(\omega) = \frac{K^2}{A^2} S_{n_s}(\omega) \qquad (6.8.37b)$$

$S_{n_s}(\omega)$ for a white noise is given by Eq. 5.8.7

$$S_{n_s}(\omega) = \eta, \quad |\omega| \leq \omega_m$$

By substituting in Eq. 6.8.37b, we get

$$S_{n_o}(\omega) = \begin{bmatrix} \dfrac{K^2}{A^2}\eta & |\omega| \leq \omega_m \\ 0 & \text{elsewhere} \end{bmatrix} \quad (6.8.38)$$

The significant difference between the power density spectrum of the output noise power in FM and PM may be seen by comparing Eqs 6.8.38 and 6.8.14. In FM, the power density has a quadratic relation with frequency, whereas in PM, it is independent of frequency as shown in Fig. 6.8.13.

The output noise power $N_o$ is given by

$$N_o = \frac{1}{\pi}\int_0^{\omega_m} S_{n_o}(\omega)d\omega = \frac{1}{\pi}\int_0^{\omega_m} \frac{K^2}{A^2}\eta\, d\omega$$

$$= \frac{2K^2\eta f_m}{A^2} \quad (6.8.39)$$

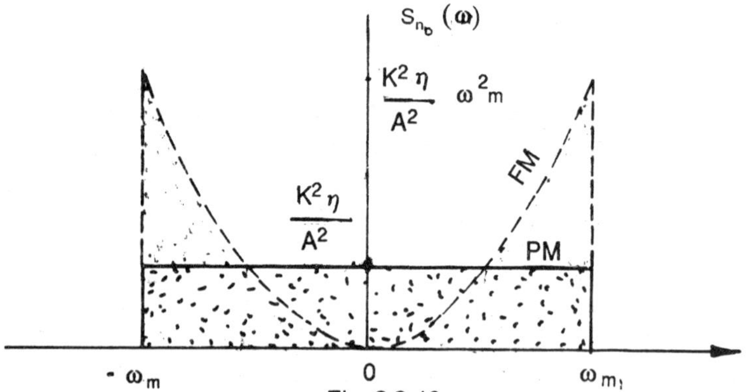

Fig. 6.8.13

From Eqs 6.8.36 and 6.8.39, the output signal to noise ratio is given by

$$\left(\frac{S_o}{N_o}\right)_{pm} = \frac{A^2 K_p^2}{2\eta f_m}\overline{f^2(t)} \quad (6.8.40)$$

Using Eqs 6.8.35 and 6.8.40, we get the figure of merit $\gamma_{pm}$ as given below:

$$\gamma_{pm} = K_p^2\,\overline{f^2(t)} \quad (6.8.41)$$

### Example 6.8.3

A single-tone modulating signal $f(t) = E_m \cos\omega_m t$ phase modulates a carrier $A\cos\omega_c t$. Show that the figure of merit $\gamma_{pm}$ is given by

$$\gamma_{pm} = \frac{1}{2}m_f^2 \quad (6.8.42)$$

**Solution**

$$f(t) = E_m \cos\omega_m t$$

$$K_p f(t) = K_p E_m \cos \omega_m t$$

m.s. values of both sides are given as

$$K_p^2 \overline{f^2(t)} = K_p^2 \frac{E_m^2}{2} \tag{6.8.43}$$

From Eq. 6.4.2a, the phase deviation $K_p E_m = \dfrac{\Delta\omega}{\omega_m} = m_f$ hence,

$$K_p^2 \overline{f^2(t)} = \frac{m_f^2}{2} \tag{6.8.44}$$

Substituting Eq. 6.8.44 in Eq. 6.8.41, we get

$$\gamma_{pm} = \frac{1}{2} m_f^2$$

Thus, the nature of noise figure dependence on the modulation index (hence bandwidth) is identical to FM system.

## Comparison of Noise Performance in Phase Modulation and Frequency Modulation

The noise performance can be compared by matching the figure of merits $\gamma_{FM}$ and $\gamma_{PM}$. For convenience, let us compare the figure of merit for the single-tone case. Thus from Eqs 6.8.18 and 6.8.42, we get

$$\frac{\gamma_{FM}}{\gamma_{PM}} = 3 \tag{6.8.45}$$

Thus, FM has a somewhat better overall noise performance for the audio band. This is also clear from Fig. 6.8.13 in which the noise power area over the band $(0, \pm\omega_m)$ in the vicinity of $\omega = 0$ is greater in PM than FM. However, as the frequency increases, the noise power in FM dominates over PM. A comparison of output noise power $N_o$ for FM and PM can be made with the help of Eqs 6.8.15 and 6.8.39. Thus

$$\frac{(N_o)_{FM}}{(N_o)_{PM}} = \frac{\omega_m^2}{3} \tag{6.8.46}$$

The ratio of the output noise powers (Eq. 6.8.46) increases as square of the modulating frequency $\omega_m$. But the ratio of figure of merits is independent of $\omega_m$ (Eq. 6.8.45). This is because the ratio of output signal powers also increases with the square function of $\omega_m$. This can be seen from Eqs 6.8.3 and 6.8.36

$$\frac{(S_o)_{FM}}{(S_o)_{PM}} = \frac{K^2 K_f^2 \overline{f^2(t)}}{K^2 K_p^2 \overline{f^2(t)}} = \frac{K_f^2}{K_p^2}$$

For an equivalent bandwidth of FM and PM, $K_f = \omega_m K_p$, and

$$\frac{(S_o)_{FM}}{(S_o)_{PM}} = \omega_m^2 \tag{6.8.47}$$

Thus the ratio of output signal powers increases three times faster than the ratio of noise powers. This is the reason why figure of merit in FM is better than in PM. However, in multichannel FDM system, PM offers some advantage over FM (Prob. 6.6). In this system, a number of voice-channels are multiplexed using SSB-SC modulation. The sub-carriers are separated by 4 kHz. The multiplexed signal forms a composite baseband signal that is used to angle-modulate a carrier signal.

## 6.9 COMPARISON BETWEEN AM AND FM

In this section we will compare amplitude modulation with frequency modulation (or angle modulation) from various aspects.

### (i) Noise Performance

We have seen that in the low-noise case (no threshold), wideband FM has a better noise performance than AM. The greater the bandwidth, better is the noise performance. Obviously, the narrowband FM has a noise performance equivalent to AM. On the contrary, in the large-noise case, AM is better than FM because the sputtering sound due to the threshold effect in FM appears at a higher $(S_i / N_i)$ ratio than in AM.

### (ii) Common Channel Interference

FM is better than AM, because of capture effect. The interfering weak signals received from the neighbouring channels are almost completely suppressed in FM, but produce distortion in AM.

### (iii) Externally Generated Noise Pulses

Let us consider a situation when the FM signal is weaker than the noise pulses produced by external sources, such as ignition systems, lightning, gallactic noise, etc. Such noise pulses rarely occur, but have extremely high amplitude for a small duration. The rise time of the receiver is normally larger than the pulse duration because of the finite bandwidth of the receiver. At the output of the receiver, the pulse width will increase. The pulse width becomes comparable with reciprocal of receiver bandwidth. Increase in the duration of pulse causes reduction in its peak magnitude by the same factor, since the area under the pulse remains unchanged. The FM receiver responds slightly to noise pulses because of its wideband RF and IF circuits. Also, the FM receiver has a limiter circuit which limits the amplitude of noise pulses so that it may not exceed the signal amplitude at the detector output. However, if the FM receiver is slightly mistuned, its ability to suppress noise pulses is highly reduced whereas, in AM such exact tuning is not essential.

### (iv) Channel Bandwidth

The wideband FM has a larger bandwidth as compared to AM because the former produces a larger number of sidebands. In a typical broadcast system, each channel bandwidth in AM is 15 kHz, whereas, in FM, it is 150 kHz. Therefore, FM has a disadvantage over AM. In a given frequency space, FM accommodates less number of channels as compared to AM, as discussed in Ex. 6.9.1(a). Hence, in a given frequency space, more number of transmitters can be accommodated using AM.

### (v) Operating Carrier Frequency

FM utilizes higher carrier frequency (above 30 MHz) because of its higher bandwidth. If FM is used at the lower carrier frequencies (HF and MF band), the number of available channels will be too less (Ex. 6.9.1b).

### (vi) Transmission Efficiency

It has been shown in Sec. 6.3 that FM has more efficiency than AM. However, the efficiency of the suppressed carrier AM is better than FM, in the sense that the latter can have 100% efficiency only for some specific values

ANGLE MODULATION SYSTEMS 419

of the modulation index $m_f$, whereas the former always has 100% efficiency. Thus, FM has better efficiency than AM, but not better than AM-SC.

## Example 6.9.1

For a modulating signal of 15 kHz, find (a) the number of channels available in an MF band (300 kHz - 3 MHz), when (i) AM is used; (ii) FM with a frequency deviation of 75 kHz is used.
(b) The maximum possible number of FM channels available in VHF band. Compare it with the number of channels available in MF band. Band gap between the channels is assumed to be zero.
*Solution*
(a) In the MF band, the frequency space available is, 3000 – 300 = 2700 kHz.
(i) When AM is used, the bandwidth of each channel = 2 × 15 = 30 kHz

Hence, the number of channels available in the MF frequency band = $\frac{2700}{30}$ = 90.

(ii) When FM is used, the bandwidth of each channel = 2 × 75 = 150 kHz.

Number of channels = $\frac{2700}{150}$ = 18.

Thus, AM has five times more number of channels as compared to FM.
(b) Frequency space available in the VHF band (30 MHz to 300 MHz) is, 300 MHz - 30 MHz = 270 MHz

Number of available FM channels = $\frac{270 \times 10^6}{150 \times 10^3}$ = 1800 channels.

This is 100 times more than those available in MF band.

## 6.10 THRESHOLD IMPROVEMENT IN DISCRIMINATORS

The threshold effect is more serious in FM as compared to AM, because in the former, the signal to noise ratio $(S_i / N_i)$ at the input of a detector, at which threshold effect starts, is higher. We have already seen that lower the threshold level $(S_i / N_i)$, better is the system, because threshold can be avoided at a comparatively lower $(S_i / N_i)$ ratio, and a small signal is needed to avoid threshold for an equivalent noise power. Hence, it is desirable to lower the threshold level in the FM receivers. The process of lowering the threshold level is known as *threshold improvement, or threshold reduction*. In this section, we will discuss the following two methods used for the improvement of the threshold.
   (i) Pre-emphasis and De-emphasis circuits, and
   (ii) FMFB (Frequency modulation with feedback).
   In the next section, we will study the FM demodulator using PLL which also provides threshold improvement.

### Pre-emphasis and De-emphasis

In the pre-emphasis and de-emphasis method, simple R-C networks are used to improve the threshold. It has been observed in Fig. 6.8.2 that the noise power density spectrum increases as a parabolic function with frequency. Also, we have observed in Chapter 1 that a message signal has a decaying Fourier spectrum with frequency. The result is that signal to noise ratio $(S_o / N_o)$ at the output of the detector becomes very low towards the higher edge of the message-band and may cause threshold effect, in spite of the fact that $(S_o / N_o)$

is large enough at lower edge of the message-band. The threshold effect may be avoided by improving $(S_o/N_o)$ at the higher edge of the message band. This is done by a simple R-C network known as *pre-emphasis circuit* (Fig. 6.10.1a) which boosts the signal amplitude of higher frequencies in the message-band before they modulate the carrier. The boosted signal at transmitter increases the $(S_o/N_o)$ ratio at the detector output.

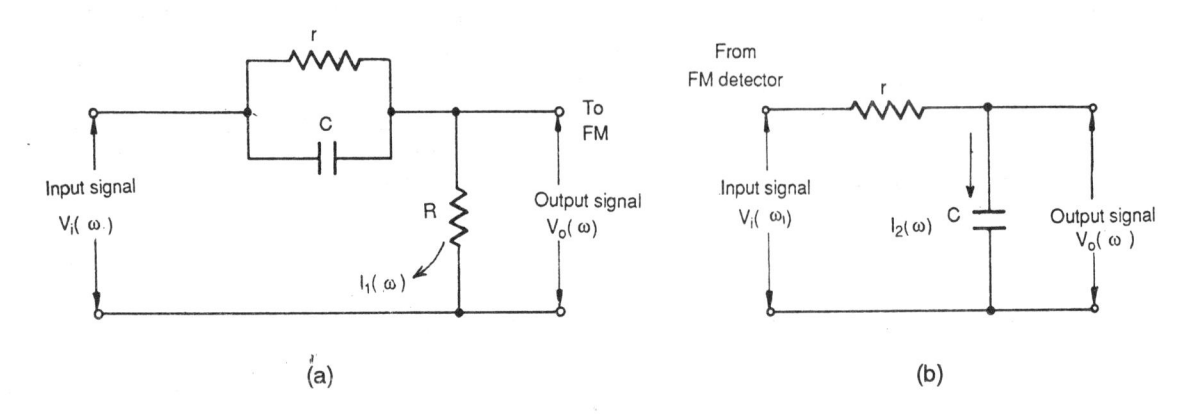

Fig. 6.10.1  (a) Pre-emphasis Circuit, (b) De-emphasis Circuit

Thus, $(S_o/N_o)$ ratio becomes large enough to improve the threshold level over the entire message-band. However, the high frequency components of the message signal reproduced by the detector are at a raised amplitude level, and therefore, amplitude distribution of the base band is disturbed. An inverse action is needed at the discriminator output to bring back the original level of high frequency components, and restore the amplitude distribution of message-band. This is done by another R-C network known as *de-emphasis circuit* (Fig. 6.10.1b). The transfer functions, $H_d(\omega)$ of the de-emphasis, and $H_p(\omega)$ of pre-emphasis circuits, have inverse relationship, so that their product is constant for the entire message band, (Ex.6.10.1), i.e.,

$$H_d(\omega) = \frac{K}{H_p(\omega)} \tag{6.10.1a}$$

or

$$H_d(\omega) H_p(\omega) = K \tag{6.10.1b}$$

The pre-emphasis circuit at the transmitter is a high pass network which behaves like a differentiator. Similarly the de-emphasis circuit at the receiver is a low-pass network which behaves like an inegrator. Thus, an FM system equipped with the facility of pre-emphasis and de-emphasis has a block diagram just like a phase modulation system shown in Fig. 6.8.12. A typical pre-emphasis and de-emphasis circuits are shown in Fig. 6.10.1(a) and (b) respectively.

Assuming $R \ll r$ in the pre-emphasis circuit, its transfer function is given by (Ex. 6.10.1)

$$H_p(\omega) = \frac{R}{r}\left(1 + j\frac{\omega}{\omega_1}\right) \tag{6.10.2}$$

where

$$\omega_1 = \frac{1}{rc} \tag{6.10.3a}$$

The transfer function $H_p(\omega)$ is plotted in 6.10.2. It is obvious that the frequency components between break frequencies $\omega_1$ and $\omega_2$ have been boosted. The second break frequency $\omega_2$ is given by

$$\omega_2 = \frac{1}{RC} \tag{6.10.3b}$$

The rate of increase in amplitude is 6 dB/octave. For the FM broadcast purpose, the lower break frequency $f_1$ is about 2.1 kHz and the higher break frequency $f_2$ is chosen to be much higher than the highest frequency term in the message band, so that $f_2$ lies outside the baseband spectral range. For audio range, $f_2$ may be taken as 30 kHz. The corresponding de-emphasis circuit has an inverse characteristic. The transfer function $H_d(\omega)$ of the de-emphasis circuit is given by (Ex. 6.10.1).

$$H_d(\omega) = \frac{1}{1 + j\left(\dfrac{\omega}{\omega_1}\right)} \tag{6.10.4}$$

This transfer function is plotted in Fig. 6.10.2. The amplitude of frequency components above $\omega_1$ are reduced at a rate of $-6\,\text{dB}$ octave and thus does the inverse action of pre-emphasis circuit to restore the amplitude distribution of the message-band.

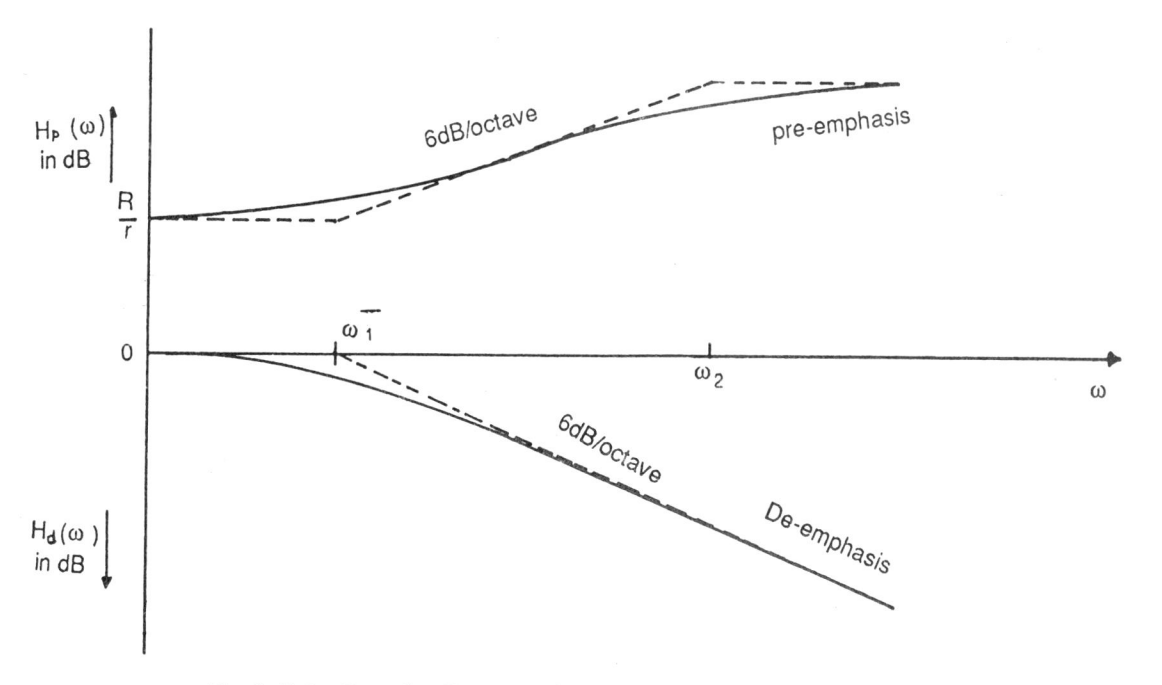

Fig. 6.10.2 Transfer Function of Pre-emphasis and De-emphasis Circuits

## Example 6.10.1

Prove that the transfer function of the pre-emphasis and de-emphasis circuits shown in Fig. 6.10.1 is given by Eqs 6.10.2 and 6.10.4 respectively, and hence show that their product is constant.

*Solution*
The transfer function $H_p(\omega)$ of pre-emphasis circuit shown in Fig. 6.10.1a is given as

$$H_p(\omega) = \frac{V_o(\omega)}{V_i(\omega)}$$

Assuming $R \ll r$, the current $I_1(\omega)$ is given by

$$I_1(\omega) = V_i(\omega)\left(\frac{1}{r} + j\omega_c\right)$$

Hence,

$$V_o(\omega) = R\left(\frac{1}{r} + j\omega_c\right)V_i(\omega)$$

The transfer function is,

$$H_p(\omega) = \frac{V_o(\omega)}{V_i(\omega)} = \frac{R}{r}(1 + j\omega c r) = \frac{R}{r}\left(1 + j\frac{\omega}{\omega_1}\right)$$

Where

$$\omega_1 = \frac{1}{rc}$$

This proves Eq. 6.10.2

Similarly, for the de-emphasis circuit shown in Fig. 6.10.1b the transfer function $H_d(\omega)$ is given by

$$H_d(\omega) = \frac{V_o(\omega)}{V_i(\omega)}$$

$$V_o(\omega) = I_2(\omega) \times \frac{1}{j\omega c} = \frac{V_i(\omega)}{r + \frac{1}{j\omega c}} \times \frac{1}{j\omega c}$$

Hence,

$$H_d(\omega) = \frac{V_o(\omega)}{V_i(\omega)} = \frac{1}{1 + j\left(\frac{\omega}{\omega_1}\right)}$$

where

$$\omega_1 = \frac{1}{rc}$$

This proves Eq. 6.10.4.
Therefore the product of transfer functions is given by

$$H_p(\omega)H_d(\omega) = \left(\frac{r}{R}\right)$$

which is constant with frequency.

## Improvement Factor (IR)

The pre-emphasis circuit does not increase the transmitted signal power because it is a passive network. The improvement in the signal to noise ratio at the detector output is because of the reduction in the noise power due to the de-emphasis circuit. The de-emphasis circuit reduces the high frequency components of the noise, and cuts down the overall noise because of its low pass filtering characteristic. The improvement in output signal to noise ratio, in effect, improves the threshold level. The extent of the improvement in signal to noise ratio depends upon the noise reduction by the de-emphasis circuit. The improvement factor (IR) is defined as

$$IR = \frac{\text{Mean output noise power without de-emphasis } (N_o)}{\text{Mean output noise power with de-emphasis } (N_o)_d}$$

Since input of the de-emphasis circuit is the same as the output of the FM detector, the noise power density at the input of the de-emphasis circuit will be the same as, given by Eq. 6.8.14. Hence, the power density at the output of the de-emphasis circuit will be given by,

$$S_{od}(\omega) = \frac{K^2 \eta \omega^2}{A^2} |H_d(\omega)|^2$$

Substituting $|H_d(\omega)|$ from Eq. 6.10.4, we get

$$S_{od}(\omega) = \frac{K^2 \eta \omega^2}{A^2} \left| \frac{1}{1 + j(\omega/\omega_1)} \right|^2 = \frac{K^2 \eta \omega^2}{A^2} \cdot \frac{\omega_1^2}{\omega^2 + \omega_1^2}$$

The noise power $(N_o)_d$ is obtained by integrating $S_{od}(\omega)$ over the frequency range 0 to $\omega_m$, i.e.,

$$(N_o)_d = \frac{1}{\pi} \int_0^{\omega_m} S_{od}(\omega) d\omega = \frac{K^2 \omega_1^2 \eta}{\pi A^2} \int_0^{\omega_m} \frac{\omega^2}{\omega^2 + \omega_1^2} d\omega$$

$$= \frac{K^2 \omega_1^3 \eta}{\pi A^2} \left[ \frac{\omega_m}{\omega_1} - \tan^{-1}\left(\frac{\omega_m}{\omega_1}\right) \right] \tag{6.10.5}$$

The noise improvement factor is given by

$$IR = \frac{N_o}{(N_o)_d} = \frac{(\omega_m/\omega_1)^3}{3\left[(\omega_m/\omega_1) - \tan^{-1}(\omega_m/\omega_1)\right]} \tag{6.10.6}$$

Thus, the improvement factor is independent of the carrier power A, and hence, the carrier power here plays no role in noise improvement. Although it is obvious that IR increases with an increase in ratio $(\omega_m/\omega_1)$, noise improvement can be achieved by increasing $\omega_m$ and by decreasing $\omega_1$. In commercial broadcast systems, $\omega_1 = 2.1$ kHz. If $\omega_m$ is taken to be 15 kHz, the improvement ratio is about 22, or 13 dB. This shows a considerable reduction in noise due to de-emphasis circuit. The pre-emphasis and de-emphasis method is quite satisfactory for the broadcast systems, with an additional advantage being that the circuit is simple and inexpensive.

## Frequency Modulation with Feedback (FMFB)

In this method, a feedback path is used for the threshold improvement, as shown in the block diagram of Fig. 6.10.3. The feedback path comprises of a voltage controlled oscillator (VCO). The mixer circuit is a voltage multiplier circuit which generates a signal at its output which is a multiplication of the two inputs of the mixer, i.e.,

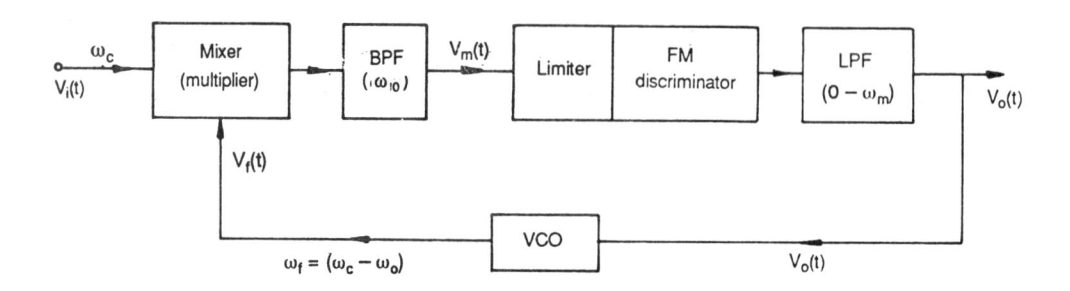

Fig. 6.10.3 Frequency Modulation with Feedback (FMFB)

$$\text{Mixer output} = v_i(t) v_f(t) \tag{6.10.7}$$

where, $v_i$ is the FM signal mixed with noise, and $v_f(t)$ is the feedback signal coming from the VCO output. The FM signal $v_i(t)$ is given by (refer Eq. 6.8.5a),

$$v_i(t) = A(t)\cos\left[\omega_c t + \psi_n(t) + \psi_s(t)\right] \tag{6.10.8}$$

where, $A(t)$ is the envelope of the carrier plus noise signal; $\psi_n(t)$, and $\psi_s(t)$, respectively are the phase deviations produced in the FM wave by the noise and message signals respectively.

### Voltage Controlled Oscillator (VCO)

Voltage controlled oscillator is an electronic sine wave oscillator. Its frequency is controlled by an externally applied voltage. Thus, VCO is a kind of frequency modulator. VCO is available in the form of IC chip. A VCO can be adjusted to generate a sine wave of frequency $\omega_c$ in the absence of any external control voltage. Now, when an external control voltage is applied, the frequency of VCO departs from $\omega_c$. The amount of frequency change per unit input voltage is defined as the *sensitivity* of the VCO, denoted by $G_o$. i.e.,

$$G_o = \frac{d\omega_i}{dv} \text{ rad/v} \tag{6.10.8}$$

where $d\omega_i$ is the change in the instantaneous frequency for a small change $dv$ in the external input voltage. Therefore, the frequency change for an external control voltage $v_o(t)$ will be $G_o v_o(t)$, and the instantaneous frequency of VCO output becomes

$$\omega_i = \omega_c + G_o v_o(t) \tag{6.10.9a}$$

The corresponding phase angle of the VCO output voltage is

$$\phi(t) = \int \omega_i \, dt = \omega_c(t) + G_o \int v_o(t) \, dt \tag{6.10.9b}$$

Hence, the VCO output voltage is given as

$$V_{vco} = C \cos\left[\omega_c t + G_o \int_{-\infty}^{t} v_o(t)\, dt\right] \qquad (6.10.10)$$

where $C$ is amplitude of VCO output voltage.

## Output of the Discriminator With, and Without Feedback

We will now see that a demodulation process can be achieved with feedback (FMFB). The demodulated output voltage is similar to an open-loop (without feedback) discriminator output, except that the output of FMFB is reduced. In the absence of a control voltage, the VCO is offset from $\omega_c$ by an amount $\omega_o$, i.e., the VCO is adjusted to have a frequency $(\omega_c - \omega_o)$. The output of the VCO then is given by (using Eq. 6.10.10)

$$v_f(t) = C \cos\left[(\omega_c - \omega_o) t + G_o \int_{-\infty}^{t} v_o(t)\, dt\right] \qquad (6.10.11)$$

The mixer output produces the sum and difference frequencies of its input signals $v_i(t)$, and $v_f(t)$. The bandpass filter, tuned to $\omega_o$, picks up only the difference frequency, and rejects the sum frequency. Thus, the output of the bandpass filter is given by

$$v_m(t) = D\cos\left[\omega_o t + \psi_s(t) + \psi_n(t) - G_o \int_{-\infty}^{t} v_o(t)\, dt\right] \qquad (6.10.12)$$

The amplitude $D$ has been arbitrarily chosen because it has no significance in FM demodulation. The output $v_o(t)$ of FMFB is given by,

$$v_o(t) = K\left[\frac{d\psi_s(t)}{dt} + \frac{d\psi_n(t)}{dt} - G_o v_o(t)\right] \qquad (6.10.13)$$

where $K$ is a constant of proportionality.
Solution of Eq. 6.10.13 yields

$$v_o(t) = \frac{K}{1 + KG_o} \frac{d}{dt}(\psi_s + \psi_n) \qquad (6.10.14)$$

The output of conventional discriminator (without feedback) is proportional to a change in the instantaneous frequency of the input signal $v_i(t)$ given by Eq. 6.10.8. Thus, in an open-loop discriminator

$$v_o(t) = K\frac{d}{dt}\left[\psi_s(t) + \psi_n(t)\right] \qquad (6.10.15)$$

This output is similar to the output voltage of FMFB given in Eq. 6.10.14, except that the amplitude of the FMFB output is reduced by a factor $(1+KG_o)$. Thus, the demodulation process is identical in both the cases. Since the amplitude reduction is identical for the noise and message signals, $S_o / N_o$ ratio remains same in both cases.

## Bandwidth of FMFB

The bandwidth $B_p$ of the FM signal at the input of the FMFB discriminator is less, as compared to the bandwidth $B_c$ of conventional open-loop FM discriminator, i.e.,

$$B_p < B_c \qquad (6.10.16)$$

This can be verified as follows. In FMFB the input to the discriminator is received from a bandpass filter tuned to $\omega_o$ and having a bandwidth $B_p$. The output of the bandpass filter is given by Eq. 6.10.12, which may be rewritten as follows (using Eq. 6.10.14)

$$v_m(t) = D \cos\left[\omega_o t + \frac{1}{1+KG_o}(\psi_s + \psi_n)\right] \qquad (6.10.17)$$

While comparing it with the conventional FM discriminator input as given by Eq. 6.10.8, we find that the frequency deviation $(\Delta\omega)_p$ in FMFB is reduced by a factor $(1+KG_o)$, as against the deviation $(\Delta\omega)$ of the conventional discriminator, i.e.,

$$(\Delta\omega)_p = \frac{1}{1+KG_o}(\Delta\omega) \qquad (6.10.18)$$

Thus, the frequency deviation in the FMFB is reduced by a factor $(1+KG_o)$. By Carson's rule, the bandwidth of the FM signal is determined by its frequency deviation. Therefore, the bandwidth $B_p$ of the FMFB is less than that of the conventional FM case. For example, CCITT recommends a frequency deviation of 75 kHz. If we take a modulating signal of the bandwidth 15 kHz, the bandwidth $B_c$ (conventional FM) is given as,

$$B_c = 2(\Delta f + f_m)$$
$$= 2(75 + 15) = 180 \text{ kHz}$$

Whereas, the bandwidth $B_p$ for FMFB is given as,

$$B_p = 2\left(\frac{\Delta f}{1+KG_o} + f_m\right)$$

Assuming $KG_o = 9$, we get

$$B_p = 2\left(\frac{75}{10} + 15\right) = 45 \text{ kHz}$$

Thus, the FMFB bandwidth $B_p$ is reduced by a factor of 4; i.e.,

$$B_p = \frac{1}{4} B_c$$

### Example 6.10.2

Prove that

$$B_p = \frac{[m_f/(1+KG_o)] + 1}{m_f + 1} B_c \qquad (6.10.19)$$

*Solution*

By Carson's rule, the bandwidth of an FM signal at the input of a conventional discriminator is,

$$B_c = 2(\Delta\omega + \omega_m) = 2\omega_m\left(\frac{\Delta\omega}{\omega_m} + 1\right) = 2\omega_m(m_f + 1) \text{ radians}$$

or

$$B_c = 2f_m(m_f + 1) \text{ Hz} \qquad (6.10.20a)$$

Similarly, for the case of FMFB,

$$B_p = 2\left(\frac{\Delta\omega}{1 + KG_o} + \omega_m\right) = 2\omega_m\left(\frac{m_f}{1 + KG_o} + 1\right) \text{ radians}$$

or

$$B_p = 2\left(\frac{m_f}{1 + KG_o} + 1\right)f_m \text{ Hz} \qquad (6.10.20b)$$

Comparing this with Eq. 6.10.20a, we find the relation given in Eq. 6.10.19.

### Threshold Extension in FMFB

In low noise case, the signal to noise ratio at the output of the FMFB remains same as in the case of conventional discriminator. However, in the large noise case, the threshold level in FMFB is extended due to a reduction in the frequency of spike generation. This is because the frequency of spike occurrence is proportional to the bandwidth of the FM signal at the input of a discriminator (Eq. 6.8.30). The bandwidth for the FMFB case $(B_p)$ is less than the bandwidth $B_c$ for the case of conventional discriminator. Hence, the frequency of spike generation is less in FMFB. Thereby, due to the reduction in bandwidth, the threshold is extended. With reduction in the bandwidth $B_p$ by a factor of 4, a threshold improvement of 6 dB may be obtained. In practice, FMFB can be designed to provide threshold improvement of 5.7 dB. This is significant improvement for the low power FM systems.

## 6.11 PHASE LOCKED LOOP FM DEMODULATOR

Phase locked loop (PLL) is a negative feedback system. In this, an error signal generated is equal to the phase difference between the input and feedback signal.

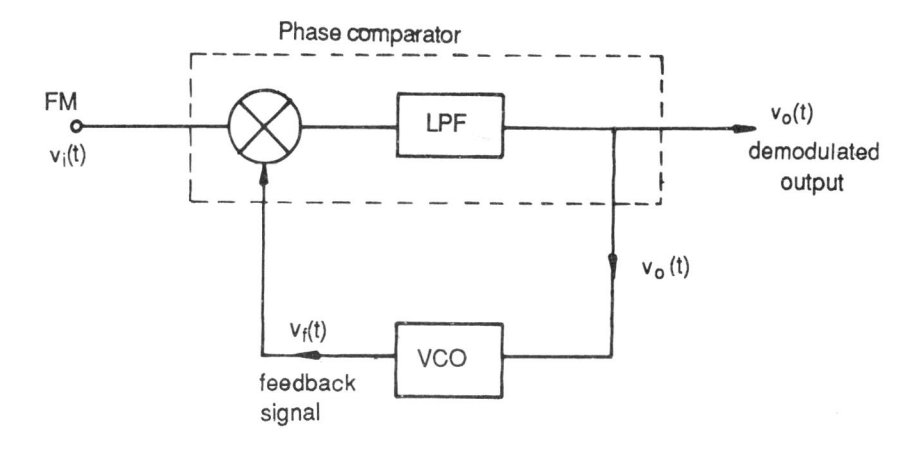

Fig. 6.11.1 Phase Locked Loop (First order)

It can demodulate FM signals and provide threshold improvement properties that are similar to those of FMFB detectors. PLL circuits are available in the form of inexpensive IC chips and can be used as a self-contained FM demodulator with an improved threshold. As shown in Fig. 6.11.1, the PLL system has two main components:

(i) Phase comparator; and (ii) VCO. The comparator comprises of a voltage multiplier and a low pass filter (LPF). The filter is designed in such a way that it selects only the difference-frequency components, and rejects the sum frequency components produced by the multiplier. The VCO is discussed in the previous section as a part of the FMFB. The PLL system is similar to the FMFB system (Fig 6.10.3) except with the following differences:

(i) *PLL has no BPF and discriminator;* thus PLL is self-contained demodulator, rather than just a method for threshold improvement of the conventional demodulator like FMFB, or de-emphasis.
(ii) when a control signal $v_o(t)$ is zero, VCO in PLL is adjusted to generate the same frequency as the unmodulated carrier; as against an offset from carrier, as in FMFB.

## Initial Equilibrium

Initially, the modulating signal $f(t)$ is zero. We will assume that initially when the control voltage $v_o(t)$ is zero, VCO has been adjusted to satisfy the following two conditions:
  (a) The frequency of VCO has been set precisely at an unmodulated carrier frequency $\omega_c$.
  (b) The VCO output $v_f(t)$ is in phase quadrature with the unmodulated carrier wave.

This establishes an initial equilibrium state in which the comparator output is zero when $v_o(t)$ is zero.

## FM Demodulation

Let us assume that the FM signal at the PLL input is given by

$$v_i(t) = A \sin\left[\omega_c t + K_f \int_{-\infty}^{t} f(t)\, dt\right]$$

$$= A \sin\left[\omega_c t + \phi_1(t)\right] \quad (6.11.1)$$

where, $\phi_1(t)$ is the angle deviation produced by the modulating signal $f(t)$. We have

$$\phi_1(t) = K_f \int_{-\infty}^{t} f(t)\, dt \quad (6.11.2)$$

The FM signal produces an output $v_o(t)$ which, in turn, will increase the frequency of VCO. Let the VCO output (which acts as feedback) be given by (using Eq. 6.10.10),

$$v_f(t) = B \cos\left[\omega_c t + G_o \int_{-\infty}^{t} v_o(t)\, dt\right]$$

$$= B \cos\left[\omega_c t + \phi_2(t)\right]$$

Where,

$$\phi_2(t) = G_o \int_{-\infty}^{t} v_o(t)\, dt \tag{6.11.3}$$

The FM signal $v_i(t)$ and the VCO output $v_f(t)$ are applied at the input of the phase comparator, which produces an output consisting of only the difference frequencies. It can be seen that the phase comparator output is given by,

$$v_o(t) = ABG_m \sin[\phi_1(t) - \phi_2(t)]$$
$$= ABG_m \sin[\psi_e(t)] \tag{6.11.4}$$

where $G_m$ is the gain of the multiplier, and $\psi_e(t)$ is the phase error defined by

$$\psi_e(t) = \phi_1(t) - \phi_2(t) \tag{6.11.5}$$

The initial equilibrium is disturbed when $f(t)$ modulates the carrier and $v_o(t)$ is generated. An increase in the frequency of the FM carrier generates a positive $v_o(t)$ at PLL output, which in turn increases the frequency of VCO. The equilibrium is reestablished when the VCO frequency is increased to an extent that it becomes equal to the frequency of the input FM signal, but no longer in phase quadrature. In this case, the error signal becomes zero, i.e.,

$$\psi_e(t) = 0 \tag{6.11.6a}$$

or

$$\phi_1(t) = \phi_2(t) \tag{6.11.6b}$$

The PLL is said to be phase locked when the phase error $\psi_e(t)$ is equal to zero. Substituting $\phi_2$ from Eq. 6.11.3, we get

$$\phi_1(t) = G_o \int_{-\infty}^{t} v_o(t)\, dt \tag{6.11.7a}$$

Differentiating both sides,

$$\frac{d\phi_1(t)}{dt} = \frac{d}{dt}\left[G_o \int_{-\infty}^{t} v_o(t)\, dt\right]$$
$$= G_o v_o(t) \tag{6.11.7b}$$

The time derivative of $\phi_1(t)$ is equal to the frequency change (denoted by $\omega$) in the FM signal, i.e.,

$$\frac{d\phi_1(t)}{dt} = \omega$$

and hence,

$$\omega = G_o v_o(t)$$

or

$$v_o(t) = \frac{\omega}{G_o} \tag{6.11.8a}$$

## 430 COMMUNICATION SYSTEMS: Analog and Digital

The frequency change $\omega$ in the FM wave is proportional to the modulating signal $f(t)$, i.e., $\omega = K_f f(t)$. Therefore Eq. 6.11.8a becomes,

$$v_o(t) = (K_f / G_o) f(t) \tag{6.11.8b}$$

Thus, the output $v_o(t)$ of the PLL is proportional to the modulating signal $f(t)$, and frequency demodulation is achieved.

*Phase error* $\psi_e(t)$ : The phase error $\psi_e(t)$ is time dependent. The output of PLL $v_o(t)$ varies sinusoidally with $\psi_e$. Fig. 6.11.2a shows the sinusoidal variation of $v_o(t)$ with $\psi_e(t)$. However, if we convert the input sinusoidal signal to a square wave before application to the multiplier, the output $v_o(t)$ will have a piecewise linear variation as shown in Fig. 6.11.2b. The linear variation in the region $-\frac{\pi}{2}$ to $\frac{\pi}{2}$ is represented by the line PQ, which may be defined as

$$v_o(t) = \psi_e \tan \theta$$

where $\theta$ is the slope of the line PQ. Since

$$\tan \theta = \frac{AB G_m}{\pi / 2}$$

Hence,

$$v_o(t) = \frac{AB G_m}{\pi / 2} \psi_e, \qquad -\frac{\pi}{2} \leq \psi_e \leq \frac{\pi}{2} \tag{6.11.9}$$

Similarly, the relation between $v_o(t)$ and $\psi_e$ can be obtained for other regions of $\psi_e$.

## First Order PLL

The block diagram shown in Fig. 6.11.1 is a first order PLL because it can be represented mathematically by a first order differential equation. This will be illustrated in the discussions to follow. For convenience, we will assume the piecewise-linear phase variation shown in Fig. 6.11.2b. However, the result will be equally valid for a sinusoidal variation also.

Differentiating Eq. 6.11.5, we get

$$\frac{d\psi_e}{dt} = \frac{d\phi_1}{dt} - \frac{d\phi_2}{dt} \tag{6.11.10a}$$

(for convenience, the argument $(t)$ is ignored)
From Eq. 6.11.3,

$$\frac{d\phi_2}{dt} = G_o v_o$$

hence,

$$\frac{d\psi_e}{dt} + G_o v_o = \frac{d\phi_1}{dt} \tag{6.11.10b}$$

By putting the value of $v_o$ from Eq. 6.11.9, we get

$$\frac{d\psi_e}{dt} + \frac{\psi_e}{\tau} = \frac{d\phi_1}{dt}, \qquad |\psi_e| \leq \frac{\pi}{2} \tag{6.11.11}$$

ANGLE MODULATION SYSTEMS 431

where

$$\tau = \frac{\pi}{2ABG_m G_o} \tag{6.11.12}$$

Alternatively, we can put the value of $\psi_e$ into Eq. 6.11.10b from Eq. 6.11.9. We now get a relation between $\phi_1$ and $v_o(t)$ as follows,

$$\frac{dv_o}{dt} + \frac{v_o}{\tau} = \frac{1}{G_o \tau} \frac{d\phi_1}{dt}, \qquad |\psi_e| \leq \frac{\pi}{2} \tag{6.11.13}$$

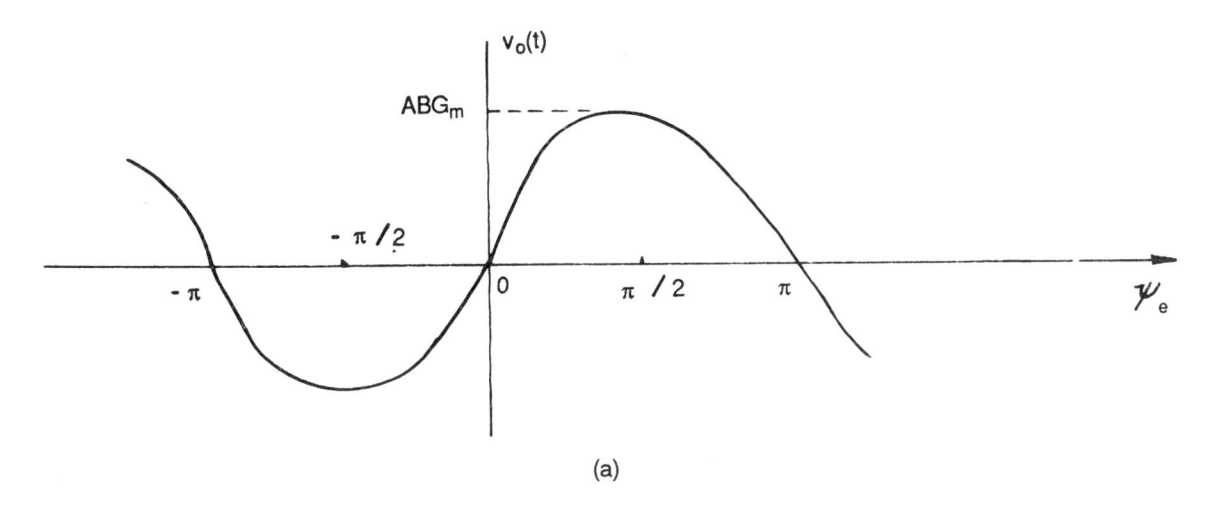

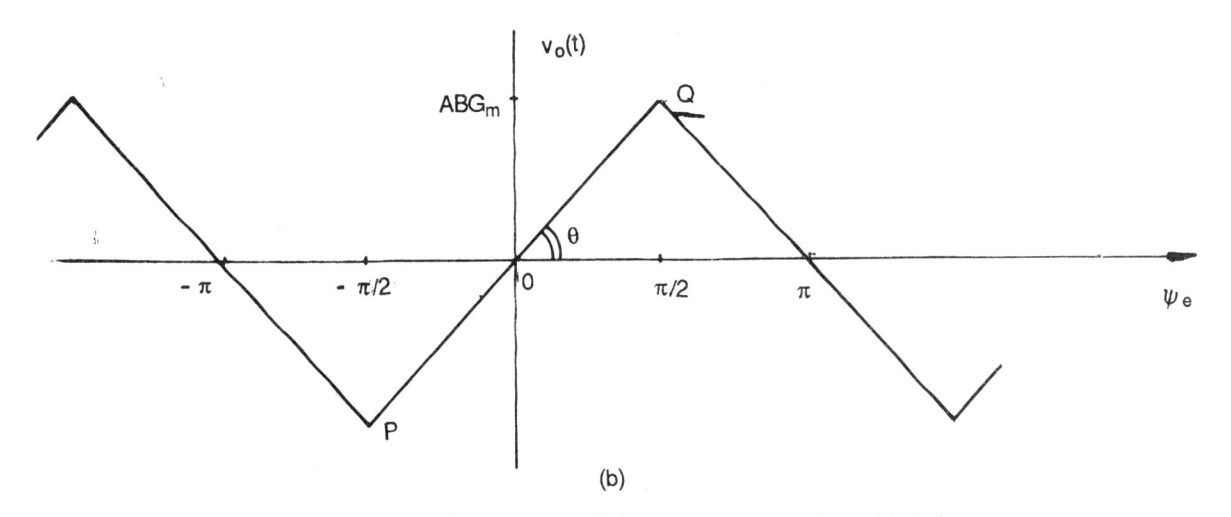

Fig. 6.11.2 (a) Sinusoidal Variation (b) Piecewise-linear Variation

Equations 6.11.11 and 6.11.13 are differential equations of the first order. For sinusoidal characteristic, the PLL differential equation will be

$$\frac{d\psi_e}{dt} + \frac{\sin \psi_e}{\tau_s} = \frac{d\phi_1}{dt} \tag{6.11.14}$$

where $\tau_s = \dfrac{1}{ABG_o G_m}$. For a small $\psi_e$, Eq. 6.11.14 becomes identical to the piecewise linear case given by Eq. 6.11.11, as $\sin \psi_e = \psi_e$. However, the time $\tau$ is modified due to difference in slopes at the origin in Figs 6.11.2a and b.

## Solution of Differential Equation

The solution of the differential equations subject to the proper initial conditions will give the performance of the PLL. Initially, when the input and feedback signals are in phase quadrature and have identical frequencies, the phase error $\psi_e = 0$. Then, for an abrupt change in carrier frequency by an amount $\omega$, $\left(\omega = \dfrac{d\phi_1}{dt}\right)$ the solution of Eq. 6.11.11 yields

$$\psi_e = \omega\tau(1 - e^{-t/\tau}), \qquad -\dfrac{\pi}{2} \le \psi_e \le \dfrac{\pi}{2}$$

Thus, $\psi_e$ grows with the time constant $\tau$ and reaches a new steady state value (when $t \to \infty$) given by

$$\psi_e = \omega\tau, \qquad -\dfrac{\pi}{2} \le \psi_e \le \dfrac{\pi}{2} \qquad (6.11.15)$$

or

$$\dfrac{\psi_e}{\tau} = \omega$$

From Eq. 6.11.9 and 6.11.12, we get $\dfrac{\psi_e}{\tau} = v_o(t) G_o$

Hence,

$$v_o(t) G_o = \omega$$

or

$$v_o(t) = \dfrac{\omega}{G_o} \qquad (6.11.16)$$

Thus, the solution gives a relationship defined by Eq. 6.11.8a.

## Operating Range of PLL

Equation 6.11.15 is valid only for the linear region PQ in Fig. 6.11.2b, i.e., for $|\psi_e| \le \dfrac{\pi}{2}$. The frequency change $\omega$ of the input carrier corresponding to $|\psi_e| = \dfrac{\pi}{2}$, is defined as operating range $R$ of the PLL. In Eq. 6.11.15, putting $\psi_e = \dfrac{\pi}{2}$ and $\omega = R$, we get

$$R = \dfrac{\pi}{2\tau} \qquad (6.11.17a)$$

It can been seen that for a sinusoidal variation of $\psi_e$ (Fig. 6.11.2a), the range is given by

$$R = \frac{1}{\tau_s} \tag{6.11.17b}$$

Thus the range behaviour of PLL depends on the time constant $\tau$ (or $\tau_s$), which in turn, depends on $G_o$, $G_m$, and the amplitudes $A,B$ of the PLL input signals.

The larger the range $R$, the greater is the speed of the PLL response to frequency change $\omega$. The range may be increased by making $\tau$ smaller. In the limit $\tau \to 0$, $\psi_e$ remains close to the origin of Fig. 6.11.2, and PLL behaves like a fast-acting loop. The fast-acting loop keeps the phase deviation $\phi_1(t)$ close to the phase of VCO and error $\psi_e$ is negligibly small. Thus, the range $R$ determines the speed of PLL.

## Bandwidth

The PLL behaves like a low pass filter. This can be verified by taking the Laplace transform (denoted by $L$) of the Eq. 6.11.11.
Let
$$L[\psi_e(t)] \equiv \psi_e(s)$$
and
$$L\left[\frac{d\phi_1}{dt}\right] = L[\omega(t)] \equiv \Omega(s)$$

Then from Eq. 6.11.11, we have

$$\psi_e(s) = \frac{\Omega(s)}{s + \frac{1}{\tau}} \tag{6.11.18}$$

The transfer function $H(S)$ of the PLL is given by

$$H(S) = \frac{\psi_e(s)}{\Omega(s)} = \frac{1}{s + \frac{1}{\tau}} \tag{6.11.19}$$

The transfer function represents a single pole in $s$-plane, accordingly PLL is termed as a *first order PLL*. Putting $s = j\omega$ in Eq. 6.11.19, we get

$$H(\omega) = \frac{\tau}{1 + j\omega\tau} \tag{6.11.20}$$

which represents the transfer function of a low pass filter with a 3dB cut off frequency $\omega_l$ given by

$$\omega_l = \frac{1}{\tau} \tag{6.11.21}$$

Therefore, PLL will produce distortion at its output if the modulating frequency $\omega_m$ is greater than $\omega_l$, just like a low pass RC filter. The distortion may be reduced by keeping $\omega_l$ much higher than the expected highest modulating term $\omega_m$. In other words, we can say that the bandidth of PLL is given by.

$$(BW)_{PLL} = \frac{1}{\tau} \tag{6.11.22}$$

## Instability in PLL

The first order PLL has been represented by Eq. 6.11.11 for a linear variation of $\psi_e$ with a positive slope $\theta$ (Fig. 6.11.2b). If $\theta$ is negative, Eq. 6.11.11 becomes

$$\frac{d\psi_e}{dt} - \frac{\psi_e}{\tau} = \frac{d\phi_1}{dt}$$

By solving the above equation, it can be seen that the PLL equilibrium may be unstable (Prob. 6.7)

## Limitation of the First Order PLL

Both bandwidth and range are controlled by a single parameter $\tau$ in first order PLL. Therefore, we cannot vary the range and bandwidth independently. If $\tau$ is adjusted to change the range, then the bandwidth also changes, and vice-versa. This is an undersirable feature, and a remedy to this is the second order PLL.

## The Second Order PLL

The second order PLL is obtained by deliberately introducing another filter of transfer function $H_p(s)$ in the first order PLL, as shown in Fig. 6.11.3. The filter $H_p(s)$ has distinct features as compared to the filter incorporated in the phase comparator. While the former will have an influence on the PLL performance, the latter has no such influence. For convenience in the discussion, the angle modulated signals encountered in the Fig. 6.11.3 are referred to by their phase angles in $s$-domain. The VCO has been characterized by a transform $G_o/s$ to represent the integration of Eq. 6.11.3. From Eq. 6.11.5, we have

$$\psi_e(s) = \phi_1(S) - \frac{G_m G_o \psi_e(s) H_p(s)}{s} \qquad (6.11.23)$$

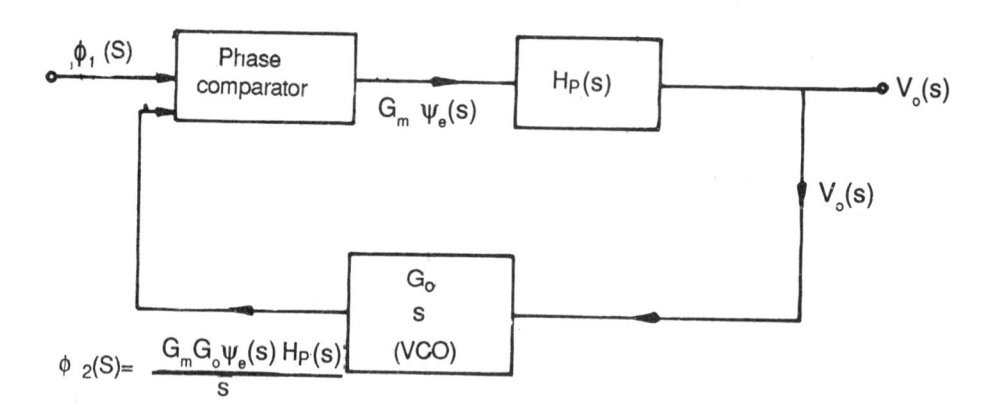

Fig. 6.11.3  PLL (Second Order)

Assume a proportional-plus-integral filter $H_p(s)$ characterized by

$$H_p(s) = 1 + \frac{b}{s} \qquad (6.11.24)$$

where $b$ is a constant. Substituting $H_p(s)$ in Eq. 6.11.23 and solving for $\psi_e(s)$, we get

$$\psi_e(s) = \frac{s\,\Omega(s)}{s^2 + \dfrac{s}{\tau} + \dfrac{b}{\tau}}; \qquad \psi(t) \le \frac{\pi}{2} \tag{6.11.25}$$

where

$$s\phi_1(s) = L\left[\frac{d\phi_1(t)}{dt}\right] = L[\omega] = \Omega(s),$$

and

$$\tau = \frac{1}{G_m G_o}$$

The transform of the output voltage $v_o(t)$ is given by

$$v_o(s) = G_m \psi_e(s) H_p(s) \tag{6.11.26a}$$

From Eqs 6.11.24, 6.11.25, and 6.11.26a, we have

$$V_o(s) = \frac{(s+b)\,G_m\,\Omega(s)}{s^2 + (s/\tau) + (b/\tau)} \tag{6.11.26b}$$

Equations 6.11.25 and 6.11.26b represent the second order differential equations having two poles in $s$-planes. Therefore, Fig. 6.11.3 represents a second order PLL.

The steady state response of the second order PLL may be obtained by using *final value theorem*. According to this theorem, a steady state value ($t \to \infty$) of a signal $f(t)$ may be obtained from its Laplace transform $F(s)$ in the limit $s \to 0$, i.e.,

$$\lim_{t \to \infty} f(t) = \lim_{s \to 0}[s F(s)] \tag{6.11.27}$$

Applying the final value theorem to Eqs 6.11.25 and 6.11.26b, the steady state values for $\psi_e$ and $v_o(t)$ can be determined.

From the basic definition of the Laplace transform, we have

$$\Omega(s) = L(\omega) = \int_0^\infty \omega e^{-st}\,dt = \frac{\omega}{s}$$

Putting this value of $\Omega(s)$ in Eq. 6.11.25, and applying final value theorem, we get

$$\psi_e(\infty) = \lim_{s \to 0}[s\psi_e(s)] = 0 \tag{6.11.28}$$

Similarly, from Eq. 6.11.26b,

$$v_o(\infty) = \lim_{s \to 0}[s V_o(s)] = G_m \omega \tau = \frac{\omega}{G_o} \tag{6.11.29}$$

It is observed that the output steady state response $v_o$ is identical to the first order PLL (refer Eq. 6.11.16), and, hence, provides an equivalent demodulation of the FM signals. But, unlike the first order PLL (refer 6.11.15), the steady state phase error $\psi_e$ in the second order PLL is zero. Therefore, the second order PLL settles back to its starting point after any transient displacement from its initial equilibrium.

Further Eq. 6.11.26b relating to the output $v_o(s)$ and frequency change $\Omega(s)$ represents the transfer function of a low pass filter. Therefore, by a proper choice of cut-off frequency, the PLL bandwidth may be adjusted to get the proper response for the entire baseband. On the other hand, Eq. 6.11.26 relating to the phase error $\psi_c(s)$, and $\Omega(s)$ represents a band pass filter; hence, the operating point of the phase comparator does not respond to the modulating signal. The operating point always remains close to the initial equilibrium point.

## (S/N) ratio at the Output of a PLL

Let us consider that an FM signal $A \cos[\omega_c t + \phi_1(t)]$ is applied to the input of PLL. The demodulated signal is given by Eq. 6.11.8.

$$v_o(t) = \frac{\omega}{G_o} = \frac{1}{G_o}\left(\frac{d\phi_1}{dt}\right)$$

Thus, the demodulated output signal of PLL is identical to the conventional discriminator output given as $v_o(t) = K\frac{d\phi_1}{dt}$ except that the constant $K$ is replaced by $(1/G_o)$. However, the low pass filtering property of PLL provides better performance under large noise conditions. The low pass filtering property of PLL is identical with an R-C network (Fig. 6.11.4) having a time constant $(RC)$ equal to the $\tau$ of PLL. The low pass filter behaves like a baseband filter of a conventional discriminator, provided its cut-off frequency is adjusted to pass the entire modulation band. Above threshold (i.e. low noise case), the spikes are rare and the performance of PLL is identical with a conventional discriminator. But below threshold (i.e., large noise case), the performance of PLL is better because the number of spikes generated are less which results a reduction in threshold. Typically PLL improves the threshold level by 3dB for $m_f = 12$. The spikes generated are reduced because most of the spikes generate a doublet at the PLL output provided the range $R$ is restricted.

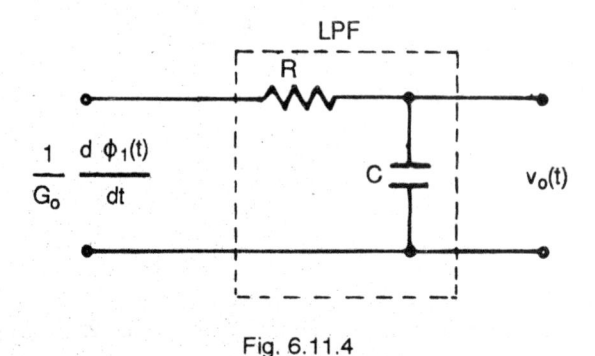

Fig. 6.11.4

## PROBLEMS

6.1. Derive an expression for an FM signal when a carrier $A \sin \omega_c t$ is being modulated by a signal
  (a) $E_m \cos \omega_m t$;
  (b) $E_m \sin \omega_m t$
  (c) Draw the wave forms of the FM and PM waves for the case (a).

6.2 A sinusoidal voltage with an amplitude of 100 volts and a frequency of 100 MHz is frequency-modulated by a sinusoidal signal of 20 kHz to generate a frequency deviation of 80 kHz. Find the amplitude of the carrier frequency and all the sidebands up to fourth present in the modulated signal. Draw the spectrum.

ANGLE MODULATION SYSTEMS    437

6.3  A carrier wave $B\cos\omega_c t$ is frequency-modulated by a periodic square wave of period $T$ and amplitude unity.

(a)  show that the FM wave is given by

$$\varphi_{FM}(t) = B \sum_{n=-\infty}^{\infty} a_n \cos(\omega_c + n\omega_o)t$$

where

$$\omega_o = \frac{2\pi}{T}$$

and

$$a_n = \frac{1}{2}\left[ Sa\left\{\frac{\pi}{2}\frac{\Delta\omega}{\omega_o} - n\right\} + (-1)^n Sa\left\{\frac{\pi}{2}\frac{\Delta\omega}{\omega_o} + n\right\}\right]$$

(b)  Draw the FM and PM waveforms.

6.4  Show that noise and signal powers at the output of an FM discriminator can be calculated independently.

6.5  Show that the output signal to noise ratio in an FM system is related to the AM system as follows:

$$\frac{[(S_o/N_o)]_{FM}}{[(S_o/N_o)]_{AM}} = 3\left(\frac{AK_f}{\omega_m}\right)^2$$

6.6  $N$-voice channels, each of bandwidth $f_m$ are multiplexed using SSB and then transmitted by FM: (i) without emphasis circuits; (ii) with emphasis circuits (equivalent to PM). The transfer function of a pre-emphasis is $|H_p(\omega)|^2 = \omega^2 \tau^2$, where, $\tau$ is the circuit constant. Show that for $Nth$ channel, the noise power $N_o$ at the output in case (i) is three times that of case (ii).

6.7  Show that the first order PLL with a negative slope $\theta(t)$ causes unstable operating point.

6.8  An Armstrong FM modulator is required in order to transmit an audio signal of bandwidth 50 Hz to 15 kHz. The narrowband (NB) phase modulator used for this purpose utilizes a crystal controlled oscillator to provide a carrier frequency $f_{c1} = 0.2$ MHz. The output of the NB phase modulator is multiplied by $n_1$ by a multiplier and passed to a mixer with a local oscillator frequency $f_{c2} = 10.925$ MHZ. The desired FM wave at the transmitter output has a carrier frequency $f_c = 90$ MHz, and a frequency deviation $\Delta f = 75$ kHz, which is obtained by multiplying the mixer output frequency with $n_2$ using another multiplier. Find $n_1$ and $n_2$. Assume that NBFM produces deviation of 25 Hz for the lowest baseband signal.

6.9  An FM signal, modulated to a depth of 8, generates a bandwidth of 180 kHz. Calculate the frequency deviation.

6.10 A single-tone FM signal is given by

$$e_{FM} = 10 \sin(16\pi \times 10^6\, t + 20 \sin 2\pi \times 10^3\, t) \text{ volts}$$

Find the modulation index, modulating frequency deviation, carrier frequency, and the power of the FM signal.

6.11. An FM radio link has a frequency deviation of 30 kHz. The modulating frequency is 3 kHz. Calculate the bandwidth needed for the link. What will be the bandwidth if the deviation is reduced to 15 kHz?

438  COMMUNICATION SYSTEMS: Analog and Digital

6.12 (a) "A differentiator circuit behaves like an FM slope detectors." justify.
   (b) State two limitations of a balanced-slope detector.
   (c) The linear characteristic deviation in a balanced-slope detector is $\frac{3B}{2}$, where $2B$ is the 3 dB bandwidth of each resonant circuit. Justify.

6.13 (a) Evaluate $(S_o/N_o)$ of FM and AM, and show that FM behaves as AM for $m_f \leq 0.5$.
   (b) Show that the noise output power density in a phase modulation receiver is uniform over the baseband.
   (c) Show that an increase in the carrier amplitude A in FM has a noise-quenching effect.

6.14 A carrier voltage $10\cos 8\pi 10^6 t$ is angle modulated by a modulating signal $5\cos 30\pi.10^3 t$.
   (a) Determine the bandwidth for frequency modulation assuming $K_f = 15$ kHz per volt.
   (b) Assuming the same bandwidth, find $K_p$ for phase modulation.
   (c) Determine the change in the bandwidth for frequency and phase modulation if the modulating signal becomes $5\cos 10\pi.10^3 t$.

6.15 A carrier wave $20\cos 8\pi.10^6 t$ is frequency-modulated by a modulating signal $2\cos(2\pi.10^3 t) + \cos(3\pi.10^3 t) + 5\cos(8\pi.10^3 t)$. Calculate the bandwidth. Assume $K_f = 40$ kHz per volt.

6.16 Explain spike generation and threshold effect in FM.

6.17 Prove the Eq. 6.5.5.

6.18 Show that the function $v(t)$ given below is a combination AM-FM signal:
   (a) $v(t) = \cos(2\pi.10^6 t) + 0.02\cos\left[2\pi(10^6 + 10^3)t\right]$.
   (b) $v(t) = \cos\omega_c t + 0.3\cos\omega_m t \sin\omega_c t$.

6.19 Show that a low pass filter can be used as a discriminator.

6.20 A wideband FM system uses a carrier wave of amplitude 10V and frequency 100 MHz. The modulating signal has a bandwidth of 5 kHz and mean square value of 50. The frequency sensitivity $K_f$ is 250 Hz/volt and maximum carrier frequency deviation is 75 kHz. The uniform noise power density on the channel is $10^{-5}$. Find,
   (a) the transfer function of the band pass filter (ideal) at the receiver input.
   (b) the signal to noise power ratio at the output of the demodulator.
   (c) the signal to noise ratio at the output, if the modulating signal is transmitted using AM and compare it with that obtained in FM.

# SEVEN
# PULSE MODULATION SYSTEMS

**INTRODUCTION**

In analog modulation systems, some parameter of sinusoidal carrier is varied in accordance with the instantaneous value of the modulating signal. In pulse modulation systems, the carrier is no longer a continuous signal but consists of a pulse train, some parameter of which is varied in accordance with the instantaneous value of the modulating signal. There are two types of pulse modulation systems:
(i) Pulse Amplitude Modulation (PAM)
(ii) Pulse Time Modulation (PTM)
In PAM, amplitude of the pulses of the carrier pulse train is varied in accordance with the modulating signal; whereas, in PTM, the timing of the pulses of the carrier pulse train is varied. There are two types of PTM.
(i) Pulse Width Modulation (PWM) or
Pulse Duration Modulation (PDM) or
Pulse Length Modulation (PLM)
(ii) Pulse Position Modulation (PPM)
In PWM, the width of pulses of the carrier pulse train is varied in accordance with the modulating signal; whereas, in PPM, the position of pulses of the carrier pulse train is varied.

According to the sampling theorem, if a modulating signal is bandlimited to BHz (i.e., there are no frequency components beyond BHz in the frequency spectrum of the modulating signal), the sampling frequency must be at least 2 BHz and, hence, the frequency of the carrier pulse train must also be at least 2 BHz.

The spectral range occupied by the basic signal is called the *baseband frequency range* or, simply, *baseband*. Hence the basic signal is sometimes called as *baseband signal*. When the basic signal is transmitted without a frequency translation, the system is known as *baseband system*.

## 7.1 PULSE AMPLITUDE MODULATION

Figure 7.1.1. explains the principle of PAM. A baseband signal $f(t)$ is shown in Fig. 7.1.1(a) and carrier pulse

440 COMMUNICATION SYSTEMS: Analog and Digital

train $f_c(t)$ is shown in Fig. 7.1.1(b). The frequency of the carrier pulse train is decided by the sampling theorem. A pulse amplitude modulated signal $f_m(t)$ is shown in Fig. 7.1.1(c). It can be seen that the amplitude of the pulses in Fig. 7.1.1(c) depends on the value of $f(t)$ during the time of the pulse.

In Fig. 7.1.1(a), the baseband signal $f(t)$ is shown to have only a positive polarity. In practice, however, we can have a baseband signal with a positive as well as negative polarity. But, in such a case, the modulated pulses will also be of positive as well as negative polarities. As the transmission of such bipolar pulses is inconvenient; a clamping circuit is used so that we always have a baseband signal with only the positive polarity as shown in Fig. 7.1.1(a).

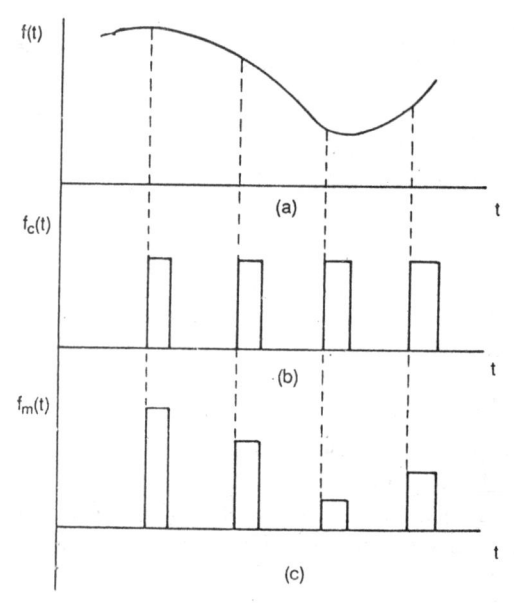

Fig. 7.1.1 (a) Pulse Amplitude Modulation-Baseband Signals $f(t)$ (b) Carrier Pulse train $f_c(t)$ (c) Pulse Amplitude Modulated Signal

There are two methods of getting the pulse amplitude modulated waveform :
(i) Natural sampling or shaped-to sampling,
(ii) Flat-top sampling.

### (i) Natural Sampling

This is explained in Fig. 7.1.2. Here, the amplitude of the carrier pulse train is adjusted to 1, the duration of the pulses is $\tau$, and they are separated by $T_s$. The PAM signal as shown in Fig. 7.1.2(d) is obtained by multiplying $f(t)$ of Fig. 7.1.2 (a) and $f_c(t)$ of Fig.7.1.2(b) in the multiplier of Fig. 7.1.2(c). It may be seen from Fig. 7.1.2(d) that the tops of the pulse-amplitude modulated pulses are not flat but they follow the natural waveform of the modulating signal $f(t)$ during the respective pulse intervals and, hence, the name *natural sampling* is given to this method.

The Fourier series of a periodic train of pulses can be given as

$$v(t) = \frac{A\tau}{T_o} + \frac{2A\tau}{T_o} \sum_{n=o}^{\infty} C_n \cos \frac{2\pi nt}{T_o} \tag{7.1.1}$$

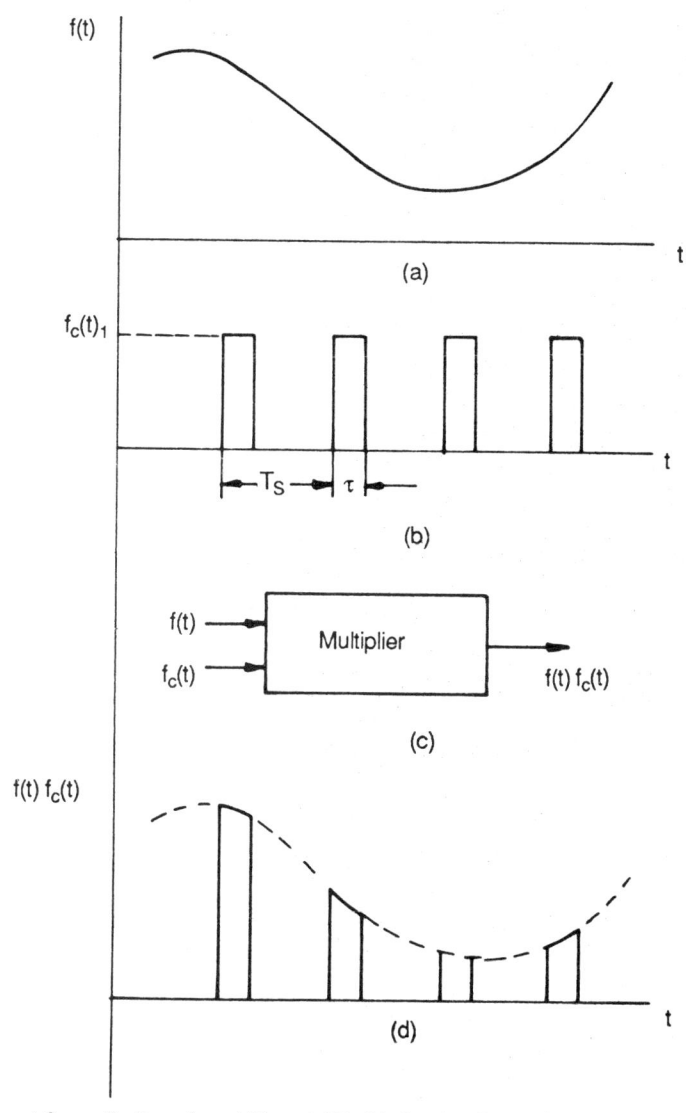

Fig. 7.1.2(a) Natural Samplig-Baseband Signal $f(t)$ (b) Carrier Pulse Train $fc(t)$ (c) Multiplier (d) PAM Signal at Output of Multiplier

where
$A$ is the amplitude of the pulse
$\tau$ the duration of the pulse
$T_o$ the period of the pulse train
and
$C_n$ is given by

$$C_n = \frac{\sin(n\pi\tau/T_o)}{n\pi\tau/T_o}$$

For the carrier pulse train of Fig. 7.1.2(b), we have

$$v(t) = f_c(t)$$
$$A = 1$$

and
$$T_o = T_s$$

Therefore Eq. (7.1.1) becomes

$$f_c(t) = \frac{\tau}{T_s} + \frac{2\tau}{T_s}\left[C_1 \cos 2\pi \frac{t}{T_s} + C_2 \cos 2 \times 2\pi \frac{t}{T_s} + \cdots\right] \qquad (7.1.2)$$

The constant $C_n$ is given by

$$C_n = \frac{\sin(2\pi\tau/T_s)}{2\pi\tau/T_s} \qquad n = 1, 2, 3, \ldots$$

The output of the multiplier is then

$$f(t)f_c(t) = \frac{\tau}{T_s}f(t) + \frac{2\tau}{T_s}\left[f(t)C_1 \cos 2\pi \frac{t}{T_s} + f(t) C_2 \cos 2 \times 2\pi \frac{t}{T_s} + \cdots\right]$$

Let us choose $T_s = \dfrac{1}{2 f_M}$ (by sampling theorem)

where $f_M$ = maximum frequency component in $f(t)$
We then get

$$f(t)f_c(t) = \frac{\tau}{T_s}f(t) + \frac{2\tau}{T_s}[f(t)C_1 \cos 2\pi(2f_M)t$$
$$+ f(t) C_2 \cos 2\pi(4 f_M)t + \cdots] \qquad (7.1.3)$$

If we don't consider the multiplying factor, then the first term on the r.h.s. of Eq. 7.1.3 is the baseband signal $f(t)$ itself. The second term is a product of $f(t)$ and a sinusoid of frequency $2f_M$. Now, $f(t)$ is bandlimited to $f_M$. Hence, the multiplication in the second term will yield the frequency spectrum given by the sum and difference frequency terms. Thus the frequency spectrum of the second term is from $(2f_M - f_M)$ to $(2f_M + f_M)$ i.e., from $f_M$ to $3f_M$. Similarly, the frequency spectrum of the third term is from $(4f_M - f_M)$ to $(4f_M + f_M)$; i.e., from $3f_M$ to $5f_M$ and so on. The magnitude plots of the spectral densities of $f(t)$ and $fc(t)$ are shown in Figs 7.1.3(a) and 7.1.3(b), respectively. If the sampled signal is passed through an ideal low pass filter with a cut off frequency $f_M$, then the output will be the baseband signal $f(t)$.

## (ii) Flat-top Sampling

The electronic circuitry needed to perform natural sampling is somewhat complicated because the pulse-top shape is to be maintained. These complications are reduced by flat-top sampling, shown in Fig. 7.1.4e. In this, the tops of the pulses are flat. Thus the pulses have a constant amplitude within the pulse interval. The

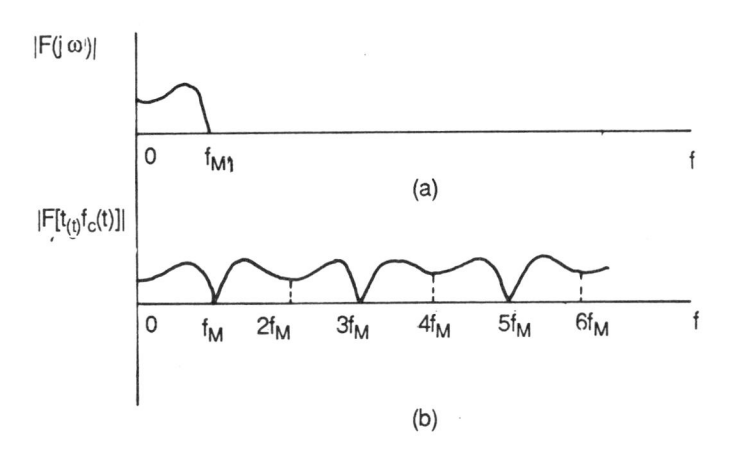

Fig. 7.1.3(a) Magnitude plot of Spectral Density of $f(t)$
(b) Magnitude plot of Spectral Density of $f(t)$ and $f_c t$

constant amplitude of the pulse can be chosen at any value of $f(t)$ within the pulse interval. In Fig. 7.1.4(e), the value at the beginning of the pulse is chosen. By making the value of the pulse amplitude constant within the pulse interval, some distortion is introduced as there is a deviation from the actual value of $f(t)$ during the pulse interval. This is discussed below:

### Spectrum of a Flat-top Sampled Signal

The flat-top sampled signal $f_m(t)$ of Fig. 7.1.4(e) may be considered as a convolution of the impulse sampled signal $f_s(t)$ of Fig. 7.1.4(a), and non-periodic pulse $p(t)$ of width $\tau$ and height 1 of Fig. 7.1.4(c). The spectrums of $f_s(t)$ and $p(t)$ are shown in Fig. 7.1.4(b) and Fig. 7.1.4(d), respectively. The spectrum of $f_m(t)$ is shown in Fig. 7.1.4(f), which is obtained by multiplying $F_{s(\omega)}$ with $P_{(\omega)}$. As the $P_{(\omega)}$ value is different at different frequencies, the shape of $F_M(\omega)$ is not similar to $F_{s(\omega)}$ which shows that a distortion will be introduced if the signal is recovered by an ideal low pass filter of a cut-off frequency $\omega_M$.

### A PAM Modulator Circuit

A PAM modulator circuit is shown in Fig. 7.1.5. This circuit is a simple emeitter follower. In the absence of the clock signal, the output follows the input. The modulating signal is applied as the input signal. Another input to the base of the transistor is the clock signal. The frequency of the clock signal is made equal to the desired carrier pulse train frequency. The amplitude of the clock signal is so chosen that the high level is at ground ($O\ V$), and the low level is at some negative voltage which is sufficient to bring the transistor in the cut-off region. Thus when the clock signal is high, the circuit behaves as an emitter follower, and the output follows the input modulating signal. When the clock signal is low, the transistor is cut-off and the output is zero. Thus the output waveform, shown in Fig. 7.1.5 is the desired pulse amplitude modulated signal.

### 7.1.1 Demodulation of PAM Signals

Demodulation of natural sampled signal can be done with the help of an ideal low pass filter with a cut-off frequency $\omega_M$. But, for this, the pulse-top shape is to be maintained after transmission. This is very difficult

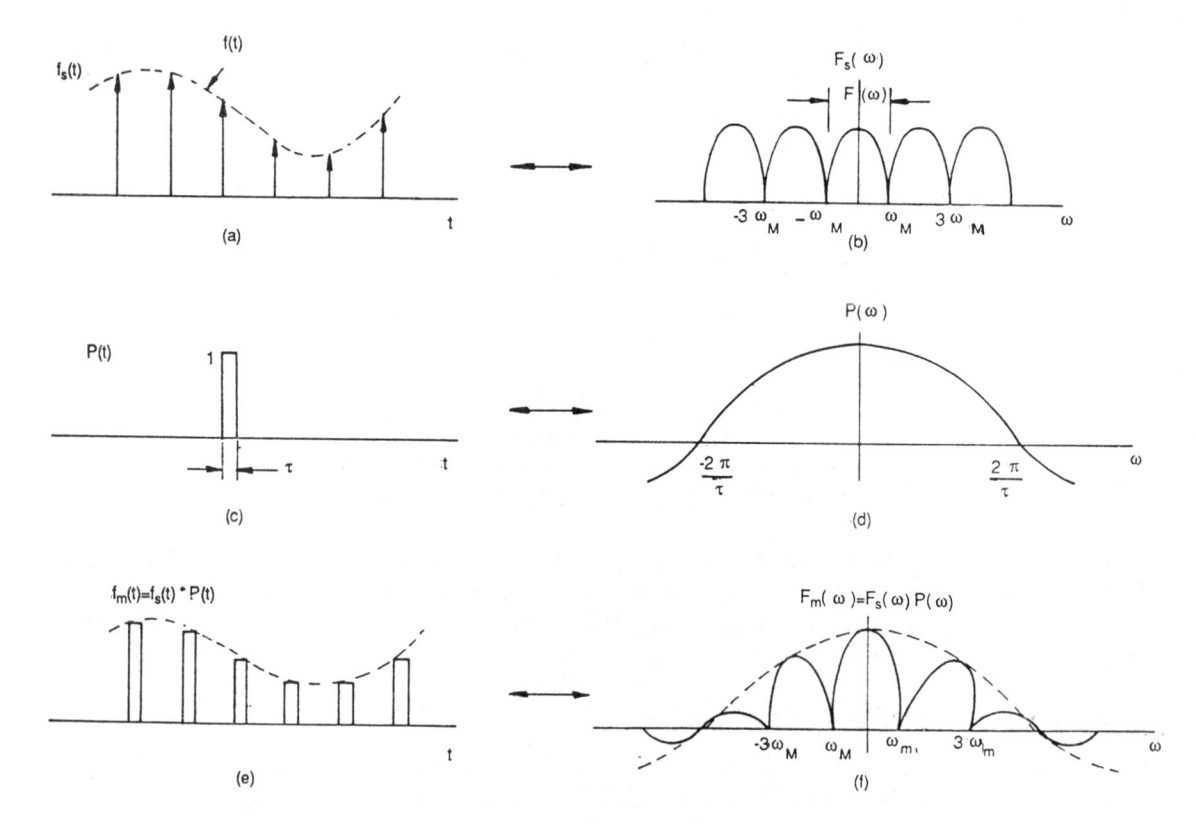

**Fig.7.1.4** Flat Top Sampling (a) Impulse Sampled Signal $f_s(t)$ (b) Spectrum of $f_s(t)$ (c) Non-periodic pulse $P_c t$ of Width $\tau$ and Height 1 (d) Spectrum of $p(t)$ (e) Flat-top Sampled PAM signal $f_m(t)$ (f) Spectrum of $f_m(t)$

due to the transmitter and receiver noise. Therefore, normally, flat-top sampling is preferred over natural sampling.

There are two demodulation methods for the flat-top sampled signal:

*(i) Using an Equalizer*

If the flat-top sampled signal is passed through an ideal low pass filter, the spectrum of the output will be $F(\omega)P(\omega)$. The time function of the output is somewhat distorted due to the multiplying factor $P_{(\omega)}$. If the low pass filter output is passed through a filter having a transfer function $1/P_{(\omega)}$ over the range $0 - \omega_M$, the spectrum at the output of this filter will be $F(\omega)P(\omega) \cdot \dfrac{1}{P(\omega)} = F(\omega)$, and, hence, the original time function $f(t)$ will be recovered. The filter with a transfer function $\dfrac{1}{P(\omega)}$ is known as equalizer. The overall arrangement is shown in Fig. 7.1.1.1(a).

The combination of an ideal LPF and equalizer is known as composite filter. The transfer function $H(\omega)$ of this composite filter is shown in Fig. 7.1.1.1(b). (It may be noted that the transfer function of the equalizer outside $\omega_M$ can be chosen according to the convenience of design). $H(\omega)$ is given by

PULSE MODULATION SYSTEMS 445

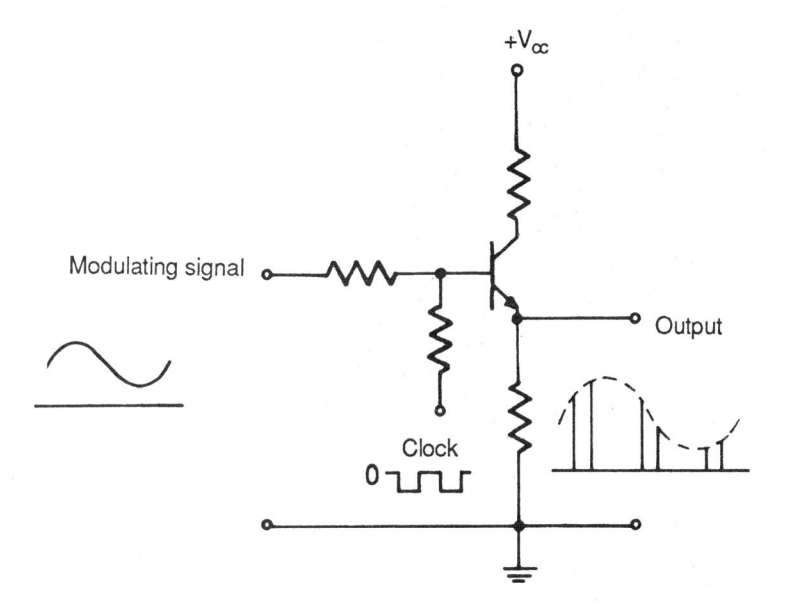

Fig.7.1.5 A PAM Modulator

(a)

(b)

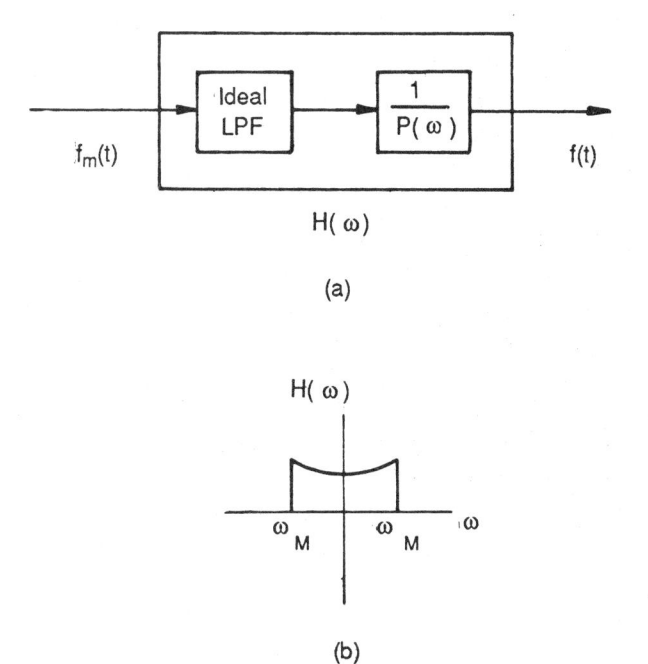

Fig. 7.1.1.1. Demodulation of Flat-Top Sampled PAM Signal Using Equalizer (a) Composite Filter with Transfer Function $H(\omega)$ (b) The Desired Characteristic of $H(\omega)$

$$H(\omega) = \frac{1}{P(\omega)}, \ |\omega| < \omega_M$$

$$= 0 \text{ otherwise}$$

*(ii) Using Holding Circuit*
In this method, the received signal is passed through a holding circuit and a LPF, shown in Fig. 7.1.1.2(a).

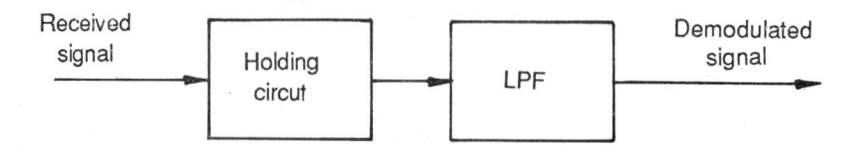

(a) Building Blocks of Demodulator

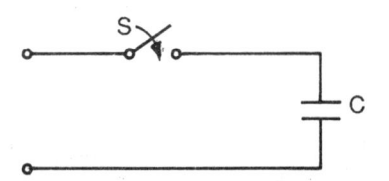

(b) Zero Order Holding Circuit

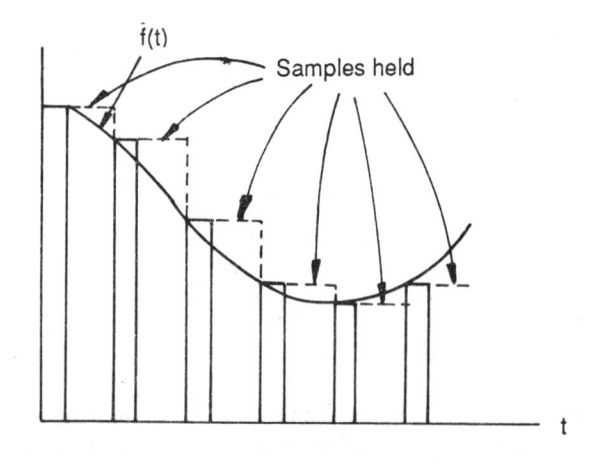

(c) Output of Holding Circuit

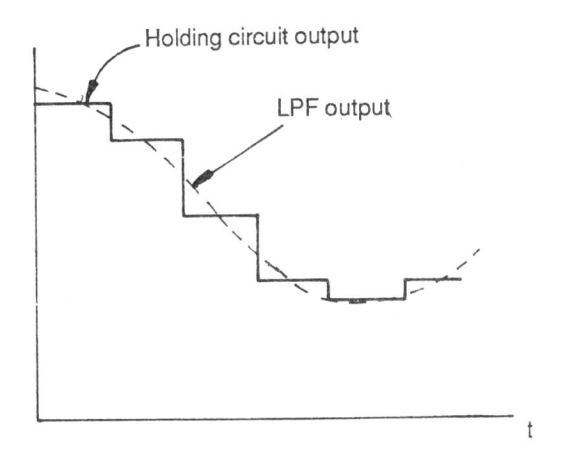

**(d) Output of LPF**

Fig. 7.1.1.2  Demodulation of Flat-Top Sampled PAM Signal

Figure 7.1.1.2(b) shows a simple holding circuit. The switch S closes after the arrival of the pulse, and it opens at the end of the pulse. The capacitor C gets charged to the pulse amplitude value, and it holds this value during the interval between the two pulses. Thus the sampled values are held as shown in Fig. 7.1.1.2(c). The holding circuit output is smoothened in LPF as shown in Fig. 7.1.1.2(d). It can be seen that some distortion is introduced because of the holding circuit. The circuit of Fig. 7.1.1.2(b) is known as *zero order* holding circuit, which considers only the previous sample to decide the value between the two pulses. The first order holding circuit considers the previous two samples; the second order holding circuit considers the previous three samples; and so on. As the order of the holding circuit increases, the distortion decreases at the cost of the circuit complexity. The amount of permissible distortion decides the order of the holding circuit.

## A PAM Demodulator Circuit

A PAM demodulator circuit is shown in Fig. 7.1.1.3. It is just an envelope detector followed by a low pass filter. The diode and R-C combination work as the envelope detector. This is followed by a second order OP-AMP low pass filter to have a good filtering characteristic. Thus, for the received pulse amplitude modulated signal as the input signal, the desired demodulated signal shown in Fig. 7.1.1.3 is the output.

### 7.1.2 Time Division Multiplexed PAM System

Normally, in a PAM system, the duration of the pulse ($\tau$) is much less than the time period of pulses $T_s$ i.e., $\tau << T_s$, as shown in Fig. 7.1.2.1(a). Thus, no information is being transmitted through the system for most of the time. The time space $T_s - \tau$ can be utilized to transmit information from other signals. In Fig. 7.1.2.1(b), the signal numbers 2, 3, and 4 are transmitting information with the help of samples numbered 2, 3, and 4, respectively. This is along with the samples numbered 1 of the signal number 1. The time period $T_s$ is equally divided between the four signals, thus allocating a time slot of $\frac{T_s}{4}$ to each signal. The duration of time slot is such that $\frac{T_s}{4} > \tau$. Thus there is a guard time $\frac{T_s}{4} - \tau$ between all successive sampling pulses, ensuring that there is less cross-talk between signals. (More about cross-talk in Sec. 7.1.3 and 7.1.4) The arrangement by

(a)

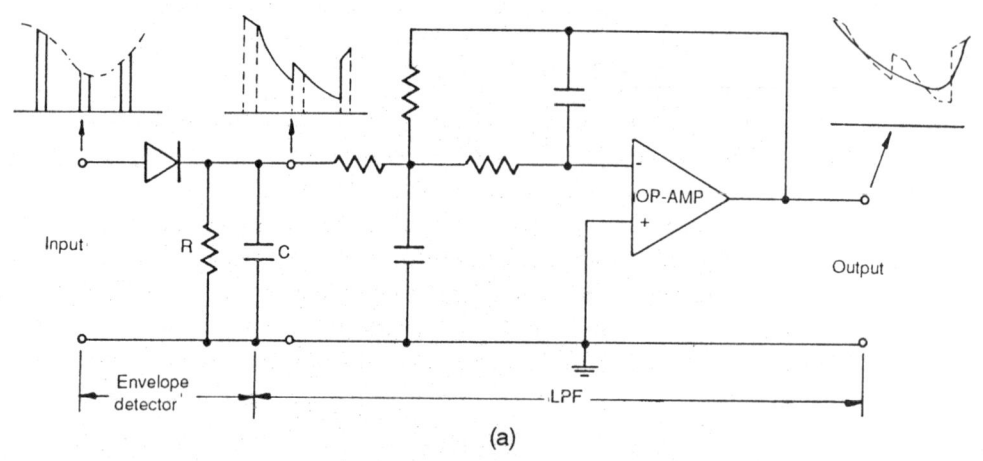

Fig. 7.1.1.3 A PAM Demodulator

(a)

(b)

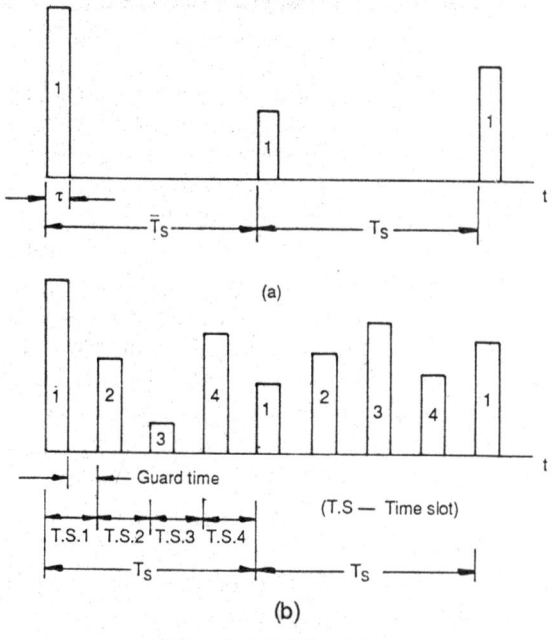

Fig. 7.1.2.1 (a) (b)

which the information from more than one signal is transmitted in this manner is known as Time Division Multiplexing (TDM). A Time Division Multiplexed PAM System is shown in Fig. 7.1.2.2 which transmits information from $n$ signals. The switch 1 and switch 2, respectively known as commutator and decommutator, are synchronized electronic switches which rotate at the same speed of $2f_M$ rotations per second. The commutator samples and combines the samples, while the decommutator separates the samples belonging to individual signals.

Synchronization is the most crucial operation in any TDM system. Thus, for example, if the commutator is at position 2, the decommutator must also be in position 2. To provide synchronization, a synchronizing pulse is transmitted in every frame (time interval between two successive samples of the same signal, i.e., $T_s$). Thus to multiplex $n$ channels, $n + 1$ time slots are provided in a frame; $n$ for channels and 1 for the synchronizing pulse. The synchronizing pulse is chosen in such a way that it is easily distinguishable. For this purpose, one of its properties is adjusted in such a way that it is never attained by the other pulses. In the case of PAM, its amplitude is made larger than the amplitudes of all the other pulses. Figure 7.1.2.2 shows the shcematic diagram of a time division multiplexed PAM system.

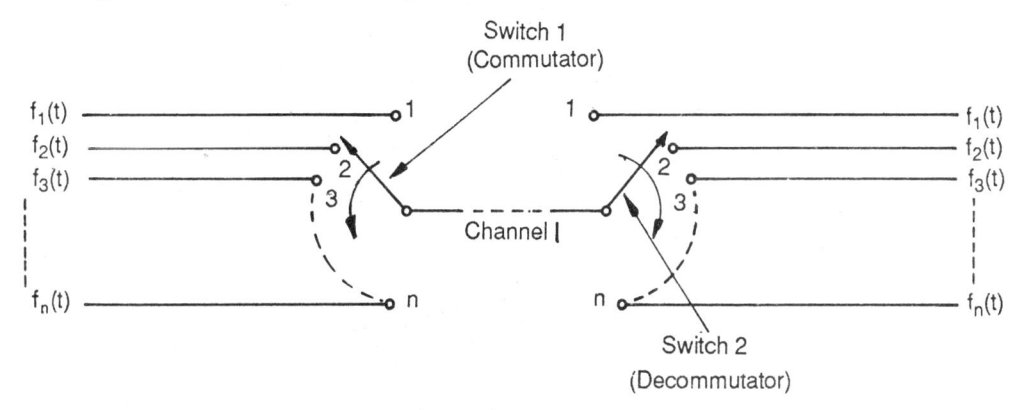

Fig. 7.1.2.2 A Time Division Multiplexed PAM System

In the above discussion, it is assumed that all signals are bandlimited to the same frequency, thereby allowing the same sampling frequency for all signals. Section 8.7 deals with a situation where this is not so.

## Comparison between FDM and TDM

There are two methods of multiplexing, viz FDM (Frequency Division Multiplexing) and TDM (Time Division Multiplexing). The basic difference between them is that in FDM the frequency scale is shared by different signals; whereas, in TDM the time scale is shared by different signals, as shown in Fig. 7.1.2.3. The bandwidth requirement of AM/SSB and PAM is $nf_m$ Hz, and that of AM/DSB and PAM/AM is $2nf_m$ Hz. Thus it can be said that the bandwidth requirement of the FDM and TDM system is the same.

TDM is superior to FDM in the following ways:

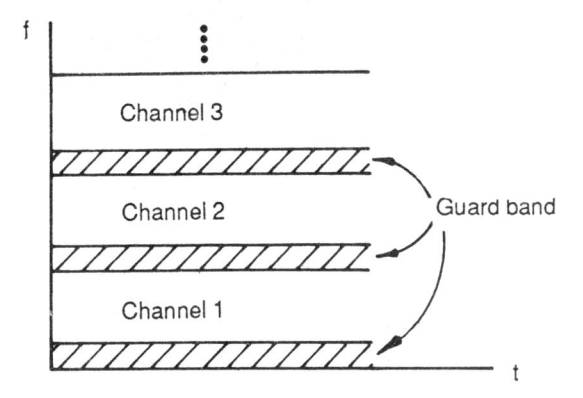

Fig. 7.1.2.3 (a) FDM

**450** COMMUNICATION SYSTEMS: Analog and Digital

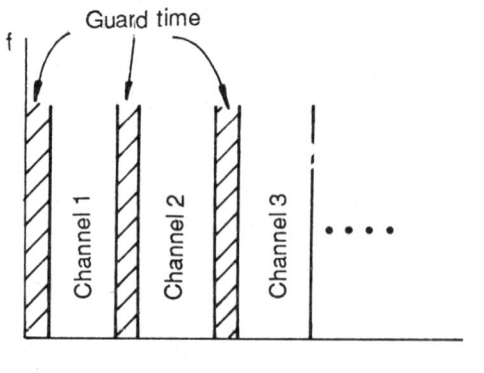

Fig.7.1.2.3 (b)TDM

(i) In the FDM system, different carriers are to be generated for different channels. Also, as each channel occupies a different frequency band, different bandpass filters are required. On the other hand, in TDM system, all the channels require identical circuits, consisting of simple synchronous switches, gates, and a low pass filter. Thus the circuitry needed in the TDM system is much simple than the one needed in the FDM system.

(ii) The non-linearities in the various amplifiers of an FDM system produce harmonic distortion and, hence, they introduce interference within the channels. Therefore, the non-linearity requirements of an FDM system are much more stringent than those for a single channel. On the other hand, in a TDM system, the signals from different channels are allotted different time slots and they are not applied to the system simultaneously. Hence, the non-linearity requirements of a TDM system are the same as those for a single channel. Thus the TDM system is relatively immune to interference within the channels (interchannel cross-talk) as compared to the FDM system.

### 7.1.3 Cross-talk due to HF cut-off of the Channel

All communication channels are of limited bandwidths. Hence a communication channel can be represented by an $RC$ lowpass filter as shown in Fig. 7.1.3.1(a) whose upper cut-off frequency is

$$f_c = \frac{1}{2\pi RC}$$

Now, when a pulse is applied to this channel, the output of the channel will be distorted as shown in Fig. 7.1.3.1(b). This is due to the HF limitations of the channel. It can be seen that the signal pulse allotted to the time slot 1 extends into the time slot 2, resulting in cross-talk. On an average, we can assume that the multiplexed signals are equally strong. Hence, during the time slot 2, the area corresponding to the signal, i.e. $A_2$ (not shown in the figure) will be the same as $A_1$. We can define the cross-talk factor, $K$, as the ratio of the cross-talk signal to the desired signal. Thus

$$K \equiv \frac{A_{12}}{A_2} = \frac{A_{12}}{A_1} \qquad (7.1.3.1)$$

The pulse in time slot 1 is almost rectangular (the figure here is exaggerated for clarity). Hence,

$$A_1 \approx V\tau \qquad (7.1.3.2)$$

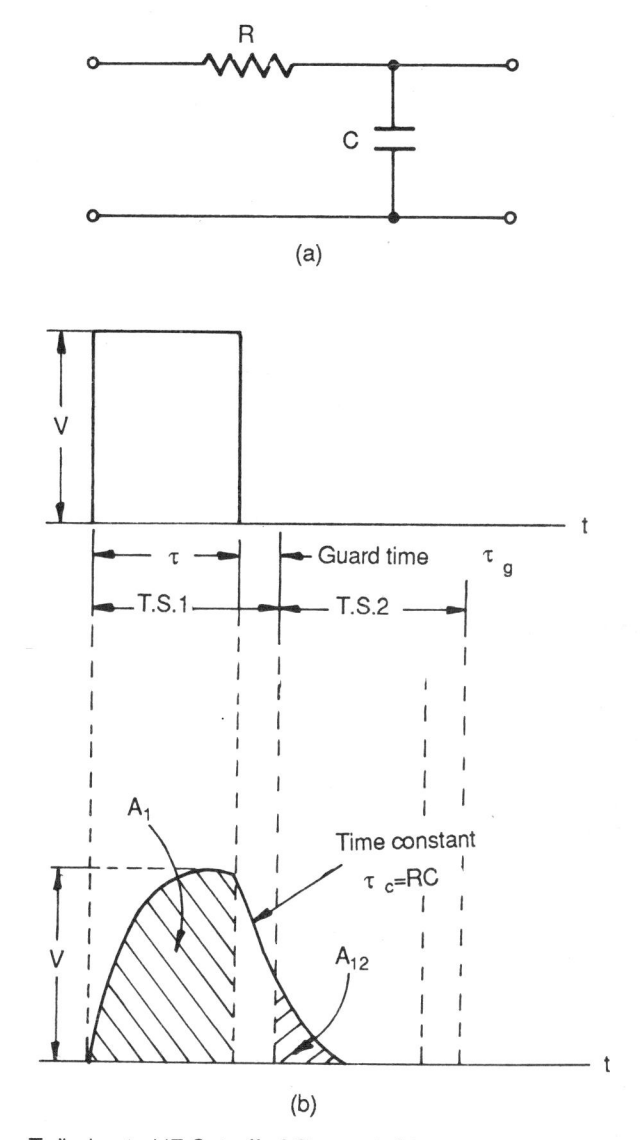

Fig 7.1.3 Cross Talk due to HF Cut-off of Channel; (a) Low pass Filter Representation of the Channel (b) Output of Channel of (a) for a Pulse Input

Now,

$$A_{12} = V\tau_c \, e^{-\tau_g/\tau_c}(1 - e^{-\tau/\tau_c}) \qquad (7.1.3.3)$$

The time constant $\tau_c$ must be much smaller than $\tau$ in order to have minimum cross-talk. Therefore, $\tau_c \ll \tau$, and Eq. 7.1.3.3 becomes

$$A_{12} \approx V\tau_c \, e^{-\tau_g/\tau_c} \qquad (7.1.3.4)$$

From Eqs (7.1.3.1), (7.1.3.2) and (7.1.3.4), we get

$$K = \frac{\tau_c}{\tau} e^{-\tau_g/\tau_c} \tag{7.1.3.5}$$

The cross-talk factor $K$ should be as low as possible. Equation 7.1.3.5 suggests that $\tau_g$ should be much larger than $\tau_c$. Of course, there are other considerations as well. For example, a large $\tau_g$ will result in less number of channels that can be multiplexed, and/or a reduced signal strength.

This type of cross-talk is restricted to the neighbouring channel because $\tau_c$ is very small, and hence the pulse ends in the neighbouring time slot only.

### 7.1.4 Cross-talk due to LF Cut-off of the Channel

Just as the channel has an upper cut-off frequency, it also has a lower cut-off frequency. Hence the channel can be represented by an $RC$ high pass filter as shown in Fig. 7.1.4.1(a), whose lower cut-off frequency is

$$f_1 = \frac{1}{2\pi RC}$$

Now, when a pulse is applied to this channel, the output of the channel will be distorted as shown in Fig. 7.1.4.1(b) due to the LF limitations of the channel. In this case, the time constant $\tau_c$ should be much greater than $\tau$ to reduce cross-talk; i.e. $\tau_c \gg \tau$.

Now,

$$\Delta V = V(1 - e^{-\tau/\tau_c}) \approx \frac{V\tau}{\tau_c} \text{ as } \tau_c \gg \tau$$

Hence

$$A_{12} \approx \Delta V \tau = \frac{V\tau^2}{\tau_c}$$

Also, as in the previous case,

$$A_2 \approx A_1 \approx V\tau$$

Therefore, the cross-talk factor in this case is

$$K \equiv \frac{A_{12}}{A_2} = \frac{A_{12}}{A_1} = \frac{V\tau^2/\tau_c}{V\tau} = \frac{\tau}{\tau_c}$$

In this case, $\tau_c$ is very large and, hence, the pulse extends to many time slots. Thus this type of cross-talk extends to more than one channel. In other words, a channel gets affected not only by the previous channel but also by some more channels prior to that.

### 7.1.5 Transmission of PAM Signals

If the PAM signals are to be transmitted directly, say over a pair of wires, then no further signal processing is necessary. However, if they are to be transmitted through the space using an antenna, they must first be amplitude or frequency or phase modulated by a high frequency carrier, and only then they can be transmitted. The overall system would be then known as PAM-AM or PAM-FM or PAM-PM respectively. At the receiving

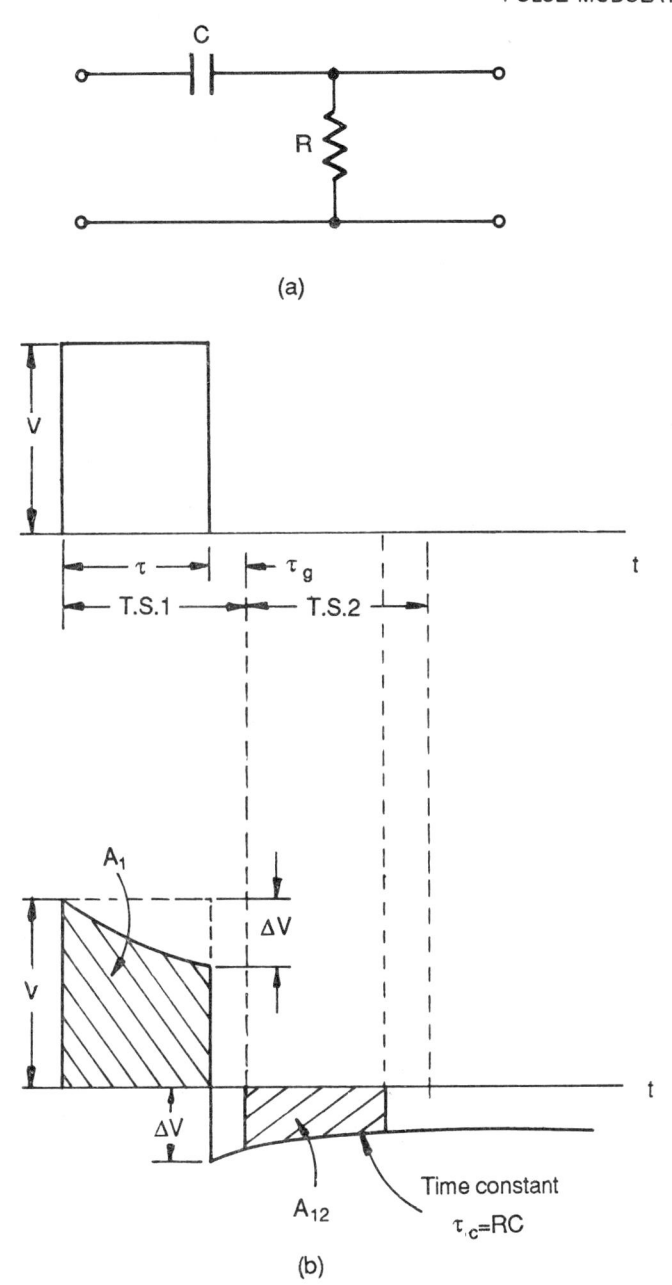

Fig. 7.1.4.1 Cross-Talk due to LF Cut-off of the Channel;
(a) High Pass Filter Representation of the Channel
(b) Output of the Channel of (a) for a Pulse Input

side, AM or FM or PM detection is first employed to get the PAM signal, and then the baseband signal is recovered from it.

### 7.1.6 Bandwidth of PAM Signals

Let us assume that we have to transmit $n$ signals, each bandlimited to $f_M$ Hz. Then, for each signal, we have to take $2f_M$ samples per seecond. Thus, in all, we have to transmit $2nf_M$ samples per second.

Now, according to the sampling theorem, a continuous signal band limited to B Hz can be transmitted by 2B samples per second. Conversely, it can be stated that 2B samples per second define a continuous signal band limited to B Hz. Therefore, the bandwidth of the PAM system of $n$ signals, each band limited to $f_M$ Hz, will be $nf_M$ Hz, because we are transmitting $2nf_M$ samples per second.

### 7.1.7 S/N Ratio of PAM System

The noise performance of PAM is identical to AM-SC signal. The figure of merit $v$ is unity. It has already been shown that the bandwidth of PAM/AM is $nf_M$, where, $n$ is the number of messages multiplexed. Therefore, the bandwidth per message is $f_M$. Thus, the transmission of sampled signal message reproduces a bandwidth at the receiver that is equivalent to the message $f(t)$. In other words, the transmission of a sampled signal is equivalent to the direct transmission of $f(t)$. The signal and noise power at the transmitter and receiver is expected to be identical. It has been shown in chapter 2 that the power of a sampled signal and the baseband signal $f(t)$ is the same. (refer Prob. 2.2)
Hence,

$$S_o = S_i = \overline{f^2(t)}$$

The bandwidth of sampled signal per message is same as that of the baseband signal $f(t)$.
Hence, for white noise

$$N_o = N_i = \eta f_M$$

where $f_M$ is the bandwidth of $f(t)$, and $\eta/2$ is the noise power per unit bandwidth. The figure of merit is given by

$$(v)_{PAM} = \frac{(S_o/N_o)}{(S_i/N_i)} = 1$$

## 7.2 PULSE TIME MODULATION

The two types of PTM systems, namely PWM and PPM, are shown in Fig. 7.2.1. The Fig. 7.2.1(a) is the baseband signal $f(t)$ whereas the Fig. 7.2.1(b) is the carrier pulse train $f_c(t)$. The Fig. 7.2.1(c) is the PWM signal where the width of each pulse depends on the instantaneous value of the baseband signal at the sampling instant. The Fig. 7.2.1(d) is the PPM signal where the shift in the position of each pulse depends on the instantaneous value of the baseband signal at the sampling instant. It can be seen that in PWM, the information about the baseband signal lies in the trailing edge of the pulse, whereas, in PPM, it lies in both the edges of the pulse. (Although, basically it lies in the leading edge, but since the width of the pulse is same always, the trailing edge also carries the same information.)

### 7.2.1 Generation of PTM Signals

*Indirect Method:* The scheme of generation of the PTM signals is shown in Fig. 7.2.1.1. Firstly, the flat-topped PAM signals are generated as explained in Sec.7.1 [Fig. 7.2.1.1(a)]. The synchronized ramp waveform shown in Fig. 7.2.1.1(b) is generated during each pulse interval. These two signals are added as shown in Fig. 7.2.1.1(c), and the sum is applied to a comparator circuit whose reference level is shown by a broken line in Fig. 7.2.1.1(c).

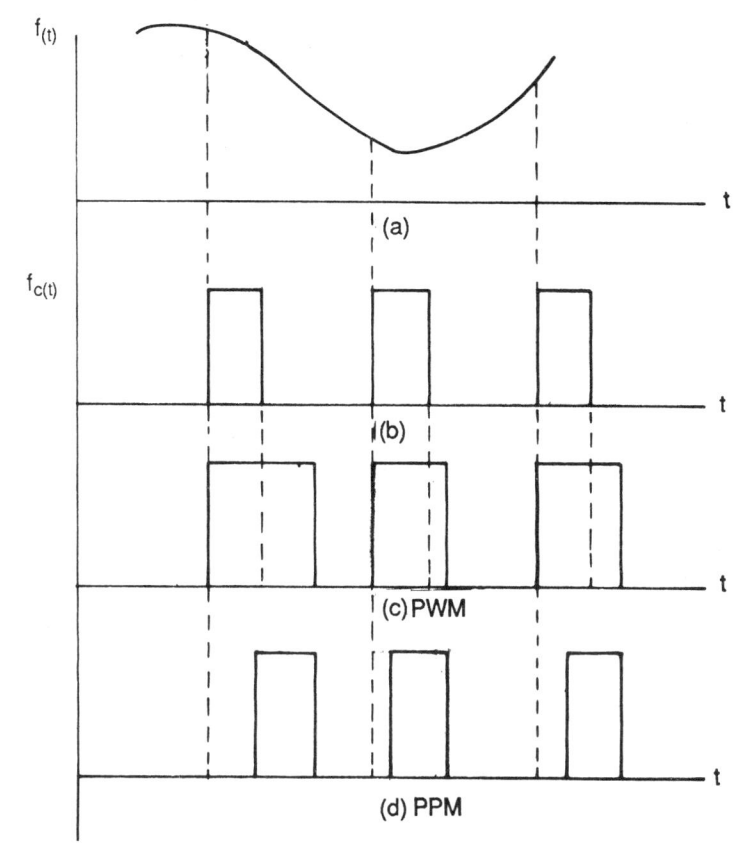

Fig.7.2.1  Pulse Time Modulation (a) Baseband Signal f(t),(b) Carrier Pulse Train $f_c(t)$,(c) Pulse Width Modulated Signal,(d) Pulse Position Modulated Signal.

The second crossing of the comparator reference level by the waveform of Fig. 7.2.1.1(c) is used to generate the pulses of constant amplitude and width as shown in Fig. 7.2.1.1(d), giving the desired PPM waveform. The leading edge of the synchronized ramp of Fig. 7.2.1.1(b) is used to start a pulse, and the trailing edge of the PPM waveform of Fig. 7.2.1.1(d) is used to terminate the pulse, as shown in Fig. 7.2.1.1(e), giving the desired PWM waveform.

The ramp amplitude is so adjusted that it is slightly greater that the maximum variation in amplitude of the PAM signals. The comparator reference level is such that it always intersects the sloping portion of the waveform of Fig. 7.2.1.1(c).

*Direct Method:* In the direct method, the PTM waveforms are generated without generating the PAM waveform. Here, the baseband signal $f(t)$ of Fig. 7.2.1.2(a) and a ramp signal of Fig. 7.2.1.1(b), occurring at the sampling instants, are added to give the waveform of Fig. 7.2.1.2(c). This is compared in a comparator whose reference level is shown in Fig. 7.2.1.2(c) by a broken line.

The PPM (Fig. 7.2.1.2d) and PWM (Fig.7.2.1.2e) waveforms are then obtained in the same manner as explained in the indirect method of generation of PTM signals.

If we want to modulate the leading edge of pulses in the PWM waveform, the ramp waveform shown in Fig. 7.2.1.3(a) should be used. For modulating both the edges of the pulses in the PWM waveform, the ramp waveform shown in Fig. 7.2.1.3(b) needs to be used.

456 COMMUNICATION SYSTEMS: Analog and Digital

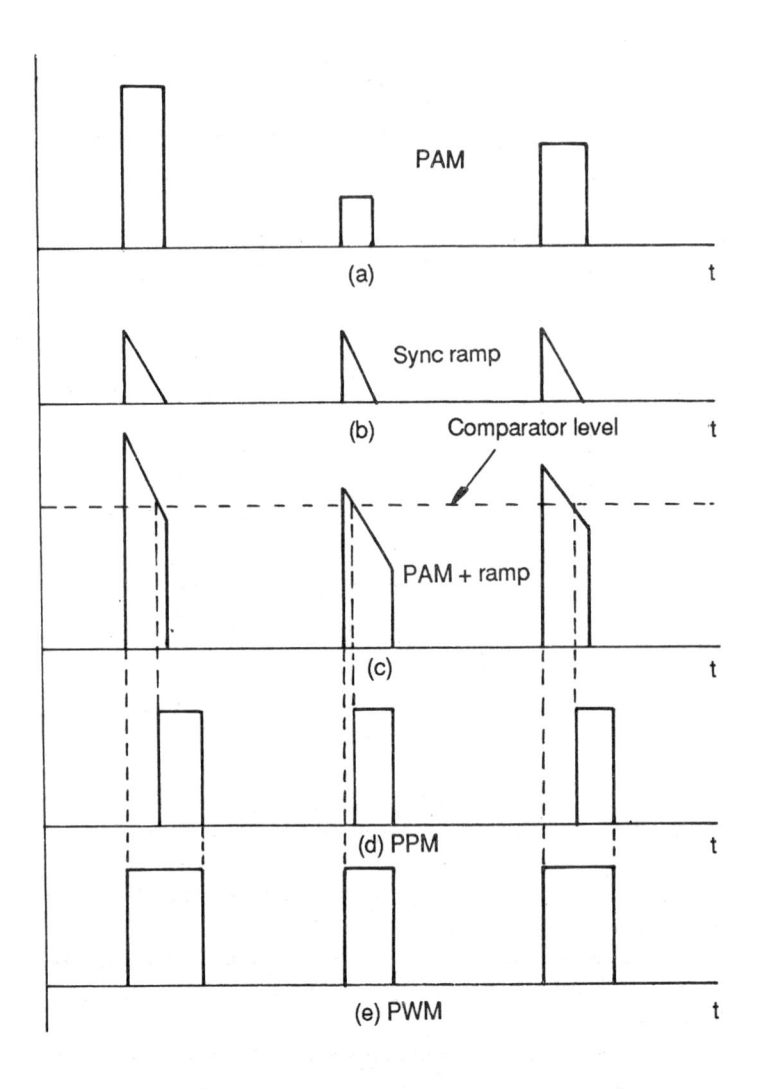

Fig.7.2.1.1 Indirect Method of Generation of PTM Signals (a) PAM Signal,
(b) Synchronized Ramp, (c) PAM + Synchronized Ramp, (d) PPM Signal, (e) PWM Signal

*A PWM Modulator Circuit:* A PWM modulator circuit is shown in Fig. 7.2.1.4. The clock signal of the desired frequency is applied as shown, from which the negative trigger pulses are derived with the help of a diode and an $R_1 - C_1$ combination which works as a differentiator. These negative trigger pulses are applied to the pin no.2 of the 555 timer which is working in the monostable mode. They decide the starting time of the PWM pulses. The end of the pulses depends on an $R_2 - C_2$ combination, and on the signal at pin no.5, to which the modulating signal is applied. Therefore, the width of the pulses depends upon the value of the modulating signal, and thus the output at pin no.3 is the desired pulse width modulated signal.

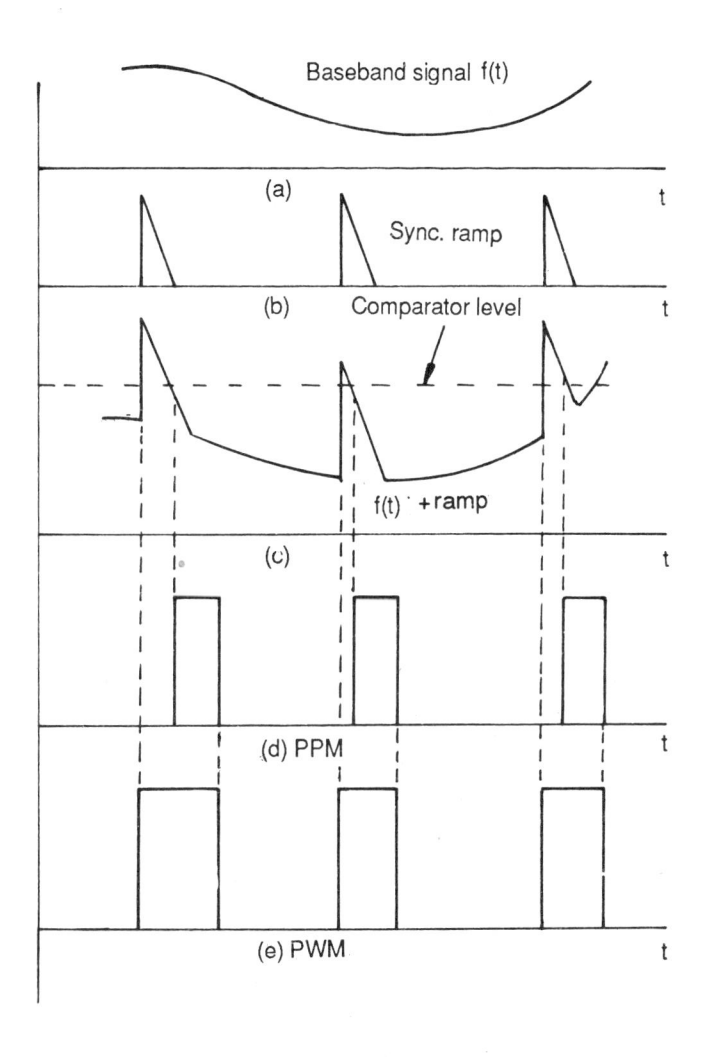

PULSE MODULATION SYSTEMS 457

Fig. 7.2.1.2 Direct Method of Generation of PTM Signals (a) Baseband Signal $f(t)$ (b) Synchronized Ramp (c) $f(t)$ + Synchronized Ramp, (d) PPM Signal (e) PWM Signal

*A PPM Modulator Circuit:* A PPM modulator circuit is shown in Fig. 7.2.1.5. The PWM signal (which is obtained as shown in Fig. 7.2.1.4) is applied to pin no.2 through the diode and $R_1$-$C_1$ combination. Thus, the input to pin no.2 is the negative trigger pulses which correspond to the trailing edges of the PWM waveform. The 555 timer is working in a monostable mode and the width of the pulse is constant (governed by an $R_2$-$C_2$ combination).

The negative trigger pulses decide the starting time of the output pulses and, thus, the output at pin no.3 is the desired pulse position modulated signal.

### 7.2.2 Demodulation of PTM Signals

The PWM waveform of Fig. 7.2.2.1(a) is used to generate a ramp waveform as shown in Fig. 7.2.2.1(b). The

(a)  (b)

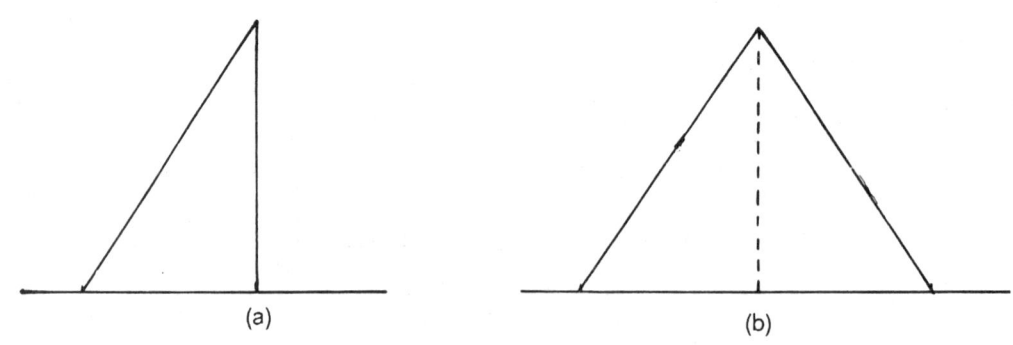

Fig.7.2.1.3(a) Synchronized Ramp for Leading Edge Modulation of PAM Signal
(b) Synchronized Ramp for Both Edge Modulation of PWM Signal

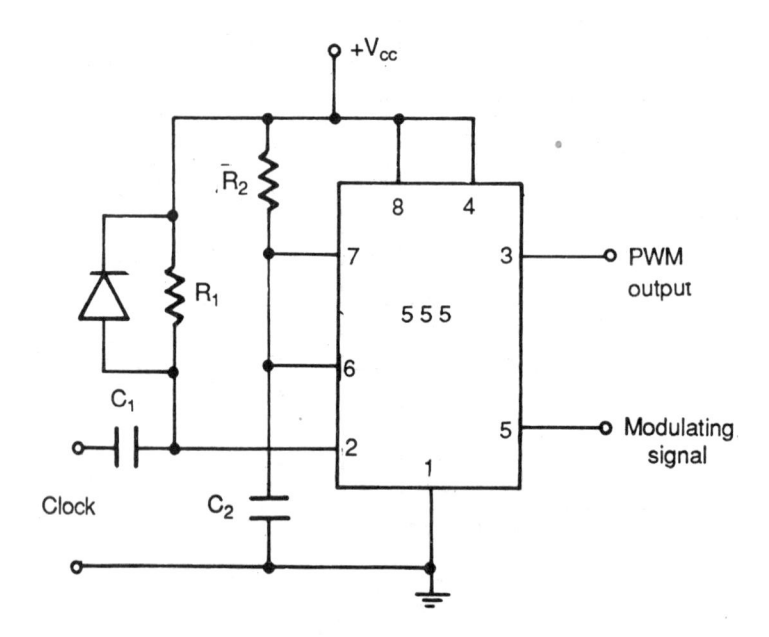

Fig.7.2.1.4  A PWM Modulator

leading edges of the PWM pulses start the ramp of same slope, and the trailing edges of the PWM pulses terminate the ramp. The height attained by the ramp is, therefore, proportional to the width of the PWM pulses. The height attained by the ramp is sustained for some time, thus creating a porch, after which the voltage returns to its initial level.

A similar type of synchronized ramp is shown in Fig. 7.2.2.2(b). This is generated with the help of the PPM pulses shown in Fig.7.2.2.2(a). Here, the ramp is initiated at the beginning of the time slot, and it is terminated by the leading edge of the PPM pulse. Thus, the height attained by the ramp is proportional to the displacement of the leading edge of the PPM pulses from the beginning of the time slot. Here too, the height attained by the ramp is sustained for some time, thus creating a porch, and then it is returned to the initial level.

The remaining procedure for both PWM and PPM is same. A sequence of locally generated pulses of a fixed amplitude are added to the synchronized ramp on the porch as shown in Figs 7.2.2.1(c) and 7.2.2.2(c). The

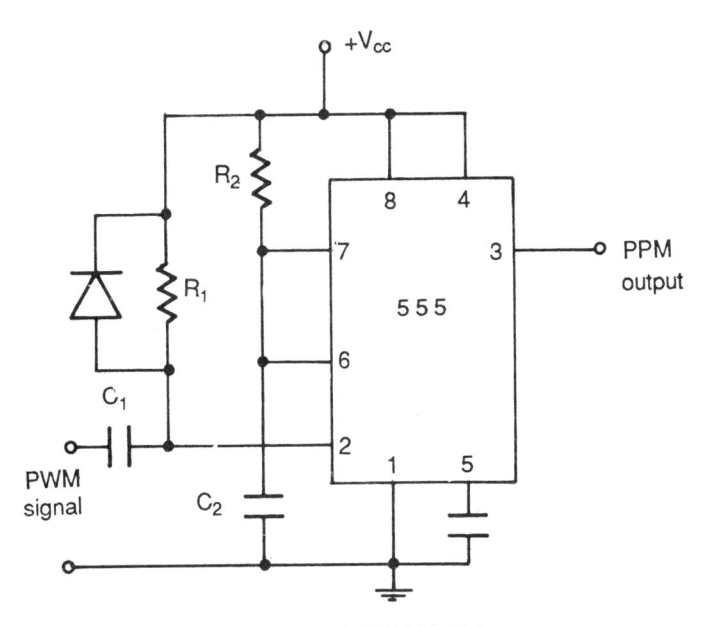

Fig.7.2.1.5 A PPM Modulator

lower portions in these two waveforms are clipped by a clipping circuit, with the clipping level adjusted in such a way that it never crosses the ramp. The output of the clipper is a PAM waveform (Fig.7.2.2.1d and Fig. 7.2.2.d), from which the baseband signal can be recovered as explained in Sec. 7.1.1.

*A PWM Demodulator Circuit:* A PWM Demodulator circuit is shown in Fig. 7.2.2.3. The transistor $T1$ works as an inverter. Hence, during the time interval $A - B$, when the PWM signal is high, the input to the transistor $T2$ is low. Therefore, during this time interval, the transistor $T2$ is cut-off and the capacitor $C$ gets charged through an $R-C$ combination. During the time interval $B - C$, when the PWM signal is low, the input to the transistor $T2$ is high, and it gets saturated. The capacitor $C$ then discharges very rapidly through $T2$. The collector voltage of $T2$ during the interval $B - C$ is then low. Thus the waveform at the collector of $T2$ is more or less a saw-tooth waveform whose envelope is the modulating signal. When this is passed through a second-order OP-AMP low pass filter, we get the desired demodulated output.

*A PPM Demodulator Circuit:* A PPM demodulator circuit is shown in Fig.7.2.2.4. This utilizes the fact that the gaps between the pulses of a PPM signal contain the information regarding the modulating signal. During the gap $A - B$ between the pulses, the transmitter is cut-off, and the capacitor $C$ gets charged through the $R$-$C$ combination. During the pulse duration $B - C$, the capacitor discharges through the transistor, and the collector voltage becomes low. Thus, the waveform at the collector is approximately a saw-tooth waveform whose envelope is the modulating signal. When this is passed through a second order OP-AMP low pass filter, we get the desired demodulated output.

### 7.2.3 Bandwidth of PTM Signals

The bandwidth can be estimated by observing the spectrum of the PTM signals. The spectral analysis for such waves is complicated and hence will not be treated in this text. We will discuss only the qualitative features of the spectrum. The one-sided spectrum of a PWM signal is shown in Fig. 7.2.3.1. Assume that the modulating signal is a single-tone sinusoid of the frequency $\omega_m$, and the sampling frequency is $\omega_s$. The PWM spectrum has the following frequency components

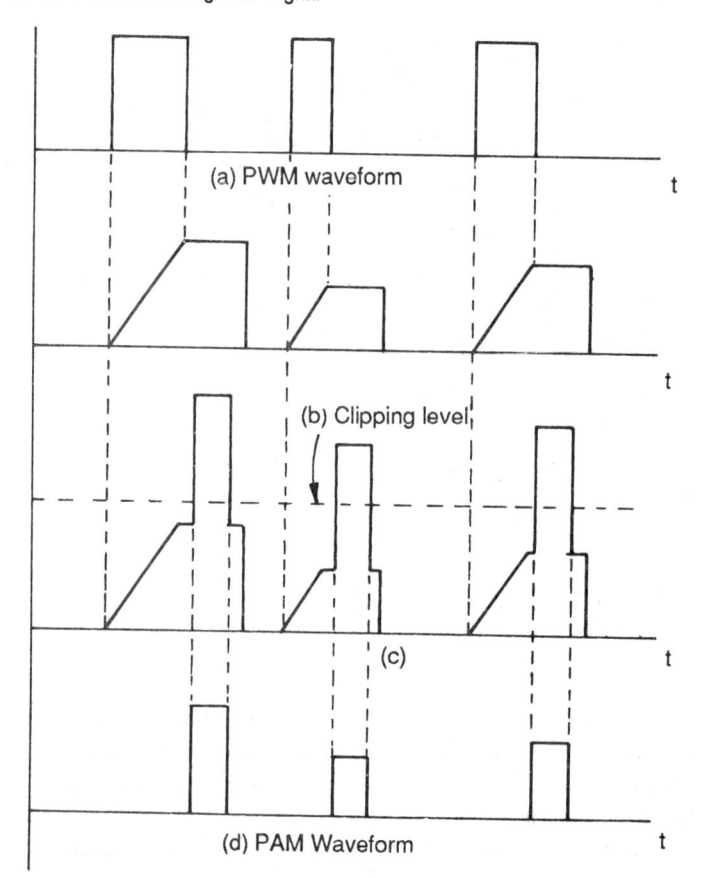

Fig.7.2.2.1 Demodulation of PWM Signals (a) PWM Waveform
(b) Ramp Waveform with Porch; (c) Ramp Waveform with locally Generated Pulse on Porch;
(d) PAM Waveform

(i) A dc component at $\omega = 0$, which represents the average value of the pulses.

(ii) The modulating frequency $\omega_m$.

(iii) The harmonics of the sampling frequency $\omega_s$.

(iv) Sidebands spaced by $\omega_m$ centered around each harmonic of $\omega_s$.

The presence of harmonics of $\omega_s$ are due to the contribution of the unmodulated pulse train which may be taken as the carrier of the PDM wave. Each harmonic of $\omega_s$ is associated with the sidebands of an FM type. The sidebands of each $\omega_s$ extend to infinity outward, but with a decaying magnitude. However, the useful message band is available in a band 0-$\omega_m$ and hence a low pass filter can be used to recover the message from PWM. But the output of LPF is distored due to the presence of cross-modulation terms that lie in the baseband. The lower sidebands of $\omega_s$ may extend to lie in the message baseband to cause distortion. This can be prevented by restricting the maximum excursion of the trailing edge of the PWM pulse.

The spectrum of a naturally sampled PPM wave for a single tone modulating signal has a form similar to that of a PDM wave, with the only difference that it contains a component proportional to the derivative of the modulating signal in place of modulating component itself. Therefore, the PPM detection can be achieved

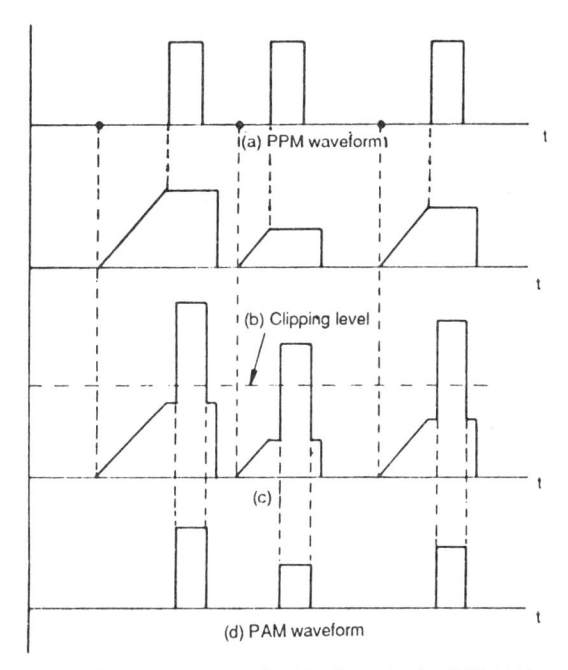

Fig. 7.2.2.2 Demodulation of PPM Signals (a) PPM Wave-form;
(b) Ramp Wave-form with Porch (c) Ramp wave-form with Locally Generated Pulse on Porch
(d) PAM Wave-form

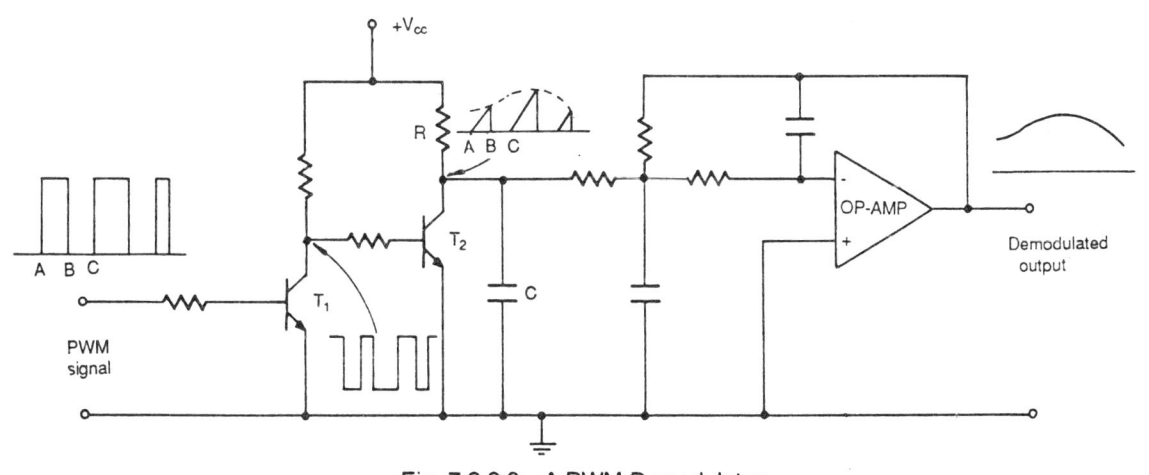

Fig. 7.2.2.3  A PWM Demodulator

by an LPF followed by an integrator. An alternative detection method is to convert PPM into PWM, and then pass it through an LPF. This provides greater signal amplitude with less distortion in the receiver.

### 7.2.4  S/N Ratio of PTM Systems

In PPM signal, the sample values of the modulating signal $f(t)$ are transmitted in terms of the pulse position

462  COMMUNICATION SYSTEMS: Analog and Digital

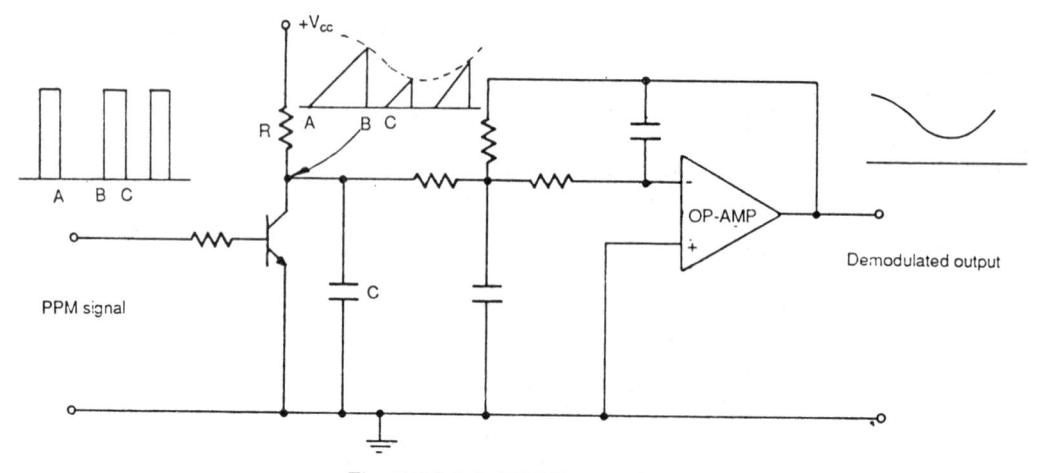

Fig. 7.2.2.4 A PPM Demodulator

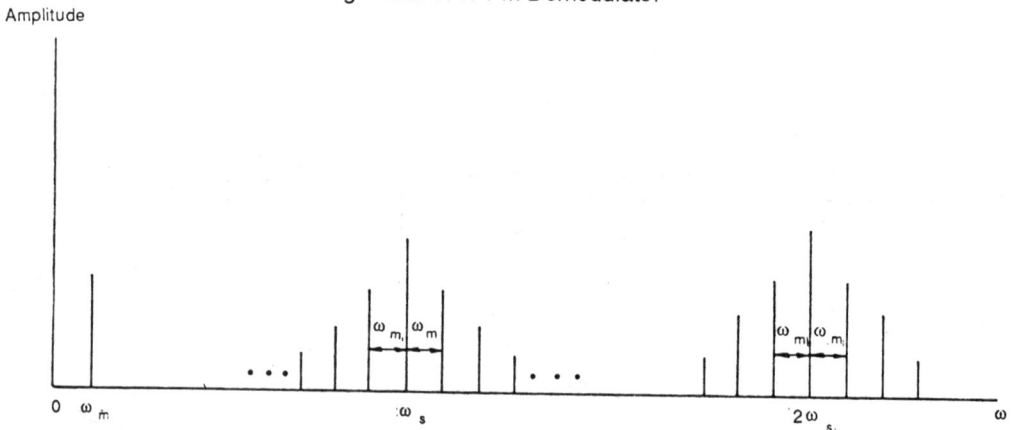

Fig.7.2.3.1 One-sided Spectrum of PWM Signal

over a channel of bandwidth BHz. This finite bandwidth of the channel causes dispersion in a received PPM pulse, (refer Fig. 2.3.7).The resultining trapezoidal pulse is shown in Fig. 7.2.4.1 (a). The rise time of the pulse is given as

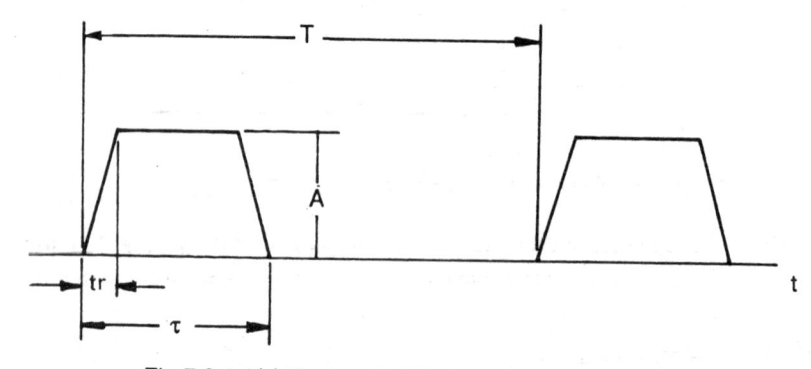

Fig.7.2.4.1(a) Trapezoidal Pulses due to Dispersion

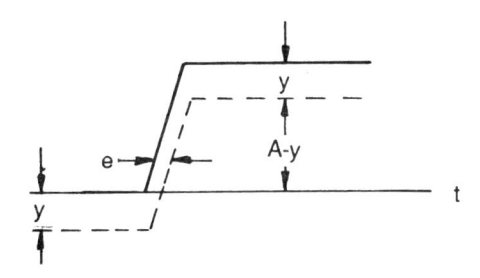

Fig. 7.2.4 (b) Shift in Pulse of (a) Due to Noise

$$t_r \approx \frac{1}{B} \qquad (7.2.4.1)$$

The position of the pulse is sensitive to any additive noise. If a noise signal $y$ is added at any instant, the position is shifted by '$e$', as shown in Fig. 7.2.4.1(b). It is obvious that

$$\frac{A}{y} = \frac{t_r}{e} \qquad (7.2.4.2)$$

As $y$ varies randomly, $e$ also varies randomly. The mean squared (m.s) values of $y$ and $e$ are related as

$$\frac{\overline{e^2}}{\overline{y^2}} = \left(\frac{t_r}{A}\right)^2 \qquad (7.2.4.3)$$

Therefore, for an additive noise $n(t)$, the m.s. value of $e$ is given as

$$\overline{e^2} = \left(\frac{t_r}{A}\right)^2 \overline{n^2(t)} \qquad (7.2.4.4)$$

The change in the position of the $m$th pulse is proportional to the value of $m$ th sample of modulating signal $f(t)$. Let us denote the change in the position by $x_m$ and the $m^{th}$ sample by $f(mT)$, each sample being spaced by $T$. Then,

$$x_m = K_1 f(mT) \qquad (7.2.4.5)$$

where $K_1$ is a proportionality constant.

Now, when noise is added, the change in the position of the $m^{th}$ pulse becomes $x_m + e_m$, where $e_m$ is random and is defined by Eq. 7.2.4.4. Let us denote this changed position of the m$^{th}$ pulse by $x'_m$, then

$$x'_m = x_m + e_m = K_1 f(mT) + e_m \qquad (7.2.4.6)$$

At the receiver, the pulse positions are converted back into samples, which are then passed through an LPF for detection. It can be seen that the output of the LPF at the receiver is given by

$$x'(t) = K_1 f(t) + e(t)$$

Therefore, the useful message in this output is

$$S_o(t) = K_1 f(t)$$

and the noise signal in the output is

$$n_o(t) = e(t)$$

The m.s. value provides the output signal and noise power as follows:

$$S_o = K_1^2 \overline{f^2(t)} \text{ and } N_o = \overline{n_o^2(t)} = \overline{e^2(t)}$$

The m.s. value of $e(t)$ is same as the m.s. value of its samples (Prob. 2.2). Therefore,

$$\overline{e^2(t)} = \overline{e_m^2} = \left(\frac{t_r}{A}\right)^2 \overline{n^2(t)}$$

and

$$N_o = \left(\frac{t_r}{A}\right)^2 \overline{n^2(t)}$$

Therefore,

$$\frac{S_o}{N_o} = \frac{K_1^2}{\overline{n^2(t)}} \left(\frac{A}{t_r}\right)^2 \overline{f^2(t)}$$

$$= K_1^2 B^2 \overline{f^2(t)} \frac{A^2}{\overline{n^2(t)}}$$

Where

$$B = \frac{1}{t_r}$$

Let the duty ratio '$d$' of the pulse be given by

$$d = \frac{\tau}{T}$$

The power contained in the input PPM wave of the amplitude '$A$' is given by

$$S_i = dA^2$$

The input noise power is given by, $N_i = \overline{n^2(t)}$
Therefore,

$$\frac{S_o/N_o}{S_i/N_i} = \frac{K_1^2}{d} \overline{f^2(t)} B^2 \qquad (7.2.4.7)$$

Thus, the $S/N$ ratio improvement is proportional to the square of the channel bandwidth. A similar result can be drawn for the PWM signal after evaluating its noise figure. It has been observed that the noise performance of PWM is inferior to PPM. This is true because the PPM system preserves all the signal information at the terminating instant of the pulses, and yet it avoids a considerable loss of power which a PWM system expends during the pulse. PWM, is, thus, less efficient than PPM with regard to the transmitter power utilization.

## PROBLEMS

7.1 Explain natural and flat-top sampling. Compare the two.
7.2 Explain the methods for demodulation of PAM signals.
7.3 Explain the cross-talk in PAM due to the HF and LF limitations of the channel. Which one of the two affects more than one channel and why?
7.4 Explain how PPM and PWM signals are generated:
   (i) From PAM signals;
   (ii) Directly. How are these detected?
7.5 Draw and explain the following circuits:
   (i) PAM modulator and demodulator.
   (ii) PWM modulator and demodulator.
   (iii) PPM modulator and demodulator.
7.6 What are FDM and TDM? Compare the two.
7.7 Two signals bandlimited to 3 and 5 kHz, are to be time devision multiplexed. Find the maximum permissible interval between two successive samples.

# EIGHT
# PULSE CODE MODULATION

## INTRODUCTION

The pulse modulation systems are not completely digital, as the amplitude, width, or position of the pulses transmitted may vary continuously in accordance with baseband signal variations. PAM is the simplest pulse modulation system. For this reason, the PAM signals are used to get a complete digital system. The PAM signals are first quantized and then coded (usually in binary code), thus, giving rise to a pulse code modulation (PCM) system. The operations of quantization and coding are explained below:

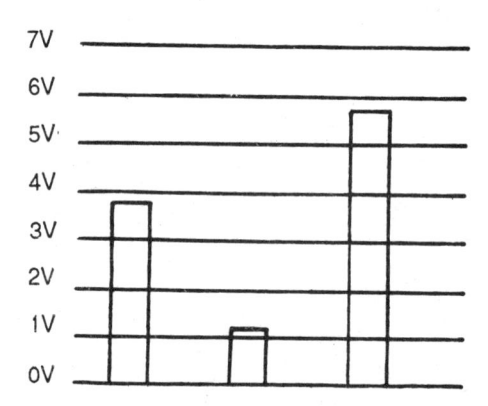

Fig. 8.1 Quantization Levels

Let us consider a PAM signal whose amplitude varies from $-0.5$ V to 7.5 V. This range is divided into eight levels known as *quantization levels* as shown in Fig. 8.1. Pulses having values from $-0.5$ V to 0.5 V are approximated (quantized) to a value 0V, and the pulses having values from 0.5 V to 1.5 V are approximated to a value 1 V, and so on. It is assumed that, as per the probability theory, the pulses having exact values as 0.5 V, 1.5V, etc., will never occur. Thus, any pulse can be approximated to one of the values of quantization levels 0V, 1V, etc. For example, the pulses A, B, and C of amplitudes 3.8 V, 1.2 V, and 5.7 V, respectively (Fig. 8.1.), will be approximated to the values 4 V, 1V and 6 V, respectively. These, then are known as quantized pulses. These quantized pulses are then encoded. The most popular code is the binary code shown in Table 8.1. The three quantized pulses having values 4V, 1V and 6V will be encoded as 100, 001, and 110

PULSE CODE MODULATION 467

respectively. These encoded values are combinations of binary digits (bits) 0 and 1. These can be transmitted easily.

Table 8.1

| Quantization Level | Binary Code |
|---|---|
| 0 | 000 |
| 1 | 001 |
| 2 | 010 |
| 3 | 011 |
| 4 | 100 |
| 5 | 101 |
| 6 | 110 |
| 7 | 111 |

For example, the presence of a pulse may be represented by 1 and its absence by 0. The transmitted signals for the pulses $A$, $B$, and $C$ will be as shown in Fig. 8.2. Note that there is a time slot allotted to each bit, a portion of which is guard time. Similarly, a time slot is allotted to each word which includes the guard time.

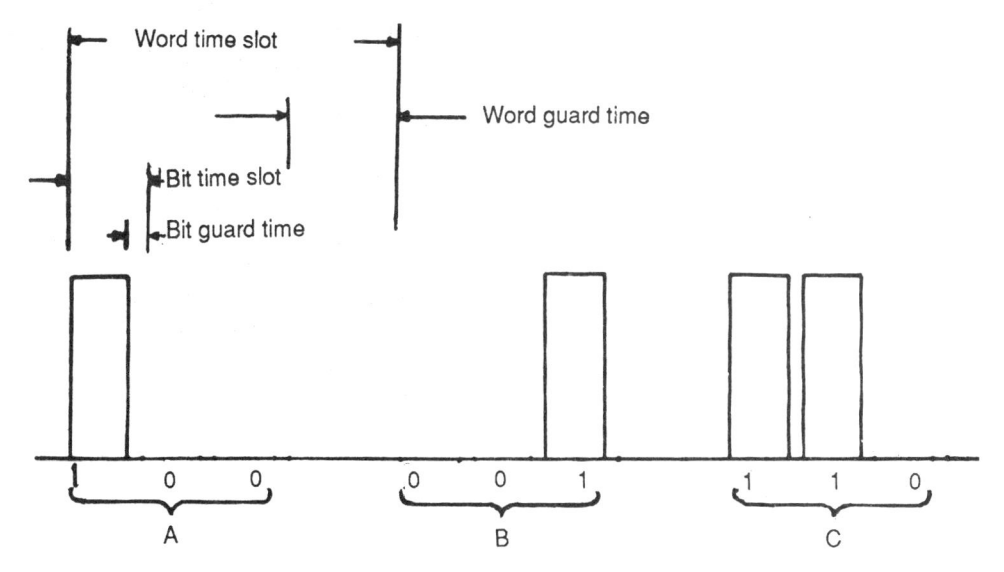

Fig. 8.2  Transmitted Signal for Quantized Pulses of Fig. 8.1

The major advantage of the PCM system is that the information does not lie in any property of the pulse, but it lies in the presence or absence of the pulse. Thus, even if noise distorts the pulse, it makes no difference so long as the decision regarding the presence or absence of the pulse is correct. Figure 8.3 explains this. The transmitted pulses are shown in Fig. 8.3(a). Two cases of received pulses are shown in Fig. 8.3(b) and 8.3(c). Figure 8.3(b) explains a situation where although the received pulses are distorted due to noise, there is no error of decision. [When the voltage in a time slot crosses the threshold level (which is at the centre of 0 and 1 level), it is received as a 1, and when the voltage in a time slot does not cross the threshold level, it is received as a 0.]  Figure 8.3(c) explains a situation where errors occur. In the time slot $X$, the voltage does not cross the threshold level and, hence, a 0 is received (an error); and in the time slot Y, the voltage crosses the threshold level and, hence, a 1 is received (an error). There is no error in the time slot Z. Thus 101 is received as 011. But the probability of the occurrence of such errors is extremely small. Moreover, in an extremely noisy condition, the difference between the 0 and 1 levels can be increased, resulting in a reduced effect of noise. Hence, practically, error-free transmission is possible. The only error that is possible is due to quantization, as there is no way to know whether a 4V signal is a result of a 3.8 V signal or a 4.3 V signal. However, the quantization error can be minimized by reducing the step size.

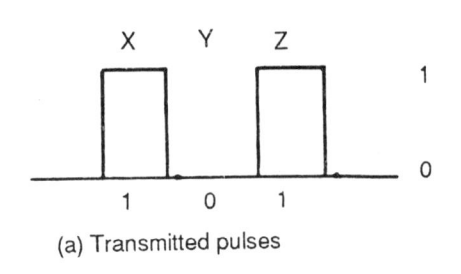

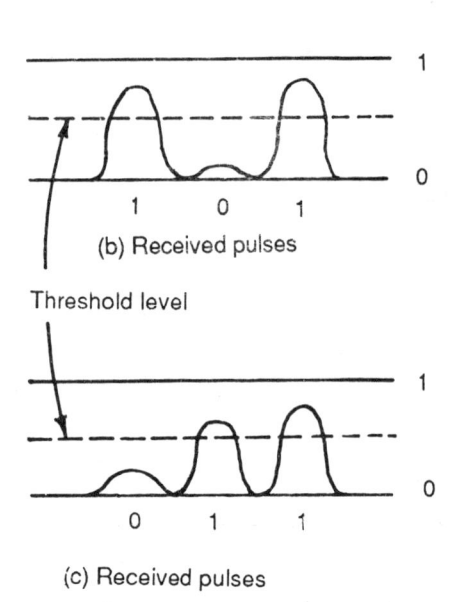

Fig. 8.3 (a) (b) and (c) Effect of Noise on PCM Pulses

## 8.1 PCM SYSTEM

The block diagram of a PCM system is shown in Fig. 8.1.1. Figure (a) shows a PCM transmitter. The baseband signal is sampled at Nyquist rate by the sampler. The sampled pulses are then quantized in the quantizer. The encoder encodes these quantized pulses into bits which are then transmitted over the channel. Figure (b) shows a PCM receiver. The first block is again a quantizer. But this quantizer is different from the transmitter quantizer because it has to take a decision about the presence or absence of a pulse. Thus, there are only two quantization levels. The output of the quantizer goes to the decoder which is an A/D converter that performs the inverse operation of the encoder. The decoder output is a sequence of quantized pulses. The original baseband signal is reconstructed in the holding circuit and LPF (Low Pass Filter) as explained in Sec. 7.1.1.

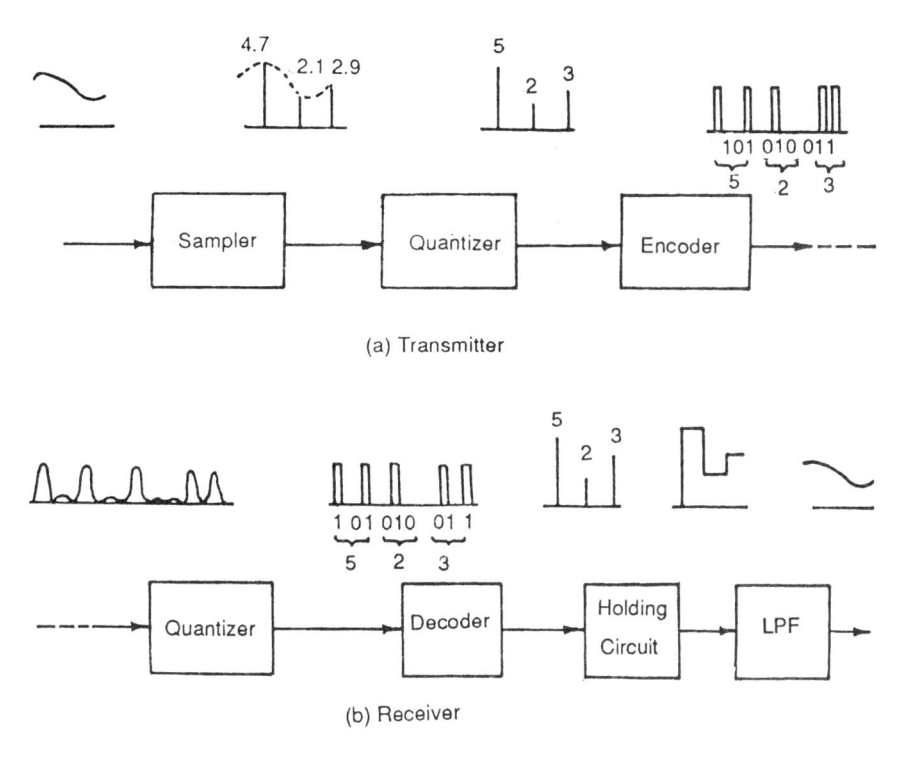

Fig. 8.1.1(a) and (b) A PCM System

## 8.2 INTERSYMBOL INTERFERENCE

As the PCM channels are band limited, the received waveforms are distorted and they extend to the next time slot, resulting in error in the determination of received bits. The situation is analogous to cross-talk in the PAM system (Sec. 7.1.4 and 7.1.5). In PAM, the adjacent time slots are different channels and hence the term cross-talk is appropriate. However, in PCM, the adjacent time slots are more generally symbols in code representation of a single quantized sample; hence, the term *intersymbol interference* is used for PCM. For example, in an 8-bit PCM signal, the probability of an adjacent time slot being a symbol in code representation of a single quantized sample is seven times greater than its being the symbol for the next channel. (For first seven bits, the adjacent bit corresponds to code representation of a single quantized sample, whereas, for the eighth bit, the adjacent bit corresponds to the next channel.)

## 8.3 EYE PATTERNS

Both the intersymbol interference and the noise cause errors to occur. A cathode ray oscilloscope can be used to give an indication of the performance of a PCM system. In this method, the received bit stream is applied to the vertical deflection plates, and the time base frequency is made equal to the bit rate so that a sweep lasts one time slot duration. When the received bit stream is ideal, as shown in Fig. 8.3.1(a), the CRO pattern will be like the one shown in Fig.8.3.1(b). If the bit stream is distorted as shown in Fig 8.3.1(c), the CRO pattern will be as shown in Fig. 8.3.1(d). The pattern of Fig. 8.3.1(d) is very much similar to the human eye, the central portion of the pattern being the opening of the eye. If the signal is further distorted, as shown in Fig. 8.3.1(e), the CRO pattern will be as shown in Fig. 8.3.1(f). In this case, the eye is further closed. Thus, an observation of the eye pattern gives an idea about the distortion in the system. The more is the opening of the eye, the less is the distortion and vice-versa.

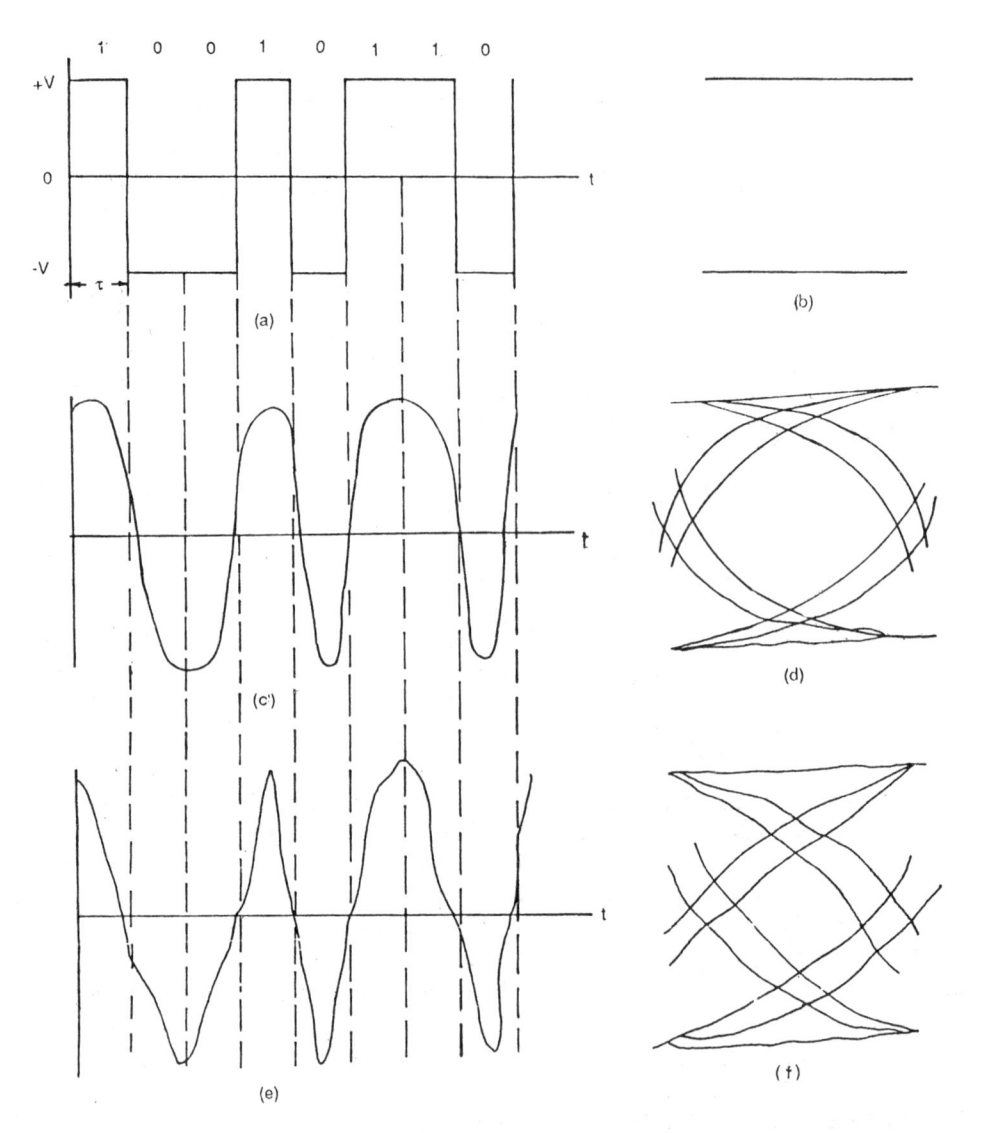

Fig. 8.3.1 Eye Patterns (a) Received Bit Stream (ideal)
(b) CRO Pattern for (a) (c) Received Bit Stream (less distortion) (d) CRO Pattern for (c)
(e) Received Bit Stream (more distortion) (f) CRO Pattern for (e)

## 8.4 EQUALIZATION

The intersymbol interference causes distortion. These distortions can be reduced by designing a proper equalizer. If the frequency response of the channel $H_c(\omega)$ is completely known, then, an equalizer is designed whose frequency response $H_e(\omega)$ is the inverse of $H_c(\omega)$. Equalizing filters are inserted between the receiving filter and the A/D converter.

An ordinary equalizer has a problem. Normally, we don't have a prior knowledge regarding telephone line connection; as such, any two subscribers can be connected. Thus, the channels are connected randomly between the transmitter and receiver. So we have to adjust the equalizer filter manually by observing the eye pattern. In an adaptive equalizer, this is done automatically by using feed-back technique.

## 8.5 COMPANDING

We have seen in the introductory portion of this chapter that the quantization error depends upon the step size. When the steps are uniform in size, the small amplitude signals will have a poorer signal to quantization noise ratio than the large amplitude signals, because in both the cases the denominator (quantization noise) is the same; whereas, the numerator is small for small amplitude signal, and large for large amplitude signals. As we have to use a fixed number of quantization levels, the only way to have a uniform signal to quantization noise ratio is to adjust the step size in such a way that the ratio remains constant. Thus, the step size should be small for small amplitude signals and large for large amplitude signals.

The effect of an adaptive step size may also be achieved in a more feasible way by distorting the signal before quantization. An inverse distortion has to be introduced at the receiver to make the overall transmission distortionless. This is explained in Fig. 8.5.1. The output is enhanced more at low amplitudes than at high amplitudes (see the dashed curve.) This output is then applied to the quantizer. Thus, the low amplitude signal will carry more quantization levels than the undistorted signal (see the solid line). A signal transmitted through a non-linear network with the characteristic shown by the dashed curve will have its extremities compressed. Hence, such a network is known as a compressor. At the receiver side, an inverse operation is to be performed to recover the original signal. This is achieved by an expander, connected between the decoder and holding circuit, whose characteristic is shown by the dotted curve in Fig. 8.5.1. The combination of compressor and expander is known as *compander,* which performs the companding operation.

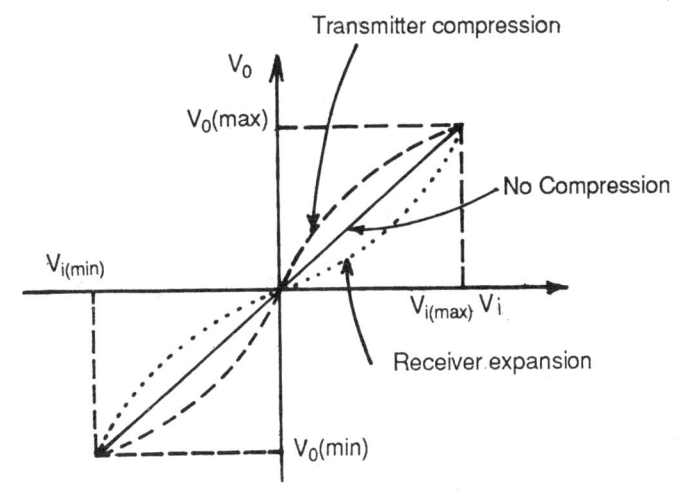

Fig. 8.5.1 Companding

## 8.6 SYNCHRONOUS TIME DIVISION MULTIPLEXING

As in the case of PAM, time division multiplexing is possible in PCM also. The method is the same as explained in Sec. 7.1.2, except that here each sample is coded into several bits. The multiplexing is possible in two ways:

(i) Bits are taken, one by one from each channel sample code. When the first bits from all channel samples, are taken, the commutator takes the second bits from all channel samples, and so on.

(ii) All code bits of the first channel samples are taken followed by all code bits of the second channel samples and so on. In this method, the desired commutator speed is less than that required in the first method.

A synchronizing bit is added at the end of each frame for synchronization between the commutator and decommutator. The signal that is to be time division multiplexed is bandlimited to the same frequency, resulting in the same sampling frequency for all the channels and, hence the name *synchronous time division multiplexing*.

## 8.7 ASYNCHRONOUS TIME DIVISION MULTIPLEXING-PULSE STUFFING

When the signals to be time division multiplexed are bandlimited to different frequencies, their sampling frequencies are also different, and hence they cannot be multiplexed by the method given in Sec.8.6. Such asynchronously sampled signals are multiplexed by a technique called *pulse stuffing*.

To multiplex the asynchronous signals, it is necessary to have a device which can store and reproduce data at different speeds. Such a storage device is known as *elastic store*. A tape recorder is an example of elastic storage device. In the tape recorder, the playback speed can be adjusted as desired by adjusting the speed with which the tape moves during feedback, and thus, different recording and playback speeds are obtained. The procedure of the asynchronous time division multiplexing is as follows:

Different signals are sampled at different sampling frequencies (because they are bandlimited to different frequencies). The samples are stored on different storage devices (such as the tape recorder). It may be noted that the recording rate of each storage device is different due to the different sampling frequencies. While transmitting these signals, each storage device is played back at different speeds in such a way that the output sample rate of each device is the same. Hence, these signals can now be synchronously time division multiplexed, and then transmitted. At the receiver, this process is reversed to recover each signal.

A major problem associated with this procedure can be explained by considering an example of two signals bandlimited to 4 kHz, and 5 kHz, respectively. The corresponding sampling frequencies will be 8 kHz, and 10 kHz, respectively, and thus the word duration will be 125 $\mu s$, and 100 $\mu s$, respectively.

Let us assume that the above two signals are recorded on two separate storage devices for a duration of 1 second. The first storage device will, then, have 8000 words of the first signal recorded on it, whereas the second storage device will have 10000 words of the second signal recorded on it. To time division multiplex these signals, each storage unit is played back at same word rate. (Notice that the playback speed of the second device should be slower than the first device by twenty per cent to have the same word rate.) The first 8000 words can now be multiplexed without any trouble. However, during the multiplexing of the last 2000 words of the second signal, there is no contribution from the first signal. Hence, these 2000 time slots of the first signal are filled with a highly improbable sequence of digits (such as all 0's, or all 1's) to indicate that actually no message is being transmitted during these time slots. As the above method requires that the pulses be stuffed into spaces provided for empty time slots, it is known as *pulse stuffing*.

## 8.8 BANDWIDTH OF THE PCM SYSTEM

Assume that there are $n$ channels, each bandlimited to $f_m$, to be time division multiplexed. Let $N$ be the length of the PCM code so that there are $2^N = M$ quantization levels. The bandwidth of the PCM system depends on the bit duration (bit time slot), which may be calculated as follows:

$$\text{sampling frequency} = 2f_m$$

and

$$\text{sampling period} = \frac{1}{2f_m}$$

# PULSE CODE MODULATION

As there are $n$ channels and $N$ bits per sample, one bit separating each channel or as a supervisory signal, and one synchronizing bit, the total number of bits/sampling period

$$= nN + n + 1$$
$$= n(N+1) + 1$$

Therefore, the bit duration $= \dfrac{\text{sampling period}}{\text{total number of bits}}$

Hence,

$$T_b = \frac{1}{[n(N+1)+1]2f_m} \text{ sec} \qquad (8.8.1)$$

where, $T_b$ is the bit duration.

For evaluating the bandwidth, it is assumed that 1's and 0's occur alternatively. Hence the bit stream in PCM is equivalent to a square wave of the pulse width $T_b$. As discussed in Sec. 1.2, the practical bandwidth of such a square wave signal is

$$BW = \frac{1}{T_b} \qquad (8.8.2)$$

By using Eqs 8.8.1 and 8.8.2, the bandwidth of the PCM system becomes

$$BW = [n(N+1)+1]2f_m \text{ Hz} \qquad (8.8.3)$$

Now, if $N >> 1$ and $n >> 1$ (as is the practical situation)
Then,

$$BW \approx 2nNf_m \text{ Hz}$$

## Example 8.8.1

24 telephone channels, each bandlimited to 3.4 kHz, are to be time division multiplexed by using PCM. Calculate the bandwidth of the PCM system for 128 quantization levels and an 8 kHz sampling frequency.

*Solution*

Given

$$n = 24$$

and

$$M = 128$$

Therefore,

$$2^N = 128$$

or

$$N = \log_2 128 = 7$$

By putting $2f_m = 8000$ Hz in Eq. 8.8.3, we get

$$BW = [24(7+1)+1]\,8000 \text{ Hz}$$
$$= 1.544 \text{ MHz}$$

On the other hand, the approximate value of $BW$, as given by equation 8.8.4, is

$$BW \approx 24 \times 7 \times 8000 \text{ Hz}$$
$$= 1.299 \text{ MHz}$$

which is approximately 15% less than the actual value.

*Note*: For comparison, if the same number of channels are frequency division multiplexed by using an SSB modulation, the required bandwidth, assuming 4 kHz per channel, will be

$$BW = 24 \times 4 = 96 \text{ kHz}$$

This example clearly shows the tremendous bandwidth requirement of the PCM system.

## 8.9 NOISE IN PCM SYSTEMS

There are two major sources of noise in a PCM system:
  (i) Transmission noise introduced outside the transmitter, and
  (ii) Quantization noise introduced in the transmitter.

The transmission noise has no effect so long as the peak noise amplitude is less than half the pulse size. When the peak noise amplitude is greater than half the pulse size, a bit error occurs, i.e. a symbol 1 is received as a symbol 0, and vice-versa, giving rise to the probability of error $Pe$, which is discussed in detail in Chapter 9. In this section, only the quantization noise is discussed.

In the PCM transmitter, a quantized value of the sample is encoded instead of its actual value. Hence, an error occurs. As the difference between the actual value and the quantized value of the sample is random, this difference, or error, may be viewed as noise due to quantization.

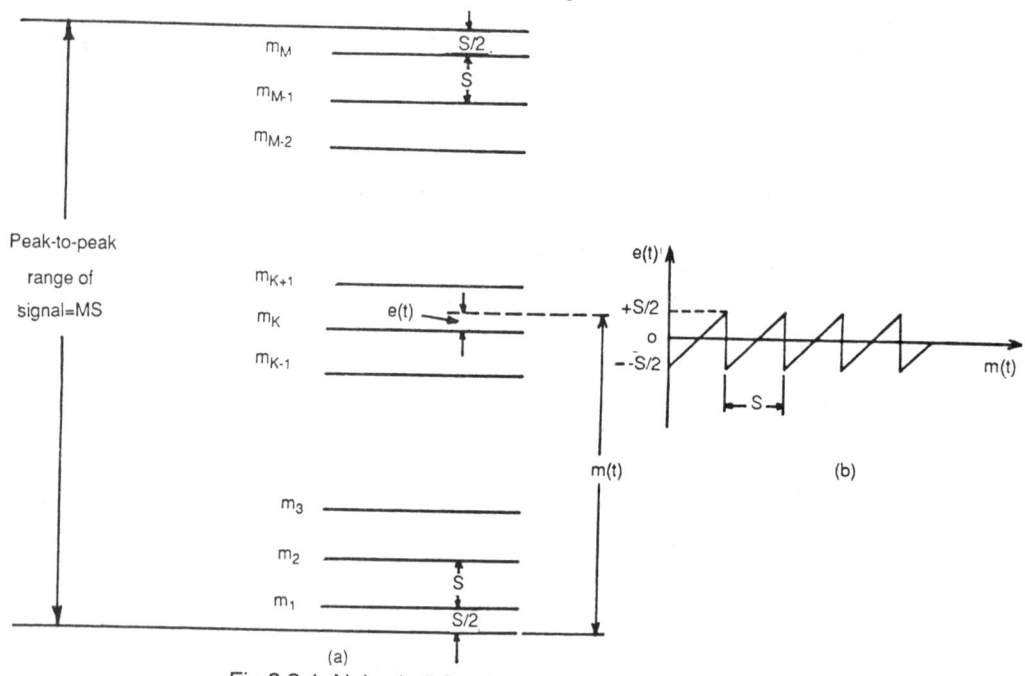

Fig.8.9.1. Noise in PCM System (a) Quantization Levels
(b) Error Voltage as a Function of Message Signal

Let the message signal be $m(t)$ and let there be $M$ equal voltage intervals, each having a magnitude of $S$ volts. At the centre of each voltage interval, there are quantization levels $m_1, m_2, \cdots m_m$, as shown in Fig. 8.91(a). The dashed level represents the actual sample value of the message signal $m(t)$ at a time $t$. Let $m(t)$ be closest to the quantization level $m_k$. Then the quantized output will be $m_k$. The quantization error is, then,

PULSE CODE MODULATION 475

$$e = m(t) - m_k$$

Figure 8.9.1(b) gives the error voltage $e(t)$ as a function of the instantaneous value of the signal $m(t)$.

Let $f(m)\,dm$ be the probability that $m(t)$ lies in the voltage range $\left(m - \dfrac{dm}{2}\right)$ to $\left(m + \dfrac{dm}{2}\right)$. Then, the mean square quantization error, or quantization noise, is

$$\overline{e^2} = Nq = \int_{m_1 - \frac{s}{2}}^{m_1 + \frac{s}{2}} f(m)(m - m_1)^2\, dm + \int_{m_2 - \frac{s}{2}}^{m_2 + \frac{s}{2}} f(m)(m - m_2)^2\, dm + \cdots + \int_{m_M - \frac{s}{2}}^{m_M + \frac{s}{2}} f(m)(m - m_M)^2\, dm \tag{8.9.1}$$

If the number of quantization levels $M$ is large (as is the usual case), it can be assumed that the probability density function $f(m)$ is constant within each quantization range, because in this case the step size $S$ is very small as compared to the peak to peak range $MS$ of the message signal.

Now, let

$f(m) = f^{(1)}$, in the first term of Eq. 8.9.1
$f(m) = f^{(2)}$, in the second term of Eq. 8.9.1
$\vdots$
$f(m) = f^{(M)}$, in the last term of Eq. 8.9.1

Equation 8.9.1 then becomes

$$Nq = f^{(1)} \int_{m_1 - \frac{s}{2}}^{m_1 + \frac{s}{2}} (m - m_1)^2\, dm + f^{(2)} \int_{m_2 - \frac{s}{2}}^{m_2 + \frac{s}{2}} (m - m_2)^2\, dm + \cdots + f^{(M)} \int_{m_M - \frac{s}{2}}^{m_M + \frac{s}{2}} (m - m_M)^2\, dm \tag{8.9.2}$$

Now, let $x = m - m_k$. Therefore, $dx = dm$, and the range of integration of all the terms in Eq. 8.9.2 become $-\dfrac{S}{2}$ to $+\dfrac{S}{2}$. Hence Eq. 8.9.2 becomes

$$Nq = f^{(1)} \int_{-\frac{s}{2}}^{+\frac{s}{2}} x^2\, dx + f^{(2)} \int_{-\frac{s}{2}}^{+\frac{s}{2}} x^2\, dx + \cdots + f^{(M)} \int_{-\frac{s}{2}}^{+\frac{s}{2}} x^2\, dx$$

$$= \left[f^{(1)} + f^{(2)} + \cdots + f^{(M)}\right] \int_{-\frac{s}{2}}^{+\frac{s}{2}} x^2\, dx$$

$$= \left[f^{(1)} + f^{(2)} + \cdots + f^{(M)}\right] \frac{S^3}{12}$$

$$= \left[f^{(1)} S + f^{(2)} S + \cdots + f^{(M)} S\right] \frac{S^2}{12} \tag{8.9.3}$$

Now, $f^{(1)}S$ is the probability that $m$ lies in the fist quantization range, $f^{(2)}S$ is the probability that $m$ lies in the second quantization range, and so on. Hence, the parantheses on the RHS of Eq. 8.9.3 is the probability that $m$ lies in the entire range of the signal. Hence,

$$f^{(1)}S + f^{(2)}S + \cdots + f^{(M)}S = 1$$

Hence, Eq. 8.9.3 becomes

$$Nq = \frac{s^2}{12} \tag{8.9.4}$$

The mean square value of the output signal is equal to the mean square value of the quantized samples.

$$\therefore \quad S_o = \overline{m_k^2} = \frac{1}{M}\left[\left(\frac{s}{2}\right)^2 + \left(\frac{3s}{2}\right)^2 + \left(\frac{5s}{2}\right)^2 + \cdots + \left\{\frac{(2M-1)s}{2}\right\}^2\right]$$

or

$$S_o = \frac{s^2}{4M}\left[1^2 + 3^2 + 5^2 + \cdots + (2M-1)^2\right]$$

$$\approx \frac{s^2}{4M} \cdot \frac{4M^3}{3} \quad\quad \text{For large } M$$

$$= \frac{s^2 M^2}{3} \tag{8.9.5}$$

Using Eqs 8.9.4 and 8.9.5, we get the output signal to quantization noise ratio as

$$\frac{S_o}{Nq} = \frac{S_o}{N_o} = \left(\frac{s^2 M^2}{3}\right)\bigg/\left(\frac{s^2}{12}\right) = 4M^2 \tag{8.9.6}$$

(Assuming $N_q = N_o$)

To calculate the noise figure, we must find the input signal to noise ratio $S_i/N_i$. This is calculated as follows:

Let the mean square value of noise $= \sigma_n^2$
Therefore,
$$N_i = \sigma_n^2 \tag{8.9.7}$$

Let us assume that 0 is represented by a 0-volt level and 1 by $A$-volt level. Assuming an equal probability for 0 and 1, the average signal power is

$$S_i = \frac{A^2}{2}$$

The value $A$ is chosen in such a way that it is much larger than the noise $\sigma_n$. Let us say

$$A = K \cdot \sigma_n$$

where

$$K = \text{constant}$$

(By choosing a large value of $K$, the probability of error is reduced.)
Therefore,

$$S_i = \frac{A^2}{2} = \frac{K^2 \sigma_n^2}{2} \tag{8.9.8}$$

From Eqs 8.8.7 and 8.8.8, we get

$$\frac{S_i}{N_i} = \frac{K^2 \sigma_n^2}{2\sigma_n^2} = \frac{K^2}{2} \qquad (8.9.9)$$

From Eqs 8.9.6 and 8.9.9, we get

Noise figure $\qquad F = \dfrac{S_i/N_i}{S_o/N_o} = \dfrac{K^2/2}{4M^2} = \dfrac{K^2}{8M^2} \qquad (8.9.10)$

## 8.10 DELTA MODULATION

By the delta modulation technique, an analog signal can be encoded into bits. Hence, in one sense a delta modulation (DM) is also PCM.

The block diagram of a DM system is shown in Fig. 8.10.1. The pulse generator produces a pulse train $P_i(t)$ of positive pulses. The modulator receives $P_i(t)$ and $\Delta(t)$, the output of the difference amplifier. The modulator output $P_o(t)$ is the input pulse train $P_i(t)$ multiplied by $+1$ or $-1$ depending upon the polarity of $\Delta(t)$. $P_o(t)$ is a positive pulse, if $\Delta(t)$ is positive and it is negative pulse if $\Delta(t)$ is negative. The magnitude of $\Delta(t)$ plays no role in deciding $P_o(t)$. Moreover, according to probability theory, the probability of $\Delta(t)$ being exactly zero is zero, and hence $\Delta(t)$ is always either positive or negative. The output of the modulator $P_o(t)$ is applied to an integrator whose output is $\tilde{m}(t)$. The input signal $m(t)$ and the integrator output $\tilde{m}(t)$ are compared in a difference amplifier whose output is $\Delta(t) = m(t) - \tilde{m}(t)$.

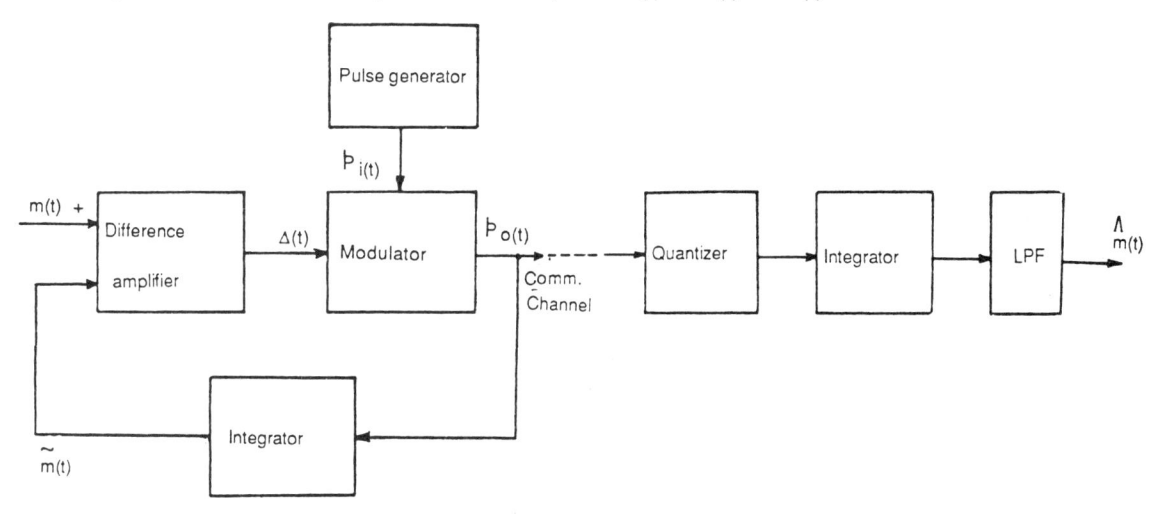

Fig. 8.10.1  DM System

Figure 8.10.2 explains the operation of a delta modulator. The initial values of $m(t)$ and $\tilde{m}(t)$ have been assumed arbitrarily. At the time $t_1$ of the first pulse in $P_i(t)$, the situation is such that $\Delta(t)$ is positive. Hence, the first pulse in $P_o(t)$ is positive. In the same way, the pulses in $P_o(t)$ are either positive or negative depending upon whether $\Delta(t)$ is positive or negative. (For example, at the time $t_6$, $\Delta(t)$ is negative and, hence, $P_o(t)$ is a negative pulse.) The waveform $\tilde{m}(t)$ approaches $m(t)$ in the form of a staircase, and then, closely follows it. Thus $\tilde{m}(t)$ is an approximation to the input signal $m(t)$.

**478** COMMUNICATION SYSTEMS: Analog and Digital

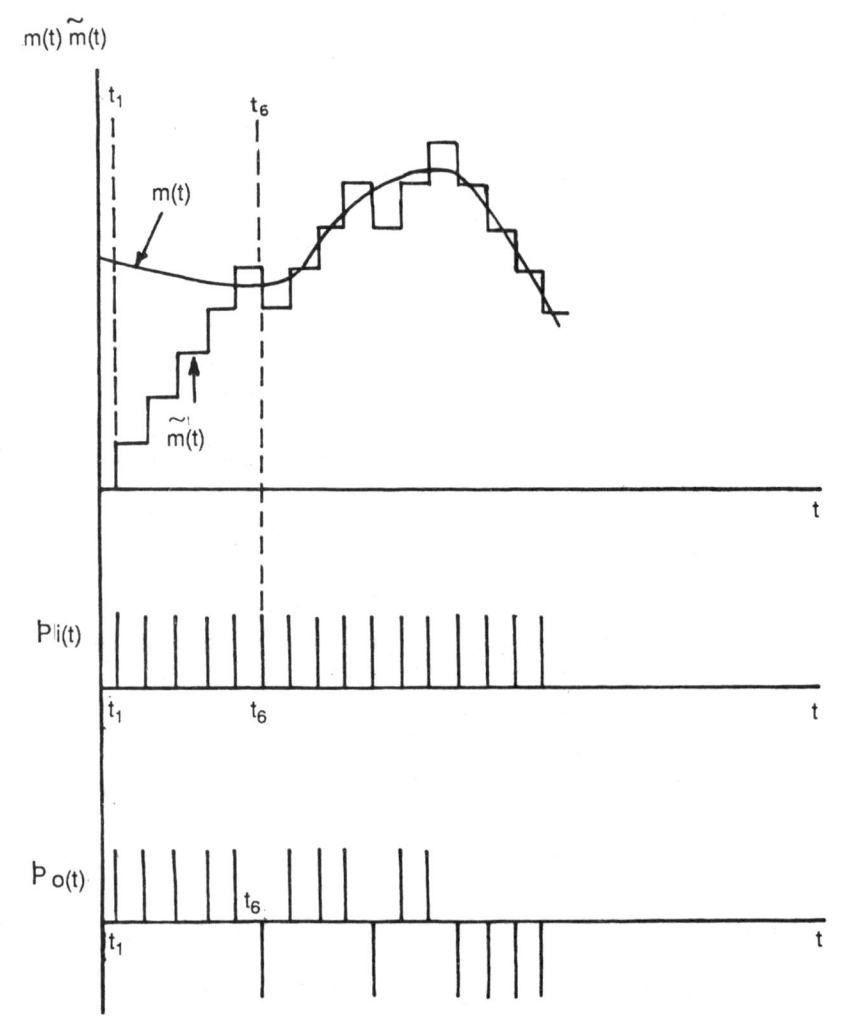

Fig. 8.10.2 Wave-forms in the DM System

The waveform $P_o(t)$ is transmitted. At the receiver side, the quantizer takes a decision whether the received pulse is positive or negative. Hence, assuming no error, the output of the quantizer is the same as the waveform $P_o(t)$ and is fed to an integrator, whose output takes the form of the waveform $\tilde{m}(t)$. The low pass filter (LPF), then, smoothens the output of the integrator, and gives a waveform $\hat{m}(t)$ which is similar to the signal $m(t)$.

As the information regarding the difference signal $\Delta(t) = m(t) - \tilde{m}(t)$ is transmitted in this method, it is known as delta modulation.

## 8.11 LIMITATIONS OF DM

The waveform $\tilde{m}(t)$ needs to closely follow the waveform $m(t)$, only then the recovered waveform $\hat{m}(t)$ resembles $m(t)$. Figure 8.11.1(a) shows a situation where the waveform $\tilde{m}(t)$ is unable to follow $m(t)$ because the slope of $m(t)$ is greater than the slope of $\tilde{m}(t)$. Same is true with Fig. 8.11.1(b), except that in this case the slopes are negative, and the slope of $m(t)$ is more negative than the slope of $\tilde{m}(t)$. In both the cases, the recovered waveform will be distorted. The DM system is, then, said to be slope overloaded.

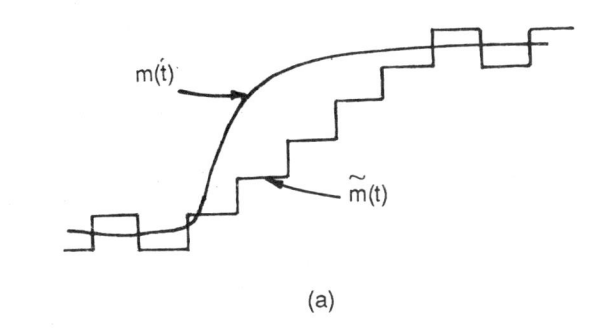

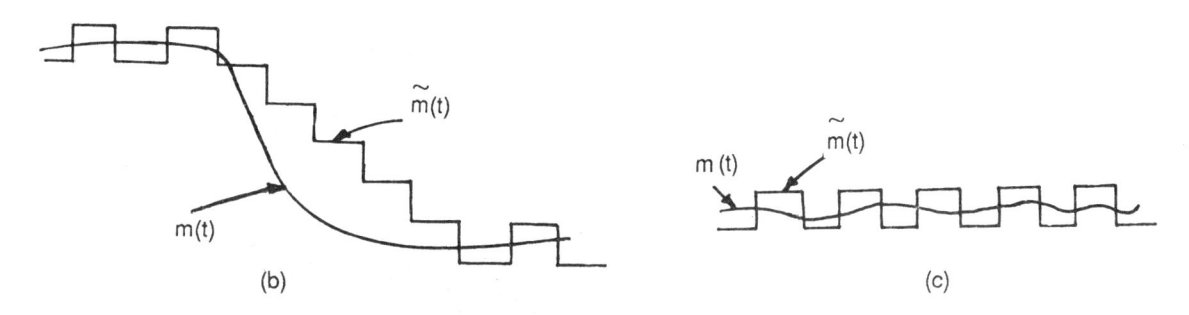

Fig. 8.11.1 Limitations of DM (a) Slope Overload (positive) (b) Slope Overload (negative) (c) Slow Varying Signal

In Fig. 8.11.1(c), the variations in $m(t)$ are such that they are within the step size. Hence, the waveform $\tilde{m}(t)$ is like a square wave. This will be recovered as d.c, whereas, the original signal $m(t)$ is not dc. Thus, in this case also distortion is resulted and the noise is known as *granular* noise.

## 8.12 ADAPTIVE DELTA MODULATION

The limitations of DM discussed in Sec. 8.11 can be overcome by suitably changing the step size. Slope-overload can be overcome if the step size is increased in such a way that the magnitude of the slope of $\tilde{m}(t)$ becomes greater than the magnitude of the slope of $m(t)$, and when the signal variations are less than the step size, the step size may be reduced to take care of the situation.

A DM system which adjusts its step size is known as the *Adaptive Delta Modulation* (ADM) system. An ADM system is shown in Fig. 8.12.1.

On the transmitter side, a variable gain amplifier is used before the integrator with $P_o(t)$ as its input. The gain of this amplifier depends on the gain control input, which is obtained by integrating $P_o(t)$ in an RC-network, and then passing the integrator output through a square law device. Under the slope-overload condition, $P_o(t)$ is a long sequence of either positive or negative pulses. The RC integrator integrates these pulses. Thus, the output of this integrator is either of a large positive or negative value. The square law device output is of a large positive value, irrespective of whether the input is positive, or negative. Thus, the gain control input of the variable gain amplifier is large and its gain increases. Hence the step size increases, which can then take care of the slope-overload (positive or negative).

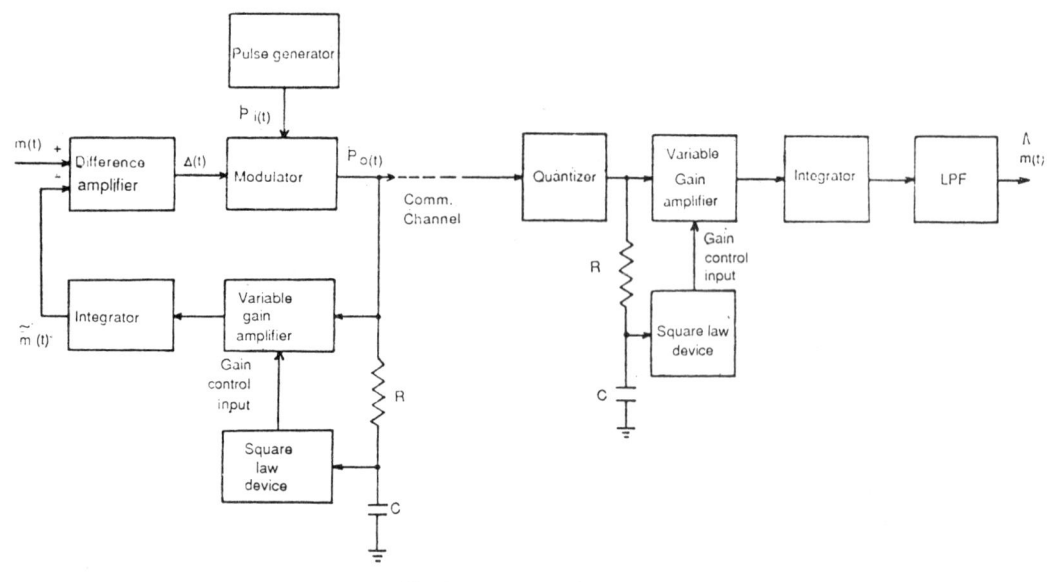

Fig.8.12.1 ADM System

When the signal variations are within the step size, $P_o(t)$ is a sequence of alternate positive and negative pulses. The RC integrator output in this case is zero and, hence, the gain Control input of the variable gain amplifier is also zero. The gain of the variable gain amplifier decreases, resulting in a reduced step size, which takes care of the situation.

On the receiver side, the output of the quantizer is fed to a variable gain amplifier whose gain control input is derived from an RC integrator and a square law device. Thus, an adaptive adjustment of the step size is obtained at the receiver, resulting in an undistorted reception of the transmitted signal.

## 8.13 COMPARISON BETWEEN PCM AND DM

DM needs a simple circuit as compared to PCM. But the signal to quantization noise ratio is less in DM than in PCM, because, in the latter case the maximum possible error due to quantization is $\frac{S}{2}$ whereas, it is $S$ in the former. Moreover, it has been found experimentally that, for voice transmission, the bit rate needed by PCM, assuming an 8 kHz sampling rate and the 7 quantization levels (which is the standard practice), is 56 kilobits per second, whereas, the bit rate needed by DM for same quality of voice transmission is much higher than 56 kilobits per second. Thus, for a good quality transmission, the bandwidth needed by DM is more than PCM. On the other hand, if the bandwidth conservation is the main criteria (at the cost of quality of transmission), then DM is preferable over PCM because in case of slightly substandard quality of transmission, the bandwidth needed by DM is less than PCM. Therefore, the use of DM can be recommended for the following two situations:

(i) When the bandwidth conservation is desirable at the cost of the quality of transmission.
(ii) When a simple circuitry is of utmost importance, and the allowable bandwidth is large.

## 8.14 DELTA OR DIFFERENTIAL PCM (DPCM)

In DPCM, instead of quantizing each sample, the difference between two successive samples is quantized, encoded, and transmitted as in PCM. This is particularly useful in voice transmission, because in this case two successive samples do not differ much in amplitude. Thus, the difference signal is much less in amplitude than the actual sample and, hence, less number of quantization levels are needed. Therefore, the number of bits per code is reduced, resulting in a reduced bit-rate. Thus, the bandwidth required in this case is less than the one required in PCM. The disadvantage of DPCM is that the modulator and demodulator circuits are more complicated than those in PCM.

## 8.15  S-ARY SYSTEM

In a binary system, pulses with one of the two possible levels are used. In an $S$-ary system, the pulses are allowed to take one of $S$ possible levels ($S > 2$). Each level corresponds to a distinct input symbol. For example, in a quarternary system ($S = 4$), the levels may be 0, 1V, 2V, and 3V, and the respective codes may be 0, 1, 2 and 3. If there are $M$ quantization levels, then we need $\log_S M$ symbols to represent a sample. For example, for $M = 64$ and $S = 4$, we need $\log_4 64 = 3$ symbols to represent a sample. As a comparison, in the binary system, we need $\log_2 64 = 6$ symbols to represent a sample. As the number of symbols needed to represent a sample is less, the $S$-ary system needs less bandwidth than the binary system.

The disadvantage of the $S$-ary system is that for a given probability of error, the needed transmitter power is more because noise is more effective due to a large number of voltage levels.

It can be shown that, for $S \gg 2$ and $P_e \ll 1$, the bandwidth is reduced by $\dfrac{1}{\log_2 S}$, and the transmitter power is to be increased by a factor of $\dfrac{S^2}{\log_2 S}$.

The circuitry for the $S$-ary system is more complicated because the receiver has to decide on one of the $S$-levels using $S-1$ comparators, or level slicers.

## PROBLEMS

8.1  "Pulse modulation systems are not digital; whereas, pulse-code modulation is." Justify.

8.2  Explain the difference between *cross-talk* and *intersymbol interference*.

8.3  What is quantization error? How does it depend upon the step size? Suggest some methods to overcome the difficulties encountered when the modulating signal amplitude swing is large.

8.4  Explain synchronous an asynchronous time division multiplexing of PCM signals.

8.5  Describe delta modulation systems. What are its limitations? How can they be overcome?

8.6  8 channels, each bandlimited to 5 kHz, are to be time division multiplexed. Each sample is coded into a 6-bit word. Find the output rate in bits/sec and the required bandwidth.

8.7  A 6-bit single channel PCM system gives an output of 60 kilo-bits per second. Determine the highest possible modulating frequency for the system.

8.8  A voice frequency signal bandlimited to 3 kHz is transmitted with the use of the DM system. The pulse repetition frequency is 30,000 pulses per second, and the step size is 40 mV. Determine the maximum permissible speech signal amplitude to avoid a slope overload.

# NINE

# DATA TRANSMISSION

## INTRODUCTION

The output of a PCM system is a string of 1's and 0's. If they are to be transmitted over copper wires, they can be directly transmitted as two voltage levels +V and − V. But if they are to be transmitted through space using antenna, some form of modulation has to be used. Amplitude, frequency, and phase modulations are commonly used. As the modulating signal consists of only two levels, the modulation techniques are known as Amplitude Shift Keying (ASK), Frequency Shift Keying (FSK), and Phase Shift Keying (PSK), respectively.

## 9.1 AMPLITUDE SHIFT KEYING (ASK)

Let $v_{(t)}$ be the binary signal to be transmitted. A 1 is represented by $v_{(t)} = +V$ and a 0 is represented by $v_{(t)} = -V$. In the ASK method, the binary signal $v_{(t)}$ [as shown in Fig. 9.1.1 (a)] is transmitted as the signal $v_{ASK}(t)$ (as shown in Fig. 9.1.1.(b)). The waveform $v_{ASK}(t)$ is $v_{ASK}(t) = u(t) A \cos \omega_o t$, where, $u(t) = 1$ for $v_{(t)} = +V$ and $u(t) = 0$ for $v_{(t)} = -V$. The signal $A \cos \omega_o t$ is the carrier signal. It can be seen that the carrier signal is on when $v_{(t)} = +V$ and it is off when $v_{(t)} = -V$. Hence this method is also known as on-off keying (OOK). As will be seen later, this method does not give a satisfactory value for probability of error. Hence ASK is rarely used.

## 9.2 FREQUENCY SHIFT KEYING (FSK)

In this method, the binary signal $v(t)$ is used to generate a waveform

$$v_{FSK}(t) = A \cos(\omega_o \pm \Omega)t \qquad (9.2.1)$$

The plus sign is applied when $v_{(t)} = +V$, and the minus sign is applied when $v_{(t)} = -V$. Thus, the frequency of the transmitted signal is high for a 1, and low for a 0 (as shown in Fig. 9.2.1).

A straightforward way to detect an FSK signal is to use a suitable filter of sharp cut-off. Although this method looks simple, it does not give optimum performance in the presence of noise.

A synchronous (coherent) method of detection of FSK is shown in Fig. 9.2.2. This method gives optimum performance in the presence of noise and will be discussed later. The disadvantage is that, being a coherent method, two synchronous local carriers of angular frequencies, $(\omega_o + \Omega)$ and $(\omega_o - \Omega)$ are required.

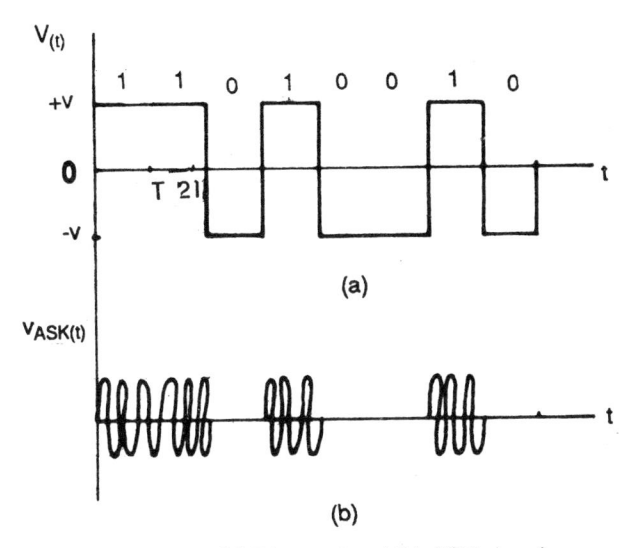

Fig. 9.1.1 (a) Binary signal (b) ASK signal

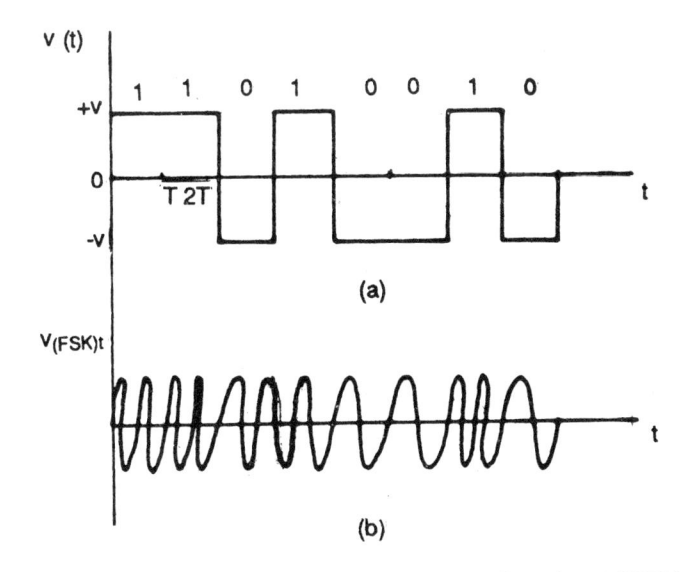

Fig. 9.2.1(a) Binary Signal (b) FSK Signal and DPSK

When $v_{FSK}(t) = A \cos(\omega_o + \Omega)t$, the input 1 to the difference amplifier is

$$A \cos(\omega_o + \Omega)t \cos(\omega_o + \Omega)t = A\left[\frac{\cos 2(\omega_o + \Omega)t + 1}{2}\right]$$

and the input 2 to the difference amplifier is

$$A \cos(\omega_o + \Omega)t \cos(\omega_o - \Omega)t = A\left[\frac{\cos 2\omega_o t + \cos 2\Omega t}{2}\right]$$

**484** COMMUNICATION SYSTEMS: Analog and Digital

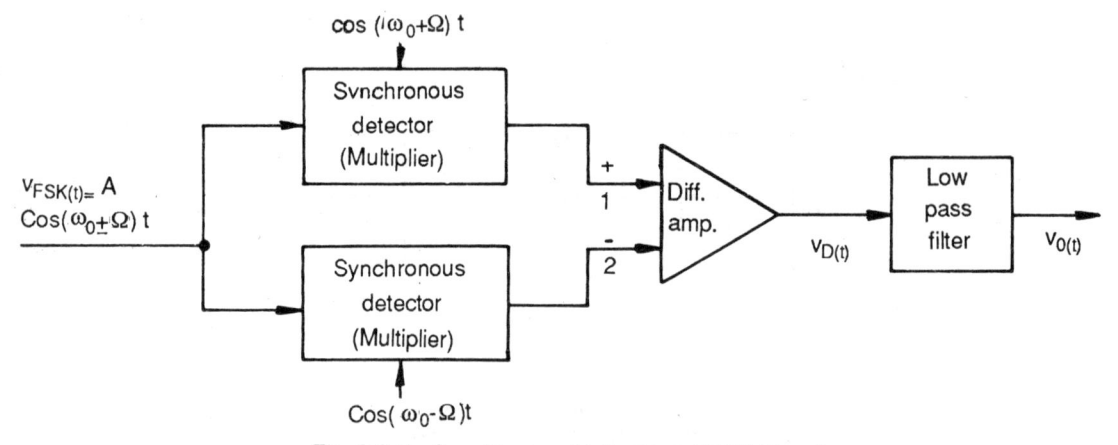

Fig. 9.2.2 Synchronous Detection of FSK Signal

Hence, the output of the difference amplifier is

$$v_{D(t)} = A\left[\frac{\cos 2(\omega_o + \Omega)t + 1}{2}\right] - A\left[\frac{\cos 2\omega_o t + \cos 2\Omega t}{2}\right]$$

$$= \frac{A}{2} + \frac{A}{2}[\cos 2(\omega_o + \Omega)t - \cos 2\omega_o t - \cos 2\Omega t] \quad (9.2.2)$$

Similarly, it can be shown that when

$$v_{FSK}(t) = A\cos(\omega_o - \Omega)t$$

the output of the difference amplifier is

$$v_{D(t)} = -\frac{A}{2} + \frac{A}{2}[\cos 2\omega_o t + \cos 2\Omega t - \cos 2(\omega_o - \Omega)t] \quad (9.2.3)$$

The low pass filter separates the dc terms in Eq. 9.2.2 and 9.2.3. Thus if $v_o(t)$ is $+A/2$, the decision is that a 1 has been received and if $v_o(t)$ is $-A/2$, the decision is that a 0 has been received. For easy separation of the d.c terms, the bit interval $T$ should be such that, in this interval, the lowest frequency term $\cos 2\Omega t$ (because normally $\omega_o \gg \Omega$) includes many cycles. Hence, it is needed that $2\Omega T \gg 2\pi$, i.e., $\Omega T \gg \pi$.

## 9.3 PHASE SHIFT KEYING (PSK)

In this method, the binary signal $v(t)$ is used to generate a waveform

$$v_{PSK}(t) = A\cos[\omega_o t + \phi(t)] \quad (9.3.1)$$

Two values of $\phi(t)$ are chosen to represent 1 or 0. Normally, for $v(t) = +V(a1), \phi(t)$ is chosen as 0 and for $v(t) = -V(a0), \phi(t)$ is chosen as $\pi$. Equation 9.3.1 may then be written as

$$v_{PSK}(t) = \frac{v(t)}{(V)} A\cos\omega_o t = \pm A\cos\omega_o t \quad (9.3.2)$$

Thus, $v_{PSK}(t) = +A\cos\omega_o t \qquad$ for $v(t) = +V(a1)$

and $\quad v_{PSK}(t) = -A\cos\omega_o t \qquad$ for $v(t) = -V(a0)$

Figure 9.3.1(a) shows the binary signal $v(t)$ and Fig. 9.3.1(b) shows the corresponding $v_{PSK}(t)$ signal.

Figure 9.3.2 shows a PSK system. In this, the waveform of Eq. 9.3.2 is generated by applying the waveform $v(t)$ and the carrier $\cos \omega_o t$ to a balanced modulator, which is a multiplier. The received signal is $\pm A \cos (\omega_o t + \theta)$, where $\theta$ is a phase angle depending upon the effective length of the path between transmitter and receiver. The detection is synchronous (coherent); hence, a synchronous local carrier is

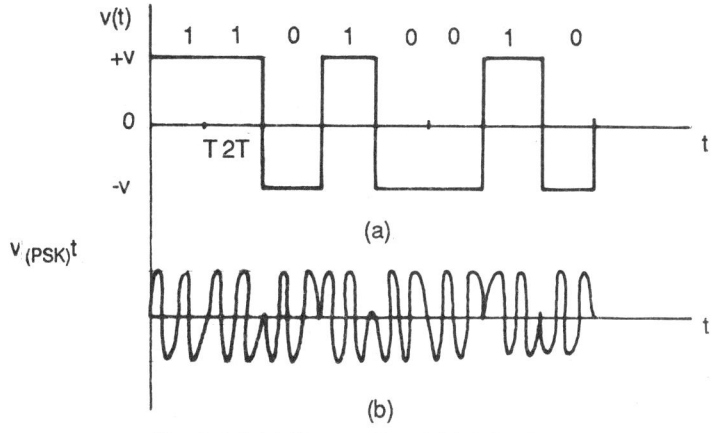

Fig. 9.3.1 (a) Binary Signal (b) PSK Signal

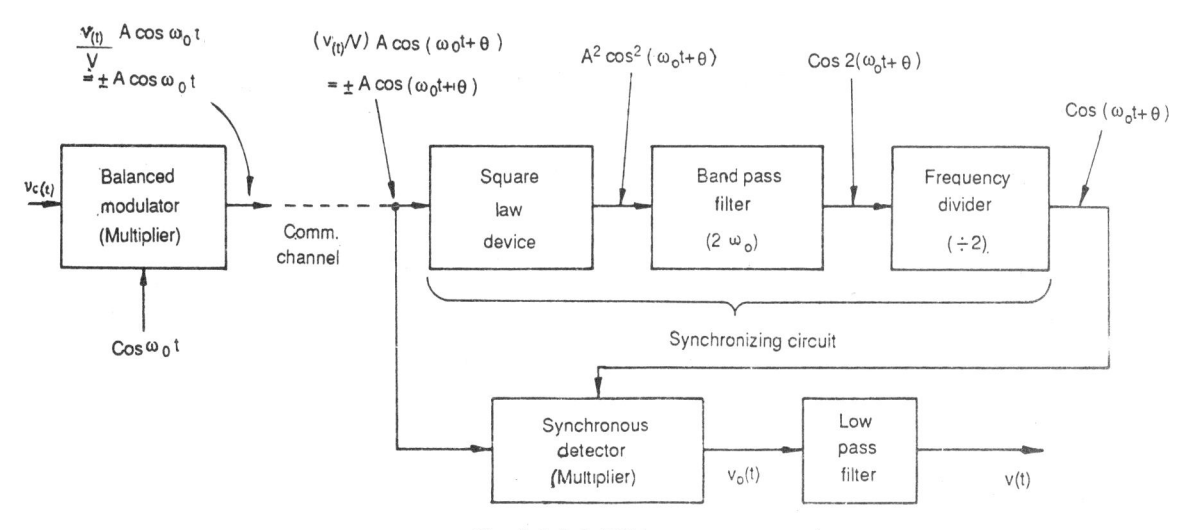

Fig. 9.3.2 A PSK system

necessary which is generated in the synchronizing circuit as shown in Fig. 9.3.2. The output of the squaring circuit is

$$A^2 \cos^2(\omega_o t + \theta) = \frac{A^2}{2}[1 + \cos 2(\omega_o t + \theta)]$$

The bandpass filter tuned to an angular frequency $2\omega_o$ separates the term $\cos 2(\omega_o t + \theta)$ which after frequency division, gives the desired $\cos(\omega_o t + \theta)$ term. In the synchronous detector, the received signal is multiplied by the locally recovered carrier. The output of the synchronous detector is

$$v_o(t) = \frac{v(t)}{V} A \cos(\omega_o t + \theta) \cos(\omega_o t + \theta)$$

$$= \frac{v(t) A}{2V}[1 + \cos 2(\omega_o t + \theta)] \quad (9.3.3)$$

The low pass filter separates the d.c term $v(t)$ in Eq. 9.3.3. For an easy separation of the dc term, the bit interval should be such that, in this interval, the term $\cos 2(\omega_o t + \theta)$ includes many cycles. Hence, it is needed that $\omega_o T \gg \pi$.

## 9.4 DIFFERENTIAL PHASE SHIFT KEYING (DPSK)

PSK has the disadvantage that it needs a complicated synchronizing circuit at the receiver. The DPSK method avoids the necessity of the synchronizing circuit at the receiver.

|  |  | b'(t) | 1 | 1 | 0 | 1 | 0 | 0 | 1 | 0 |
|---|---|---|---|---|---|---|---|---|---|---|
| I | b(t) |  | 1 | 1 | 1 | 0 | 0 | 1 | 0 | 0 | 1 |
|   | phase |  | 0 | 0 | 0 | $\pi$ | $\pi$ | 0 | $\pi$ | $\pi$ | 0 |
| II | b(t) |  | 0 | 0 | 0 | 1 | 1 | 0 | 1 | 1 | 0 |
|   | phase |  | $\pi$ | $\pi$ | $\pi$ | 0 | 0 | $\pi$ | 0 | 0 | $\pi$ |

Fig. 9.4.1 A DPSK Receiver (a) Binary Message and Auxiliary Message Streams

Let $b'(t)$ be the binary message to be transmitted as shown in Fig. 9.4.1(a). An auxiliary message stream $b(t)$ is generated from $b'(t)$ by using a logic circuit. The first bit in $b(t)$ is arbitrary. The subsequent bits in $b(t)$ are determined on the basis of the rule that when $b'(t)$ is a 1, $b(t)$ does not change its value and, when $b'(t)$ is a 0, $b(t)$ changes its value. Figure 9.4.1(a) shows two possible bit streams $b(t)$ and the respective phases. In the first bit stream, the initial bit is a 1 and, in the second bit stream, the initial bit is a 0. Figure 9.4.1(b) shows a DPSK signal generator. The logic circuit which generates the bit stream $b(t)$ is an XNOR gate whose output is a 1 when both the inputs are same, and a 0 when the inputs are different. Corresponding to the bit stream, a waveform $v(t)$ is generated, where $v(t) = +V$ for $b(t) = 1$ and $v(t) = -V$ for $b(t) = 0$. This waveform is applied to a balanced modulator whose carrier input is $A \cos \omega_o t$. The balanced modulator is a multiplier whose output is

$$v_{DPSK}(t) = \frac{v(t)}{V} A \cos \omega_o t = \pm A \cos \omega_o t \quad (9.4.1)$$

and is transmitted over the channel.

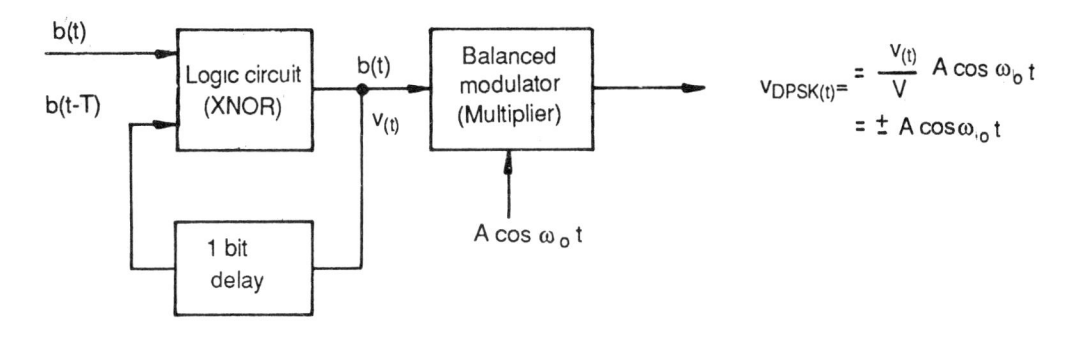

Fig. 9.4.1 (b) Block Schematic of DPSK Transmitter

$$v_{DPSK(t)} = \frac{v_{(t)}}{V} A \cos \omega_0 t = \pm A \cos \omega_0 t$$

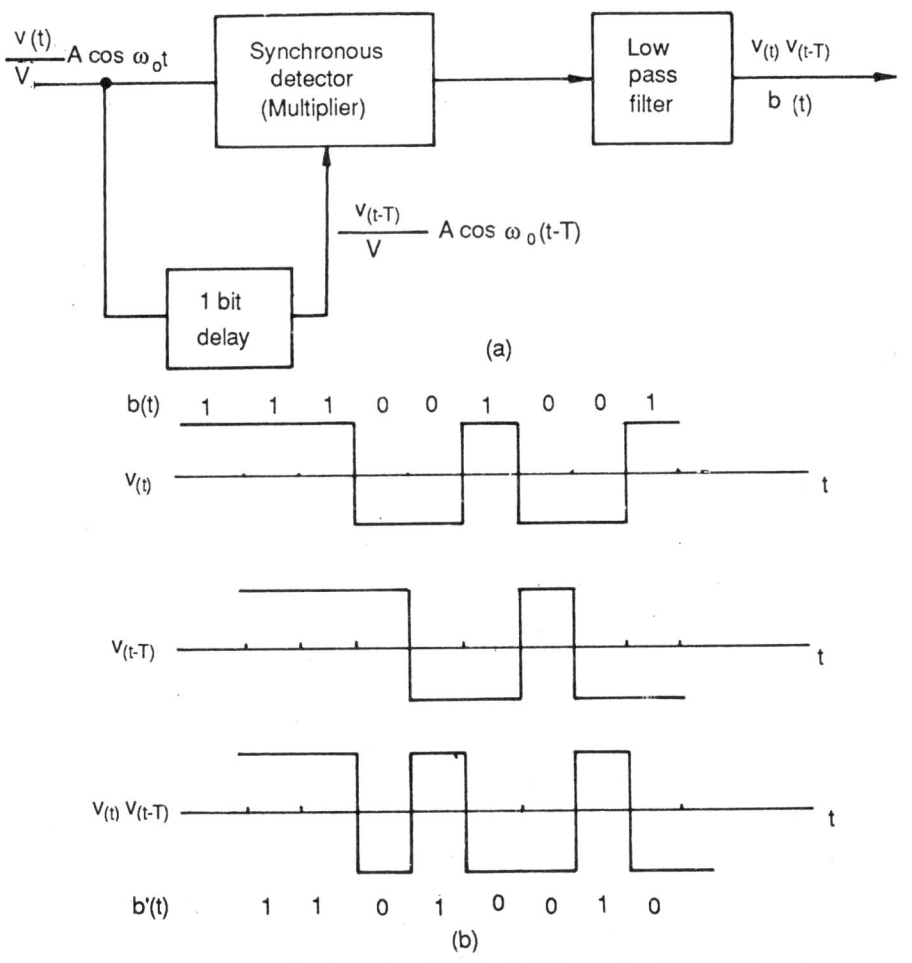

Fig. 9.4.2 A DPSK Receiver (a) Block Schematic of DPSK Receiver (b) Wave-form Showing Recovery of Binary Message

A method for the reception of DPSK signal is shown in Fig. 9.4.2(a). Here, the received signal is applied to a synchronous detector, to which another input is the received signal delayed by 1 bit. The output of the synchronous detector is

$$\frac{v(t)}{V} A \cos \omega_o t \frac{v(t-T)}{V} A \cos \omega_o (t - T)$$

$$= v(t) v(t - T) \frac{A^2}{2V^2} \left[ \cos \omega_o T + \cos 2\omega_o \left( t - \frac{T}{2} \right) \right] \quad (9.4.2)$$

The first term on the RHS of Eq. 9.4.2 is the waveform $v(t) v(t - T)$ with a multiplying constant. The second term is the double frequency carrier term which can be removed by using the low pass filter. The product $v(t) v(t - T)$ represents the original message stream $b'(t)$ [see Fig. 9.4.2(b)]. Equation 9.4.2 indicates that, for the signal output to be as large as possible, the bit interval $T$ should be so selected that $\cos \omega_o T = \pm 1$. Thus, the carrier frequency $\omega_o / 2\pi$ should be so selected that the bit interval $T$ is an integer of half cycles in duration.

The advantage of DPSK is that no synchronous carrier is necessary at the receiver, and the disadvantage is that an error occurs in pairs, because one erroneous bit affects the decision in two successive bit intervals. Hence, in DPSK, the error rate is greater than in PSK.

## 9.5 BASEBAND SIGNAL RECEIVER

Let the binary encoded PCM baseband signal be a sequence of voltage levels +V and − V representing a 1 and a 0, respectively. In a bit interval, a sample value is taken at a particular time and, if it is positive, the decision is that a 1 has been received, and if it is negative, the decision is that a 0 has been received. If, at the sampling instant, the noise voltage is of larger magnitude than V, and is of opposite polarity, an error will occur. Such a situation is shown in Fig. 9.5.1, where, at the sampling instant $t = t_1 + \Delta t_1$, the sampled value is negative, whereas, the bit transmitted is a 1.

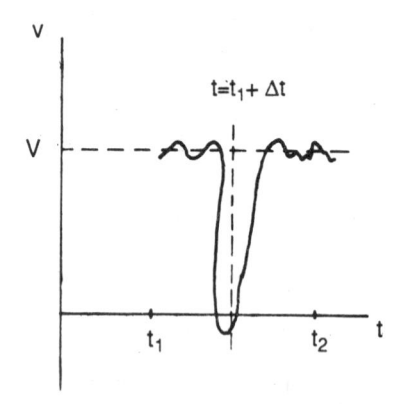

Fig. 9.5.1 Error Due to Noise at Sampling Instant in a PCM System

The probability of error can be reduced if we are able to find the sampling instant at which the sample voltage due to signal is emphasized as against the sample voltage due to noise. Such a receiver is shown in Fig. 9.5.2. The noise $n(t)$ is assumed to be a white Gaussian of power spectral density $\eta/2$. The signal $S(t)$ and the noise $n(t)$ are applied to an integrator. Just before the bit interval, the capacitor $C$ is discharged by a brief closing of the dump switch SW1. The sample is taken at the output of the integrator at the end of the

DATA TRANSMISSION    489

bit interval (i.e., at $t = T$) by a brief closing of the sample switch SW2. The receiver of Fig. 9.5.2, is known as integrate and dump, the term dump referring to the abrupt discharge of the capacitor after each sampling.

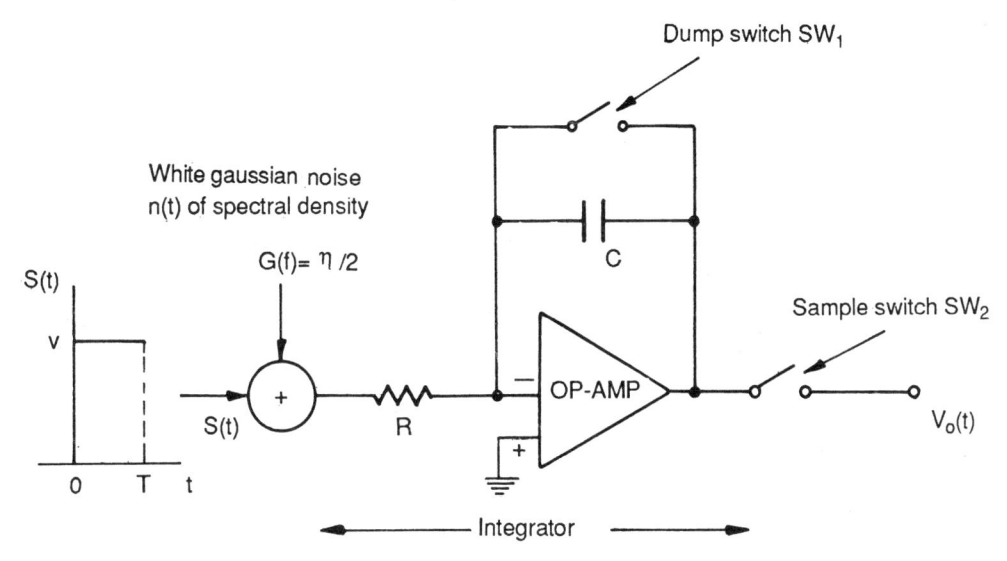

Fig. 9.5.2   Integrate and Dump Receiver

The output is

$$v_o(T) = \frac{1}{\tau}\int_0^T [s(t) + n(t)]dt \qquad \text{where } \tau = RC$$

$$= \frac{1}{\tau}\int_0^T s(t)dt + \frac{1}{\tau}\int_0^T n(t)dt \qquad (9.5.1)$$

The sample voltage due to the signal is

$$s_o(T) = \frac{1}{\tau}\int_0^T V\,dt = \frac{VT}{\tau} \qquad (9.5.2)$$

The sample voltage due to the noise is

$$n_o(T) = \frac{1}{\tau}\int_0^T n(t)dt \qquad (9.5.3)$$

The variance of $n_o(T)$ is

$$\sigma_o^2 = \overline{n_o^2(T)} = \frac{\eta T}{2\tau^2} \qquad (9.5.4)$$

[$n_o(T)$ has a Gaussian probability density.]

The output $S_o(t)$ shown in Fig. 9.5.3(a), is a ramp voltage and at the end of bit interval, the magnitude of $S_o(t)$ is maximum $(+(VT/\tau)$ or $-(VT/\tau)$ depending upon whether the bit is a 1 or a 0); whereas, the noise $n_o(t)$ shown in Fig. 9.5.3(b), has a random value at the end of the bit interval. Thus, at the end of the bit interval, the sample voltage due to signal is emphasized as compared to the sample voltage due to noise. Hence, the end of the bit interval is selected as the sampling instant.

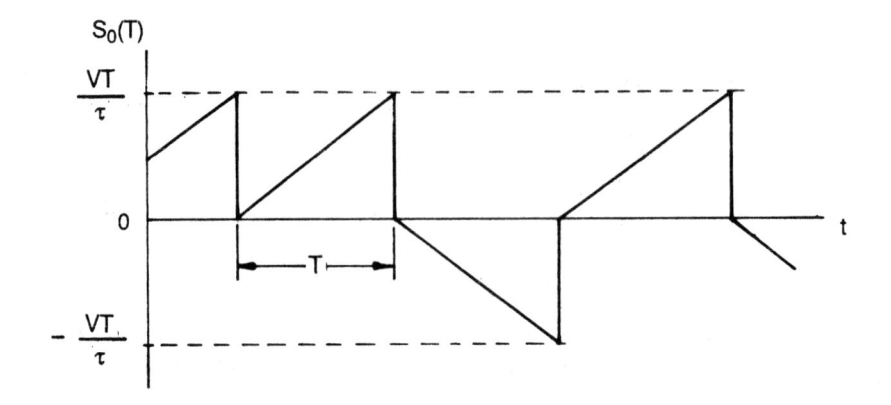

Fig. 9.5.3 (a) Sample Voltage due to Signal

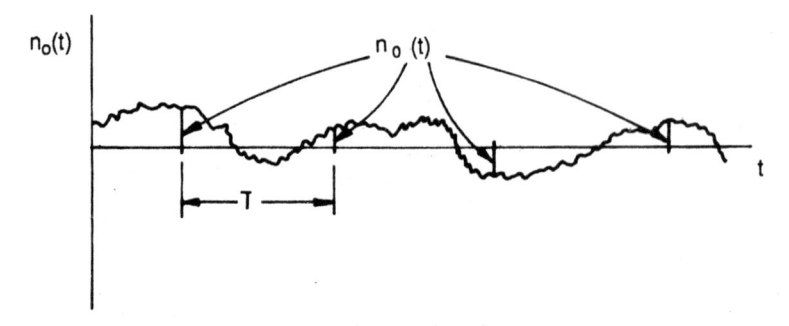

Fig. 9.5.3 (b) Sample Voltage due to Noise

The figure of merit of the receiver is the signal to noise ratio given by

$$\frac{[S_o(T)]^2}{[n_o(T)]^2} = \frac{(VT/\tau)^2}{\eta T/2\tau^2}$$

$$= \frac{2V^2 T}{\eta} \qquad (9.5.5)$$

## 9.6 PROBABILITY OF ERROR

The probability density of the noise sample $n_o(T)$ is Gaussian as shown in Fig. 9.6.1. Hence,

$$f[n_o(T)] = \frac{e^{-n_o^2(T)/2\sigma_o^2}}{\sqrt{2\pi\sigma_o^2}}$$

Since, at the end of bit interval, $S_o T = -\dfrac{VT}{\tau}$ (for a 0), an error will occur only when $n_o(T) > \dfrac{VT}{\tau}$. Hence, the probability of error in this case is given by the shaded area (1) in Fig. 9.6.1. Hence,

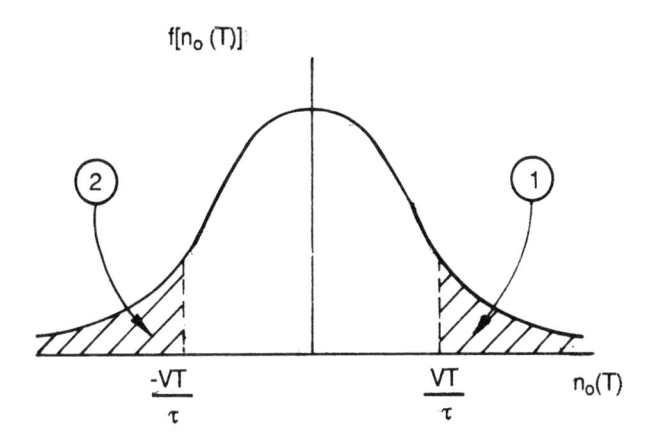

Fig. 9.6.1 Probability Density of Noise Sample $n_o(T)$

$$Pe = \int_{VT/\tau}^{\infty} f[n_o(T)]\,dn_o(T)$$

$$= \int_{VT/\tau}^{\infty} \frac{e^{-n_o^2(T)/2\sigma_o^2}}{\sqrt{2\pi\sigma_o^2}}\,dn_o(T)$$

substituting $x \equiv n_o(T)/\sqrt{2}\,\sigma_o$ and using Eq. 9.5.4, we get

$$Pe = \frac{1}{2}\frac{2}{\sqrt{\pi}} \int_{x=V\sqrt{T/\eta}}^{\infty} e^{-x^2}\,dx$$

$$= \frac{1}{2}\mathrm{erfc}(V\sqrt{T/\eta})$$

$$= \frac{1}{2}\mathrm{erfc}\left(\frac{V^2 T}{\eta}\right)^{\frac{1}{2}}$$

$$= \frac{1}{2}\mathrm{erfc}\left(\frac{Es}{\eta}\right)^{\frac{1}{2}} \qquad (9.6.1)$$

Also, at the end of the bit interval, $S_o(T) = \dfrac{VT}{\tau}$ (for a 1). Hence, an error will occur if $n_o(T)$ is negative and is of a greater magnitude than $\frac{VT}{\tau}$. This error probability is given by the shaded area (2) in Fig. 9.6.1. Because of the symmetry of the Gaussian curve, both the shaded areas are equal, and hence the error probability given by Eq. 9.6.1 is quite general.

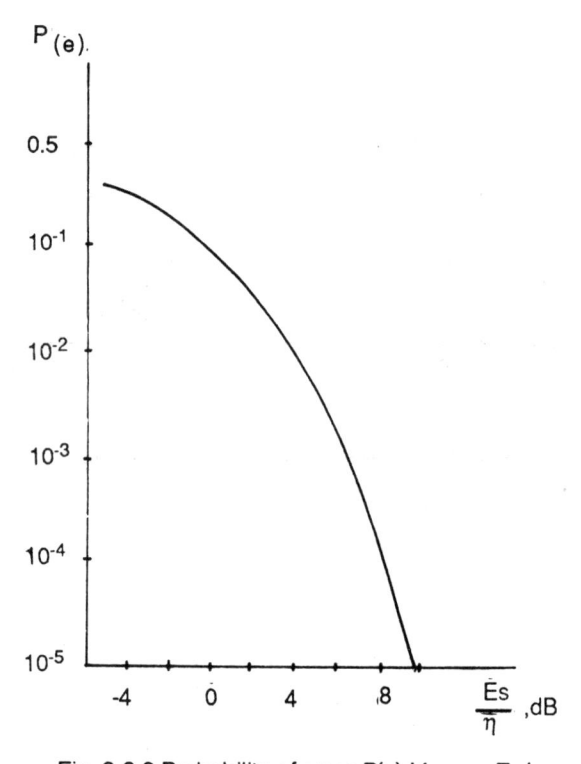

Fig. 9.6.2 Probability of error P(e) Versus $Es/\eta$

The graph of $P(e)$ versus $\frac{Es}{\eta}$ in dB is shown in Fig. 9.6.2. Note that $P(e)$ decreases rapidly as $\frac{Es}{\eta}$ increases. The maximum value of $P(e)$ is 1/2. Thus, even if the signal is entirely lost in the noise, the receiver cannot be wrong more than half the time on the average.

## 9.7 THE OPTIMUM FILTER

In the receiver of Fig. 9.5.2, the signal is passed through a filter (i.e., the integrator). It is important to know

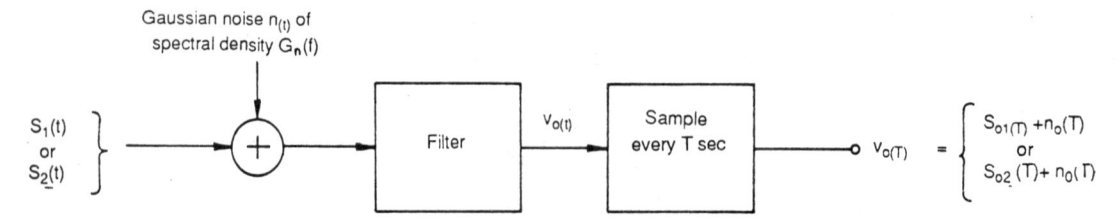

Fig. 9.7.1 Generalized Receiver for a Binary Coded PCM

## DATA TRANSMISSION

whether the integrator is an *optimum filter* which gives the minimum probability of error. Hence, a more general discussion of optimum filter is necessary.

Figure 9.7.1 shows a generalized receiver for a binary coded PCM. If the input is $S_1(t)$, the output is $v_o(T) = S_{o1}(T) + n_o(T)$ and, if the input is $S_2(T)$, the output is $v_o(T) = S_{o2}(T) + n_o(T)$. In the absence of noise, the output would be $S_{o1}(T)$ or $S_{o2}(T)$. Thus, the decision in the presence of noise regarding the transmitted signal depends upon whether $v_o(t)$ is closer to $S_{o1}(T)$ or $S_{o2}(T)$. The decision boundary is, therefore, $\dfrac{S_{o1}(T) + S_{o2}(T)}{2}$.

Let us consider a case where $S_{o1}(T) > S_{o2}(T)$ and $S_2(T)$ is transmitted. An error will occur if at the sampling instant, the noise $n_o(T)$ is positive and larger than

$$\frac{1}{2}[S_{o1}(T) + S_{o2}(T)] - S_{o2}(T) \quad \text{or} \quad \frac{S_{o1}(T) - S_{o2}(T)}{2}$$

Thus, an error will occur if

$$n_o(T) \geq \frac{S_{o1}(T) - S_{o2}(T)}{2} \tag{9.7.1}$$

Hence, the probability of error is

$$Pe = \int_{\frac{S_{o1}(T) - S_{o2}(T)}{2}}^{\infty} \frac{e^{-n_o^2(T)/2\sigma_o^2}}{\sqrt{2\pi\sigma_o^2}} dn_o(T)$$

$$= \frac{1}{2} \frac{2}{\sqrt{\pi}} \int_{\frac{S_{o1}(T) - S_{o2}(T)}{2\sqrt{2}\sigma_o}}^{\infty} e^{-x^2} dx \quad \text{assuming } x \equiv n_o(T)/\sqrt{2}\,\sigma_o$$

$$= \frac{1}{2} \operatorname{erfc}\left[\frac{S_{o1}(T) - S_{o2}(T)}{2\sqrt{2}\sigma_o}\right] \tag{9.7.2}$$

$P(e)$ decreases as $\left[\dfrac{S_{o1}(T) - S_{o2}(T)}{2\sqrt{2}\sigma_o}\right]$ increases (see Fig. 9.6.2). The optimum filter is the filter which maximizes the ratio

$$\gamma = \frac{S_{o1}(T) - S_{o2}(T)}{\sigma_o} \tag{9.7.3}$$

The transfer function $H(f)$ of this optimum filter must be calculated. Note that $\gamma$ is proportional only to the difference signal and not to the individual signals. Let the difference signal be

$$p(t) = S_1(t) - S_2(t) \tag{9.7.4}$$

The corresponding output signal of the filter then is

$$p_o(t) = S_{o1}(t) - S_{o2}(t) \tag{9.7.5}$$

Let $P(f)$ and $P_o(f)$ be the Fourier transforms of $P(t)$ and $P_o(t)$, respectively. Then,

$$P_o(f) = H(f)P(f) \tag{9.7.6}$$

where $H(f)$ is the transfer function of the filter.
Now,

$$P_o(T) = \int_{-\infty}^{\infty} P_o(f) e^{j2\pi fT} df$$

$$= \int_{-\infty}^{\infty} H(f) P(f) e^{j2\pi fT} df \tag{9.7.7}$$

Let $Gn(f)$ and $Gn_o(f)$ be the power spectral densities of $n(t)$ and $n_o(t)$, respectively. Then

$$Gn_o(f) = |H(f)|^2 Gn(f) \tag{9.7.8}$$

The normalized output noise power, i.e. the noise variance $\sigma_o^2$, is then (using Parseval's Theorem)

$$\sigma_o^2 = \int_{-\infty}^{\infty} Gn_o(f) df = \int_{-\infty}^{\infty} |H(f)|^2 Gn(f) df \tag{9.7.9}$$

From Eqs 9.7.7 and 9.7.9, we get

$$\gamma^2 = \frac{P_o^2(T)}{\sigma_o^2} = \frac{\left| \int_{-\infty}^{\infty} H(f) P(f) e^{j2\pi fT} df \right|^2}{\int_{-\infty}^{\infty} |H(f)|^2 Gn(f) df} \tag{9.7.10}$$

We shall now make use of "Schwartz inequality". It states that

$$\frac{\left| \int_{-\infty}^{\infty} X(f) Y(f) df \right|^2}{\int_{-\infty}^{\infty} |X(f)|^2 df} \leq \int_{-\infty}^{\infty} |Y(f)|^2 df \tag{9.7.11}$$

The equality sign applies when

$$X(f) = KY^*(f)$$

where $K$ is an arbitrary constant and $Y^*(f)$ is the complex conjugate of $Y(f)$. Now, if

DATA TRANSMISSION 495

and
$$X(f) \equiv \sqrt{Gn(f)}\, H(f) \tag{9.7.12}$$

$$Y(f) \equiv \frac{1}{\sqrt{Gn(f)}} P(f) e^{j2\pi fT} \tag{9.7.13}$$

Equation 9.7.10 becomes

$$\left.\begin{aligned}\gamma^2 &= \frac{\left|\int_{-\infty}^{\infty} X(f)\, Y(f)\, df\right|^2}{\int_{-\infty}^{\infty} |X(f)|^2\, df} \leq \int_{-\infty}^{\infty} |Y(f)|^2\, df \\ &= \int_{-\infty}^{\infty} \frac{|P(f)|^2}{Gn(f)}\, df\end{aligned}\right\} \tag{9.7.14}$$

$\gamma^2$ will attain its maximum value when

$$X(f) = KY^*(f)$$

From Eqs 9.7.12 and 9.7.13, we find that the optimum filter which gives maximum $\gamma^2$ has a transfer function

$$H(f) = K \frac{P^*(f)}{Gn(f)} e^{-j2\pi fT} \tag{9.7.15}$$

and $\gamma^2_{max}$ is [from Eq. 9.7.14]

$$\gamma^2_{max} = \int_{-\infty}^{\infty} \frac{|P(f)|^2}{Gn(f)}\, df \tag{9.7.16}$$

[Note that, for mathematical convenience, $\gamma^2$ maximized in place of $\gamma$.]

## 9.8 THE MATCHED FILTER

When the input noise is white, the optimum filter is known as a *matched filter*. In this case $Gn(f) = \eta/2$, and Eq. 9.7.15 becomes

$$H(f) = K \frac{P^*(f)}{\eta/2} e^{-j2\pi fT} \tag{9.8.1}$$

The impulse response of this filter is

$$h(t) = F^{-1}[H(f)]$$

$$= \frac{2k}{\eta} \int_{-\infty}^{\infty} P^*(f)\, e^{-j2\pi fT}\, e^{j2\pi ft}\, df$$

$$= \frac{2k}{\eta} \int_{-\infty}^{\infty} P^*(f) e^{-j2\pi f(t-T)} df$$

$$= \frac{2k}{\eta} \int_{-\infty}^{\infty} P(f) e^{-j2\pi f(t-T)} df$$

$$\left[ \because \int_{-\infty}^{\infty} P^*(f) df = \int_{-\infty}^{\infty} P(f) df \right.$$

$$\left. \text{Since } h^*(t) = h(t) \text{ for physical realizability of the filter.} \right]$$

$$= \frac{2K}{\eta} p(T-t)$$

$$= \frac{2K}{\eta} [S_1(T-t) - S_2(T-t)] \tag{9.8.2}$$

$$[\because p(t) \equiv S_1(t) - S_2(t)]$$

The probability of error of a matched filter can be found as follows:
In the case of matched filter

$$G_n(f) = \eta/2$$

Hence, Eq. 9.7.16 becomes

$$\gamma^2_{max} = \frac{2}{\eta} \int_{-\infty}^{\infty} |P(f)|^2 df \tag{9.8.3}$$

From Parseval's theorem, we have

$$\int_{-\infty}^{\infty} |P(f)|^2 df = \int_{-\infty}^{\infty} p^2(t) dt = \int_{0}^{T} p^2(t) dt \tag{9.8.4a}$$

$$[\because p(t) \text{ persists only for a time } T]$$

Hence,

$$\gamma^2_{max} = \frac{2}{\eta} \int_{0}^{T} [S_1(t) - S_2(t)]^2 dt \tag{9.8.4b}$$

$$= \frac{2}{\eta} \int_{0}^{T} [S_1^2(t) + S_2^2(t) - 2S_1(t) S_2(t)] dt$$

$$= \frac{2}{\eta}[E_{S1} + E_{S2} - 2E_{S12}] \qquad (9.8.5)$$

where $E_{S1}$ and $E_{S2}$ are energies in $S_1(t)$ and $S_2(t)$ respectively; and $E_{S12}$ is the joint energy in the product waveform $\sqrt{S_1(t)S_2(t)}$.

Now, if
$S_2(t) = -S_1(t)$, so that both $S_1(t)$ and $S_2(t)$ have the same energy, we get

$$Es_1 = Es_2 = -Es_{12} = Es$$

Hence Eq. 9.8.5 becomes

$$\gamma^2_{max} = \frac{8Es}{\eta} \qquad (9.8.6)$$

Rewriting Eq. 9.7.2 using Eq. 9.7.5, we get

$$Pe = \frac{1}{2} erfc\left[\frac{p_o(T)}{2\sqrt{2}\,\sigma_o}\right]$$

$$= \frac{1}{2} erfc\left[\frac{p_o^2(T)}{8\sigma_o^2}\right]^{\frac{1}{2}}$$

$$= \frac{1}{2} erfc\left[\frac{1}{8}\gamma^2\right]^{\frac{1}{2}} \qquad (9.8.7)$$

*Pe* is minimum when $\gamma^2$ is maximum.

Hence,

$$(Pe)\min = \frac{1}{2} erfc\left[\frac{1}{8}\gamma^2_{max}\right]^{\frac{1}{2}}$$

$$= \frac{1}{2} erfc\left[\frac{1}{8}\frac{8Es}{\eta}\right]^{\frac{1}{2}}$$

$$= \frac{1}{2} erfc\left[\frac{Es}{\eta}\right]^{\frac{1}{2}} \qquad (9.8.8)$$

The term $Es$ in Eq. 9.8.8 signifies that the error probability depends only on the signal energy, and not on the signal waveshape.

## 9.9 CORRELATOR

The correlator is a coherent system of signal reception as shown in Fig. 9.9.1.

The received signal plus noise $v_1(t)$ is multiplied by a locally generated signal $S_1(t) - S_2(t)$. The output of the multiplier is passed through an integrater whose output is sampled at $t = T$. Immediately after each sampling, all energy storing elements of the integrator are discharged. Since we are correlating the received signal and noise with the local signal $[S_1(t) - S_2(t)]$, the receiver is called a correlator.

**498** COMMUNICATION SYSTEMS: Analog and Digital

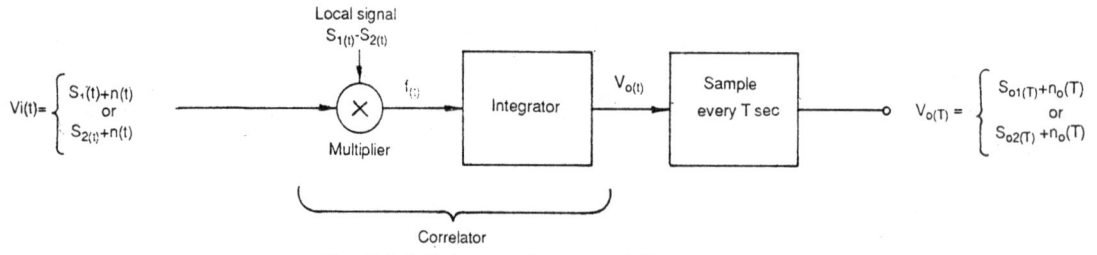

Fig. 9.9.1 Coherent System of Signal Reception

The output signal and noise are

$$S_o(T) = \frac{1}{\tau}\int_0^T S_i(t)[S_1(t) - S_2(t)]dt \qquad (9.9.1)$$

$$n_o(T) = \frac{1}{\tau}\int_0^T n(t)[S_1(t) - S_2(t)]dt \qquad (9.9.2)$$

where

$$S_i(t) = S_1(t) \text{ or } S_2(t)$$

and $\tau$ is the constant of the integrator.

Let us now compare the correlator with the matched filter. The output of the matched filter can be found by using the convolution integral. Thus

$$v_o(t) = \int_{-\infty}^{\infty} v_i(\lambda) h(t-\lambda) d\lambda = \int_0^T v_i(\lambda) h(t-\lambda) d\lambda \qquad (9.9.3)$$

where $h(t)$ is the impulse response of the matched filter and the limit of the integral in Eq. 9.9.3 is 0 to $T$, because the bit extends only over that interval.

From Eq. 9.8.2, we get

$$h(t-\lambda) = \frac{2k}{\eta}[S_1(T-t+\lambda) - S_2(T-t+\lambda)] \qquad (9.9.4)$$

substituting Eq. 9.9.4 into 9.9.3, we get

$$v_o(t) = \frac{2k}{\eta}\int_0^T v_i(\lambda)[S_1(T-t+\lambda) - S_2(T-t+\lambda)]d\lambda$$

or

$$S_o(t) + n_o(t) = \frac{2k}{\eta}\int_0^T [S_i(\lambda) + n(\lambda)][S_1(T-t+\lambda) - S_2(T-t+\lambda)]d\lambda$$

Substituting $t = T$, we get

$$S_o(T) + n_o(T) = \frac{2k}{\eta}\int_0^T S_i(\lambda)[S_1(\lambda) - S_2(\lambda)]d\lambda$$

$$+ \frac{2k}{\eta} \int_o^T n(\lambda)[S_1(\lambda) - S_2(\lambda)]d\lambda$$

Hence,

$$S_o(T) = \frac{2k}{\eta} \int_o^T S_i(\lambda)[S_1(\lambda) - S_2(\lambda)]d\lambda \tag{9.9.5}$$

and

$$n_o(T) = \frac{2k}{\eta} \int_o^T n(\lambda)[S_1(\lambda) - S_2(\lambda)]d\lambda \tag{9.9.6}$$

Thus, we see that $S_o(T)$ and $n_o(T)$ for correlator [Eqs 9.9.1 and 9.9.2] and for matched filter [Eqs 9.9.5 and 9.9.6] are identical. Hence, the performance of the correlator and the matched filter are identical.

## 9.10 PROBABILITY OF ERROR IN ASK

In ASK, $S_1(t) = A \cos \omega_o(t)$ and $S_2(t) = 0$
Thus,
$$p(t) = S_1(t) - S_2(t) = A \cos \omega_o t$$
Now,
$$\gamma_{max}^2 = \frac{2}{\eta} \int_o^T [S_1(t) - S_2(t)]^2 \, dt \quad \text{[from Eq. 9.8.4b]}$$

$$= \frac{2}{\eta} \int_o^T A^2 \cos^2 \omega_o t \, dt$$

$$= \frac{A^2 T}{\eta} \tag{9.10.1}$$

Hence,

$$Pe = \frac{1}{2} erfc \left[ \frac{1}{8} \gamma_{max}^2 \right]^{\frac{1}{2}}$$

$$= \frac{1}{2} erfc \left[ \frac{A^2 T}{8\eta} \right]^{\frac{1}{2}}$$

$$= \frac{1}{2} erfc \left[ \frac{Es}{4\eta} \right]^{\frac{1}{2}} \tag{9.10.2}$$

$$\left[ \because E\hat{s} = \frac{A^2 T}{2} \right]$$

## 9.11 PROBABILITY OF ERROR IN FSK

In FSK,
$$S_1(t) = A\cos(\omega_o + \Omega)t$$
and
$$S_2(t) = A\cos(\omega_o - \Omega)t$$
Hence,
$$p(t) = S_1(t) - S_2(t)$$
$$= A[\cos(\omega_o + \Omega)t - \cos(\omega_o - \Omega)t]$$

Now,
$$\gamma^2_{max} = \frac{2}{\eta}\int_0^T [p(t)]^2 dt$$

$$= \frac{2}{\eta}\int_0^T A^2[\cos(\omega_o + \Omega)t - \cos(\omega_o - \Omega)t] dt$$

$$= \frac{2A^2 T}{\eta}\left[1 - \frac{\sin 2\Omega T}{2\Omega T} + \frac{1}{2}\frac{\sin[2(\omega_o + \Omega)T]}{2(\omega_o + \Omega)T}\right.$$

$$\left. - \frac{1}{2}\frac{\sin[2(\omega_o - \Omega)T]}{2(\omega_o - \Omega)T} - \frac{\sin(2\omega_o T)}{(2\omega_o T)}\right] \quad (9.11.1)$$

If we make the following assumptions,
$$\omega_o T \gg 1 \text{ and } \omega_o \gg \Omega$$
which are usually encountered in practical systems, the last three terms in Eq. 9.11.1 can be neglected. Hence,

$$\gamma^2_{max} = \frac{2A^2 T}{\eta}\left[1 - \frac{\sin 2\Omega T}{2\Omega T}\right] \quad (9.11.2)$$

The quantity $\gamma^2_{max}$ attains its maximum value when $\Omega$ is so selected that
$$2\Omega T = 3\pi/2$$
For this value of $\Omega$, we have
$$\gamma^2_{max} = 2.42 \frac{A^2 T}{\eta} \quad (9.11.3)$$
Hence,
$$P_e = \frac{1}{2} erfc\left[\frac{1}{8}\gamma^2_{max}\right]^{\frac{1}{2}}$$

$$\approx \frac{1}{2} erfc\left[0.3\frac{A^2 T}{\eta}\right]^{\frac{1}{2}}$$

$$= \frac{1}{2} erfc \left[ 0.6 \frac{Es}{\eta} \right]^{\frac{1}{2}} \qquad (9.11.4)$$

$$\left[ \because Es = \frac{A^2 T}{2} \right]$$

## 9.12 PROBABILITY OF ERROR IN PSK

In PSK,
$$S_1(t) = A \cos \omega_o t$$
and
$$S_2(t) = -A \cos \omega_o t$$

As in this case, $S_1(t) = -S_2(t)$, Eq. 9.8.8 is applicable here. Hence

$$Pe = \frac{1}{2} erfc \left( \frac{Es}{\eta} \right)^{\frac{1}{2}} \qquad (9.12.1)$$

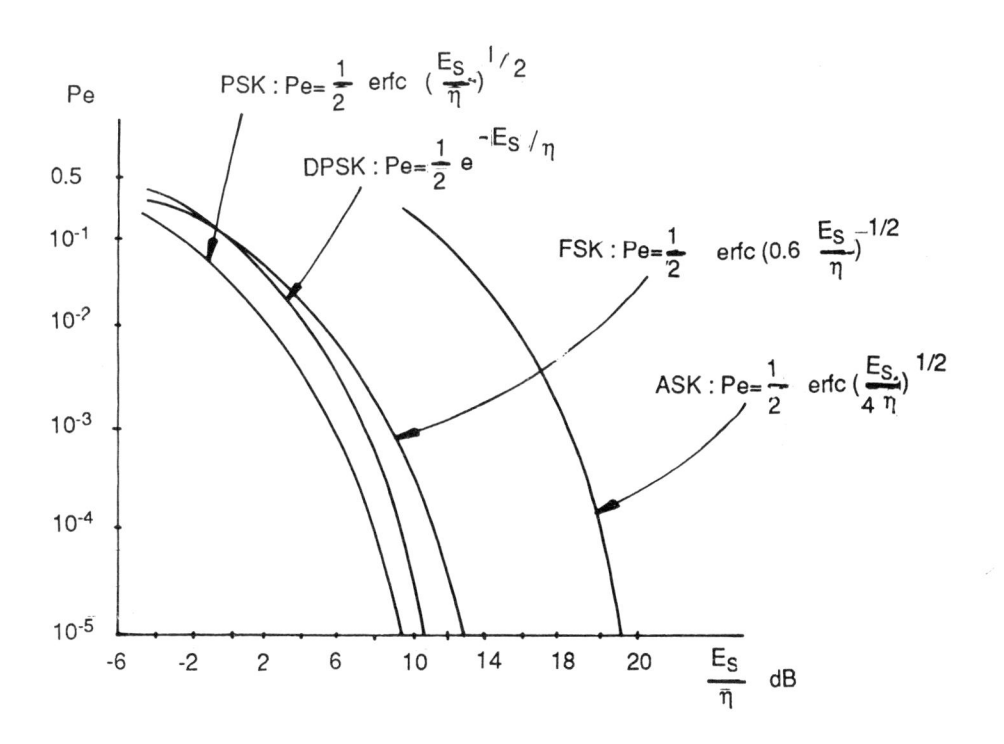

Fig. 9.14.1 Probability of Error P(e) for ASK, FSK, PSK and DPSK

## 9.13 PROBABILITY OF ERROR IN DPSK

The calculation of $Pe$ in the case of DPSK is beyond the scope of this book. The result is

$$Pe = \frac{1}{2} e^{-Es/\eta} \qquad (9.13.1)$$

## 9.14 COMMENTS ON Pe OF DIFFERENT METHODS

Figure 9.14.1 gives $Pe$ curves for different methods. As ASK gives the maximum $Pe$ for all signal to noise ratios $Es/\eta$, it is rarely used. PSK is the best method because, in this case, we have $S_1(t) = -S_2(t)$. FSK does not give as good a performance as PSK, because, in this case, $S_1(t) \neq S_2(t)$. The reason for a greater value of $Pe$ in DPSK than in PSK for all signal to noise ratios $Es/\eta$ is that, in the former method, one erroneous bit affects decision in two successive bit intervals.

## PROBLEMS

9.1 Explain frequency shift keying. Describe coherent detection of FSK signals. What should be the relationship between bit-rate and frequency-shift for a better performance?

9.2 Explain PSK and DPSK; compare the two.

9.3 Explain the working of an integrate and dump baseband signal receiver. Find its output signal to noise ratio.

9.4 What are optimum and matched filters? Find their transfer functions.

9.5 What is a correlator? Show that the performance of the correlator and matched filter are identical.

# TEN
# INFORMATION THEORY

## INTRODUCTION

Information theory is a branch of probability theory, which can be applied to the study of the communication systems. In general, communication of information is statistical in nature and the main aim of information theory is to study the simple ideal statistical communication models. Information theory was invented by communication scientists while they were studying the statistical structure of electrical communication equipments.

Communication systems deal with the flow of some sort of information in some network. The information may be electrical signals, words, pictures, music etc. There are three basic blocks of a communication system:
(i) Transmitter or source.
(ii) Channel or transmission network, which conveys the communique from transmitter to receiver.
(iii) Receiver or destination.

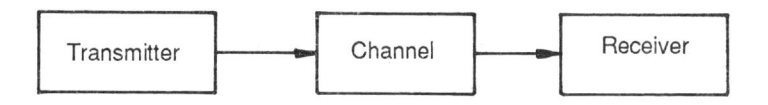

Fig.10.1 A Communication System

Figure 10.1 shows the simplest form of a communication system. In practice, generally, there are a number of transmitters and receivers with a complex network. In such cases, it is desirable to study the distribution of information in the system. Therefore some sort of transmission efficiency is to be defined which will lead to the most efficient transmission.

When the communique is readily measurable, such as an electric current, the study of the communication system is relatively easy. But, when the communique is *information*, the study becomes rather difficult. How to define the measure for an amount of information? And having defined a suitable measure, how can it be applied to improve the communication of information? Information theory answers these questions.

## 10.1  UNIT OF INFORMATION

The communication system considered here are of statistical nature; i.e. the performance of the system can never be described in a deterministic sense. It is always described in statistical terms. Thus the most significant feature of the communication system shown in Fig. 10.1 is its unpredictability or uncertainty. The transmitter transmits at random any one of the pre-specified messages. The probability of transmitting each individual message is known. When the communication system model is statistically defined, we are able to describe its *overall* or *average* performance. Thus our quest for an amount of information is virtually a search for a statistical parameter associated with a probability scheme. The parameter should indicate a relative measure of uncertainty relevant to the occurrence of each message in the message ensemble.

The principle of improbability — *"if a dog bites a man, it's no news, but if a man bites a dog, it's a news"* —helps us in this regard. The probability of a dog biting a man is quite high, so this is not a news, i.e. very little amount of information is communicated by the message *"a dog bites a man"*. On the other hand, the probability of a man biting a dog is extremely small, so this becomes a news, i.e., quite an amount of information is communicated by the message *"a man bites a dog"*. Thus, we see that there should be a sort of inverse relationship between the probability of an event and the amount of information associated with it. The more the probability of an event, the less is the amount of information associated with it, and vice-versa. Thus,

$$I(x_j) = f\left[\frac{1}{p(x_j)}\right] \tag{10.1.1}$$

where $x_j$ is an event with a probability $p(x_j)$ and the amount of information associated with it is $I(x_j)$.

Now, let there be another event $y_K$ such that $x_j$ and $y_k$ are independent. Hence, the probability of the joint event is $p(x_j, y_K) = p(x_j) p(y_K)$ with associated information content

$$I(x_j, y_K) = f\left[\frac{1}{p(x_j, y_K)}\right] = f\left[\frac{1}{p(x_j) p(y_K)}\right] \tag{10.1.2}$$

The total information $I(x_j, y_k)$ must be equal to the sum of individual informations $I(x_j)$ and $I(y_k)$ where, $I(y_K) = f\left[\frac{1}{p(y_K)}\right]$.

Thus, it can be seen that the function on RHS of Eq. 10.1.2 must be a function which converts multiplication into addition. Logarithm is one such function. Thus,

$$\begin{aligned} I(x_j, y_K) &= \log \frac{1}{[p(x_j) p(y_K)]} \\ &= \log \frac{1}{p(x_j)} + \log \frac{1}{p(y_K)} \\ &= I(x_j) + I(y_K) \end{aligned}$$

Hence the basic equation defining the amount of information (or self information) is

$$I(x_j) = \log \frac{1}{p(x_j)} = -\log p(x_j) \tag{10.1.3}$$

Different units of information can be defined for different bases of logarithms. When the base is '2' the unit is a *bit*, when the base is '*e*', the unit is '*nat*', and when the base is '*10*', the unit is *decit* or *Hartley*. The base 2, or binary system, is of particular importance. Hence, when no base is mentioned, it is to be assumed as 2 and the unit of information as bit.

Table 10.1.1 Conversion of Information Units

| Unit | Bits (base 2) | Nats (base $e$) | Decits (Base 10) |
|---|---|---|---|
| Bits (base 2) | — | 1 bit = $\dfrac{1}{\log_2 e}$ = 0.6932 nat | 1 bit = $\dfrac{1}{\log_2 10}$ = 0.3010 decit |
| Nats (base $e$) | 1 nat = $\dfrac{1}{\ln 2}$ = 1.4426 bits | — | 1 nat = $\dfrac{1}{\ln 10}$ = 0.4342 decit |
| Decits (base 10) | 1 decit = $\dfrac{1}{\log_{10} 2}$ = 3.3219 bits | 1 decit = $\dfrac{1}{\log_{10} e}$ = 2.3026 nats | — |

$\log_2 2 = 1$     $\log_2 e = 1.4426$     $\log_2 10 = 3.3219$
$\ln 2 = 0.6932$     $\ln e = 1$     $\ln 10 = 2.3026$
$\log_{10} 2 = 0.3010$     $\log_{10} e = 0.4342$     $\log_{10} 10 = 1$

## 10.2 ENTROPY

A communication system is not only meant to deal with a single message, but with all possible messages. Hence, although the instantaneous information flows corresponding to individual messages from the source may be erratic, we may describe the source in terms of average information per individual message, known as *entropy* of the source.

Let there be $M$ different messages $m_1, m_2 ... m_M$, with their respective probabilities of occurrences $P_1, P_2, ... P_M$. Let us assume that in a long time interval, $L$ messages have been generated. Let $L$ be very large so that $L \gg M$; then, the number of messages $m_1 = P_1 L$.

The amount of information in message $m_1 = \log \dfrac{1}{P_1}$. Thus the total amount of information in all $m_1$ messages

$$= P_1 L \log \dfrac{1}{P_1}.$$

The total amount of information in all $L$ messages will then be

$$I_t = P_1 L \log \dfrac{1}{P_1} + P_2 L \log \dfrac{1}{P_2} + ... + P_M L \log \dfrac{1}{P_M}$$

The average information per message, or entropy, will then be

$$H \equiv \dfrac{I_t}{L} = p_1 \log \dfrac{1}{P_1} + P_2 \log \dfrac{1}{P_2} + ... + P_M \log \dfrac{1}{P_M}$$

$$= \sum_{k=1}^{M} p_k \log \dfrac{1}{p_k} = -\sum_{k=1}^{M} p_k \log p_k \tag{10.2.1}$$

If there is only a single possible message, i.e., $M = 1$ and $p_k = P_1 = 1$, then

$$H = p_1 \log \frac{1}{p_1} = 1 \log \frac{1}{1} = 0$$

Thus, it can be seen that in the case of a single possible message, the reception of that message conveys no information.

On the other hand, let there be only one message out of the $M$ message having a probability 1 and all others 0. In that case

$$H = \sum_{k=1}^{M} P_K \log \frac{1}{P_K}$$

$$= p_1 \log \frac{1}{p_1} + \lim_{p \to 0} \left[ p \log \frac{1}{p} + p \log \frac{1}{p} + \ldots \right]$$

$$= 1 \log \frac{1}{1} + 0$$

$$= 0$$

Thus, if all the probabilities are zero except for one, which ought to be unity the entropy is zero. In all the other cases, the entropy is greater than 0 as can be seen from Eq. 10.2.1.

For a binary system ($M = 2$), the entropy is

$$H = p_1 \log \frac{1}{p_1} + p_2 \log \frac{1}{p_2}$$

Let

$$p_1 = p, \text{ then } p_2 = 1 - p_1 = 1 - p = q$$

Hence,

$$H = p \log \frac{1}{p} + (1-p) \log \frac{1}{(1-p)} \tag{10.2.2}$$

$$= p \log \frac{1}{p} + q \log \frac{1}{q} = H(p) = H(q)$$

A plot of $H$, as a function of $p$, as in Eq. 10.2.2 is shown in Fig. 10.2.1.

The condition for maximum entropy and its value can be found as follows:
Differentiating Eq. 10.2.2 w.r.t. $p$ yields

$$\frac{dH}{dp} = 0 = -\ln 2 - \log p + \ln 2 + \log(1-p)$$

i.e.,     $\log p = \log(1-p)$

i.e.,     $p = 1 - p$

i.e.,     $p = \frac{1}{2}$

Fig. 10.2.1

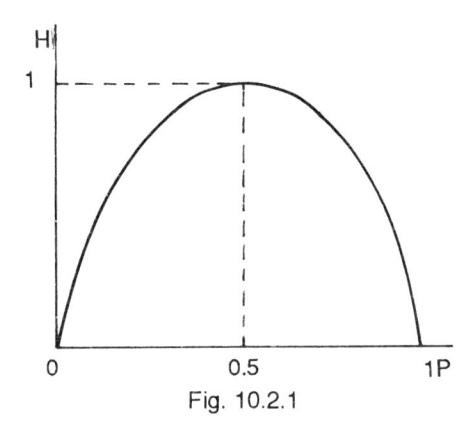

Also, we have

$$\frac{d^2H}{dp^2} = -\frac{1}{p} - \frac{1}{1-p} < 0$$

Hence, $H$ has a maxima at $p = \frac{1}{2}$

The maximum value of $H$ can be found from Eq. 10.2.2 by putting $p = \frac{1}{2}$ in it. Thus,

$$H_{max} = H\bigg|_{p=\frac{1}{2}} = \frac{1}{2}\log 2 + \frac{1}{2}\log 2 = 1 \text{ bit / message}$$

We have seen that for the binary case ($M = 2$), the entropy is maximum when $p = \frac{1}{2}$, i.e., when both the messages are equally likely. Similarly, it can be shown that for an $M$'ary case, the entropy is maximum when all the messages are equally likely. In this case, the maximum entropy is then

$$H_{max} = \sum_{k=1}^{M} p_k \log \frac{1}{p_k}$$

$$= \sum \frac{1}{M} \log M \text{ since } p_k = \frac{1}{M}$$

or

$$H_{max} = \log M \text{ bits / message} \tag{10.2.3}$$

since there are $M$ terms in the summation.

The important properties of entropy can now be summarised as follows:

(i) $\log M \geq H(x) \geq 0$

(ii) $H(x) = 0$ if all probabilities are zero, except for one, which must be unity

(iii) $H(x) = \log M$ if all the probabilities are equal so that $p(x)_i = p_i = \frac{1}{M}$ for all $i$'s

Now, let us examine $H$ under different cases for $M = 2$:

Case I: $p_1 = 0.01, p_2 = 0.99, H = 0.08$

**508** COMMUNICATION SYSTEMS: Analog and Digital

Case II:   $p_1 = 0.4, p_2 = 0.6, H = 0.97$
Case III:   $p_1 = 0.5, p_2 = 0.5, H = 1.00$

In case I, it is very easy to guess whether the message $m_1$ with a probability $p_1$ (= 0.01) will occur, or the message $m_2$ with a probability $p_2$ (= 0.99) will occur. (Most of the time message $m_2$ will occur.) Thus, in this case, the uncertainty is less. In case II, it is somewhat difficult to guess whether $m_1$ will occur or $m_2$ will occur as their probabilities are nearly equal. Thus, in this case, the uncertainty is more. In case III, it is extremely difficult to guess whether $m_1$ or $m_2$ will occur, as their probabilities are equal. Thus, in this case, the uncertainty is maximum. We have seen that the entropy is less when uncertainty is less and it is more when uncertainty is more. Thus, we can say that *entropy is a measure of uncertainty*.

## 10.3 RATE OF INFORMATION

If a message source generates messages at the rate of $r$ messages per second, the rate of information $R$ is defined as the average number of bits of information per second. Now, $H$ is the average number of bits of information per message. Hence,

$$R = rH \text{ bits/sec.} \tag{10.3.1}$$

Let us consider two sources of equal entropy $H$, generating $r_1$ and $r_2$ message per second, respectively. The first source will transmit the information at a rate $R_1 = r_1 H$ and the second source will transmit the information at a rate $R_2 = r_2 H$. Now, if $r_1 > r_2$, then $R_1 > R_2$. Thus in a given period, more information is transmitted from the first source than the second source, placing greater demands on the communication channel. Hence, the source is not described by its entropy alone but also by its rate of information.

Sometimes, $R$ is referred to as bits/sec entropy, and $H$ is referred to as bits/message entropy.

### Example 10.3.1

An event has six possible outcomes with the probabilities $p_1 = 1/2, p_2 = 1/4, p_3 = 1/8, p_4 = 1/16, p_5 = 1/32, p_6 = 1/32$. Find the entropy of the system. Also find the rate of information if there are 16 outcomes per second.
*Solution*
The entropy $H$ is

$$H = \sum_{k=1}^{6} p_k \log \frac{1}{p_k}$$

$$= \frac{1}{2} \log 2 + \frac{1}{4} \log 4 + \frac{1}{8} \log 8 + \frac{1}{16} \log 16 + \frac{1}{32} \log 32 + \frac{1}{32} \log 32$$

$$= \frac{31}{16} \text{ bits/message}$$

Now, $r = 16$ outcomes/s

Hence, the rate of information $R$ is

$$R = rH$$

$$= 16 \times \frac{31}{16}$$
$$= 31 \text{ bits/s}$$

### Example 3.3.2

A continuous signal is bandlimited to 5 kHz. The signal is quantized in 8 levels of a PCM system with the probabilities 0.25, 0.2, 0.2, 0.1, 0.1, 0.05, 0.05, and 0.05. Calculate the entropy and the rate of information.

*Solution*

The signal should be sampled at a frequency $5 \times 2 = 10$ kHz (Sampling theorem). Each sample is then quantized to one of the eight levels. Looking at each quantized level as a message, we get

H = – (0.25 log 0.25 + 0.2 log 0.2 + 0.2 log 0.2 + 0.1 log 0.1 + 0.1 log 0.1 + 0.05 log 0.05 + 0.05 log 0.05 + 0.05 log 0.05)

= 2.74 bits/message

As the sampling frequency is 10 kHz, the message rate = 10,000 messages/s. Hence, the rate of information is

$$R = rH = 10,000 \times 2.74 = 27,400 \text{ bits/s}.$$

## 10.4 JOINT ENTROPY AND CONDITIONAL ENTROPY

So far we have considered only a single probability scheme (known as one-dimensional probability scheme) and its associated entropy. This probability scheme may pertain either to the transmitter or to the receiver. Thus we can study the behaviour of either the transmitter, or the receiver. But, to study the behaviour of a communication system, we must simultaneously study the behaviour of the transmitter and the receiver. This gives rise to the concept of a two-dimensional probability scheme. The results of a one-dimensional probability scheme may be extended to a two-dimensional probability scheme, which may further be extended to study any finite-dimensional probability scheme.

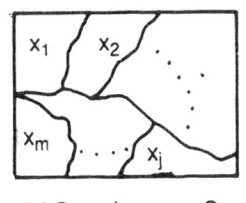

(a) Sample space $S_1$    (b) Sample space $S_2$    (c) Sample space $S = S_1 S_2$

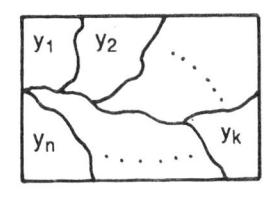

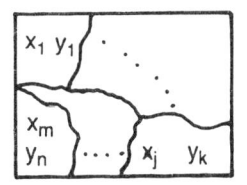

Fig. 10.4.1

Let there be two finite discrete sample spaces $S_1$, and $S_2$, and let their product space be $S = S_1 S_2$ (See Fig. 10.4.1)

Let
$$[X] = [x_1 \ x_2 \ldots x_m]$$
and
$$[Y] = [y_1 y_2 \ldots Y_n]$$

be the sets of events in $S_1$, and $S_2$, respectively. Each event $x_j$ of $S_1$ may occur in conjunction with any event

$j_K$ of $S_2$. Hence, the complete set of events in $S = S_1 S_2$ is

$$[XY] = \begin{bmatrix} x_1 y_1 & x_1 y_2 & \cdots & x_1 y_n \\ x_2 y_1 & x_2 y_2 & \cdots & x_2 y_n \\ \cdots & \cdots & \cdots & \cdots \\ x_m y_1 & x_m y_2 & \cdots & x_m y_n \end{bmatrix}$$

Thus, we have three sets of complete probability Schemes

$$P(X) = [P(x_j)]$$
$$P(Y) = [P(y_k)]$$
$$P(XY) = [P(x_j, y_k)]$$

[A probability scheme $(x_j)$ is said to be complete when $\sum_j P(x_j) = 1$]

We have three complete probability schemes and naturally there will be three associated entropies.

$$H(X) = -\sum_{j=1}^{m} p(x_j) \log p_{(x_j)} \qquad (10.4.1)$$

where

$$p(x_j) = \sum_{k=1}^{n} p(x_j, y_k)$$

$$H(Y) = -\sum_{k=1}^{n} p(y_k) \log p(y_k) \qquad (10.4.2)$$

where

$$p(y_k) = \sum_{j=1}^{m} p(x_j, y_k)$$

$$H(XY) = -\sum_{j=1}^{m} \sum_{k=1}^{n} p(x_j, y_k) \log p(x_j, y_k) \qquad (10.4.3)$$

$H(X)$ and $H(Y)$ are marginal entropies of $X$ and $Y$ respectively, and $H(XY)$ is the joint entropy of $X$ and $Y$.

The conditional probability $P(X/Y)$ is given by

$$p(X/Y) = \frac{p(X,Y)}{p(Y)}$$

We know that $y_k$ may occur in conjuction with $x_1, x_2 \ldots x_m$. Thus,

$$[X/y_k] = \begin{bmatrix} \dfrac{x_1}{y_k} & \dfrac{x_2}{y_k} & \cdots & \dfrac{x_m}{y_k} \end{bmatrix}$$

and the associated probability scheme is

$$P(X/y_k) = [p(x_1/y_k)\ p(x_2/y_k) \ldots p(x_m/y_k)]$$
$$= \left[\dfrac{p(x_1, y_k)}{p(y_k)}\ \dfrac{p(x_2, y_k)}{p(y_k)} \ldots \dfrac{p(x_m, y_k)}{p(y_k)}\right] \quad (10.4.4.)$$

Now,

$$P(x_1, y_k) + p(x_2, y_k) + \ldots p(x_m, y_k) = P(y_k)$$

Thus the sum of elements of the matrix given by Eq. 10.4.4 is unity. Hence the probability scheme defined by Eq. 10.4.4 is complete. Therefore, an entropy may be associated with it. Thus,

$$H(X/y_k) = -\sum_{j=1}^{m} \dfrac{p(x_j, y_k)}{p(y_k)} \log \dfrac{p(x_j, y_k)}{p(y_k)}$$
$$= -\sum_{j=1}^{m} p(x_j/y_k) \log p(x_j/y_k)$$

We may take the average of this conditional entropy for all admissible values of $y_k$ in order to obtain a measure of an average conditional entropy of the system.

$$H(X/Y) = \overline{H(X/y_k)}$$
$$= \sum_{k=1}^{n} p(y_k) H(X/y_k)$$
$$= -\sum_{k=1}^{n} p(y_k) \sum_{j=1}^{m} p(x_j/y_k) \log p(x_j/y_k)$$
$$= -\sum_{j=1}^{m} \sum_{k=1}^{n} p(y_k)\, p(x_j/y_k) \log p(x_j/y_k)$$
$$= -\sum_{j=1}^{m} \sum_{k=1}^{n} p(x_j, y_k) \log p(x_j/y_k) \quad (10.4.5)$$

Similarly, it can be shown that

$$H(Y/X) = -\sum_{j=1}^{m} \sum_{k=1}^{n} p(x_j, y_k) \log p(y_k/x_j) \quad (10.4.6)$$

$H(X/Y)$ and $H(Y/X)$ are *average conditional entropies*, or simply *conditional entropies*.

Thus, in all, there are five entropies associated with a two-dimensional probability scheme. They are: $H(X)$, $H(Y)$, $H(X,Y)$, $H(X/Y)$, and $H(Y/X)$.

Now, let $X$ represent a transmitter, and $Y$ a receiver. The following interpretations of the different entropies for a two-port communication system can be derived:

$H(X)$ : Average information per character at the transmitter, or entropy of the transmitter.

$H(Y)$ : Average information per character at the receiver, or entropy of the receiver.

$H(X,Y)$ : Average information per pair of the transmitted and received characters, or average uncertainty of the communication system as a whole.

$H(X/Y)$ : A received character $y_k$ may be the result of transmission of one of the $x'_j s$ with a given probability. The entropy associated with this probability scheme, when $y_k$ covers all received symbols, i.e., $\overline{H(X/y_k)}$ is the conditional entropy $H(X/Y)$; a measure of information about the transmitter, where it is known that $Y$ is received.

$H(Y/X)$ : A transmitted character $x_j$ may result in the reception of one of the $y_k's$ with a given probability. The entropy associated with this probability scheme, when $x_j$ covers all transmitted symbols, i.e., $\overline{H(Y/x_j)}$ is the conditional entropy $H(Y/X)$ ; a measure of information about the receiver, where it is known that $X$ is transmitted.

$H(X)$ and $H(Y)$ give indications of the probabilistic nature of the transmitter, and the receiver, respectively. $H(X/Y)$ indicates how well one can recover the transmitted symbols, from the received symbols; i.e. it gives a measure of equivocation. $H(Y/X)$ indicates how well one can recover the received symbols from the transmitted symbols; i.e., it gives a measure of error, or noise. The meaning of $H(X/Y)$ and $H(Y/X)$ will be more clear after we study mutual information.

The relationship between the different entropies is found as follows:

$$H(XY) = -\sum_{j=1}^{m}\sum_{k=1}^{n} p(x_j, y_k) \log p(x_j, y_k)$$

$$= -\sum_{j=1}^{m}\sum_{k=1}^{n} p(x_j, y_k) \log \left[p(x_j/y_k) p(y_k)\right]$$

$$= -\sum_{j=1}^{m}\sum_{k=1}^{n} p(x_j, y_k) [\log p(x_j/y_k) + \log p(y_k)]$$

$$= -\sum_{j=1}^{m}\sum_{k=1}^{n} [p(x_j, y_k) \log p(x_j/y_k) + p(x_j, y_k) \log p(y_k)]$$

$$= H(X/Y) - \sum_{j=1}^{m}\sum_{k=1}^{n} p(x_j, y_k) \log p(y_k)$$

$$= H(X/Y) - \sum_{k=1}^{n}\left[\sum_{j=1}^{m} p(x_j, y_k)\right] \log p(y_k)$$

$$= H(X/Y) - \sum_{k=1}^{n} p(y_k) \log p(y_k)$$

$$= H(X/Y) + H(Y) \tag{10.4.7}$$

Similarly, it can be shown that
$$H(XY) = H(Y/X) + H(X) \tag{10.4.8}$$

## Example 10.4.1

Complete the following probability matrix in all possible ways:

$$\begin{bmatrix} 0.1 & a & 0.2 & 0.4 \\ 0.3 & 0.1 & b & 0.5 \\ c & 0.4 & 0.1 & 0.3 \\ 0.2 & 0.2 & 0.1 & d \end{bmatrix}$$

*Solution*

The given matrix is a two-dimensional matrix as it is neither a row nor a column matrix. Moreover, as the sum of all the given entries is 2.9 (which is greater than 1), it cannot be a joint-probability matrix.

As the sum of all the given entries in each row of the matrix is less than 1, it can be a conditional-probability matrix $P(Y/X)$ with the values of $a, b, c$ and $d$ so chosen that the sum of each row becomes 1. Thus, when $a = 0.3, b = 0.1, c = 0.2$, and $d = 0.5$, the given matrix is the conditional-metric $P(Y/X)$. Again, as the sum of all the given entries in the fourth column of the matrix is 1.2 (which is greater than 1), it cannot be a conditional-probability matrix $P(X/Y)$.

Thus, there is only one possible way to complete the given probability matrix and it is:

$$\begin{bmatrix} 0.1 & 0.3 & 0.2 & 0.4 \\ 0.3 & 0.1 & 0.1 & 0.5 \\ 0.2 & 0.4 & 0.1 & 0.3 \\ 0.2 & 0.2 & 0.1 & 0.5 \end{bmatrix} = P(Y/X)$$

## Example 10.4.2

A discrete source transmits messages $x_1, x_2$ and $x_3$ with the probabilities 0.3, 0.4 and 0.3. The source is connected to the channel given in Fig. 10.4.2. Calculate all the entropies.

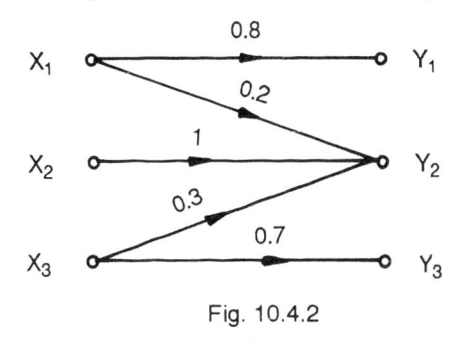

Fig. 10.4.2

*Solution*

Figure 10.5.2 gives the conditional-probability matrix $P(Y/X)$ as

## 514 COMMUNICATION SYSTEMS: Analog and Digital

$$P(Y/X) = \begin{array}{c} \\ x_1 \\ x_2 \\ x_3 \end{array} \begin{array}{c} y_1 \quad y_2 \quad y_3 \\ \begin{bmatrix} 0.8 & 0.2 & 0 \\ 0 & 1 & 0 \\ 0 & 0.3 & 0.7 \end{bmatrix} \end{array}$$

Also given is

$$P(X) = [0.3 \quad 0.4 \quad 0.3]$$

The joint probability matric $P(X,Y)$ can be obtained by multiplying the rows of $P(Y/X)$ by $p(x_1)$, $p(x_2)$ and $p(x_3)$ i.e. by 0.3, 0.4, and 0.3 respectively, giving

$$P(X,Y) = \begin{bmatrix} 0.8 \times 0.3 & 0.2 \times 0.3 & 0 \\ 0 & 1 \times 0.4 & 0 \\ 0 & 0.3 \times 0.3 & 0.7 \times 0.3 \end{bmatrix}$$

$$= \begin{array}{c} \\ x_1 \\ x_2 \\ x_3 \end{array} \begin{array}{c} y_1 \quad y_2 \quad y_3 \\ \begin{bmatrix} 0.24 & 0.06 & 0 \\ 0 & 0.4 & 0 \\ 0 & 0.09 & 0.21 \end{bmatrix} \end{array}$$

(As a check, it can be seen that the sum of all the entries in $P(X,Y)$ is 1.)

The probabilities $p(y_1)$, $p(y_2)$ and $p(y_3)$ can be obtained by adding the column of $P(X,Y)$ giving

$$p(y_1) = 0.24, \; p(y_2) = 0.06 + 0.4 + 0.09 = 0.55, \; p(y_3) = 0.21$$

The conditional-probability matrix $P(X/Y)$ can be obtained by dividing the columns of $P(X,Y)$ by $P(y_1)$, $p(y_2)$, and $p(y_3)$, respectively, giving

$$P(X/Y) = \begin{array}{c} \\ x_1 \\ x_2 \\ x_3 \end{array} \begin{array}{c} y_1 \quad y_2 \quad y_3 \\ \begin{bmatrix} 1 & 0.109 & 0 \\ 0 & 0.727 & 0 \\ 0 & 0.164 & 1 \end{bmatrix} \end{array}$$

(*Check:* The sum of all the columns of $P(X/Y)$ is 1.)

The entropies can now be calculated as follows:

$$H(X) = -\sum_{j=1}^{3} P(x_j) \log p(x_j)$$

$$= -(0.3 \log 0.3 + 0.4 \log 0.4 + 0.3 \log 0.3)$$
$$= 1.571 \text{ bits/message}$$

$$H(Y) = -\sum_{k=1}^{3} P(y_k) \log p(y_k)$$

$$= -[0.24 \log 0.24 + 0.55 \log 0.55 + 0.21 \log 0.21]$$

$$H(X, Y) = -\sum_{j=1}^{3}\sum_{k=1}^{3} p(x_j, y_k) \log p(x_j, y_k)$$

$$= 1.441 \text{ bits/message}$$

$$= -[0.24 \log 0.24 + 0.06 \log 0.06 + 0.4 \log 0.4 + 0.09 \log 0.09 + 0.21 \log 0.21]$$

$$= 2.053 \text{ bits/message}$$

$$H(X/Y) = -\sum_{j=1}^{3}\sum_{k=1}^{3} p(x_j, y_k) \log p(x_j/y_k)$$

$$= -[0.24 \log 1 + 0.06 \log 0.109 + 0.4 \log 0.727 + 0.09 \log 0.164 + 0.21 \log 1]$$

$$= 0.612 \text{ bit/message}$$

$$H(Y/X) = -\sum_{j=1}^{3}\sum_{k=1}^{3} P(x_j, y_k) \log p(y_k/x_j)$$

$$= -[0.24 \log 0.8 + 0.06 \log 0.2 + 0.4 \log 1 + 0.09 \log 0.3 + 0.21 \log 0.7]$$

$$= 0.482 \text{ bit/message}$$

[*Note*: $H(X/Y)$ and $H(Y/X)$ can also be found by using Eqs 10.4.7 and 10.4.8, respectively.]

### Example 10.4.3

A transmitter has an alphabet of four letters $[x_1\ x_2\ x_3\ x_4]$ and the receiver has an alphabet of three letters $[y_1\ y_2\ y_3]$. The joint probability matrix is

$$P(X,Y) = \begin{array}{c} \\ x_1 \\ x_2 \\ x_3 \\ x_4 \end{array} \begin{array}{c} \begin{matrix} y_1 & y_2 & y_3 \end{matrix} \\ \begin{bmatrix} 0.3 & 0.05 & 0 \\ 0 & 0.25 & 0 \\ 0 & 0.15 & 0.05 \\ 0 & 0.05 & 0.15 \end{bmatrix} \end{array}$$

Calculate all the entropies.

*Solution*

The probabilities of the transmitter symbols are found by a summation of rows and the probabilities of the receiver symbols are found by a summation of columns of the matrix $P(X,Y)$.
Thus,

$$P(X) = [0.35\ \ 0.25\ \ 0.2\ \ 0.2]$$

and

$$P(Y) = [0.3\ \ 0.5\ \ 0.2]$$

The entropies can now be calculated as follows:

$$H(X) = -\sum_{j=1}^{4} P(x_j) \log P(x_j)$$

$$= [0.35 \log 0.35 + 0.25 \log 0.25 + 0.2 \log 0.2 + 0.2 \log 0.2]$$
$$= 1.96 \text{ bits/message}$$

$$H(Y) = -\sum_{k=1}^{3} P(y_k) \log P(y_k) = -[0.3 \log 0.3 + 0.5 \log 0.5 + 0.2 \log 0.2]$$
$$= 1.49 \text{ bits/message}$$

$$H(X, Y) = -\sum_{j=1}^{4}\sum_{k=1}^{3} P(x_j, y_k) \log p(x_j, y_k)$$
$$= -[0.3 \log 0.3 + 0.05 \log 0.05 + 0.25 \log 0.25$$
$$+ 0.15 \log 0.15 + 0.05 \log 0.05 + 0.05 \log 0.05$$
$$+ 0.15 \log 0.15]$$
$$= 2.49 \text{ bits/message}$$

$$H(X/Y) = H(X, Y) - H(Y)$$
$$= 2.49 - 1.49$$
$$= 1.00 \text{ bit/message}$$

$$H(Y/X) = H(X, Y) - H(X)$$
$$= 2.49 - 1.96$$
$$= 0.53 \text{ bit/message}$$

## 10.5 MUTUAL INFORMATION

We are interested in the transfer of information from a transmitter through a channel to a receiver. Prior to the reception of a message, the state of knowledge at the receiver about a transmitted signal $x_j$ is the probability that $x_j$ would be selected for transmission. This is *a-priori* probability $p(x_j)$. After the reception and selection of the symbol $y_k$, the state of knowledge concerning $x_j$ is the conditional probability $p(x_j / y_k)$ which is also known as *a-posteriori* probability. Thus, before $y_k$ is received, the uncertainty is

$$-\log p(x_j)$$

After $y_k$ is received, the uncertainty becomes

$$-\log p(x_j / y_k)$$

The information gained about $x_j$ by the reception of $y_k$ is the net reduction in its uncertainty, and is known as *mutual information* $I(x_j, y_k)$.
Thus,

$I(x_j, y_k)$ = initial uncertainty − final uncertainty

$$= -\log p(x_j) - [-\log p(x_j / y_k)]$$
$$= \log \frac{p(x_j / y_k)}{p(x_j)}$$

$$= \log \frac{p(x_j, y_k)}{p(x_j)p(y_k)}$$

$$= \log \frac{p(y_k/x_j)}{p(y_k)}$$

$$= I(y_k; x_j)$$

Thus, we see that mutual information is symmetrical in $x_j$ and $y_k$ i.e.,

$$I(x_j; y_k) = I(y_k; x_j)$$

[*Note*: Self-information may be treated as a special case of mutual information when $y_k = x_j$. Thus,

$$I(x_j; x_j) = \log \frac{p(x_j/x_j)}{p(x_j)} = \log \frac{1}{p(x_j)}$$

$$= I(x_j) \,]$$

The average of mutual information i.e., the entropy corresponding to mutual information, is given by

$$I(X; Y) = \overline{I(x_j; y_k)}$$

$$= \sum_{j=1}^{m} \sum_{k=1}^{n} p(x_j, y_k) \, I(x_j; y_k)$$

$$= \sum_{j=1}^{m} \sum_{k=1}^{n} p(x_j, y_k) \log \frac{p(x_j/y_k)}{p(x_j)}$$

$$= \sum_{j=1}^{m} \sum_{k=1}^{n} p(x_j, y_k) \, [\log p(x_j/y_k) - \log p(x_j)]$$

$$= - \sum_{j=1}^{m} \sum_{k=1}^{n} p(x_j, y_k) \log p(x_j)$$

$$\quad - \left[ - \sum_{j=1}^{m} \sum_{k=1}^{n} p(x_j, y_k) \log p(x_j/y_k) \right]$$

$$= - \sum_{j=1}^{m} \left[ \sum_{k=1}^{n} p(x_j, y_k) \right] \log p(x_j) - H(X/Y)$$

$$= - \sum_{j=1}^{m} p(x_j) \log p(x_j) - H(X/Y)$$

$$= H(X) - H(X/Y) \tag{10.5.1}$$
$$= H(X) + H(Y) - H(X, Y) \text{ using Eq. 10.5.7} \tag{10.5.2}$$
$$= H(Y) - H(Y/X) \text{ using Eq. 10.5.8} \tag{10.5.3}$$

$I(X;Y)$ does not depend on the individual symbols $x_j$ or $y_k$; it is a property of the whole communication system. On the other hand $I(x_j; y_k)$ depends on the individual symbols $x_j$ or $y_k$. In practice, we are interested in the whole communication system and not in individual symbols. For this reason, although $I(X;Y)$ is an average mutual information, it is mostly referred as mutual information. Since $I(X;Y)$ indicates a measure of the information transferred through the channel, it is also known as *transferred information* or *transinformation* of the channel.

Equations 10.5.1 and 10.5.3 explain the meaning of conditional entropies $H(X/Y)$ and $H(Y/X)$ respectively. Equation 10.5.1 states that the transferred information $I(X,Y)$ is equal to the average source information minus the average uncertainty that still remains about the messages. In other words, $H(X/Y)$ is the average additional information needed at the receiver after reception in order to completely specify the the message sent. Thus $H(X/Y)$ gives the information lost in the channel. This is also known as equivocation. Equation 10.5.3 states that the transferred information $I(X;Y)$ is equal to the receiver entropy minus that part of the receiver entropy which is not the information about the source. Thus $H(Y/X)$ can be considered as the noise entropy added in the channel. Thus, $H(Y/X)$ is a measure of noise or error due to the channel.

## Example 10.5.1

Find the mutual information for the channel given in (A) Ex. 10.4.2, and (B) Ex.10.4.3.
*Solution*
(A) $I(X;Y) = H(X) - H(X/Y) = 1.571 - 0.612 = 0.959$ bit / message
As a check,
$I(X;Y) = H(Y) - H(Y/X) = 1.441 - 0.482 = 0.959$ bit / message
(B) $I(X;Y) = H(X) - H(X/Y) = 1.96 - 1.00 = 0.96$ bit / message
As a check,
$I(X;Y) = H(Y) - H(Y/X) = 1.49 - 0.53 = 0.96$ bit / message

### 10.5.1 Noise-free Channel

Let us consider the communication channel shown in Fig. 10.5.1.1. It is known as a noise-free channel.

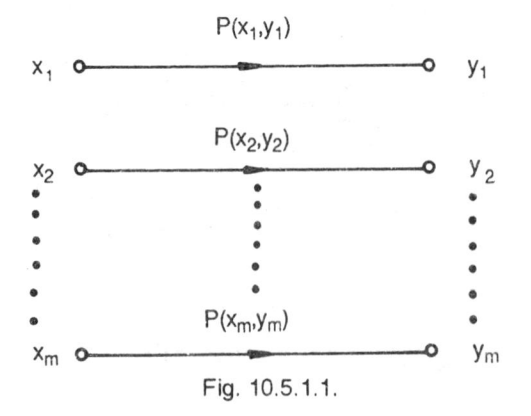

Fig. 10.5.1.1.

In such channels, there is a one-to-one correspondence between input and output, i.e., each input symbol is received as one and the only one output symbol. Also, $n = m$. A discrete noise-free channel is shown in Fig. 10.5.1.1. The joint probability matrix $P(X,Y)$ is of the diagonal form

$$[P(X,Y)] = \begin{bmatrix} P(x_1,y_1) & 0... & 0 & 0 \\ 0 & P(x_2,y_2) & 0... & 0 \\ ...... & ............... & ... & ... \\ 0 & 0... & 0 & p(x_m,y_m) \end{bmatrix} \qquad (10.5.1.1)$$

and the channel probability matrices $[P(Y/X)]$ and $[P(X/Y)]$ are unity-diagonal matrices.

$$[P(Y/X)] = [P(X/Y)] = \begin{bmatrix} 1 & 0 & 0 & ... & 0 & 0 \\ 0 & 1 & 0 & ... & 0 & 0 \\ ... & ... & .. & ... & ... & ... \\ 0 & 0 & 0 & ... & 0 & 1 \end{bmatrix} \qquad (10.5.1.2)$$

From Eq. 10.5.1.1, it can be seen that

$$H(X,Y) = -\sum_{j=1}^{m}\sum_{k=1}^{m} P(x_j, y_k) \log p(x_j, y_k)$$

$$= -\sum_{j=1}^{m} P(x_j, y_j) \log p(x_j, y_j)$$

Since
$$p(x_j, y_k) = 0 \qquad \text{for } j \neq k$$

Also, from Eq. 10.5.1.1, it can be seen that
$$p(x_j, y_j) = p(x_j) = p(y_j)$$

Hence,
$$H(X,Y) = H(X) = H(Y)$$

From Eq. 10.5.1.2, we get
$$H(Y/X) = H(X/Y) = -m(1 \log 1) = 0$$

(Because there are $m$ unity terms in the matrix and the remaining terms are zero.)
Thus, for a noise-free channel,

$$I(X;Y) = H(X) - H(X/Y) = H(X) = H(Y) = H(X,Y) \qquad (10.5.1.3)$$

### 10.5.2 Channel with Independent Input and Output

In such channels, there is no correlation between the input and output symbols.

Let us consider the channel shown in Fig. 10.5.2.1(a). The joint probability matrix in this case is

$$[P(X,Y)] = \begin{array}{c} \\ x_1 \\ x_2 \\ \\ x_m \end{array} \begin{array}{c} y_1 \quad y_2 \quad y_n \\ \begin{bmatrix} P_1 & P_1 \cdots & P_1 \\ P_2 & P_2 \cdots & P_2 \\ ... & ... & ... \\ P_m & P_m \cdots & P_m \end{bmatrix} \end{array} \qquad (10.5.2.1)$$

**520** COMMUNICATION SYSTEMS: Analog and Digital

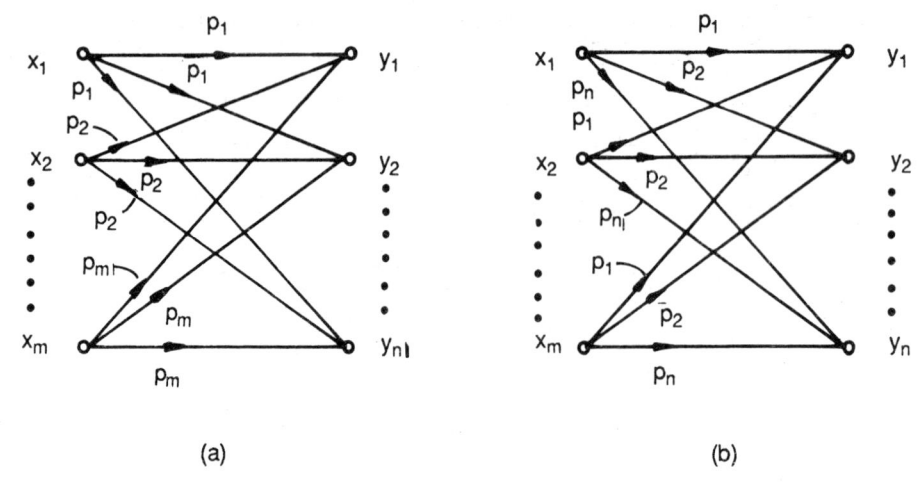

Fig. 10.5.2.1 (a) and (b)

It can be seen from Eq. 10.5.2.1 that

$$\sum_{j=1}^{m} p_j = \frac{1}{n}$$

$$p(x_j) = np_j \quad j = 1,2,\ldots, m$$

$$p(y_k) = \frac{1}{n} \quad k = 1,2,\ldots, n$$

and
$$p(x_j, y_k) = p_j$$

Hence,
$$p(x_j, y_k) = p(x_j)\, p(y_k) \tag{10.5.2.2}$$

Equation 10.5.2.2 shows that $x_j$ and $y_k$ are independent for all $j$ and $k$, i.e. input and output are independent for the channel shown in Fig. 10.5.2.1(a).

From Eq. 10.5.2.2, we get

$$\frac{p(x_j, y_k)}{p(y_k)} = p(x_j) \text{ or } p(x_j/y_k) = p(x_j) \tag{10.5.2.3}$$

and

$$\frac{p(x_j, y_k)}{p(x_j)} = p(y_k) \text{ or } p(y_k/x_j) = p(y_k) \tag{10.5.2.4}$$

Now,

$$H(Y/X) = -\sum_{j=1}^{m}\sum_{k=1}^{n} p(x_j, y_k) \log p(y_k/x_j)$$

$$= -\sum_{j=1}^{m}\sum_{k=1}^{n} p(x_j) p(y_k) \log p(y_k)$$

Thus,

$$H(Y/X) = -\sum_{k=1}^{n}\left[\sum_{j=1}^{m}p(x_j)\right]p(y_k)\log p(y_k)$$

$$= -\sum_{k=1}^{n}p(y_k)\log p(y_k) \qquad \text{Since } \sum_{j=1}^{m}p(x_j) = 1$$

$$= H(Y)$$

Similarly, it can be shown that
$$H(X/Y) = H(X)$$

Hence, for the channel shown in Fig. 10.5.2.1(a), we have

$$\left.\begin{array}{l}I(X;Y) = H(X) - H(X/Y) = H(X) - H(X) = 0\\ I(X;Y) = H(Y) - H(Y/X) = H(Y) - H(Y) = 0\end{array}\right\} \qquad (10.5.2.5)$$

Equations 10.5.2.5 state that no information is transmitted through the channel shown in Fig.10.5.2.1(a).

The joint probability matrix for the channel shown in Fig. 10.5.2.1(b) is

$$[P(X,Y)] = \begin{array}{c} \\ x_1 \\ x_2 \\ \vdots \\ x_m \end{array}\begin{array}{c}y_1 \quad y_2 \quad \cdots \quad y_n\\ \begin{bmatrix}P_1 & P_2 & \cdots & P_n\\ P_1 & P_2 & \cdots & P_n\\ \cdots & \cdots & \cdots & \cdots\\ P_1 & P_2 & \cdots & P_n\end{bmatrix}\end{array} \qquad (10.5.2.6)$$

It can be seen from Eq. 10.5.2.6 that

$$\sum_{k=1}^{n}P_k = \frac{1}{n}$$

$$p(x_j) = \frac{1}{n} \qquad j = 1, 2, \ldots, m$$

$$p(y_k) = nP_k \qquad k = 1, 2, \ldots, n$$

and
$$p(x_j, y_k) = P_k$$

Hence,
$$p(x_j, y_k) = p(x_j)p(y_k) \qquad (10.5.2.7)$$

Equation 10.5.2.7 shows that $x_j$ and $y_k$ are independent for all $j$ and $k$, i.e., input and output are independent for the channel shown in Fig. 10.5.2.1(b). Following the same procedure, it can be shown that in this case also $I(X;Y) = 0$; i.e., no information is transmitted through the channel shown in Fig. 10.5.2.1(b).

Hence it can be said that in the case of a channel with an independent input and output, no information is transmitted through it, i.e., $I(X;Y) = 0$.

[*Note*: A channel is said to be with an independent input and output when the joint probability matrix satisfies at least one of the following conditions:
(i) Each row, consists the same element.
(ii) Each column consists the same element.]

## Example 10.5.2.1

Find the mutual information for the channel shown in Fig. 10.5.2.2.

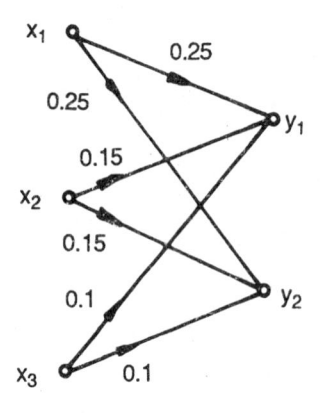

Fig. 10.5.2.2

*Solution*

The joint probability matrix for the channel shown in Fig. 10.5.2.2 is

$$[P(X, Y)] = \begin{array}{c} x_1 \\ x_2 \\ x_3 \end{array} \begin{array}{cc} y_1 & y_2 \\ \begin{bmatrix} 0.25 & 0.25 \\ 0.15 & 0.15 \\ 0.1 & 0.1 \end{bmatrix} \end{array}$$

From $P(X,Y)$, we get

$$P(x_1) = 0.25 + 0.25 = 0.5; P(x_2) = 0.15 + 0.15 = 0.3;$$
$$P(x_3) = 0.1 + 0.1 = 0.2$$

and

$$P(y_1) = P(y_2) = 0.25 + 0.15 + 0.1 = 0.5$$

Hence,

$$H(X) = -\sum_{j=1}^{3} p(x_j) \log p(x_j)$$

$$= -[0.5 \log 0.5 + 0.3 \log 0.3 + 0.2 \log 0.2]$$

$$= 1.485 \text{ bits/message}$$

$$H(Y) = -\sum_{k=1}^{2} p(y_k) \log p(y_k)$$

$$= -[0.5 \log 0.5 + 0.5 \log 0.5]$$

= 1 bit/message

$$H(X, Y) = -\sum_{j=1}^{3}\sum_{k=1}^{2} p(x_j, y_k) \log p(x_j, y_k)$$

$$= -[0.25 \log 0.25 + 0.25 \log 0.25 + 0.15 \log 0.15 + 0.15 \log 0.15 + 0.1 \log 0.1 + 0.1 \log 0.1]$$

$$= 2.485 \text{ bits/message}$$

Hence,
$$I(X;Y) = H(X) + H(Y) - H(X, Y) = 1.485 + 1 - 2.485 = 0$$

The channel shown in Fig. 10.5.2.2 is with an independent input and output (as is clear from its joint probability matrix), and the answer $I(X;Y) = 0$ confirms that the mutual information in such a case is zero.

## 10.6 CHANNEL CAPACITY

The mutual information $I(X;Y)$ indicates a measure of the average information per symbol transmitted in the system. A suitable measure for efficiency of transmission of information may be introduced by comparing the actual rate and the upper bound of the rate of information transmission for a given channel. Shannon has introduced a significant concept of *channel capacity* defined as the maximum of mutual information. Thus, the channel capacity $C$ is given by

$$C = \max I(X;Y) = \max [H(X) - H(X/Y)] \qquad (10.6.1)$$

The transmission efficiency or channel efficiency is defined as

$$\eta = \frac{\text{actual transinformation}}{\text{maximum transinformation}}$$

or

$$\eta = \frac{I(X;Y)}{\max I(X;Y)} = \frac{I(X;Y)}{C} \qquad (10.6.2)$$

The *redundancy* of the channel is defined as $R = 1 - \eta = \dfrac{C - I(X;Y)}{C}$ \qquad (10.6.3)

### 10.6.1 Noise-free Channel

For a noise-free channel, the mutual information is given by Eq. 10.5.1.3. Thus,

$$I(X; Y) = H(X)$$

Hence, the channel capacity in this case is

$$C = \max I(X;Y) = \max H(X)$$

But, according to Eq. 10.2.3

$$\max H(X) = \log M \text{ bits/message}$$

where $M$ is the total number of messages.
Hence, for a noise-free channel,

$$C = \log M \text{ bits/message} \qquad (10.6.1.1)$$

## 10.6.2 Symmetric Channel

A symmertic channel is defined as the one for which (i) $H(Y/x_j)$ is independent of $j$; i.e. the entropy corresponding to each row of $[P(Y/X)]$ is the same, and (ii) $\sum_{j=1}^{m} P(y_k/x_j)$ is independent of $k$, i.e., the sum of all the columns of $[P(Y/X)]$ is the same.

It can be seen that a channel is symmetric if the rows and columns of the channel matrix $D = P(Y/X)$ are independently identical; except for permutations. If $D$ is a square matrix, then, for a symmetric channel, the rows and columns are identical, except for permutations. The following examples will make the concept of symmetric channel clear:

(a) $P(Y/X) = \begin{bmatrix} \frac{1}{2} & \frac{1}{4} & \frac{1}{4} \\ \frac{1}{4} & \frac{1}{2} & \frac{1}{4} \\ \frac{1}{4} & \frac{1}{4} & \frac{1}{2} \end{bmatrix}$ This is a symmetric channel as the rows and columns are identical, except for permutations $\left(\text{each contains one } \frac{1}{2} \text{ and two } \frac{1}{4}\right)$.

(b) $P(Y/X) = \begin{bmatrix} \frac{1}{2} & \frac{1}{4} & \frac{1}{4} \\ \frac{1}{4} & \frac{1}{4} & \frac{1}{2} \\ \frac{1}{2} & \frac{1}{4} & \frac{1}{4} \end{bmatrix}$ This is not a symmetric channel, as although the rows are identical, except for permutations, the columns are not.

(c) $P(Y/X) = \begin{bmatrix} \frac{1}{3} & \frac{1}{6} & \frac{1}{3} & \frac{1}{6} \\ \frac{1}{6} & \frac{1}{3} & \frac{1}{6} & \frac{1}{3} \end{bmatrix}$ This is a symmetric channel, as each row contains two $1/3$ and two $1/6$, and each column contains one $1/3$ and one $1/6$.

(d) $P(Y/X) = \begin{bmatrix} \frac{1}{3} & \frac{1}{6} & \frac{1}{3} & \frac{1}{6} \\ \frac{1}{3} & \frac{1}{3} & \frac{1}{6} & \frac{1}{6} \end{bmatrix}$ This is not a symmetric channel, as, although the rows are identical except for permutations, the columns are not.

(e) $P(Y/X) = \begin{bmatrix} 0.4 & 0.6 \\ 0.3 & 0.7 \\ 0.6 & 0.4 \\ 0.7 & 0.3 \end{bmatrix}$ This is not a symmetric channel, as although the columns are identical except for permutations, the rows are not.

[Note that since the rows of the above matrices are complete probability schemes, the sum of each row in each matrix is unity.]
For a symmetric channel,

$$I(X;Y) = H(Y) - H(Y/X)$$

$$= H(Y) - \sum_{j=1}^{m} H(Y/x_j) \, p(x_j)$$

$$= H(Y) - A \sum_{j=1}^{m} p(x_j)$$

where $A = H(Y/x_j)$ is independent of $j$ and hence is taken out of the summation sign.
Also,

$$\sum_{j=1}^{m} p(x_j) = 1,$$

Hence,

$$I(X;Y) = H(Y) - A \tag{10.6.2.1}$$

The channel capacity of a symmetric channels is

$$\begin{aligned} C &= \max \, I(X;Y) \\ &= \max \, [H(Y) - A] \\ &= \max \, [H(Y)] - A \end{aligned}$$

or

$$C = \log n - A \tag{10.6.2.2.}$$

where $n$ is total number of receiver symbols.
Since

$$\max H(Y) = \log n$$

### 10.6.3  Binary Symmetric Channel (BSC)

The most important case of a symmetric channel is the Binary Symmetric Channel (BSC). In this case $m = n = 2$, and the channel matrix is $D = [P(Y/X)] = \begin{bmatrix} p & 1-p \\ 1-p & p \end{bmatrix} = \begin{bmatrix} p & q \\ q & p \end{bmatrix}$

BSC can be represented graphically as shown in Fig. 10.6.3.1

### Example 10.6.3.1

For a BSC shown in Fig. 10.6.3.1, find the channel capacity for (i) $p = 0.9$; (ii) $p = 0.6$.

$$\begin{aligned} C &= \log n - A \\ &= \log 2 - H(Y/x_j) \\ &= \log 2 - \left[ -\sum_{j=1}^{2} p(y_k/x_j) \log p(y_k/x_j) \right] \\ &= \log 2 + p \log p + (1-p) \log (1-p) \\ &= 1 - (p \log p + q \log q) \end{aligned}$$

**526** COMMUNICATION SYSTEMS: Analag and Digital

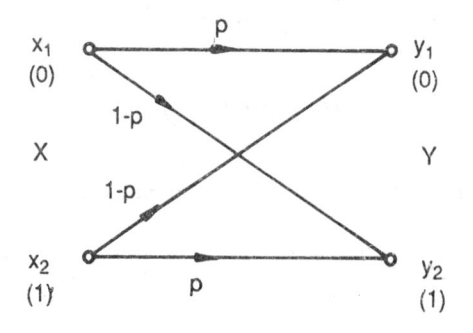

Fig. 10.6.3.1

$$= 1 - H(p)$$
$$= 1 - H(q)$$

(i) For $\quad p = 0.9,$
$$C = 1 + 0.9 \log 0.9 + 0.1 \log 0.1$$
$$= 0.531 \text{ bit/message}$$

(ii) For $\quad p = 0.6,$
$$C = 1 + 0.6 \log 0.6 + 0.4 \log 0.4$$
$$= 0.029 \text{ bit/message}$$

### 10.6.4 Cascaded Channels

Sometimes channels are to be cascaded for some reasons. Let us consider the case of two cascaded Binary Symmetric Channels as shown in Fig. 10.6.4.1. The analysis of these cascaded channels is as follows :

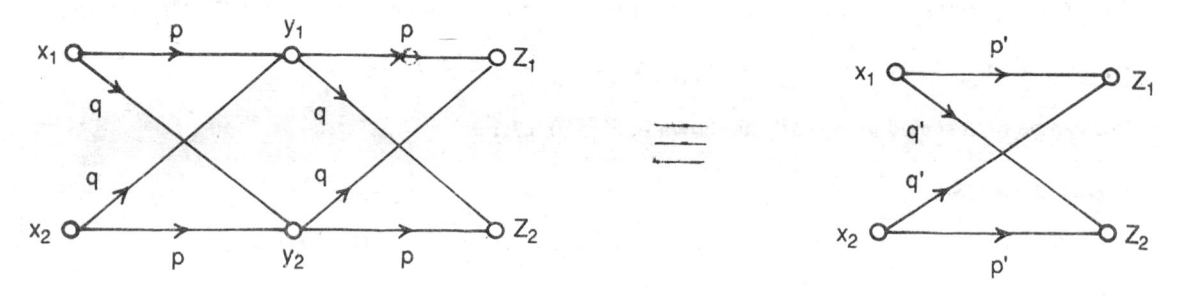

Fig. 10.6.4.1

The message from $X_1$, reaches $Z_1$ in two ways : $X_1 - y_1 - z_1$ and $x_1 - y_2 - z_1$. The respective path probabilities are $p.p$ and $q.q$.
Hence,
$$p' = p^2 + q^2 = (p + q)^2 - 2pq = 1 - 2pq$$

Similarly, the message from $x_1$ reaches $z_2$ in two ways : $x_1 - y_1 - z_2$ and $x_1 - y_2 - z_2$. The respective path probabilities are $p.q$ and $q.p$.

Hence,
$$q' = pq + qp = 2pq$$

[As a check, it can be seen that $p' + q' = (1 - 2pq) + (2pq) = 1$]

Thus, the channel matrix of the cascaded channel is

$$P(Z/X) = \begin{bmatrix} 1-2pq & 2pq \\ 2pq & 1-2pq \end{bmatrix} = \begin{bmatrix} p' & q' \\ q' & p' \end{bmatrix} \quad (10.6.4.1)$$

Thus, the cascaded channel is equivalent to a single Binary Symmetric Channel with error probability = $2pq$. We know that the channel capacity of a BSC is given by

$$C = 1 - H(q)$$

Hence, the channel capacity of cascaded channel is

$$C = 1 - H(q^1) = 1 - H(2pq) \quad (10.6.4.2)$$

For $0.5 > q > o$, $2pq$ is greater than $q$. Hence, the channel capacity of two cascaded BSC's is less than a single BSC, as expected.

### 10.6.5 Binary Erasure Channel (BEC)

A Binary Erasure Channel (BEC) has two inputs (0,1) and three outputs (0,y,1) as shown in Fig.10.6.5.1. BEC is also very important. Here, 0 and 1 are transmitted, and they are received as 0, y and 1. The symbol y indicates that, due to noise, no deterministic decision can be made as to whether the received symbol is a 0 or a 1. In other words, the symbol y indicates that the output is erased. Hence, the name Binary Erasure Channel. In practice, whenever decision is in favour of y, i.e., whenever deterministic decision in favour of 0 or 1 is not possible, the receiver requests the transmitter for re-transmission till the decision is taken either in favour of 0 or in favour of 1.

For BEC, the channel matrix is

$$D = [P(Y/X)] = \begin{bmatrix} p & q & 0 \\ 0 & q & p \end{bmatrix}$$

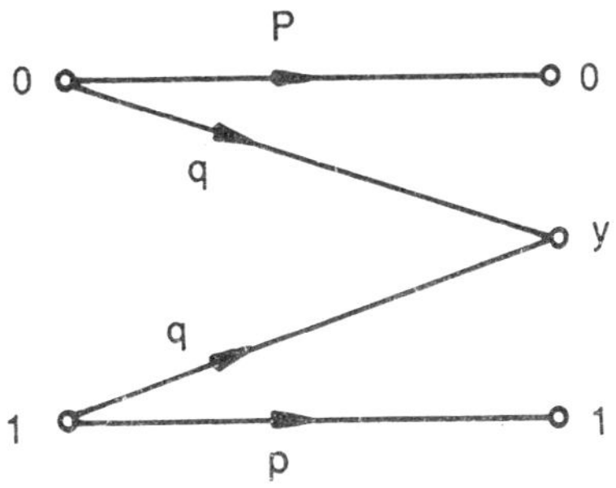

Fig. 10.6.5.1

Let us assume that $p(0) = \alpha$ and $p(1) = 1 - \alpha$ at the transmitter. Hence;

$$H(X) = \alpha \log \frac{1}{\alpha} + (1-\alpha) \log \frac{1}{(1-\alpha)}$$

Now, since $p(x_1) = p(0) = \alpha$ and $p(x_2) = p(1) = (1 - \alpha)$, the joint probability matrix $P(X,Y)$ can be found by multiplying the rows of $p(Y/X)$ by $\alpha$, and $(1 - \alpha)$, respectively.

Hence,

$$[P(X, Y)] = \begin{bmatrix} \alpha p & \alpha q & 0 \\ 0 & (1-\alpha)q & (1-\alpha)p \end{bmatrix}$$

The summation of the columns give

$$p(y_1) = \alpha p, \; p(y_2) = \alpha q + (1-\alpha)q = q, \; p(y_3) = (1-\alpha)p$$

The conditional probability matrix $P(X/Y)$ can be found by dividing the columns of $P(X,Y)$ by $p(y_1)$, $p(y_2)$, and $p(y_3)$, respectively. Thus,

$$P(X/Y) = \begin{bmatrix} \dfrac{\alpha p}{\alpha p} & \dfrac{\alpha q}{q} & 0 \\ \dfrac{0}{\alpha p} & \dfrac{(1-\alpha)q}{q} & \dfrac{(1-\alpha)p}{(1-\alpha)p} \\ & & \end{bmatrix}$$

$$= \begin{bmatrix} 1 & \alpha & 0 \\ 0 & 1-\alpha & 1 \end{bmatrix}$$

Now,

$$H(X/Y) = -\sum_{j=1}^{2}\sum_{k=1}^{3} p(x_j, y_k) \log p(x_j/y_k)$$

$$= -[\alpha p \log 1 + \alpha q \log \alpha + (1-\alpha)q \log(1-\alpha) + (1-\alpha)p \log 1]$$

$$= -q[\alpha \log \alpha + (1-\alpha)\log(1-\alpha)]$$

$$= qH(X)$$

$$= (1-p)H(X) \qquad \text{since } q = 1-p$$

In this case, the mutual information and channel capacity are

$$I(X, Y) = H(X) - H(X/Y)$$
$$= H(X) - (1-p)H(X)$$
$$= pH(X)$$

and

$$C = \max I(X; Y)$$
$$= \max [pH(X)]$$
$$= p \max [H(X)]$$
$$= p \qquad \text{since max. } [H(X)] = 1$$

### 10.6.6 Repetition of Signals

The concept of BEC is used in the following way:
The transmitted signal is repeated at the channel input to increase the channel efficiency as shown in Fig. 10.6.6.1. In this case, the acceptable output signals are $y_1 = 00$ and $y_2 = 11$. The outputs $y_3 = 01$ and $y_4 = 10$ are discarded (erased) and request for re-transmission is made as in BEC.

The channel matrix is

$$P(Y/X) = \begin{array}{c} \\ x_1 \\ x_2 \end{array} \begin{array}{c} y_1 \quad y_2 \quad y_3 \quad y_4 \\ \begin{bmatrix} p^2 & q^2 & pq & pq \\ q^2 & p^2 & pq & pq \end{bmatrix} \end{array}$$

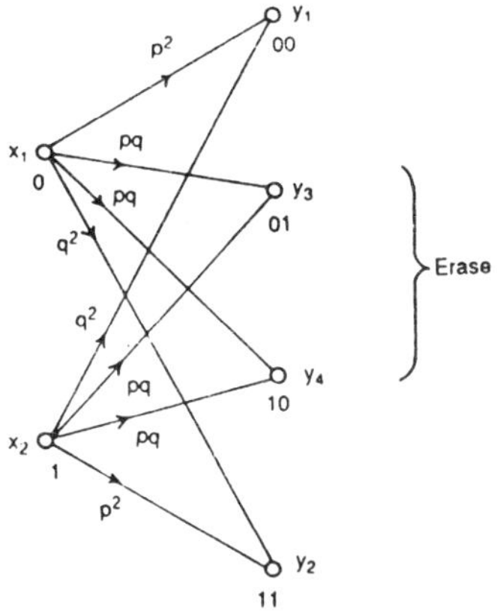

Fig. 10.6.6.1

Let us assume $p(x_1) = p(x_2) = 0.5$
Therefore,

$$P(Y, X) = \begin{array}{c} \\ x_1 \\ x_2 \end{array} \begin{array}{c} y_1 \qquad y_2 \qquad y_3 \qquad y_4 \\ \begin{bmatrix} p^2/2 & q^2/2 & pq/2 & pq/2 \\ q^2/2 & p^2/2 & pq/2 & pq/2 \end{bmatrix} \end{array}$$

Hence,

$$p(y_1) = \frac{p^2 + q^2}{2} = p(y_2)$$

and

$$p(y_3) = pq = p(y_4)$$

Therefore,

$$H(Y) = p(y_1) \log \frac{1}{p(y_1)} + p(y_2) \log \frac{1}{p(y_2)}$$

$$+ p(y_3) \log \frac{1}{p(y_3)} + p(y_4) \log \frac{1}{p(y_4)}$$

$$= (p^2 + q^2) \log \left( \frac{2}{p^2 + q^2} \right) + 2 pq \log \left( \frac{1}{pq} \right) \qquad (10.6.6.1)$$

$$H(Y/X) = p^2 \log \left( \frac{1}{p^2} \right) + q^2 \log \left( \frac{1}{q^2} \right) + pq \log \left( \frac{1}{pq} \right) + pq \log \left( \frac{1}{pq} \right) \qquad (10.6.6.2)$$

$$I(X;Y) = H(Y) - H(Y/X)$$

$$= (p^2 + q^2) \log \left( \frac{2}{p^2 + q^2} \right) + 2 pq \log \left( \frac{1}{pq} \right) - \left[ p^2 \log \left( \frac{1}{p^2} \right) + q^2 \log \left( \frac{1}{q^2} \right) + 2pq \log \left( \frac{1}{pq} \right) \right]$$

$$= (p^2 + q^2) \log \left( \frac{2}{p^2 + q^2} \right) + p^2 \log p^2 + q^2 \log q^2$$

**530** COMMUNICATION SYSTEMS: Analog and Digital

$$= (p^2 + q^2)\left[1 + \log\left(\frac{1}{p^2+q^2}\right)\right] + p^2 \log p^2 + q^2 \log q^2$$

$$= (p^2 + q^2)\left[1 + \left(\frac{p^2+q^2}{p^2+q^2}\right)\log\left(\frac{1}{p^2+q^2}\right) + \left(\frac{p^2}{p^2+q^2}\right)\log p^2 + \left(\frac{q^2}{p^2+q^2}\right)\log q^2\right]$$

$$= (p^2 + q^2)\left[1 + \left(\frac{p^2}{p^2+q^2}\right)\log\left(\frac{1}{p^2+q^2}\right) + \left(\frac{q^2}{p^2+q^2}\right)\log\left(\frac{1}{p^2+q^2}\right)\right.$$
$$\left. + \left(\frac{p^2}{p^2+q^2}\right)\log p^2 + \left(\frac{q^2}{p^2+q^2}\right)\log q^2\right]$$

$$= (p^2 + q^2)\left[1 + \left(\frac{p^2}{p^2+q^2}\right)\log\left(\frac{p^2}{p^2+q^2}\right) + \left(\frac{q^2}{p^2+q^2}\right)\log\left(\frac{q^2}{p^2+q^2}\right)\right]$$

$$= (p^2 + q^2)\left[1 - H\left(\frac{q^2}{p^2+q^2}\right)\right] \tag{10..6.6.3}$$

Thus, the channel is now equivalent to a BSC with error probability $q' = \frac{q^2}{p^2+q^2}$. Since $q' < q$, the mutual information $I(X,Y)$ is greater than the original value (i.e. when there is no repetition of signals) $1 - H(q)$.

### 10.6.7 Binary Channel

Although it is easy to analyse a BSC, in practice, we come across binary channels with non-symmetric structures. A binary channel is shown in Fig. 10.6.7.1. The channel matrix is

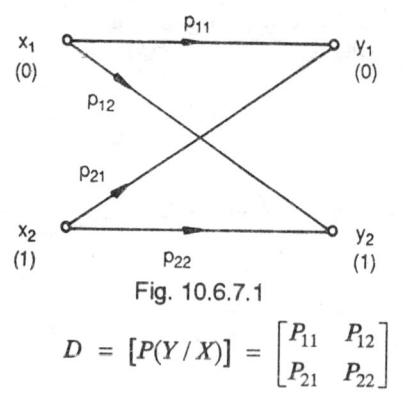

Fig. 10.6.7.1

$$D = [P(Y/X)] = \begin{bmatrix} P_{11} & P_{12} \\ P_{21} & P_{22} \end{bmatrix}$$

To find the channel capacity of a binary channel, the auxiliary variables, $Q_1$, and $Q_2$, are defined by

$$[P][Q] = -[H]$$

or

$$\begin{bmatrix} P_{11} & P_{12} \\ P_{21} & P_{22} \end{bmatrix}\begin{bmatrix} Q_1 \\ Q_2 \end{bmatrix} = \begin{bmatrix} P_{11}\log P_{11} + P_{12}\log P_{12} \\ P_{21}\log P_{21} + P_{22}\log P_{22} \end{bmatrix} \tag{10.6.7.1}$$

The channel capacity is then given by

$$C = \log(2^{Q_1} + 2^{Q_2}) \qquad (10.6.7.2)$$

[The proof of Eq. 10.6.7.2 is too complicated to be given in this book.]

### Example 10.6.7.1

Find the mutual information and channel capacity of the channel shown in Fig. 10.6.7.2. Given $p(x_1) = 0.6$, $p(x_2) = 0.4$

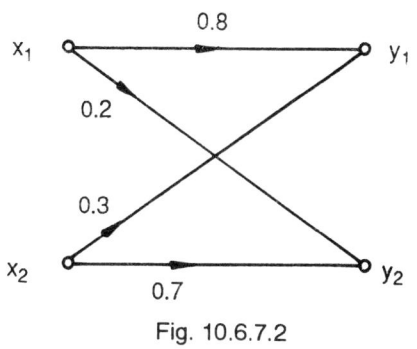

Fig. 10.6.7.2

*Solution*

The channel matrix is
$$D = [P(Y/X)] = \begin{bmatrix} 0.8 & 0.2 \\ 0.3 & 0.7 \end{bmatrix}$$

The joint probability matrix is obtained by multiplying the rows of $P(Y/X)$ by $p(x_1)$ and $P(x_2)$ respectively.

$$[P(X, Y)] = \begin{bmatrix} 0.8 \times 0.6 & 0.2 \times 0.6 \\ 0.3 \times 0.4 & 0.7 \times 0.4 \end{bmatrix} = \begin{bmatrix} 0.48 & 0.12 \\ 0.12 & 0.28 \end{bmatrix}$$

$p(y_1)$ and $p(y_2)$ are obtained by summing the columns of $[P(X,Y)]$

$$P(y_1) = 0.48 + 0.12 = 0.6$$
$$P(y_2) = 0.12 + 0.28 = 0.4$$

The matrix $[P(X/Y)]$ is obtained by dividing the columns of $\{P(X,Y)\}$ by $P(y_1)$, and $p(y_2)$, respectively

$$[P(X/Y)] = \begin{bmatrix} 0.48/0.6 & 0.12/0.4 \\ 0.12/0.6 & 0.28/0.4 \end{bmatrix} = \begin{bmatrix} 0.8 & 0.3 \\ 0.2 & 0.7 \end{bmatrix}$$

$H(X)$ and $H(X/Y)$ can be found as

$$H(X) = -\sum_{j=1}^{2} p(x_j) \log p(x_j)$$

$$= -[0.6 \log 0.6 + 0.4 \log 0.4]$$

$$= 0.971 \text{ bit/message}$$

$$H(X/Y) = -\sum_{j=1}^{2}\sum_{k=1}^{2} p(x_j, y_k) \log p(x_j/y_k)$$

$$= -[0.48 \log 0.8 + 0.12 \log 0.3 + 0.12 \log 0.2 + 0.28 \log 0.7]$$

$$= 0.786 \text{ bit/message}$$

Hence,

Now,
$$I(X;Y) = H(X) - H(X/Y) = 0.971 - 0.786 = 0.185 \text{ bit/message}$$

$$P_{11} = 0.8; P_{12} = 0.2; P_{21} = 0.3 \text{ and } P_{22} = 0.7$$

Hence Eq. 10.6.7.1 becomes

$$\begin{bmatrix} 0.8 & 0.2 \\ 0.3 & 0.7 \end{bmatrix} \begin{bmatrix} Q_1 \\ Q_2 \end{bmatrix} = \begin{bmatrix} 0.8 \log 0.8 + 0.2 \log 0.2 \\ 0.3 \log 0.3 + 0.7 \log 0.7 \end{bmatrix}$$

Solving for $Q_1$ and $Q_2$ yields :

$$Q_1 = -0.6568 \text{ and } Q_2 = -0.9764$$

Hence Eq. 10.6.7.2 becomes

$$C = \log(2^{-0.6568} + 2^{-0.9764})$$

$$= \log(0.633 + 0.513)$$

$$= \log 1.146$$

$$= 0.2 \text{ bit/message}$$

## 10.7 SHANNON'S THEOREM

This theorem is concerned with the rate of information transmission over a communication channel. The term communication channel covers all the features and component parts of the transmission system which introduce noise, or limit the bandwidth. Shannon's theorem says that it is possible, in principle, to device a means whereby a communication system will transmit information with an arbitrarily small probability of error provided that the information rate $R$ is less than or equal to a rate $C$, the channel capacity. The statement of Shannon's theorem is:

Given a source of $M$ equally likely messages, with $M \gg 1$, which is generating information at a rate $R$. Given a channel with a channel capacity $C$. Then, if $R \leq C$, there exists a coding technique such that the output of the source may be transmitted over the channel with a probability of error of receiving the message which may be made arbitrarily small.

The important feature of the theorem is that it indicates that for $R \leq C$, error free transmission is possible in the presence of noise.

There is a negative statement associated with Shannon's theorem. It is: Given a source of $M$ equally likely messages, with $M \gg 1$, which is generating information at a rate $R$. Given a channel with a channel capacity $C$. Then, if $R > C$, the probability of error is close to unity for every possible set of $M$ transmitter signals.

The negative theorem shows that, if the information rate $R$ exceeds a specific value $C$, the error probability approaches unity, as $M$ increases.

Thus, the channel capacity $C$ of a communication channel is its very important characteristic. It decides the maximum permissible rate at which an errorfree transmission of information is possible through it.

## 10.8 CONTINUOUS CHANNEL

Till now we have confined our discussion only to discrete channels. A number of communication systems use continuous sources and, thus, use the channel continuously. AM, FM, PM are examples of systems using continuous channel.

The information theory concept of discrete channels can be extended to continuous channels. The definitions of different entropies in the discrete case were based on the concept of different averages. In a similar way, we may define different entropies in case of continuous distributions. If $p(x)$ is the probability density function associated with the signal $x(t)$, then the entropy of the source is

$$H(x) = E[-\log p(x)] = -\int_{-\infty}^{\infty} p(x) \log p(x)\, dx \qquad (10.8.1)$$

In a similar way, the different entropies associated with two-dimensional random variables with a joint density $p(x,y)$ and marginal densities $p_1(x)$ and $p_2(y)$ may be defined as

$$H(x) = E[-\log p_1(x)] = -\int_{-\infty}^{\infty} p_1(x) \log p_1(x)\, dx \qquad (10.8.2)$$

$$H(y) = E[-\log p_2(y)] = -\int_{-\infty}^{\infty} p_2(y) \log p_2(y)\, dy \qquad (10.8.3)$$

$$H(x, y) = E[-\log p(x, y)] = -\int_{-\infty}^{\infty}\int_{-\infty}^{\infty} p(x, y) \log p(x, y)\, dx\, dy \qquad (10.8.4)$$

$$H(x/y) = E[-\log p(x/y)] = -\int_{-\infty}^{\infty}\int_{-\infty}^{\infty} p(x, y) \log p(x/y)\, dx\, dy \qquad (10.8.5)$$

$$H(y/x) = E[-\log p(y/x)] = -\int_{-\infty}^{\infty}\int_{-\infty}^{\infty} p(x, y) \log p(y/x)\, dx\, dy \qquad (10.8.6)$$

All the definitions here are contingent upon the existence of the corresponding integrals.

In the discrete case, all the entropies involved are positive quantities because the probability of occurrence of an event is always positive and less than 1. In the continuous case, however,

$$\int_{-\infty}^{\infty} p(x)\, dx = 1 \qquad (10.8.7)$$

and

$$\int_{-\infty}^{\infty}\int_{-\infty}^{\infty} p(x,y)\, dx\, dy = 1 \qquad (10.8.8)$$

Hence, the density functions $p(x)$ and $p(x,y)$ need not be less than 1 for all values of the random variable. This may lead to a negative entropy. For this reason, the concept of self-information can no longer be associated with $H(x)$, as in a discrete case. We call $H(x)$ as the entropy function, but $H(x)$ no longer indicates the average self-information of the source. The same is true for the other entropies. Thus, it follows that the individual entropies may be negative. But it does not mean that the concept of entropy is useless for the continuous system. Because, it is transinformation $I(x,y)$ and not the entropies such as $H(x)$, $H(y)$, $H(x,y)$, $H(x/y)$ and $H(y/x)$ that have direct interpretation as far as the information processed through the channel is concerned. It will be shown that $I(x;y)$ of a continuous system is non-negative. For a continuous case, $I(x;y)$ is defined as

$$I(x;y) = \int_{-\infty}^{\infty} \int_{-\infty}^{\infty} p(x,y) \log \frac{p(x,y)}{p_1(x) p_2(y)} dx\, dy \qquad (10.8.9)$$

Note that, as before, we have

$$I(x;y) = H(x) + H(y) - H(x,y) = H(x) - H(x/y) = H(y) - H(y/x)$$

Equation 10.8.9 can be written as

$$I(x;y) = -\int_{-\infty}^{\infty} \int_{-\infty}^{\infty} p(x,y) \log \frac{p_1(x)}{p(x/y)} dx\, dy$$

Hence,

$$I(x;y) \geq -\int_{-\infty}^{\infty} \int_{-\infty}^{\infty} p(x,y) \left[ \frac{p_1(x)}{p(x/y)} - 1 \right] \log e\, dx\, dy$$

or

$$I(x;y) \geq \left[ -\int_{-\infty}^{\infty} \int_{-\infty}^{\infty} p_2(y) p_1(x) \log e\, dx\, dy + \int_{-\infty}^{\infty} \int_{-\infty}^{\infty} p(x,y) \log e\, dx\, dy \right]$$

Now,

$$\int_{-\infty}^{\infty} \int_{-\infty}^{\infty} p(x,y)\, dx\, dy = 1$$

and

$$\int_{-\infty}^{\infty} \int_{-\infty}^{\infty} p_2(y) p_1(x)\, dx\, dy$$

$$= \int_{-\infty}^{\infty} p_1(x) \left[ \int_{-\infty}^{\infty} p_2(y)\, dy \right] dx$$

$$= \int_{-\infty}^{\infty} p_1(x) [1]\, dx$$

$$= 1$$

Hence,
$$I(x;y) \geq \log e - \log e$$
or
$$I(x;y) \geq 0$$

Hence, the transinformation of a continuous system is non-negative.

It may be recalled that transinformation is the difference between *a-priori* and *a-posteriori* expectations of the systems; and, although these expectations can be negative individually, their magnitudes are such that their difference is never negative.

### Example 10.8.1

A random variable has a density function as shown in Fig. 10.8.1. Find the corresponding entropy.

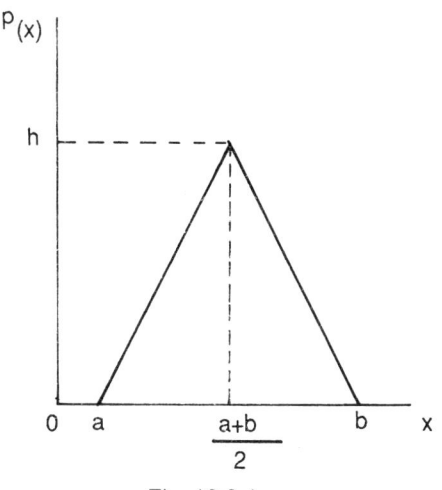

Fig. 10.8.1

*Solution*

$$p(x) = \frac{2h}{b-a}(x-a) \qquad a \leq x \leq \frac{a+b}{2}$$

$$p(x) = \frac{2h}{b-a}(b-x) \qquad \frac{a+b}{2} \leq x \leq b$$

Hence,

$$H(x) = -\int_{a}^{\frac{a+b}{2}} \frac{2h}{(b-a)}(x-a) \ln \frac{2h}{(b-a)}(x-a)\,dx$$

$$-\int_{\frac{a+b}{2}}^{b} \frac{2h}{(b-a)}(b-x) \ln \frac{2h}{(b-a)}(b-x)\,dx$$

This can be solved by using the following equation:

$$\int x \ln(\lambda x)\,dx = \frac{x^2}{2}\ln(\lambda x) - \frac{x^2}{4}$$

Thus,

$$H(x) = -\frac{2h}{(b-a)}\left[\frac{(x-a)^2}{2}\ln\left\{\frac{2h}{(b-a)}(x-a)\right\} - \frac{(x-a)^2}{4}\right]_{a}^{\frac{a+b}{2}}$$

$$-\frac{2h}{(b-a)}\left[-\left\{\frac{(b-x)^2}{2}\ln\left\{\frac{2h}{(b-a)}(b-x)\right\} - \frac{(b-x)^2}{4}\right\}\right]_{\frac{a+b}{2}}^{b}$$

$$= \frac{h(b-a)}{2}\left[-\ln h + \frac{1}{2}\right]$$

Now, since

$$\frac{h(b-a)}{2} = \int_{-\infty}^{\infty} p(x)\,dx = 1$$

Hence,

$$H(x) = -\ln h + \frac{1}{2}$$

It may be seen that

$$H(x) > 0 \quad \text{for } h < \sqrt{e}$$
$$H(x) = 0 \quad \text{for } h = \sqrt{e}$$
$$H(x) < 0 \quad \text{for } h > \sqrt{e}$$

Thus, we see that for $h > \sqrt{e}$, the entropy is negative.

## 10.9 CAPACITY OF A GAUSSIAN CHANNEL: SHANNON-HARTLEY THEOREM

The noise characteristics of channels encountered in practice is generally Gaussian. (Channels with Gaussian noise characteristics are known as Gaussian channels.) Moreover, the results obtained for a Gaussian channel often provide a lower bound on the performance of a system with the non-Gaussian channel. Thus, if a particular encoder-decoder is used with a Gaussian channel, giving an error probability $Pe$, then, with a non-Gaussian channel, another encoder-decoder can be designed for which the error probability will be less than $Pe$. Hence, the study of a Gaussian channel is very important.

For a Gaussian channel,

$$p(x) = \frac{1}{\sqrt{2\pi\sigma^2}} e^{-x^2/2\sigma^2} \tag{10.9.1}$$

Hence,

$$H(x) = -\int_{-\infty}^{\infty} p(x) \log p(x)\,dx$$

But
$$-\log p(x) = \log \sqrt{2\pi\sigma^2} + \log e^{x^2/2\sigma^2}$$
Hence,
$$H(x) = \int_{-\infty}^{\infty} p(x) \log \sqrt{2\pi\sigma^2}\, dx + \int_{-\infty}^{\infty} p(x) \log e^{x^2/2\sigma^2}$$

This may be evaluated to yield
$$H(x) = \log \sqrt{2\pi\, e\sigma^2} \text{ bits/message} \quad (10.9.2)$$

Now, if the signal is bandlimited to $\omega$ Hz, then it may be uniquely specified by taking $2\omega$ samples per second (sampling theorem). Hence the rate of information transmission is
$$R(x) = 2\omega\, H(x)$$
$$= 2\omega \log\left(\sqrt{2\pi\, e\sigma^2}\right)$$
$$= \omega \log\left(\sqrt{2\pi\, e\sigma^2}\right)^2$$

or
$$R(x) = \omega \log(2\pi\, e\sigma^2) \quad (10.9.3)$$

If $p(x)$ is a bandlimited Gaussian noise with an average noise power $N$, then we have
$$R(n) = R(x) = \omega \log(2\pi\, eN) \quad (\because \sigma^2 = N) \quad (10.9.4)$$

Now, consider a continuous source transmitting information over a noisy channel. If the received signal is composed of a transmitted signal $x$ and a noise $n$, then the joint entropy (in bit/s basis) of the source and noise is given by
$$R(x, n) = R(x) + R(n/x)$$

Assuming that the transmitted signal and noise are independent (as is the practical situation),
$$R(x, n) = R(x) + R(n) \quad (10.9.5)$$

Since the received signal $y$ is the sum of the transmitted signal $x$ and the noise $n$, we may equate
$$H(x, y) = H(x, n)$$
or
$$H(y) + H(x/y) = H(x) + H(n)$$
or
$$R(y) + R(x/y) = R(x) + R(n) \quad (10.9.6)$$

The rate at which the information is received from a noisy channel is
$$R = R(x) - R(x/y)$$

By using Eq. 10.9.6 we get,
$$R = R(y) - R(n) \text{ bits/s} \quad (10.9.7)$$

The channel capacity in bit/sec is
$$C = \text{Max}\,[(R)] \text{ bits/s}$$

or
$$C = \text{Max}\,[R(y) - R(n)]\,\text{bits/s} \qquad (10.9.8)$$

Since $R(n)$ is assumed to be independent of $x(t)$, maximizing $R$ requires maximizing $R(y)$. Let a transmitted signal be limited to an average power $S$ and the noise on the channel be white Gaussian with an average power $N$ within the bandwidth $\omega$ of the channel. The received signal will now have an average power $(S+N)$. $R(y)$ is maximum when $y(t)$ is also a Gaussian random process because noise is assumed to be Gaussian. Thus, the entropy from Eq. 10.9.4 on a per-second basis is

$$R(y) = \omega \log\,[2\pi\,e(S+N)]\,\text{bits/s}$$

While the entropy of the noise is given by

$$R(n) = \omega \log\,[2\pi\,eN]\,\text{bits/s}$$

The channel capacity may now be obtained directly since $R(y)$ has been maximized. Thus,

$$\begin{aligned}C &= \max[R(y) - R(n)] \\ &= \omega \log[2\pi\,e[S+N]] - \omega \log[2\pi\,eN] \\ &= \omega \log\left[\frac{S+N}{N}\right]\end{aligned}$$

or

$$C = \omega \log\left[1 + \frac{S}{N}\right]\,\text{bits/s} \qquad (10.9.9)$$

Equation 10.9.9 is the famous *Shannon-Hartley theorem*, which is complementary to Shannon's theorem and applies to a Gaussian noise channel. The statement of the Shannon-Hartley theorem is given below:

The channel capacity of a white-bandlimited Gaussian channel is

$$C = \omega \log\left[1 + \frac{S}{N}\right]\,\text{bits/s}$$

where $\omega$ is the channel bandwidth, $S$ is the average signal power, and $N$ is the average noise power.

If $\eta/2$ is the two-sided power spectral density of noise in watts/Hz, then

$$N = \eta\omega$$

and
$$C = \omega \log\left[1 + \frac{S}{\eta\omega}\right]\,\text{bits/s} \qquad (10.9.10)$$

### Example 10.9.1

Calculate the bandwidth of the picture (video) signal in a television. The following are the available data:
(i) number of distinguishable brightness levels = 10;
(ii) the number of elements per picture frame = 300,000;
(iii) picture frames transmitted per second = 30; and
(iv) S/N required = 30 dB.

*Solution*

The information per picture element = $\log 10 = 3.32$ bits/element. The information per picture frame $300,000 \times 3.32 = 996,000$ bits/picture frame. As 30 picture frames are transmitted per second, the information rate is

$$R = 996,000 \times 30 = 29.9 \times 10^6 \,\text{bits/s}$$

To transmit information at this rate, the channel capacity $C$ must be such that
$$C \geq R$$
Now,
$$C = \omega \log\left(1 + \frac{S}{N}\right)$$
$$= \omega \log(1 + 1000)$$
$$\left(\because \frac{S}{N} = 30 \text{ dB} = 1000\right)$$

Hence, we must have
$$\omega \log(1001) \geq 29.9 \times 10^6$$
or
$$\omega \geq 3.02 \times 10^6 \text{ Hz}$$

Thus, the minimum bandwidth required to transmit the picture (video) signal in a Television is approximately 3 MHz.

## 10.10 BANDWIDTH*N S/N* TRADE-OFF

Equation 10.9.9 shows that a noiseless channel $(S/N = \infty)$ has an infinite capacity. On the other hand, the channel capacity does not become infinite as the bandwidth approaches infinity because, with an increase in bandwidth, the noise power also increases. Thus, for a fixed signal power, and in the presence of white Gaussian noise, the channel capacity approaches an upper limit with increasing bandwidth. This is explained below:

We have,
$$C = \omega \log\left(1 + \frac{S}{N}\right)$$
$$= \omega \log\left(1 + \frac{S}{\eta\omega}\right)$$
$$= \frac{S}{\eta} \frac{\eta\omega}{S} \log\left(1 + \frac{S}{\eta\omega}\right)$$
$$= \frac{S}{\eta} \log\left(1 + \frac{S}{\eta\omega}\right)^{\eta\omega/s}$$

Now,
$$\lim_{x \to 0} (1 + x)^{1/x} = e$$

If $x = \dfrac{S}{\eta\omega}$, then as $\omega$ approaches infinity, $x$ approaches zero

Hence,
$$\lim_{\omega \to \infty} \left(1 + \frac{S}{\eta\omega}\right)^{\frac{\eta\omega}{s}} = e$$

Hence,
$$\lim_{\omega \to \infty} C = \frac{S}{\eta} \log e = 1.44 \frac{S}{\eta} = R \max \qquad (10.10.1)$$

Let us now consider the trade-off between the bandwidth and $S/N$ ratio:
Let
$$\frac{S}{N} = 15 \text{ and } \omega = 5 \text{ kHz}$$
Then,
$$C = \omega \log\left(1 + \frac{S}{N}\right)$$
$$= 5 \log(1 + 15)$$
$$= 20 \text{ k bits/s}$$

If the $S/N$ ratio is increased to 31, the bandwidth for the same channel capacity can be found from
$$C = 20 = \omega \log(1 + 31)$$
or
$$\omega = \frac{20}{\log 32} = \frac{20}{5} = 4 \text{ kHz}$$

With a 4 kHz bandwidth, the noise power will be $\frac{4}{5}$ times the noise power at 5 kHz. Thus, the signal power will have to be increased by a factor $\frac{4}{5} \times \frac{31}{15} = 1.65$. In other words, a 20% reduction in the bandwidth (5 kHz to 4 kHz) requires a 65% increase in the signal power. Thus, to decrease the bandwidth, the signal power has to be increased. Similarly, it can be shown that to decrease the signal power, the bandwidth must be increased.

### Example 10.10.1

A Gaussian channel has 1 MHz bandwidth. Calculate the channel capacity if the signal power to noise spectral density ratio $(S/\eta)$ is $10^5$ Hz. Also find the maximum information rate.

*Solution*
$$C = \omega \log\left(1 + \frac{S}{N}\right)$$
$$= \omega \log\left(1 + \frac{S}{\eta \omega}\right)$$
$$= 10^6 \log\left(1 + \frac{10^5}{10^6}\right)$$
$$= 13800 \text{ bits/s}$$

The maximum information rate is given by Eq. 10.10.1.

Hence,
$$R\max = 1.44 \frac{S}{\eta}$$
$$= 1.44 \times 10^5$$
$$= 144000 \text{ bits/s}$$

## PROBLEMS

10.1 What is entropy? Show that the entropy is maximum when all the messages are equi-probable. Assume $M = 3$.

10.2 The voice frequency modulating signal of a PCM system is quantized in 16 levels with following probabilities.

$p_1 = p_2 = p_3 = p_4 = 0.1, p_5 = p_6 = p_7 = p_8 = 0.05,$
$p_9 = p_{10} = p_{11} = p_{12} = 0.075$ and $p_{13} = p_{14} = p_{15} = p_{16} = 0.025$

Calculate the entropy and information rate. Assume $f_m = 3$ kHz.

10.3 In a message conveyed through a long sequence of dots and dashes, the probability of occurrence of a dash is one third of that of a dot. The duration of a dash is three times that of a dot. If the dot lasts for 10 m sec and the same time is allowed between symbols, determine the following :

(a) the information in dot and dash,

(b) average information in the dot-dash code,

(c) average information rate.

10.4 Find the transferred information for the channels shown in Fig. Prob. 10.4.

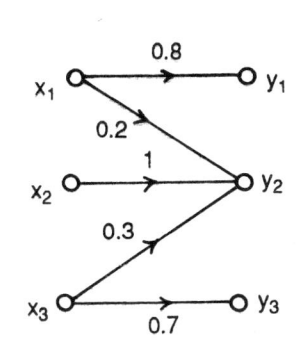

$P(x_1) = 0.2, P(x_2) = 0.5, P(x_3) = 0.3$

(a)

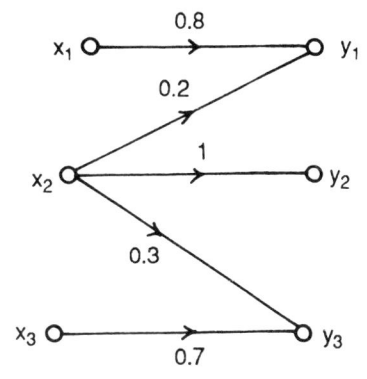

$P(y_1) = 0.2, P(y_2) = 0.5, P(y_3) = 0.3$

(b)

Fig. Prob. 10.4

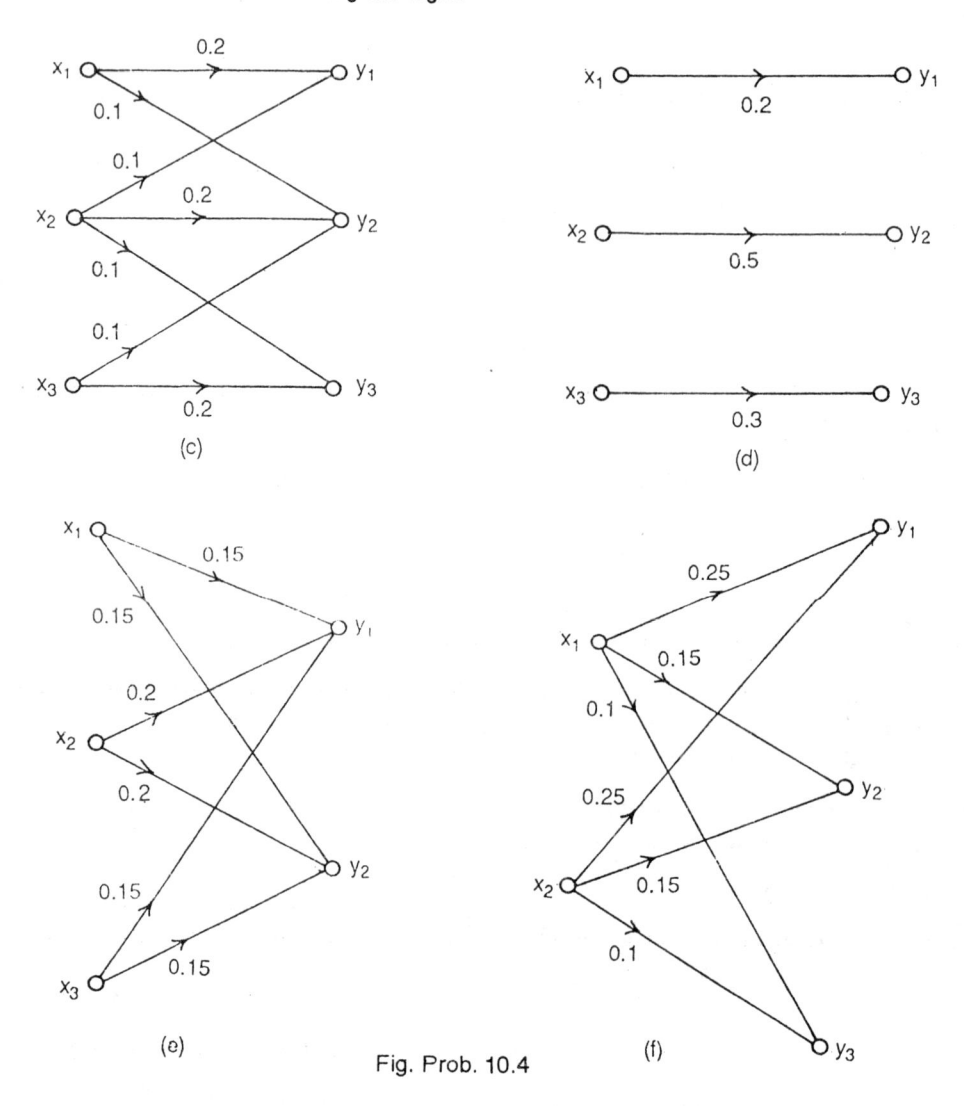

**10.5** Find the channel capacity for the channels shown in Fig. Prob. 10.5.

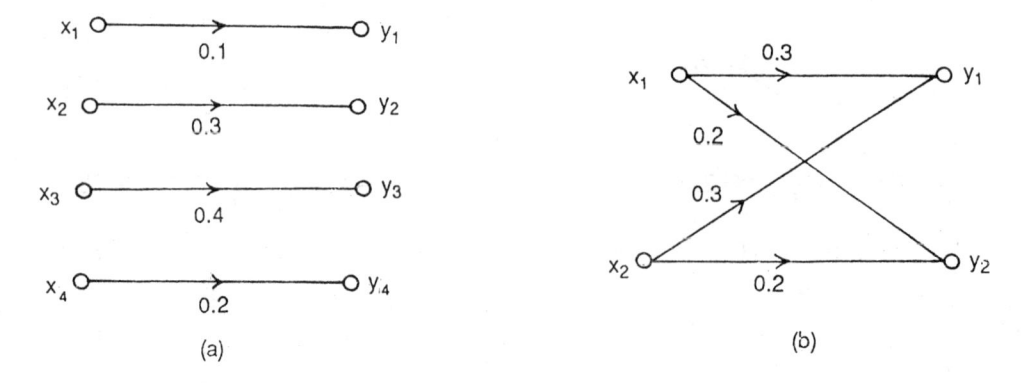

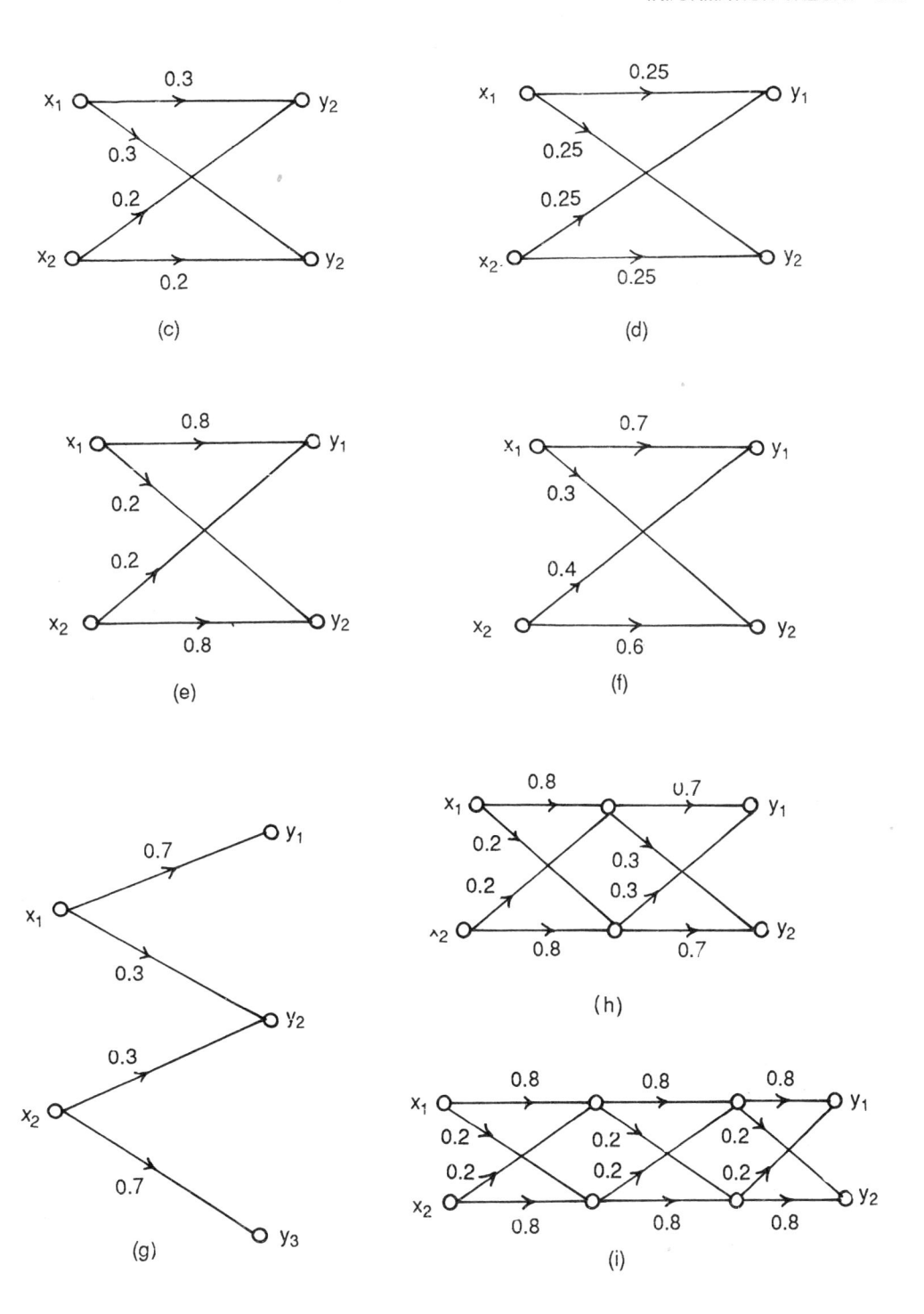

Fig. Prob. 10.5

10.6 Find the entropy corresponding to the random variabe whose density function is shown in Fig. Prob.10.6.

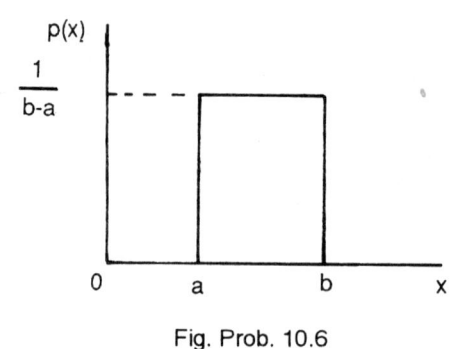

Fig. Prob. 10.6

10.7 In a fascimile transmission of a picture, there are about $2.25 \times 10^6$ picture elements per frame. For good reproduction, 12 brightness levels are necessary. Assume that all these levels are equi-probable. Calculate the channel bandwidth required to transmit 1 picture every 3 minutes. Assume a signal to noise power ratio of 30 dB.

10.8 A communication system has $\frac{S}{N} = 31$ and BW = 3 kHz. Find the allowable percentage reduction in signal power if BW is increased to 4 kHz.

# ELEVEN
# CODING

## INTRODUCTION

Coding offers the most significant appliction of the information theory. The main purpose of coding is to improve the efficiency of the communication system in some sense.

Coding is a procedure for mapping a given set of messages $[m_1 \, m_2 ... m_N]$ into a new set of encoded messages $[C_1 \, C_2 ... C_N]$ in such a way that the transformation is one-to-one; i.e. for each message, there is only one encoded message. This is source coding. Moreover, generally, by coding one seeks to improve the efficiency of transmission. It is, possible to device codes for a special purposes (such as secrecy, or minimum probability of error) without relevance to the efficiency of transmission. This is channel coding. It is also possible to resort to codes that do not have a one-to-one association. However, in this chapter we will study only the one-to-one codes, which (i) improve some sort of transmission efficiency (sec. 11.1-11.3) and (ii) reduce probability of error by detecting and correcting errors (sec. 11.4-11.6)

(a) Without encoder-decoder

(b) With encoder- decoder

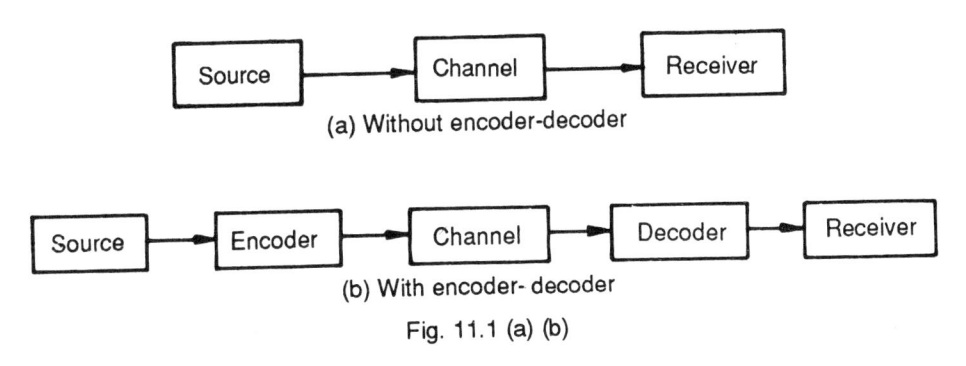

Fig. 11.1 (a) (b)

Figure 11.1(a) shows a communication system without the encoder-decoder. In Fig. 11.1(b), the messages are first encoded by the encoder and, then, transmitted via channel. At the receiving end, the received messages are first decoded in the decoder and, then, the original messages are recovered.

The following terminology is associated with coding:

(a) Letter, symbol, or character: Any individual member of the alphabet set.
(b) Message, or word: A finite sequence of letters of an alphabet.
(c) Length of the word: The number of letterrs in a message.
(d) Coding, encoding, or enciphering: A procedure for associating words constructed from a finite alphabet of a language with the given words of another language in a one-to-one manner.
(e) Decoding, or deciphering: The inverse operation of assigning words of the second language corresponding to the given words of the first language.
(f) Uniquely decipherable, or separable encoding and decoding: In this operation, the correspondence of all possible sequences of words between the two languages is one-to-one when there is no space between the words.
(g) Irreducibility or prefix property: When no encoded words can be obtained from each other by the addition of more letters, the code is said to be irreducible or of a prefix property.

When a code is irreducible, it is also uniquely decipherable; but the reverse is not true. This will be clear from the following examples:

Let $C_1 = 0$, $C_2 = 10$ and $C_3 = 110$. This code is irreducible as an addition of a 0, or a 1 to any of the code words does not produce other code words. The same code is uniquely decipherable as any received message string can be uniquely decoded. For example, if we receive

$$0110101000101101 1010$$

it is uniquely decoded as

$$C_1 C_3 C_2 C_2 C_1 C_1 C_2 C_3 C_3 C_2$$

Thus, it can be said that when a code is irreducible, it is also uniquely decipherable.

Now, let $C_1 = 0$, $C_2 = 01$ and $C_3 = 011$. This code, too, is uniquely decipherable, as any received message string can be uniquely decoded. For example, if we receive

$$00011011010010101011$$

it is uniquely decoded as

$$C_1 C_1 C_3 C_3 C_2 C_1 C_2 C_2 C_2 C_3$$

But the same code is not irreducible, as codes $C_2$, and $C_3$, are formed by adding 1 to codes $C_1$, and $C_2$, respectively. Thus, it can be said that when a code is uniquely decipherable, it is not necessarily irreducible. [*Note*: The English language is not uniquely decipherable. For example, if *FACTOR* is received, it may be decoded either as a single word FACTOR, or as a combination of two words *FACT* and *OR* .]

## 11.1 CODING EFFICIENCY

Let $M$ be the number of symbols in an encoding alphabet. Let there be $N$ messages $[m_1, m_2 ... m_N]$ with the probabilities $[P_{(m_1)}, P_{(m_2)} \cdots P_{(m_N)}]$. Let $n_i$ be the number of symbols in the $i$ th message. The average length of the message or the average length per code word is then given by

$$\overline{L} = \sum_{i=1}^{N} n_i p(n_i) \text{ letters/message} \qquad (11.1.1)$$

$\overline{L}$ should be minimum to have an efficient transmission. Coding efficiency, then, can be defined as

$$\eta = \frac{\overline{L} \min}{\overline{L}} \qquad (11.1.2)$$

Now, let $H(x)$ be the entropy of the source in bits/message. Also, let $\log M$ be the maximum average information associated with each letter in bits/letter.

CODING 547

Hence, the ratio $\dfrac{H(x)}{\log M}$ having a unit $\dfrac{\text{bits/message}}{\text{bits/letter}}$ or letters/message, gives the minimum average number of letters per message.

or

$$\frac{H(x)}{\log M} = \overline{L}\ \min$$

Hence, the coding efficiency is

$$\eta = \frac{\overline{L}\ \min}{\overline{L}} = \frac{H(X)}{\overline{L} \log M} \qquad (11.1.3)$$

and

$$\text{redundancay} = 1 - \eta \qquad (11.1.4)$$

Let us consider an example

$$[M] = [m_1\ m_2\ m_3\ m_4]$$

$$[P(M)] = \left[\frac{1}{2}\ \frac{1}{4}\ \frac{1}{8}\ \frac{1}{8}\right]$$

Without coding and considering a one-to-one correspondance (i.e., a noiseless channel), the efficiency is

$$\eta = \frac{I(X;Y)}{C} = \frac{H(X)}{\log N}$$

$$= \frac{-\left[\dfrac{1}{2}\log\dfrac{1}{2} + \dfrac{1}{4}\log\dfrac{1}{4} + \dfrac{1}{8}\log\dfrac{1}{8} + \dfrac{1}{8}\log\dfrac{1}{8}\right]}{\log 4}$$

$$= \frac{7/4}{2}$$

$$= 87.5\%$$

Now, let us use a binary code for coding. Let the code letters be 0 and 1. Thus $M$ is 2. One possible code is given in the following table:

| Message | code | length of code |
|---|---|---|
| $m_1$ | $C_1 = 00$ | $n_1 = 2$ |
| $m_2$ | $C_2 = 01$ | $n_2 = 2$ |
| $m_3$ | $C_3 = 10$ | $n_3 = 2$ |
| $m_4$ | $C_4 = 11$ | $n_4 = 2$ |

For this code,

$$\overline{L} = \sum_{k=1}^{4} n_k p(k) = \left(2 \times \frac{1}{2}\right) + \left(2 \times \frac{1}{4}\right) + \left(2 \times \frac{1}{8}\right) + \left(2 \times \frac{1}{8}\right) = 2\ \text{Letters/message}$$

$$\eta = \frac{H(X)}{\bar{L} \log M} = \frac{7/4}{2 \log 2} = \frac{7/4}{2} = 87.5\%$$

Thus, we see that this coding procedure is not improving the efficiency. Let us consider another code:

| message | code | length of code |
|---|---|---|
| $m_1$ | $C_1 = 0$ | $n_1 = 1$ |
| $m_2$ | $C_2 = 10$ | $n_2 = 2$ |
| $m_3$ | $C_3 = 110$ | $n_3 = 3$ |
| $m_4$ | $C_4 = 111$ | $n_4 = 3$ |

In this case,

$$\bar{L} = \sum_{k=1}^{4} n_k p(k)$$

$$= \left(1 \times \frac{1}{2}\right) + \left(2 \times \frac{1}{4}\right) + \left(3 \times \frac{1}{8}\right) + \left(3 \times \frac{1}{8}\right)$$

$$= \frac{7}{4} \text{ Letters/message}$$

$$\eta = \frac{H(X)}{\bar{L} \log M} = \frac{7/4}{7/4 \times \log 2} = \frac{7/4}{7/4} = 100\%$$

Thus, we can say that the second coding technique is better as compared to the first. Let us now try to find out why the second method is giving the best result.

In the first method, the probability of 0 is given by

$$P(0) = \frac{\sum_{k=1}^{4} p_k C_{k0}}{\sum_{k=1}^{4} p_k n_k}$$

where $C_{k0} \rightarrow$ number of 0's in $k$th coded message.
Hence,

$$P(0) = \frac{\left(\frac{1}{2} \times 2\right) + \left(\frac{1}{4} \times 1\right) + \left(\frac{1}{8} \times 1\right) + \left(\frac{1}{8} \times 0\right)}{2}$$

as $\sum_{k=1}^{4} p_k n_k$ is already found to be equal to 2.
Hence,

$$P(0) = \frac{11/8}{2} = \frac{11}{16}$$

Similarly,

$$P(1) = \frac{\sum_{k=1}^{4} p_k C_{k1}}{\sum_{k=1}^{4} p_k n_k}$$

where $C_{K1} \to$ number of 1's in $k$th coded message.
Hence,

$$P(1) = \frac{\left(\frac{1}{2} \times 0\right) + \left(\frac{1}{4} \times 1\right) + \left(\frac{1}{8} \times 1\right) + \left(\frac{1}{8} \times 2\right)}{2} = \frac{(5/8)}{2} = \frac{5}{16}$$

$P(1)$ can also be found out in following way:

$$P(0) + P(1) = 1$$

$\therefore \quad P(1) = 1 - P(0) = 1 - \frac{11}{16} = \frac{5}{16}$

In the second method,

$$P(0) = \frac{\sum_{k=1}^{4} p_k C_{k0}}{\sum_{k=1}^{4} p_k n_k} = \frac{\left(\frac{1}{2} \times 1\right) + \left(\frac{1}{4} \times 1\right) + \left(\frac{1}{8} \times 1\right) + \left(\frac{1}{8} \times 0\right)}{\left(\frac{1}{2} \times 1\right) + \left(\frac{1}{4} \times 2\right) + \left(\frac{1}{8} \times 3\right) + \left(\frac{1}{8} \times 3\right)} = \frac{7/8}{7/4} = \frac{1}{2}$$

and

$$P(1) = 1 - P(0) = 1 - \frac{1}{2} = \frac{1}{2}$$

Thus, we can see that the coding efficiency is more when $P(0) = P(1)$; and it is less when the two probabilities are different. This is because when both $P(0)$ and $P(1)$ are equal, the information content is maximum. As the difference in $P(0)$ and $P(1)$ increases, the information content decreases and, hence the coding efficiency also decreases. It can be seen that the code $C_1 = 111, C_2 = 110, C_3 = 10$ and $C_4 = 0$ gives 66.67% efficiency, as in this case $P(0) = \frac{4}{21}$ and $P(1) = \frac{17}{21}$. While coding, the following general rule may be followed:

Encode a message with a high probability in a short code word. Only then the average length of the code word $\overline{L}$ will decrease resulting in an increase in efficiency.

## 11.2 SHANNON-FANO CODING

This method of coding is directed towards constructing reasonably efficient separable binary codes. Let $[X]$ be the ensemble of the messages to be transmitted, and $[P]$ be their corresponding probabilities. The sequence $C_k$ of binary numbers of the length $n_k$ associated to each message $x_k$ should fulfil the following conditions:

(1) No sequences of employed binary numbers $C_k$ can be obtained from each other by adding more binary digits to the shorter sequence (prefix property).
(2) The transmission of an encoded message is reasonably efficient; i.e., 1 and 0 appear independently, with almost equal probabilities.

The actual procedure of the Shannon-Fano coding is as follows:

The messages are first written in the order of non-increasing probabilities. The message set then is partitioned into two most equi-probable subsets $[X_1]$ and $[X_2]$. A 0 is assigned to each message contained in one subset, and a 1 to each message contained in the other subset. The same procedure is repeated for the subsets of $[X_1]$ and $[X_2]$ i.e., $[X_1]$ will be partitioned into two subsets $[X_{11}]$ and $[X_{12}]$, and $[X_2]$ will be partioned into two subsets $[X_{21}]$ and $[X_{22}]$. The codewords in $[X_{11}]$ will start with 00, $[X_{12}]$ will start with 01; $[X_{21}]$ will start with 10, and $[X_{22}]$ will start with 11. The procedure is continued until each subset contains only one message. Note that each digit 0 or 1 in each partioning of the probability space appears with a more or less equal probability and is independent of the previous or subsequent partioning. Hence, $P(0)$ and $P(1)$ are also more or less equal.

### Example 11.2.1

Apply the Shannon-Fano coding procedure for the following message ensemble:

$$[X] = [x_1 \quad x_2 \quad x_3 \quad x_4 \quad x_5 \quad x_6 \quad x_7 \quad x_8]$$
$$[P] = [1/4 \quad 1/8 \quad 1/16 \quad 1/16 \quad 1/16 \quad 1/4 \quad 1/16 \quad 1/8]$$

Take $M = 2$

**Solution**

| Message | Probability | Encoded message | Length ($n_i$) |
|---------|-------------|-----------------|----------------|
| $x_1$   | 0.25        | 0 0             | 2              |
| $x_6$   | 0.25        | 0 1             | 2              |
| $x_2$   | 0.125       | 1 0 0           | 3              |
| $x_8$   | 0.125       | 1 0 1           | 3              |
| $x_3$   | 0.0625      | 1 1 0 0         | 4              |
| $x_4$   | 0.0625      | 1 1 0 1         | 4              |
| $x_5$   | 0.0625      | 1 1 1 0         | 4              |
| $x_7$   | 0.0625      | 1 1 1 1         | 4              |

$$\bar{L} = \sum_{k=1}^{8} P_k n_k = \left(\tfrac{1}{4} \times 2\right) + \left(\tfrac{1}{8} \times 3\right) + \left(\tfrac{1}{16} \times 4\right) + \left(\tfrac{1}{16} \times 4\right)$$
$$+ \left(\tfrac{1}{16} \times 4\right) + \left(\tfrac{1}{4} \times 2\right) + \left(\tfrac{1}{16} \times 4\right) + \left(\tfrac{1}{8} \times 3\right)$$

or

$$\bar{L} = 2.75 \text{ letters/message}$$

$$H(X) = -\sum_{k=1}^{8} P_k \log P_k$$

$$= -\left[\tfrac{1}{4} \log \tfrac{1}{4} + \tfrac{1}{8} \log \tfrac{1}{8} + \tfrac{1}{16} \log \tfrac{1}{16} + \tfrac{1}{16} \log \tfrac{1}{16} \right.$$
$$\left. + \tfrac{1}{16} \log \tfrac{1}{16} + \tfrac{1}{4} \log \tfrac{1}{4} + \tfrac{1}{16} \log \tfrac{1}{16} + \tfrac{1}{8} \log \tfrac{1}{8}\right]$$

$$= 2.75 \text{ bits/message}$$

$$\log M = \log 2 = 1 \text{ bits/letter}$$

Therefore, $$\eta = \frac{H(X)}{\overline{L} \log M} = \frac{2.75}{2.75 \times 1} = 100\%$$

## Example 11.2.2

Apply the Shannon-Fano coding procedure for the following message ensemble:

$$[X] = [\ x_1 \quad x_2 \quad x_3 \quad x_4 \quad x_5 \quad x_6 \quad x_7\ ]$$
$$[P] = [\ 0.4 \quad 0.2 \quad 0.12 \quad 0.08 \quad 0.08 \quad 0.08 \quad 0.04\ ]$$

Take $M = 2$

### Solution

For the first partitioning, there are two ways

(i) $[X_1] = [x_1\ x_2]$, $[X_2] = [\ x_3 \quad x_4 \quad x_5 \quad x_6 \quad x_7]$
(ii) $[X_1] = [x_1]$, $[X_2] = [\ x_2 \quad x_3 \quad x_4 \quad x_5 \quad x_6 \quad x_7]$

(i)

| Message | Probability | Encoded message | Length |
|---|---|---|---|
| $x_1$ | 0.4  | 0 0     | 2 |
| $x_2$ | 0.2  | 0 1     | 2 |
| $x_3$ | 0.12 | 1 0 0   | 3 |
| $x_4$ | 0.08 | 1 0 1   | 3 |
| $x_5$ | 0.08 | 1 1 0   | 3 |
| $x_6$ | 0.08 | 1 1 1 0 | 4 |
| $x_7$ | 0.04 | 1 1 1 1 | 4 |

$$\overline{L} = \sum_{k=1}^{7} p_k n_k = (0.4 \times 2) + (0.2 \times 2) + (0.12 \times 3) + (0.08 \times 3)$$
$$+ (0.08 \times 3) + (0.08 \times 4) + (0.04 \times 4)$$
$$= 2.52 \text{ letters / message}$$

(ii)

| Message | Probability | Encoded message | Length |
|---|---|---|---|
| $x_1$ | 0.4  | 0          | 1 |
| $x_2$ | 0.2  | 1 0 0      | 3 |
| $x_3$ | 0.12 | 1 0 1      | 3 |
| $x_4$ | 0.08 | 1 1 0 0    | 4 |
| $x_5$ | 0.08 | 1 1 0 1    | 4 |
| $x_6$ | 0.08 | 1 1 1 0    | 4 |
| $x_7$ | 0.04 | 1 1 1 1    | 4 |

$$\overline{L} = \sum_{k=1}^{7} p_k\ n_k = (0.4 \times 1) + (0.2 \times 3) + (0.12 \times 3) + (0.08 \times 4) + (0.08 \times 4) + (0.08 \times 4) + (0.04 \times 4)$$

= 2.48 letters/message

Thus, it can be seen that the second method is better as it gives a lower value for $\bar{L}$.
Now,

$$H(X) = -\sum_{k=1}^{7} p_k \log p_k = -[(0.4 \log 0.4) + (0.2 \log 0.2) + (0.12 \log 0.12) + (0.08 \log 0.08)$$
$$+ (0.08 \log 0.08) + (0.08 \log 0.08) + (0.04 \log 0.04)]$$
$$= 2.42 \text{ bits/message}$$

Hence, efficiency in the second method is

$$\eta = \frac{H(x)}{\bar{L} \log M} = \frac{2.42}{2.48 \log 2} = 97.6\%$$

Example 11.2.2 shows that sometimes the Shannon-Fano method is ambiguous. The ambiguity arises due to the availability of more than one schemes of partitioning. Moreover, as $M$ tends to increase, this method is not suitable and the formation of $M$ approximately equi-probable groups is rather difficult and with little choice. Example 11.2.3 will make this point clear.

### Example 11.2.3

Solve Ex.11.2.2 for $M = 3$.
*Solution*
Let the encoding alphabet be $-1\ 0\ 1$

| Message | Probability | Encoded message | Length |
|---------|-------------|-----------------|--------|
| $x_1$ | 0.4 | $-1$ | 1 |
| $x_2$ | 0.2 | $0\ -1$ | 2 |
| $x_3$ | 0.12 | $0\ 0$ | 2 |
| $x_4$ | 0.08 | $1\ -1$ | 2 |
| $x_5$ | 0.08 | $1\ 0$ | 2 |
| $x_6$ | 0.08 | $1\ 1\ -1$ | 3 |
| $x_7$ | 0.04 | $1\ 1\ 0$ | 3 |

$$\bar{L} = \sum_{k=1}^{7} p_k n_k = (0.4 \times 1) + (0.2 \times 2) + (0.12 \times 2) + (0.08 \times 2) + (0.08 \times 2) + (0.08 \times 3) + (0.04 \times 3)$$
$$= 1.72 \text{ letters/message}$$

Hence,

$$\eta = \frac{H(X)}{\bar{L} \log M} = \frac{2.42}{1.72 \times \log 3} = 88.7\%$$

## 11.3 HUFFMAN CODING

The Huffman coding method leads to the lowest possible value of $\bar{L}$ for a given $M$, resulting in a maximum

$\eta$. Hence, it is also known as the minimum redundancy code, or optimum code. The procedure is as follows:
(1) $N$ messages are arranged in an order of non-increasing probability.
(2) The probabilities of $[N - K[M - 1)]$ least likely messages are combined, where k is the highest integer that gives a positive value to the bracket, and the resulting $[K(M - 1) + 1]$ probabilities are re-arranged in a non-increasing manner. This step is called 'reduction'. The reduction procedure is repeated as often as necessary, by taking $M$ terms every time, until there remain $M$ ordered probabilities. It may be noted that by combining $[N - K(M - 1)]$, and not $M$, terms in the first reduction, it is ensured that there will be exactly $M$ terms in the last reduction.
(3) Encoding begins with the last reduction, which consists of exactly $M$ ordered probabilities. The first element of the encoding alphabet is assigned as the first digit in the codewords for all source messages associated with the first probability of the last reduction. Similarly, the second element of the encoding alphabet is assigned as the second digit in the codewords for all source messages associated with the second probability of last reduction and so on.

The same procedure is repeated for the second from last reduction, to the first reduction, in that order.

## Example 11.3.1

Solve Ex. 11.2.2 by the Huffman method.

| Message | Probability | First reduction | Second reduction | Third reduction | Fourth reduction | Fifth reduction |
|---|---|---|---|---|---|---|
| $x_1$ | 0.4 | 0.4 | 0.4 | 0.4 | 0.4 | 0.6  1<br>0.4  0 |
| $x_2$ | 0.2 | 0.2 | 0.2 | 0.24<br>0.2  1 | 0.36  1<br>0.24  0 | |
| $x_3$ | 0.12 | 0.12<br>0.12 | 0.16<br>0.12  1<br>0.12  0 | 0.16  0 | | |
| $x_4$ | 0.08 | 0.08  1 | | | | |
| $x_5$ | 0.08 | 0.08  0 | | | | |
| $x_6$ | 0.08  1 | | | | | |
| $x_7$ | 0.04  0 | | | | | |

First digit from left (1)
Second digit from left (1)
Third digit from left (0)
Fourth digit from left (1)

Code-word $C_4 = 1101$

### Solution

As per step (2) of procedure, last two terms should be combined in the first reduction.

*Explanation of construction of the codeword $C_4$ for the message $X_4$*: The dashed-line shows the path for deciding the codeword $C_4$ for the message $X_4$. The '1' encountered in the fifth reduction becomes the first digit from left in $C_4$. The '1' encountered in the fourth reduction becomes the second digit from left in $C_4$. The '0' encountered in third reduction becomes the third digit from left in $C_4$. No digit is encountered in the second reduction. The '1' encountered in the first reduction becomes thee fourth digit from left. Thus, the codeword $C_4$ is $C_4$ = 1101. In the same way, other codewords can be formed.

|        | Code | Length |
|--------|------|--------|
|        | $C_1 = 0$ | 1 |
|        | $C_2 = 111$ | 3 |
|        | $C_3 = 101$ | 3 |
|        | $C_4 = 1101$ | 4 |
|        | $C_5 = 1100$ | 4 |
|        | $C_6 = 1001$ | 4 |
|        | $C_7 = 1000$ | 4 |

$$\overline{L} = \sum_{k=1}^{7} p_k\, n_k = (0.4 \times 1) + (0.2 \times 3) + (0.12 \times 3) + (0.08 \times 4) + (0.08 \times 4) + (0.08 \times 4) + (0.04 \times 4)$$

= 2.48 letters/message

This is the same as the one obtained in the second answer of Example 11.3.2 which gives the maximum eficiency.

### Example 11.3.2

Solve Ex. 11.3.1 for $M = 3$.

*Solution*

As per step (2) of procedure, last three terms should be combined in the first reduction.
Let the encoding alphabet be -1 0 1

The construction of the codeword $C_1 = 01 - 1$ is explained by the dashed path. Similarly, other codewords can be formed.

| Message | Code-word | Probability | First reduction | Second reduction |
|---------|-----------|-------------|-----------------|------------------|
| $x_1$ | $C_1 = 1$ | 0.4 | 0.4 | 0.4 \| 1 |
|       |           |     |     | 0.4 \| 0 |
| $x_2$ | $C_2 = -1$ | 0.2 | 0.2 | 0.2 \| -1 |
|       |           |     | 0.2 \| 1 | First digit from left (0) |
| $x_3$ | $C_3 = 0\,0$ | 0.12 | 0.12 \| 0 |  |
| $x_4$ | $C_4 = 0\,-1$ | 0.08 | 0.08 \| -1 | Second digit from left (1) |
| $x_5$ | $C_5 = 0\,1\,1$ | 0.08 \| 1 |  |  |
| $x_6$ | $C_6 = 0\,1\,0$ | 0.08 \| 0 |  |  |
| $x_7$ | $C_7 = 0\,1\,-1$ | 0.04 \| -1 | Third digit from left (-1) |  |

Code-word $C_7 = 0\,1\,-1$

$$\overline{L} = \sum_{k=1}^{7} p_k\, n_k = (0.4 \times 1) + (0.2 \times 1) + (0.12 \times 2) + (0.08 \times 2) + (0.08 \times 3) + (0.08 \times 3) + (0.04 \times 3)$$

$$= 1.5 \text{ letters/message}$$

Hence,
$$\eta = \frac{H(x)}{\overline{L} \log M}$$

$$= \frac{2.42}{1.5 \log 3}$$

$$= 99.5\%$$

It can be seen that $\eta$ obtained by the Huffman method is much better than that obtained by the Shannon-Fano method. (See Ex.11.2.3)

## 11.4 ERROR-CONTROL CODING

The probability of error for a particular signalling scheme is a function of signal-to-noise ratio at the receiver input and the information rate. In practical systems, the maximum signal power and the bandwidth of the channel are restricted to some fixed values. Also, the noise power spectral density $\eta/2$ is fixed for a particular operating environment. With all these constraints, it is often not possible to arrive at a signalling scheme which will yield an acceptable probability of error for a given application. The only practical alternative for reducing the probability of error is the use of error-control coding. The channel encoder systematically adds digits to the transmitted message digits. Although these additional digits convey no new information, they make it possible for the channel decoder to detect, and correct errors in the information bearing digits. The overall probability of error is reduced due to error detection and/or correction.

Error-control codes are divided into two categories: Block codes and Convolutional codes. In block codes, a group of $r$ check bits are derived from a block of $k$ information bits. The structure of block codes is such that the information bits are followed by the check bits. The check bits are then used to verify the information bits at the receiver. In convolutional codes, the check bits are continuously interleaved with information bits. The check bits verify information bits not only in the block immediately preceding them, but in other blocks as well. The main difference between block codes and convolutional codes is that in the block code, a block of $n$ digits generated by the encoder in a particular time unit depends only on the block of $k$ input message digits within that time unit; whereas, in the convolutional code, a block of $n$ code digits generated by the encoder in a time unit depends not only on the block of $k$ message digits within that time unit, but also on the preceding $(N-1)$ blocks of messages $(N > 1)$.

## 11.5 BLOCK CODES

In block codes (also known as arithmetic codes, or group codes), each block of $k$ message bits is encoded into a block of n bits $(n > k)$, as shown in Fig. 11.5.1. The check bits are derived from the message bits and are added to them. The $n$-bit block of a channel encoder output is called a codeword and the codes (or coding schemes) in which the message bits appear at the beginning of a codeword, are called *systematic codes*.

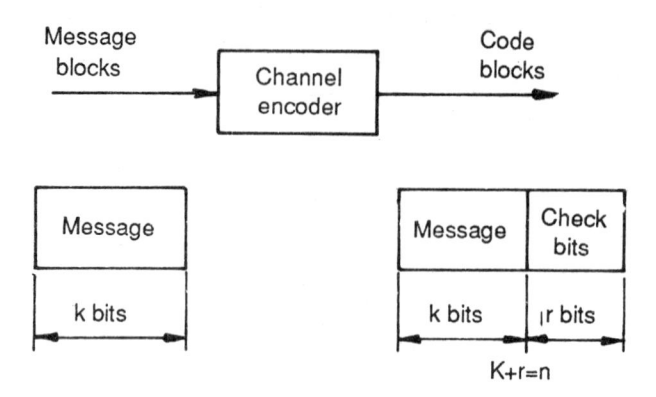

```
            Message                            Code
            blocks        ┌─────────┐          blocks
         ──────────────▶  │ Channel │ ──────────────▶
                          │ encoder │
                          └─────────┘

         ┌──────────┐              ┌──────────┬───────┐
         │          │              │          │ Check │
         │ Message  │              │ Message  │ bits  │
         │          │              │          │       │
         ├──────────┤              ├──────────┼───────┤
         │  k bits  │              │  k bits  │ r bits│
         └──────────┘              └──────────┴───────┘
                                         K+r=n
```

Fig. 11.5.1

### 11.5.1 Parity Check Codes

The simplest possible block code is when the number of check bits is one. These are known as parity check codes. When the check bit is such that the total number of 1's in the codeword is even, it is an *even parity check code*, and when the check bit is such that the total number of 1's in the codewords is odd, it is an *odd parity check code*.

The following example explains the parity check code:

| Message | Code for even parity |          | Code for odd parity |          |
|---------|----------------------|----------|---------------------|----------|
|         | message              | checkbit | message             | checkbit |
| 010011  | 010011               | 1        | 010011              | 0        |
| 101110  | 101110               | 0        | 101110              | 1        |

If a single error occurs in a received message, it can be immediately detected; although the position of the erroneous bit cannot be determined. Thus, with this code, though a signle error can be detected, it cannot be corrected.

### 11.5.2 Study of Binary Code Space

In this section certain important concepts such as the weight of a code, Hamming distance, etc., are introduced.

The *weight* of a codeword is defined as the number of non-zero components in it. For example,

| Code word | Weight |
|-----------|--------|
| 010110    | 3      |
| 101000    | 2      |
| 000000    | 0      |

The *Hamming distance* between two code words is defined as the number of components in which they differ.
For example,
let
$$U = 1010$$
$$V = 0111$$
$$W = 1001$$

Then, $D(U, V)$ = distance between $U$ and $V$ = 3
Similarly,
$$D(U, W) = 2$$
and
$$D(V, W) = 3$$
Mathematically, the Hamming distance, can be defined as
$$D(U, V) = \sum_{k=1}^{n} (\alpha_k \oplus \beta_k) \quad (10.5.2.1)$$
where
$$U = \alpha_1 \alpha_2 \alpha_3 \ldots \alpha_n$$
$$V = \beta_1 \beta_2 \beta_3 \ldots \beta_n$$
($\alpha$'s and $\beta$'s are binary digits 0 or 1).

The notation $\oplus$ means modulo – 2 addition, for which the rules are
$$0 \oplus 0 = 0$$
$$0 \oplus 1 = 1$$
$$1 \oplus 0 = 1$$
$$1 \oplus 1 = 0$$

Then, for $U$ = 1010 and $V$ = 0111 Eq.10.5.2.1 gives
$$D(U, V) = [(1 \oplus 0) + (0 \oplus 1) + (1 \oplus 1) + (0 \oplus 1)] = [1 + 1 + 0 + 1] = 3$$

The *minimum distance* of a block code is defined as the smallest distance between any pair of codewords in the code.

Now, let us consider a block code with a minimum distance two. If a single error occurs, a word will be erroneously received as a meaningless word, i.e. the word does not exist in the codebook. Thus, in such a set-up any single error can be detected, but it cannot be corrected. This will be clear from the following example:

Let us consider a block code of two digits with a minimum distance two. Two codebooks are possible. They are 00, 11 and 01, 10. Let our codebook be 01, 10. Now, with a single error, 01 may be received either as 00 or as 11. Let us suppose that it is received as 00. Since 00 is not in our codebook, an error has been detected. But a decision cannot be taken as to whether 01 or 10 was transmitted, as both are at equal distance from 00. Hence, the error cannot be corrected. If we have a codebook of a minimum distance three, the single error can be corrected as the distance of the erroneous word is 1 from only one codeword, and more than 1 from all other codewords. For example, if 000, 111 is our codebook, and if 001 is received, a decision can be taken that 000 is received since the distance between 000 and 001 is one, whereas, the distance between 111 and 001 is two.

By extending above ideas, the following data are given by Hamming:

| *Minimum distance* | *Descripton of Coding* |
| --- | --- |
| 1 | Error cannot be detected |
| 2 | Single error detection |
| 3 | Single error correction |
| 4 | Single errorr correction plus double error detection |
| 5 | Double error correction |
| 6 | Double error correction plus triple error detection |

In general, if $n$ is the minimum distance of a block code, then

(i) $\dfrac{n-1}{2}$ errors can be corrected if $n$ is odd

(ii) $\dfrac{n-2}{2}$ errors can be corrected and $\dfrac{n}{2}$ errors can be detected if $n$ is even.

It is interesting to find the number of maximum possible codewords of length $n$ and a minimum distance $d$. This is given by $B(n,d)$ and Hamming got the following values:

$$B(n, 1) = 2^n$$

$$B(n, 2) = 2^{n-1}$$

$$B(n, 3) = 2^m \leq \frac{2^n}{n+1}$$

$$B(n, 4) = 2^m \leq \frac{2^{n-1}}{n}$$

$$B(n, 2k) = B(n-1, 2k-1)$$

$$B(n, 2k+1) = 2^m \leq \frac{2n}{1 + \binom{n}{2} + \binom{n}{2} + \ldots + \binom{n}{k}}$$

The equality in $B(n, 3) \leq \frac{2^n}{n+1}$ is valid when $\frac{2^n}{n+1}$ is an integer. Such codes are referred to as *close packed codes*. These are obtained by selecting $n = 2^k - 1$ where $k$ is a positive integer. For example, $k = 2,3,4$ gives $n = 3,7,15$; resulting in close packed codes $B(3, 3)$, $B(7, 3)$, $B(15, 3)$ respectively. Thus,

$$B(3, 3) = \frac{2^3}{3+1} = 2$$

$$B(7, 3) = \frac{2^7}{7+1} = 16$$

$$B(15, 3) = \frac{2^{15}}{15+1} = 2048$$

when $n \neq 2^k - 1$, the number of maximum possible code words is found out from $B(n, 3) < \frac{2^n}{n+1}$. Thus, for $n = 5$,

$$B(5, 3) = 2^m \leq \frac{2^5}{5+1}$$

or

$$2^m \leq 5.33, \text{ giving } m = 2$$

Hence,
$$B(5, 3) = 2^2 = 4$$

For $n = 6$,
$$B(6, 3) = 2^m \leq \frac{2^6}{6+1} \text{ or } 2^m \leq 9.14. \text{ Thus } m = 3$$

Hence,
$$B(6, 3) = 2^3 = 8$$

The codebooks $B(5,3)$ and $B(6,3)$ may be found as

$$\underbrace{B(5,3)}$$  $$\underbrace{B(6,3)}$$
00000    000000  100110
01101    010101  110011
10110    111000  011110
11011    101101  001011

### 11.5.3 Linear Block Codes

If each of the $2^k$ codewords of a systematic code can be expressed as linear combinations of $k$ linearly independent code-vectors, the code is called a *linear block code* or *systematic linear block code*.

There are two steps in the encoding procedure for linear block codes: (1) The information sequence is segmented into message blocks of $k$ successive information bits. (2) Each message block is transformed into a larger block of $n$ bits by an encoder according to some pre-determined set of rules. The $n$-$k$ additional bits are generated from linear combinations of the message bits. The encoding operations can be described with the help of matrices. Let a message block be a row vector

$$D = [d_1 d_2 \ldots d_k]$$

where each message bit can be a 0 or a 1. Thus, we have $2^k$ distinct message blocks. Each message block is transformed into a code word $C$ of length $n$ bits

$$C = [C_1 C_2 \ldots C_n]$$

by the encoder, and there are $2^k$ distinct codewords. It may be noted that there is one unique codeword for each distinct message block. This set of $2^k$ codewords, also known as code-vectors, is called an $(n,k)$ *block code*. The rate efficiency of this code is $\frac{k}{n}$.

In a systematic linear block code, the first $K$ bits of the codeword are the message bits, i.e.

$$c_i = d_i, i = 1, 2, \ldots k \qquad (11.5.3.1)$$

The last $n$-$k$ bits in the codeword are check bits generated from $k$ message bits according to some pre-determined rule:

$$\left. \begin{array}{l} C_{k+1} = p_{11} d_1 \oplus p_{21} d_2 \oplus \ldots \oplus p_{k,1} d_k \\ C_{k+2} = p_{12} d_1 \oplus p_{22} d_2 \oplus \ldots \oplus p_{k,2} d_k \\ \vdots \\ C_n = p_{1,n-k} d_1 \oplus p_{2,n-k} d_2 \oplus \ldots \oplus p_{k,n-k} d_k \end{array} \right\} \qquad (11.5.3.2)$$

The coefficients $p_{i,j}$ in Eq.11.5.3.2 are 0's and 1's so that $C_k$'s are 0's and 1's. The additions in Eq. 11.5.3.2 are modulo-2 additions. Equations 11.5.3.1 and 11.5.5.2 can be combined to give a matrix equation:

$$[C_1 C_2 \ldots C_n] = [d_1 d_2 \ldots d_k] \begin{bmatrix} 1000 & \cdots & 0 & \vdots & p_{11} & p_{12} & \cdots & p_{1,n-k} \\ 0100 & \cdots & 0 & \vdots & p_{21} & p_{22} & \cdots & p_{2,n-k} \\ 0010 & \cdots & 0 & \vdots & p_{31} & p_{32} & \cdots & p_{3,n-k} \\ \cdots & \cdots & \cdots & & \cdots & \cdots & & \cdots \\ 0000 & \cdots & 1 & \vdots & p_{k,1} & p_{k,2} & \cdots & p_{k,n-k} \end{bmatrix}_{k \times n} \qquad (11.5.3.3)$$

or
$$C = DG \tag{11.5.3.4}$$
where $G$ is the $K \times n$ matrix on the RHS of Eq. 11.5.3.3. It is called the generator matrix of the code and is used in encoding operation. It has the form
$$G = [I_k \vdots P]_{k \times n} \tag{11.5.3.4a}$$
where $I_k$ is the identity matrix of the order $K$ and $P$ is an arbitrary $K \times (n-k)$ matrix. The matrix $P$ completely defines the $(n,k)$ block code. The selection of a $P$ matrix is an important step in the design of an $(n,k)$ block code because then the code generated by $G$ achieves certain desirable properties such as the ease of implementation, ability to correct errors, high rate efficiency, etc.

### Example 11.5.3.1

The generator matrix for a (6,3) block code is given below. Find all code-vectors of this code.

$$G = \begin{bmatrix} 100 & \vdots & 110 \\ 010 & \vdots & 011 \\ 001 & \vdots & 111 \end{bmatrix}$$

*Solution*

The message block size $k$ for this code is 3, and the length of the code-vector $n$ is 6. The code-vector for the message block $D = [\,1\,0\,1\,]$ is given by

$$C = DG = [101] \begin{bmatrix} 100\ 110 \\ 010\ 011 \\ 001\ 111 \end{bmatrix}$$

$$= [\,1\,0\,1\,0\,0\,1\,]$$

Other code-vectors can be found out in the same way. The code-vectors of this code are:

| Message | Code-Vector |
|---|---|
| 0 0 0 | 0 0 0   0 0 0 |
| 0 0 1 | 0 0 1   1 1 1 |
| 0 1 0 | 0 1 0   0 1 1 |
| 0 1 1 | 0 1 1   1 0 0 |
| 1 0 0 | 1 0 0   1 1 0 |
| 1 0 1 | 1 0 1   0 0 1 |
| 1 1 0 | 1 1 0   1 0 1 |
| 1 1 1 | 1 1 1   0 1 0 |

The parity check matrix $H$ is associated with each $(n,k)$ block code and is given by

$$H = \begin{bmatrix} p_{11} & p_{21} & \cdots & p_{k1} & \vdots & 100 & \cdots & 0 \\ p_{12} & p_{22} & \cdots & p_{k2} & \vdots & 010 & \cdots & 0 \\ \cdots & \cdots & \cdots & \cdots & \vdots & \cdots & \cdots & \cdots \\ p_{1,n-k} & p_{2,n-k} & \cdots & p_{k,n-k} & \vdots & 000 & \cdots & 1 \end{bmatrix}$$

$$= [P^T \vdots I_{n-k}]_{(n-k) \times n} \tag{10.5.3.5}$$

where $P^T$ is the transpose of matrix $P$. The parity check matrix can be used to verify whether a codeword $C$ is generated by the matrix $G$. The rule for verification is:
$C$ is a code word in the $(n,K)$ block code generated by $G$ if, and only, if

$$CH^T = 0 \tag{10.5.3.6}$$

where $H^T$ is the transpose of $H$ and is also given by

$$H^T = \left[\frac{P}{I_{n-k}}\right] \tag{10.5.3.7}$$

The parity check matrix $H$ is used in the decoding operation as follows:

Consider a linear $(n,k)$ block code with a generator matrix $G = [I_k \vdots p]$ and a parity check matrix $H = [P^T \vdots I_{n-k}]$. Let $C$ and $R$ be the transmitted and received code-vectors, respectively, in a noisy communication system. The vector $R$ is the sum of the transmitted code-vector $C$, and an error vector $E$, i.e.,

$$R = C + E \tag{11.5.3.8}$$

The function of the receiver is to decode $C$ from $R$, and the message block $D$ from $C$. The receiver does the decoding operation by determining an $(n-k)$ vector $S$ (known as error syndrome of $R$) defined as

$$S = RH^T \tag{11.5.3.9}$$

Equation 11.5.3.9 can be rewritten as

$$S = (C + E) H^T$$
$$= CH^T + EH^T$$
$$= EH^T \text{ since } CH^T = 0$$

Thus, the syndrome of the received vector is zero if $R$ is a valid codevector. If errors occur in transmission, the syndrome $S$ of the received vector is non-zero. Moreover, $S$ is related to $E$ and the decoder uses $S$ to detect and correct errors.

### Example 11.5.3.2

Consider a (7,4) block code generated by

$$G = \begin{bmatrix} 1 & 0 & 0 & 0 & \vdots & 1 & 1 & 0 \\ 0 & 1 & 0 & 0 & \vdots & 0 & 1 & 1 \\ 0 & 0 & 1 & 0 & \vdots & 1 & 0 & 1 \\ 0 & 0 & 0 & 1 & \vdots & 1 & 1 & 1 \end{bmatrix}$$

$$\quad\quad\quad I_4 \quad\quad\quad\quad P$$

Explain how the errors syndrome $S$ helps in correcting a single error.

*Solution:*
The parity check matric $H$ for the given code is

$$H = [P^T \vdots I_{7-4}]$$

$$= \begin{bmatrix} 1 & 0 & 1 & 1 & \vdots & 1 & 0 & 0 \\ 1 & 1 & 0 & 1 & \vdots & 0 & 1 & 0 \\ 0 & 1 & 1 & 1 & \vdots & 0 & 0 & 1 \end{bmatrix}$$

For a message block $D = [1\ 1\ 0\ 1]$, the code-vector $C$ is given by

$$C = DG = [1\ 1\ 0\ 1 \vdots 0\ 1\ 0]$$

For this code-vector, the syndrome $S$ is given by

$$S = CH^T = [0\ 0\ 0]$$

If the third bit from left of the code-vector $C$ suffers an error in transmission, the received vector $R$ will be

$$R = [1\ 1\ 1\ 1\ 0\ 1\ 0]$$
$$= [1\ 1\ 0\ 1\ 0\ 1\ 0] \oplus [0\ 0\ 1\ 0\ 0\ 0\ 0]$$
$$= C \oplus E$$

and the syndrome of $R$ is

$$S = RH^T$$

or

$$S = [1\ 1\ 1\ 1\ 0\ 1\ 0] \begin{bmatrix} 1 & 1 & 0 \\ 0 & 1 & 1 \\ 1 & 0 & 1 \\ 1 & 1 & 1 \\ 1 & 0 & 0 \\ 0 & 1 & 0 \\ 0 & 0 & 1 \end{bmatrix} = [101] = EH^T$$

It can be noted that the syndrome vector $S$ for an error in the third bit is equal to the third row of $H^T$. It can also be noted that, for this code, a single error in the $i^{th}$ bit of $C$ leads to a syndrome vector that is equal to the $i^{th}$ row of $H^T$.

Thus, single errors can be corrected at the receiver by comparing $S$ with the rows of $H^T$ and correcting the $i^{th}$ received bit if $S$ is equal to the $i^{th}$ row of $H^T$. When the syndrome vector $S$ is zero, it is concluded that no error has occurred.

**Selection of $H^T$:**

First $K$ rows of $H_T$ (i.e., the matrix $P$) should be chosen to be distinct. Moreover, none of them can be all 0's, since it corresponds to no-error condition. Also, none of them can be any row of $I_{n-k}$ (i.e. with a single 1 in it), since the last $n-k$ rows of $H^T$ constitute the identity matrix $I_{n-k}$.

### 11.5.4 Hamming's Single Error Correcting Code

We know that when a single error occurs, say in the $i^{th}$ bit of the code word, the syndrome of a received vector is equal to the $i^{th}$ row of $H^T$. Hence, if $n$ rows of $n \times (n-k)$ matrix $H^T$ are chosen to be distinct, then the syndrome of all single errors will be distinct, and we can correct the single errors. Once $H^T$ is chosen, the generator matrix $G$ can be obtained by using Eqs 11.5.3.4 and 11.5.3.5.

Each row in $H^T$ has $(n-k)$ entries. Each one of these entries could be a 0, or a 1. Hence, we can have $2^{n-k}$ distinct rows of $(n-k)$ entries from which $2^{n-k} - 1$ distinct rows of $H^T$ can be selected. (It is to be kept in mind that the row with all 0's cannot be selected.) Since the matrix $H^T$ has $n$ rows, the condition for all of them to be distinct is

$$2^{n-k} - 1 \geq n$$

or

$$(n - k) \geq \log_2(n + 1)$$

or

$$n \geq k + \log_2(n + 1) \tag{11.5.4.1}$$

Thus the minimum size $n$ for the codeword can be determined. (Note that $n$ has to be an integer.)

### Example 11.5.4.1

Design a block code with a minimum distance of three and a message block size of eight bits.

*Solution*

Since the minimum distance is three, only a single error correcting code is desired.
From Eq. 11.5.4.1 we have

$$n \geq 8 + \log_2(n + 1)$$

The smallest value of $n$ that satisfies the above inequality is $n = 12$. Thus, a (12,8) block code is needed. $H^T$ will be of the size $12 \times (12-8)$. The last four rows of $H^T$ will be a $4 \times 4$ identity matrix. The first eight rows are arbitrarily chosen, with the restriction that none of the rows have all 0's or single 1 in it, and that they are distinct. One choice for $H^T$ is as follows

$$H^T = \left[\frac{P}{I_{n-k}}\right] = \begin{bmatrix} 0 & 0 & 1 & 1 \\ 0 & 1 & 0 & 1 \\ 0 & 1 & 1 & 0 \\ 0 & 1 & 1 & 1 \\ 1 & 0 & 0 & 1 \\ 1 & 0 & 1 & 0 \\ 1 & 0 & 1 & 1 \\ 1 & 1 & 0 & 0 \\ \cdots & \cdots & \cdots & \cdots \\ 1 & 0 & 0 & 0 \\ 0 & 1 & 0 & 0 \\ 0 & 0 & 1 & 0 \\ 0 & 0 & 0 & 1 \end{bmatrix}$$

The generator matrix for this code is

$$G = [I_k \vdots P] \begin{bmatrix} 1 & 0 & 0 & 0 & 0 & 0 & 0 & 0 & \vdots & 0 & 0 & 1 & 1 \\ 0 & 1 & 0 & 0 & 0 & 0 & 0 & 0 & \vdots & 0 & 1 & 0 & 1 \\ 0 & 0 & 1 & 0 & 0 & 0 & 0 & 0 & \vdots & 0 & 1 & 1 & 0 \\ 0 & 0 & 0 & 1 & 0 & 0 & 0 & 0 & \vdots & 0 & 1 & 1 & 1 \\ 0 & 0 & 0 & 0 & 1 & 0 & 0 & 0 & \vdots & 1 & 0 & 0 & 1 \\ 0 & 0 & 0 & 0 & 0 & 1 & 0 & 0 & \vdots & 1 & 0 & 1 & 0 \\ 0 & 0 & 0 & 0 & 0 & 0 & 1 & 0 & \vdots & 1 & 0 & 1 & 1 \\ 0 & 0 & 0 & 0 & 0 & 0 & 0 & 1 & \vdots & 1 & 1 & 0 & 0 \end{bmatrix}$$

The (12,8) block code can now be generated by using the equation

$$C = DG$$

### 11.5.5 Cyclic Codes

Cyclic codes form a subclass of linear block codes. They are important for two reasons: First, encoding and syndrome calculations can be easily implemented by using simple shift registers with feedback connections. Second, the mathematical structure of these codes is such that it is possible to design codes having useful error-correcting properties.

An $(n,k)$ linear block code $C$ is called a cyclic code if it satisfies the following property: If an $n$-tuple

$$v = (v_o, v_1, \ldots, v_{n-1}) \tag{11.5.5.1}$$

is a code-vector of $C$, then the $n$-tuple

$$v^{(1)} = (v_{n-1}, v_o, v_1, \ldots, v_{n-2})$$

which is obtained by shifting $v$ cyclically one place to the right, is also a code-vector of $C$. From the above definition it is clear that

$$v^{(i)} = (v_{n-i}, v_{n-i+1}, \ldots, v_o, v_1, \ldots, v_{n-i-1}) \tag{11.5.5.2}$$

is also a code-vector of $C$.
An example of cyclic code:

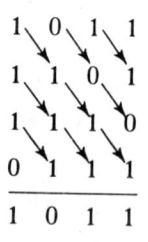

```
1 0 1 1
1 1 0 1
1 1 1 0
0 1 1 1
───────
1 0 1 1
```

It can be seen that the code 1 0 1 1, 1 1 0 1, 1 1 1 0, 0 1 1 1 is obtained by a cyclic shift of $n$-tuple 1011 ($n=4$). The code obtained by rearranging the four words is also a cyclic code. Thus 1011, 1110, 1101, 0111 are also cyclic codes. (All 0's or all 1's can be words of any cyclic code as all shifts result in the same word.)

The code word $v$ can be represented by a code polynomial as

$$V(x) = v_o + v_1 x + v_2 x^2 + \ldots + v_{n-1} x^{n-1} \tag{11.5.5.3}$$

# CODING

The coefficients of the polynomial are 0's and 1's, and they belong to a binary field which satisfies the following rules of addition and multiplication:

$$0 + 0 = 0 \qquad 0.0 = 0$$
$$0 + 1 = 1 \qquad 0.1 = 0$$
$$1 + 0 = 1 \qquad 1.0 = 0$$
$$1 + 1 = 0 \qquad 1.1 = 1$$

(It can be seen that + means modulo-2 addition previously denoted by $\oplus$.)

Also, $x^2 = x.x; x^3 = (x^2).x = x.x.x$; etc

Now, we will state a theorem (without giving its proof) which is very useful for a cyclic code generation.

*Theorem*

If $g(x)$ is a polynomial of the degree $(n-k)$ and is a factor of $x^n + 1$, then, $g(x)$ generates an $(n,k)$ cyclic code in which the code polynomial $V(x)$ for a data vector $D = (d_0, d_1, d_2, \ldots d_{k-1})$ is generated by

$$V(x) = D(x) g(x) \qquad (11.5.5.4)$$

## Example 11.5.5.1

The generator polynomial of a (7,4) cyclic code is $g(x) = 1 + x + x^3$. Find the 16 codewords of this code.

*Solution*

Consider the message vector

$$D = (d_0 \ d_1 \ d_2 \ d_3) = (0101)$$

The message polynomial is

$$D(X) = (0)1 + (1)x + (0)x^2 + (1)x^3$$
$$= x + x^3$$

The code polynomial $V(x)$ is given by

$$V(x) = D(x) g(x)$$
$$= (x + x^3)(1 + x + x^3)$$
$$= x + x^2 + x^3 + x^4 + x^4 + x^6$$

Now,

$$x^4 + x^4 = (1+1)x^4 = (0)x^4 = 0$$

Hence,

$$V(x) = x + x^2 + x^3 + x^6$$
$$= (0)1 + (1)x + (1)x^2 + (1)x^3$$
$$+ (0)x^4 + (0)x^5 + (1)x^6$$

Hence the code-vector $V$ is

$$V = (0\ 1\ 1\ 1\ 0\ 0\ 1)$$

In the same way all code-vectors can be found out.

| Message | Code-Vector |
|---|---|
| 0 0 0 0 | 0 0 0 0 0 0 0 |
| 0 0 0 1 | 0 0 0 1 1 0 1 |
| 0 0 1 0 | 0 0 1 1 0 1 0 |
| 0 0 1 1 | 0 0 1 0 1 1 1 |
| 0 1 0 0 | 0 1 1 0 1 0 0 |
| 0 1 0 1 | 0 1 1 1 0 0 1 |
| 0 1 1 0 | 0 1 0 1 1 1 0 |
| 0 1 1 1 | 0 1 0 0 0 1 1 |
| 1 0 0 0 | 1 1 0 1 0 0 0 |
| 1 0 0 1 | 1 1 0 0 1 0 1 |
| 1 0 1 0 | 1 1 1 0 0 1 0 |
| 1 0 1 1 | 1 1 1 1 1 1 1 |
| 1 1 0 0 | 1 0 1 1 1 0 0 |
| 1 1 0 1 | 1 0 1 0 0 0 1 |
| 1 1 1 0 | 1 0 0 0 1 1 0 |
| 1 1 1 1 | 1 0 0 1 0 1 1 |

It can be seen that there are seven code-vectors with three 1's in each of them, which satisfy Eq. 11.5.5.2. Similarly, there are seven code-vectors with four 1's in each of them, which also satisfy Eq.11.5.5.2. Apart from these fourteen code-vectors, the remaining two code-vectors are all 0's and all 1's, which can be words of any cyclic code. Thus, the above code is a cyclic code.

## 11.6 CONVOLUTIONAL CODES

As already mentioned, in convolutional codes the message bit stream is encoded in a continuous fashion; rather than in piecemeal as in block codes. Convolutional codes are easily generated with a *shift register* shown in Fig. 11.6.1.

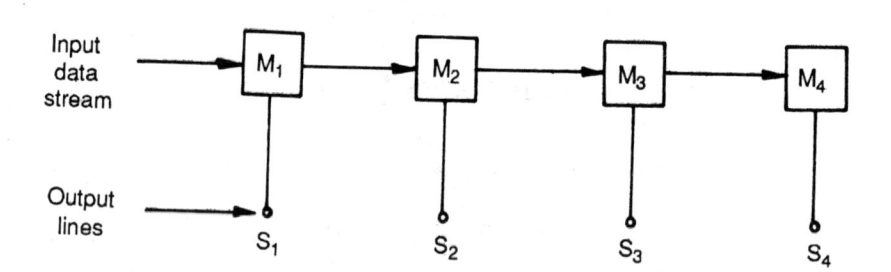

Fig.11.6.1.(a)   A Four-stage Shift Register

Storage (memory) devices, such as flip flops, connected in cascade, form a shift register. Each flip flop is capable of storing one bit. A four-bit shift register is shown in Fig. 11.6.1(a). $M_1, M_2, M_3$ and $M_4$ are memory devices. A stream of binary data is applied to $M_1$ in MSB (Most Significant Bit) first fashion. $S_1, S_2, S_3$ and $S_4$ are outputs taken from $M_1, M_2, M_3$ and $M_4$ respectively. $M_1$ stores the most recent bit of input data stream

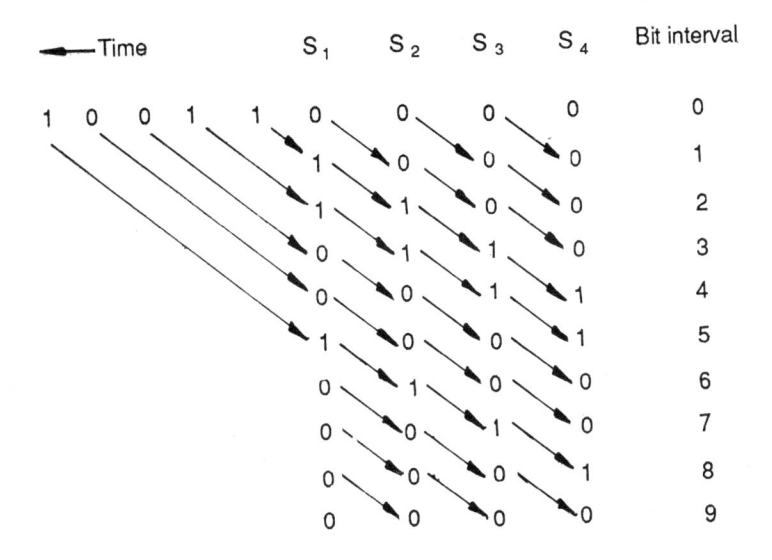

```
                                S₁   S₂   S₃   S₄   Bit interval
    ←—Time
                                 0    0    0    0      0
     1   0   0   1   1   0       1    0    0    0      1
                         1       1    0    0           2
                         0       1    1    0    0      3
                         0       0    1    1    0      4
                         1       0    0    1    1      5
                         0       1    0    0    1      6
                         0       0    1    0    0      7
                         0       0    0    1    0      8
                         0       0    0    0    1      9
                                 0    0    0    0
```

Fig.11.6.1(b)  Successive States of Stages for the Input Train 11001.

and indicates its state on the output line $S_1$. Therefore, the output $S_1$ is the same as the MSB of the input data stream. After one bit interval, the bit stored in $M_1$ shifts one stage to the right, i.e., to $M_2$. Thus, the output $S_2$ of $M_2$ is the same as $S_1$, i.e., the input bit stream with one bit interval delay. In this way, the input bit stream appears at every output line with an increased delay.

The operation of the shift register of Fig.11.6.1(a) is explained in Fig.11.6.1(b). Here, it is assumed that initially the shift register is clear, i.e., all memories are storing zeroes. A five-bit input data stream is applied to the shift register and this figure traces the path of the data stream through the register. The input data stream is 11001, but, for convenience of decoding, the train is represented against a reversed time scale, i.e., MSB which is the bit on the extreme left of the input data stream, enters the register first. With each succeeding bit, the contents of each memory device are shifted into the next device. At the ninth bit interval, the register returns to its clear state after allowing the input data stream to pass through it.

## 11.6.1  Encoder for Convolutional Code

An encoder for a convolution code is shown in Fig. 11.6.1.1. In this case,

$K$ = no. of shift registers
   = 3

$v$ = no of modulo-2 adders
   = no. of bits in the code-block
   = 3

$L$ = length of input data stream
   = 4

The outputs $v_1$, $v_2$ and $v_3$ of the adders are

**568** COMMUNICATION SYSTEMS: Analog and Digital

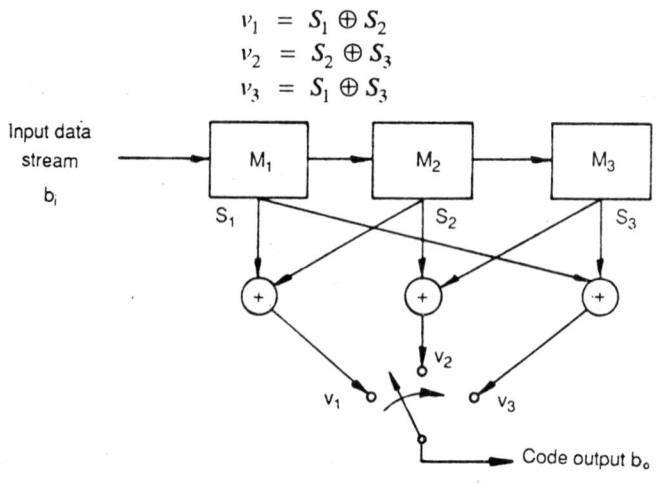

$$v_1 = S_1 \oplus S_2$$
$$v_2 = S_2 \oplus S_3$$
$$v_3 = S_1 \oplus S_3$$

Fig. 11.6.1.1. Encoder for Convolutional Code

It is assumed that, initially, the shift register is clear. The operation of the encoder is explained for the input data stream of a four-bit sequence

$$m = 1101$$

This is entered in the shift register from MSB. Thus, at the first-bit interval, $S_1 = 1, S_2 = 0, S_3 = 0$. Now, $v_1, v_2$ and $v_3$ can be found from Eq. 11.6.1.1. Thus

$$v_1 = 1 \oplus 0 = 1, v_2 = 0 \oplus 0 = 0, v_3 = 1 \oplus 0 = 1$$

Hence, the output at the first-bit interval is 101. Similarly, at the second bit interval, $S_1 = 1, S_2 = 1, S_3 = 0$. Thus, $v_1 = 1 \oplus 1 = 0, v_2 = 1 \oplus 0 = 1, v_3 = 1 \oplus 0 = 1$. Hence, the output at second-bit interval is 011.

In the same manner, outputs at other bit intervals can be found out. Since $L = 4$ and $k = 3$, the register resets at seventh ($L + k = 4 + 3 = 7$) bit interval. The output at each bit interval consists of $v$ bits (in this case $v = 3$). Thus, for each message, there are $v(L+k)$ bits in the output codeword. Notice that each message bit remains in the shift register for $k$-bit intervals. Hence, each input bit has an influence on the $k$ groups of $v$ bits; i.e., on $vk$ output bits.

Table 11.6.1.1 gives the coded output bit stream for all input data streams for the encoder shown in Fig. 11.6.1.1. The MSB column of input data stream is such that it is divided into two subsets (eight 0's and eight 1's), resulting in two subsets of the first code block of three bits in the coded output bit stream (eight 000 and eight 101). Each of these two subsets of the MSB column is further divided into two subsets (four 0's and four 1's) in the second MSB column, resulting in two subsets of second code block of three bits in the coded output bit stream (four 000-four101 and four110-four 011). In the same way, each subset is further divided into two subsets, till there is only one code block of three bits in each subset. Thus, it is possible to construct a code tree shown in Fig. 11.6.2.1, from the table 11.6.1.1, if the input data stream is entered from the MSB in the convolutional code encoder. On the other hand, it is not possible to construct such code tree if the input data stream is entered from the LSB, as successive division in two subsets is not possible if we start from the LSB column. Hence, in the convolutional encoder, the input data stream is entered from the MSB and not from the LSB.

### 11.6.2 Decoding a Convolutional Code

*The Code Tree*: Figure 11.6.2.1 shows the code tree for the encoder of Fig. 11.6.1. This is derived from Table 11.6.1.1. The starting point on the code tree is at the extreme left and corresponds to the situation before the arrival of the first message bit. The first message bit may be either a 0, or a 1. When an input bit is 0, the upward path is taken, and when it is 1, the downward path is taken. The same rule is followed at each junction or node. The path through the tree shown by the dashed line is for input message 1101. The code for the input message

1101 can be found by reading the bits encountered from the entrance to exit of the tree along the dashed path. Thus, the desired code is 101 011 101 110 110 011 000, same as in Table 11.6.1.1. Codes for other messages can be found out with the help of an appropriate path on the code tree. Note that any path through the tree

Table 11.6.1.1 Convolutional Code for Fig. 11.6.1.1

| Input data stream | Coded output bit stream |||||||
|---|---|---|---|---|---|---|---|
| 0000 | 000 | 000 | 000 | 000 | 000 | 000 | 000 |
| 0001 | 000 | 000 | 000 | 101 | 110 | 011 | 000 |
| 0010 | 000 | 000 | 101 | 110 | 011 | 000 | 000 |
| 0011 | 000 | 000 | 101 | 011 | 101 | 011 | 000 |
| 0100 | 000 | 101 | 110 | 011 | 000 | 000 | 000 |
| 0101 | 000 | 101 | 110 | 110 | 110 | 011 | 000 |
| 0110 | 000 | 101 | 011 | 101 | 011 | 000 | 000 |
| 0111 | 000 | 101 | 011 | 000 | 101 | 011 | 000 |
| 1000 | 101 | 110 | 011 | 000 | 000 | 000 | 000 |
| 1001 | 101 | 110 | 011 | 101 | 110 | 011 | 000 |
| 1010 | 101 | 110 | 110 | 110 | 011 | 000 | 000 |
| 1011 | 101 | 110 | 110 | 011 | 101 | 011 | 000 |
| 1100 | 101 | 011 | 101 | 011 | 000 | 000 | 000 |
| 1101 | 101 | 011 | 101 | 110 | 110 | 011 | 000 |
| 1110 | 101 | 011 | 000 | 101 | 011 | 000 | 000 |
| 1111 | 101 | 011 | 000 | 000 | 101 | 011 | 000 |

passes through only as many nodes ($L$) as there are bits in the input message. The node corresponds to the point where alternate paths are possible depending on the next message bit being 1 or 0.

*Decoding in the Presence of Noise: Exhaustive Search Method* In the absence of noise, the codeword will be received as transmitted. Hence, it is easy to reconstruct the original message. But due to noise, the word that is received is not the one transmitted. Decoding in the presence of noise is done in the following manner (the procedure is explained for $k = 3, L = 4, v = 3$).

The first message bit has an effect on the first $kv = 9$ bits. From the code tree of Fig. 11.6.2.1, it is clear that there are eight possible combinations of the first nine digits which are acceptable codewords. All these combinations are compared with the first nine bits of the received word, and the path corresponding to the combination giving a minimum discrepancy is accepted as the correct path.

If the path goes upwards at the first node $A$, then the first message bit is taken as 0, and if the path goes downwards, then the first message bit is taken as 1. Say, the path is downwards (as shown by $A\ B$ in Fig.11.6.2.1). Thus, it is concluded that the first message bit is 1. Now, we are at node $B$. The second message bit will have an effect on the next nine bits for which, again, there are eight possible ways. Using the same procedure, the direction of the path at the node $B$, and hence, the second message bit is decided. In the same way, all the message bits are decided and the received word is decoded.

It may be seen that the probability of error decreases exponentially with $K$. Hence, $K$ should be made as large as possible. But, on the other hand, the decoding of each bit requires an examination of the $2^k$ branch sections (in our case $2^k = 8$, as $k = 3$) of the code tree. Hence, with a large $k$, the decoding procedure becomes lengthy. Another method known as *sequential decoding* is manageable even for a large $k$.

*Sequential Decoding*
The main advantage of sequential decoding is that it avoids the lengthy process of examining every branch of the $2^k$ possible branches of the code tree while decoding a single message bit. In this method, at the arrival

**570** COMMUNICATION SYSTEMS: Analog and Digital

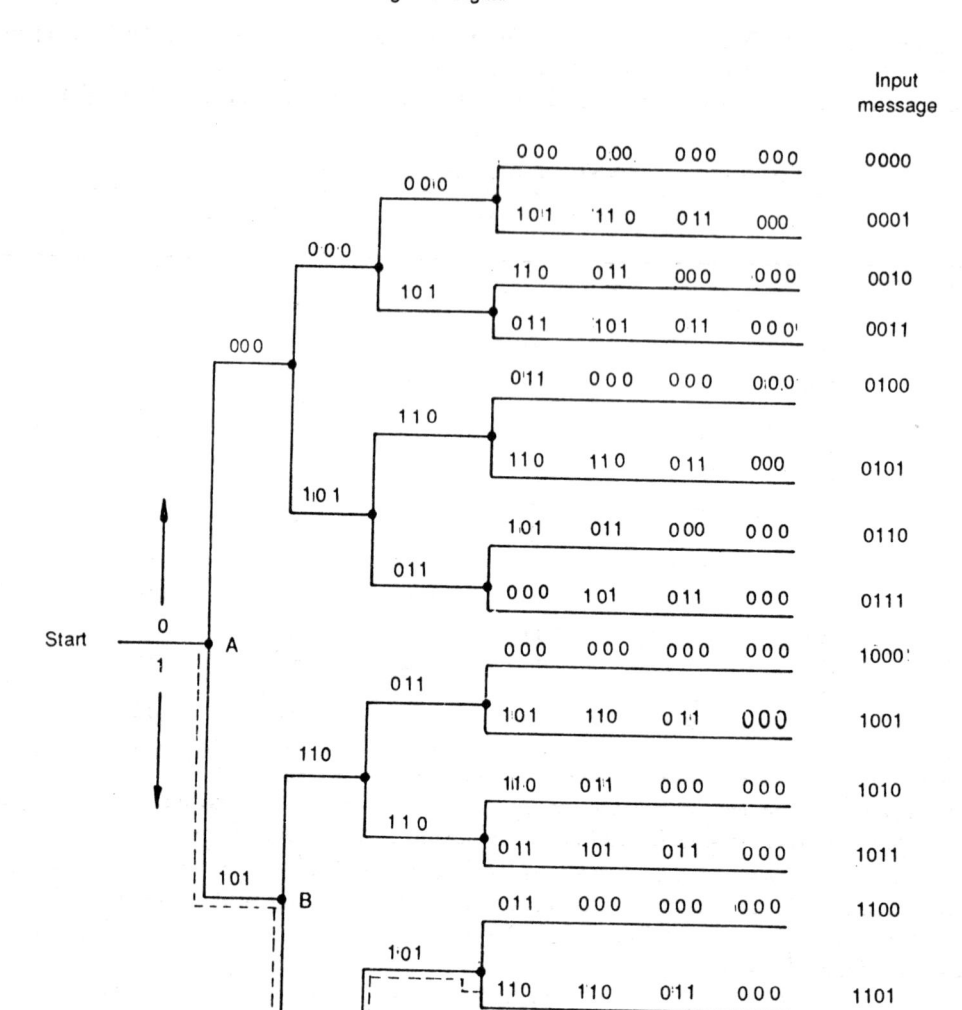

Fig.11.6.2.1 Code Tree for the Encoder of Fig.11.6.1.1.

of a $v$-bit code block, the encoder compares these bits with the code blocks of the branches diverging from the starting node. The encoder follows the branch whose code block gives lesser discrepencies with the received code block. The same procedure is repeated at each node.

Figure 11.6.2.2. illustrates how a decoder decides that it has taken a wrong turn. Let $P(e)$ be the probability that a received bit is in error. Then, the total number of errors $d(\ell) = v\ell P(e)$, where $v$ is the number of bits in a code block, and $\ell$ the number of nodes traversed. Then, if a plot $\ell$ Vs d($\ell$) is plotted, the ideal plot is a straight line $d(\ell) = v\ell P(e)$; and the correct path curve oscillates about this line within reasonable limits. If the decoder takes a wrong turn at any node, the total number of errors will increase rapidly after that node. When it crosses the discard level, the decoder judges that it has made an error, and it retraces to the previous node and takes the alternate turn. If it is still on the incorrect path, it will again retrace the path and follow

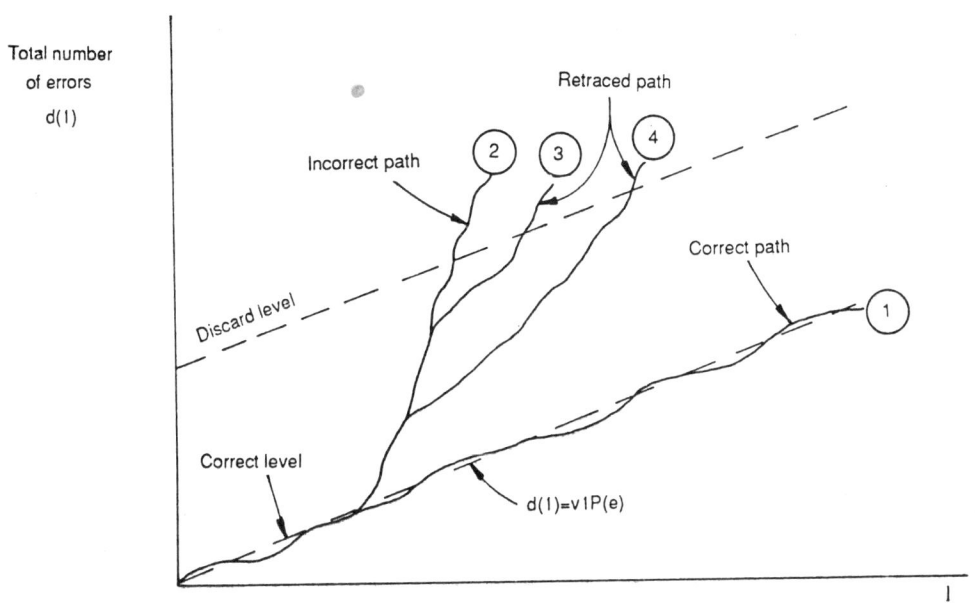

Fig. 11.6.2.2

the same procedure. After few retraces, the decoder will finally follow the correct path. In Fig.11.6.2.2, (2) is the incorrect path; (3) and (4) are retraced paths; and (1) is the correct path. Thus, the sequential decoder operates mostly on short code blocks and reverts to a trial-and-error search over long code blocks only when it judges that an error has been made. The end result is that the sequential decoding may generally be accomplished with much less computation than the exhaustive search method.

## PROBLEMS

11.1 Give the desired properties of a code. Distinguish between irreducible and separable codes.

11.2 What is coding efficiency? Show that the coding efficiency is maximum when $P(0) = P(1)$.

11.3 Apply Shannon-Fano coding procedure for $M = 2$ and $M = 3$.
Given
$$[x] = [x_1\ x_2\ x_3\ x_4\ x_5\ x_6\ x_7\ x_8]$$
$$[P] = [0.1\ 0.25\ 0.15\ 0.05\ 0.15\ 0.1\ 0.05\ 0.15]$$

11.4 Solve problem 11.3 by using the Huffman coding procedure.

11.5 The generator matrix for a (7,4) block code is
$$G = \begin{bmatrix} 1 & 0 & 0 & 0 & \vdots & 0 & 1 & 1 \\ 0 & 1 & 0 & 0 & \vdots & 1 & 0 & 1 \\ 0 & 0 & 1 & 0 & \vdots & 1 & 1 & 0 \\ 0 & 0 & 0 & 1 & \vdots & 1 & 1 & 1 \end{bmatrix}$$

Find all the codewords of the code.

11.6 The generator polynomial of a (6,3) cyclic code is
$$g(x) = 1 + x^2$$
Find all the codewords of the code.

11.7 The encoder for a convolutional code is shown below:

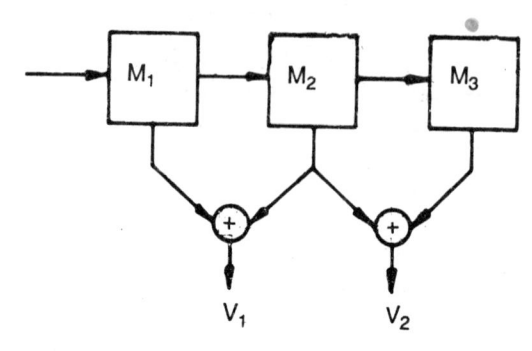

Fig. Prob.11.7

Find all the codewords for a 4-bit input data.

11.8 Construct a code tree for problem 11.7. Explain the exhaustive search and sequential decoding methods.

# ANSWERS TO SELECTED UNSOLVED PROBLEMS

## CHAPTER 1

(1.6) $f(t) = \dfrac{1}{T} \displaystyle\sum_{n=-\infty}^{\infty} e^{jn\omega_0 t}, \ (-\infty < t < \infty)$.

(1.10) (a) $F(\omega) = \dfrac{1}{(a+j\omega)^2}$; (b) $-\dfrac{2}{\omega^2}$

(1.11) (a) $\sigma\sqrt{2\pi}\ e^{-\sigma^2\omega^2/2}$

(1.16) $Sa^2(\omega + 2\pi.10^6) + Sa^2(\omega - 2\pi.10^6)$

(1.17) $\dfrac{\pi}{10} \displaystyle\sum_{n=-\infty}^{\infty} Sa\left(\dfrac{n\pi}{20}\right) \delta(\omega - 4\pi)$

(1.18) $j10\pi\big[\delta(\omega + 2\pi.10^6 t) - \delta(\omega - 2\pi.10^6 t)\big]$
$+ j4\pi\big[\delta(\omega + 2\pi.10^3 t + 2\pi.10^6.t) + \delta(\omega + 2\pi.10^3 t - 2\pi.10^6 t)$
$- \delta(\omega - 2\pi.10^3 t + 2\pi.10^6 t) - \delta(\omega - 2\pi.10^3 t - 2\pi.10^6 t)\big]$
$- 3\pi[\,\delta(\omega + 2\pi.10^2 t + 2\pi.10^6 t) - \delta(\omega + 2\pi 10^2 t - 2\pi.10^6 t)$
$- \delta(\omega - 2\pi.10^2 t + 2\pi\ 10^6 t) + \delta(\omega - 2\pi.10^2 t - 2\pi.10^6 t)]$

(1.19) $10^{-6} Sa\big[10^{-6}(\omega - \omega_c)\big] + 10^{-6} Sa\big[10^{-6}(\omega + \omega_c)\big]$; where $\omega_c$ is carrier frequency.

(1.20) $\dfrac{2A}{1-\omega^2} \cos\left(\dfrac{\omega\pi}{2}\right)$

(1.21) $1, \dfrac{\pi}{3}$ (b) $1, \dfrac{\pi}{6}$

## CHAPTER 2

(2.7) (a) $A^2/4$ (b) $\left(B^2 + \dfrac{1}{2}\right)$

(2.10) $\psi_r(\omega) = \begin{cases} 36/\omega^2; |\omega| < 2 \\ 0 \quad \text{elsewhere}; \end{cases}$ $E_i = 4.5;\ E_o = \dfrac{36}{\pi}\tan^{-1}(2)$

(2.11) $A^2$

(2.12) 14.5

(2.14) (iii) No

(2.15) 40.5

(2.16) 75

(2.17) 0.5

## CHAPTER 3

(3.2) (i) 0.012  (ii) 0.004
(3.3) (i) 0.0054  (ii) 0.2353

(3.5) $\dfrac{3}{32}, \dfrac{5}{16}$

(3.6) (a) $\dfrac{1}{15}$     (b) $\dfrac{4}{15}$

(c) $f_1(x) = \begin{cases} \dfrac{1}{5} & \text{for } x = 1 \\ \dfrac{4}{5} & \text{for } x = 2 \end{cases}$   $f_{2(y)} = \begin{cases} 0 & \text{for } y = 0 \\ \dfrac{1}{3} & \text{for } y = 1 \\ \dfrac{2}{3} & \text{for } y = 2 \end{cases}$

(3.7) (a) $\dfrac{1}{28}$     (b) $\dfrac{15}{56}$     (c) $\dfrac{13}{21}$

(3.8) $\dfrac{17}{5}, \dfrac{5}{3}, \dfrac{57}{5}$

(3.9) $\dfrac{133}{28}, \dfrac{8}{7}, \dfrac{200}{21}$

(3.11) $\dfrac{5}{16},\ 2.5,\ 1.118$

(3.12) 0.8754
(3.13) 0.636

# CHAPTER 4

(4.8)  2.76
(4.9)  (a) $19.3 \times 10^{-18}$ (b) $4.825 \times 10^{-18}$
(4.10) 2.041 dB
(4.11) 1.067

# CHAPTER 5

(5.1)  Carrier power = 200 W; power of each sideband = 8 W
(5.2)  159235.67, 15.92, 557.32, 0.2, 0.01, 0.22
(5.3)  84 W, 42 W
(5.4)  796178, 50 V, (796178 ± 796), 17.5 V, 796178 ± 159.2, 7.5 V, 1612.5 W, 362.5 W
(5.5)  0.96
(5.6)  10.335 KW
(5.7)  760375, 25 V, 758375, 25 V, 75 V
(5.8)  0.88, 2.78 KW
(5.9)  201.34 W
(5.10) (i) 0.4, (ii) 14.8%, (iii) 5.75 KW, (iv) 1.08 KW
(5.11) (i) 50%, (ii) 5.625 KW, (iii) 83%, (iv) 1K, 1.125 K
(5.14) (a) $H(\omega) = G_{20\,kHz}(\omega + 2 \times 10^5) + G_{20\,kHz}(\omega - 2 \times 10^5)$, (b) 33 dB, 33 dB.
(5.15) $\left[\cos(2\pi.10^3 t) + \cos(4\pi.10^3 t)\right]\cos(40\pi.10^3.t)$
$\pm \left[\sin(2\pi.10^3 t) + \sin(4\pi.10^3 t)\right]\sin(40\pi.10^3.t)$
(5.17) (i) (a) $H(\omega) = G_{2.5\,kHz}(\omega + 2.5 \times 10^3 + 2 \times 10^5) + G_{2.5\,kHz}(\omega - 2.5 \times 10^3 - 2 \times 10^5)$
(b) 33 dB, 33 dB
(ii) (a) $H(\omega) = G_{2.5kHz}(\omega + 2.5 \times 10^3 - 2 \times 10^5) + G_{2.5kHz}(\omega - 2.5 \times 10^3 + 2 \times 10^5)$
(b) 33 dB, 33 dB

# CHAPTER 6

(6.7)  64, 48.
(6.8)  70 kHz.
(6.9)  20, 1 kHz, 20 kHz, 8 MHz, 50 W
(6.10) 66 kHz, 36 kHz
(6.13) (a) 240 kHz, (b) 1.5 kHz, (c) 160 kHz, 7.5 kHz.
(6.14) 408 kHz

## CHAPTER 7

(7.7) 50 $\mu$sec

## CHAPTER 8

(8.6) 6,00,000 bits/sec, 570 kHz
(8.7) 5 kHz
(8.8) 63.7 mV

## CHAPTER 10

(10.2) 3.842 bits/message, 23052 bits/sec
(10.3) (a) 1.414 bits, 3 bits  (b) 1.405 bits/message
 (c) 112.4 bits/sec
(10.4) (a) 0.273 bit/message  (b) 0.924 bit/message
 (c) 0.42 bit/message  (d) 1.4855 bits/message
 (e) 0  (f) 0
(10.5) (a) 2 bits/sec  (b) 0
 (c) 0  (d) 0
 (e) 0.2781 bit/sec  (f) 0.067 bit/sec
 (g) 0.7 bit/sec  (h) 0 0.041 bit/sec
 (i) 0.035 bit/sec
(10.6) $-\log(b-a)$.
(10.7) 5 kHz
(10.8) 47.2%

## MULTIPLE CHOICE QUESTIONS

### Chapter 1  Signal Analysis

1. The sampling of a function $f(t)\ \sin(2\pi f_0 t)$ starts from a zero crossing. The signal can be detected if sampling time $T$ is

   (a) $T = \dfrac{1}{2f_0}$  (b) $T > \dfrac{1}{2f_0}$

   (c) $T < \dfrac{1}{2f_0}$  (d) $T \le \dfrac{1}{2f_0}$

2. A function is sampled at Nyquist rate $f_s = 2f_0$. The function can be recovered from its samples only if it is a
   (a) sine wave of frequency $f_0$;
   (b) Triangular wave of fundamental frequency $f_0$,
   (c) periodic square wave of fundamental frequency $f_0$,
   (d) a unit step function.

ANSWERS TO SELECTED UNSOLVED PROBLEMS 577

3. A signal $f(t) = \cos 8\pi t + 0.5 \cos 4\pi t$ is instaneously sampled. The maximum allowable value of sampling interval $T_s$ in sec. is
   (a) $\frac{1}{4}$ sec,
   (b) $\frac{1}{8}$ sec,
   (c) $\frac{1}{14}$ sec,
   (d) $\frac{1}{8\pi}$ sec.

4. The value of integral $\int_{-\infty}^{\infty} \delta(t) \cos t \, dt$ is equal to
   (a) zero,
   (b) one,
   (c) infinity,
   (d) undefined.

5. Solution of integral $\int_{-1}^{2} (t^3 + 3) \delta(t + 2) dt$ is equal to
   (a) $-5$,
   (b) $-2$,
   (c) zero,
   (d) 11

6. Frequency domain of a periodic triangular function is a
   (a) discrete sampling function,
   (b) descrete sync function,
   (c) continuous sampling function,
   (d) continuous sampling square function.

7. An impulse function consists, of
   (a) entire frequency range with same relative phase,
   (b) infinite bandwidth with linear phase variations,
   (c) pure d.c.
   (d) large d.c. along with weak harmonics.

8. Sinusoidal function is used as basic function in electrical communication because,
   (a) convenience,
   (b) response to linear system is sine wave,
   (c) both a and b,
   (d) none.

9. Laplace transform has the following merit ever fourier transform,
   (a) former is more informative,
   (b) latter is more informative
   (c) former converges for more functions,
   (d) latter has exponential decay.

10. The derivative of an ideal step function is
    (a) an impulse function,
    (b) zero,
    (c) sync function,
    (d) undefined.

11. A trigonometric series has,
    (a) single sided spectrum,
    (b) double sided spectrum,
    (c) may have both,
    (d) none

12. A signum function is,
    (a) zero for $t$ greater than zero,
    (b) zero for $t$ less than zero,
    (c) unity for $t$ greater than zero,
    (d) $2u(t) - 1$.

13. A sine wave generator starts at $t = 0$. It consists of
    (a) a pure sine wave,
    (b) a pure cosine wave,
    (c) both,
    (d) sine wave plus other frequencies.

14. Following is true.
    (a) fourier series is unique,
    (b) fourier transform is unique,
    (c) both are unique,
    (d) none is unique.

578   COMMUNICATION SYSTEMS: Analog and Digital

15. Following is not a reason of distortion in communication system.
    a. insufficient channel bandwidth
    b. random variations in channel
    c. external Interference
    d. none.

## CHAPTER 2   Linear Systems

16. Let function $A = \sin \omega_0 t$ and $\sin(\omega_0 + \phi)$. The power of two signals; $P_A$ and $P_B$ is related as
    (a) $P_B = \phi P_A$
    (b) $P_B = \dfrac{1}{\phi} P_A$
    (c) $P_B = \tan \phi P_A$
    (d) $P_B = P_A = P$

17. Let $f_1(t) = G_1(t) + 3$; $f_2(t) = G_2(t) + 2$. If $G_1(t)$ and $G_2(t)$ are uncorrelated, the correlation between $f_1(t)$ and $f_2(t)$ is,
    (a) 5
    (b) 6
    (c) 1
    (d) zero.

18. The unit impulse response of a linear system is sampling square function. Its transfer function is a
    (a) Gate function
    (b) square of Gate function
    (c) triangular function
    (d) sampling square function.

19. Transfer function of a linear system is $100\, e^{-j5\omega}$. The system is a,
    (a) distortionless amplifier,
    (b) distortionless attenuator,
    (c) amplifier with phase distortion,
    (d) attenuator with phase distortion.

20. A linear system is characterized by, $H(\omega) = A e^{-b\omega^3}$. The system is physically
    (a) unrealizable,
    (b) realiazable,
    (c) depends on the value of $b$
    (d) depends on the value of $A$.

21. An energy signal has $F(\omega) = 5$. Its energy density spectrum is
    (a) 10
    (b) 5
    (c) 25
    (d) 1

22. Highest value of the autocorrelation of a function $10 \sin 10 \pi t$ is equal to,
    (a) 50
    (b) 100
    (c) zero
    (d) depends on time $t$.

23. The autocorrelation of a sampling function is a.
    (a) triangular function
    (b) square of sampling function
    (c) signum function
    (d) Gate function

24. Two function $f_1(t)$ and $f_2(t)$ with correlation of 5 has average power of 2 and 5 respectively. The power of $f_1(t) + f_2(t)$ is
    (a) 7
    (b) 3
    (c) 12
    (d) 17

25. A rectangular pulse is passed through an LPF. The response is a
    (a) triangular function
    (b) trapezoidal function
    (c) sampling function
    (d) none

# CHAPTER 3    Probability and Random Signal Theory

26. Which of the following is incorrect?
    (a) $A - B = A\overline{B}$
    (b) $\overline{AB} = \overline{A}\,\overline{B}$
    (c) $A\overline{A} = 0$
    (d) $AA = A$
27. Pick the odd man out,
    (a) stochastic variable
    (b) stochastic function
    (c) random variable
    (d) random experiment.
28. Pick the odd man out
    (a) binomial distribution
    (b) normal distribution,
    (c) uniform distribution
    (d) Rayleigh distribution.
29. Which of the following is incorrect?
    (a) $P(v) = 1$
    (b) $P(\overline{A}) = P(A) - 1$
    (c) $0 \leq P(A) \leq 1$
    (d) If A and B are mutually exclusive, then $P(A+B) = P(A)+P(B)$
30. The total area under the probability distribution curve is
    (a) 1
    (b) 0
    (c) depends on the nature of the distribution
    (d) none of the above
31. The spectral density of white noise
    (a) varies with frequency
    (b) varies with bandwidth
    (c) varies with amplitude of the signal
    (d) is constant
32. The theoretical power of white noise is
    (a) zero
    (b) finite
    (c) infinite
    (d) depends on frequency of the signal
33. The stationary process has
    (a) ensemble average equal to time average
    (b) all the statistical properties dependent on time
    (c) all the statistical properties independent of time
    (d) zero variance
34. Events A and B are statistically independent if
    (a) A and B occur simultaneously
    (b) A and B occur at different times
    (c) occurance of A includes occurrence of B
    (d) none of the above
35. Pick the odd man out
    (a) expectation
    (b) variance
    (c) standard deviation
    (d) Tchebycheff's inequality

## State True or False

36. The classical approach for probability theory does not explain the situation when the number of outcomes of an experiment is small
    (a) True
    (b) False
37. Mutually exclusive events are also statistically independent
    (a) True
    (b) False
38. The probability that a continuous random variable takes on a particular value is zero.
    (a) True
    (b) False
39. Unit of variance is same as that of the random variable.
    (a) True
    (b) False

40. If a random process is ergodic then it is also stationary.
    (a) True
    (b) False

## CHAPTER 4  Noise

41. An amplifier having noise figure of 20 dB and available power gain of 15 dB is followed by a mixer circuit having noise figure of 9 dB. The overal noise figure as referred to input in dB is
    (a) 11.07
    (b) 10.44
    (c) 21.52
    (d) 0.63
42. Thermal noise is independent of
    (a) bandwidth
    (b) Temperature
    (c) center frequency
    (d) Boltzman's constant
43. Transistor $T_1$ operates at 20 kHz and $T_2$ operates at 200 Hz. The flicker noise is
    (a) more in $T_1$
    (b) more in $T_2$
    (c) equal in both
    (d) depends on bias
44. A triode has transconductance equal to 25 $\mu v$. The equivalent noise resistance is
    (a) IM ohms
    (b) IK ohms
    (c) 10 K ohms
    (d) 100 K ohms
45. Johnson noise is
    (a) always white
    (b) white for all practical purposes
    (c) never white
    (d) depends on temperature
46. The spectrum of the white noise and an impulse function is similar in following respects
    (a) both have similar magnitude spectrum
    (b) both has similar phase spectrum
    (c) both have similar magnitude and phase spectrum
    (d) they have nothing similar
47. Parallel combination of a resistance $R$ and a capacitance $C$ develops a noise voltage source at its common terminal. The rms value of the voltage varies;
    (a) proportional to $R$
    (b) inversely proportional to $C$
    (c) inversely proportional to square root of $C$
    (d) proportional to $RC$
48. A noise voltage source has a resistance of 10 ohms. Its power density spectrum is $0.26 \times 10^{-5}$. The corresponding available power density is
    (a) $2.6 \times 10^{-5}$
    (b) 0.025
    (c) $26 \times 10^{-5}$
    (d) $6 \times 10^{-8}$
49. A narrowband noise shows
    (a) amplitude modulation only
    (b) frequency modulation only
    (c) both AM and FM
    (d) none
50. A narrowband noise source $n(t)$ has symmetrical spectrum and has power density spectrum $0.2 \times 10^{-6}$ The power density of quadrature component is
    (a) $0.2 \times 10^{-6}$
    (b) $0.1 \times 10^{-6}$
    (c) $0.4 \times 10^{-6}$
    (d) $0.05 \times 10^{-6}$

## CHAPTER 5  Amplitude Modulation

51. The positive RF peaks of an AM voltage rise to a maximum value of 12 V and drop to a minimum value of 4 V. The modulation index assuming single tone modulation is
    (a) 3
    (b) $\frac{1}{3}$
    (c) $\frac{1}{4}$
    (d) $\frac{1}{2}$

52. The most suitable method for detecting a modulated signal $(2.5 + 5 \cos \omega_m t) \cos \omega_c t$ is
    (a) envelope detector  (b) synchronous detector
    (c) ratio detector  (d) both a and b
53. The main advantage of superheterodyne receiver is,
    (a) simple circuit  (b) better tracking
    (c) improvement in selectivity and sensitivity  (d) better alignment
54. The received signal frequency at any time of a superheterodyne receiver having IF = 456 kHz, is 1 MHz. The corresponding image signal is
    (a) within its medium band
    (b) outside the medium band
    (c) depends on modulation index
    (d) depends on modulating frequency
55. The resonant frequency of an RF amplifier is 1 MHz and its bandwidth is 10 kHz. The Q-factor will be
    (a) 10  (b) 100
    (c) 0.01  (d) 0.1
56. A plot of modulation index verses carrier amplitude yields a
    (a) horizontal line  (b) verticle line
    (c) parabola  (d) hyperbola
57. A carrier is amplitude modulated to a depth of 40%. The increase in power is
    (a) 40%  (b) 20%
    (c) 16%  (d) 8%
58. Following is not the purpose of modulation
    (a) multiplexing  (b) effective radiation
    (c) Narrowbanding  (d) increase in signal power
59. An AM wave is given by $e_{AM} = 10(1 + 0.4 \cos 10^3 t + 0.3 \cos 10^4 t \cos 10^6 t$
    The modulation index of the envelope is
    (a) 0.4  (b) 0.5
    (c) 0.3  (d) 0.9

## CHAPTER 6  Angle Modulation

60. Armstrong F.M. transmitter performs frequency multiplication in stages
    (a) to increase the overall S/N ratio,(b) to reduce bandwidth
    (c) to find the desired value of carrierfrequency as well as frequency deviation,
    (d) for convenience.
61. Limiter is not essential in the following detector
    (a) Foster–Seeley  (b) balanced slope
    (c) ratio  (d) none
62. Figure of merit is always unity in
    (a) SSB-SC  (b) AM
    (c) FM  (d) All the three
63 The output $V_R$ of the ratio detector is related with the output $V_F$ of similar Foster–seeley discriminator as follows:
    (a) $V_F = V_R$  (b) $V_F > V_R$
    (c) $V_F = 0.51 V_R$  (d) $V_F = 2V_R$

582 COMMUNICATION SYSTEMS: Analog and Digital

64. Which one is an advantage of AM over FM
    (a) FM is more immune to noise
    (b) FM has better fidelity
    (c) Probability of noise spike generation is less in AM
    (e) FM has wide bandwidth
65. The message carrying efficiency is best in
    (a) FM
    (b) AM
    (c) AM-SC
    (d) Phase modulation
66. Following is not advantage of FM over AM
    (a) noise immunity
    (b) fidelity
    (c) capture effect
    (d) sptuttering effect
67. The modulating frequency in freque ncy modulation is increased from 10 kHz to 20 kHz. The bandwidth is
    (a) doubled
    (b) halved
    (c) increases by 20 kHz
    (d) increased tremendously
68. A narrowband FM does not have the following feature
    (a) it has two sidebands
    (b) Both sidebands are equal in amplitude
    (c) both sidebands have same phase difference with respect to carrier
    (d) it does not show amplitude variations
69. In time division multiplexing, the FM detector has
    (a) more
    (b) less
    (c) equal
    (d) unknown noise contribution as compared to phase modulation
70. In a single tone FM discriminator $(S_o / N_o)$ is
    (a) proportional to deviation
    (b) proportional to cube of deviation
    (c) inversely proportional to deviation
    (d) proportional to square of deviation
71. Assuming other parameters unchanged, if the modulating frequency is halved in a modulating systems, the modulation index is doubled. The modulation system is
    (a) AM
    (b) FM
    (c) phase modulation
    (d) angle modulation

## CHAPTER 7  Pulse Modulation Systems

72. The maximum permissible distance between two samples of a 2 kHz signal is
    (a) 1000 $\mu$ sec
    (b) 500 $\mu$ sec
    (c) 250 $\mu$ sec
    (d) None of the above
73. Pick the odd man out
    (a) PWM
    (b) PPM
    (c) PDM
    (d) PLM
74. The main advantage of TDM over FDM is that it
    (a) needs less power
    (b) needs less bandwidth
    (c) needs simple circuitry
    (d) gives better S/N ratio
75. The PWM needs
    (a) more power than PPM
    (b) more samples per second than PPM
    (c) more bandwidth than PPM
    (d) none of the above
76. The PAM signal can be detected by
    (a) bandpass filter
    (b) bandstop filter
    (c) high pass filter
    (d) low pass filter

## State True or False

77. The guard time between pulses increases transmission efficiency
    (a) True  (b) False
78. Noise can be reduced by increasing sampling rate
    (a) True  (b) False
79. Adaptive Delta Modulation is preferred over Delta Modulation because its step size changes as per the requirement
    (a) True  (b) False

## CHAPTER 8   Pulse Code Modulation

80. In PCM, the quantization noise depends on
    (a) sampling rate  (b) number of quantization levels
    (c) signal power  (d) none of the above
81. Which of the following modulation is digital in nature
    (a) PAM  (b) PPM
    (c) DM  (d) none of the above
82. Which of the following modulation is analog in nature
    (a) PCM  (b) DPCM
    (c) DM  (d) none of the above
83. In PCM, if the number of quantization levels is increased from 4 to 64, then the bandwidth requirement will approximately be
    (a) 3 times  (b) 4 times
    (c) 8 times  (d) 16 times
84. Quantization noise occurs in
    (a) PAM  (b) PPM
    (c) DM  (d) none of the above
85. Companding is used in PCM to
    (a) reduce bandwidth  (b) reduce power
    (c) increase $S/N$ ratio  (d) get almost uniform $S/N$ ratio
86. Pulse stuffing is used in
    (a) synchronous TDM  (b) asynchronous TDM
    (c) any TDM  (d) none of the above
87. The main advantage of PCM is
    (a) less bandwidth  (b) less power
    (c) better $S/N$ ratio  (d) possibility of multiplexing
88. The main disadvantage of PCM is
    (a) large bandwidth  (b) large power
    (c) complex circuitry  (d) quantization noise
89. The main advantage of DM over PCM is
    (a) less bandwidth  (b) less power
    (c) better $S/N$ ratio  (d) simple circuitry
90. In a DM system, the granular noise occurs when modulating signal
    (a) increases rapidly  (b) decreases rapidly
    (c) changes within the step size  (d) has high frequency component

## CHAPTER 9 Data Transmission

91. Which of the following gives maximum probability of error
    (a) ASK
    (b) FSM
    (c) PSK
    (d) DPSK
92. Which of the following gives minimum probability of error
    (a) ASK
    (b) FSK
    (c) PSK
    (d) DPSK

**State True or False**

93. In DPSK no synchronous carrier is needed at the receiver
    (a) True
    (b) False
94. In FSK no synchronous carrier is needed at the receiver
    (a) True
    (b) False
95. Probability of error in DPSK is less than PSK
    (a) True
    (b) False
96. Amplitude shift keying is also known as on off keying
    (a) True
    (b) False
97. Correlator is a coherent system of signal reception
    (a) True
    (b) False

## CHAPTER 10 Information Theory

98. A given source will have maximum entropy if the messages produced are :
    (a) two in number
    (b) mutually exclusive
    (c) statistically independent
    (d) equiprobable
99. Entropy gives
    (a) amount of information
    (b) rate of information
    (c) measure of uncertainity
    (d) probability of message
100. Which of the following is incorrect
    (a) $H(y/x) = H(x,y) - H(x)$
    (b) $I(x,y) = H(x) - H(y/x)$
    (c) $H(x,y) = H(x/y) + H(y)$
    (d) $I(x,y) = H(y) - H(y/x)$

**State True or False**

101. Unit of mutual information is bits/message
    (a) True
    (b) False
102. More is probability, less is information
    (a) True
    (b) False
103. The entropy $H(x/y)$ gives a measure of noise
    (a) True
    (b) False
104. Entropy of a continuous channel can be negative
    (a) True
    (b) False
105. Decit is unit of information rate
    (a) True
    (b) False
106. The message "weather is cloudy" carries no information
    (a) True
    (b) False

ANSWERS TO SELECTED UNSOLVED PROBLEMS 585

107. The statement "a head or a tail will appear in a coin tossing experiment" carries no information
    (a) True (b) False
108. Channel capacity is a measure of information rate:
    (a) True (b) False

## CHAPTER 11  Coding

### State True or False

109. When a code is irreducible, it is also separable.
    (a) True (b) False
110. Huffman code is also known as maximum redundancy code
    (a) True (b) False
111. English language is not uniquely dicipherable
    (a) True (b) False
112. In binary system, the coding efficiency increases as $P(0)$ appreaches 0.5.
    (a) True (b) False
113. A code with Hamming distance 4 is capable of double error correction
    (a) True (b) False
114. A code with Hamming distance 2 is not capable of error correction
    (a) True (b) False
115. Error cannot be corrected with the help of parity check code
    (a) True (b) False
116. Cyclic code is a subclass of convolutional code
    (a) True (b) False
117. Exhaustive search method of decoding a convolutional code is preferred over sequential decoding method
    (a) True (b) False

## ANSWERS TO MULTIPLE CHOICE QUESTIONS

| (1) c | (2) a | (3) b | (4) b | (5) c |
| --- | --- | --- | --- | --- |
| (6) d | (7) a | (8) c | (9) c | (10) d |
| (11) a | (12) d | (13) d | (14) b | (15) d |
| (16) d | (17) d | (18) c | (19) a | (20) a |
| (21) c | (22) a | (23) d | (24) d | (25) b |
| (26) b | (27) d | (28) a | (29) b | (30) a |
| (31) d | (32) d | (33) c | (34) d | (35) d |
| (36) b | (37) b | (38) a | (39) b | (40) a |
| (41) b | (42) c | (43) b | (44) d | (45) b |
| (46) a | (47) c | (48) d | (49) c | (50) b |
| (51) d | (52) c | (53) c | (54) a | (55) b |
| (56) d | (57) d | (58) d | (59) b | (60) c |
| (61) c | (62) a | (63) c | (64) c | (65) c |
| (66) d | (67) c | (68) c | (69) b | (70) d |

| | | | | | | | | | |
|---|---|---|---|---|---|---|---|---|---|
| (71) | b | (72) | c | (73) | b | (74) | c | (75) | a |
| (76) | d | (77) | b | (78) | b | (79) | a | (80) | b |
| (81) | c | (82) | d | (83) | a | (84) | c | (85) | d |
| (86) | b | (87) | c | (88) | a | (89) | d | (90) | c |
| (91) | a | (92) | c | (93) | a | (94) | b | (95) | b |
| (96) | a | (97) | a | (98) | d | (99) | c | (100) | b |
| (101) | a | (102) | a | (103) | b | (104) | a | (105) | b |
| (106) | b | (107) | a | (108) | b | (109) | a | (110) | b |
| (111) | a | (112) | a | (113) | b | (114) | a | (115) | a |
| (116) | b | (117) | b | | | | | | |

# REFERENCES

1. Abramson N., *Information theory and coding*, McGraw-Hill, 1963.
2. Alley C.L. and Atwood K.W., *Electronic engineering*, John Wiley, 1973.
3. Ash R., *Information theory*, John Wiley, 1965.
4. Das J., Mullick S.K. and Chatterjee P.K., *Principles of Digital Communication*, Wiley Eastern, 1986.
5. Davenport W.B. and Root W.L., *An introduction to the theory of random signals and noise*, McGraw-Hill, 1958.
6. Filler W., *An introduction to probability theory and its applications*, Vol. I, John Wiley, 1968.
7. Hancock J.C., *Introduction to the principles of communication theory*, McGraw-Hill, 1963.
8. Haykins, *Communication systems*, Wiley Eastern, 1982.
9. Lathi B.P., *Communication Systems*, Wiley Eastern, 1983.
10. Lee Y.W., *Statistical theory of communication*, John Wiley, 1960.
11. Miller G.M., *Modern electronic communication*, Prentice-Hall, 1978.
12. Mithal G.K., *Radio engineering*, Khanna Publisher, 1985.
13. Papoulis A., *Probability, random variables and stochastic processes*, McGraw-Hill, 1965.
14. Reza F.M., *An introduction to information theory*, McGraw-Hill, 1961.
15. Roden M.S., *Analog and digital communication systems*, Prentice Hall, 1979.
16. Rosie A.M., *Information and communication theory*, Van Nostrand, 1973.
17. Schwartz M., *Information transmission, modulation and noise*, McGraw-Hill, 1959.
18. Schwartz M., Bennet W.R. and Stein S., *Communication systems and techniques*, McGraw-Hill, 1966.
19. Shanmugham. K.S., *Digital and analog communication systems*, John Wiley, 1979.
20. Spiegel M.R., *Theory and problems of probability and statistics*, McGraw-Hill, 1975.
21. Stein S. and Jones J.J., *Modern communication principles*, McGraw-Hill, 1967.
22. Taub H. and Schilling D.L., *Principles of communication systems*, McGraw-Hill, 1971.
23. Terman F.E., *Electronic and radio engineering*, McGraw-Hill, 1955.
24. Thomas J.B., *An introduction to statistical communication theory*, John, Wiley, 1969.
25. Viterbi A.J., and Omura J.K., *Principles of digital communication and coding*, McGraw-Hill, 1979.
26. Wozencraft J.M., and Jacobs I.M., *Principles of communication engineering*, John Wiley, 1965.

# INDEX

(DSB-SC) 244
 Sensitivity 323
 0V 466

## A

Asynchronous time division multiplexing 472
Active filters 99
Adaptive Delta Modulation 479
Adjacent channel selectivity 329
Aliasing effect 78
AM 242, 284, 304
AM Receiver 322
AM-SC 243
Amplified and delayed AVC 330
Amplitude distortion 92
Amplitude shift keying 482
Analog modulation 240
Analytic signal 265
Angle-modulation 242, 348
Aperiodic signals 87
Armstrong method 384, 385
Associative laws 131
Atmospheric 193
Attenuation 91
Autocorrelation 122, 233
Available power 211

## B

Balanced modulator 256, 312, 485
Band limited signal 75
Bandpass signals 83
Bandwidth 250
Bar chart 151
Baseband signal 240
Bayes' theorem 146
Binary channel 530
Binary code 466
Binary erasure channel 527

Binary symmetric channel 525
Binomial distribution 181
Bit 504
Block codes 555
Broadcast system 284

## C

Capacitive reactance tube 383
Cascade amplifier 227
Cascaded channels 526
Cascaded stages 225
Causal signals 88
Causality condition 94
Central moment 177
Chopper-type (switching) modulator 251
Code polynomial 564
Coding 545
Coherence 119
Coherent detection 245
Collector modulation 307
Companding 471
Compatible single sideband 277
Commutative laws 130
Complement laws 130
Conditional distributions 164
Conditional probability 139
Continuous channel 533
Continuous probability distributions 155
Continuous random variable 154
Continuous spectrum 23
Correlation 118
Correlator 497
Costa's receiver 247
Cross-talk factor 450

## D

Damped sinusoidal waveform 51
De-emphasis 388, 419

Decibel 91
Decit 504
Delayed AVC 330
Delta function 28
Delta modulation 477
Delta or differential PCM (DPCM) 480
Demodulation 244
Demorgan's laws 131
Density function 155, 163
Detection 244
Deterministic signals 87
Diagonal clipping 315
Difference laws 131
Differential phase shift keying 486
Dirac delta function 31
Dirichlet's conditions 3
Discrete probability distributions 150
Discrete random variable 150
Discrete spectrum 9
Dispersion 103
Distant sets 129
Distortionless 91
Distribution 183
Distribution function 152
Distributive laws 130
Donald duck voice effect 274
Double spotting 330

### E

Elastic store 472
Empty set 129
Energy density spectrum 106
Energy signal 104
Entropy 505
Envelope detectors 313, 338
Equalization 470
Equivalent noise temperature 216
Error-control coding 555
Even parity 556
Expectation 169
Expected value 169
Experiment 149
Eye Patterns 469

### F

Feeder 321
Fidelity 324
Filter method 268
Filters 95
Flat-top sampling 442
Fluctuation noise 193

FM demodulators 388
FM modulators 373
FM Transmitter 384
FMFB 419, 423
Foster-Seely 388
Fourier series 1
Fourier series expansion of a periodic function 6
Fourier transform 23
Frequency conversion 253
Frequency convertors 326
Frequency discriminators 388
Frequency division multiplexing 449
Frequency domain 11
Frequency modulation 242
Frequency shifting property 45
Frequency spectrum 9
Friss formula 226
Front-end circuits 327
Function 90

### G

Ganged capacitor 327
Gate function 13, 64, 107
Gaussian distribution 183
Gaussian pulse 60
Granular noise 479
Guard time 467

### H

Hamming distance 556
Harmonic generator 320
Heterodyning 322
High-level modulation 318
Hilbert transform 262
Histogram 151
Homodyne detection 245

### I

Ideal filters 96
Identical sets 129
Idempotent laws 130
Identity laws 130
Image signal 325
Image signal rejection 329
Impulse function 28, 30
Impulse-noise 407
Incoherent function 119
Independent random variables 164
Indeterministic function 194
Information theory 503
Instantaneous frequency 349

# INDEX

Interception 322
Intermediate frequency 324
Intersymbol interference 469
Inverse Hilbert transform 262
Inverse Fourier transform 23

## J

Joint distributions 160
Joint probabilities 139
Joint probability function 160, 163

## L

Laplace transform 23
Large noise case 406
Line spectrum 9
Linear system 88
Linear modulation 372
Linear block codes 559
Linear diode detector 314
Low-level modulation 318

## M

Marginal density functions 163
Marginal distribution function 163
Marginal probability function 161
Master oscillator 319
Mean 169
Miller capacitance 376
Minimum distance 557
Mixer stage 326
Modulated class C amplifier 306
Modulation 240
Modulation 45, 240, 242
Modulation index 303, 357
Moment 177
Moment generating function 178
Multiplexing 240
Multiplication theorem 141
Mutual information 516
Mutually exclusive events 133

## N

Narrowband FM 355
Narrowband noise 233
Narrowband FM 364
Narrowbanding 241
Nat 504
Natural sampling 440
NBFM 385
Negative frequencies 5
Neper 91

Neutralization 321
Noise bandwidth 209
Noise 193
Noise figure 216, 219, 476
Noise impulse 406
Noise resistance 229
Noise temperature 214, 216
Non linear 88
Non-causal 88
Non-linear circuits 256
Non-linear modulation 372
Non-periodic 87
Normal distribution 183
Normalized random variable 176
Null set 129
Nyquist interval 78

## O

Odd symmetry 10
On-Off keying 482
Overmodulated AM 246

## P

Paley-Wiener criterion 93, 94
Parameter variation method 373
Parameters 119
Parity check code 556
Parity check matrix 560
Parseval's power theorem 110
Parseval's theorem 105, 496
Partially suppressed carrier system 246
Partition noise 197
Period 87
Periodic 87
Periodic function 6
Phase deviation 358
Phase distortion 92
Phase locked loop 248
Phase modulation 242
Phase shift keying 484
Phase-shift-method 268
Phasing method 268
Phasor diagram 358
Phasor representation 250
Pilot carrier 246
Plate modulation 307
PLL 419, 427
Poisson distribution 182
Porterior approach 133
Power density signals 104, 109
Power density spectrum 111

## INDEX

Pre-emphasis 388, 419
Pre-envelope 260, 265
Prefix property 546
Probability 132
Probability density function 155
Probability function 151
Proper subset 129
Proportional-plus-integral filter 434
Pulse amplitude modulation 439
Pulse code modulation 466
Pulse duration modulation 439
Pulse length modulation 439
Pulse modulation systems 239, 439
Pulse position modulation 439
Pulse stuffing 472
Pulse time modulation 439
Pulse width modulation 439
Pulse-triplet 407

## Q

Quadrature null effect 246
Quantization 466
Quantization error 474
Quantization levels 474
Quantization noise 474
Quantizer 468

## R

Random 150
Random pulse 50
Random experiments 132
Random function 150
Random signal 87
Random variable 150, 183
Rate of information 508
Ratio detector 388
Rayleigh's energy theorem 105
Reactance tubes 373
Reactance tube modulator 381
Reactor modulator 373
Receiver 323, 388
Resistor noise 198
Ring modulator 255
Rise time 99

## S

Sample space 129
Sampling function 16
Sampling square function 60
Sampling theorem 75, 81

Saturable reactor modulator 373
Scanning parameters 119
Schottky formula 196
Schwartz inequality 494
Selection process 322
Selectivity 323
Sensitivity 323
Separable encoding 546
Sequential decoding 569
Set 128
Shannon-Hartley theorem 536
Shannon's theorem 532
Signal 87
Signal analysis 1
Signal to noise ratio 221
Significant sidebands 364
Signum function 37
Simple AVC 330
Single-tuned discriminator 388
Singular function 34
Singularity functions 28
Sine integral 101
Slope detectors 388
Slope overload 478
Smooth noise 406
Spectral amplitudes 4
Spectral density 105
Spectral density function 24
Spike noise 407
Sputtering sound 407
Square law detectors 313
Squarer 248
SSB-SC 258
Stagger-tuned discriminator 388
Standard deviation 174
Standardized random variable 176
Statistically independent 141
Stochastic function 150
Stochastic variable 150
Subset 129
Superheterodyne 323
Superheterodyne Receiver 324
Symmertic channel 524
Symmetry property 41
Synchronizing circuit 485
Synchronous detection 245
Synchronous time division multiplexing 471
System 88
System function 89, 90
System bandwidth 99
Systematic codes 555

## T

Tchebycheff's inequality   179
The optimum filter   492
Temperature   216
Threshold effect   342, 407
Threshold improvement   419
Time constant R-C   315
Time differentiation   55
Time division multiplexing   449
Time domain   11
Time shifting property   43
Time-invariant   88
Time-varying   89
Transfer function   90
Transferred information   518
Transinformation   518
Transmitters   373
Trapezoidal function   56
TRF   322
TRF receiver   322
Triangular RF pulse   58
Trigonometric Fourier series   3
Triplet   407

## U

Uncertainty   508
Uncorrelated   119
Union or sum   129
Unit impulse response   88
Unit step function   31, 38
Universal set U   129

## V

VCO   373
Venn diagrams   129
Vestigial-Sideband   277

## W

Weight of a code   556
Wideband FM   355, 361